Managing Strategi ge

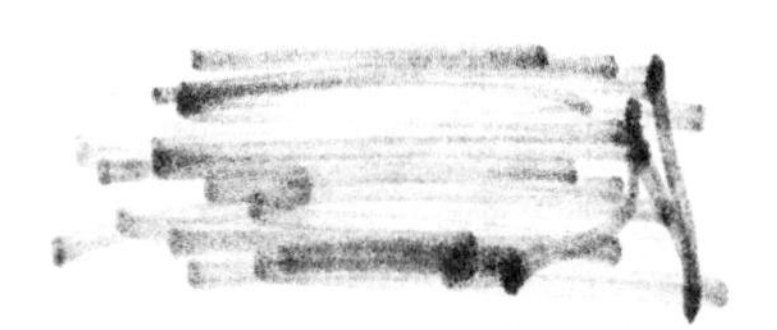

MANAGING STRATEGIC INNOVATION AND CHANGE

A Collection of Readings

MICHAEL L. TUSHMAN
Graduate School of Business
Columbia University

PHILIP ANDERSON
The Amos Tuck School
Dartmouth College

New York Oxford
OXFORD UNIVERSITY PRESS
1997

Oxford University Press

Oxford New York
Athens Auckland Bangkok Bogotá Bombay
Buenos Aires Calcutta Cape Town Dar es Salaam
Delhi Florence Hong Kong Istanbul Karachi
Kuala Lumpur Madras Madrid Melbourne
Mexico City Nairobi Paris Singapore
Taipei Tokyo Toronto

and associated companies in
Berlin Ibadan

Published by Oxford University Press, Inc.
198 Madison Avenue, New York, New York 10016

Library of Congress Cataloging-in-Publication Data
Managing strategic innovation and change : a collection of readings/
edited by Michael L. Tushman and Philip Anderson.
p. cm.
Includes bibliographical references and index.
ISBN 0-19-510010-7—ISBN 0-19-510011-5 (pbk.)
1. Industrial management.
2. Technological innovations.
3. Creative ability in business.
4. Organizational change.
I. Tushman, Michael.
II. Anderson, Philip, 1956– .
HD31.M29425 1996 658.4'06—dc20 96-2633

9 8 7 6 5 4 3 2 1
Printed in the United States of America
on acid-free paper

PREFACE

Why are we economically better off than our parents were? The economist Josef Schumpeter asked this question more than fifty years ago, and arrived at a powerful insight: the prime driver of economic progress is technological innovation. The economics of his day held perfect competition to the end of policy. Schumpeter pointed out, however, that people were better off in the 1940s than they had been at the turn of the century because of technical advances, not because more industries had come to approach the perfectly competitive ideal. Modern life is a triumph of innovation, not antitrust policy; the refrigerator, the air conditioner, the radio, the closed-body automobile, synthetic rubber, and a host of other breakthroughs created a better life and ultimately made possible a mass consumer society. As we approach the twenty-first century, Schumpeter's insight holds more firmly than ever. The goal of both societies and organizations should be fostering the fastest rate of innovation possible, since technical progress is a key factor elevating the economic well-being of a people.

Those innovations that shape our lives are created in organizations. Yet, managing innovation over time seems to be quite difficult. Suppose ten years ago you had made a list of the most successful companies in the world with the highest growth prospects. Your list today would look quite different; many stars have fallen, while many others have risen with astonishing speed. The common denominator in most cases is this: the most successful firms are those that have been able to systematically exploit innovation. In a few cases, historically successful firms have forged the future and reshaped their competitive environments. In most cases, however, those historically successful firms have not taken advantage of technological opportunities they themselves often invent. Rather, fundamentally new products and services are often commercialized by new firms—firms that are able to capture the value that technological breakthroughs create.

Managing innovation to create value is a complex, cross-functional, historically dependent endeavor. It is not the same as managing research and development (R&D), or "high-tech management," or new-product development, or implementing change. It certainly involves more than enhancing creativity; managers very often tell us that there already exist within their companies more great ideas than they know how to implement. The key themes of managing innovation are reflected in the way we have organized this book:

1. The kinds of innovation critical to success vary over time in a recurring cycle. Again and again, we find that industries go through long periods of incremental technological change, punctuated by occasional technological discontinuities, major breakthroughs that push forward the state of the art in an industry's core technologies by an order of magnitude. Each discontinuity inaugurates an era of ferment, a period of rapid technological change in which different designs often clash as a new technology supplants its predecessor. This culminates in a dominant design that evolves into the standard architecture expressing the original, crude breakthrough idea.

With the appearance of a standard, incremental change reigns again until the next discontinuity. The technical and competitive challenges that firms encounter differ greatly over this cycle, so there is no "one best way" to manage innovation at all times. Furthermore, history matters to the firm, not just to the industry. Faced with technological change, organizations often find that their previous successes are the very things that hinder adaptation. An innovative firm must evolve, but evolution takes place through a process of branching, where the new must spring from, yet ultimately break free of the old.

2. Management of innovation is an organizational problem. The architecture of an organization—its formal structure, its competencies, its job and career structure, its culture, and its power—determines its capacity to nurture, sustain, and exploit innovation. There is no one best way to organize, but absent an effective organizational weapon, brilliant ideas, good timing, and incisive strategies very seldom lead to successful innovation.
3. Technology is central to competitive strategy, although strategic management encompasses much more than just innovation. It is difficult to capture profits from innovation without an explicit strategy for exploiting the competitive openings that new technologies provide. Research and development create valuable options and opportunities, but these are converted to survival, growth, and profits only through strategic leadership and organization execution capabilities.
4. Research and development is often the heart of a firm's capacity for technological innovation. However, management of innovation is a cross-functional challenge. Firms only capture the full value of their technological advances when each of their functional areas brings its strength to bear in support of an innovation. Coordinating highly capable functional areas is difficult, but disciplinary strength both within and outside R&D is an essential element of successful innovation.
5. Successful innovators must excel at managing linkages and interfaces between organizations. Within the firm, effective cross-linking of strong functional areas must occur within the context of teams and overlapping organizing structures. The global enterprise must also effect strong internal linkages between country-based subunits. Vertical linkages up and down a hierarchy should allow innovation to bubble up from many sources, glued together by a strategic vision and coherence that only top management can provide. Managing the firm's intellectual capital is an exercise in linking the various sources of a firm's knowledge to the problems posed by the changing markets it serves.
6. Innovation is managed by leaders and teams with multiple competencies. Effective managers of innovation are able to develop organizational architectures that produce innovations, manage the organizational changes that always accompany innovation, and manage the dual requirements of operating in both the short and long terms simultaneously. Organizational architectures, strategies, cross-functional competencies, and linking systems are tools in the hands of executives. Innovation requires enterprises to manage the dual challenge of efficiency and adaptiveness, to create change without a debilitating degree of chaos. Visionary leaders and strong executive teams surmount these problems; structures and systems alone do not.

How can managers master this set of core issues? Experience alone is insufficient. It must be set against (and used to challenge) a conceptual took kit drawn from research that speaks to pragmatic problems while transcending individual cases. This book is a multidisciplinary approach to helping you build your inventory of useful mental models. We have drawn on a variety of disciplines and research programs worldwide. Such research-anchored models accelerate your learning, and enhance your ability to penetrate below the surface of things, to distinguish symptoms from problems and incidents from principles.

A strength of this book is that it draws from

a range of disciplines that contribute different, if converging insights. Your conceptual took kit must be broad, because innovation poses different demands at different points in an historical cycle. You cannot focus only on generating breakthrough innovations, lest you falter when the competitive environment shifts toward a focus on incremental, evolutionary advances. You cannot focus solely on "high technology," because many of the challenges you face require solutions that do not draw on leading-edge technical capabilities. Although innovation and entrepreneurship are often closely linked, innovation management is a key issue for small and large organizations, venture start-ups and firms with decades of tradition, services, and manufacturing enterprises.

A common thread is that most of the readings take a dynamic approach to thinking about processes that play out over time. We begin the book by putting forward a view of innovation that is sensitive to history, timing, cycles, and the constant tension between environmental pressures for stability and change. The introduction and overview section begins with our own view of how firms manage innovations over time, focusing on the notion of managing streams of innovations that play out across different eras in a technology cycle. The introduction continues with a look at the rise and fall of the Swiss watch industry, a rich case study that illustrates the impact of technology shifts on entire production networks cutting across individual firms.

Section II introduces you to a way of thinking about innovation as an historical, cyclical process. We introduce the notion of "punctuated equilibrium," a pattern of change in which long periods of routine evolution alternate with short bursts of rapid transformation. We then focus on how standards and dominant designs emerge, and explore when and how patient, incremental advance is the key to competitive success. This section concludes with a look at how new technologies substitute for old, against a backdrop of institutional resistance to change.

In Section III, we view managing innovation as a problem in creating organizational architectures to facilitate the development and implementation of innovations. Innovations are the products of organizational action, so it is necessary to understand how adaptive organizations are built and managed. We introduce a congruence model that lies at the root of our approach to managing innovation. This methodology has been used in helping managers around the world in diagnosing innovation problems and, in turn, managing innovation and change. Then, we examine how core competencies, career patterns, culture, power, and organizational structure all interact, jointly determining whether the organization can achieve and maintain a high degree of fit with a fast-changing environment. In keeping with our theme of processes that play out over time, we conclude by examining how congruence itself not only creates success, but creates a set of rigidities that can inhibit future innovation.

Section IV deals with the interface between technological innovation and strategy at the business unit level. The readings examine how technical capabilities and strategic advantages evolve over time, and how firms capture value from the innovations they pioneer. Technology strategy is operationalized by a firm's choice of innovation projects to pursue. The last reading in this section introduces the notion that we should value and rank these projects by treating them as "call options" on the future.

Section V shifts the spotlight to several roles three key functional areas—R&D, marketing, and operations—play in the innovation process. How do effective R&D organizations operate, and how should R&D be linked to strategy? How do market-driven organizations ensure that innovation efforts address customer needs while transcending the customer's own short-term vision of what he wants? How can organizations balance the need for efficiency and quality with the demands placed on operations by streams of innovation?

Section VI looks at the problem of weaving together organizational components to create an innovation engine that is more effective than the sum of its parts. We examine this problem at three different levels. First, we focus on effecting cross-functional linkages through managing teams and working the boundaries between functions. Second, we explore the connections between organizational units, as opposed to functional disciplines. These may be geographically separated divisions, levels in a hierarchy, or "network" forms of organization. Third, we investigate how successful firms form and administer ties with other organizations, either strategic partners or quasi-independent ventures spun out of the parent.

In Section VII, we investigate how top executives manage innovation and change. We look at the way effective leaders steer the organization through cycles of stability and reorientation, asking how top teams implement genuinely new organizational designs. We conclude by examining how chief executive officers at Xerox and Alcoa have communicated new visions and values, and how they have adapted their organizations' architectures to fast-changing environments. Both organizations have been informed by ideas and methods discussed in our book.

This book is designed to meet the needs of managers and future managers. We have built on the ideas in these readings to design MBA courses and executive programs on managing innovation at Columbia University, Dartmouth's Tuck School of Business, MIT, Cornell University, California Institute of Technology, INSEAD, and Chalmers University. We have also employed them in consulting engagements for a variety of firms around the globe, both large and small. We are particularly grateful for input and support from managers at Ciba-Geigy, Bristol Myers Squibb, BOC, Ericsson, Pfizer, Thompson, Grand Met, and the American Electronics Association. Managers and students looking for a broad, yet integrated approach should find that this book opens a window to a diverse set of ideas that are seldom brought together in one place. By combining both classic articles and recent, cutting-edge works, we have endeavored to provide a view of the subject that is at once timely, yet of enduring value.

ACKNOWLEDGMENTS

It would be impossible to thank everyone whose ideas and constructive criticism have contributed to this book. Both the Columbia University Graduate School of Business and the Amos Tuck School at Dartmouth College have generously underwritten our research and the teaching. The Innovation and Entrepreneurship and Strategy Research Centers at Columbia have provided direct support, for which we are grateful. Charles O'Reilly, Jeff Pfeffer, and David Nadler have been companions in building, extending, and employing these ideas on managing innovation and change. Without the imagination and hard work of Joy Glazener, this book would not exist, and Cynthia Wiegand's careful proofreading saved us from many an error. We also would like to express our appreciation to Herb Addison and Ken MacLeod at Oxford University Press for helping shepherd this work to completion. Over the years, we have worked with several thousand executives and MBAs, and we have learned a great deal from them. Their feedback has constantly challenged us to reorganize and rethink the way we present this subject. The present set of readings culminates a long process of learning by doing that only their patience and goodwill has made possible.

M. T., *New York*
P. A., *Hanover, N.H.*
June 1996

CONTENTS

Contributors, xiii

I Introduction and Overview, 1

1. Technology Cycles, Innovation Streams, and Ambidextrous Organizations: Organization Renewal Through Innovation Streams and Strategic Change, 3
   Michael L. Tushman, Philip C. Anderson, and Charles O'Reilly

2. Technological Discontinuities and Flexible Production Networks: The Case of Switzerland and the World Watch Industry, 24
   Amy Glasmeier

II Innovation over Time and in Historical Context, 43

Technology Cycles

3. Managing Through Cycles of Technological Change, 45
   Philip C. Anderson and Michael L. Tushman

4. Exploiting Opportunities for Technological Improvement in Organizations, 53
   Marcie J. Tyre and Wanda J. Orlikowski

Dominant Design

5. The Panda's Thumb of Technology, 68
   Stephen Jay Gould

6. Strategic Maneuvering and Mass-Market Dynamics: The Triumph of VHS over Beta, 75
   Michael A. Cusumano, Yiorgos Mylonadis, and Richard S. Rosenbloom

Era of Incremental Change

7. Managing Product Families: The Case of the Sony Walkman, 99
   Susan Sanderson and Mustafa Uzumeri

8. The Technology-Product Relationship: Early and Late Stages, 121
   Ralph Gomory

Technological Substitution and Innovation in Context

9. Gunfire at Sea: A Case Study of Innovation, 129
   Elting Morison

10. How Established Firms Respond to Threatening Technologies, 141
    Arnold C. Cooper and Clayton G. Smith

III Organization Architectures and Managing Innovation, 157

11. A Congruence Model for Organization Problem Solving, 159
    David A. Nadler and Michael L. Tushman

12. The Role of Core Competencies in the Corporation, 172
    C. K. Prahalad

13. Managing Professional Careers: The Influence of Job Longevity and Group Age, 183
    Ralph Katz

14. Using Culture for Strategic Advantage: Promoting Innovation Through Social Control, 200
    Charles O'Reilly III and Michael L. Tushman

15. Understanding Power in Organizations, 217
    Jeffrey Pfeffer

16. Dynamic Tension in Innovative, High Technology Firms: Managing Rapid Technological Change Through Organizational Structure, 233
    Claudia Bird Schoonhoven and Mariann Jelinek

17. Core Capabilities and Core Rigidities: A Paradox in Managing New Product Development, 255
    Dorothy Leonard-Barton

IV Technology and Business Strategy, 271

18. Technology Strategy: An Evolutionary Process Perspective, 273
    Robert A. Burgelman and Richard S. Rosenbloom

19. Capturing Value from Technological Innovation: Integration, Strategic Partnering, and Licensing Decisions, 287
    David J. Teece

20. Managing R&D as a Strategic Option, 307
    Graham R. Mitchell and William F. Hamilton

V Managing Functional Competencies, 317

Research and Development

21. Tech Talk: How Managers Are Stimulating Global R&D Communication, 319
Arnoud De Meyer

22. Japan's Management of Global Innovation: Technology Management Crossing Borders, 331
Kiyonori Sakakibara and D. Eleanor Westney

23. Research That Reinvents the Corporation, 342
John Seely Brown

Marketing

24. Marketing and Discontinuous Innovation: The Probe and Learn Process, 353
Gary S. Lynn, Joseph G. Morone, and Albert S. Paulson

25. Developing New Product Concepts via the Lead User Method: A Case Study in a "Low-Tech" Field, 376
Cornelius Herstatt and Eric von Hippel

Operations/Manufacturing

26. Managing Flexible Automation, 385
Paul S. Adler

27. A Perspective on Quality Activities in American Firms, 402
Noriaki Kano

VI Managing Linkages, 417

Cross-Functional Linkages

28. Organizing and Leading "Heavyweight" Development Teams, 419
Kim B. Clark and Steven C. Wheelwright

29. Making Teamwork Work: Boundary Management in Product Development Teams, 433
Deborah Gladstein Ancona and David F. Caldwell

30. Redundant, Overlapping Organization: A Japanese Approach to Managing the Innovation Process, 443
Ikujiro Nonaka

Organizational Linkages

31. Managing Innovation in the Transnational Corporation, 452
Christopher A. Bartlett and Sumantra Ghoshal

32. The Logic of Global Business: An Interview with ABB's Percy Barnevik, 477
William Taylor

33. Toward Middle-Up-Down Management: Accelerating Information Creation, 494
Ikujiro Nonaka

34. Managing Intellect, 506
James Brian Quinn, Philip Anderson, and Sydney Finkelstein

Extra-Organizational Linkages and Venturing

35. Engines of Progress: Designing and Running Entrepreneurial Vehicles in Established Companies; Raytheon's New Product Center, 1969–1989, 524
Rosabeth Moss Kanter, Jeffrey North, Lisa Richardson, Cynthia Ingols, and Joseph Zolner

36. Entering New Businesses: Selecting Strategies for Success, 541
Edward B. Roberts and Charles A. Berry

37. The Use of Alliances in Implementing Technology Strategies, 556
Yves Doz and Gary Hamel

VII Executive Leadership and Managing Innovation and Change, 581

38. Convergence and Upheaval: Managing the Unsteady Pace of Organizational Evolution, 583
Michael L. Tushman, William H. Newman, and Elaine Romanelli

39. Implementing New Designs: Managing Organizational Change, 595
David A. Nadler and Michael L. Tushman

40. Vision, Values, Milestones: Paul O'Neill Starts Total Quality at Alcoa, 607
Peter J. Kolesar

41. The CEO as Organizational Architect: An Interview with Xerox's Paul Allaire, 631
Robert Howard

Index, 643

CONTRIBUTORS

Paul S. Adler	Graduate School of Management, USC
Deborah Gladstein-Ancona	Sloan School of Management, MIT
Philip Anderson	Amos Tuck School, Dartmouth College
Christopher A. Bartlett	Graduate School of Business, Harvard University
Charles A. Berry	Pilkington Bros., PLC
John Seely Brown	Palo Alto Research Center, Xerox
Robert A. Burgelman	Graduate School of Business, Stanford University
David F. Caldwell	School of Business, Univ. of California-Santa Clara
Kim B. Clark	Graduate School of Business, Harvard University
Arnold C. Cooper	Krannert School of Management, Purdue University
Michael A. Cusumano	Sloan School of Management, MIT
Arnoud De Meyer	Insead
Yves Doz	Insead
Sydney Finkelstein	Amos Tuck School, Dartmouth College
Sumantra Ghoshal	London Business School
Amy Glasmeier	Penn State University, Dept. of Geography
Ralph Gomory	Sloan Foundation

Stephen Jay Gould	Harvard University
Gary Hamel	London Business School
William F. Hamilton	Wharton School
Cornelius Herstatt	Arthur D. Little Company, Zurich
Robert Howard	Boston Consulting Group
Cynthia Ingols	Management Consulting Firm-Corporate Classrooms, Cambridge, MA
Mariann Jelinek	William and Mary College
Noriaki Kano	Faculty of Engineering, Dept. of Management Science, Tokyo Rika Daigaku
Rosabeth Moss Kanter	Graduate School of Business, Harvard University
Ralph Katz	School of Management, Northeastern University
Peter J. Kolesar	Graduate School of Business, Columbia University
Dorothy Leonard-Barton	Graduate School of Business, Harvard University
Gary S. Lynn	Department of Management and Engineering Management, Stevens Institute of Technology
Graham R. Mitchell	GTE Corporation
Elting Morison	MIT, deceased
Joseph G. Morone	Rensselaer Polytechnic Institute
Yiorgos Mylonadis	University of Pennsylvania
David A. Nadler	Delta Consulting Group
William H. Newman	Graduate School of Business, Columbia University
Ikujiro Nonaka	Hito Tsubashi University
Jeffrey North	Staples, Inc., Framingham, MA
Charles O'Reilly	Graduate School of Business, Stanford University

WANDA J. ORLIKOWSKI	Sloan School of Management, MIT
ALBERT S. PAULSON	Rensselaer Polytechnic Institute
JEFFREY PFEFFER	Graduate School of Business, Stanford University
C.K. PRAHALAD	School of Business, University of Michigan
JAMES BRIAN QUINN	Amos Tuck School, Dartmouth College
LISA RICHARDSON	Bain
EDWARD B. ROBERTS	Sloan School of Management, MIT
ELAINE ROMANELLI	School of Management, Georgetown University
RICHARD S. ROSENBLOOM	Graduate School of Business, Harvard University
KIYONORI SAKAKIBARA	London Business School, England
SUSAN SANDERSON	School of Management, Rensselaer Polytechnic Institute
CLAUDIA BIRD SCHOONHOVEN	Amos Tuck School, Dartmouth College
CLAYTON G. SMITH	University of Notre Dame
WILLIAM TAYLOR	Fast Company, Boston, MA
DAVID J. TEECE	HAAS School of Business, Berkeley, CA
MICHAEL TUSHMAN	Graduate School of Business, Columbia University
MARCIE J. TYRE	Sloan School of Management, MIT
MUSTAFA UZUMERI	Auburn University
ERIC VONHIPPEL	Sloan School of Management, MIT
D. ELEANOR WESTNEY	Sloan School of Management, MIT
STEVEN C. WHEELWRIGHT	Graduate School of Business, Harvard University
JOSEPH ZOLNER	Harvard Graduate School of Education

CHAPTER ONE

INTRODUCTION AND OVERVIEW

Managing innovation would be relatively easy if it were the executive's only concern. It is difficult because managers must juggle conflicting forces. On the one hand, firms that do not adapt and innovate often fail. On the other hand, being adaptive may create inefficiency, threatening survival in highly competitive environments. Usually, this dual challenge requires an organization to embrace inconsistent internal divisions, some concerned with short-term needs and others focused on long-term requirements. Furthermore, this balancing act takes place in a shifting environment, where sometimes radical innovation prevails, and at other times incremental change is the dominant theme.

This first section introduces the concept of innovation streams, patterns of innovation that respond to these conflicting pressures. Managing innovation streams allows the organization to surmount the different challenges posed at different points of an industry's passage through technology cycles. The second reading illustrates these themes of cyclical change and flexible/efficient organization through a careful study of the world's watch manufacturing industry.

1

Technology Cycles, Innovation Streams, and Ambidextrous Organizations: Organization Renewal Through Innovation Streams and Strategic Change

MICHAEL L. TUSHMAN
PHILIP C. ANDERSON
CHARLES O'REILLY

THE CHALLENGE

Innovation and new product development are crucial sources of competitive advantage. After all the cost-cutting, downsizing and re-engineering, innovation and product development are levers through which firms can reinvent themselves. For example, Noika, Intel, Sony, Seiko, Corning, and Motorola have all generated sustained competitive advantage through continuous streams of incremental, architectural (the combination or linkage of existing technology in novel ways), and discontinuous innovation. More generally, across a variety of industries, these superior firms realized 49 percent of their revenues from products developed over the prior five years versus only 11 percent for mediocre performers (Deschamps and Nayak, 1995). There is a burgeoning literature on the competitive importance of innovation and the linkage between innovation and organization renewal across industries and countries (Schoonhoven et al., 1990; Morone, 1993; Hamel and Prahalad, 1994; Burgelman, 1994; Brown and Eisenhardt, 1995b; Henderson and Clark, 1990; Rosenbloom and Christensen, 1994; Utterback, 1994; Jellinek and Schoonhoven, 1990; and Tushman and Anderson, 1986).

Yet with all the attention paid to the importance of innovation and new product development, a curious puzzle should concern reflective managers. Technology and resource-rich firms often fail to compete in the very technologically turbulent environments that they helped create. Consider SSIH, the Swiss watch consortium, and Oticon, the Danish hearing aid firm. Both organizations dominated their respective worldwide markets (SSIH through the 1970s and Oticon through the 1980s), and both developed new technologies that had the capabilities to re-create their markets (e.g., quartz movements and in-the-ear [ITE] volume and tone control). Though SSIH and Oticon had the technology and the resources to innovate, it was smaller, more aggressive firms that initiated new technology in watches and hearing aids. SSIH and Oticon prospered until new industry standards (what we will call dominant designs) destroyed both SSIH's and Oticon's market positions in a matter of a few years.

In both the emergence of quartz watches and ITE hearing aids as dominant designs, it was not new technology that led to the demise of the Swiss or the Danes—indeed, SSIH and Oticon were the technology leaders. Nor was the sudden loss in market share due to lack of financial resources or governmental regulations. Rather, the sudden demise of SSIH and the huge losses at Oticon were rooted in organization complacency and inertia. These pathologies of sustained success led to SSIH's and Oticon's inability to renew themselves through proactively initiating streams of innovation. Innovation streams are patterns of innovations that simultaneously build on and extend prior products (e.g., mechanical watches, behind-the-ear [BTE] hearing aids) *and* destroy those very products that account for a firm's historical success (e.g., quartz watches, ITE hearing aids). These innovation streams shape and reshape markets.

SSIH and Oticon are not unique. The stultifying, innovation-numbing effects of success are a global phenomenon. Consider the recent economic performance of firms listed in Figure 1.1. Each firm on the list, whether American, French, Japanese, Dutch, German, or Swiss, dominated their respective market for years, only to be rocked by economic crisis as their markets shifted. This paradoxical pattern in which winners, with all their competencies and assets, often become losers is found across industries and countries (e.g., Rosenbloom and Christensen, 1994; Henderson and Clark, 1990). It seems that building core competencies and managing through continuous improvement are not sufficient for sustained competitive advantage. Worse, under a remarkably frequent set of conditions, building on core competencies (e.g., the Swiss building on precision mechanics) and engaging in continuous, incremental improvement actually traps the organization in its distinguished past and leads to catastrophic failure as technologies and, in turn, markets shift. Core competencies often turn into core rigidities (Barton, 1992).

Every firm in Figure 1.1, like SSIH or Oticon, was out-innovated by more nimble, foresightful (though not more resource-rich) competitors. This success paradox, these competence-based pathologies, is driven by historically constrained managerial action (often inaction) and organizational processes in the context of changing technological and market opportunities, not by the invisible hand of the market nor by governmental or public policy. Thus the troubles at General Motors, IBM, Philips, or Siemens were not rooted in public policy issues, but rather by these firms' inability to either take advantage of technological opportunities, or to directly shape the nature of technological change in their respective markets (like the Swiss and Oticon). These firms were unable to build and extend their existing competencies to develop innovations that would create new markets (as Starkey did with ITE hearing aids) and/or rewrite the competitive rules in existing markets (as Seiko did with quartz watches).

IBM	Syntex	Smith Corona
Kodak	Philips	Siemans
Sears	Volkswagen	Fuji Xerox
General Motors	SSIH	Zenith TV
	Oticon	

Fig. 1.1. The paradox of success.

This success paradox is not deterministic; core competencies need not become core rigidities. Some organizations are capable of moving from strength to strength, extending and/or replacing competencies and proactively moving to shift bases of competition through streams of innovation. These firms are able to develop incremental innovation as well as innovations that alter industry standards, substitute for existing products, and/or reconfigure products to fundamentally different markets. For example, in the watch industry, Seiko was able to not only compete in mechanical watches (the historically dominant technology), but also was willing to experiment with the quartz and tuning fork movements. With this technological experimentation, Seiko managers made the bold, proactive decision to substitute their quartz watch for their existing mechanical watches. The switch to quartz movements led to fundamentally different competitive rules in the watch industry. Similarly, Starkey (the U.S. hearing aid company) was able to move beyond BTE hearing aids to ITE hearing aids by simply reconfiguring existing hearing aid components. This seemingly minor innovation led to a new industry standard (as ITE designs substituted for BTE designs) and different industry rules (focusing on sound quality and fashion).

Seiko and Starkey are not unique. Nokia, Sony, HP, Ericsson, and Motorola are all firms that have been able to shift from today's strength to tomorrow's strength (e.g., Barton, 1992; Jellinek and Schoonhoven, 1990; Foster, 1987; Morone, 1993; Brown and Eisenhardt, 1995b). These organizations are able to develop and migrate competencies such that old competencies provide a platform for building new, often fundamentally different, competencies. Those firms that are able to sustain competitive advantage over time are able to shape technology cycles through creating streams of innovation. These streams include incremental, competence-enhancing, innovation (e.g., thinner mechanical watches), architectural or configurational innovation (e.g., Starkey's ITE hearing aid), as well as fundamentally new, competence-destroying, innovation (e.g., Seiko's quartz movement substituting for mechanical movements). By actively managing streams of innovation, senior teams increase the probability that their firm will be able to shape industry standards (that provide the assumptions around which incremental innovation occurs), take advantage of fundamentally new markets for existing technology, and proactively introduce substitute products which, as they cannibalize existing products, create new markets and competitive rules (see also Hurst, 1995; Hamel and Prahalad, 1994; Utterback, 1994).

OUR CORE IDEAS

We build on research and practice to develop ways out of these competency traps. We provide concepts and tools in helping managers manage for today's innovation requirements even as they simultaneously build both technical and organizational competencies to develop innovation that will shape tomorrow's competitive requirements. While managers cannot know in advance what innovations will achieve market success, in managing simultaneously for today and tomorrow, managers build organizations that are systematically more "lucky" than the competition. We introduce our core ideas here and provide detail in the next section.

Building on the technology life cycle literature, we introduce the idea of *innovation streams*—patterns of innovation that are required for sustained competitive advantage. Innovation streams focus attention away from innovations in isolation, toward patterns of fundamentally different innovation as a market unfolds. Further, a focus on innovation streams calls attention to products as made up of a set of subsystems, each of which has its own innovation stream. Thus watches are made up of a set of subsystems (e.g., energy source, oscillation device, face, etc.), each of which was transformed between 1970 and 1984. Sustained competitive advantage is gained from managing innovation streams for each subsystem (Abernathy and Clark, 1985; Tushman and Rosenkopf, 1992).

Innovation streams are driven by shifts in the underlying *technology cycle*. Technology cycles call managerial attention to the importance of industry standards (e.g., the quartz movement, Windows operating system, the QWERTY keyboard) as well as technological discontinuities (e.g., batteries replacing springs as the energy source in watches) as critical junctures in a market's evolution. Dominant designs and technological discontinuities are demarcation points between fundamentally different competitive arenas. Dominant designs and technological discontinuities are not technically determined; rather, they are windows of competitive opportunity where managerial action can shape market rules and subsequent innovation patterns (e.g., Tushman and Anderson, 1986; Anderson and Tushman, 1990; Utterback, 1994; Baum and Korn, 1995).

If sustained competitive advantage is built by simultaneously creating fundamentally different types of innovation (e.g., mechanical *and* quartz watches at Seiko, BTE, *and* ITE hearing aids at Starkey), managers must be capable of building and managing organizations that can simultaneously create both incremental and discontinuous innovation. *Ambidextrous organizations* have multiple organization architectures to concurrently nurture these diverse innovation requirements. We link innovation streams to multiple, inconsistent, internally contradictory organization architectures that must co-exist within a single organization (i.e., at Seiko, the development of the quartz watch was done in a different facility, with different managers, roles, structures, and culture than the mechanical watch facility). Such ambidextrous organizations are capable of operating simultaneously for the short and long term, for today and tomorrow, for both incremental and discontinuous innovation. Such dual organizations build in the experimentation, improvisation, and luck associated with small organizations, along with the efficiency, consistency, and reliability associated with larger organizations (e.g., Eisenhardt and Tabrizi, 1995; Imai et al., 1985).

Ambidextrous organizations are, however, inherently unstable. Most often, the older, larger, more traditional units sabotage those more entrepreneurial units—usually today's efficiency (and incremental innovation) kills tomorrow's architectural (e.g., Oticon's inability to initiate the ITE hearing aid) and/or discontinuous innovation (e.g., SSIH's inability to initiate the quartz movement). Given these dynamics, senior management teams must both create the *internal diversity* associated with ambidextrous organizations as well as *provide balance* between the needs of today's innovation demands with that of tomorrow's innovation possibilities.

In such internally inconsistent organizations, the senior management team must provide clear roles and responsibilities for the contrasting units and be a force for integration and balance. A clear, common, shared competitive vision is a powerful tool to bind internally inconsistent units together (e.g., Morone, 1993; Collins and Porras, 1994). Without experimental, entrepreneurial units there can be no future, yet without efficient, consistent units there can be no present. These dualities bring internal conflict and politics. These necessary tensions can only be managed by a senior team that can both build ambidextrous organizations and can articulate a clear, common vision within which these tensions make sense.

Ambidextrous organizations create the intra-organizational diversity that provides senior management with data, insight, and opportunity to take strategic action to proactively shape technology cycles and, in turn, market rules. For example, internal experimentation at Seiko gave senior managers the knowledge to make a choice between further funding mechanical watches and/or shifting to the quartz (or tuning fork) movements. At critical junctures in a market's evolution, particularly at the shaping of industry standards and at technological discontinuities, internal experimentation provides senior managers with alternatives from which those managers can act to proactively shape industry standards (as Seiko and Starkey did), create new markets (e.g., Starkey's move into the fashion market), and/or initiate product substitutes (e.g., Seiko's substituting quartz for mechanical watches).

But such strategic decisions *must be coupled with strategic organization change*. Managers can only rewrite industry rules if they are also willing to rewrite their organization's rules. For example, the proactive innovation moves at both Seiko and Starkey were executed by sweeping changes in strategy, structure, competencies, and processes. Where Seiko and Starkey initiated proactive strategic change coupled with their strategic innovation, SSIH and Oticon were forced to make much more costly and traumatic reactive strategic change.

Proactive strategic innovation and change and proactive moves to rewrite market rules always face entrenched resistance from the status quo. Such consequential decisions are often clouded by the firm's prior success or are made proactively but poorly implemented. Strategic innovation always ruptures stable political equilibria, embedded competencies, cultures, and organization processes. Thus managing innovation streams requires not only building ambidextrous organizations, but also organizations capable of *executing systemwide change,* even in the face of prior success. Managing strategic innovation is as much about managing change as it is about managing technology.

We discuss, then, developing streams of innovation, building ambidextrous organizations, the role of the senior management team in building and integrating this diversity, and senior management's role in managing large system change associated with strategic innovation. These are all crucial competencies for sustained competitive advantage; for building from today's to tomorrow's competitive strength.

TECHNOLOGY CYCLES AND INNOVATION STREAMS

Technology Cycles. Much is written on the relative importance of incremental versus discontinuous innovation and on the relative importance of market pull versus technology push innovation (e.g., Morone, 1993; Deschamps and Nayak, 1995; Gomory, 1989). In competitive, technology-intensive global markets, competitive advantage can only be built through a combination of different types of innovation—through the creation of not only product substitutes, but also architectural innovation as well as through continuous, incremental innovation (Abernathy and Clark, 1985; Iansiti and Clark, 1994; Brown and Eisenhardt, 1995a; Sanderson and Uzumeri, 1995). It is the ability to produce streams of different kinds of innovation that drives sustained competitive advantage. To better understand the structure and flow of innovation streams, we need to understand technology cycles. Clarifying technology cycles helps untangle the relative timing and importance of incremental, architectural, and discontinuous innovation.

Technology cycles are composed of technological discontinuities (i.e., quartz and tuning fork movements in watches) that trigger periods of technological and competitive ferment. These turbulent innovation periods are closed with the emergence of an industry standard or dominant design (e.g., quartz movement dominated both the mechanical escapement and tuning fork mechanisms) (Tushman and Anderson, 1986; Anderson and Tushman, 1990; and Utterback, 1994). For example, in early radio transmission, continuous wave transmission was a technological discontinuity that threatened to replace spark-gap transmission. Continuous-wave transmission initiated competition not only between continuous-wave transmission and spark-gap transmission, but also between three variants of continuous wave transmission (alternating wave, arc, and vacuum tube transmission) (Aitken, 1985). This period of technological ferment led to vacuum tube transmission as the dominant design in radio transmission (Rosenkopf and Tushman, 1994). The emergence of a dominant design ushers in a period of incremental as well as architectural technological change, which at some point, is broken by the next substitute product (i.e., electronic typewriters replaced electric typewriters, which had previously replaced mechanical typewriters). This subsequent technological discontinuity then triggers the next wave of technological variation, selection, and retention (see Figure 1.2).

Technology cycles are seen most directly in simple, or nonassembled products (e.g., skis, tennis racquets, glass, chemicals). For example, in crop fungicides, Ciba-Geigy's propiconazol (Tilt) was a new chemical entity that challenged Bayer's and BASF's products. Propiconazol triggered competition between chemical entities as well as between a vast number of formulations within propiconazol. Ciba eventually created its EC 250 version, which became the industry standard in crop fungicides. More recently, Ciba Crop Protection division has initiated several product substitutes (including genetically engineered seeds) to cannibalize and replace propiconazol. These fundamentally new crop protection products will initiate the next technology cycle in the crop protection market.

In more complex assembled products (e.g., watches) and systems (e.g., radio, voice mail), these technology cycles apply at the subsystem level. For example, watches are assembled products made up

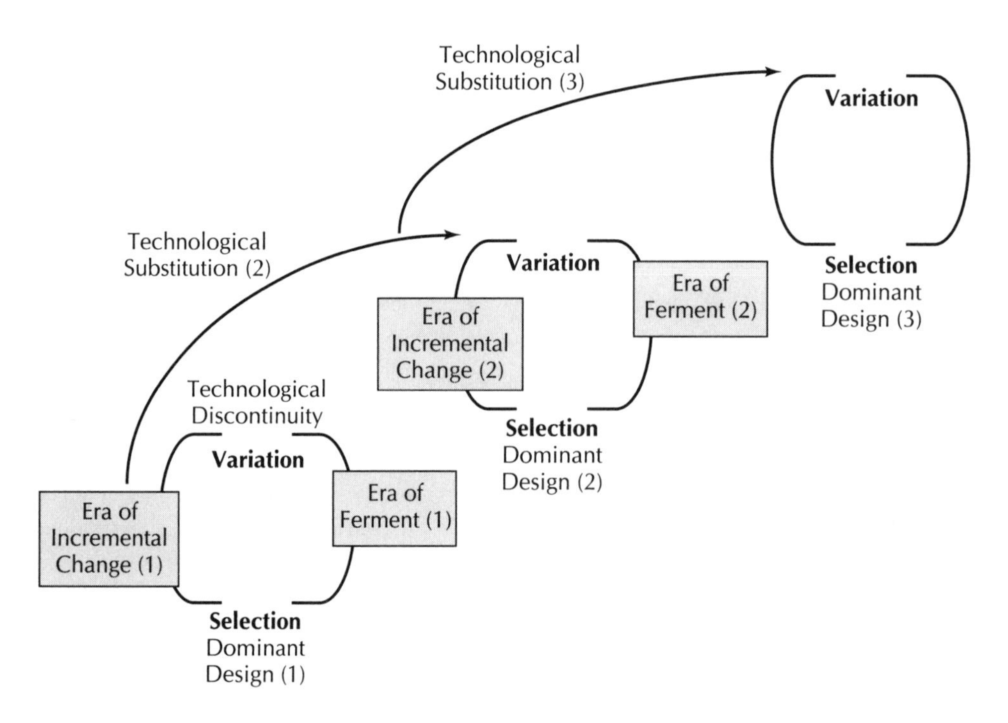

Fig. 1.2. Technology cycles over time.

of at least four subsystems: energy source, oscillation device, transmission, and display. Each of these subsystems has its own technology cycle. For example, the pin-lever escapement became the dominant design in watch oscillation in the late nineteenth century. Escapements became better and better through incremental changes in the same fundamental design until the late 1960s. Between 1968 and 1972, escapements were threatened by both tuning fork and quartz oscillation. This period of technological competition between escapements, tuning fork and quartz movement, ended with the emergence of quartz oscillation as the dominant design in the oscillating subsystem (see Landes, 1983). As with mechanical escapements, the emergence of quartz movements as the dominant design led, in turn, to numerous incremental improvements in the quartz movement and sharp decreases in innovation in tuning fork and escapement oscillation devices.

Note that in the watch industry, between 1970 and 1985, every subsystem of the watch, from energy source to face, was transformed through its own technology cycle of technical variation, selection of a dominant design, and subsequent period of incremental change. While watches are a unique product, they do provide a generic illustration of complex products as made up of interconnected subsystems, each of which has its own technology cycle. Further, each subsystem shifts in relative strategic importance as the industry evolves. For example, where oscillation was the key strategic battlefield through the early 1970s, once the quartz movement became the dominant design, the locus of strategic innovation shifted to the face, energy, and transmission subsystems. Similar dynamics of subsystem and linkage technology cycles have been documented in a variety of industries (see Tushman and Rosenkopf, 1992; Henderson and Clark, 1990; Henderson, 1995; Noble, 1984; Hughes, 1983; Van de Ven and Garud, 1994; David, 1985; Morone, 1993).

Technology cycles are made up of alternating periods of technological variation, competitive selection, and retention (or convergence). *Technological discontinuities* initiate technology cycles; they are rare, unpredictable events triggered by scientific advance (e.g., battery technology for watches) or through the unique combination of existing technology (e.g., Sony's Walkman, or continuous aim gunfire). Technological discontinuities rupture existing incremental innovation patterns and spawn a period of technological ferment in which competing technological variants vie for market acceptance (e.g., the competition between tuning fork, quartz, and escapement oscillation, or the competition between MAC, UNIX, OS/2, and Windows operating systems). Such periods of technological ferment involve competition between existing versus new technology (e.g., escapement versus tuning fork and quartz; AC versus DC power systems) as well as competition between variants of the new technology (e.g., 60, 120, 240 cycles AC power). These eras of ferment are confusing, uncertain, and costly to customers, suppliers, vendors, and regulatory agencies (see Figure 1.3).

From such periods of variation, or eras of technological ferment, a *single dominant design* emerges. For example, numerical control dominated record playback in numerical control machine tools, 60 cycle AC power dominated 120 cycle power (in the United States), fan beam dominated pencil beam in CT scanning, high strength magnetic fields dominated low strength fields in MRI imaging, VHS format dominated Beta in VCR, and Windows format dominated MAC, UNIX, and OS/2 in operating systems (Noble, 1984; Morone, 1993; Hughes, 1983; Cusumano et al., 1992). Absent strong patents, competing firms switch to the new standard or risk getting locked out of the market (see Figure 1.4).

How do dominant designs emerge? Except for the most simple, nonassembled products (e.g., cement), the closing on a dominant design is not technologically driven. Rather, dominant designs emerge out of competition between alternative technological trajectories initiated and pushed by competitors, alliance groups, and governmental regulators, each with their own political, social, and economic agendas. Because no technology can dominate on all possible dimensions of merit, the closing on a dominant design takes place not through the invisible hand of the market and not through natural selection, but through competi-

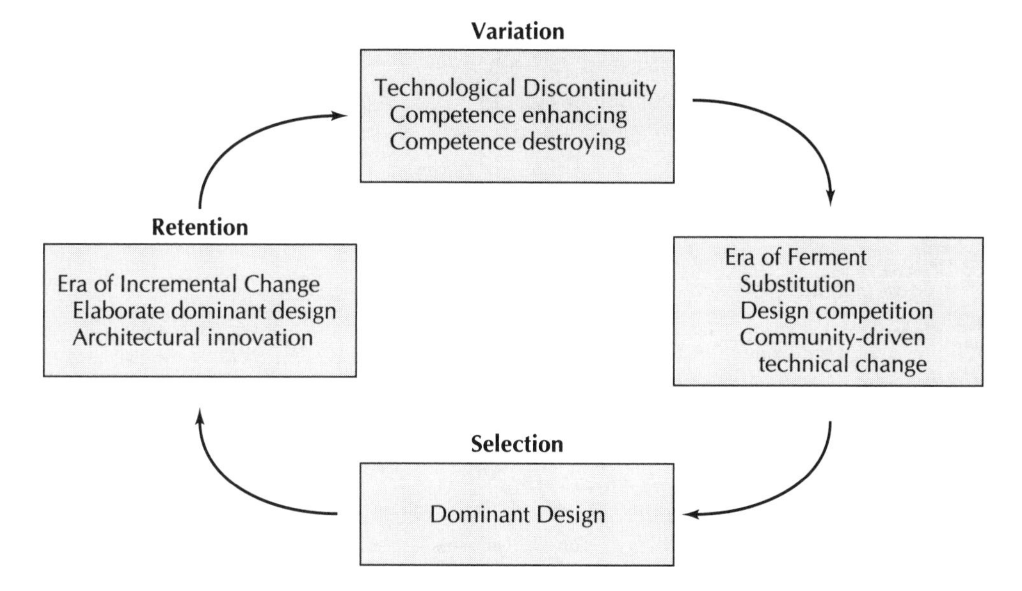

Fig. 1.3. A technology cycle.

- Model T
- AC Power System
- DC-3
- IBM 360
- Smith Model 5
- CMA
- VHS, 1/2 inch
- Windows (operating system)
- Quartz movement
- Watches

Dominant designs are driven by technological, market, and regulatory forces

Windows of opportunity (competitive selection)

Fig. 1.4. Examples of dominant designs.

tive/political competition between the alternative technological variants (Henderson, 1995; Hughes, 1983; Noble, 1984; Baum and Korn, 1995; Tushman and Rosenkopf, 1992).

Dominant designs are watershed events in a technology cycle. Where before the dominant design technology progress is driven by competition between alternative technological trajectories, after the dominant design subsequent technological change is driven by the logic of the selected technology itself (see Figure 1.3). The closing on a dominant design shifts product innovation from major product variation to major process innovation and, in turn, to incremental innovation—building, extending, and continuously improving the dominant design (Abernathy and Utterback, 1978). These periods of incremental innovation lead to profound advances in the now standard product (Hollander, 1965; Florida and Kenney, 1990; Tushman and Anderson, 1986). In contrast, the consequences of betting on the "wrong" design are devastating—particularly if that subsystem is a core subsystem (e.g., IBM's losing control of the microprocessor and operating system of PCs to Intel and Microsoft, respectively).

The emergence of a dominant design also permits the development of product platforms and families (Morone, 1993; Sanderson and Uzumeri, 1994; Meyer and Utterback, 1993). A product family is a set of products built from the same fundamental technology platform. These product families share traits, architecture, components, and interface standards. For example, once Sony closed on the WM-20 platform for its Walkman, it was then able to generate more than thirty incremental versions within the same family. Indeed, over a ten-year period, Sony was able to develop four Walkman product families and over 160 incremental versions of those four fundamental families. Such sustained attention to technological discontinuities at the subsystem level (e.g., flat motor and miniature battery), closing on a few standard platforms, and incremental product proliferation helped Sony control industry standards in its product class and outperform Japanese, American, and European competitors (Sanderson and Uzumeri, 1994).

These periods of incremental innovation are, in turn, broken by subsequent technological discontinuities. *Subsequent product or process technological discontinuities* trigger the next cycle of technological variation, selection, and incremental change (see Figures 1.2 and 1.5). For example, in Ciba's Crop Protection Division, Wolfgang Samo hosted the development of fundamentally new chemical fungicides (competence enhancing), as well as biologically engineered seeds (competence destroying). Samo's point of view was that if Ciba did not substitute for its own successful product line, BASF or Bayer would. Creating substitute products and/or processes and cannibalizing one's product line prior to the competition is an impor-

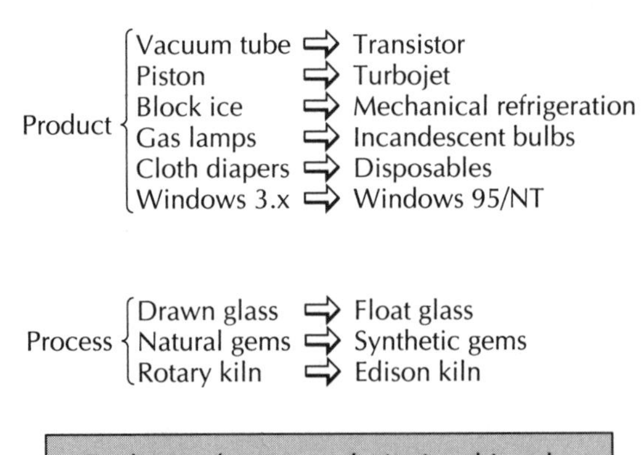

Fig. 1.5. Product process substitution.

tant source of competitive advantage (Foster, 1986; Iansiti and Clark, 1994; and Morone, 1993).

Finally, architectural innovations are those innovations that affect how a given set of core subsystems are linked together (Henderson and Clark, 1990). These innovations (i.e., smaller fans, watches, hearing aids, or disk drives) reconfigure the same core technology and take the reconfigured product to fundamentally different markets (e.g., Starkey's move into the fashion hearing aid market, Honda's early move to smaller motorcycles, or the migration of disk drive technology from mainframes to personal computers). While architectural innovation may be technologically simple and associated with substantial economic returns, these shifts are frequently missed by incumbent firms (Henderson and Clark, 1990; Henderson, 1995; Rosenbloom and Christensen, 1994).

Technology cycles are composed of periods of technological ferment, the closing on a dominant design, leading to a period of incremental innovation, which is, in turn, broken by the next technological discontinuity and subsequent technology cycle and/or by the next architectural innovation. For example, in typewriters, the Smith Model 5, the IBM Selectric, and word processing reflect the march of dominant designs over an eighty year period (Utterback, 1994). These cycles apply for both product subsystems as well as for linking technologies. These patterns hold across product classes—the only difference between high (e.g., minicomputer and low (e.g., cement) technological industries is the length of time between the closing on a standard and the subsequent discontinuity. Thus from an innovation perspective, what lies behind the familiar S-shaped product life cycle curve are fundamentally different innovation requirements. Eras of ferment require discontinuous product variants and dominant designs are associated with fundamental process innovation, while eras of incremental change are associated with streams of incremental and architectural innovation (Utterback, 1994; Anderson and Tushman, 1994) (see Figure 1.2).

Innovation Streams. Given the nature of technology cycles, the roots of sustained competitive advantage lie in a firm's ability to proactively initiate incremental, architectural, as well as discontinuous innovation. Management must develop the diverse competencies and organizational capabilities to shape and take advantage of dominant designs (e.g., Windows versus OS/2), shape architectural innovation, as well as introduce substitute products prior to the competition (e.g., Windows 95 versus IBM's WARP). Thus the route to sustained competitive advantage is not through either incremental (market driven) innovation or through discontinuous (technology driven) innovation, but through the capabilities of producing streams of innovation over time (see Figure 1.6).

Innovation streams have very different characteristics over time and require diverse and shifting technological, marketing and manufacturing, and organizational capabilities. Opening a new product class (e.g., crop protection products, CT scanners, or portable phones) is the most turbulent, whitewater part of the innovation stream. This era of ferment is associated with major product variation as producers offer their variants to new markets (e.g., multiple early crop protection agents, multiple CT variants, and multiple portable phone variants). These multiple variants are produced within firms as well as between firms (e.g., Ciba versus BASF; Motorola versus Ericsson; and EMI versus GE) (see Figure 1.6).

This turbulent, uncertain period is closed with the emergence of a dominant design. The closing on a dominant design is associated with the innovation stream shifting away from product variation toward major process innovation. Dominant designs result from firms betting on a product variant and then initiating major process innovation to drive down cost (e.g., Motorola's 1983 commercial portable phone, IBM's 360 design; GE's fan beam CT technology; or Ciba's bet on EC-250. The success of these bets can only be in retrospect.

After the closing on an industry standard, the innovation stream is more stable; there ensues a relatively long period of incremental and/or generational innovation (e.g., Sony's four Walkman generations and 160 incremental innovations) (see Figure 1.6). Periods of incremental innovation lead,

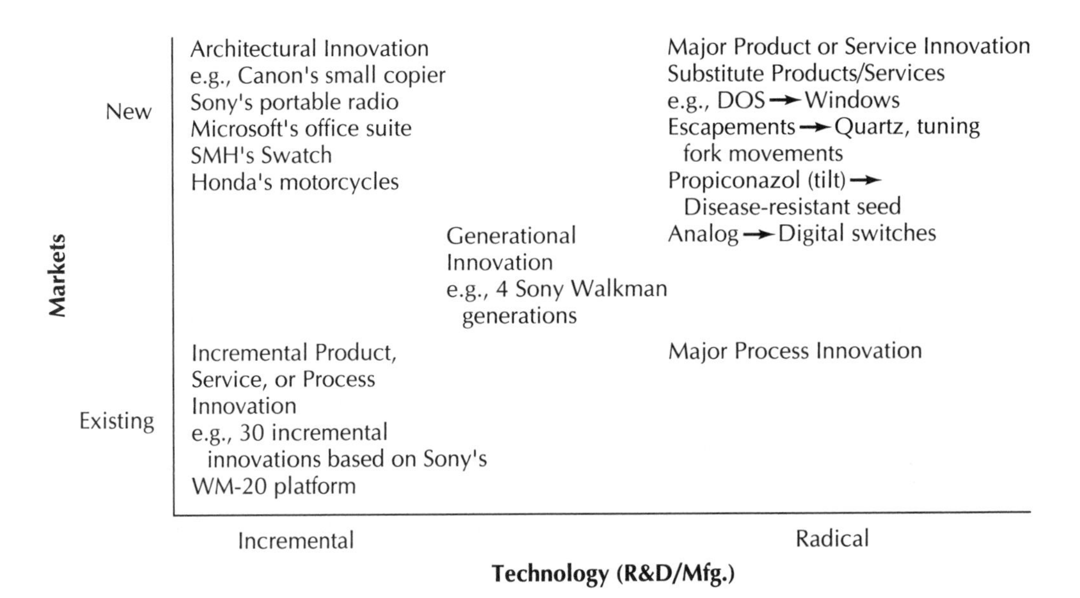

Fig. 1.6. Innovation streams.

in turn, to either architectural innovation, taking the standard product to fundamentally new markets (e.g., Starkey taking its ITE hearing aids to more fashion-conscious European markets) and/or to developing product substitutes (e.g., Seiko developing the quartz and tuning fork movements or Ciba developing fungicide-resistant seeds). Product substitutes initiate the next cycle of technological variation, dominant design, and era of incremental change.

Innovation streams direct managerial attention away from a particular innovation or a particular innovation orientation (e.g., market push versus technology pull; incremental versus discontinuous), toward streams of contrasting innovations that must be produced within a business unit over time (e.g., within Ciba's Crop Protection division). Innovation streams call attention to maintaining control over core product subsystems, and to dominant designs, architectural innovation, and product substitution as windows of opportunity where proactive managerial action can shape technological evolution and, in turn, bases of competition. Finally, through such proactive management of technology cycles and innovation streams managers have the opportunity to build on mature technologies (e.g., mechanical movements), which provide the base from which the new technology (e.g., quartz movement) can emerge. Thus the dying technology provides the compost, which allows its own seeds, its own variants, to grow and thrive.

BUILDING AMBIDEXTROUS ORGANIZATIONS: ORGANIZATION ARCHITECTURES AND INNOVATION STREAMS

Technology cycles and the nature of innovation streams indicate that sustained competitive advantage is built on simultaneously operating in multiple modes: managing incremental, architectural, as well as discontinuous innovation, managing for short-term efficiency and long-term innovation. If sustained competitive advantage is built via creating different types of innovation contemporaneously, managers must be capable of building organizations that can handle such innovative and strategic diversity (e.g., Rosenbloom and Christensen, 1994; Iansiti and Clark, 1994;

Morone, 1993). Executive teams must be capable of building ambidextrous organizations; organizations with multiple, internally inconsistent architectures (Weick, 1979; Burgelman, 1994).

Much is written on strategic intent and industry foresight (e.g., Hamel and Prahalad, 1994). While a firm's strategic intent and innovation strategy may be well articulated, the ability of the senior team to develop organizational capabilities to actually execute strategy may be the more scarce strategic capability. For example, where disk drive manufacturers knew the different markets for their disk drive technologies, very few incumbents were able to take their technologies across the mainframe, mini-computer, and PC computer markets (Rosenbloom and Christensen, 1994). Management teams must be able not only to craft strategic intents, but also to directly couple their strategic intents to organizational architectures (Tushman and O'Reilly, in press).

Managers have hardware and software tools in building organizational architectures (Nadler and Tushman, 1989, in press; Nadler et al., 1992). Hardware tools include organization structures, systems, and rewards as well as work processes and flows. Software tools include the firm's human resource capabilities, its culture, norms, and social networks, as well as the characteristics and competencies of the senior team (see Figure 1.7). Anchored by the unit's strategy, objectives, and vision, management teams build social and technical systems, internally congruent organization arrangements, human resources, culture, and work processes that can execute the unit's strategic intent more rapidly than the competition (Tushman and O'Reilly, in press). Performance shortfalls are

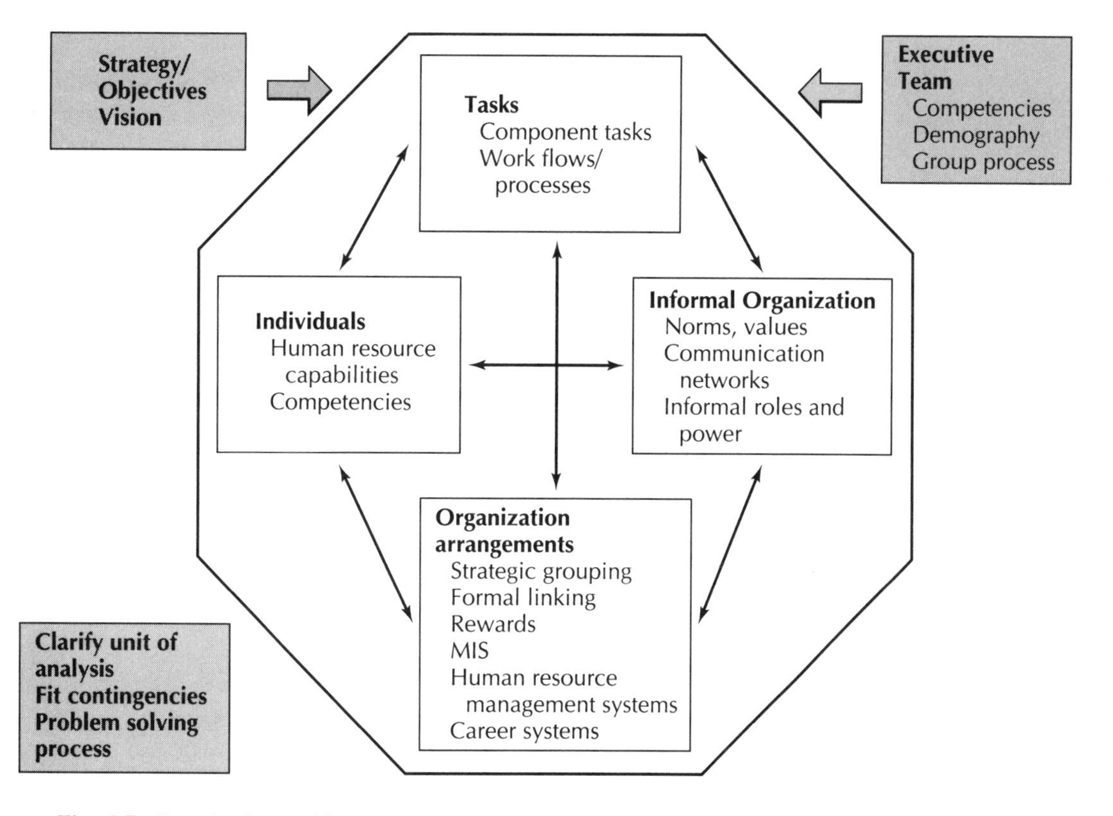

Fig. 1.7. Organization architecture: A congruence model of organizations. (*Source:* Adapted from Tushman and Nadler, 1994.)

caused either by internal inconsistencies within the technical and social systems and/or by inappropriate strategies.

Organizational architectures for incremental innovation are fundamentally different than for discontinuous innovation. What organizational architecture is appropriate for incremental innovation? Continuous, incremental improvement in both the product and associated processes, and high volume throughput associated with incremental innovation requires organizations with relatively structured roles and responsibilities, centralized procedures, efficiency-oriented cultures, highly engineered work processes, strong manufacturing and sales capabilities, and demographically more homogeneous, older, and experienced human resources (Tushman and Nadler, 1986; Eisenhardt and Tabrizi, 1995). These efficiency-oriented units celebrate continuous improvement and the elimination of variability and have relatively short time horizons. These units have well-developed knowledge systems, have learned through continuous, incremental improvement, are highly inertial, and often have glorious histories (e.g., SSIH, U.S. Steel, IBM, Philips) (see Figure 1.8). Neither discontinuous nor architectural innovation is a natural outcome of efficiency-oriented organizations (e.g., Henderson and Clark, 1990).

What organizational architecture is appropriate for discontinuous innovation? In dramatic contrast to incremental innovation, discontinuous innovation emerges from entrepreneurial, skunkworks types of organizations. These entrepreneurial units are relatively small, have loose, decentralized structures, experimental cultures, loose, jumbled work processes, strong entrepreneurial and technical competencies, and a relatively young and heterogeneous human resource profile. These entrepreneurial units generate the experiments, the failures, and the variation from which the senior team can learn about the future. These units explicitly build new experience bases, knowledge systems, and networks to break from the organization's history. They generate the variants from which the senior team can make bets on both possible dominant designs and technological discontinuities (see Figure 1.8) (Burgelman, 1994; Nonaka, 1988; Eisenhardt and Tabrizi, 1995). In contrast to the larger, more mature, efficiency-oriented units, these small entrepreneurial units are inefficient, rarely profitable, and have no established histories.

Architectural innovation does not require different core technologies, but rather takes existing technologies and links these technologies in novel ways. For example, Starkey took extant hearing aid technology, repackaged this technology into a smaller space, and created the ITE hearing aid market. Architectural innovation is not built on new

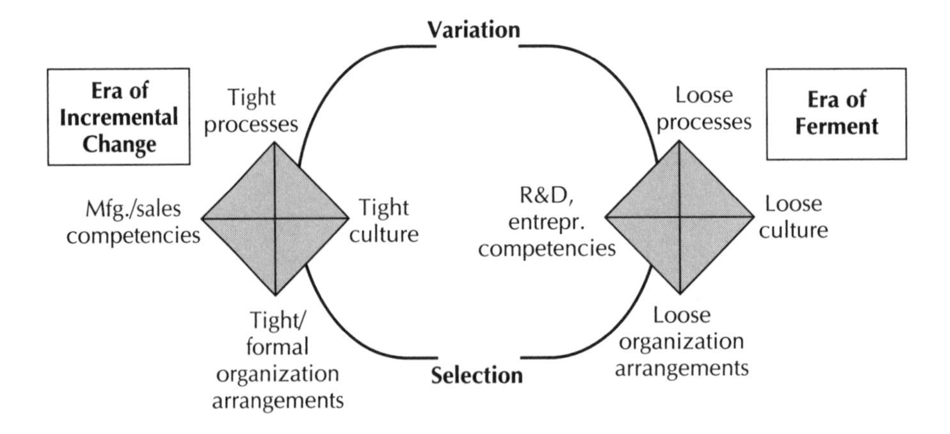

Fig. 1.8. Organization architectures and technology cycles.

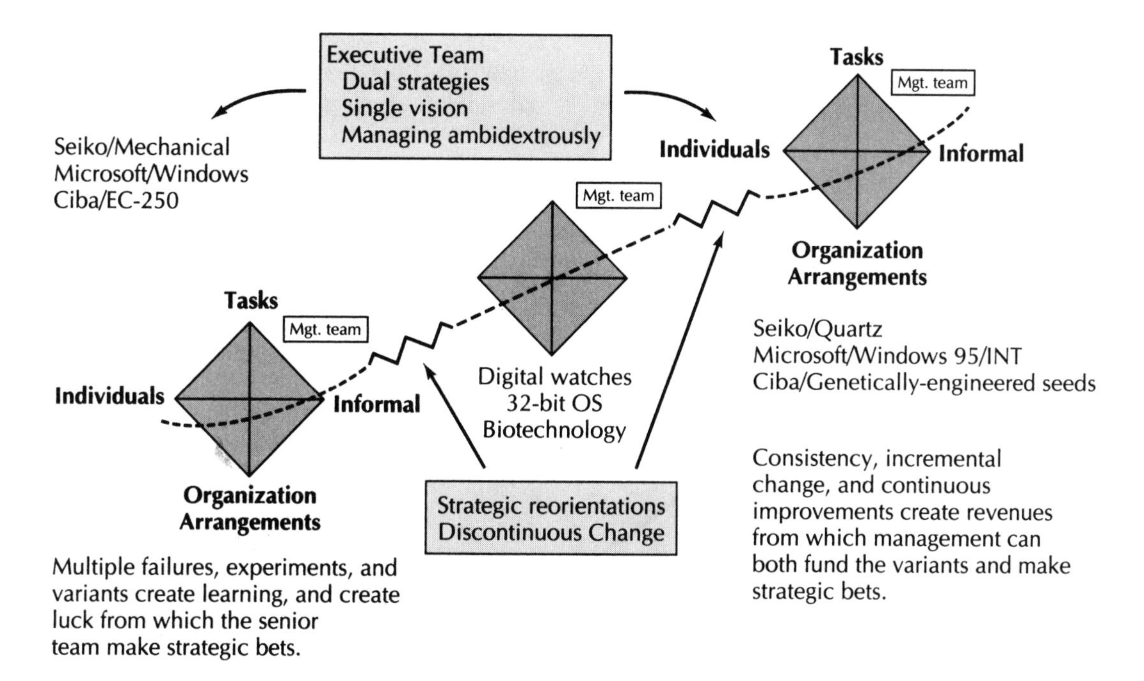

Fig. 1.9. Organization architectures. Managing ambidextrously: Creating both consistency and luck.

technical breakthroughs, but rather on integrating competencies from both efficiency and entrepreneurial units and in building distinct organizational architectures to bring this new product to new markets (see also Iansiti and Clark's [1994] discussion of supercomputers and Brown and Eisenhardt's [1995b] discussion of computers and workstations). While technologically simple, architectural innovations are often not initiated by incumbents because of the difficulties in developing organizational linking competencies (e.g., Henderson and Clark, 1990; Henderson, 1995).

Innovation streams require organizations to be simultaneously capable of producing incremental, architectural, as well as discontinuous, innovations. But these different innovation types require fundamentally different organization architectures within a simple business unit. For example, Samo at Ciba Crop Protection cannot afford an either/or innovation strategy. He must continue to support incremental innovation in his mature propiconazol product even as he works to develop fundamental product substitutes (e.g., a biologically engineered seed that will never get ill). Samo's challenge is to build an ambidextrous Crop Protection division—a division that is capable of simultaneously hosting fundamentally different organizational architectures (see Figure 1.9). More generally, the senior management's challenge is to build into a single organization multiple integrated organizational architectures. A single organization must host multiple cultures, structures, processes, and human resource capabilities required to be incrementally innovative (new and improved propiconazol) while at the same time create products that might cannibalize the existing product line (e.g., a new seed).

Ambidextrous organizations have built-in contradictions. These cultural, structural, demographic, and process contradictions are necessary if the organization is to be able to produce streams of innovations. Yet these necessary internal contradictions create conflict and dissensus between the different organization units—between those historically profitable, large, efficient, older, cash-generating units versus the young, entrepreneurial, risky, cash-absorbing units. Because the power, resources, and traditions of organizations are usually anchored in the older, more traditional units, these units usually work to ignore, trample, or otherwise kill the entrepreneurial units.

The certainty of today's incremental advance often works to destroy the potential of tomorrow's architectural and/or discontinuous advance. For example, the Swiss watch producers had all the quartz and tuning fork technology, but chose to destroy those variants and reinvest in mechanical movements. Oticon had both the idea and technology to move into ITE hearing aids, yet it was incapable of innovating until after Starkey had taken this new market. Similarly, IBM and Xerox had the software, microprocessor, and imaging competencies within their firms, but chose to reinvest in mainframe and reprographic technology. Given these inertial, defensive, and political dynamics, the management team must not only protect and legitimize the entrepreneurial units, but also keep these units physically, culturally, and structurally separate from those more mature units (e.g., Cooper and Smith, 1992). But this is difficult; in industry after industry, those major firms, those with the competencies to unleash architectural and/or discontinuous innovation, chose to close off internally generated learning and reinvest in their past; core capabilities over and over become core rigidities (Barton, 1992; Utterback, 1994; Kearns and Nadler, 1992; Foster, 1986; Morone, 1993; Hamel and Prahalad, 1994; and Rosenbloom and Christensen, 1994).

BUILDING AMBIDEXTROUS MANAGEMENT TEAMS AND MANAGING STRATEGIC INNOVATION AND CHANGE

Ambidextrous Management Teams. As innovation streams require internally inconsistent organization architectures, it is the senior management team that must balance today's incremental innovation demands with tomorrow's architectural and discontinuous innovation requirements. The senior team must be able to host and reconcile the dual, paradoxical organization requirements of ambidextrous organizations (Quinn and Cameron, 1988). This internal diversity provides the senior team with knowledge and competencies for both today and tomorrow. These insights provide the learning from which managers can develop foresight to alter industry rules.

Yet ambidextrous organizations are unstable. Senior teams must themselves have the internal heterogeneity and competencies to handle these dual innovation demands (e.g., Ancona and Nadler, 1992; Eisenhardt, 1989; Nonaka, 1988; and Eisenhardt and Tabrizi, 1995). The senior team must be able to host the diverse roles, structures, processes, and cultures of the different organization architectures. For example, Samo at Ciba Crop Protection must encourage one manager to continue to champion and innovate in propiconazol even as Samo encourages another product champion to develop a substitute for propiconazol (see also De Castro's dual strategy at Data general [Kidder, 1982]). These senior teams must provide the push, drive, and direction for incremental innovation even as they dream and challenge another piece of the organization to re-create the future.

Senior management must not only create and manage highly differentiated organizations (that is, organizations for both today and tomorrow), but also highly integrated organizations (Iansiti and Clark, 1994; Lawrence and Lorsch, 1967). Beyond organization differentiation, the senior team must provide the formal rules, regulations, roles, and rewards as well as informal networks, norms, and values to achieve integration (Brown and Eisenhardt, 1995a). But formal integration is insufficient. The senior team can also use competitive visions to achieve organization integration in ambidextrous organizations (see Collins and Porras, 1994). Clear, simple visions, like Motorola's "best in class in portable communication devices," or Ciba's "most dominant crop protection competitor, worldwide," permit the senior team to celebrate incremental innovation and efficiency along with discontinuous change, experimentation, and learning through failure. It is only through such clarity of vision that the senior team can support and embrace internally contradictory organization architectures associated with ambidextrous organizations and still be consistent.

Competitive vision provides the strategic axiom from which management teams can balance the paradoxical requirements of strategic innovation. Clear, emotionally engaging visions create a point of clarity within which organizations can be simultaneously incremental and discontinuous, short-term and long-term, efficient and innovative, centralized and decentralized (Collins and Porras, 1994; Nonaka, 1988; and Imai et al., 1985). Thus in the mid-1980s, Seymour Cray and John Rollwagen's vision for Cray was "we make the world's fastest computer." Within such a vision, Cray was able to hold onto its classic Cray Strategy of custom-made supercomputers as well as move into high-volume, low-cost supercomputers. An image of a management team managing streams of innovation is, then, that of a juggler juggling multiple balls simultaneously. As with jugglers, most management teams can do one thing well (i.e., juggle one ball). It is the rare management team that can articulate a clear, compelling vision and simultaneously host multiple organization architectures (i.e., multiple balls) without sounding confused or, worse, hypocritical.

Strategic Innovation and Discontinuous Organization Change. With ambidextrous organizations managers build into their business units the capability of managing in multiple time frames; managing simultaneously for today's innovation requirements and tomorrow's innovation possibilities. Entrepreneurial units provide learning-by-doing data, insight, and luck regarding possible dominant designs, architectural innovation and product substitution. These entrepreneurial units provide a range of experiences from which the senior team can learn about possible futures. In contrast, more mature units drive sustained incremental innovation and more short-term learning. With these diverse types of innovation and organization learning taking place within the firm, the senior management team has the raw data from which to make strategic decisions—when to try and create a dominant design (and what product variant to bet on), when to initiate an architectural innovation, and/or when to introduce a product substitute that might cannibalize the existing product line (McGrath et al., 1992). For example, Samo and his senior team at Ciba Crop Protection must understand the options for both continued development of propiconazol and the genetically engineered seed, and make a decision when, where, and how to initiate the substitute product.

Dominant designs, architectural innovation, and product substitution events are windows of strategic opportunity in which managerial action can shape and/or reinvent a product class (Tushman and Rosenkopf, 1992; Hamel and Prahalad, 1994). During eras of ferment, management can move to shape the closing on a dominant design; during eras of incremental change, management can act to substitute for its existing product line. For example, IBM's bet on the 360 mainframe series fundamentally shaped the evolution of that product class just as Seiko's move on the quartz movement fundamentally changed the watch product class. While one can know dominant designs and successful product substitutes ex post facto (e.g., VHS over Beta, quartz over escapements), management teams cannot know the "right" decisions on either dominant designs or substitution events ex ante. Yet through building dual organizational capabilities, the management team maximizes the probability that it will have both the "luck" and the expertise from which to make industry shaping decisions proactively versus reactively (e.g., Peters, 1990; Cohen and Levinthal, 1990). Further, while "correct" strategic innovation bets can only be known in retrospect, managerial action within the firm, with collaborators, alliance partners, and governmental agencies can affect the ultimate closing on an industry standard and/or the success of the product substitute (see Teece, 1987; McGrath et al., 1992).

At the closing on a dominant design, innovation within the firm shifts from major product variation to major process innovation and, in turn, to sustained incremental innovation (Abernathy, 1978; Abernathy and Clark, 1985). At product substitution events and for architectural innovation, strategic innovation shifts from incremental innovation to major product or process innovation. As strategic innovation requirements shift at these junctures, so too must the dominant organizational capabilities. Those organizational architectures, structures, roles, cultures, processes, and competencies so appropriate during eras of ferment are no longer appropriate during eras of incremental change, just as those organizational architectures so appropriate during eras of incremental change are no longer appropriate during eras of ferment (see Figure 1.9). For example, in Ciba's Crop Protection division, the organization architecture used so effectively to produce, promote, and sell mature propiconazol will have to be completely reconfigured if Samo's team decides to introduce the disease-resistant seed.

At these strategic junctures, shifts in a firm's innovation stream can only be executed with concurrent, discontinuous organization change. Sharp shifts in the innovation stream can only be executed through sharp organizational shifts. Managers can only attempt to rewrite their industry's rules if they are willing to rewrite their organization's rules (Anderson and Tushman, 1994; Romanelli and Tushman, 1994; Utterback, 1994; and Morone, 1993). For example, IBM's 360 decision in mainframes was coupled with sweeping shifts in IBM's structure, controls, systems, and culture. Bold strategic moves and/or great technology left uncoupled to revised organizational capabilities leads to underperformance. For example, Sony's superior Beta technology format was not able to counter JVC's combination of an adequate VHS technology coupled with brilliant organizational capabilities and strategic alliances (e.g., with Matsushita, Cusumano et al., 1992; Teece, 1987). Shifts in innovation streams must, then, be tightly coupled with shifts in organization structures, competencies, cultures, and processes (see also Rosenbloom and Christensen, 1994). Strategic innovation is as much rooted in organization architectures as it is in technological prowess.

Strategic innovation requires fundamentally different organizational architectures. Yet the need for discontinuous organization change runs headlong into internal forces for inertia. Patterns in organization evolution across industries and countries suggest that these contrasting forces cannot be adjudicated through incremental organization change. Incremental change benefits today's organization even as it stunts the move to tomorrow's organization (Tushman et al., 1986; Romanelli and Tushman, 1994). Organization renewal and shifts in innovation streams can be executed through strategic reorientation—discontinuous and concurrent shifts in strategy, structure, competencies and processes. These frame-breaking organizational changes are often initiated by revised senior teams (see Romanelli and Tushman, 1994; Miller, 1990; and Meyer et al., 1990).

Our model of business unit evolution is one of long periods of incremental change punctuated by discontinuous, frame-breaking, organization change (see Figure 1.10). Organizations can move from today's strength to tomorrow's strength through strategic reorientations. Strategic reorientations are coupled to shifts in the innovation stream; coupled to moves on a dominant design, architectural innovation, and product substitution.

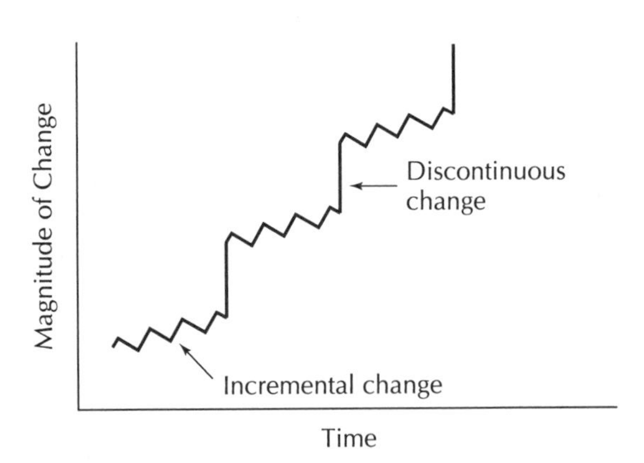

Fig. 1.10. Patterns in organization evolution.

For example, the shift to the 360 series at IBM, Seiko's shift to the quartz movement, and Starkey's shift to ITE hearing aids were all coupled with proactive strategic reorientations. If strategic reorientations are not done proactively (as in the IBM 360, Seiko, and Starkey examples), they will have to be done reactively (as with Burroughs in mainframes, SSIH in watches, and Oticon in hearing aids). Reactive reorientations (often called turnarounds) are more risky than proactive reorientations, because they must be implemented under crisis conditions and under considerable time pressure, which hinders the firm's ability to learn (Tushman and O'Reilly, in press; Hamel and Prahalad, 1994; Hurst, 1995; and Rosenbloom and Christensen, 1994).

As innovation streams shift over time, so too must organizational architectures. The management team must develop not only its ability to conceptualize strategic reorientations, but also its ability to execute the strategic changes associated with going from a given present state to a fundamentally different future state (Nadler et al., 1995; Tushman and O'Reilly, in press). The senior team must build its capabilities to manage not only incremental organization change, but also discontinuous organization change (Hurst, 1995).

The management challenge for discontinuous change is fundamentally different from managing incremental change. Discontinuous change must be initiated and directed by the senior team, must be driven by an integrated change agenda, and must be rapidly implemented—driven by the senior team's clear vision and committed actions. These strategic changes are often coupled with shifts in the senior team and within middle management. Further, the management of strategic orientations must attend to the politics of the change, individual resistance to change, and maintaining control during the transition period (Nadler et al., 1995; Tushman and O'Reilly, in press; Nadler and Tushman, 1989). If implemented through incremental change methods, strategic reorientations run the risk of being sabotaged by the politics, structures, and competencies of the status quo (e.g., Virany et al., 1992; Kearns and Nadler, 1992). For example, in Ciba's Crop Protection division, the transition from fundamentally different fungicides to EC-50 (its bet on a dominant design) was executed through sweeping changes in the division and through a new fungicide team. In contrast, breakthrough innovation at Xerox in the late 1970s and early 1980s was not coupled with corresponding organizational shifts. The politics of stability held Xerox hostage to its past (Smith and Alexander, 1990; Kearns and Nadler, 1992).

MANAGING AMBIDEXTROUSLY: EMBRACING CONTRADICTIONS AND ORGANIZATION LEARNING

Managing streams of innovation is about managing dualities: managing and embracing not efficiency versus innovation, not tactical versus strategic, not large versus small, not today versus tomorrow, but efficiency *and* innovation, tactical *and* strategic, large *and* small, today *and* tomorrow. Managing innovation streams is about consistency and control as well as variability, learning by doing, and the creation of luck. It is the crucial role of the senior team to embrace these contradictions and take advantage of the tensions and synergies that emerge from juggling multiple competencies simultaneously (see also Hurst, 1995; Collins and Porras, 1994. It is the role of vision and strategy to bind these dualities, these paradoxical requirements together. Only through clarity of vision can these multiple contradictions be reconciled and integrated and innovation streams executed.

While success seems to be hazardous, it is possible to get and stay on top. While relatively rare, Corning's success in ceramics, Ciba's success in crop protection, and Motorola's success in mobile phones indicate that it is possible to build sustained competitive advantage and to move from strength to strength (Morone, 1993; Brown and Eisenhardt, 1995b). We have emphasized building executive teams and leadership throughout the organization that can simultaneously manage the strategies, structures, competencies, work processes, and cultures for both short-term efficiency as well as create the conditions for tomorrow's strategic innovation. Thus we have talked not about organization structure or culture for innovation, but about structures and cultures for strategic innovation and change. Organizations with dual capabilities are able to maximize the probability that the firm will be more effective in the short term *and* be able to make proactive and informed strategic bets. It is through such dual, ambidextrous organizations that managers can build the capabilities to be most effective in the short run and create the conditions to initiate and implement strategic innovation. It is through such dual organizational capabilities that managers create both the expertise and luck from which they can shape their firm's future.

Sustained competitive advantage is about senior team and organization learning. The senior team must learn from its own behavior and from the organization's responses to their actions. First order learning is based on continuous improvement, consistency, balance, and congruence between task demands and people, organization arrangements, and culture. This learning mode is about getting better and better in the execution of today's strategy. Second order learning, in contrast, is dedicated to destroying today's business, cannibalizing today's product, and moving out from the organization's past. This learning mode is created by building experimentation and variants into the system. These variants, created by contrasting structures, people, and cultures, generate the data and expertise from which managers can then make strategic innovation bets. These strategic innovation bets must then be implemented.

Managing strategic innovation is about managerial action; action driven by clear strategies, objectives, and vision, action disciplined by systematic data gathering, and action that eventually renews the organization by proactive strategic innovation and change. Strategic innovation is not about quick fixes or fashionable techniques—it is not about re-engineering, speed, strategic intent, decentralization, or empowerment. It is about a comprehensive approach to managerial problem solving and action based on an integrative problem solving framework, and an understanding of the linkages between innovation streams, executive teams, and organization evolution.

Finally, strategic innovation is about implementation. Strategy, vision, and innovation are empty words—indeed, the world is full of great visions, strategies, technologies, and innovations that never were implemented, or were implemented well after the competition. As vision, strategy, and innovation streams are really in their execution, we have highlighted the importance of effectively managing strategic change. As innovation always is associated with organizational changes, managers must focus on managing change within their units, change across their organizations, as well as with organizations outside the focal organization. We have coupled the ideas of managing innovation streams with the dynamics of managing change—managing politics, control, and individual resistance to change.

Even if periods of incremental change do contain the seeds of their own destruction, organizations can move from strength to strength. Through proactive strategic change, senior teams can manage the rhythm by which as each strength dies it gives way to the next. Prior organizational competencies provide a platform so that the next phase of an organization's evolution does not start from ground zero. Organizations can then renew themselves through a series of proactive strategic reorientations anchored by a common vision. These strategic reorientations are coupled to bets on dominant designs, architectural innovation, and/or product substitution. Like a dying vine, the prior period of incremental change provides the compost for its own seeds, its own variants, to thrive in the subsequent period of incremental change.

Our approach to strategic innovation evokes several images of managers involved in the management of innovation streams. Managers can be seen as simultaneously organization architects/social engineers, as network builders/politicians, as well as artists/jugglers (see Figure 1.11). The manager as engineer/architect focuses on social engineering, using the tools of strategy, structures, human resource competencies, and cultures to build

The Manager As:	Role:
☐ Architect/Engineer	☐ Building fit, consistency and congruence of structures, human resources, and cultures to execute critical tasks in service of strategy, objectives, and vision. ☐ Continuous improvement.
☐ Politician/Network Builder	☐ Managing strategic change by building networks and shaping coalitions down, across, ups, and outside the manger's unit.
☐ Artist	☐ Building in contradictory strategies, structures, competencies, and cultures in service of both incremental and discontinous innovation, as well as integrating these contradictions through vision and strategy.

Fig. 1.11. Managerial metaphors in managing strategic innovation and change.

robust organizations to get today's work and tomorrow's innovation accomplished. Yet at junctures in a product class, shifts in the innovation stream must be coupled with discontinuous organization change. These change requirements run counter to the dynamic conservatism found in successful firms. Therefore, the second managerial image focuses on the manager as a network builder and politician, building cliques and coalitions in service of innovation and change. Finally, we have focused on managers as artists—building in the tensions and contradictions of managing both for today as well as for tomorrow, even as these managerial artists integrate these contradictions into an internally consistent whole. These multiple managerial images and associated competencies seem to be necessary to manage innovation streams; these managerial and, in turn, organizational competencies seem to be necessary to achieve and sustain competitive advantage in turbulent product classes.

REFERENCES

Abernathy, W.J. *The Productivity Dilemma*. Baltimore: Johns Hopkins Press, 1978.

Abernathy, W. and Clark, K. "Innovation: Mapping the Winds of Creative Destruction." *Research Policy*, 1985. 14, 3–22.

Abernathy, W. and J. Utterback. "Patterns of Industrial Innovation." *Technology Review*, 1978, 2, 40–47.

Aitken, H. *The Continuous Wave*. Princeton: Princeton University Press, 1985.

Ancona, D. and D. Nadler. "Top Hats and Executive Teams." *Sloan Management Review*, 1989, 31 (1), 19–28.

Anderson, P. and M. Tushman. "Technological Discontinuities and Dominant Designs: A Cyclical Model of Technological Change." *Administrative Science Quarterly*, 1990, 35, 604–633.

Barton, L. "Core Capabilities and Core Rigidities: A Paradox in Managing New Product Development. *Strategic Management Journal*, 1992, 13, 111–125.

Baum, J. and H. Korn. "Dominant Designs and Population Dynamics in Telecommunications Services." *Social Science Research*, 1995, 24 (2), 97–135.

Brown, S. and K. Eisenhardt. "Accelerating Adaptive Processes: Product Innovation in the Global Computer Industry." Administrative Science Quarterly [ASQ] 1995a, 40, 84–110.

Brown, S. and K. Eisenhardt. "Product Development: Past Research, Present Findings and Future Directions." *Academy of Management Review*, 1995b, 20 (2).

Burgelman, R. "Fading Memories: A Process Theory of Strategic Business Exit." *Administrative Science Quarterly*, 1994.

Cohen, W. and D. Levinthal. "Absorptive Capacity: A New Perspective on Learning and Innovation." *Administrative Science Quarterly*, 1990, 35 (1), 128–152.

Collins, J. and J. Porras. *Built to Last*. New York: Harper Business, 1994.

Cooper, A. and C. Smith. "How Established Firms Respond to Threatening Technologies." *Academy of Management Executive*, 1992, 6 (2), 55–70.

Cusumano, M., Y. Mylonadis, and R. Rosenbloom.

"Strategic Maneuvering and Mass Market Dynamics: The Triumph of VHS Over Beta." *Business History Review,* 1992, 66, 51–94.

David, P. "Clio and the Economics of Qwerty." *American Economic Review,* 1985, 75 (2), 332–337.

Deschamps, J. and P. Nayak. *Product Juggernauts.* Cambridge, Mass.: Harvard Business School Press, 1995.

Eisenhardt, K. "Making Fast Strategic Decisions in High Velocity Environments." *Academy of Management Journal,* 1989, 32 (3), 543–576.

Eisenhardt, K. and B. Tabrizi. "Acceleration Adaptive Processes." *Administrative Science Quarterly,* 1995, 40 (1), 84–110.

Florida, R. and Kenney, M. *The Breakthrough Illusion.* New York: Basic Books, 1990.

Foster, R. *Innovation: The Attacker's Advantage.* New York: Summit Books, 1986.

Gersick, C. "Pacing Strategic Change: The Case of New Ventures." *Academy of Management Journal,* 1994, 37 (1), 9–45.

Gomory, R. and R. Schmitt. "Science and Product." *Science,* 1988, 240, 1131–1204.

Hamel, G. and C. Prahalad. *Competing for the Future.* Cambridge, Mass.: Harvard Business School Press, 1994.

Henderson, R. "Of Life Cycles Real and Imaginary: The Unexpectedly Long Old Age of Optical Lithography." *Research Policy,* 1995, 24 (4), 631–643.

Henderson, R. and K. Clark. "Architectural Innovation: The Reconfiguration of Existing Product Technologies and the Failure of Established Firms." *Administrative Science Quarterly,* 1990, 35 (1), 9–30.

Hollander, S. *Sources of Efficiency.* Cambridge, Mass.: MIT Press, 1965.

Hughes, T. *Networks of Power.* Baltimore: Johns Hopkins Press, 1983.

Hurst, D. *Crisis and Renewal.* Cambridge, Mass.: Harvard Business School Press, 1995.

Iansiti, M. and K. Clark. "Integration and Dynamic Capability." *Industrial and Corporation Change,* 1994, 3 (3), 557–605.

Imai, K., I. Nonaka, and H. Takeuchi. "Managing the New Product Development Process: How Japanese Firms Learn and Unlearn." In Clark, K. et al., (eds.), *The Uneasy Alliance.* Cambridge, Mass.: Harvard Business School Press, 1985.

Jellinek, M. and C. Schoonhoven. *The Innovation Marathon.* Cambridge, Mass.: Blackwell, 1990.

Kearns, D. and D. Nadler. *Prophets in the Dark.* New York: Harper, 1992.

Kidder, T. *The Soul of a New Machine.* Boston: Little, Brown, 1982.

Landes, D. *Revolution in Time.* Cambridge, Mass.: Harvard University Press, 1983.

Lawrence, P. and J. Lorsch. *Organizations and Environments.* Cambridge, Mass.: Harvard Business School Press, 1967.

McGrath, R., I. MacMillan and M. Tushman. "The Role of Executive Team Actions in Shaping Dominant Designs: Towards Shaping Technological Progress." *Strategic Management Journal,* 1992, 13, 137–161.

Meyer, A., G. Brooks, and J. Goes. "Environmental Jolts and Industry Revolutions." *Strategic Management Journal,* 1990, 11.

Meyer, M. and J. Utterback. "Product Family and the Dynamics of Core Capability." *Sloan Management Review,* 1993, 34 (3).

Miller, D. *The Icarus Paradox: How Exceptional Companies Bring about Their Own Downfall.* New York: Harper, 1990.

Morone, J. *Winning in High Tech Markets.* Cambridge, Mass.: Harvard Business School Press, 1993.

Nadler, D., M. Gerstein, and R. Shaw. *Organization Architecture.* San Francisco: Jossey-Bass, 1992.

Nadler, D., R. Shaw, and E. Walton. *Discontinuous Change.* San Francisco: Jossey-Bass, 1995.

Nadler, D. and M. Tushman. *Strategic Organization Design.* Glenville, Ill: Scott Foresman, 1988.

Noble, D. *Forces of Production.* New York: Alfred Knopf, 1984.

Nonaka, I. "Creating Organizational Order Out of Chaos: Self Renewal in Japanese Firms." *California Management Review,* 1988, 30 (3), 57–73.

Peters, T. "Get Innovative or Get Dead." *California Management Review,* 1990, Pt. 1, 33 (1), 9–26; Pt. 2, 33 (2), 9–23.

Quinn, R. and K. Cameron. *Paradox and Transformation.* Cambridge, Mass.: Ballinger Publications, 1988.

Romanelli, E. and M. L. Tushman, "Organizational Transformation as Punctuated Equilibrium: An Empirical Test." *Academy of Management Journal,* 1994, 37, 1141–1166.

Rosenbloom, D. and C. Christensen. "Technological Discontinuities, Organization Capabilities, and Strategic Commitments." *Industry and Corporate Change,* 1994.

Rosenkopf, L. and M. Tushman. "The Coevolution of Technology and Organization." In Baum, J. and J. Singh (eds.), *Evolutionary Dynamics of Organizations.* New York: Oxford University Press, 1994.

Sanderson, S. and V. Uzumeri. "Managing Product Families." *Research Policy,* 1995, 24 (5), 761–782.

Schoonhoven, K., K. Eisenhardt, and K. Lyman. "Speeding Products to Market." *Administrative Science Quarterly,* 1990, 35 (1), 177–207.

Smith, D. and R. Alexander. *Fumbling the Future.* New York: Harper, 1990.

Teece, D. "Profiting from Technological Innovation." In D. Teece (ed.), *The Competitive Challenge.* New York: Harper & Row, 1987.

Teece, D. and G. Pisano. "Dynamic Capabilities of Firms." *Industry and Corporate Change,* 1994, 3, 537.

Tushman, M. and P. Anderson. "Technological Discontinuities and Organization Environments." *Administrative Science Quarterly,* 1986, 31, 439–465.

Tushman, M. L. and D. Nadler. "Organizing for Innovation." *California Management Review,* 1986, 28 (3), 74–92.

Tushman, M., W. Newman, and E. Romanelli. "Convergence and Upheaval: Managing the Unsteady Pace of Organizational Evolution." *California Management Review,* 1986, 29 (1), 29–44.

Tushman, M. and C. O'Reilly. *Evolution and Revolution: Mastering the Dynamics of Innovation and Change.* Cambridge, Mass.: Harvard Business School Press, in press.

Tushman, M. and E. Romanelli. "Organization Evolution: A Metamorphosis Model of Convergence and Reorientation." In L. Cummings and B. Staw (eds.), *Research in Organizational Behavior,* vol. 7. Greenwich, Conn.: JAI Press, 1985, 171–222.

Tushman, M. and L. Rosenkopf. "On the Organizational Determinants of Technological Change: Towards a Sociology of Technological Evolution." In B. Staw and L. Cummings (eds.), *Research in Organization Behavior,* vol. 14. Greenwich, Conn.: JAI Press, 1992, 311–347.

Utterback, J. *Mastering the Dynamics of Innovation.* Cambridge, Mass.: Harvard Business School Press, 1994.

Van de Ven, A. and R. Garud. "The Coevolution of Technical and Institutional Events in the Development of an Innovation." In Baum, J. and J. Singh (eds.), *Evolutionary Dynamics of Organizations.* New York: Oxford University Press, 1994.

Virany, B., M. Tushman, and E. Romanelli. "Executive Succession and Organization Outcomes in Turbulent Environments." *Organization Science,* 1992, 3 (1), 72–91.

Weick, K. *The Social Psychology of Organizing.* Reading, Mass.: Addison-Wesley, 1979.

2

Technological Discontinuities and Flexible Production Networks: The Case of Switzerland and the World Watch Industry

AMY GLASMEIER

The twentieth-century history of the Swiss watch industry illustrates how cultures and industrial production systems experience great difficulty adapting to external change at different points in time. The current emphasis on production networks—unique reservoirs of potential technological innovation realized through cooperation rather than competition among firms—lacks a detailed appreciation of historic networks, and in particular their fragile character in times of economic turmoil. While networks can and do promote innovation within an existing technological framework, historical experience suggests their fragmented, atomistic structure is subject to disorganization and disintegration during periods of technological change. An exclusive focus on "production" ignores other constraints that are powerful forces governing the reaction abilities of regions. Previous research has largely relied on a model of oligopolistic competition to explain how the Swiss lost control of the world watch industry. I conclude, on the contrary, that the Swiss experience must be understood from the standpoint of how technological change challenges previous ways of organizing production, industry, culture, and society. Technology shifts present a series of strategic turning points that industrial leaders must navigate during a period of technological change.

INTRODUCTION

The history of the Swiss watch industry is instructive as countries and regions experiment with network production systems in attempts to maintain and augment their competitiveness in a global economy. On the eve of the electronics revolution, the Swiss watch production system, centered in the mountainous Jura region, was flexible, cost effective, and extremely profitable. Both horizontally and vertically disintegrated, the Swiss system offered enormous variety while maintaining quality and timeliness of delivery. "The multiplicity of enterprises, and the competition and emulation that characterized the industry, yielded a product of superior quality known the world over for high fashion, design, and precision" [21, p. 48].

Beginning in the 1970s, when foreign compe-

This article is based on original research conducted in Japan, Hong Kong, and Switzerland. The assistance of Luc Tissot, Wendy Taillard, Pierre Rossel, and Jocelyn Tissot, of the Tissot Economic Foundation, was greatly appreciated. The Foundation and its staff were instrumental in the successful completion of this research. A full elaboration of the argument presented in this article can be found in various working papers in the Graduate Program in Community and Regional Planning, Working Paper Series. Reprinted from *Research Policy*, Vol. 20, A. Glasmeier, "Technological Discontinuities and Flexible Production Networks: The Case of Switzerland and the World Watch Industry," pp. 469–485, 1991, with kind permission from Elsevier Science B.V., Amsterdam, The Netherlands.

tition hurdled technological frontiers in watch movements, advancing from mechanical to electric, electronic, digital and finally quartz technology, the Jura's undisputed dominance ended.[1] Massive job loss and out-migration occurred as firms, unable or unwilling to adapt to new technologies, closed their doors. Today, while still world leaders in watch export value, Swiss watchmakers produce only a fraction of their pre-1970s output levels, and resources needed to invest in new product research and development are scarce [40]. In a span of less than 30 years, the world's dominant watch region yielded technological leadership (in watchmaking and micromechanics) to its Far Eastern rivals. What lessons can be learned about network production systems and technological innovation from the experience of Switzerland's watch region?

Industrial restructuring of the past 20 years has left once dominant manufacturing regions such as America's industrial heartland and Germany's Ruhr valley debilitated. Reincorporating technological innovation within production systems of deindustrialized regions has become a major concern. Even technologically vibrant regions such as Route 128, Silicon Valley, and Emilia Romagna confront uncertain figures in the current period of intense technological development and international competition. How can a region remain innovative during a period of technological change? Do network production systems offer a more flexible and permanent means of regional adaption? By examining the experience of the watch industry I hope to tie together empirical experiences of researchers asking similar questions concerning the relationship between structure of production, regional culture, technological change, the formation and maintenance of core skills, and state-led regional development in industrial hinterlands.

This article reviews the twentieth-century history of the Swiss watch industry. My purpose is to suggest that focussing solely on production transactions does not adequately explain how economic, social, and cultural conditions interact to form a complex of human relations that can remain flexible and innovative over time. While networks are quite proficient at production and innovation within an existing technological framework, disintegrated systems may neither accumulate profits nor demonstrate a collective will to make essential investments in research, marketing, and distribution in response to technological change. At a more refined level, this article suggests that the current emphasis on production networks—unique reservoirs of potential technological innovation realized through cooperation rather than competition among firms—lacks a detailed appreciation of historic networks, and in particular their fragile character in times of economic turmoil [29,30,31,32]. While networks can and do promote innovation within an existing technological system, historical experience suggests their fragmented, atomistic structure is subject to disorganization and disintegration during periods of major technological change [8,9,35].[2]

Inter-firm networks are important ingredients of technologically innovative and flexible industrial production systems. This insight is an important addition to contemporary theoretical discussions about the relationship between regional and technology development. As the case of the Swiss watch industry suggests (within a particular structure of production and organization), a highly articulated system of production, tied together by elaborate cultural, institutional, and economic relationships, exhibits flexibility and adaptability within an existing technological framework. The case study also illustrates, however, that within any production system flexibility and innovativeness are time-dependent. Therefore, given specific historical circumstances, they are vulnerable.

The lack of coordination and organization within small firm complexes has precipitated discussions about the need for governing systems to regulate small firm atomistic behavior and to replicate functions performed by the vertically integrated corporation such as R&D, management training, marketing, and distribution. As Saxenian's work suggests, network production systems are vulnerable to technological threats from the outside [30]. Calls for institution-building must be tempered with the knowledge that coordinating organizations quickly become absorbed into the regional fabric and ossify over time. During the course of two centuries the Swiss watch industry established elaborate institutions, organized technological competitions and other social events, and supported diverse organizations to bolster the industry's innovative capabilities [8]. The strength of these institutions has been and still is significant. The Swiss Watch Industry Federation (FH) was the leading organization in trade negotiations for the world watch industry [24]. Long after the Swiss lost volume leadership, the FH continued to lead GATT negotiations [27]. Market share, quotas, and tariffs are still negotiated by the FH for the world watch industry. Yet despite its worldwide reputation, in the 1960s and 1970s the FH was unable to overcome resistance to new technology. While the organization sponsored Swiss R&D, it could not force members to incorporate new technologies into existing products.

This article departs from past treatments of the watch industry's post-war experience by examining the rise of world competition through the lens of technological shifts. Previous treatments have largely relied on a model of oligopolistic competition to explain how the Swiss lost control of the world watch industry. I conclude, on the contrary, that the Swiss experience must be viewed from another angle. How do technological shifts challenge previous ways of organizing production, industry, culture, and society? Fundamental changes in technology present a series of strategic turning points that industrial leaders must navigate. The Swiss were no exception.

This article first considers the historic evolution of the Swiss watch industry during the final decades of the last century and shows its early adaptability to new production innovations within the framework of mechanical watch technology. The bulk of the remaining discussion traces the evolution of the industry—illustrating the difficulties experienced in the face of radical technological developments.

The Swiss watch industry provides an important case study of an industrial and cultural system that retained technological supremacy for two centuries and that still holds its dominant position within the earlier mechanical paradigm. The industry has, however, yielded technological leadership to foreign competitors in a major geographical shift in world production [36]. The Swiss are now followers rather than leaders of industry trends.

THE EARLY TWENTIETH-CENTURY HISTORY OF THE SWISS WATCH INDUSTRY

Historically the Swiss industry has shown surprising resilience in the face of change. At the end of the last century (1876–1900) the industry was issued a major challenge by the U.S. watch production system. America's watch manufacturers developed machinery to produce watches at high volume with low cost, low skill, and relatively high levels of precision. Watch movements were drastically simplified and more economical to produce. While hand-adjustment was still required in final assembly, the overall skill content in American watches was drastically reduced.

The Swiss response to U.S. technological challenges was decisive. Over a period of 20 years (1885–1905) they proved more than capable of making needed technical progress [7]. While the Swiss lost considerable market shares in the U.S., the country's manufacturers did not yield control of global markets [19]. Over the course of two decades, the Swiss system adopted aspects of the American system that were cost effective. The Swiss system shifted from its reliance on small-scale cottage production to an intermediate form that combined mechanization and partial vertical integration. Standard parts were mechanically manufactured at large scale in centralized factories, while flexibility was maintained in dispersed design and assembly activities. Even the more complicated parts were eventually mechanized using "versatile machines which were susceptible of all manner of adjustment, hence required some skill to operate . . ." [20, p. 40]. Thus within the existing mechanical technology system, the industry achieved new levels of profitability and international renown.

There is no doubt that by the 1910s Swiss mechanical watches dominated the world watch industry [18]. The Swiss controlled the micromechanical export industry by cost competitiveness, superior manufacturing competency, high levels of precision, and extraordinary attention to detail and style. The vertically integrated parts manufacturers achieved economies of scale through volume production. This benefit was passed on to assemblers in the form of low cost movements. In the most labor-intensive aspects of the industry, the vertically disintegrated system of assembly and case manufacture kept overhead charges low.

International Economic Chaos and the Call for Regulation

The early 1920s was a period of great instability in the watch industry.[3] Disruptions in the watch market presented the Swiss with new and different problems [18]. Significant sums of capital had been invested to meet the American manufacturing challenge. Firms were larger, and the industry represented a larger share of gross national product [21]. The severity of the crisis forced family businesses to take drastic steps simply to reduce inventory. Opportunism, price-cutting, and increased export of movements and parts further destabilized the industry [39]. This unprecedented threat resulted in a call for industry regulation, and a cartel was formed.[4]

During the 1920s various associations were created to represent the interests of industry members. The Swiss Watch Industry Federation (FH) was organized to govern both firms assembling watches from component parts and the few firms with integrated manufacturing operations. The 17 manufacturers of ébauches (watch movements) were organized into a trust EBAUCHE S.A. Manufacturers of components other than ébauches (balance wheels, assortments, hair springs) were organized into the Union des Branches Annexes de l'Horlogerie (UBAH). In the late 1920s members of the various associations agreed to set levels of output and prices, and explicit rules were designed to restrict exportation of parts [18].

When this degree of collaboration proved insufficient to control opportunistic firms, the government intervened. In conjunction with industry and banking leaders, the federal government created the massive holding company ASUAG (which included EBAUCHE S.A. as well as other leading component producers). This final merger halted the exportation of parts and components to competitor countries [18].

The Statut de l'Horlogerie and the Codification of the Swiss System

The Statut de l'Horlogerie of the early 1930s established a regulatory system that governed Swiss watch manufacturing for more than 30 years. Through a combination of cartelization and government ownership, the Swiss industry was regulated to control vertical integration, foreign sourcing, and off-shore production. Swiss manufacturers could buy only from Swiss component producers, and component producers could sell only to Swiss firms. To further limit competition, government regulated the sale of machinery. The Statut de l'Horlogerie regulated the volume of Swiss watch production by requiring permits for the construction and expansion of production facilities [15].

The resulting industry structure consisted of the parts manufacturers who sold their output to assemblers, the assemblers, and the brand name manufacturers. ASUAG could sell only to firms recognized by the Swiss government under the law. It could not export parts or technology. Manufacturers fabricated complete watches but were restricted from selling movements and other parts to assemblers—thus eliminating competition with parts suppliers. They were also restricted from setting up production in other countries. Assemblers were prohibited from establishing production outside Switzerland, and they could buy parts from non-Swiss manufacturers only if prices were 20 percent below Swiss levels. The law's greatest effects were in regulating who was allowed to produce, what could be produced, and how much could be produced. By requiring export and manufacturing permits, the government essentially held supply below world demand and ensured Swiss firms handsome profit levels.

From 1933 to 1961 the Swiss watch industry experienced considerable stability matched by handsome growth. All industry sectors enjoyed the benefits. Under the Statut de l'Horlogerie, market shares were effectively stabilized. This predictably encouraged firms to reinvest profits in new process technology. High profits earned in this period allowed firms to develop a mechanical watch manufacturing system unparalleled in efficiency.

Abandoning Industry Regulation: Instituting Industrial Change

In the early 1960s three decades of stability once again gave way to uncertainty. Foreign competition ended the Swiss monopoly on mechanical watch production and the country's quasi-monopoly on the world watch industry. The slow erosion of Swiss world export market share met with cries from industry members to change laws that had regulated the industry for 30 years.

Reasons for industry discontent were numerous. The more profitable and better run firms lobbied against the cartel arguing that it protected firms that were producing low quality watches [39,21]. Laws were also criticized for fixing the level of Swiss production at a time when other countries were making substantial inroads in the Swiss world export market share. In 1961 the Federal Assembly of the Swiss Confederation ratified a new decree eliminating the regulation of output and encouraging rationalization of the industry. The new law took effect in 1962, but it was not until the early 1970s that restrictions on watch manufacturing were entirely eliminated.

As expected, the watch industry underwent a series of unprecedented mergers. The healthier and larger establishments joined forces to match the sizes of their Far East Asian and American rivals. Within two years three firms were producing 32 percent of Swiss exports. SSIH (formed in the 1930s with the merger of Tissot and Omega to become a leading vertically integrated manufacturer) became the third largest watch manufacturer in the world (behind Timex and Seiko). In 1971 the ASUAG expanded beyond strictly component production by creating the General Watch Company, a holding organization of several brand names and component manufacturers [18]. A third holding company, Société des Garde-Temps (SGT), was created primarily to manufacture low-price and electronic watches.[5]

In addition to the three holding companies, there were a number of important groups. Rolex, although privately held, had 1972 sales estimated at 200 million Swiss francs (almost a quarter of Swiss exports by value) [17]. There were also four middle-sized groups including two subsidiaries of U.S. companies, Zenith and Bulova, and the prestige brands, Piaget, Patek Philippe and others. The remainder of the industry was made up of hundreds of small companies assembling and selling watches.

THE WORLD MARKET FOR WATCHES

At the end of the 1980s watch producers manufactured approximately 500 million watches annually worldwide (not including Eastern European production). The market for watches is made up of segments in a pyramid-like structure. The base of the pyramid consists of mass producers selling watches with a wholesale value of less than $50 (1990 dollars). These sales account for 90 percent of total volume. Although this segment of the market is dominated by Hong Kong producers, it also includes American Timex and the Swiss Swatch watch. The mid-price ($50–500) watch market segment makes up 9 percent of all watch sales. Firms marketing in this segment include well known Swiss (Tissot, Omega, Longines, Rado), Japanese (Seiko and Citizen), and specialty American name brands (Hamilton).[6] The luxury market segment comprises only 1 percent. Luxury watch prices start at about $750. There are 20 brands in this category and include such world-renowned names as Cartier, Ebel, Rolex and Patek Philippe. Some firms span more than one market segment. These include firms such as Rolex that add value to the basic steel case watch model by altering the external parts of the watch (e.g. diamond-encrusted bezel).

TABLE 2.1. Industry Structure

	Houses	Employment
Movements and parts	40	8000
Casing and bracelets	135	6500
Subcontractors	135	6000
Integrated manufacturers and assemblers	250	12000

Source: Radja [26].

Structure of the Swiss Watch Industry

The structure of the Swiss industry emerging after the collapse of the cartel maintains important vestiges of the old system. The Swiss watch industry reflects both horizontal and vertical disintegration of establishments. Although the number of individual producers has declined over time, the shares of establishments in the different industry segments have remained the same [25,26]. The watch industry consists of four levels of production: movements and parts manufacturers; case and bracelet manufacturers; subcontractors; and assemblers and integrated manufacturers. Table 2.1 lists the number of Swiss watch firms and their employment by industry level.

Parts and Assemblers

Production of parts and components passes through various channels to arrive at final assemblers. Final assemblers assemble for all price segments of the watch market under nationally and internationally known brand names. Together, parts manufacturers, subcontractors, and assemblers employ approximately 20,000 workers with a turnover of more than 3 billion dollars, or 40 percent of total industry revenues (1990). Mid-price watch assemblers include such well known names as Chopard, Century, Corum, Eterna, and Raymond Weil. The luxury end of the assembly industry includes "international names" such as Cartier, Chanel, Christian Dior, Gucci, and Dunhill. These brand labels represent individual designers that subcontract with Swiss component and case manufacturers to assemble their watches.[7]

Manufacturers

In contrast to the disintegrated structure of parts manufacturers and assemblers, the Swiss industry also includes enterprises that are vertically integrated. Manufacturers produce the entire product from movements through parts, casing, and final assembly. There are a dozen important integrated watch manufacturers. Rolex is the single largest manufacturer and enjoys 25 percent of total Swiss industry revenues. At the very high end luxury market segment are found the most prestigious integrated manufacturers, including Piaget, Patek Philippe, Audemars Piguet, Breguet, Ebel, and Blancpain. These firms produce watches that sell in the thousands of dollars and are produced in small numbers.

Movement Manufacturers

In response to economic instability in the 1930s, movement manufacturing was concentrated in a single firm (ASUAG). In the early 1980s reorganization further concentrated this activity. ASUAG was combined with a number of large previously independent manufacturers (Tissot, Rado, Longines, Omega, Certina, and Swatch) into SMH. SMH accounts for approximately 30 percent of all watch revenues and controls almost 25 percent of total watch employment. While SMH is the single largest supplier of parts and movements, there are a number of other movement producers including Fabrique D'Ebauches de Sonceboz et Ebosa, Piguet, and La Novelle Lemania et Laeger-Lecoultre. These firms cater to the luxury watch industry.

Movement production in the Swiss watch industry is the most vertically integrated. It is composed of 40 firms employing 8,000 workers with an average plant size of 200 employees. This component of production is the most technologically advanced and enjoys significant economies of scale. Parts and case makers, subcontractors, and independent manufacturers represent the majority of establishments but employ far fewer workers per firm (averaging 45 employees). Observing size of

TABLE 2.2. Firm Distribution by Size

Size Category	Number of Establishments	Number of Employees
1–9	200	1,500
10–19	130	2,000
20–49	100	4,000
50–99	70	5,000
100–199	34	5,000
200–499	32	10,000
500 plus	4	5,000

Source: Radja [26].

establishment verifies the disintegrated structure of production. Table 2.2 lists the number of firms by employee size category. The structure of the industry is essentially the same as it was in the 1970s.

GLOBAL COMPETITORS AND THE WORLD MARKET FOR WATCHES

In the early 1970s the Swiss struggled to reorient their factories while nimble competitors flooded the field. But their production system was not easy to dismantle or rearrange. Japanese, American and Hong Kong firms posed unique challenges to Swiss watchmakers. This new and rising competition and the advent of a new movement technology were both significant problems.

The Japanese industry were vertically integrated and therefore a low-cost producer [18,22]. Seiko and Citizen made major inroads in the world watch market as both component and finished watch manufacturers. The Japanese made high quality low-priced movements that were sold to firms around the world. By the early 1970s, Japanese watch companies had succeeded in capturing 14 percent of the world watch market.

The Japanese increased their share of world export markets through various means. They had lower labor costs and an undervalued currency. They were vertically integrated and employed manufacturing automation. The Japanese developed the capacity to manufacture standardized movements and watch models. Japan was also selling large volumes of movements to the U.S. and Hong Kong. Although the Japanese produced models for every price range, they targeted the lucrative middle range, undercutting Swiss competitors.

Unlike the Swiss, the Japanese had the advantage of a large protected home market. Because other watch manufacturers were effectively locked out of their domestic market, Japanese producers enjoyed artificially high domestic prices that covered fixed costs. Thus in international markets watches could be sold close to or at marginal costs.[8]

Government regulation and financial assistance (R&D grants) accelerated Japanese penetration of the world watch industry. The Japanese watch industry continued to rationalize—undergoing further vertical integration that streamlined operations and reduced inefficiencies. The government encouraged vertical integration to "minimize the proliferation of marginal watch producers and to minimize the drain on foreign reserves caused by the importation of watch machinery" [18, p. 23]. In 1978, 88 percent of Japanese production was attributable to two firms. Switzerland's leading manufacturer accounted for only 9 percent of total national Swiss production [2].

The Structure of the Japanese Watch Industry

The level of vertical integration in the Japanese watch industry is high by national standards. High levels of vertical integration are primarily associated with the small number of industry competitors (Citizen, Seiko and Casio) and individual company product development strategies [10]. Citizen and Casio are primarily electronics firms which concentrate on digital watch manufacturing [3,4]. Both companies pursued watch manufacturing to assure a market for their primary product, electronic components [8]. Seiko dominates Japan's watch industry. The company is responsible for 60 percent of the nation's output. Vertical integration was part of a conscious strategy to be the industry leader [34,14]. After World War II (like other Japanese firms such as Honda), Seiko followed a product diversification strategy built around the company's core competence, precision manufacturing.[9]

The most critical advantage of Japan's capital-intensive system is the ability to manufacture components in huge volumes at low cost. The sale of movements and watch kits cemented the industry's 1970s world volume leadership.

The U.S. Market and American Watch Manufacturers

Japan was not the only significant challenger to the Swiss watch industry. The U.S. was both the world's largest and most competitive watch market [8]. The vast majority of American demand was satisfied by domestic firms. America's two stellar watch manufacturers, Timex and Bulova, essentially controlled two-third's of the nation's market [19]. American watchmaking firms were dominant in the U.S. partly because of high tariffs that were based on the number of jewels in the watch movement and implemented to protect the domestic industry. Swiss manufacturers responded by redesigning their watches to include fewer jewels. But by redesigning watches, production was further fragmented with more models designed for the U.S. market. A bad side effect of this strategy was an inability to produce at volumes that would allow for productivity gains [27]. On the other hand, American firms enjoyed a loophole in trade policy which permitted off-shore watch assembly by low-wage laborers. Because American firms could avoid paying duties, they could sell cheaper products than the Swiss.

Bulova and Timex presented significant problems for Japanese and Swiss manufacturers. Both corporations followed the American system of mass production. Employing a combination of sophisticated production technology and labor flexibility (through internationalization of production), Bulova produced a range of products spanning all price categories. Bulova's strength was the medium price range. The company produced hundreds of different styles in its Swiss factories. An international production system maximized site-specific advantages such as skill levels, technology, and markets.[10] The company's international orientation provided important opportunities to test-market new products. By having a strong brand policy and aggressively marketing products, Bulova moved into markets worldwide. At the high end, with its aggressively marketed tuning fork technology, Bulova was unique.

Alternatively, Timex sold a product that was cheap, simplified, and standardized. It was therefore easily mass produced. The company developed highly efficient, dedicated production equipment to produce huge volumes of standardized products. Timex also engineered true interchangeability. Parts could be exchanged not only within but between plants [21]. Because the U.S. lacked skilled watch workers, Timex pursued a capital-intensive production strategy. Machines were automated to reduce human involvement to a minimum. The company designed a dramatically simplified but well-manufactured watch with a relatively long life.[11]

But Timex did not confine itself to the low-price market segment. By the early 1960s Timex had developed a low-priced higher quality jeweled watch line. In addition to its traditional and effective distribution channels (high traffic locations such as drug stores), Timex introduced its watches into jewelry stores and other, more conventional watch sales outlets. Within 20 years the company had gone from bankruptcy to control of 45 percent of the U.S. market and 86 percent of U.S. domestic watch production.[12]

The Hong Kong Industry[13]

While the Swiss were battling for market share with U.S. and Japanese firms, the Hong Kong industry emerged. From its experience as Japan's low-cost assembly location and a long standing preeminence in case and bracelet manufacturing, in less than 30 years the Hong Kong watch industry rose to become the world's volume leader. With little capital investment, Hong Kong watchmakers developed the capacity to produce thousands of watch models each year. Assembling in excess of 300 million units in 1988, Hong Kong produced more watches than any other nation. Based on value of exports, the country recently surpassed Japan to become the second largest watch producer, behind the Swiss.[14]

Early in the 1970s, with advances in diode technology, Hong Kong watchmakers moved into light-emitting digital (LED) display watches. LED watches dominated output for a short time, but declined when liquid crystal display watches emerged later in the decade. Until the early 1980s, this newer technology dominated the Hong Kong watch industry. In the early 1980s, the emergence of quartz analog watches breathed new life into the industry. Like other components of watch manufacturing, quartz analog watch production in Hong Kong relied upon foreign parts. And because it also required a higher level of capital investment, Hong Kong's production system took time to adjust.

But adjust it did. In the late 1980s quartz analog watches began to dominate the industry, and evidence of their growing importance is striking. In 1980 digital watches accounted for approximately 60 percent of the value of total watch output. Analogs made up only 8 percent, and mechanical watches accounted for the remainder. In 1984 quartz analogs and digitals each made up approximately 43 percent of total output value [13]. By 1988 quartz analog watches dominated the market, accounting for 82 percent of total output by value. Digital watches made up only 12 percent. Because of its extremely fluid industrial structure, a rapid and complete transformation of the watch industry's product mix was possible [8,13].

Hong Kong's flexibility to respond to technological change derives from the fact that its watchmaking industry is a user, not a producer, of new technology. Hong Kong has been unable to develop its own movement technology due to a lack of skills and capital investment. The "sweatshop" nature of the industry means that labor absorbs the cost of change. Watchmaking in Hong Kong is ephemeral. Like other low-wage assembly industries, Hong Kong's momentary advantage can evaporate with the slightest increase in wages. A significant portion of watch assembly is already done in mainland China. Thus, while at the moment watchmaking thrives in Hong Kong, the industry has little long-term attachment to the island.

TECHNOLOGICAL CHANGE AND INDUSTRIAL INSTABILITY

Until the 1970s, the world watch industry grew steadily, and production was shared among three countries, the U.S., Japan, and Switzerland.[15] Trade liberalization in the 1950s, coupled with GATT and U.S. tariff reductions in the 1960s, set the stage for enormous expansion of markets in the 1970s. In a span of 10 years the market for watches doubled from 230 to 450 million watches [41].

In the early 1970s world demand for watches was overwhelmingly for mechanical devices. Only 2 percent of export sales were electronic watches. But in just two decades, the structure of demand changed. The competitive terrain shifted from precision based on mechanical know-how to accuracy based on electronic engineering. By the late 1980s electronic products comprised 76 percent of world consumption—approximately 60 percent digital, and the remainder analog. While the Swiss were the first to develop electronic watch technology, competitors succeeded in commercializing it.

Science Replaces Art in Watch Manufacturing

The introduction of electronic watches in the early 1970s had a profound impact on the Swiss share of world markets (Table 2.3). In 1974 Swiss watches made up 40 percent of the world export market (by volume). Ten years later this figure had fallen to 10 percent. The loss occurred almost en-

TABLE 2.3. Export of Watch Movements and Completed Watches 1951–1980 (thousands of units)

	Japan	Switzerland
1951	31	33,549
1955	19	33,742
1960	145	40,981
1965	4,860	53,164
1970	11,399	71,437
1975	17,017	65,798
1980[a]	68,300	50,986

[a]Includes movements [21].

tirely in the high volume, low- and medium-price watch market segments.

How was it that the Swiss share of world markets fell so precipitously? The watch cartel insulated Swiss manufacturers from the effects of inter-firm competition. Enjoying (volume) control of the world market (based on mechanical devices), it was easy for firms to become myopic about external events and new technology introduced by distant competitors. Because ASUAG looked only to members of the Swiss Watch Industry Federation (FH) for market information, new developments outside Switzerland did not filter into existing information channels.

When pressured to incorporate radical technological innovations, the Swiss industry proved unprepared to commercialize new ideas. Although inventions were very frequent, industry leaders were often skeptical about the viability of new proposals—particularly if they implied a radical reorientation of existing timekeeping methods. As one leading watch family head commented on the industry's failure to capitalize on tuning fork technology: "Every day someone came to the factory door with a so-called innovation. Claims of new and different technologies were a dime a dozen. Given production pressures, problems and uncertainties, what was one to do?" As Morgan Thomas notes, "many firms attempt to screen basic science and technical knowledge relevant to the firm's mission" [37]. This skeptical complacency proved costly when Hetzel, the Swiss inventor of tuning fork technology was ignored by Swiss watch manufacturers. After he successfully commercialized his new technology in the United States, the Swiss were forced into a defensive position just to gain access to the new technology.

The Organizational Structure of the Swiss Watch Industry

The tightly articulated network surrounding watch manufacturing strengthened the status quo. The watch industry was heavily geographically concentrated in the Jura Mountain region. To the outside world towns were identified by the factories of either major manufacturers (Longines in St. Emier; Tissot in Le Locle), or by specific watch products (Bienne and SMH formerly made ASUAG watch movements).

Regional institutions were interwoven into the fabric of the industry. Educational institutions were steeped in watchmaking tradition, turning out skilled workers who spent up to four years learning to make watches from start to finish. Machine tool firms such as Dixie, the originator of the jig bore, provided equipment to parts houses and claimed world preeminence in the manufacture of precision tools. Banking institutions were deeply implicated in the fortunes of the watch industry. In the early 1970s regional banks were known to have as much as 50 percent of their loanable funds invested in family-run watch-related enterprises. And the industry made heavy investment in collective R&D laboratories. The complicated web of watch manufacturing permeated the core of the region's social, political, and economic institutions.

Watch manufacturing's fragmented production structure also presented problems. Subcontracting levels were high, and the region's dominant firms could not exercise control over the myriad component producers. Fixation with precision had lulled the region's firms into believing they were invulnerable to external forces [5,11]. Supreme precision, however, did not require a theoretical understanding of new scientific developments; rather it necessitated great attention to detail. As Pierre Rossel notes, "the region's firms were unprepared to overcome a technological paradigm shift that devalued the region's long-standing comparative advantage" [28].

Transferring a foreign technology into existing products was crippled by a manufacturing culture steeped in tradition. Rapid change was the antithesis of watch culture which rewarded patient methodical actions within an existing technological trajectory. This was the key. The transition from mechanical to electronic movement manufacturing called into question the heart of the Swiss watch industry. To say that precision metal machining no longer ruled the sacred domain of timekeeping accuracy was simply too much for the centuries-old

Swiss tradition to endure. Rather than embracing this new threat as they had done when confronted by the American system of mass production, they chose to diminish its significance—with grave consequence.

Invention Does Not Guarantee Innovation: Technological Discontinuity and the Advent of Quartz

Organizational limitations inhibited ASUAG, the major movement producer, from moving into quartz. Because its market was literally hundreds of mechanical watch assemblers, no individual firm's demand was enough to persuade ASUAG to commit to one quartz movement design. But neither did any single manufacturer have an incentive to switch technologies. And even when ASUAG recognized the importance of quartz technology, the company lacked the marketing capability to successfully sell a quartz product to its primary market, Swiss assemblers [38]. A captive producer (unable to sell movements outside Switzerland), ASUAG lacked the incentive, necessity, and the ability to develop the marketing skills to compete internationally. Simultaneously, key watch manufacturers such as SSIH were unable to decide upon a quartz model. They therefore invested in numerous efforts to develop a quartz movement.

As with other challenges, the Swiss responded initially to the new quartz threat. Convinced that quartz was a passing fancy, nonetheless the Swiss rose to confront the new menace. Setting technicians to the task, the Swiss produced the first quartz watch movement simply to show that it could be done. But the elaborate network then stopped in its tracks, confident that quartz would eventually be related (as the electric watch had been) to the status of curio.

By the time the Swiss developed an industry-wide response to quartz technology, they lagged two years behind the Japanese. Having developed the initial technology, they failed to commercialize it. The Swiss are not alone in this fate. As Hoffman notes, "an innovation may be a technical success but a commercial failure in the innovator firm but a commercial success in the imitator firm" [12]. While the Swiss could claim that they were the first to develop a quartz watch (1971), they had to buy the necessary accompanying semiconductor technology from the Americans. Increased investments in R&D could not overcome the Swiss lag in microelectronics technology. The Swiss failure to act in the face of quartz technology could perhaps be blamed on bad judgement. But the problem was more fundamental and went to the heart of its highly fragmented network system of production.

The Swiss did not anticipate that the new technology would dominate the market in such a short time. But as integrated circuit prices fell precipitously, quartz watches became increasingly affordable. Because the Japanese had been vertically integrated since the 1960s, companies such as Seiko were poised to take full advantage of manufacturing developments occurring at various stages in the watch manufacturing process—further cementing their technological lead. Simultaneously, they could cross-subsidize component manufacturing, reducing per unit prices while raising per unit performance. The late development of Swiss domestic production of integrated circuits in a free-standing enterprise could not take advantage of information passing between component producers and watch manufacturers.

Unlike the Japanese watch manufacturers who saw semiconductor technology as an end in itself, the Swiss' forays into microprocessor technology were oriented strictly toward watches. This end-market focus did not facilitate synergies between semiconductor manufacturers and a wide range of users. In contrast with Japanese watchmakers who had a commercial electronics industry to rely upon for market outlets, the Swiss watch industry had to go it alone. Given the size of the watch industry's demand for chips, it was difficult to operate a chip production facility at optimal scale. It also made investments in R&D very costly per unit of expected demand for chips. Based on the network tradition, companies such as SSIH attempted to overcome their lack of technological capacity through a joint R&D project with Battelle. FH also initiated a joint R&D project to lessen Swiss dependence on U.S. semiconductor technology. With the FH, Brown Boveri and Philips of the Netherlands formed FASELEC, a laboratory to develop Swiss semiconductor production capacity. By the mid-1960s it was hard to judge the success of the venture because operations were never made public [18,39].

Complicating matters further was the rapid development of digital display technology. The commercialization of digital watches occurred with lightening speed. Digital display technology increased demand for quartz watches, leading to further price declines. This time the competition included American semiconductor manufacturers producing their own brands. Price reductions were dramatic, and by 1975 Texas Instruments had introduced a very inexpensive digital watch in a plastic case for $19.95. Although problems with the battery momentarily resulted in high reject levels, and consumers really wanted watches with move visual appeal, it did not take long to solve these problems. Battery longevity was vastly increased, watch designs improved aesthetically, and within an astonishingly short period of time, digital watches took over a large share of the market.

The Swiss responded slowly to change in digital technology largely because when it was introduced, it was crude. Given what promised to be a reasonably long developmental period between the introduction of the quartz technology and its eventual market success, the Swiss were understandably skeptical. As Dosi characterizes this moment,

> Especially when a technological trajectory is very "powerful", it might be difficult to switch from one trajectory to an alternative one. Moreover, when some comparability is possible between the two (i.e., when they have some dimensions in common) the frontier on the alternative (new) trajectory might be far behind that on the old one with respect to some or all the common dimensions. In other words, whenever the technological paradigm changes, one has got to start (almost) from the beginning in the problem-solving activity [6, p. 154].

The quartz "problem" permeated the Swiss watch manufacturing network. Every segment of the industry was affected. The rapid development of quartz means there were now many sets of tools needed to produce cases and dials. Uncertainty in both technology and consumer preference forced the Swiss watch companies to compete in three watch markets—digital, tuning fork, and quartz. The succession of innovations and new model development resulted in excess inventory. It seemed that just as a watch was developed, it became obsolete. During this period of rapid technological change, Swiss firms (and others) were forced to take back and in many cases write down inventory—an extremely costly endeavor [39]. The problems of the industry did not become widely apparent, however, until hidden reserves were consumed, and firms were forced to reveal their weakened position.

The Limits of the Network

The experience of the Swiss watch industry is indicative of the turmoil experienced when a new technological trajectory unfolds. Signals about which direction the technology will ultimately take are filtered through networks of institutions which often have competing short-term interests. In the case of the watch industry, firms had a vested interest in mechanical watchmaking. They were receiving positive signals about their existing product, and demand was strong. Therefore suggestions about a possible technological shift seemed misplaced. While the market provides a good focusing device after a decision is taken by industry participants, it is rarely helpful in deciding *ex ante* which direction the technology will ultimately take. As Dosi suggests,

> . . . the point we wish to stress, however, is the general weakness of market mechanisms in the *ex ante* selection of technological directions especially at the initial stage of the history of an industry. This is, incidentally, one of the reasons that militates for the existence of "bridging institutions" between "pure" science and applied R&D. Even when a significant "institutional focussing" occurs, there are likely to be different technological possibilities, an uncertain process of search with different organizations, firms and individuals "betting" on different technological solutions. With different competing technological paradigms, competition does not only occur between the "new" technology and the

"old" one which it tends to substitute, but also among alternative "new" technological approaches [6, p. 87].

The introduction of the electronic watch resulted in unprecedented change in the organization of watch production. The differences between electronic and mechanical watches were dramatic. Whereas labor costs constituted as much as 70 percent of a mechanical watch, in electronic watches labor costs were very low (less than 10 percent). Another major difference was the control of technology. The Swiss effectively controlled mechanical watch technology (due to the watch statute), and Bulova controlled the tuning fork. Electronics were fundamentally different. The technology was widely available, thus increasing the likelihood of new competitors with little or no prior watchmaking experience. Given the evolution of electronics, it was almost a foregone conclusion that price declines would occur in tandem with increases in capability. Thus, even the cheapest watch could be a good watch.

Network Rigidities Hamper Industry Response

Internal industry organizational and cultural impediments hampered a rapid response to the electronic watch. For example, the production planning time horizon for mechanical watches differed radically from electronics. The manufacturing cycle was organized according to the lead time needed to manufacture tools and dies for the fabrication of a new caliber, or watch dimension. Once committed to a design, tools and dies were crafted to cut the necessary metal parts. After parts were manufactured, movements were assembled and sold. Introducing a new watch model took up to two years. With electronic watches there were fewer parts to be manufactured. Consequently the time needed to make a watch dropped dramatically. Thus, when the Swiss were faced with the need to shift to a new technology, they were already two years behind, given the differences in the manufacturing cycles.

Ironically, product variety further hampered the industry. Few factories specialized in a single caliber. Therefore, firms were unable to achieve economies of scale. And because most factories produced several calibers' parts, inventory overhead was costly. Parts were required for each caliber—resulting in huge volumes of work in process. And the manufacturing cycle had to be managed across a wide range of products from tool making to product assembly.

Manufacturers had no choice but to focus on quality to differentiate themselves from the assemblers. Moreover, marketing strategy dictated the need to produce a family of watches to preserve firm market share. Since manufacturers could not sell movements, they could not achieve sufficient economies of scale to enjoy minimum efficiencies. Low volume of output led to high prices.

The effort required to overcome technological deficiencies associated with quartz technology required an industry-wide response. Given the industry's weakened condition, no single firm could afford the costs of developing such an uncertain technology. Numerous industry associations were formed to develop the technology. This new form of collaboration created serious problems, however, because no single firm could appropriate the fruits of collective research and translate it into a competitive advantage to capture new markets. Unlike times past, when pursuit of new innovation formed the basis of market share, collective research became collective knowledge. Firms were compelled to embark upon research to create technological differentiation based on the original quartz innovation. These efforts were costly, uncertain, and occasionally unsuccessful.

Distribution

The Swiss also had to contend with a centuries-old distribution system built around the watch as a piece of jewelry. Mechanical watches were traditionally distributed through jewelry stores, and jewelers made steady profits on repair. But quartz technology threatened to change all that.

Swiss distribution outlets initially balked at

the quartz watch. Early rejection was partially attributable to awkward styling: electronic watches were bulky and unattractive [39]. But more importantly, watch distributors effectively stalled the introduction of Swiss quartz analog watches in defense of their own market for watch repair. Quartz watches were more accurate and relatively unbreakable compared with mechanical watches.

Unlike the Swiss, the Japanese did not have an age-old distribution system. Market channel conflicts did not confront Japanese quartz watch manufacturers. Indeed Japanese channel strategy selected outlets through which the benefits of quartz longevity and error-free operation were maximized. The quartz watch was easier to sell, and it was more accurate. Timing was also important. The Japanese quest for large markets occurred simultaneously with the retail revolution. Mass marketing greatly expanded the number of outlets for watches. By the 1970s consumers were more likely to buy a watch in a variety store than a jewelry shop.

By the mid-1970s the Swiss were running just to catch up. Major Japanese competitors introduced increasingly cheap, long-lived, and refined watches. They pursued a strategy of short production runs; each time improving upon previous designs and climbing the learning curve more rapidly. Because the manufacturing cycle for the electronic watch was much shorter than for the mechanical watch, the Japanese could experiment within a relatively short time period.

The final blow came when the benefits of quartz converged to produce a cheaper, smaller, thinner, stylish, and accurate woman's watch. Before quart efforts at further miniaturization, women's watches had been less accurate and more costly to manufacture than men's. Now accuracy no longer distinguished cheap from expensive watches. The entire basis of Swiss market hegemony—precision—had evaporated.

Reorganization and Rationalization

By the early 1980s the Swiss industry was in disarray. The international recession dealt the final blow to the Swiss watch industry's historic organization. Faced with operating losses and massive inventories, SSIH was eventually a victim of industry reorganization. The company could not solve the equation of low prices, wide assortment, small volume, rapid change, short delivery time, and large model series [38]. Seiko, Japan's largest watch producer, was able to respond because it had the market volume to offer a wide assortment with economical series, low prices, and short delivery. A single statistic says it all, "on the average Japan produced, under each brand name, 6 million watches in the 1970s compared with fewer than 100,000 in Switzerland" [16, p. 221].

In the early 1980s SSIH and ASUAG were forced by the banks to merge. While the national significance of the Swiss watch industry could not be abandoned, neither could industry organization be allowed to continue as it had in the past. The merged SMH Group was taken over by powerful Swiss industrialists. One of the most dramatic changes arising from the merger was the introduction of a wholly new product, the "Swatch," propelling the Swiss back into the low-priced segment of the market [2,23].[16]

SUMMARY, REFLECTIONS AND CONCLUSIONS

Over the course of the last 20 years the Swiss lost both volume market leadership and technological supremacy. Given the industry's well-tuned production system, high level of profitability, and persistent success in its traditional line of business, what precipitated this historic reversal?

Beginning in the late 1920s the industry organized as a cartel to reduce the opportunistic behavior of industry participants. The resulting structure, though highly efficient and profitable, outlived its usefulness. Following the rescission of the Statut de l'Horlogerie, the network structure of production, while efficient and flexible, was also fragmented. Faced with the need to shift from a technology based on mechanics to one based on electronics, a time-lag built into the fragmented system inhibited rapid information flow. Shifting technological systems required that institutions and other critical components of the existing system be substantially modified. But this task proved difficult. The 200-year dominance of the previous paradigm constituted an "outlook which focused the eyes and efforts of technologists, engineers, and institutions in defined directions" [6, p. 158]. Initially, the region did not have the training capacity to provide electronics engineers. These skilled workers had to be imported from outside. In the case of Swiss watches, the decades old distribution system promoted Swiss watches based on their mechanical precision. Other organizations which represented the industry, such as the FH, were still predicting mechanical watch supremacy as late as the early 1970s. Educational and technical institutions—the core of the region's production complex—took even longer to respond to the new technological regime.

Amid radical change, organizations could not form a single voice to respond. The watch industry's collective research efforts to pioneer new technology could not overcome organizational inertia and infighting that arose with the need to commercialize the new technology. Without detailed and prearranged specifications about how the benefits of research were to be distributed, institutional inertia slowed the process of change. Since no single firm could be the "first" to introduce the collectively developed innovation, each firm had to develop its own [39]. When industrial reorganization eventually occurred, efforts were insufficient to address the structural crisis. Longstanding inefficiencies embedded in the production system led many firms into bankruptcy, resulting in bank ownership of some of the region's most famous and successful firms.

As the Swiss case attests, elaborate network production systems suffer like any other organizational form in the face of unexpected technological change [36]. All prior means of governance are called into question. The peculiar advantage of decentralized systems are also potentially their greatest flaw. Extreme change often necessitates radical reorientation. In such instances the ability to respond rapidly dictates who will be the ultimate victors. Although watches represent a specific case, their longevity as a product should bring pause to pronouncements that network production systems are somehow immune to technological change.

APPENDIX—HISTORY OF WATCH TECHNOLOGY

Evolution of the Watch

Since its creation almost 300 years ago, the watch has remained remarkably the same. Up until the 1960s alterations in the watch occurred mostly to the exterior in response to fashion and consumer tastes. The internal mechanism remained stable. The advent of electronics represented the first significant departure in the internal functions of the watch. The impact of this new technological development was profound. Centuries-old traditions in the manufacture of watches were revolutionized over night. Prior to electronics, timekeeping accuracy was associated with the precision with which metal parts were cut, filed, and fitted together. The most accurate watches were made by a single craft worker who cut each part, polished each metal edge, and placed each tiny part in the watch movement. Tuning the watch to achieve accuracy required very careful attention to detail. A short description of the evolution of watches helps clarify the meaning of electronics to the industry. This section draws heavily from Knickerbocker's treatment of watch development [18].

The Standard Spring-Powered Watch

A standard mechanical watch consists of three groups of parts: the "ébauche," or movement blank; the regulating components; and other generic parts. The ébauche consists of the framework (or back bone) of the watch, the gear train, and the winding and setting mechanism. Regulating components make the movement work at a correct rate. The miscellaneous parts include the case, crystal, etc.

A mechanical watch is driven by a main spring which transfers stored energy (in the spring coil) to the gears that move the hands. The release of energy is controlled by the escapement mechanism. Although numerous escapement models were developed over time, the mechanism is constrained by the anchor fork to give up a precise amount of energy. The anchor fork rocks back and forth allowing the escapement wheel to advance in tiny increments, and these increments are converted by other gears to the watch hands.

The anchor fork moves in conjunction with a balance wheel that moves back and forth. The balance wheel is motivated by a hairspring that coils and uncoils keeping the balance wheel in motion. As the anchor fork disengages from the escapement wheel, it transmits enough power to the hairspring to coil it. As the hairspring uncoils, it rotates the balance wheel in the opposite direction. This motion rocks the anchor fork in the opposite direction, starting the next cycle of the regulating mechanism.

Not all watch movements are the same. Differences in movement quality relate to the technical composition of the parts used. Precision, ornamentation, and movement miniaturization differentiate a high from a low quality watch. Internal jeweling in watches does not reflect differences in quality. Jewels are used to reduce friction between touching metal parts. The majority of jewels used in the interior of a watch are made out of synthetic materials and do not add significant value to the watch. Within mechanical watches there is a qualitative difference between jeweled and pin lever watches. The pin lever watch contains a more simplified movement compared to a jeweled watch (an example is Timex). A pin lever watch has few moving parts and does not use jewels to reduce friction between metal parts.

The Electric Watch

In this century, the first major technological advance in watch movement manufacture was the introduction of the electric watch. The electric watch movement was made possible by World War II R&D developments in the miniaturization of motors and batteries. The electric watch was only a partial step away from the mechanical watch. The main spring and many of the components of the escapement were eliminated and replaced by current from a battery that drove a tiny balance wheel motor. The electric watch was introduced in 1957 and was available world wide by the 1960s. Because electric watches were no more accurate than most mechanical watches, they did not make major inroads in the medium- and high-price watch markets.

The Tuning Fork Watch

The second major innovation in watch technology, the tuning fork, had a profound effect on the watch industry. The tuning fork is stimulated by an electric current from a battery. The current causes the tuning fork to vibrate at 360 cycles per second. A tiny strip of metal connected to the tuning fork transfers the vibration to a set of gears, which like a conventional watch, drives the watch hands. Because the tuning fork vibrates 31 million times a day, the mechanism is arm more accurate than a mechanical watch. Tuning fork technology was invented in the early 1950s and became commercially available in the early 1960s. A women's version was eventually introduced in the early 1970s.

The Quartz Crystal Watch

The third, and most significant innovation in watch manufacturing occurred in the late 1960s with the use of quartz crystals to regulate increments of time. When electric current is passed through quartz it vibrates at very high frequency. Micro circuitry subdivides the crystal's frequency into electric pulses which drive the watch. In some cases the quartz is used to power a stepping motor, which is connected to a gear train that moves the hands. Quartz technology can also be used to stimulate a tuning fork device. In solid state watches the pulses are fed into integrated circuits that convert the pulses into minute and second time increments. This last type of watch incorporates no moving parts. The face and hands of a solid state watch are replaced with different methods to display time.

Changes to the Watch Face

Prior to the introduction of the quartz crystal watch, changes in the timekeeping mechanism were completely internal to the watch. With the advent of the quartz crystal, time could be displayed by conventional means (analog) or by digital display. Two primary displays are important: light-emitting diodes (LED) and liquid crystal display (LC). LEDs are semiconductors that emit light (much like a light bulb). Originally used in calculators, LEDs became fashionable in watches in the early 1970s. Because LEDs require considerable power they are not illuminated at all times. A push-button activates the display. In contrast, an LC display consists of a glass sandwich with a thin coating of electrically sensitive chemical between the glass plates. When a current is passed through an LC, the chemical changes its crystalline structure. The altered crystals reflect light coming from an outside source. While less power consumptive than an LED, LC's brightness and precision depend on the brightness of the external illumination.

The advent of quartz altered both internal and external features of the watch. With high levels of accuracy, the quartz watch could also incorporate numerous functions. Within an incredibly small space, a quartz watch could include multiple time-keeping mechanisms including alarms and other sophisticated functions. In combination, quartz technology revolutionized the watch industry.

Notes

1. The Appendix provides a brief review of watch technologies.

2. The use of the term "disintegration" is meant to imply chaos not the evolution of the spatial division of labor as used by Scott [32].

3. World War I created severe disruptions in the world watch market. Russia, a major Swiss market, closed its borders to international trade, while other countries raised protectionist barriers in attempts to preserve domestic industries. Demand for Swiss watches declined precipitously between 1916 and 1921.

4. It was the larger firms which had made the capital investments in equipment that wanted to inject order into the historically anarchistic industry. To recoup capital investments, the more advanced firms had to control the small firms that easily sprang up and produced cheap watches [15].

5. The SGT holding company also acquired two American watch companies, Waltham and Elgin.

6. The Hamilton Watch company is American in name only. The Swiss corporation SMH owns the brand.

7. Until recently these international brands did not operate their own manufacturing plants. Cartier (one of the largest international brands) previously subcontracted production to EBEL, a Swiss prestige manufacturer. The French firm recently opened a plant to manufacture its own products.

8. When domestic wages began to rise, the Japanese quickly shifted assembly to Hong Kong where wages were lower (creating a spatial division of labor to ensure low price and timely product delivery).

9. Today Seiko has major market share in certain types of semiconductors, micro machinery, miniature circuit board manufacturing, and small plastic and rare metal parts. The company has capitalized on the original core skill by pursuing markets and product niches in related fields.

10. For example, Accutron was made in the U.S. where technology levels were high despite lesser manual labor skills. Medium- and low-priced mechanical watches were manufactured by high-skilled Swiss workers.

11. Given that their watch was cheap, Timex made no pretense of providing after sales service. When the watch stopped running, it was simply thrown away and a new one purchased.

12. Like Bulova, Timex established international market presence and production capacity. The company had 20 plants scattered around the globe. Each market was carefully analyzed, and sales strategies were adjusted according to local customers [18].

13. In 1988 there were approximately 1,386 watch-making firms registered with the Hong Kong government. From 1983 to 1988 employment increased steadily from 25,200 to 26,444. Over 90 percent of firms employ fewer than 50 employees. The structure of the watch industry is remarkably similar to that of electronics in Hong Kong. A recent study by the Hong Kong Government Department of Industry indicated that many of the structural weaknesses evident in the watch industry are also apparent in electronics. These include lack of local brands and design capacity, a fragmented production structure, and low levels of capital investment. Hong Kong watch prices (FOB) are extraordinarily low by world standards. The average wholesale price of a watch in 1988 U.S. dollars was $3.00. Advertised wholesale prices ranged from $4.00–10.00 per watch (with a lead time of between 25 and 60 days). Even jeweled watches cost a fraction of those manufactured in Switzerland.

Orders can be as small as 100 watches, and in some cases firms have no minimum lot size.

14. Hong Kong is an established forerunner in innovation and exportation of watch parts, cases, bands, and accessories. The case and band industry is well-developed, and Hong Kong firms export finished products to major watch producing countries that include Japan, Switzerland, and the U.S. Surprisingly, even up-scale companies such as Cartier use Hong Kong watch bands. In 1984 (the latest year for which statistics are available), there were 484 case-making firms employing 9,200 workers in Hong Kong. This is almost four times the number of case producers in Switzerland.

15. We do not include Eastern Block country production in these figures. A considerable volume of watches is produced in the Soviet Union, East Germany, and other Eastern Block nations [18].

16. Swatch is a plastic watch manufactured at high volume using advanced automation and assembly-line methods. But the real innovation is in marketing the watch as a high fashion, mood-oriented product. Ownership of multiple models is stressed, and marketing is targeted toward specific age groups [1].

REFERENCES

1. J. Arbose, The Turnaround in Swiss Industry: How the Smokestack Crowd Learned Marketing, *International Management (UK)* 42 (1) (1987) 16–23.
2. *Business Month,* Up From Swatch (March 1988) 57.
3. Casio Computer, I.J. *Corporate Annual Reports* (1984–1988).
4. Citizen Watch Company (Japan), *Corporate Annual Reports* (1985–1988).
5. O. Crevoisier, M. Fragomichelakis, F. Hainard, and D. Maillat, *Know-how, Innovation, and Regional Development,* Paper presented at the 29th European Congress, Cambridge, England, September 1989.
6. G. Dosi, Technological Paradigms and Technological Trajectories, *Research Policy* 11 (1982) 147–162.
7. G. Dosi, Technological Paradigms and Technological Trajectories, in: C. Freeman (ed.), *Long Waves and the World Economy* (Butterworth, England, 1984).
8. A. Glasmeier, *The Hong Kong Watch Industry,* Final report to the Tissot Economic Foundation, Le Locle Switzerland, Working Paper 14, Graduate Program in Community and Regional Planning, University of Texas at Austin, 1989.
9. A. Glasmeier, *The Japanese Small Business Sector,* Final report to the Tissot Economic Foundation, Le Locle, Switzerland, Working Paper 16, Graduate Program in Community and Regional Planning, University of Texas at Austin, 1989.
10. A. Glasmeier and R. Pendall, *The History of the World Watch Industry,* Preliminary report to the Tissot Economic Foundation, Le Locle Switzerland, Working Paper 13, Graduate Program in Community and Regional Planning, University of Texas at Austin, 1989.
11. F. Hainard, *Savoir-faire et culture technique dans l'Arc Jurassien* (UNESCO, Université de Neuchâtel, 1988).
12. W.D. Hoffman, Market Structure and Strategies of R&D Behaviour in the Data Processing Market—Theoretical Thoughts and Empirical Findings, *Research Policy* 5 (1976) 334–53.
13. Hong Kong Government Industry Department, Industrial Profile, Hong Kong's Watches and Clocks Industry (Hong Kong, October 1985).
14. *International Management,* Hattori Inspires Seiko to a High-Tech Future, *International Management* 39 (7) (July 1984) 20–25.
15. E. Jaquet and A. Chapuis, with the co-operation of G.A. Berner and S. Guye, *Technique and History of the Swiss Watch Industry* (Spring Books, London, 1970).
16. P. Katzenstein, *Small States and World Markets: Industrial Policy in Europe* (Cornell University Press, Ithaca, New York, 1985).
17. A. Knickerbocker, *Notes on the Watch Industries of Switzerland, Japan and the United States* (abridged) (Harvard Business School, Boston, 1974).
18. A. Knickerbocker, *Notes on the Watch Industries of Switzerland, Japan and the United States* (revised) (Harvard Business School, Boston, 1976).
19. D.S. Landes, Watchmaking: A Case Study of Enterprise and Change, *Business History Review* 53 (1) (Spring 1979) 1–38.
20. D.S. Landes, *Revolution in Time: Clocks and The Making of the Modern World* (Belknap Press, Cambridge, MA, 1983).
21. D.S. Landes, Time Runs Out for the Swiss, *Across the Board* 21 (1) (January 1984) 46–55.
22. D. Maillat, *Technology: A Key Factor for Regional Development* (Georgi Publishing Company, Saint-Saphorin, 1982).
23. L. Pilarski, Can New Management Team Keep Swiss Watches Ticking? *International Management* 40 (6) (June 1985) 57–58.
24. Conversation between Tihomil Radja and Amy Glasmeier, Switzerland (August 1989).
25. Conversation between Tihomil Radja and Amy Glasmeier, Switzerland (November 1990).
26. T. Radja Unpublished tables (1990).
27. Conversation between Rene Retornaz and Amy Glasmeier, Neuchâtel, Switzerland (August 1989).

28. P. Rossel, Unpublished note, Tissot Economic Foundation, Le Locle Switzerland (1990).
29. C. Sabel, Reemergence of Regional Economies, in: J. Zeitlin and P. Hirst, *Reversing Industrial Decline? Industrial Structure and Policy in Britain and Her Competitors* (Berg, Leamington Spa, 1989).
30. A. Saxenian, *Regional Networks and the Resurgence of Silicon Valley,* Working Paper 508, Institute of Urban and Regional Development, University of California at Berkeley, 1989.
31. A. Saxenian, *The Origins and Dynamics of Production Networks in Silicon Valley,* Working Paper 516, Institute of Urban and Regional Development, University of California at Berkeley, 1990.
32. A. Scott, *Metropolis* (University of California Press, Berkeley, 1988).
33. A. Scott and M. Storper, High Technology Industry and Regional Development: A Theoretical Critique and Reconstruction, *International Social Science Journal* 112 (1989) 215–232.
34. Seiko, Hattori I.J. *Corporate Annual Report* (1985).
35. G.P.F. Steed, The Northern Ireland Linen Complex, 1950–1970, *Annals of the Association of American Geographers* 64 (3) (September 1974) 397–408.
36. M. Storper and R. Walker, *The Capitalist Imperative: Territory, Technology and Industrial Growth* (Basil Blackwell, Oxford, 1989).
37. M. Thomas, Growth and Structural Change: The Role of Technical Innovation, in: A. Amin and J. Goddard (eds.), *Technological Change, Industrial Restructuring and Regional Development* (Allen and Unwin, London, 1986).
38. Conversation between Luc Tissot and Amy Glasmeier, Zurich, Switzerland (August 1989).
39. Conversation between Luc Tissot and Amy Glasmeier, Austin, Texas (February 1990).
40. Union Bank of Switzerland, *The Swiss Watchmaking Industry,* UBS Publications on Business, Banking and Monetary Topics No. 100, Zurich (1986).
41. Union Bank of Switzerland, *Economic Survey of Switzerland* (1987).

CHAPTER TWO

INNOVATION OVER TIME AND IN HISTORICAL CONTEXT

Technology evolution, innovation, and change are dynamic processes. They play out over time and are history-bound; to understand the impact of a major innovation, it is necessary to understand what led up to it and what is likely to follow it. A technological discontinuity, such as the introduction of electronic watches, inaugurates an era, a cycle of technical evolution that often overturns existing industry structures. However, it would be a mistake to focus solely on discontinuities. Predictably, they lead to eras of ferment and technical experimentation, as new technologies substitute for older ones, and variants of the breakthrough idea vie with one another for acceptance. One variant usually emerges as an industry standard, and when it achieves dominance, an era of incremental elaboration of the standard ensues, changing the competitive landscape.

Anderson and Tushman describe and illustrate this cycle in detail, suggesting that long periods of incremental change are punctuated by discontinuities and dominant designs. Tyre and Orlikowski show that firms adapt to new technologies in a similar way, alternating short bursts of change with longer periods of gradual, routine progress. Gould highlights the critical importance of standards, using the introduction of the QWERTY typewriter keyboard layout to show how even chance factors can become "locked in" when a standard achieves enough momentum behind it. Cusumano and his coauthors investigate the strategies firms use to establish the designs they back as industry standards, illuminating their ideas with a careful examination of the Beta/VHS videocassette standard war. Sanderson and Uzumeri show how a firm builds on a standard platform, creating competitive advantage by building a web of small extensions around a basic design. Competition becomes a struggle of product family versus product family, not model versus model. Gomory's article highlights the importance of slow, patient accumulation of innovations during a period of incremental change. He argues that managers need to focus less on dramatic, radical breakthroughs, because the gradual accretion of seemingly minor advances can threaten firms that fail to keep up with the "ladder" of progress. However, when the next discontinuity arrives it is extremely difficult for established firms to adapt to it, because they have learned to excel at incremental innovation when that was called for. Morison's classic article discusses the sources of resistance to a clearly superior new technology, showing how strong values and beliefs that once were functional create institutional shortsightedness. Cooper and Smith dissect the characteristic mistakes that industry veterans make when faced with radical, superior technologies that challenge a prevailing regime.

3

Managing Through Cycles of Technological Change

Four lessons are drawn from the evidence that technology progresses in a series of cycles, hinging on discontinuities and the emergence of dominant designs.

PHILIP ANDERSON
MICHAEL L. TUSHMAN

". . . industrial mutation . . . incessantly revolutionizes the economic structure from within. This process of Creative Destruction is the essential fact about capitalism."

—JOSEF SCHUMPETER, 1942.

We are managing in what Peter Drucker has termed "the age of discontinuity." Examples of revolutionary technological changes that transform industries abound. Ceramic engine parts will replace metal engine parts in the next decade, thanks to their high strength-to-weight ratio and resistance to heat. Flat-screen displays will obsolesce today's bulky cathode-ray tubes in television screens and computer monitors. Optical disks capable of storing billions of bytes will supplant today's magnetic fixed disks for mass computer storage. Lithium batteries will supersede today's lead-acid technology.

It is precisely this sort of discontinuous change that brings about "creative destruction," the overturning of established industry structures which Schumpeter saw as the fundamental engine of capitalist progress. Building on a tradition extending back to the 1950s (see for example Strassmann, 1956, and Bright, 1964), Richard Foster (1986) argues that industry leaders become losers because they have difficulty managing technological discontinuities—movements from one technology to another with inherently higher limits.

Examples of creative destruction based on both product and process revolutions abound: the shift from vacuum tubes to semiconductors overturned the dominance of firms such as RCA and Sylvania; with the installation of new, energy-saving cement manufacturing technology, eight of the ten largest American cement makers were acquired by foreign firms between 1973 and 1980.

Managing through periods of upheaval and transformation requires that we develop a useful model of technological change. Are there predictable patterns of innovation that recur time and time again in industry after industry? Are there predictable consequences of technological discontinuities? Who pioneers discontinuous innovations? When do leaders become losers?

Reprinted from *Research/Technology Management*, May/June 1991, pp. 26–31, with kind permission from the Industrial Research Institute, Washington D.C.

Foster's depiction of technological progression through a series of S-curves suggests that technological change follows a cyclical pattern. The best-known model of technological change, the Abernathy/Utterback model, originally viewed technological progress as a single cycle, leading toward more process and less product innovation and culminating in the "productivity dilemma." Yet more recent updates of this framework in the early 1980s also conclude that technological change is cyclical—"dematurity" can in effect set the clock back and return an industry from a "specific" to a "fluid" state.

Our study of the entire history of three industries (see editorial box, next page) leads us to conclude that technology progresses in a series of cycles, hinging on technological discontinuities and the emergence of dominant designs. Here, we discuss:

- The cyclical nature of technological change.
- The influence of "competences."
- The empirical character of observed technology cycles.
- Who pioneers discontinuities and dominant designs.
- The process of "creative destruction."
- The implications of technology cycles for managers.

TECHNOLOGY CYCLES

As Foster's notion of a series of S-curves suggests, an industry evolves through a *succession* of technology cycles. Each cycle begins with a *technological discontinuity.* Discontinuities are breakthrough innovations that advance by an order of magnitude the technological state-of-the-art which characterizes an industry. They are based on new technologies whose technical limits are inherently greater than those of the previous dominant technology, along economically relevant dimensions of merit.

To illustrate, examine Figure 3.1. The manufacture of window glass has been characterized by three great discontinuities. In the 19th century, skilled artisans blew molten glass into long cylinders, which were cut with a wire and flattened into glass sheets. In 1903, the Lubbers process substituted an automatic blowing machine for the artisan. In 1917, the Colburn machine, which drew a con-

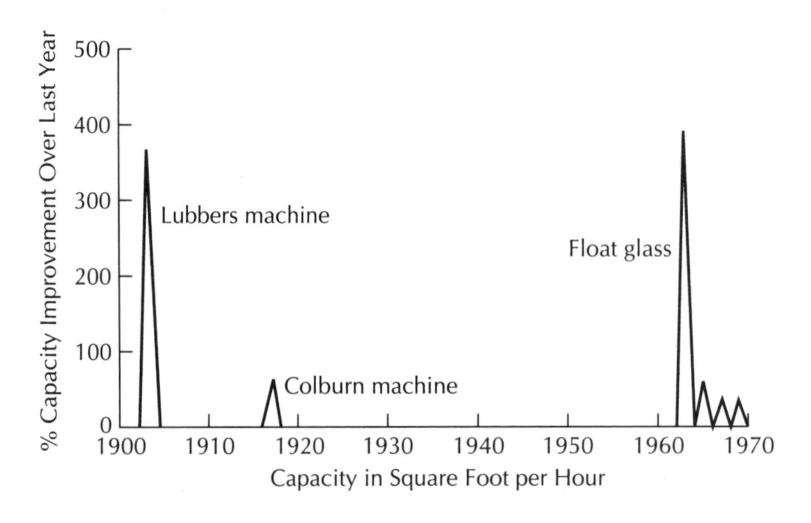

Fig. 3.1. Three great discontinuities mark the development of machinery for manufacturing window glass in the United States.

How the Study Was Conducted

Lehigh University's Center for Innovation Management Studies and Columbia University's Strategy Research Center funded our three-year investigation of the entire history of the U.S. minicomputer, cement and glass (containers, plate and windows) industries. We tracked the entry and exit of firms in these industries from their inception via directories extensively cross-checked with archival sources and data from trade associations and consultants. We also tracked a single key performance measure for each industry to empirically identify discontinuities: we focused on kiln capacity for cement, machine capacity for glass containers and flat glass, and CPU speed for minicomputers. To measure dominant design, we looked at new process installations for glass and cement and sales by model for minicomputers. A dominant design was considered to have emerged when one fundamental architecture accounted for 50 percent or more of new product sales or process installations for three straight years. Complete details of the data sources, methodology, and statistical analyses performed are contained in Anderson (1988), available from University Microfilms.

Important limitations to the study should be noted. First, due to the length of the time series examined, only three industries are included; it would be unwise to over-generalize the results. In particular, these findings may not completely apply to service industries or sectors where an oligopoly exists, where regulation is an important factor, or where strong patent positions are common. Furthermore, the study looked at only one performance measure per industry (due to limitations of historical data); in most industries, one would measure the technical frontier using several parameters. We were only able to examine survival and exit rates; the study draws no conclusion about the effect on firm performance of being first-to-innovation or first-to-standard.

—P.A. and M.T.

tinuous ribbon from a tank of molten glass, was introduced. In 1963, the Pilkington float glass process was introduced in the United States, producing a continuous ribbon by floating molten glass across a bed of molten alloy. In each case, a process with inherently higher limits redefined the state of the art, increasing machine capacity by an order of magnitude while lowering costs and improving quality.

Each technological discontinuity inaugurates a *technology cycle* (Figure 3.2). The breakthrough initiates an *era of ferment,* characterized by two processes. First, the new technology displaces its predecessor during an *era of substitution.* Though Foster argues that new technologies appear only when the old technology reaches its technical limits, often the older technology improves markedly in response to the competitive threat. Gaslight technology, for example, improved dramatically in the decade after the introduction of the Edison electric light; Apple has pushed the limits of 8-bit microcomputer technology forward dramatically since the appearance of 16-bit and 32-bit replacements for the once-dominant Apple II. Despite these improvements, Fisher and Pry (1971) demonstrate that in many cases, the substitution process proceeds with mathematical inevitability once a small initial penetration is achieved.

The second process partly overlaps the first. An era of *design competition* follows a disconti-

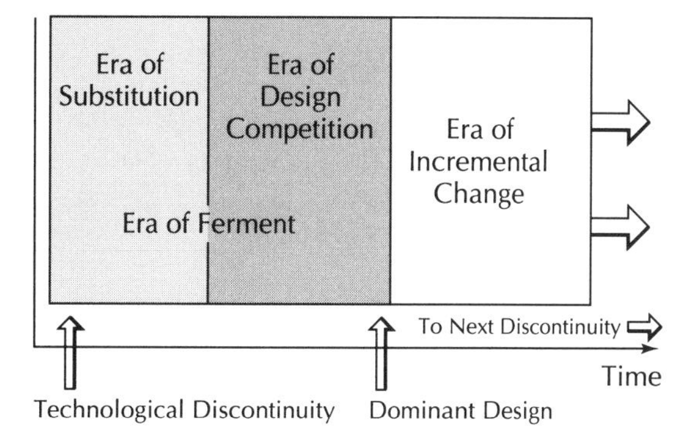

Fig. 3.2. Industries evolve through successions of technology cycles, each inaugurated by a technological discontinuity. (Reproduced with permission of copyright owner. Further reproduction prohibited.)

nuity. Radical innovations are usually crude, and are replaced by more refined versions of the initial product or process. Typically, several competing designs emerge, each embodying the fundamental breakthrough advance in a different way. Examples include the tremendous proliferation of automobile designs following Duryea's first auto, or the appearance of dozens of competing airplane models after the Wright brothers' invention.

The design competition culminates in the appearance of what Abernathy and Utterback (1978) term a "dominant design," also called a "technological guidepost" by Sahal (1981). This design is a single basic architecture that becomes the accepted market standard. Dominant designs are not necessarily better than competing designs, and they often pioneer no innovative features themselves. Rather, they represent a *combination* of features, often pioneered elsewhere, that sets a benchmark to which all subsequent designs are compared. Examples include the IBM 360 computer series, the Fordson tractor, and the Ford Model T automobile.

The emergence of a dominant design marks the end of the era of ferment and the beginning of a period of incremental change. Here, the rate of design experimentation drops sharply, and the focus of competition shifts to market segmentation and lowering costs (via design simplification and process improvement). Many scholars and R&D managers contend that it is the patient accumulation of small improvements that accounts for the bulk of technological progress. Though this may not be true in every case, there is little doubt that once a design becomes a standard, it establishes a trajectory for future technical progress and changes the basis of competition in the industry. This era of competition based on slight improvements on a standard design continues until the next technological discontinuity emerges to kick off a new technology cycle.

INFLUENCE OF "COMPETENCES"

The nature of the technology cycle is dramatically affected by the cutting dimension of *competence*. Some discontinuous innovations are *competence-destroying*. They obsolesce existing knowhow; mastery of the old technology does not imply mastery of the new. Firms must embark on a new learning curve which is essentially unaffected by the firm's existing knowhow, and technical professionals require new training. The transistor illustrates a competence-destroying product innovation; mastery of vacuum tube technology proved as much a hindrance as a help to engineers trying to understand semiconductor electronics, and the learning curve for firms struggling to master the technology was unaffected by the firm's vacuum tube knowhow. Similarly, float glass is a competence-destroying process discontinuity; a firm's knowledge of Colburn drawing technology conferred little advantage in mastering the Pilkington float glass process.

Other discontinuous innovations are *competence-enhancing*. These breakthroughs push forward the state-of-the-art by an order of magnitude, but build on existing knowhow instead of obsolescing it. Thus the turbofan jet engine is a competence-enhancing product innovation. It markedly improved engine performance, but built on existing knowhow instead of overturning it. The introduction of process control in cement kilns was a competence-enhancing process innovation. Computerization made possible enormous kilns, allowing cement manufacturers to employ their existing cement-making knowhow to make more and better cement than any human operator could produce.

Both product and process innovations may either enhance or destroy existing competences. Yet there is a fundamental difference between product and process innovations. Product innovations normally affect more links in the value chain than do process innovations. The customer must be made aware of new products; often, he is not aware of process innovations *per se*. New products often require distribution channels and suppliers different from those which serviced older products. Process innovations usually make the product better and cheaper without necessarily disrupting upstream and downstream linkages. Thus, a key factor is not only whether the core technical knowhow of an industry is disrupted by an innovation, but whether links in the value chain are overturned or reinforced by the new technology.

CHARACTERIZING THE TECHNOLOGY CYCLE

Discontinuities are generally uncommon, and their frequency varies greatly by industry. Nonetheless, they characterize both young and mature industries. We tracked 24 years of minicomputer data, over 100 years of cement industry history, and nearly 200 years of glass industry history and located only 17 discontinuities. The minicomputer industry passed through three discontinuities in a quarter-century, while the cement and glass industry experienced 50-year periods of incremental change. However, *every* industry we studied experienced at least one discontinuity since 1960, and the "mature" cement industry witnessed two.

A *single* dominant design *always* emerged following a discontinuity, except in two situations. When one discontinuity follows another very rapidly (within 3–4 years), a dominant design may not have time to emerge before the second new technology displaces the first. When several producers each patent their own proprietary process and refuse to license to others, a dominant design may not emerge. Otherwise, in every case a single product or process architecture accounted for over 50 percent of new installations. Ultimately one standard prevails; we did not observe cases where two standards coexisted or where the position of dominance rotated among several competing designs.

The original discontinuous innovation *never* became a standard. Some improved version of the initial breakthrough became the basis of a dominant design in every case. Furthermore, more often than not dominant designs lagged behind the state-of-the-art at the time they were introduced. The winner of the design competition is seldom at the industry's performance frontier; typically, the industry pushes the state-of-the-art forward during the era of ferment, then standardizes on a design that is *behind* the leading edge of the technology.

The length of the era of ferment (the lag from introduction of the new technology to establishment of a dominant design with 50 percent of the market) depends on whether the discontinuity enhances or destroys existing knowhow. It took longer for an industry to converge on a dominant design following a competence-destroying discontinuity than it took to converge on a dominant design following a competence-enhancing discontinuity. When existing knowhow is reinforced, the industry arrives at a standard relatively rapidly; when it is overturned, it takes considerably longer for the design competition to culminate in a single technological guidepost. Furthermore, when a series of discontinuities enhance the same underlying competence, the length of the era of ferment grew shorter in each successive technological cycle, bolstering the argument that the more familiar the underlying knowhow, the easier it is to reach a standard.

PIONEERS OF DISCONTINUITIES AND DOMINANT DESIGNS

A key competitive question is, when will a discontinuity overturn an industry—when will leaders become losers? Figure 3.3 summarizes our find-

Discontinuities	Product	Process
Competence Destroying	Newcomer	Veteran
Competence Enhancing	Veteran	Veteran

Dominant Designs	Product	Process
Competence Destroying	Veteran	Veteran
Competence Enhancing	Veteran	Veteran

Fig. 3.3. Veteran firms are more likely to pioneer each class of discontinuity and dominant design except the competence-destroying product innovation.

ings. Focusing on the first five firms to adopt an innovation, we observed that in general veterans—firms which competed in the industry before the discontinuity—are more likely to pioneer breakthrough innovations. This runs counter to the often-heard argument that revolutions usually come from outside an industry. It is often the case that the *initial* innovator is a newcomer to the industry, but when we look at the group of first-movers, we usually find that veterans predominate.

It is easy to understand why this is so when an innovation builds on existing knowhow. Firms which possess that knowhow—the veterans—are most likely to build on that expertise. It is also easy to understand why competence-destroying innovations are pioneered by newcomers. The new technology obsolesces what the veterans know, temporarily knocking down barriers to entry. Veterans are reluctant to adopt the new technology because it wipes out their considerable investments and forces them to change in fundamental ways. It is in this case that leaders are most likely to become losers. However, competence-destroying process innovations are typically pioneered by veterans, despite the fact that they are obsolescing their own process knowhow. We argue that veterans still are able to exploit strengths upstream and downstream in the value chain following a process discontinuity; only their core technical knowhow is overturned. As a result, veterans are willing to write off investments in existing facilities and expertise to exploit the price/performance advantage of the new technology.

Finally, dominant designs are *always* pioneered by veterans, whether or not they build on or destroy competences. The revolutionary is seldom the standard-setter. Recall that dominant designs seldom are state-of-the-art, and that industry experience is needed to understand what the market needs in a standard.

CREATIVE DESTRUCTION

Industries are characterized by waves of foundings and failures. A period when the failure rate is unusually high is often termed a "shakeout." The conventional wisdom is that overcapacity or downturns in demand cause shakeouts. By analyzing mortality rates, we found no relationship between changes in demand and failure rates. Instead, failure rates were remarkably higher during eras of ferment than in any other period. The inability to adapt to a new technical order seems to kill more firms than the inability to withstand a recession in the industry. Interestingly, only one American cement firm failed during the Great Depression; in contrast, dozens failed when confronted with the challenge of adapting to new kiln technology.

IMPLICATIONS FOR MANAGERS

The model of technology cycles provided here is one step toward developing what Foster terms "a language and a facility for talking about and directing technology." It allows managers in different industries to organize their view of the industry's technical history, and to compare the effects of various types of innovations on the industry's structure. Beyond this, we draw four principal lessons for managers from this research.

1. Expect discontinuities. They do not happen frequently, but they do occur even in mature industries, and they are watershed events. When evaluating potential discontinuities on the horizon, consider whether they would enhance or destroy fundamental competences in your industry. Consider developing competences that survive technological revolutions, such as flexible manufacturing capability or strong distribution channels.

2. When a discontinuity appears, expect an era of ferment culminating in a single dominant design (with the two exceptions noted above). Expect several designs to compete; expect one to emerge as a winner. The dominant design will seldom be a state-of-the-art architecture; it is usually introduced by industry veterans, and the time it takes to reach a design depends on whether the discontinuity is competence-enhancing or competence-destroying.

3. Realize that technological revolutions may be introduced by an industry newcomer, but the

How the Electric Auto Could Evolve

To see how the ideas developed in this research can help managers understand the probable evolution of a specific new technology, consider the predicted development of an electric automobile, a technology currently in its infancy:

A commercial electric automobile will become feasible following a breakthrough innovation, either in the power of electric motors (e.g., via superconductivity) or in power generation/storage technology (e.g., solar cells or improved batteries).

The initial electric automobile will be a crude design, which will be elaborated and altered by dozens of imitators (*unless* the innovation enjoys strong legal protection from the outset). The displacement of conventional automobiles by electric automobiles will follow a classic S-curve, whose takeoff will be greatly aided by one or two key sales (e.g., to government vehicle fleets or a major auto rental company). Rival versions of the initial breakthrough will compete for legitimacy and substantially improve the product's performance.

From the many versions of the initial innovation, one will emerge as the industry standard architecture, *de facto* or by regulation/agreement. This "dominant design" will account for over 50 percent of new electric automobile sales following its establishment.

Following the establishment of the dominant design, the competitive focus of this product class will shift from performance improvement via significant architectural variations to cost reduction, market segmentation, and development/elaboration of the infrastructure that supports electric automobiles. Improvements will be incremental, and virtually all successful models will incorporate the key features of the dominant design. This regime of continuous improvement will continue until another discontinuity overthrows this generation of electric automobiles.

If the discontinuity which paves the way for growth in this product class is a *process* discontinuity (e.g., superconducting power transmission), one should expect incumbent auto manufacturers and entrants from closely related fields (e.g., truck manufacture) to pioneer the new technology. The same would apply if the discontinuity is a component easily retrofitted to existing automobiles (e.g., a breakthrough battery). *Only* if the discontinuity involves a fundamental redefinition of automobile design (e.g., new concepts in motors, the body, power train, etc.) would we expect the process of "creative destruction" to replace today's vehicle makers with a new generation of companies spawned by the new technology.

—*P.A. and M.T.*

group of firms that adopt it earliest typically includes a majority of veterans. Only in the case of competence-destroying product discontinuities do we observe a preponderance of newcomers in the pool of first-movers. It is worthwhile to monitor potential competitors from outside an industry, particularly when you suspect that a new product technology can obsolesce existing knowhow. But more often than not, the pioneers of discontinuities are competitors you already know, not newcomers to the industry.

4. Consider the implications of the finding that technological change, not downturns in demand, is associated with shakeouts. Top management always pays attention to industry recessions and is willing to make painful cost-cutting moves when demand drops. Yet it is not this form of competition that threatens the very survival of the firm and its rivals. Maintaining the organization's ability to navigate the rapids of creative destruction brought on by technological discontinuities is the key to fulfilling management's first duty to shareholders—preserving their capital by ensuring the continuance of the enterprise. The ability to direct the firm's marketing and financial operations helps top managers improve a firm's profitability. The ability to direct process and product innovation affects not only profitability but the viability of the firm itself in a world of technological upheaval.

BIBLIOGRAPHY

Abernathy, William. 1978. *The Productivity Dilemma.* Baltimore: Johns Hopkins University Press.

Abernathy, William and Clark, Kim. 1985. "Innovation: mapping the winds of creative destruction." *Research Policy,* 14:3–22.

Abernathy, William and Utterback, James. 1978. "Patterns of industrial innovation." *Technology Review,* 2:40–47.

Abernathy, William; Clark, Kim; and Kantrow, Alan. 1983. *Industrial Renaissance.* New York: Basic Books.

Anderson, Philip. 1988. *On the Nature of Technological Progress and Industrial Dynamics.* Unpublished Ph.d. dissertation, Columbia University.

Bright, James. 1964. *Research, Development and Technological Innovation.* Homewood, Illinois: Richard D. Irwin.

Clark, Kim. 1983. "Competition, technical diversity, and radical innovation in the U.S. auto industry." In Robert Rosenbloom (Ed.), *Research on Technological Innovation, Management, and Policy:* 103–149. Greenwich, CT: JAI Press.

Fisher, J. and R. Pry. 1971. "A simple substitution model of technological change." *Technological Forecasting and Social Change,* 3:75–88.

Foster, Richard. 1986. *Innovation: the Attacker's Advantage.* New York: Summit Books.

Sahal, Devendra. 1981. *Patterns of Technological Innovation.* Reading, MA: Addison-Wesley.

Schumpeter, Josef. 1942. *Capitalism, Socialism and Democracy.* New York: Harper & Brothers.

Strassman, Paul. 1956. *Risk and Technological Innovation.* Ithaca, NY: Cornell University Press.

4

Exploiting Opportunities for Technological Improvement in Organizations

MARCIE J. TYRE
WANDA J. ORLIKOWSKI

We often hear that companies must learn to embrace change. This is particularly true of companies that are applying advanced technologies to improve their competitive position. The full advantages of such technologies cannot simply be purchased off the shelf; they are won by patiently and carefully tailoring the technology to fit a given firm's organizational and strategic context. At the same time, organizational skills, procedures, and assumptions within the firm need to be adapted to fit the new technology.[1]

Little is known, however, about how organizations actually go about modifying new process technologies, or how they adapt their own practices in response to technological change. Most of the research on this topic has assumed that users learn about and modify new technologies gradually. These assumptions have been built into our theories and images about technological adaptation—such as the familiar learning curve, which implies a highly regular accretion of improvements over time. The same assumptions are built into the prescriptions many researchers offer to management. These researchers exhort managers to "allow plenty of time" to digest new process technologies and to strive for "continuous improvement" (see Figure 4.1).

Yet most of the research on which these assumptions are based was performed at the aggregate level. Certainly, an entire firm or factory must strive for continuous improvement. But, at the level of a particular new technology, the process of learning about and modifying a new process may not be continuous at all. Indeed, our research suggests that the pattern of adaptation for an individual new technology is often a decidedly "lumpy" or episodic one (see Figure 4.2). In general, it appears that the introduction of a new technology into an operating environment triggers an initial burst of adaptive activity, as users explore the new technology and resolve unexpected problems. However, this activity is often short-lived, with effort and attention declining dramatically after the first few months of use. In effect, the technology, as well as the habits and assumptions surrounding it, tend to become "taken for granted" and built into standard operating procedures. This initiates a period of stability in which users focus attention more on regular production tasks than on further adaptation. Later on, users often refocus their attention on unresolved

Reprinted from "Exploiting Opportunities for Technological Improvement in Organizations" by Tyre, M. and W. Orlikowski, *Sloan Management Review*, (Fall 1993) pp. 13–26, by permission of the publisher.

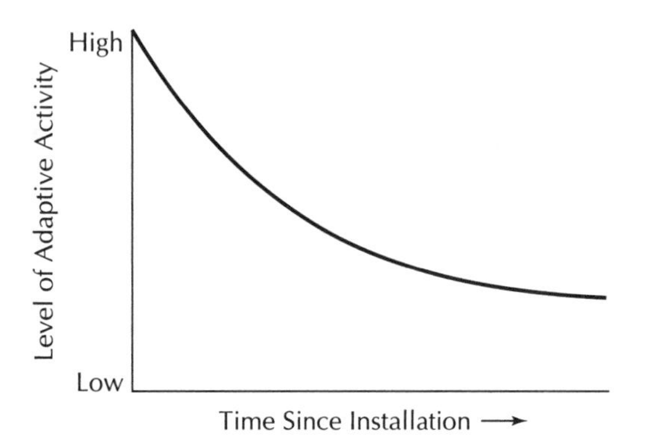

Fig. 4.1. A continuous pattern of technological adaptation.

problems or new challenges, creating additional spurts of adaptive activity. In many cases, this episodic pattern continues over time, with brief periods of adaptation followed by longer periods of relatively stable use.

In this paper, we discuss the evidence for such an episodic process of adaptation. We draw on our own research in U.S. and European companies, as well as existing research on the practices of successful Japanese companies. After presenting this evidence, we also discuss why such an episodic pattern—provided it is understood and managed—may serve as an effective and powerful way to pursue ongoing improvement of new process technologies.

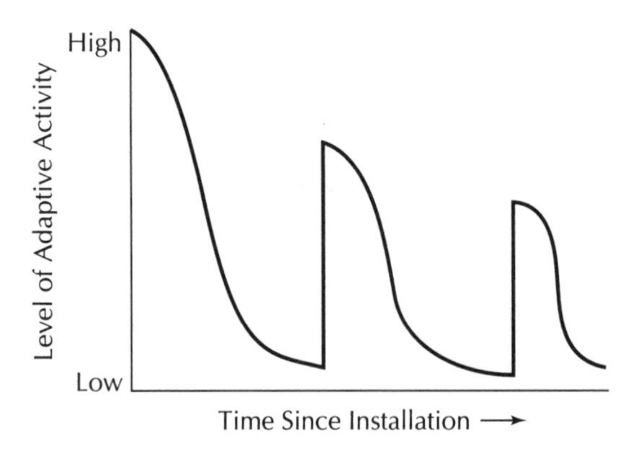

Fig. 4.2. A "lumpy" or episodic pattern of technological adaptation.

EXPLICATING THE PATTERN OF ADAPTATION OVER TIME

In a recent research study, we investigated how three manufacturing and service organizations in the United States and Europe adapted new process technologies.[2] The first site was BBA (names have been disguised), a leading manufacturer of precision metal components, where we studied forty-one projects involving the introduction of new capital equipment in European and U.S. factories. The second site was SCC, a multinational software developer of custom-built computer applications, where we followed the introduction of computer-aided software engineering (CASE) tools in three U.S. offices. The third site was Tech, a research university in the United States, where we examined modifications made to user-customizable computer tools such as text editors and electronic mail utilities.

Our main findings are consistent with the pattern described in Figure 4.2. First, we found that the installation of a new process technology was followed by an immediate and relatively brief burst of adaptation efforts. Thereafter, such efforts fell off precipitously. Thus, experimentation was more likely to occur and significant changes more apt to be implemented *immediately* following introduction than at any later time. This rapid fall-off of adaptive activity was apparently not a simple "learning curve" phenomenon, because it occurred even when outstanding problems had not been fully addressed.

However, the initial period was not the only time when important modifications were made. In each company, events sometimes triggered new episodes of intensive adaptation effort. These later episodes were also short-lived, but they were critical because they enabled users to tackle outstanding problems and to apply the additional insights gained through use over time. Thus, the cycle of intensive improvement followed by relatively stable operations tended to repeat itself.

The timing of adaptation at BBA illustrates this pattern. As shown in Figure 4.3, we found that most of the adaptations made were accomplished within a very short time after implementation—on

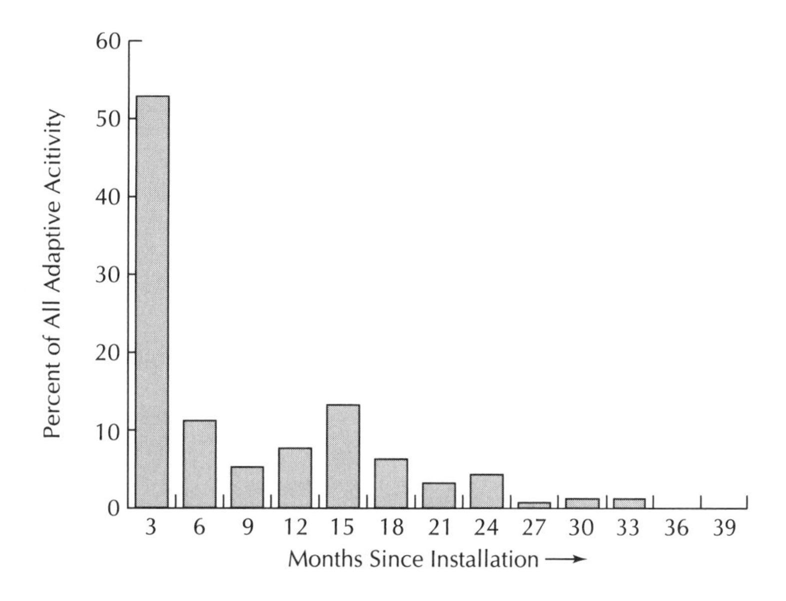

Fig. 4.3. Timing of technological adaptation at BBA.

average, 54 percent of all adaptive activity was completed in the first three months, or only 12 percent of the average total time to full integration. This pattern was remarkably consistent across all of the projects analyzed; the episode of adaptation that seemed to accompany initial implementation lasted about the same time (approximately three months) whether the project involved five people or fifty, and whether the technology was familiar or a departure from current procedures. Further, it was clear that adaptation efforts were not falling off simply because the users had resolved all problems within this period; on average, respondents reported five significant problems still outstanding at the time when initial adaptation efforts were curtailed. Indeed, most of the new technologies were not considered "production worthy" for many months.

Following the initial burst of activity, most of the technologies entered a phase of regular use as a part of the overall production process. On the other hand, participants did not completely ignore possible improvements to new technologies after the initial period of adaptation. In most projects, they regrouped and refocused attention on modifications some time later, again in a concentrated manner and for a short period (two to three months). Three-quarters of all projects at BBA showed a second spurt of adaptive activity. On average, this episode began about eleven months after the initial installation, and it accounted for an average of 23 percent of all reported adaptive activities. Further, in several of the projects, there was a third such spurt of adaptive activity about six to twelve months after the second episode.

Similar patterns emerged at the other two firms studied (see Table 4.1). At SCC, a large amount of adjustment and modification took place directly following initial installation of CASE tools into a new project site. In each project, the tools had to be fitted to the particular client organization. However, once application programmers (i.e., the users responsible for the actual production of new application software) began work on the project and started to use the CASE tools as process technology, further changes to the tools halted. These tool users required that their process technology be stable and reliable to facilitate production work. Thus, further refinement of the tools declined very sharply after the initial spurt of adaptive activity.

As at BBA, however, a significant surprise or major breakdown later in the project could turn

TABLE 4.1. Evidence of the Episodic Pattern of Technological Adaptation

	Early Focus on Adaptation Was Quickly Subsumed by Routine Use		Later on, Routine Use Was Broken by Occasional Episodes of Further Adaptation	
Site	Number of Projects Exhibiting This Pattern	Percent of All Projects	Number of Projects Exhibiting This Pattern	Percent of All Projects
BBA	34	86	31	75
SCC	4	80	1	25
Tech	33	65	49	96

users' attention back to the need for ongoing improvements. At these times, technical support personnel were reassigned to undertake a new round of adaptations.

At Tech, too, users' adaptation of their computer tools fell off abruptly soon after initial implementation. In particular, exploring or experimenting to learn about the technology virtually ceased after the first few weeks of use. Instead, users quickly settled on a computing environment and tried to maintain its stability. As on Tech employee explained, few people even thought about making changes once they had become comfortable with the software: "It's just the way I do it. . . . It's not that [further changes would be] hard, it's just that it's not worth the effort." Yet most users at Tech (forty-nine of fifty-one) noted that specific events did occasionally refocus their attention on the software and trigger further customizations; thus, further adaptations occurred, clustered in relatively brief spurts that were interspersed with periods of routine use of the technology.

In short, all three of these very different organizations displayed a distinctly discontinuous pattern in the way they adapted new process technologies. Significantly, this did not seem to be a conscious management policy in any of the companies. To the contrary, managers (and users) frequently stated that they recognized the need for continuous ongoing changes to new technologies, but that it was difficult to keep people focused on this sort of modification activity for more than a short time. Thus, once users became familiar with a new technology, it tended to become a "taken for granted" part of normal operations.

The forces for stability and routinization, however, were occasionally disrupted by events that forced—or allowed—technology users to ask new questions and to reexamine old problems. Typically, the events that created additional opportunities for adaptation were new developments that somehow interrupted routine operations. At BBA, for instance, the reported new episodes of adaptation were generally associated with events that placed new demands on existing operations and also created a pause in the normal production schedule. For example, when new machines were added to the production line where the technology was in use, they often created increased demands for high-precision or high-speed processing that had not yet been achieved. At the same time, the installation of the new machines imposed a temporary line shutdown. Users in our study often took advantage of this time to address old problems and to initiate new adaptations to their technology. Similarly, the introduction of new products or product requirements, the imposition of new production procedures, or occasional breakdowns of the new technology were also times when the need for improvement became apparent, while providing a brief and sanctioned stop in the action. At Tech, the release of new versions of computer software forced users to interrupt their normal routines; these events accounted for almost one-third (28 percent) of all later episodes of adaptation observed there. Users at Tech also reported that they occasionally returned their attention to making software modifications when existing procedures became too frustrating, or when they were exposed to new ideas for making their routines more efficient.

REGULARITY OR PATHOLOGY?

Several aspects of our findings are notable when compared to the conventional wisdom about technological adaptation. First, improvement is episodic, not continuous. That is, the initial burst of adaptation as well as later episodes are limited in duration, quickly giving way to longer periods of relatively routine operation. Second, some unusual event that interrupts normal productive operations typically triggers these periods of adaptation or at least triggers users to ask new questions. The contrary nature of these results raises a number of questions. Do these results suggest a new way of understanding the adaptation process, or are they simply the result of mismanagement at the companies we studied?

To answer these questions, we examined several detailed accounts of how some successful firms in other industries and nations absorb and modify new technologies. In particular, we were interested in Japanese firms that embrace "continuous improvement" in their overall operation. We asked whether the pattern of adaptations around a specific new technology in such firms is in fact gradual and continuous over time, or whether it reflects the episodic pattern that we observed in our data.

We discovered that the successful Japanese operations do not invite or expect continuous adaptations to specific new technologies. Instead we found a discontinuous model of adaptation that basically resembled our findings. An important difference, however, was that these Japanese firms appeared to consciously and carefully manage the timing of adaptations. Managers in these organizations apparently create *and exploit* the very episodic pattern that we have described. Specifically, the managers in these companies appear to do three things. First, they aggressively utilize the introduction period to adapt new technologies. That is, they identify and make the maximum number of modifications as early as possible. Then, following this period, they impose routine on the use of new technologies, *and* they exploit that routine for what it can teach them. Third, they consciously and periodically create new opportunities for further adaptations.

Next we describe these three aspects of technological management in more detail. We then present reasons why exploiting the episodic pattern of technological adaptation can be a particularly *effective* (and attainable) approach to learning and improving over time.

Aggressively Adapting the New Technology

The Japanese operations we examined truly exploit the initial period of technology introduction. They do not build in a great deal of extra time for debugging a new technology before moving into full production schedules. Rather, they develop very demanding, early production commitments for new technologies—and then they take steps to ensure that the new process technology will be ready. The ability to do this stems partly from the careful early design of new products and processes, well documented by Wheelwright and Clark.[3] What is less widely understood is that meeting tight production deadlines in Japanese operations also stems from "the intensity of revisions on the spot during startup."[4] At Toyota, for instance, there is "a direct engineering assault to correct [problems] in the beginning, [which] prevents the need to dribble a constant stream of engineering changes through the formal system over a long period of time."[5]

This early spurt of activity is often not apparent to outsiders. For example, U.S. observers who toured a successful Japanese electronics operation reported that "the Japanese can simply 'flick a switch'" to start up new process technologies at high yield. In realty, successful introduction required three months of intensive, exhausting effort by a team of engineers. These engineers knew that there would be problems, but they also knew that they had to resolve the problems during the brief ramp-up period. Production commitments were absolutely firm, and the company's profitability depended on meeting them.[6]

In the automobile industry, for example, Clark and Fujimoto describe the Japanese approach to managing initial ramp-up of new technologies as far more intense than normal problem-solving activities—in fact, they label it "the Japanese 'war time' approach."[7] They find that, while U.S. and European automobile manufacturers typically allow almost one year after the start of production to meet their target quality level on a new line, the Japanese allow only one to two months. The idea behind this Japanese strategy, explains Hall, is to "work through as many engineering changes as possible when a new model starts into production so that, at the close of start-up, it is ready to run smoothly."[8] This strategy is backed up with significant resource commitments; for example, at Toyota, "design engineers, manufacturing engineers, production control personnel, quality managers, and anyone else who is necessary live on the production floor during start-up. [Time losses are minimized because] engineering changes are made on the spot."[9]

Imposing and Exploiting Routine

While many Japanese firms pursue continuous improvement, this does not mean that a newly implemented technology is subjected to a constant stream of changes. Instead, before any improvement is undertaken, "it is essential that the current standards be stabilized" and institutionalized.[10] In the *kaizen* philosophy, "there can be no improvement where there are no standards."[11] This means that standards and measures must be imposed, "and it is management's job to see that everyone works in accordance with the established standards."[12]

Specifically, in the case of new process technologies, once the initial episode of debugging and modifying the technology is completed, standard operating routines are developed and enforced. Managers do not expect or allow operators to introduce modifications on an informal, everyday basis.

Jaikumar provides a vivid example; in his sample of flexible manufacturing system (FMS) users in Japan, operations were so smooth that production planning took only one hour per week and unexpected downtime was virtually nil. With such predictability in operations, there was no need and "no allowances for in-the-line, people-intensive adjustments."[13] Even the appropriate recovery routine for each identifiable failure mode had been codified and provided to operators.[14]

This does not mean that technology users in such firms neglect ongoing improvements—only that the changes they actually implement are tightly controlled. Operators in these firms perform considerable off-line experimentation, but they do not make unauthorized changes to production technology.[15] Except for urgent corrections, they are permitted to make adaptations to the technology or production practices only at designated times.[16]

This emphasis on routinization is echoed in nonmanufacturing operations. In Japan's new software factories, there is rigid adherence to existing standards and little allowance for individuals to make changes during the production of new software systems.[17] Managers in Hitachi's software works, for example, realized that creating standardized procedures not only helped them to improve programming efficiency and productivity but also "helped them identify best practices within the company and within the industry for dissemination to all projects and departments."[18]

Periodically Creating New Opportunities for Adaptation

Descriptions of technological adaptation in these Japanese firms reveal that, at the level of a given technology, continuous change really involves repeated cycles of change and stability. Imposing the discipline of routine procedures ensures that the timing of further changes is carefully managed. Whenever possible, the managers batch modifications together systematically and implement them in one intensive episode of adaptation.

Very often these episodes are timed to coincide with other major changes, such as product-model changeovers, releases of new software versions, or yearly factory shutdowns. Hall explains that the most efficient manufacturing companies "generally plan engineering changes one to two months before the effective date. The effective dates of most changes are timed to occur at schedule-change times."[19] Similarly, in Japanese software factories, development methods, tools, and products are generally held steady during a given project but are periodically and systematically upgraded or revised. A review of Hitachi's software factory shows that Hitachi managers knew that the procedures they initially created would not be perfect; they "clearly recognized that these procedures and standards would have to evolve as personnel and technology changed, and they made provisions to revise performance standards annually."[20]

In short, by creating distinct episodes of adaptation, managers provide the operation with both the benefits of routine and the vitality of ongoing change.

WHY A DISCONTINUOUS PATTERN OF CHANGE CAN BE EFFECTIVE

One way of interpreting the evidence from our research and the Japanese studies is to suggest that effective companies take the naturally "lumpy" pattern of adaptation and exploit it. That is, managers in the firms cited achieve maximum benefit from their adaptation efforts by carefully managing both spurts of adaptation and periods of routine operation. Indeed, there is evidence that such a *dis*continuous pattern of modification can yield important benefits. First, there appears to be a natural surge of energy at the start of projects, which smart managers exploit fully. Second, managers can enhance both learning and efficiency goals by imposing (and using) periods of routine operation in between periods of rapid change. And third, by revisiting the adaptation agenda at intervals, managers can make problems more tractable and can render change more attractive and more manageable. Next we discuss each of these issues.

Utilizing the Initial Window of Opportunity

The period immediately following initial introduction of a new technology provides a special "window of opportunity" for adaptation. At the start of the project, the level of energy is high. The novelty of the situation helps people to focus on the new technology and to see it as a distinct and malleable tool. As Weick argues, "The point at which technology is introduced is the point at which it is most susceptible to influence. Beginnings are of special importance because they constrain what is learned about the technology and how fast it is learned."[21]

In the three organizations we studied, we noticed powerful organizational forces that underscored the importance of beginnings. In many cases, the motivation to change and to achieve targets was highest at the start of the introduction period. Challenging project objectives had a catalyzing effect at the start that often faded over time. For example, when one of the BBA plants began using an advanced precision grinding cell, the plant manager explained that "grinding all five faces [is] *the* key objective in this project. " Productivity improvements were viewed as less challenging and were simply assumed. Yet eighteen months later, when users still had not achieved five-face grinding, the motivating power of these objectives had faded. One engineer simply dismissed the earlier objective, commenting, "We only tried doing all five faces on this machine as an experiment. It was sort of an add-on that did not work." Thus, in this case, as in others we observed, people slowly lost sight of their aggressive original objectives as they became accustomed to the level of performance actually achieved.

Another powerful factor is that modifications and improvements are easiest to implement at the beginning of the introduction efforts. Successful implementation means that, over time, the technology becomes increasingly integrated into the production process. The new technology gets physically interconnected with the rest of the production process, and users learn to rely on it for production needs. Thus, after the initial period, further adaptation threatens to disrupt the physical flow of goods or services. Later modifications also threaten to destroy the routines and procedures for using the technology that users establish over time. For example, one engineer at BBA explained that it was hard to get production people to agree to further adaptation because "now the operators depend on the machine—it's built in, they don't want to change." Even at Tech, where there were few system constraints on the changes that individuals made to their personal computing environments, users admitted that their own routines or habits tended to constrain further change. One user stated, "I got a set of custom [settings] from [a colleague] about four years ago. Now they're ingrained."

Another reason why beginnings are so potent is that, as time goes on, key project personnel often move to other assignments and are not available to help with continued fine-tuning and modification of the new technology. Even when team members are not reassigned, teams tend to lose enthusiasm. At a BBA plant in Germany, an engineering manager commented, "It's easy to get plant engineers to start working on large projects, but it's extremely difficult to keep attention focused on the details over time."

Research on human behavior suggests that these tendencies are not just a function of mismanagement or short-term thinking in Western companies but rather are normal aspects of human task performance. Psychological studies have shown that people's motivation to engage in effortful problem solving is partly a function of time: with extended exposure to a given phenomenon, people tend to become less alert and to notice fewer details than they do initially.[22] Familiarity also makes people less willing to invest time and effort in difficult problem solving.[23] Sustaining such motivation over long periods of time is difficult.

Finally, identifying and resolving problems during the initial introduction period is often easier because there are fewer competing demands on people's time. Later on, production issues tend to dominate the list of priorities, simply because urgent problems have the power to remove attention from more important, but less urgent, issues. One project manager we interviewed stated, "The basic operating fact is that you need to produce good parts every day. [Even if] these people [the project team members] see a problem or get an idea and want to try it, some days you just can't."

Some project managers in our study imagined that they could continue to modify the technology gradually, "as we went along," but found that production and adaptation did not mix well. One project engineer explained:

> On this project, we tried to mix production and engineering work. But once we really got into production, time to do important engineering work was squeezed out by everyday work with the machine and operators. The sheer volume of work made it impossible to search very far for new solutions, or to examine and test ideas before they went on line. There was plenty of money, but no time—we only had Saturdays for testing new solutions. The result? Lots of grey hair!

The difficulty of simultaneously tackling production and adaptation was not unique to Western firms. One Japanese engineer faced with this problem is quoted as saying:

> The production quota was high at that time. Even when team members gathered at the operation room on time to do an experiment, we were often kept waiting until the production quota was filled. Experiments often started at midnight. These circumstances deteriorated the efficiency of the work significantly. Consequently, it took longer to fix the problems than we had expected. This experience taught us the importance of finding the potential bottlenecks as soon as possible. The later the problems are found, the more it becomes difficult to solve them.[24]

Learning from Routine Use

Once an initial set of modifications has been identified and implemented, there are several reasons to hold the technology relatively constant for a period of time. First, there are obvious efficiency benefits in allowing the operation to run for some time in a stable fashion without frequent, disruptive changes. Hayes and Clark have shown that plants not subject to high levels of "confusion" from frequent changes in their tasks and technology have superior growth in productivity relative to "higher confusion" environments.[25] In his study of Toyota, Hall argues that "dribbling a constant stream of changes" into the operating system can seriously compromise operating effectiveness.[26]

Second, routine operations are an important test bed; if constant and uncontrolled changes are being made, it will be very difficult to assess the effectiveness of previous modifications.[27] Especially in "noisy," complex environments, where signals are misleading or difficult to interpret, problem solvers need to observe the system over relatively long periods of time before defining further adaptations.[28]

Finally, periods of routine operation allow users to explore a new technology and to learn from their own reactions.[29] Through extended use, they can identify features that are inconvenient or that cannot meet the evolving daily demands of a particular operating environment.[30] Thus, as Imai explains, progress can occur only when one "institutionalizes [a given] improvement as a new practice to improve on."[31]

Reopening the Window of Opportunity

There are several reasons why it makes sense to return to the adaptation agenda in short but intensive spurts, rather than to try for a more gradual pattern of change. First, short but intensive episodes of adaptive activity exploit "economies of scale" in problem solving. Many of the problems that affect a new technology require diverse resources. Managers, engineers, and users must create a cross-functional team, call in outside experts, set up experimental apparatus, develop prototypes, and so on. Frequently, regular operations must come to a halt. Gathering these resources only *once* to attack a number of issues is obviously more efficient than gathering them repeatedly as issues arise.

Changes are also less disruptive to ongoing operations if they are bundled together than if they are trickled out piecemeal to operations.[32] Indeed, some problems simply cannot be solved—or cannot be solved effectively—unless they are examined as part of a complex set of interdependent issues. When problems are dealt with en masse, instead of one by one, these interdependencies can be identified and utilized.

There are also motivational economies of scale associated with brief, intensive spurts of adaptive activity. A team (or even an individual) may be more easily motivated to devote attention and energy when the goal is large and obviously significant (resolving many problems) than when it is small and apparently unimportant (solving just one issue). Similarly, the rewards and satisfaction that come from resolving a significant set of issues can motivate team and individual efforts.

Finally, episodic cycling between short spurts of change and longer periods of regular use makes it possible to revisit problems or issues as knowledge is gained with experience. Multiple cycles make it possible to respond to changes in the operating environment that occur long after initial implementation of a new system—whether these are exogenous developments or shifts in users' preferences and expectations.[33] When users move occasionally into an adaptation mode, problems that eluded their understanding during one phase can be reframed and perhaps approached more successfully during another phase.

MANAGING ATTENTION AND EFFORT OVER TIME

These arguments suggest that a "lumpy" pattern of adaptation around a specific new technology may not be ineffective. Instead, relatively long periods of routine use, coupled with occasional but intensive episodes of adaptation, may be a powerful combination. Reaping the joint benefits of short episodes of adaptation and longer periods of regular use, however, is not automatic. It requires explicit management of the attention and effort applied in both phases.

Managing cycles of change and routine use of technologies demands a way of thinking that is unfamiliar to many managers. Users in all three of our research sites noted that managers very seldom took explicit action to create episodes of adaptation. For example, managers seldom intervened in a specific project to raise performance expectations, nor did they require regular project audits that might have focused attention on persistent performance shortfalls. Indeed, our interviews with managers showed that, while they were concerned with the way their organizations were introducing and using new process technologies, they were not sure how to manage the process. None of the people we interviewed seemed aware of the need to create or take advantage of discrete windows of opportunity for technological adaptation.

Our findings suggest that managers must consider how to create opportunities for adaptation, how to utilize those opportunities, and how to exploit periods of regular use of technologies for generating new insights and ideas. We cannot offer a recipe for doing this. However, we can suggest some ideas for moving toward a more conscious management of opportunities for technological improvement.

Creating Opportunities for Adaptation

Sometimes events outside of managers' direct control create new challenges or expose users to new ideas. At BBA, the imposition of new products and additional equipment often created renewed opportunities to focus on problems with process technology. However, we found few cases where managers actively created new opportunities, even though they could have. For example, requesting post-project audits of new process technology is one way of helping users to refocus on original project objectives and to compare them with current operations. Managers can also precipitate "minicrises" within the operating environment that push users to stretch existing capabilities—for example, managers can declare a target of zero defects during a given week. Taking a different approach, managers can inject new resources for problem solving on a temporary basis. For example, during a period when orders are low, plant managers can call a "problem-solving day" when no production takes place, and each person suggests and works at implementing improvements.

Rotating people between assignments is another way to generate opportunities for change. We know that new project team members tend to bring fresh ideas and pose questions that existing team members have ceased asking.[34] Additional team members also represent resources that are often needed to undertake modification activities. For example, we noticed that at BBA the addition of a new engineer into the factory environment sometimes triggered new episodes of adaptation. Likewise, existing team members who spend time in another setting or in a different function may return with fresh ideas to put into practice or a different perspective for assessing performance.

In creating windows of opportunity, it is important to realize that simply knowing that there are problems (or that better alternatives exist) is not enough. Somehow management must provide the incentive to undertake improvements, the resources necessary to accomplish this, and the chance to halt normal production rhythms for at least a short time. New product introductions, new quality requirements, or special assignments are events that often provide these—but not always. If operations are too chaotic, or if additional requirements place too great a burden on existing resources, it will be difficult to focus on potential new improvements at the same time.

Besides the specific actions taken, the way these actions are framed is also important. Employees may interpret a new or unexpected event as a threat (which produces insecurity and rigidity), when they could see it as a welcome opportunity for change.[35] The words, rewards, and actions managers use all affect such framing. To recognize a new development as an opportunity, employees must believe that the situation holds the potential for meaningful gain, and that they will have access to the competencies and resources necessary to develop that potential.[36] Consider, for example, the post-project audit. This can easily become just one more onerous reporting requirement, in which employees spend considerable effort justifying actions already taken. Yet audits can present opportunities for real exploration if managers frame them differently. Managers can ask users to skip the economic justification, asking them instead to identify three unfilled objectives and to outline plans for improvement.

An interesting question is whether new windows of opportunity must be surprising or unexpected events. Some organizations seem to turn very regular events, such as yearly plant shutdowns, new product introductions, scheduled technology upgrades, or maintenance reviews into successful episodes of adaptation.[37] Yet there are also many cases in which the element of surprise is important for jolting people out of entrenched routines and assumptions. In a well-known study, Meyer showed how a potentially crippling doctors' strike led some hospitals to undertake new organizational experiments, to restructure operations, and to redefine internal power relationships.[38] Other examples are also common; an extreme case was the San Francisco earthquake in 1989. Following the quake, some computer users discovered they had lost the information they had saved on disks. Although experiencing a loss, they were also released from past constraints, and some used this opportunity to reconfigure their workspaces both electronically and physically. Whether such a disruptive event is necessary to create new opportunities for adaptation may depend on how deeply entrenched existing routines have become.

Exploiting Windows of Opportunity for Adaptation

At least three factors influence how an organization exploits opportunities for adaptation: its capability to act rapidly during a limited time window; the knowledge to select and undertake useful adaptations; and the choice of objectives to guide activities during this period.

The ability to act rapidly following a surprise, disruption, or halt in normal operations is critical if windows of opportunity are inherently brief. Yet this ability is rare; it requires that the operation be organized for fast response. One important aspect of fast response is what Bohn has called "information turn-around time"; that is, the time required to collect data (by observing regular operations or by running special tests), analyze them, and make decisions about the next steps.[39] If test results are obsolete before action is taken, then modifications will be both slow and often misdirected. One plant engineer we interviewed explained: "We could work on this [mold] to improve it, but we'd have to send it out to the lab [for evaluation after each trial], and the lab lead time is one week. You can't develop a process like that! So, we just decided to consider this part 'done'."

Similarly, how support functions respond is critical to maximizing the limited time for adaptation. Very often, modifications require input from technicians, development engineers, systems analysts, programmers, maintenance personnel, and so on. If users find it difficult and time consuming to get the attention of support personnel and if weeks elapse before specialists respond to users' requests, then the opportunities to undertake modifications will dissolve rapidly.

One of the Italian project teams we studied created an innovative solution to this problem. Plant managers recognized the need to respond quickly to the problems arising with a new process technology. They created a special workstation near the new process, called *pronto intervento*, where dedicated maintenance personnel and other resources (such as extra tooling and machine parts) were available to deal with problems immediately. Similarly, rapid response from external part or tool vendors, combined with smooth functioning of internal purchasing and receiving units, can be critical.

It is important to note that the organization's capability to respond rapidly is highly systemic. If project deadlines are habitually allowed to lengthen because of outstanding problems, than key personnel will be perpetually busy and unable to assist with modifications. Worse, these commitments will be unpredictable, and the whole system will suffer increasingly from lags and unresponsiveness.[40]

This illustrates the self-reinforcing quality of rigid project deadlines and limited windows of opportunity: when project deadlines are respected, key personnel are free to make improvements in other parts of the operation. When episodes of adaptation are short but intense, such personnel can devote their full attention to their work, yet still be able to return to other projects in a short time.

Similarly, strict deadlines can help focus energy on needed adaptations. When tasks are bounded by definite deadlines, it often helps focus people's attention and increases the chance that they will work actively on the problems. Especially when people are working in groups, deadlines help to keep individual efforts aligned and to maintain motivation.[41] Awareness of tight time limits helps groups to assess their progress and to develop new approaches when existing ones are not working.[42]

Of course, deadlines can also become dysfunctional if they are unrealistic or too rigid to accommodate unanticipated contingencies. Remembering that new windows of opportunity can be opened in the future for further work may help managers set more reasonable deadlines and expectations for the period immediately following installation of a new technology.

Perhaps the most serious problem that distracts personnel from attending to technological adaptation is the need for continuous fire fighting just to maintain normal operations. When operations are out of control to begin with, introducing new (and advanced) technology only creates confusion and chaos.[43] Similarly, introducing technology that is itself very immature can swamp the operation with crises and divert attention from genuine improvement efforts. One of the engineers involved in a disappointing project at BBA explained: "There were so many machine and quality problems at first that we had to change so many things. . . . Once we got into these problems, I couldn't do any more [significant improvements], only try to attend to the little problems. The result was a lot of frustration." The more that problems can be resolved *before* introducing the new technology, the more users can exploit windows of opportunity for making improvements, rather than simply for keeping their heads above water.[44]

Besides availability of physical and support resources, users need considerable technical capability if they are to take advantage of opportunities to improve operating technologies. Thus, technical training for users of new process technologies is critical. However, many conventional training programs ignore the fact that opportunities for learning, like technological adaptation, may also be episodic or cyclical over time. Traditionally, user training occurs before installation of a new process technology. Most systems development processes schedule user training as one of the last phases before handing off the system to the users. Certainly, some basic-level skills are needed at installation. However, just as all problems with a new technology do not show up immediately, so users cannot absorb all that they need to understand at the outset. As they gain experience with the new technology, they gain insights and increase their "absorptive capacity" for further formal training.[45] Increased comfort with the new technology and a greater understanding of their own operating requirements allow users to better exploit educational opportunities on more advanced topics.

Moving away from the "one-shot" training approach would be particularly valuable in the case of software. Frequently, users simply cannot explore or appreciate the complex features and functionality without significant experience with the technology. One company we know has specifically implemented a two-tier training program for its personal computer users. The company does not regard the second tier as optional training, but as the continuation of the training course begun six months earlier.

Other users are another source of insight into how to adapt new process technologies. For instance, at Tech, other computer tool users suggested a number of software modifications.[46] However, physical and organizational boundaries often prevent users from visiting other sites or from borrowing ideas from other users. This blocks new opportunities for adaptation and prevents informed choices.

Finally, the objectives organizations apply are perhaps the most critical determinant of whether organizations exploit opportunities for technological adaptation. Too often, managers view the period immediately following initial installation of a new process technology as the time to get the system up and running smoothly. Thus, the objective is to resolve problems that will interfere with full assimilation of the new technology. An alternative and more aggressive approach is to view this period as a time to surface as many problems as possible. The reason is that if problems are not identified early, it may be difficult to focus on (or even to recognize) them later. Yet such unresolved issues may compromise performance of the technology over the long term. With such an alternative approach, the objective of the initial startup period is not just a working technology, but also a new understanding of how the technology can be most fully exploited in a given operating environment.

Ogawa, who studied new technology introductions in the Japanese steel industry, argues that many managers hold a mistaken view of the test period.[47] They assume that the purpose of this period is to enable operators to get accustomed to the new equipment, to set operating parameters, and to get product samples approved by customers. Ogawa suggests that the best managers also see the test period as a time to surface all major problems with the new technology. They reason that the new equipment should be placed under unusually severe demands as early as possible. Ogawa calls this the "rapid max" strategy. Under "rapid max," the new system is quickly but temporarily brought up to maximum operating rates during the start-up and test period. The point with "rapid max" is not to reach stabilization rapidly, since this can lead to stabilization of suboptimal routines, or even to "false stabilization." Rather, the objective is to experience rapidly a full range of issues and to uncover problems that would be difficult to deal with later.

Employing a "rapid max" strategy in new process start-ups would be a bold move for many managers. The approach employs the same counterintuitive logic as the *kanban* system, which tells managers to decrease inventory in order to make their problems more obvious. Indeed, if we recognize that experience brings with it inevitable forces for routinization, the "rapid max" strategy appears to be a powerful learning tool that can contribute to long-term effectiveness.

Exploiting Periods of Regular Use of Process Technology

We have argued that periods of sustained regular use can be complementary to episodes of adaptation. Periods of regular use provide data on how the technology is working, on whether previous changes are yielding positive results, and on what new problems or opportunities need to be addressed. Thus the key to utilizing periods of regular use is careful observation of operations and collection of data about them.

This is not a simple task. In many companies, users of technology perceive a conflict between production goals and data-gathering activities. One user at Tech remarked that she often runs into a certain problem with her software, but is typically "too busy" to carefully log the circumstances surrounding the error as a base for future analysis.

This indicates that managers should make data collection a part of users' regular responsibilities. For example, Jaikumar reports that, in Japanese FMS installations, operators spend almost one-third of their time observing system behavior, examining statistics on system performance, or running tests to generate new data about the system.[48]

Managers could also make data gathering easier with a number of automated methods. For example, if software users are too busy to log errors and the circumstances surrounding them, an automatic "log-it" function could capture relevant data when problems occur. Or users could employ an electronic mail message system to send ideas and remarks about system problems to a central person, who could than collate and categorize comments. Additionally, automatic tracking of manufacturing operations (e.g., as part of statistical process control systems) gathers performance variations over time. Users could cull existing records of repairs, service calls, or engineering change orders for data on problems and opportunities. At the very least, users could keep individual electronic journals of observations or just a file of ideas (including things they like or do not like). Whatever the mechanism, useful data should include both well-structured information on predetermined topics (e.g., level of defects, nature of defects), as well as unstructured observations and ideas from users as they interact with the new technology.

Nonelectronic innovations can be equally useful. For example, one of our colleagues keeps a file of ideas from students or others on ways to improve his course. Instead of ignoring a suggestion because he is too busy to deal with it, he simply jots it down and files it. Once a year (when he revises his syllabus over the summer), he reviews the contents and acts on any ideas that have come up repeatedly.

Framing the data-gathering effort by setting appropriate objectives is also important. During periods of regular operation, managers need to help users focus on testing existing solutions, discovering unresolved problems, and increasing their understanding of the technology—not just on getting product out the door. This ensures that the next time users find an opportunity to focus on adaptations, they will be able to investigate problems at a deeper level and to tackle more challenging types of change. In this way, the repeated cycle of adaptation could offer the chance for something better than just ongoing modification. It could create opportunities to identify increasingly subtle problems or to set increasingly challenging objectives for the technology (and its users) over time.

CONCLUSION

We have described a central paradox that affects the implementation and later use of new process technologies. On the one hand, ongoing adaptation is an important success factor for implementing and using many new process technologies—and such adaptation takes time and experience. On the other hand, the more experience that users gain with a new technology, the more they rely on established routines and habits. Over time, their sharp focus on the technology as a separate and malleable object fades; thus, both the technology and the way it is used are eventually taken for granted.

Given these tendencies, is it possible to pursue ongoing improvement while enjoying the benefits of stable, routine operations? We suggest that it is, but that achieving both objectives requires careful management of time and attention. At the level of a specific technology, there are important benefits to applying adaptation efforts in an uneven, episodic manner, rather than on a gradual, continuous basis. The episodic pattern described here allows users to rely on stable production routines most of the time, but also provides discrete "windows of opportunity" to reexamine and change those routines. Short but intensive episodes of adaptation enable team members to devote their attention to adaptation efforts without undue distraction from—or interference with—ongoing operations. Such brief, intensive periods of adaptation also make it possible to exploit economies of scale by attending to many small problems at once.

Unfortunately, many companies neither recognize nor manage these episodic cycles of change and stability. Managers typically exhort employees to seek out problems and to pursue improvements on a continuous basis. Yet we suggest that managers could exploit a more discontinuous pattern to good effect by: (1) aggressively exploiting the opportunities for change that accompany the initial introduction of a new technology into the organization; (2) mining subsequent periods of regular, routine use for new data and new insights into technological problems and opportunities; and (3) periodically creating and utilizing new opportunities for further adaptation. This last point may be the most challenging. It involves both focusing users' attention on the need for change and providing the resources and capabilities needed to act quickly—before the window of opportunity closes once again.

Note

The research reported in this paper was partly funded by the International Center for Research on the Management of Technology at the MIT Sloan School of Management and the MIT Leaders for Manufacturing Program. The authors gratefully acknowledge this support.

REFERENCES

1. D.A. Leonard-Barton, "Implementation as Mutual Adaptation of Technology and Organization," *Research Policy* 17 (1988):251–267; and A.H. Van de Ven, "Central Problems in the Management of Innovation," *Management Science* 32 (1986): 590–607.
2. M. Tyre and W. Orlikowski, "Windows of Opportunity: Temporal Patterns of Technological Adaptation in Organizations," *Organization Science* 5 (forthcoming, 1994).
3. S.C. Wheelwright and K.B. Clark, *Revolutionizing Product Development: Quantum Leaps in Speed, Efficiency and Quality* (New York: Free Press, 1992).
4. R.W. Hall, *Zero Inventories* (Homewood, Illinois: Dow Jones-Irwin, 1983), p. 197.
5. Ibid., p. 199.
6. N.S. Langowitz, "Plus Development Corp (A)"

(Boston: Harvard Business School, Case No. 9-687-001, 1986).
7. K.B. Clark and T. Fujimoto, *Product Development Performance* (Boston: Harvard Business School Press, 1991), p. 202.
8. Hall (1983), p. 60.
9. Ibid.
10. M. Imai, *Kaizen: The Key to Japan's Competitive Success* (New York: McGraw-Hill Publishing Company, 1986), p. 63.
11. Ibid., p. 74 See, also: P.S. Adler, "Time-and-Motion Regained," *Harvard Business Review*, January-February 1993, pp. 97–108.
12. Imai (1986), p. 75.
13. R. Jaikumar, "Postindustrial Manufacturing," *Harvard Business Review*, November-December 1986, p. 76.
14. K.B. Clark, R. Henderson, and R. Jaikumar, "A Perspective on Computer-Integrated Manufacturing Tools" (Boston: Harvard Business School, Working Paper No. 88-048, 1989).
15. R.J. Schonberger, *Japanese Manufacturing Practices* (New York: Free Press, 1982).
16. Hall (1983); and Schonberger (1982).
17. M.A. Cusumano, *Japan's Software Factories* (New York: Oxford University Press, 1991).
18. M.A. Cusumano, "Shifting Economies: From Craft Production to Flexible Systems and Software Factories," *Research Policy* 21 (1992):468.
19. Hall (1983), p. 200.
20. Cusumano (1992), p. 468.
21. K. Weick, "Technology as Equivoque," *Technology and Organizations*, ed. P.S. Goodman et al. (San Francisco: Jossey-Bass, 1990), pp. 21–22.
22. D. Newtson, "Attribution and the Unit of Perception of Ongoing Behavior," *Journal of Personality and Social Psychology* 28 (1973): 1, 28–38; and A.W. Kruglanski and T. Freund, "The Freezing and Unfreezing of Lay Interferences: Effects on Impressional Primacy, Ethnic Stereotyping, and Numerical Anchoring," *Journal of Experimental Social Psychology* 19 (1983): 448–468.
23. E.J. Langer and L. Imber, "When Practice Makes Imperfect: The Debilitating Effects of Overlearning," *Journal of Personality and Social Psychology* 37 (1979):2014–2025.
24. H. Ogawa, "Information Flow and Learning in New Process Development: Construction Project in the Steel Industry" (Cambridge, Massachusetts: MIT Sloan School of Management, unpublished thesis, 1991).
25. R.H. Hayes and K.B. Clark, "Exploring the Sources of Productivity Differences at the Factory Level," *The Uneasy Alliance*, ed. Clark et al. (Boston: Harvard University Press, 1985).
26. Hall (1983).
27. Imai (1986).
28. D. Levinthal and J. March, "A Model of Adaptive Organizational Search," *Journal of Economic Behavior and Organization* 2 (1981): 307–333.
29. Weick (1990); and A. Etzioni, "Humble Decision Making," *Harvard Business Review*, July-August 1989, pp. 122–126.
30. E. von Hippel and M.J. Tyre, "How Learning by Doing Is Done: Problem Identification in Novel Process Equipment, " *Research Policy* (forthcoming).
31. Imai (1986), p. 62.
32. Hall (1983); and Clark and Fujimoto (1991).
33. Von Hippel and Tyre (forthcoming).
34. R. Katz, "The Effects of Group Longevity on Project Communication and Performance," *Administrative Science Quarterly* 27 (1982): 81–104.
35. A.D. Meyer, "Adapting to Environmental Jolts," *Administrative Science Quarterly* 27 (1982):551–537; J.E. Dutton and S.E. Jackson, "Categorizing Strategic Issues: Links to Organizational Action," *Academy of Management Review* 12 (1987): 76–90; and Weick (1990).
36. S.E. Jackson and J.E. Dutton, "Discerning Threats and Opportunities," *Administrative Science Quarterly* 33 (1988):370–387.
37. Hall (1983); and Jaikumar (1986).
38. Meyer (1982).
39. R.E. Bohn, "Learning by Experimentation in Manufacturing" (Boston: Harvard Business School, Working Paper No. 88-001, 1988).
40. M.G. Bradac, D.E. Perry, and L.G. Votta, "Prototyping a Process Monitoring Experiment" (Baltimore, Maryland: Proceedings of the Fifteenth International Conference on Software Engineering, May 1993).
41. J.R. Hackman, *Groups That Work (and Those That Don't)* (San Francisco: Jossey-Bass, 1990).
42. C.J. Gersick, "Time and Transition in Work Teams: Toward a New Model of Group Development," *Academy of Management Journal* 31 (1988):1, 9–31.
43. Hayes and Clark (1985).
44. M.J. Tyre and O. Hauptman, "Effectiveness of Organizational Response Mechanisms to Technological Change in the Production Process," *Organization Science* 3 (1992):301–320.
45. W.M. Cohen and D.A. Levinthal, "Absorptive Capacity: A New Perspective on Learning and Innovation," *Administrative Science Quarterly* 35 (1990):128–152.
46. W. Mackay, "Users and Customizable Software: A Co-adaptive Phenomenon" (Cambridge Massachusetts: MIT Sloan School of Management, unpublished Ph.D. thesis, 1990).
47. Ogawa (1991).
48. Jaikumar (1986).

5

The Panda's Thumb of Technology

STEPHEN JAY GOULD

The brief story of Jephthah and his daughter (Judg. 11:30–40) is, to my mind and heart, the saddest of all biblical tragedies. Jephthah makes an intemperate vow, yet all must abide by its consequences. He promises that if God grant him victory in a forthcoming battle, he will sacrifice by fire the first living thing that passes through his gate to greet him upon his return. Expecting (I suppose) a dog or a goat, he returns victorious to find his daughter, and only child, waiting to greet him "with timbrels and with dances."

Handel's last oratorio, *Jephtha,* treats this tale with great power (although his librettist couldn't bear the weight of the original and gave the story a happy ending, with angelic intervention to spare Jephthah's daughter at the price of her lifelong chastity). At the end of part 2, while all still think that the terrible vow must be fulfilled, the chorus sings one of Handel's wonderful "philosophical" choruses. It begins with a frank account of the tragic circumstance:

> How dark, O Lord, are thy decrees! . . .
> No certain bliss, no solid peace,
> We mortals know on earth below.

Yet the last two lines, in a curious about-face, proclaim (with magnificent musical solidity as well):

> Yet on this maxim still obey:
> WHATEVER IS, IS RIGHT

This odd reversal, from frank acknowledgment to unreasonable acceptance, reflects one of the greatest biases ("hopes" I like to call them) that human thought imposes upon a world indifferent to our suffering. Humans are pattern-seeking animals. We must find cause and meaning in all events (quite apart from the probable reality that the universe both doesn't care much about us and often operates in a random manner). I call this bias "adaptationism"—the notion that everything must fit, must have a purpose, and in the strongest version, must be for the best.

The final line of Handel's chorus is, of course, a quote from Alexander Pope, the last statement of the first epistle of his *Essay on Man,* published just thirteen years before Handel's oratorio. Pope's text contains (in heroic couplets to boot) the most striking paean I know to the bias of adaptationism. In my favorite lines, Pope chastises those people who may be unsatisfied with the senses that nature bestowed upon us. We may wish for more acute vision, hearing, or smell, but consider the consequences.

> If nature thunder'd in his op'ning ears
> And stunn'd him with the music of the spheres
> How would he wish that Heav'n had left him still
> The whisp'ring zephyr, and the purling rill!

And my favorite couplet, on olfaction:

> Or, quick effluvia darting thro' the brain,
> Die of a rose in aromatic pain.

What we have is best for us—whatever is, is right.

By 1859, most educated people were prepared to accept evolution as the reason behind similarities and differences among organisms—thus accounting for Darwin's rapid conquest of the intellectual world. But they were decidedly not ready to acknowledge the radical implications of Darwin's proposed mechanism of change, natural selection, thus explaining the brouhaha that the *Origin of Species* provoked—and still elicits (at least before our courts and school boards).

Darwin's world is full of "terrible truths," two in particular. First, when things do fit and make sense (good design of organisms, harmony of ecosystems), they did not arise because the laws of nature entail such order as a primary effect. They are, rather, only epiphenomena, side consequences of the basic causal process at work in natural populations—the purely "selfish" struggle among organisms for personal reproductive success. Second the complex and curious pathways of history guarantee that most organisms and ecosystems cannot be designed optimally. Indeed, to make the statement even stronger, imperfections are the primary proofs that evolution has occurred, since optimal designs erase all signposts of history.

This principle of imperfection has been the main theme of these essays for several years. I call it the panda principle to honor my favorite example, the panda's false thumb, subject of an old essay (*Natural History,* November 1978) that reemerged as the title to one of my books. Pandas are the herbivorous descendants of carnivorous bears. Their true anatomical thumbs were, long ago during ancestral days of meat eating, irrevocably committed to the limited motion appropriate for this mode of life and universally evolved by mammalian Carnivora. When adaptation to a diet of bamboo required more flexibility in manipulation, pandas could not redesign their thumbs but had to make do with a makeshift substitute—an enlarged radial sesamoid bone of the wrist, the panda's false thumb. The sesamoid thumb is a clumsy, suboptimal structure, but it works. Pathways of history (commitment of the true thumb to other roles during an irreversible past) impose such jury-rigged solutions upon all creatures. History inheres in the imperfections of living organisms—thus we know that they had a different past, converted by evolution to their current state.

We can accept this argument for organisms (we know, after all, about our own appendixes and aching backs). But is the panda principle more pervasive? Is it a general statement about all historical systems? Will it apply, for example, to the products of technology? We might deem it irrelevant to the manufactured objects of human ingenuity—and for good reason. After all, constraints of genealogy do not apply to steel, glass, and plastic. The panda cannot shuck its digits (and can only build its future upon this inherited ground plan), but we can abandon gas lamps for electricity and horse carriages for motor cars. Consider, for example, the difference between organic architecture and human buildings. Complex organic structures cannot be reevolved following their loss; no snake will redevelop front legs. But the apostles of so-called post-modern architecture, in reaction to the sterility of so many glass-box buildings of the international style, have juggled together all the classical forms of history in a cascading effort to rediscover the virtues of ornamentation. Thus, Philip Johnson could place a broken pediment atop a New York skyscraper and raise a medieval castle of plate glass in downtown Pittsburgh. Organisms cannot recruit the virtues of their lost pasts.

Yet I am not so sure that technology is exempt from the panda principle of history, for I am now sitting face to face with the best example of its application. Indeed, I am in most intimate (and striking) contact with this object—the typewriter keyboard.

I could type before I could write. My father was a court stenographer, and my mother is a typist. I learned proper eight-finger touch-typing when I was about nine years old and still endowed with small hands and weak, tiny pinky fingers. I was thus, from the first, in a particularly good position to appreciate the irrationality of the distribution of letters on the standard keyboard, called QWERTY by all aficionados in honor of the first six letters on the top letter row.

Clearly, QWERTY makes no sense (beyond the whiz and joy of typing QWERTY itself). More than 70 percent of English words can be typed with the letters DHIATENSOR, and these should be on the most accessible second, or home, row—as they were in a failed competitor to QWERTY introduced as early as 1893. But in QWERTY, the most common English letter, E, requires a reach to the top row, as do the vowels U, I, and O (with O struck by the weak fourth finger), while A remains in the home row but must be typed with the weakest finger of all (at least for the dexterous majority of right handers)—the left pinky. (How I struggled with this as a boy. I just couldn't depress that key. I once tried to type the Declaration of Independence and ended up with: th t ll men re cre ted equ l.)

As a dramatic illustration of this irrationality, consider the keyboard of an ancient Smith-Corona upright, identical with the one (my Dad's original) that I use to type these essays (a magnificent machine—no breakdown in twenty years and a fluidity of motion unmatched by any manual typewriter since). After more than half a century of use, some of the most commonly struck keys have been worn right through the surface into the soft pad below (they weren't solid plastic in those days). Note that E, A, and S are worn in this way—and note also that all three are either not in the home row or are struck with the weak fourth and pinky fingers in QWERTY.

This claim is not just a conjecture based on idiosyncratic personal experience. Evidence clearly shows that QWERTY is drastically suboptimal. Competitors have abounded since the early days of typewriting, but none have supplanted or even dented the universal dominance of QWERTY for English typewriters. The best-known alternative, DSK, for Dvorak Simplified Keyboard, was introduced in 1932. Since then, virtually all records for speed typing have been held by DSK, not QWERTY, typists. During the 1940s, the U.S. Navy, ever mindful of efficiency, found that the increased speed of DSK would amortize the cost of retraining typists within ten days of full employment. (Mr. Dvorak was not Anton of the *New World Symphony,* but August, a professor of education at the University of Washington, who died disappointed in 1975. Dvorak was a disciple of Frank B. Gilbreth, pioneer of time and motion studies in industrial management.)

Since I have a special interest in typewriters (my affection for them dates to those childhood days of splendor in the grass and glory in the flower), I have wanted to write such an essay for years. But I never had the data I needed until Paul A. David, Coe Professor of American Economic History at Stanford University, kindly sent me his fascinating article, "Understanding the Economics of QWERTY: The Necessity of History" (in *Economic History and the Modern Economist,* edited by W.N. Parker. New York: Basil Blackwell Inc., 1986, pp. 30–49). Virtually all the nonidiosyncratic data in this essay come from David's work, and I thank him for this opportunity to satiate an old desire.

The puzzle of QWERTY's dominance resides in two separate questions: Why did QWERTY ever arise in the first place? And why has QWERTY survived in the face of superior competitors?

My answers to these questions will invoke analogies to principles of evolutionary theory. Let me, then, state some ground rules for such a questionable enterprise. I am convinced that comparisons between biological evolution and human cultural or technological change have done vastly more harm than good—and examples abound of this most common of all intellectual traps. Biological evolution is a bad analogue for cultural change because the two systems are so very different for three major reasons that could hardly be more fundamental.

First, cultural evolution can be faster by orders of magnitude than biological change at its maximal Darwinian rate—and questions of timing are of the essence in evolutionary arguments. Second, cultural evolution is direct and Lamarckian in form: the achievements of one generation are passed by education and publication directly to descendants, thus producing the great potential speed of cultural change. Biological evolution is indirect and Darwinian, as favorable traits do not descend to the next generation unless, by good fortune, they arise as products of genetic change. Third, the basic topologies of biological and cultural change are completely different. Biological evolution is a system of constant divergence without any subsequent joining of branches. Lineages once distinct, are separate forever. In human history, transmission across lineages is, perhaps, the major source of cultural change. Europeans learned about corn and potatoes from Native Americans and gave them smallpox in return.

So, when I compare the panda's thumb with a typewriter keyboard, I am not attempting to derive or explain technological change by biological principles. Rather, I ask if both systems might not record common, deeper principles of organization. Biological evolution is powered by natural selection, cultural evolution by a different set of principles that I understand but dimly. But both are systems of historical change. There must be (perhaps I now only show my own bias for intelligibility in our complex world) more general principles of structure underlying all systems that proceed through history—and I rather suspect that the panda principle of imperfection might reside among them.

My main point, in other words, is not that typewriters are like biological evolution (for such an argument would fall right into the nonsense of false analogy), but that both keyboards and the panda's thumb, as products of history, must be subject to some regularities governing the nature of temporal connections. As scientists, we must believe that general principles underlie structurally related systems that proceed by different overt rules. The proper unity lies, not in false applications of these overt rules (like natural selection) to alien domains (like technological change), but in seeking the more general rules of structure and change themselves.

The Origin of QWERTY: True randomness has limited power to intrude itself into the forms of organisms. Small and unimportant changes, unrelated to the working integrity of a complex creature, may drift in and out of populations by a process akin to throwing dice. But intricate structures, involving the coordination of many separate parts, must arise for an active reason—since the bounds of mathematical probability for fortuitous association are soon exceeded as the number of working parts grows.

But if complex structures must arise for a reason, history may soon overtake the original purpose—and what was once a sensible solution becomes an oddity or imperfection in the altered context of a new future. Thus, the panda's true thumb permanently lost its ability to manipulate objects when carnivorous ancestors found a better use for this digit in the limited motions appropriate for creatures that run and claw. This altered thumb then becomes a constraint imposed by past history upon the panda's ability to adapt in an optimal way to its new context of herbivory. The panda's thumb, in short, becomes an emblem of its different past, a sign of history.

Similarly, QWERTY had an eminently sensible rationale in the early technology of typewriting but soon became a constraint upon faster typing as advances in construction erased the reason for QWERTY's origin. The key (pardon the pun) to QWERTY's origin lies in another historical vestige easily visible on the second row of letters. Note the sequence: DFGHJKL—a good stretch of the alphabet in order, with the vowels E and I removed. The original concept must have simply arrayed the letters in alphabetical order. Why were the two most common letters of this sequence removed from the most accessible home row? And why were other letters dispersed to odd positions?

Those who remember the foibles of manual typewriters (or, if as hidebound as yours truly, still use them) know that excessive speed or unevenness of stroke may cause two or more keys to jam near the striking point. You also know that if you don't reach in and pull the keys apart, any subsequent stroke will produce a repetition of the key leading the jam—as any key subsequently struck will hit the back of the jammed keys and drive them closer to the striking point.

These problems were magnified in the crude

technology of early machines—and too much speed became a hazard rather than a blessing, as key jams canceled the benefits of celerity. Thus, in the great human traditions of tinkering and pragmatism, keys were moved around to find a proper balance between speed and jamming. In other words—and here comes the epitome of the tale in a phrase—QWERTY arose in order to slow down the maximal speed of typing and prevent jamming of keys. Common letters were either allotted to weak fingers or dispersed to positions requiring a long stretch from the home row.

This basic story has gotten around, thanks to short takes in *Time* and other popular magazines, but the details are enlightening, and few people have the story straight. I have asked nine typists who knew this outline of QWERTY's origin and all (plus me for an even ten) had the same misconception. The old machines that imposed QWERTY were, we thought, of modern design—with keys in front typing a visible line on paper rolled around a platen. This leads to a minor puzzle: key jams may be a pain in the butt, but you see them right away and can easily reach in and pull them apart. So why QWERTY?

As David points out, the prototype of QWERTY, a machine invented by C.L. Sholes in the 1860s, was quite different in form from modern typewriters. It had a flat paper carriage and did not roll paper right around the platen. Keys struck the paper invisibly from beneath, not patently from the front as in all modern typewriters. You could not view what you were typing unless you stopped to raise the carriage and inspect your product. Keys jammed frequently, but you could not see (and often did not feel) the aggregation. Thus, you might type a whole page of deathless prose and emerge only with a long string of E's.

Sholes filed for a patent in 1867 and spent the next six years in trial-and-error efforts to improve his machine. QWERTY emerged from this period of tinkering and compromise. As another added wrinkle (and fine illustration of history's odd quirks), R joined the top row as a last-minute entry, and for a somewhat capricious motive according to one common tale—for salesmen could then impress potential buyers by smooth and rapid production of the brand name TYPE WRITER, all on one row. (Although I wonder how many sales were lost when TYPE EEEEEE appeared after a jam!)

The Survival of QWERTY: We can all accept this story of QWERTY's origin, but why did it persist after the introduction of the modern platen roller and frontstroke key? (The first typewriter with a fully visible printing point was introduced in 1890). In fact, the situation is even more puzzling. I though that alternatives to keystroke typing only became available with the IBM electric ball, but none other than Thomas Edison filed a patent for an electric print-wheel machine as early as 1872, and L.S. Crandall marketed a writing machine without typebars in 1879. (Crandall arranged his type on a cylindrical sleeve and made the sleeve revolve to the required letter before striking the printing point.)

The 1880s were boom years for the fledgling typewriter industry, a period when a hundred flowers bloomed and a hundred schools of thought contended. Alternatives to QWERTY were touted by several companies, and both the variety of printing designs (several without typebars) and the improvement of keystroke typewriters completely removed the original rationale for QWERTY. Yet during the 1890s, more and more companies made the switch to QWERTY, which became an industry standard by the early years of our century. And QWERTY has held on stubbornly, through the introduction of the IBM Selectric and the Hollerith punch card machine to that ultimate example of its nonnecessity, the microcomputer terminal (Apple does offer a Dvorak option with the touch of a button but emblazons QWERTY on its keyboard and reports little use of this high-speed alternative).

To understand the survival (and domination to this day) of drastically suboptimal QWERTY, we must recognize two other commonplaces of history, as applicable to life in geological time as to technology over decades—contingency and incumbency. We call a historical event—the rise of mammals or the dominance of QWERTY—contingent when it occurs as the chancy result of a long string of unpredictable antecedents, rather than as a necessary outcome of nature's laws. Such contingent events often depend crucially upon choices from a distant past that seemed tiny and trivial at the time. Minor perturbations early in the game can nudge a process into a new pathway, with cascading consequences that produce an outcome vastly different from any alternative.

Incumbency also reinforces the stability of a pathway once the little quirks of early flexibility push a sequence into a firm channel. Suboptimal politicians often prevail nearly forever once they gain office and grab the reins of privilege, patronage, and visibility. Mammals waited 100 million years to become the dominant animals on land and only got a chance because dinosaurs succumbed during a mass extinction. If every typist in the world stopped using QWERTY tomorrow and began to learn Dvorak, we would all be winners, but who will bell the cat or start the ball rolling? (Choose your cliché, for they all record this evident truth.) Stasis is the norm for complex systems; change, when it happens at all, is usually rapid and episodic.

QWERTY's fortunate and improbable ascent to incumbency occurred by a concatenation of circumstances, each indecisive in itself, but all probably necessary for the eventual outcome. Remington had marketed the Sholes machine with its QWERTY keyboard, but this early tie with a major firm did not secure QWERTY's victory. Competition was tough, and no lead meant much with such small numbers in an expanding market. David estimates that only 5,000 or so QWERTY machines existed at the beginning of the 1880s.

The push to incumbency was complex and multifaceted, dependent more upon the software of teachers and promoters than upon the hardware of improving machines. Most early typists used idiosyncratic hunt-and-peck, few-fingered methods. In 1882, Ms. Longley, founder of the Shorthand and Typewriter Institute in Cincinnati, developed and began to teach the eight-finger typing that professionals use today. She happened to teach with a QWERTY keyboard, although many competing arrangements would have served her purposes as well. She also published a do-it-yourself pamphlet that was widely used. At the same time, Remington began to set up schools for typewriting using (of course) its QWERTY standard. The QWERTY ball was rolling but this head start did not guarantee a place at the summit. Many other schools taught rival methods on different machines and might have gained an edge.

Then a crucial event in 1888 probably added the decisive increment to QWERTY's small advantage. Longley was challenged to prove the superiority of her eight-finger method by Louis Taub, another Cincinnati typing teacher, who worked with four fingers on a rival non-QWERTY keyboard with six rows, no shift action, and (therefore) separate keys for upper and lower case letters. As her champion, Longley engaged Frank E. McGurrin, an experienced QWERTY typist who had given himself a decisive advantage that, apparently, no one had thought of before. He had memorized the QWERTY keyboard and could therefore operate his machine as all competent typists do today—by what we now call touch-typing. McGurrin trounced Taub in a well-advertised and well-reported public competition.

In public perception, and (more importantly) in the eyes of those who ran typing schools and published typing manuals, QWERTY had proved its superiority. But no such victory had really occurred. The tie of McGurrin to QWERTY was fortuitous and a good break for Longley and for Remington. We shall never know why McGurrin won, but reasons quite independent of QWERTY cry out for recognition: touch-typing over hunt-and-peck, eight fingers over four fingers, the three-row letter board with a shift key versus the six-row board with two separate keys for each letter. An array of competitions that would have tested QWERTY were never held—QWERTY versus other arrangements of letters with both contestants using eight-finger touch-typing on a three-row keyboard or McGurrin's method of eight-finger touch-typing on a non-QWERTY three-row keyboard versus Taub's procedure to see whether the QWERTY arrangement (as I doubt) or McGurrin's method (as I suspect) had secured his success.

In any case, the QWERTY steamroller now

gained crucial momentum and prevailed early in our century. As touch-typing by QWERTY became the norm in America's typing schools, rival manufacturers (especially in a rapidly expanding market) could adapt their machines more easily than people could change their habits—and the industry settled upon the wrong standard.

If Sholes had not gained his tie to Remington, if the first man who decided to memorize a keyboard had used a non-QWERTY design, if McGurrin had a bellyache or drank too much the night before, if Longley had not been so zealous, if a hundred other perfectly possible things had happened, then I might be typing this essay with more speed and much greater economy of finger motion.

But why fret over lost optimality. History always works this way. If Montcalm had won a battle on the Plains of Abraham, perhaps I would be typing *en français*. If a portion of the African jungles had not dried to savannas, I might still be an ape up a tree. If some comets had not struck the earth (if they did) some 60 million years ago, dinosaurs might still rule the land, and all mammals would be rat-sized creatures scurrying about in the dark corners of their world. If *Pikaia*, the only chordate of the Burgess Shale, had not survived the great sorting out of body plans after the Cambrian explosion, mammals might not exist at all. If multicellular creatures had never evolved after five-sixths of life's history had yielded nothing more complicated than an algal mat, the sun might explode a few billion years hence with no multicellular witness to the earth's destruction.

Compared with these weighty possibilities, my indenture to QWERTY seems a small price indeed for the rewards of history. For if history were not so maddening quirky, we would not be here to enjoy it. Streamlined optimality contains no seeds for change. We need our odd little world, where QWERTY rules and the quick brown fox jumps over the lazy dog.*

*I must close with a pedantic footnote, lest nonaficionados be utterly perplexed by this ending. This quirky juxtaposition of uncongenial carnivores is said to be the shortest English sentence that contains all twenty-six letters. It is, as such, *de rigueur* in all manuals that teach typing.

6

Strategic Maneuvering and Mass-Market Dynamics: The Triumph of VHS over Beta

MICHAEL A. CUSUMANO
YIORGOS MYLONADIS
RICHARD S. ROSENBLOOM

This article explores the evolution of a dynamic mass market and the strategic maneuvering to establish a product standard among firms that commercialized the videocassette recorder (VCR) for household use. The VCR was only one of several consumer electronics products (others include televisions, radios, stereos, audio tape recorders, and miscellaneous items ranging from digital watches to calculators) whose basic technology and initial applications came from within the United States or Europe. In each case, Japanese firms mastered the essentials of consumer-oriented product design and then went on to develop superior capabilities in mass production and mass distribution. As a result, during the 1970s and 1980s Japanese industry came to dominate the global consumer electronics business. In the U.S. market, for example, of an estimated $30 billion in sales for 1986, American firms accounted for merely 5 percent, compared to nearly 100 percent of U.S. sales in the 1950s.[1]

After its first appearance in the early 1970s, the VCR surpassed color television to become the largest single consumer electronics product in terms of sales by the early 1980s. One format, the U-Matic, developed primarily by the Sony Corporation, soon emerged as the dominant design for professional and educational uses, replacing other kinds of video players and recorders. By the mid-1970s, variations of this machine embodying more integrated electronics and narrower ($^1/_2$-inch) tape resulted in two formats designed exclusively for home use: the Betamax, introduced in 1975 by Sony, and the VHS (Video Home System), introduced in 1976 by the Victor Company of Japan (Japan Victor or JVC) and then supported by JVC's parent company, Matsushita Electric, as well as the majority of other firms in Japan, the United States, and Europe.[2] Despite their common ancestry and technical similarities, Beta and VHS machines remained incompatible, because they used different tape-handling mechanisms and cassette sizes, as well as coding schemes for their video signals that varied just enough so that tapes were not interchangeable.

Beta was the first compact, inexpensive, reliable, and easy-to-use VCR; it accounted for the majority of VCR production during 1975–77 and enjoyed steadily increasing sales until 1985.

Nonetheless, it fell behind the VHS in market share during 1978 and steadily lost share thereafter. By the end of the 1980s, Sony and its partners had ceased producing Beta models, with Sony promoting another similar but incompatible standard using a smaller (8mm) tape, primarily for home movies (see Tables 6.1 and 6.2). The outlines of this competition have been discussed before, both in English and in Japanese.[3] This study examines how and why the VCR rivalry unfolded as it did.

The literatures on both management and economics contain discussions of the strategic challenges that a new large-scale industry poses to innovators and later entrants. Of importance to this story, given the particular characteristics of the VCR product and market, are the roles of first movers versus other technological pioneers and later entrants. The first movers—the first firms to commercialize a new technology—often benefit from superior technology and reputation, which they may sustain through greater experience or a head start in patenting. Being first often provides a unique opportunity to shape product definitions, forcing followers to adapt to a standard or to invest in order to differentiate their offerings.[4] The first movers may also exploit opportunities for the early acquisition of scarce critical resources, as exotic as specialized production equipment or as mundane as retail shelf space; they can accumulate above-average profits if they enjoy a de facto monopoly position, as occurred in the early days of the industrial video recorder used by television stations (invented and commercialized by Ampex), the mainframe business computer (commercialized most successfully by IBM), and the plain-paper copier (commercialized by Xerox).[5] Rather than to inventors, however, the largest payoffs may actually go to the firms that lead in creating the necessary systems and investments for successful mass production and mass distribution.[6]

With technologies and markets that require years to develop, being the inventor or first mover

TABLE 6.1. Beta-VHS Annual Production and Cumulative Shares, 1975–1988

	Beta Format			VHS Format		
Year	(A)	(B)	(C)	(A)	(B)	(C)
1975	20	20	100/100	—	—	—
1976	175	195	61/64	110	110	39/36
1977	424	619	56/58	339	449	44/42
1978	594	1,213	40/48	878	1,327	60/52
1979	851	2,064	39/44	1,336	2,663	61/56
1980	1,489	3,552	34/39	2,922	5,585	66/61
1981	3,020	6,572	32/35	6,478	12,063	68/65
1982	3,717	10,289	28/32	9,417	21,480	72/68
1983	4,572	14,861	25/30	13,645	35,125	75/70
1984	6,042	20,903	20/26	23,464	58,589	80/74
1985	3,387	24,290	8/20	40,977	99,566	92/80
1986	1,106	25,396	4/16	29,553	129,119	96/84
1987	669	26,065	2/13	39,767	168,886	98/87
1988	148	26,213	0.3/11	44,761	213,647	99.7/89
			8mm FORMAT			
1984	10	10				
1985	566	576				
1986	1,051	1,627				
1987	1,351	2,978				
1988	1,531	4,509				

Units: (A) = annual production in thousands of units; (B) = cumulative production in thousands of units; (C) = share of total VHS and Beta production/share of total VHS and Beta cumulative production in percent.

Sources: For 1976–83, *Nikkei Business* (in Japanese), 27 June 1983; for 1981–83, *Nihon Keizai Shimbun* (Japan Economic Journal, in Japanese), 21 Dec. 1984; for 1975 and 1985–88, and 8mm format, JVC, Public Relations Dept.

TABLE 6.2. VCR Production and Format Shares, 1975–1984 (percent)

	1975	1976	1977	1978	1979	1980
Beta Group						
Sony	100	56	51	28	24	22
Others	—	5	5	12	15	11
Subtotal	100	61	56	40	39	34
VHS Group						
Matsushita	—	29	27	36	28	29
JVC	—	9	15	19	22	18
Others	—	1	2	5	11	19
Subtotal	—	39	44	60	61	66
	1981	1982	1983	1984	...	1989
Beta Group						
Sony	18	14	12	9		
Sanyo	9	10	8	6		
Toshiba	4	4	4	3		
Others	1	1	2	2		
Subtotal	32	28	25	20		0
VHS Group						
Matsushita	28	27	29	25		
JVC	19	20	16	17		
Hitachi	10	10	11	15		
Sharp	7	7	9	9		
Mitsubishi	3	3	3	4		
Sanyo	—	3	4	5		
Others	2	2	2	5		
Subtotal	68	72	75	80		100

Sources: Same as Table 2 plus Yoichi Yokomizo, "VCR Industry and Sony" (MS Thesis, MIT, Sloan School of Management, 1986).

in commercialization may not be as useful as coming into the market second or third, as long as the rapid followers have comparable technical abilities, which usually result from having been among the pioneers who participated in developing the technology for commercial applications.[7] These firms, which, along with the inventors, are also technological pioneers, may follow the first mover quickly enough to neutralize its advantages while still exploiting the benefits that come from being a leader in creating the set of complementary assets in manufacturing, marketing, and distribution needed for market dominance.[8] For example, rapid followers who are also pioneers should be able to copy the best features of the first product while adding others to differentiate their offerings. They may have better information about buyer preferences after watching early consumer reactions and have more time to plan for manufacturing, distribution, licensing, or the use of complementary products and services. Follower pioneers and later entrants may also exploit investments made by the first mover, such as in solving engineering and manufacturing problems (if the solutions become public knowledge) or in educating buyers in the use of a new product (as occurred with the video recorder and the personal computer). They may benefit as well from the mistakes or inflexibility of the first mover as the market develops and the technology changes.[9]

In a mass consumer market, the time required to create a dominant standard may be so great that first-mover advantages are minimal, especially for products subject to what economists and others have termed "bandwagon" effects and "network externalities." The bandwagon effect refers to situations where early sales or licensing of a particular product lead (either accidentally or deliberately) to rising interest in that product. A momentum builds up that encourages other potential licenses, distributors, and customers to support the product that seems most likely to become the industry standard, regardless of whether it is technically superior, cheaper, or "better" in other ways than alternatives. The support for one standard over another can become especially dynamic and self-reinforcing if, for reasons apart from the main product itself (such as the need for and relative availability of a complementary product like software programs for computers or prerecorded tapes for VCRs), customers perceive value in owning the standard that becomes the most commonly available in the industry. Network externalities refer to whether or not there is a usage pattern that depends on such a complementary product, as well as to how and how much customers use it with the main product.[10]

While a market is unfolding, both early and later entrants can maneuver to establish a sustainable winning position before the game is decided. Each has particular advantages and disadvantages associated with the timing of decisions and the extent of commitments. Each can affect, at least in part, whether or not support for its standard occurs and how much it continues. In the case of the VCR, the potential global market measured hundreds of millions of units. Its very scale created a window of opportunity lasting a few years, during which firms with comparable engineering and manufacturing capabilities could challenge Sony, the first mover in refining the technology for consumers as well as in making preparations to exploit the mass market. As demand grew at rates outstripping the supply capabilities of Sony or any one producer, rapid followers who were also technological pioneers stimulated the occurrence of a first bandwagon that affected the formation of alliances for production and distribution. The emergence of demand for a complementary product—prerecorded tapes (usually movies)—set off a second bandwagon in the 1980s, as retail outlets for tape rental chose to focus on stocking tapes in the format being adopted by a majority of users, even though Sony's original format still enjoyed substantial acceptance. Of particular interest to historians, economists, and students of management strategy is how the initial moves of the main rivals shaped their long-term competitive positions as well as their eventual success or failure in this market.

INVENTORS, PIONEERS, AND STANDARD-SETTERS

Magnetic video recording technology was created in the United States, but numerous European and Japanese companies competed and collaborated in the 1960s and 1970s to adapt the technology to the requirements of a mass market. Ampex Corporation, a small California company, invented a video recorder for broadcasting applications in 1956.[11] This came after several years of competition with Radio Corporation of America (RCA) to use magnetic tape (as earlier used in audio tape recorders) to record television signals, and freed the broadcast industry from a reliance on live performances or on a clumsy system of film recording. In the late 1950s, Sony, JVC, and Matsushita, as well as several other Japanese firms, began studying and improving on the $50,000-plus Ampex machine, employing novel recording-head mechanisms and solid-state electronic circuits, as well as other product and process innovations, which allowed them to miniaturize the video recorder and to reduce its price dramatically.

Design technology for video recording had been difficult for Ampex to master but proved more difficult to protect from a select handful of companies that had made audio tape recorders and then invested in the development of video recording. Although Ampex retained control of important patents, Japanese firms challenged these in Japanese courts and also explored ways to invent around them. By the mid-1960s, several firms in Japan, along with Ampex in the United States and Philips in Europe, had accumulated considerable expertise in video recording design and manufacture.

Despite a series of products through the 1960s that did not appeal to consumers because of high prices, poor pictures quality, bulky housings, and inconvenient reel-to-reel formats, the Japanese pioneers continued to improve their machines until, in 1971, Sony succeeded in designing a cassette model with $^3/_4$ inch-wide tape. This machine, called the U-Matic, was still too large and expensive for regular home use. Nonetheless, it found a market among schools and other institutions, and it embodied the core design concepts that served as the basis for both the Beta and VHS formats.[12] In conjunction with an agreement to adopt Sony's U-Matic as a standard for institutional machines, three Japanese firms that later competed for the home video standard—Sony, Matsushita, and JVC—signed a cross-licensing agreement for video recording patents in 1970.[13] Philips did not join this group and pursued its own distinctive VCR design.

Although engineers and managers recognized that a standard format would be better for consumers and producers (who would benefit from expansion of the market), agreement on a single home video format proved impossible to reach. In fact, Sony's experience with the U-Matic had made its engineers particularly reluctant to cooperate in establishing or refining a new standard. As early as 1970, Sony had appeared ready to introduce a smaller machine that used a more sophisticated (azimuth) recording system and that might have proved popular with consumers. Since Matsushita and JVC were not yet ready to mass produce this type of machine, the U-Matic ended up as a compromise design, requiring a wide tape and a large cassette. The compromise thus forced Sony, by agreeing to support what became the industry standard for institutional machines, to miss a potential opportunity to enter even earlier into the home market.[14]

Utilizing nearly two decades of experience with video recorder design, engineering, and manufacturing, Sony and JVC both proceeded to develop 1/2 inch-wide tape. VCRs for the home and introduced them in 1975 and 1976. Meanwhile, other companies, including Ampex, RCA, Matsushita, Toshiba, Sanyo, and Philips, introduced or experimented with alternative formats. Unlike the Sony and JVC designs, both of which resembled the effective U-Matic design, the other VCRs were based on distinctive design concepts that proved to be inferior to Beta and VHS.

In addition, just as Sony's Betamax was essentially a miniaturization of the U-Matic but with a more advanced recording technique, the VHS closely resembled the U-Matic (and thus the Betamax), even though the recording format, tape-handling mechanisms, and cassette sizes remained different. Accordingly, it proved difficult for Sony and JVC, and the firms that carried their machines, to differentiate their products through basic features. Hence, neither Beta nor VHS could gain a technological advantage in design or manufacturing that could be sustained long enough to gain a dominant market position. Sony did establish an advantage in reputation if not in actual design and manufacturing skills because of its unique history as an innovator in home video and as primary inventor of the U-Matic. As discussed later in this article, however, Sony's first-mover role and strategic initiatives did not result in a sustainable advantage. Its chief competitors also had superb technical skills, and domination of the huge global market required cooperation with other firms in mass production, licensing, and distribution of both hardware and software. It was by no means certain, however, that the VHS—which came to market after Betamax and was backed by a small firm (JVC) with limited manufacturing and distribution capabilities—would prove superior in the global marketplace.

THE GLOBAL MASS MARKET

Demand for a novel consumer-electronics product can rise rapidly as masses of new customers appear each year. In home video, for example, everyone with a television set was a potential customer. In contrast, professional video had been a very limited market. Machines for broadcast use were expensive and complex, and the number of buyers equaled the number of television stations—hundreds, not millions, in the United States, Japan, and Europe combined. As a result, one firm was able to supply most of the new and replacement demand for many years. For example, Ampex had produced approximately 75 percent of all video recorders in use worldwide in 1962, and it was able to dominate the broadcast market for two decades after its invention of the video recorder in 1956.[15]

The Beta and VHS models opened up a true mass market, allowing video recorders to parallel and then in the early 1980s to pass color television sets to become Japan's (and the world's) top consumer electronics product in production value.[16] The vast size and worldwide structure of this new demand made it nearly impossible for any one firm to accommodate it. Annual production of home videocassette recorders in Japan exceeded one million as early as 1978, having commenced only in 1975, and continued to double each year until 1981. Japanese firms exported 53 percent of the video recorders they produced in 1977 and approximately 80 percent from 1979 onward. The top export destination was the United States during 1976–79, but European exports consumed a larger share during 1980–82, as VCR sales boomed with the increasing availability of prerecorded tapes (see Table 6.3).[17] Europe was probably a more favorable market in which to promote the use of software than the United States because of the smaller number of television stations and available broadcast programs.

Thus, the characteristics of home video—the market's "mass" and global nature, as well as the product's technical complexity—meant that efficient mass production capacity, broad distribution channels, and clear market preferences would require years to emerge. An early mover into the market had no guarantee of a sustainable advantage from simply being first, but needed an effective strategy to capitalize on its position. The need for strategic action was especially strong because other pioneers, after observing customer reactions to the initial product offering, had the option of moving in with a comparable product, lower prices, better features, or superior distribution. In fact, Matsushita was known for competing in that manner: monitoring a broad range of technical developments and gradually building up in-house skills while waiting for Sony, JVC, or other innovative consumer-electronics firms to introduce a new product. Matsushita would then enter the market six months to a year later with a similar but lower-priced version, usually manufactured more efficiently because of Matsushita's mass production skills and willingness to invest to achieve scale economies where they proved useful. The scale of Matsushita manufacturing reflected broad distribution guaranteed through an enormous domestic sales network, which marketed products under brand names that included Panasonic, Technics, National, and Quasar. Matsushita also could schedule large production runs because of its willingness to sell finished products to original equipment manufacturers (OEMs) in Japan and abroad for sale under their labels.[18]

THE ARGUMENT

A VCR by itself is worthless. Users can employ it only in conjunction with a complementary product, the videotape cassette, that is designed to conform to the interface specification of the VCR. This is a common characteristic of contemporary information technologies, such as the personal computer (PC) and its software programs, compact disc (CD) players and discs, or TV receivers and broadcast signals. Interface standards for innovative products of this sort can be established by various means: government regulation (the Federal Communications Commission for television), formal agreement among a large number of producers of the primary product (CD players), or implicit acceptance by producers reflecting the market power of a sponsor (IBM PC).

In the case of the VCR, since no single producer or coalition was strong enough to impose a worldwide standard, and since repeated efforts to bring producers to an agreement failed, the marketplace set the standard. Furthermore, the existence of a "network externality" had two important consequences. First, given rival products of approximately equal cost and capabilities, buyers will

TABLE 6.3. Japanese VCR Exports, 1975–1983 (value in billions of yen, production in thousands of units)

				Exports by Region/Total Exports (%)		
	Value	Units	Export (%)	N. America	Europe	Other
1976	31	139	48	75	17	8
1977	66	402	53	85	8	7
1978	126	973	73	60	28	12
1979	222	1,671	78	46	33	21
1980	444	3,444	78	32	42	26
1981	854	7,355	84	34	44	22
1982	1,080	10,661	82	27	52	21
1983	1,261	15,237	80	41	38	21

Source: Nomura Management School, "VTR Sangyo noto" [VTR industry note] (Tokyo, 1984), 43.

tend to choose the one that has been chosen, or appears likely to be chosen, by a greater number of other buyers. Second, this creates a dynamic system with a "positive feedback": the perceived benefit of choosing a given standard increases as more buyers choose it, thus increasing the probability of purchase by others not yet in the marketplace. An early lead in this sort of contest, however achieved, may become self-reinforcing.

In the drama of the VCR standardization battle, there were three sets of principal players: 1) the main protagonists, Sony, JVC, and Philips, sponsors of the three principal rival formats and major producers of the core product, the VCR; 2) the remaining consumer electronics producers, each of whom would adopt one of the standard formats for production and/or distribution; and 3) the producers and distributors of an important complementary product, prerecorded software.

As it played out, the crucial battle was between Beta and VHS, Sony and JVC. (Although Philips held on to a different standard in Europe for a decade, it never posed a serious challenge to the other two). The facts are simple: Beta reached the market first, took 58 percent of the market in 1975–77, and fell behind VHS in 1978. For the next six years, sales of Beta-format VCRs increased every year, even as its share of the worldwide market fell every year. After being outsold four-to-one by VHS in 1984, Beta began a rapid decline to extinction (see Tables 6.1 and 6.2).

The figures show how quickly the VHS format turned a slight early lead in sales into a dominant position. Chance events might have produced that early lead, and, as the theory suggests, that might be enough to explain the outcome. The thesis of this article, however, is that the early lead and the eventual outcome reflect the deliberate actions of the main players. Strategic maneuvering by the principal protagonists in 1975–77 led to an alignment of producers of the core product and to the exploitation of mass production and distribution capabilities sufficient to account for the early dominance of VHS sales. In a second phase of rivalry, in the 1980s, the strategic alignment of producers of complementary products reinforced the VHS advantage and hastened the demise of Beta, which might otherwise have survived as a second format.

EMERGENCE OF THE VCR STANDARD

A three-year period, from mid-1974 to 1977, proved decisive in determining the outcome of the standardization battle that would rage on for another decade. At the start of this period, diversity characterized the positions of the world's largest consumer electronics companies with respect to home video, a market that remained wholly speculative in 1974. VCR designs based on six different incompatible formats were in late stages of development at rival companies, and three of the majors, Hitachi, Sharp, and Zenith, had no commitments at all to home video development. By mid-1977, the pattern had changed sharply, as all ten of the biggest firms were marketing home VCRs, and the industry had divided into three "families," supporting either Sony's Beta, JVC's VHS, or the Philips format. The line-ups, and data about each firm's color TV sales and prior VCR commitments, are identified in Table 6.4.

The decisive factors in the standards battle were few. First, of the six designs being developed around the world in 1974, four were significantly flawed and destined to fail. The Philips N-1500, Sanyo-Toshiba V-Code, and Matsushita VX designs were marketed vigorously yet fell short, despite the introduction of improved second-generation models in each case. RCA's VCR design never got past the prototype stage, since management abandoned the project after seeing the Betamax. Although a later Philips model, the V-2000, had many fine technical features, it proved complex and costly to manufacture and was introduced too late to capture a viable market share. Like RCA, Philips also had a video disc system under development, which distracted management attention away from the VCR; JVC and other Japanese firms also had disc systems under development but concentrated on refining and marketing their VCR machines.

Because of the common technical heritage in

TABLE 6.4. Home-Video Families and World Color TV Shares, 1976–1977

			1976 World Color TV Sales	
Company	Format	1974 VCR Commitments	Rank	Share (%)
Sony	Beta	Betamax prototype	3	7.4
Sanyo	"	V-Code in Japan	5	6.2
Toshiba	"	V-Code in Japan	6	5.8
Zenith	"	none	4	6.4
Total Beta				25.8
Matsushita	VHS	VX-100 prototype	1	12.7
Hitachi	"	none	7	5.6
RCA	"	Selectavision prototype	8	5.2
Sharp	"	none	10	3.1
Total VHS				26.6
Philips	Philips	N-1500 in Europe	2	11.5
Grundig	"	N-1500 in Europe	9	3.8
Total Philips				15.3

Source: For color TV sales: Harvard Business School, "The Television Set Industry in 1979" (Boston, Mass., Case no. 9380-191, 1980).

the U-Matic, the Beta and VHS designs were closely comparable in cost and performance. Sony had a clear lead primarily in timing; it would take JVC roughly two more years to match the stage that Sony had achieved by late 1974. But moving first was not sufficient, in itself, to win the prize in this market; how Sony moved and what its principal rivals did also mattered. In retrospect, as Akio Morita, then Sony's president, later acknowledged, he and Masaru Ibuka, then chairman, made a "mistake" and "should have worked harder to get more companies together in a 'family' to support the Betamax format."[19] JVC, in the number two position, did "try harder" and was more effective at forming alliances in support of VHS.

JVC's more effective campaign to form an alliance behind VHS produced a coalition that matched the Beta family in global market power. JVC and its principal ally (and parent), Matsushita, followed that with strategic commitments that gained a decisive edge in market share for VHS, beginning in 1978. Matsushita exploited its generic skills in mass production and substantial previous experience in VCR manufacture by establishing production capacity for the VHS that exceeded the combined capacities of all other Japanese VCR producers. JVC, meanwhile, moved aggressively to bring leading European consumer electronics firms into the VHS family, almost preempting that market from Beta.

STRATEGIC ALIGNMENT OF PRIMARY PRODUCERS

A set of assumptions that proved to be in conflict shaped Sony's strategy for commercializing the Betamax. Sony's leaders believed that the Beta design was good enough to be a winner, and they knew that they were ahead of their rivals in VCR development. But they also understood that no producer, on its own, could establish a VCR format, however good the design, as a recognized global standard. Thus, Sony set out to interest other VCR pioneers in adopting the Beta format, concentrating especially on winning the allegiance of Matsushita, its most formidable rival. But two premises hampered their ability to recruit allies. As Japan's leading developer of video technology, Sony believed that it should not have to delay commercialization of the Betamax in order to cooperate, and probably compromise, on the development of an industry standard with other firms. Sony managers and engineers felt that their earlier willingness to compromise on the U-Matic had been a competitive error. Consequently, Sony went ahead and began manufacturing preparations for the Betamax in the fall of 1974, before approaching other firms to discuss the prospect of their adopting the Sony machine as an industry standard (see Appendix A).

Furthermore, Sony was reluctant to build VCRs for its licensees. Sony had always been uniquely innovative with consumer products that incorporated advanced electronics. Its management had never before agreed to ship Sony products to other companies for distribution under their labels, preferring to build up the Sony name and reputation and to avoid sharing the benefits of Sony innovations with too many levels of distributors. For example, Sony developed and marketed Japan's first audio-tape recorder (1950), stereo audio system for broadcasting (1952), transistorized radio (1955), transistorized video-tape recorder (1958), and transistorized micro-television (1959), as well as unique products such as the Trinitron television, whose picture-tube technology did not follow the industry standard established by RCA.[20] Thus, although Sony managers realized that they would have to license the Beta format to ensure its widest distribution, they were unwilling to compromise on their standard or to help potential licensees with OEM shipments.

Sony first demonstrated the Betamax to representatives of RCA, an American video pioneer, in September 1974. At the same time, Sony began talking to JVC and Matsushita, its U-Matic partners, about "joint development" of a home video format. But Sony did not manage these relationships well. When it approached the other firms, Sony had already begun tooling up for the Betamax, signaling to prospective partners a commitment to proceed with mass production irrespective of their support. Sony thus acted as a true first mover, perhaps believing that its lead in the market would convince other firms to follow. At the same time, having begun manufacturing preparations also made Sony less flexible, because altering the design of its machine would require expensive changes in manufacturing equipment.

The 1974 discussions with RCA accomplished one of Sony's objectives by persuading RCA to kill its own VCR development program, but they also brought to light the most vulnerable aspect of the initial Beta design, its limited playing time. RCA had given two hundred of its own VCRs to U.S. customers in a market test during early 1974 and concluded that a minimum two-hour playing time was necessary for commercial success.[21] RCA executives knew from the Betamax demonstration that their efforts to develop VCR technology had been far surpassed by the innovative Japanese, and they terminated their own program. But they decided to wait for further progress in the technology, especially for longer playing times, before making a commitment to market a particular VCR.

When Sony demonstrated the Betamax to Matsushita and JVC in December 1974, Matsushita also questioned the adequacy of a one-hour playing time.[22] These negative reactions to the Betamax then convinced managers at JVC that a successful machine would have to offer at least two hours of playing time and strengthened their commitment to the VHS, whose development had always proceeded on that assumption. JVC now joined RCA and Matsushita in declining to adopt the Beta format.[23]

Sony managers eventually realized that they were not in a strong bargaining position and decided to modify the Betamax for two-hour recordings. Sony postponed further licensing negotiations, losing valuable time and opportunities to continue attempts at enlisting licensees. In particular, when Hitachi, another major producer of consumer electronics products, showed an interest in July 1975 in licensing the Betamax, Sony managers refused, insisting that the Betamax was not yet perfected and thus not available for licensing.[24] It seems that Sony managers were still primarily interested in persuading Matsushita to adopt the Beta standard, rather than Hitachi; they knew by this time that JVC was working on a competing format, which, because of JVC's position as a Matsushita subsidiary, Matsushita was likely to support if Sony did not make a special effect to persuade them to adopt the Beta format.

Moreover, Sony sought partners who could quickly manufacture VCRs on their own rather than requiring Sony to provide complete machines. Sony chairman Akio Morita was unequivocal about this strategy, declaring early in 1976 that "Sony is not an OEM manufacturer."[25] In this regard, Matsushita, which had a large manufacturing capability for VCRs based on previous unsuccessful products, was a better fit than Hitachi, which had made only a few broadcast-use VCRs through a subsidiary and needed an OEM relationship before it could establish in-house production.[26]

Sony resumed seeking partners as soon as it revised the Betamax to play for two hours. Top executives from Sony and Matsushita met again in March 1976 to discuss adopting Beta as the common standard. In July, Sony demonstrated the latest machine to Matsushita, JVC, Hitachi, Sharp, Mitsubishi, Toshiba, and Sanyo and also appealed to Japan's Ministry of International Trade and Industry (MITI) for support. MITI officials tried to negotiate a settlement and favored Sony in these discussions, since it already had a machine in the market. Toshiba and Sanyo eventually agreed to back Beta, but the other firms decided to wait for the VHS, which JVC announced publicly in September 1976.[27]

In contrast to Sony, JVC followed a strategy aimed at forming as large a group as possible, aggressively pursuing both licensing and OEM agreements, including exports.[28] Management first established a group of adherents in Japan who could boost JVC's manufacturing and marketing capabilities—before completing the design and its own preparations for manufacture. JVC initiated this process in the spring of 1975, shortly after Sony's initial demonstration of the Betamax, and by the end of 1976 had lined up Hitachi, Mitsubishi, and Sharp, in addition to Matsushita. JVC also proposed an OEM relationship to Matsushita, which turned it down because JVC did not have enough capacity to supply Matsushita's huge distribution network and also because Matsushita was capable of producing the VHS machine on its own within a few months.[29] In addition, JVC agreed to provide machines to Hitachi, whereas Sony would not; JVC began shipments to Hitachi in December 1976. In January and February 1977, JVC also began supplying VCRs to Sharp and Mitsubishi, which Hitachi had helped to recruit.[30]

As a second step, toward the end of 1976, JVC moved to establish a footing in the U.S. market by negotiating with RCA. The U.S. company rejected this offer for an OEM relationship because of JVC's small production capacity.[31] Yet, rather than giving up on OEM agreements outside Japan, JVC turned toward European firms, which would be satisfied with smaller quantities than RCA needed. JVC pursued these European alliances far more actively and effectively than any other VHS or Beta producer, even after establishing a large production base and gaining worldwide recognition for its brand name (see Table 6.5).

In addition, to entice other firms to support VHS, JVC was willing to let other companies participate in refining the standard, such as moving from two hours to longer recording times or adding new features. JVC also provided considerable assistance in manufacturing and marketing.[32] Yet another important difference from Sony proved to be style: JVC managers approached prospective partners in an exceedingly "polite and gentle" manner, and encouraged them to adopt as the common VCR standard "the best system we are all working on," rather than the VHS per se.[33] One outcome of JVC's approach was that prospective manufacturing partners truly believed they would have some stake in the future evolution of VHS features.[34] Allowing partners to share in development also improved the VHS in ways that JVC might not have pursued itself. For example, after JVC exhibited the VHS prototype to Matsushita in spring 1975, Matsushita provided technical feedback that sped the completion of the new VCR.[35] Matsushita also took the lead in increasing recording and playback time after consulting with RCA.

JVC also strengthened the position of the VHS family by moving aggressively to line up European distribution. Philips, the leader in the consumer electronics market in Europe, still commanded less than 25 percent of the market for color television in the region. With its German ally, Grundig, the

TABLE 6.5. Group Alignments (1983–1984)

Japan	U.S.	Europe
	VHS Group (40)	
JVC	Magnavox (Ma)	Blaupunkt (Ma)
Matsushita	Sylvania (Ma)	Zaba (J)
Hitachi	Curtis Mathes (Ma)	Nordmende (J)
Mitsubishi	J.C. Penny (Ma)	Telefunken (J)
Sharp	GE (Ma)	SEL (J)
Tokyo Sanyo	RCA (H)	Thorn-EMI (J)
Brother (Mi)	Sears (H)	Thomson-Brandt (J)
Ricoh (H)	Zenith (J)*	Granada (H)
Tokyo Juki (H)		Hangard (H)
Canon (Ma)		Sarolla (H)
Asahi Optical (H)		Fisher (T)
Olympus (Ma)		Luxer (Mi)
Nikon (Ma)		
Akai Trio (J)		
Sansui (J)		
Clarion (J)		
Teac (J)		
Japan Columbia (H)		
Funai		
	Beta Group (12)	
Sony	Zenith (S)*	Kneckerman (Sa)
Sanyo	Sears (Sa)	Fisher (Sa)
Toshiba		Rank (To)
NEC		
General (To)		
Aiwa		
Pioneer (S)		
	V-2000 (7)	
		Philips
		Grundig
		Siemens (G)
		ITT (G)
		Loewe Opta (G)
		Korting (P)
		B&O (P)

Note: Suppliers indicated by initial (J = JVC, Ma = Matsushita, H = Hitachi, Mi = Mitsubishi, T = Tokyo Sanyo, S = Sony, To = Toshiba, Sa = Sanyo, P = Philips, G = Grundig).
*In spring 1984, Zenith switched from the Beta group to VHS.
Source: Nomura Management School, "VTR Sangyo noto"; and JVC, Public Relations Dept.

TABLE 6.6. VCR Sales by Country and Format (1983)

	Unit Sales (millions)	VHS %	Beta %	V-2000 %
USA	4.1	75	25	0
Japan	3.4	70	30	0
Britain	2.3	74	24	2
W. Germany	1.5	60	20	20
France	0.4	70	20	10
Italy	0.2	60	20	20
Above Totals	11.9	72	25	3

Source: Nomura Management School, "VTR Sangyo noto," 5.

number two producer, Philips was producing home VCRs based on its 1972 technology, now outmoded by the Beta and VHS innovations. Most of the other European consumer electronics firms had earlier marketed VCRs produced by Philips and Grundig, but by 1975 all of them had dropped the product. In contrast to RCA's reaction to the Japanese innovations, Philips determined to surpass the new designs with an innovative machine, for which they launched development in 1975. Meanwhile, Philips and Grundig persisted with the old design, upgraded in 1977 to provide two-hour recordings. The Philips V-2000 reached the market in 1980 but, despite impressive technical features, it was too expensive and too late (see Table 6.6).

JVC exploited this opportunity to recruit Telefunken, Thomson, Thorn, Nordmende, and other strong European brands into the VHS family. Moving quickly with its Japanese partners, JVC had defined the technical standards for a PAL (the European color standard) VCR in 1977. JVC's readiness to supply machines on an OEM basis as well as to help firms prepare for manufacturing in Europe, plus the evident superiority of VHS over the current Philips offering, won commitments in rapid order from the remaining major European firms.[36]

The marketing clout wielded by the rival families is worth close analysis, because all the participants understood that VCRs would be sold as adjuncts to television and audio equipment. A rough proxy for market power in that industry in the mid-1970s was a company's share of the color television receiver market. At one level, the rivals appear evenly balanced. Among the world's top ten consumer electronics companies, the VHS and Beta groups were evenly matched, each selling slightly more than one-quarter of the color sets sold in 1976 (see Table 6.4). whereas Philips and Grundig together accounted for less than one-sixth. But the

VHS family was more successful in gaining the allegiance of smaller brands. Hence, within each of the three major geographic markets, VHS started out with a market share advantage. The VHS family—Matsushita, JVC, Hitachi, Sharp, and Mitsubishi—accounted for nearly 60 percent of color TV sales in Japan in 1976, compared to only 37 percent for Sony, Toshiba, and Sanyo. In the U.S. market, the VHS brands, led by RCA, had a 49 percent share of color TV sales in 1976, compared to only 41 percent for Zenith, Sony, Sears, and the rest of the Beta family. And by 1978, almost all the European brands not committed to the Philips format adopted VHS, leaving Beta in a minority position.

In 1975 and 1976, all the world's leading consumer electronics producers entered the home video market. Those that had bet wrong on video development, choosing an inferior design approach, or electing not to invest at all, reversed their positions and adopted one of the three contending formats. In the course of these two years, JVC, by adroit maneuvering (and with a major boost from Matsushita), transformed the structure of the rivalry to establish a standard format for home VCRs. In mid-1975, Sony had stood out in a field of diverse contenders, including rival VCRs as well as potential alternatives such as videodisc. Its Beta design was the only format both ready for market and capable of performing at the level required for a mass market. By mid-1977, VHS could challenge it from a position of parity, both in product cost and functionality and in the market power of the VHS family.

PRODUCT DIFFERENTIATION

Did the market performance of VHS result from differentiating features, price, or quality? A comparison of models introduced during 1975–85 by Sony, JVC, and Matsushita, the major home VCR producers, indicates some differences in all three dimensions (see Appendixes B, C, and D). In general, however, at no time did either format establish more than a transient advantage in features, prices, or picture quality.

For example, although Sony's initial models played for one hour and VHS machines two hours, Sony increased its machine's capacity to two hours merely five months after JVC entered the market and several months before Matsushita appeared (see Table 6.7). Sony offered more low-priced models until 1980, when Sanyo introduced inexpensive Beta models. Nevertheless, Matsushita quickly surpassed Sony in share once it entered the VHS market in 1977, and the VHS standard was dominant worldwide by the end of 1978. Beta and VHS offered basic models at similar prices; the VHS group included more brand names, yet Sony led in the introduction of most new features even as it was losing market share to the VHS group. Between 1977 and 1983, Sony was the first company to offer wireless remote control, half-speed and one-third speed machines, multi-function machines (scan, slow, and still), high fidelity (hi-fi) sound, and a one-unit movie camera (camcorder). But, as can be seen in Table 6.8, Matsushita or JVC usually matched Sony's new features within a few months, and sometimes more quickly. JVC also was first with several innovations, such as slow/still functions, a portable VCR, and stereo recording (which Matsushita marketed at the same time).

Differences in picture quality are more difficult to assess, but VHS did not have a reputation

TABLE 6.7. Recording-Playing Time Comparison

Year/Month	Beta	VHS
1975/5	1 hr. (Sony)	
1976/10		2 hr. (JVC)
1977/3	2 hr. (Sony)	
1977/10		4 hr. (Matsushita)
1978/10	3 hr. (Sony)	
1979/3	4.5 hr. (Sony)	
1979/8		6 hr. (Matsushita)
1979/8		4 hr. (JVC)
1979/12		6 hr. (JVC)
1982/3	8 hr. (Sony)	
1982/9	5 hr. (Sony)	

Note: Some of the longer playing times for Beta were achieved with thinner tape, not new machine models.

Source: Itami Hiroyuki, *Nihon no VTR sangyo: naze sekai o seiha dekita no ka* [Japan's VTR industry: why it was able to dominate the world] (Tokyo, 1989), 208; JVC, Public Relations Dept. (See Appendix C).

TABLE 6.8. Special Effects Comparison (Sony and Matsushita)

	Introduction Date (Year/Month)		
	Sony	Matsushita	JVC
Wireless remote	1977/3*	1977/6	1979/6
1/2-Speed machine	1977/3*	1977/6	1979/8
Slow/still	1979/3	1978/7	1977/12*
Portable VCR	1978/9	1980/2	1978/2*
1/3-Speed machine	1979/3*	1979/8	1979/12
Scan/slow/still	1979/3*	1980/6	1979/8
Stereo recording	1980/7	1979/8*	1979/8*
Hi-fi	1983/4*	1983/5	1983/11
One-unit camera-recorder	1983/7*	1985/1	1984/3

*Marks the first to introduce the feature.
Source: Yokomizo, "VCR Industry and Sony"; and Appendixes B, C, and D.

as being superior to Beta, and the truth may indeed have been the opposite.[37] In addition, physical differences existed in the machine weights and cassette sizes, but it remains unclear how these affected the course of events, except that the smaller Beta cassette made it more difficult for Sony to increase recording or playing time simply by putting more tape into its cassettes.[38]

The key issue here is that Beta machines still might have survived as an alternative format used for high-quality recording of broadcast programs off the air or for home movies (the market niche Sony has exploited with 8mm camcorders). To have achieved this with Beta, Sony would have had to distinguish its VCR through special effects or features that made it especially convenient or superior to VHS in performance. Yet, as with basic features and prices, Sony was unable to differentiate Beta models for a significant length of time because of the technical skills and initiatives of JVC and Matsushita, as well as those of their partners in the VHS group.

It also seems that Matsushita was able to counter Sony in the Japanese and U.S. markets by utilizing its huge engineering and manufacturing resources to offer a product line with more combinations of features and prices. Compared to Sony, Matsushita introduced both less and more expensive VCRs between 1978 and 1981 and manufactured about twice the number of model types Sony produced during the same time period (see Appendixes B and D). Other marketing measures helped VHS firms overcome Sony's image for high quality and reliability; for example, RCA and Matsushita (which marketed Panasonic and Quasar brands in the United States) both offered an extended labor warranty for their machines.

MASS PRODUCTION AND MASS DISTRIBUTION

By 1978, the VHS family had gained a significant edge in manufacturing capability, as well as in market power. Both the Beta and VHS machines were complex to manufacture, compared to other consumer electronics products such as radios, televisions, or audio equipment, in particular because they required high precision for machining the heads and sophisticated assembly skills for building the tape-handling mechanism and other components. The difficulty of designing and then mass-producing an inexpensive VCR kept Ampex and RCA from entering this segment of the market in the 1970s, even though both designed home VCR prototypes in their laboratories.[39] Philips, in addition to difficulties with product reliability, also had to price its VCRs 20 to 30 percent higher than VHS and Beta machines.[40]

Both Sony and JVC mastered the problems of mass production engineering and manufacturing benefiting from experiences gained through earlier video recorder production. They also relied on integrated development teams for the Beta and VHS projects that brought together members with both design and operations backgrounds. JVC, which had less experience making VCRs than Sony, paid special attention to making its VCR easy to manufacture and service by creating a relatively simple, low-cost design with fewer components and assembly steps than the Betamax—characteristics that also appealed to companies wishing to license a VCR for in-house manufacturing. In contrast, although Sony had the manufacturing expertise to produce the Betamax economically, potential licensees appeared concerned over their ability to mass produce the Beta design.[41]

Matsushita also made low-cost production a major priority as it modified the VHS design and prepared its own plants. The company spent at least fourteen months studying manufacturing issues before formally adopting the VHS standard in January 1977. Matsushita engineers knew what problems to expect, because they had accumulated invaluable experience producing earlier VCR machines, including a cartridge model once made in a plant with 1,200 workers and a monthly capacity of 10,000 units, as well as the VX cassette model, which Matsushita had made in 1976 before switching to the VHS.[42] Matsushita not only emphasized a reduction in parts but also invested in manufacturing automation and scheduled large production runs, anticipating that its vast distribution system would enable it to sell a great number of VCRs.[43] Matsushita's ability to deliver low-priced VCRs with an increasing variety of features also helped it undercut Sony prices and win contracts to supply machines to overseas distributors—arrangements that further increased Matsushita's scale of operations and ability to justify additional investments in product development and automation.[44]

Managers at Matsushita believed that the manufacturer who would dominate the world market would be the company that captured the largest share of the U.S. market, where the major VCR distributors were likely to be RCA and Zenith, the leaders in color television sales.[45] Sony moved first after developing a two-hour model by establishing a relationship with Zenith, after having been rebuffed by RCA. RCA intended to lead in the market for home video players but wanted lower-priced machines as well as a longer recording time. Meanwhile, Matsushita took a strong interest in RCA's distribution resources. These mutual interests brought RCA and Matsushita together in negotiations for an OEM agreement after discussions broke down between RCA and JVC, which did not have the manufacturing capacity to supply RCA with the volume of machines it wanted.

As RCA managers pondered which Japanese producer with which to link up, they reconsidered the issue of tape length. In February 1977, apparently to the astonishment of Matsushita executives, RCA requested a VCR that "could record a football game." This meant a recording time of at least three hours. Rather than ending the negotiations, Matsushita launched an intensive effort to double playing time from two to four hours by using the approach Sony had taken to double the playing time of its one-hour machine: halving the width of each recording track (called the track pitch) as well as slowing the recording speed. Matsushita put seventy engineers on this project alone and achieved the increase in playing time in merely two months; it then set up production capacity for 10,000 units per month within six months. By the end of March 1977, Matsushita had an agreement to supply RCA with approximately 50,000 four-hour VCRs by year's end.[46]

A large part of the VHS advantage came from the sheer ability to deliver more VHS machines than Beta producers could make early on in the competition. Even in 1978, because of Matsushita's massive capacity, the VHS group accounted for approximately 66 percent of the total Japanese VCR production capacity of 191,000 units per month (see Table 6.9). Matsushita—not JVC—thus proved instrumental in winning over RCA and pushing the VCR competition toward the areas where Sony was weakest: low prices and mass distribution, as well as longer playing and recording times. JVC personnel opposed a doubling of the playing time, arguing that this constituted a "bastardization" of the VHS (that is, a compromise in picture quality), and they refrained from collabo-

TABLE 6.9. VCR Monthly Production Capacity (1978) (thousands of machines, average monthly capacity)

VHS Group		Beta Group	
100	Matsushita	45	Sony
20	JVC	10	Toshiba
6	Hitachi	10	Sanyo
126	VHS Total	65	Beta Total

Source: Itami, *Nihon no VTR sangyo*, 220.

rating with Matsushita in pursuing this feature. JVC eventually built a two-speed (two- and four-hour) machine in August 1977, primarily to satisfy its OEM partners, but not until July 1979 did it introduce such a machine commercially under the JVC brand name.[47] JVC, which had about one-tenth the sales volume of Matsushita, also took six months to build a machine with four-hour play and twelve months to achieve a monthly capacity of 10,000 units.[48]

Most important, the nature of competition changed as a result of Matsushita's alliance with RCA. First, momentum clearly built up for VHS in the U.S. market, as General Electric, Sylvania, Magnavox, and Curtis Mathes scrambled to join this group in 1977, under the rationale that the format RCA supported would probably become the dominant machine in the American market.[49] U.S. distributors initially had been indifferent to the choice of standards and appeared to be waiting for clearer market signals before selecting a format. Second, because of the longer playing time, Matsushita and its distributors, and later other firms in the VHS group, were able to establish an image of the Beta machine as deficient with respect to this basic feature. Sony increased the Betamax's playing time to three hours in October 1978, but not until March 1979, a year and a half after Matsushita introduced the four-hour VHS, did Sony introduce a 4.5-hour machine (see Table 6.7).

Thus, by spring 1977 Matsushita was able to plan a large-scale entry into the worldwide VCR market and to begin exploiting its skills and investments in low-cost manufacturing and mass distribution. These assets, in turn, helped RCA, which had brand recognition as well as extensive distribution channels, to offer reliable products at low prices. The effective Matsushita-RCA combination then damaged Sony's competitive position in both the U.S. and Japanese markets, not only because Sony's market share and distinctiveness declined. Shortly after RCA's announcement of a reduction in prices to undercut Sony in August 1977, Zenith demanded a renegotiation of its OEM agreement with Sony, to whom it was paying $100 more for Beta machines than RCA paid Matsushita for VHS machines.[50] After a lag of more than two months, Sony and Zenith responded by matching RCA's prices.[51] Yet these moves portended a difficult future: Sony would now play the game on terms that Matsushita and RCA had set, and play it poorly. In fact, Sony had trouble matching the prices of both Matsushita and JVC in the low end of the VCR market between 1979 and 1981 (see Fig. 6.1). Sanyo took over as the primary supplier of the lowest-priced Beta machines, but it did not have the range of alliances or the distribution channels to which Matsushita had access.

STRATEGIC ALIGNMENT FOR COMPLEMENTARY PRODUCTS

Of the three principal functions of the VCR—namely, "time-shifting" (recording broadcast programs for later viewing), making and viewing home movies, and playing prerecorded cassette programs—only in the last one did the greater availability of VHS prove to be a significant factor for consumers. Blank cassettes used for time-shifting and movies were readily available for both machines. The format did represent a potential constraint on the sharing of these tapes among households, once recorded, but such use remained small. On the other hand, users quickly perceived that prerecorded tapes were more available in VHS than in Beta, and that difference appeared very salient to users intending to rent or buy programs.

Until the early 1980s, that difference did not matter much in the marketplace. The VCR was broadly perceived to be a niche product, appealing primarily to certain demographic segments. In 1980 and 1981, with VCR ownership in only 5 to 10 percent of television households in most advanced countries, forecasts typically projected a leveling of demand at penetration levels of 15 to 30 percent in the late 1980s.[52] Users gave little evidence of interest in prerecorded tapes. In the United States in the late 1970s, three-quarters of all VCR owners bought no prerecorded tapes.[53] In 1983, several years after the beginning of the tape-rental business, 40 percent of VCR owners never used such

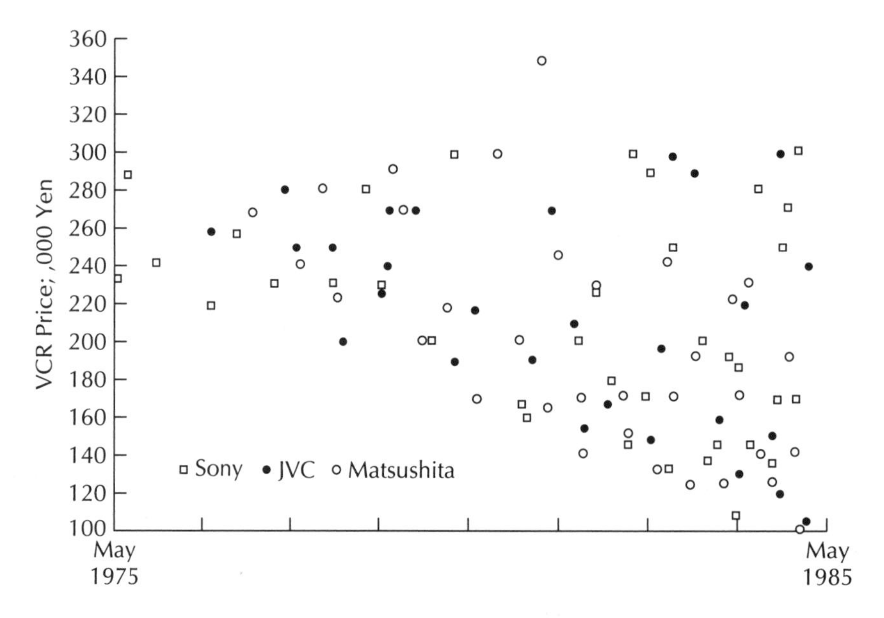

Fig. 6.1. VCR price summary comparison. (*Source:* Appendixes B, C, D.)

tapes and only 8 percent identified them as "important."[54] With a small installed base of players and low consumer interest, producers and distributors of programs had little incentive to invest heavily in prerecorded tapes and video rental stores.

All that changed in the mid-1980s. Confounding the forecasts, the VCR turned into a mass-market product, reaching 30 percent of American homes by 1985, five years ahead of most forecasts, and still climbing. Sales and rentals of prerecorded cassettes began to grow exponentially, doubling each year from 1982 to 1986. Although at least one leading U.S. firm concluded in 1982 that tape rentals would not be accepted by U.S. consumers and that the economics of the rental business would not support a large industry, entrepreneurs flocked to open rental stores in every neighborhood.[55]

Europe stood at the leading edge of this change. VCRs began to achieve mass-market penetration in Europe earlier than elsewhere, apparently due to the availability of fewer broadcast channels there. In 1983, when penetration had reached 10 percent in the United States and 12 percent in Japan, it was 29 percent in the United Kingdom and still growing. Because TV set rental was a common practice in Britain, extended readily to VCRs, the practice of renting programs on tape was a natural adjunct. The linkages formed by JVC and Hitachi with Thorn and Granada, the leading British TV-rental operations, led those distributors to emphasize the VHS format in tape rental as well. Program producers and distributors, observing the preponderance of European brands adopting VHS, tended to emphasize it over Beta and Philips formats. One pioneer in tape production, Magnetic Video, in 1980 had three times as much capacity in Europe for VHS production as for either Beta or V-2000.[56]

In the United States, aggressive steps by RCA in the late 1970s contributed significantly to the momentum behind the VHS standard, which still did not overtake Beta decisively until the mid-1980s. Because of its ambitious videodisc venture, RCA had well-developed ideas about the consumer market for recorded video programming. To promote its VCR in 1978, RCA developed an important alliance with Magnetic Video Corporation of America (MV). MV was a leader in prerecorded video (primarily used then for education and training) and was the first to offer feature films on cassette. RCA supplied two MV program cassettes free with each VCR in 1978, along with a membership in the MV "club." MV, which soon found most of its growth coming in the VHS format, expanded capacity to enable it to duplicate 2.4 VHS tapes for every Beta tape. Matsushita facilitated this by developing equipment for high-speed duplication and by rapidly making low-cost decks available to MV and others. When the British firm, Granada, began opening rental shops in the United States in 1980, it offered only VHS players and cassettes.

Sony matched most of these moves, but with a lag and with less effect. In 1979 Sony linked up with Video Corporation of America (VCA), but this firm continued to promote VHS as well. Sony also proved less effective than Matsushita in supplying equipment for duplication of tapes in the Beta format. As a consequence of these and other moves, by 1980 the VHS format clearly dominated Beta in the channels for prerecorded tapes. According to one estimate, VHS then accounted for 70 to 90 percent of the revenues of cassette dealers in the United States.[57]

As the mass market began to grow in subsequent years, VHS sustained and multiplied this initial advantage. The greater abundance of VHS program material gave buyers greater incentive to choose VHS players, which then led tape distributors to stock more VHS tapes, in a reinforcing pattern. By 1984, contrary to most forecasts made as recently as 1980 or 1981, the sale and rental of prerecorded tapes was a billion-dollar business in the United States, dominated by the VHS format.[58] When Zenith, the leading U.S. color TV brand, switched from Beta to VHS in 1984, the end was in sight for the Beta format.

CONCLUSIONS

The VCR story provides a classic example of the dynamics possible in a standardization contest affected by bandwagons and complementary products. The evidence cited here also shows this case to be an important illustration of how strategic maneuvering can harness the dynamic power of a special marketplace—the mass-consumer market—to make a winner out of a second mover with extensive technological skills but a weak starting position in manufacturing and distribution capabilities.

In April 1975, Sony enjoyed what looked like an insurmountable lead. Its Betamax, already on the market in Japan, was clearly superior to VCRs being offered by major rivals—Matsushita, Sanyo, Toshiba, and Philips. The company had a lustrous reputation globally as an innovator and leader in consumer electronics. JVC, a minor factor in the industry, was still struggling to perfect VHS prototypes that seemed to offer few evident technological advantages. Matsushita was struggling with its poorly received VX product. Two years later, though Beta still enjoyed a lead, JVC, supported by Matsushita, set in motion the fundamental forces that would continually erode, and then extinguish, Beta's share of a massive global market.

In retrospect, it is possible to identify the key events and to "explain" the outcome in terms of a few factors. But as events were unfolding, the implications of each strategic move must have been more difficult to discern. Each of the key protagonists acted in a way that made sense in context. Sony's behavior followed patterns that had brought it great success over two decades. JVC, the undergo, could not reasonably have been less humble or flexible in its relationships. Matsushita, along with Toshiba, Sanyo, and Philips, were actually failed first movers, since they introduced unsuccessful VCRs at nearly the same time as the Betamax. Matsushita, however, exhibited its usual mixture of caution and flexibility. Had the market grown more slowly, as nearly all observers expected, Sony might have been able to respond more effectively to its early mistakes and to the actions of its key competitors.

A few important moves made the difference. JVC created a winning alignment of VCR producers in Japan by the way its managers conducted the formation of alliances, showing versatility and humility, whereas Sony pressed commitment and reputation. The alliance with the giant Matsushita brought huge added benefits. Matsushita's management waited until VHS seemed likely to be a viable alternative to Betamax before abandoning its own VX model and then quickly switched over to the new machines, investing massively in capacity in advance of demand while pushing the product technology to meet RCA's requirements of a longer recording time. JVC completed the sweep by moving ahead of Sony to enlist a huge number of European partners behind VHS.

JVC's early success in aligning itself with Matsushita and other Japanese producers allowed the company to gain a decisive edge in the race for distribution rights. Sony's reluctance to be an OEM supplier, and its underestimation of the threat from VHS, left Beta in a minority position for potential market power in North America and Western Europe as well as in Japan. As the theories discussed in this article suggest, once VHS took the lead, it became more and more likely that it would continue to gain share year after year. The final contest, among producers and distributors of video software, accelerated this process. Even without the growing importance of software, the outcome probably would have been the same in the long run. Nonetheless, the dominance of VHS in tape-rental channels hastened the demise of Beta and made certain it would not survive even as a second format.

Louis Pasteur said that "chance favors the prepared mind." Chance no doubt played a role in the dynamic growth of the VCR industry and the eventual success of VHS. But the alliances that JVC formed for production and distribution and the timely strategic commitments of its ally, Matsushita, proved to be the decisive factors in the triumph of VHS over Beta.

Notes

1. Michael L. Dertouzos, Richard K, Lester, and Robert M. Solow, *Made in America: Regaining the Productivity Edge* (Cambridge, Mass., 1989), 216–18.

2. Betamax is a trademark of the Sony Corporation. VHS is a trademark of the Victor Company of Japan (JVC).

3. In English see, for example, James Lardner, *Fast Forward: Hollywood, the Japanese, and the VCR Wars* (New York, 1987); and P. Ranganath Nayak and John M. Ketteringham, *Breakthroughs!* (New York, 1986); in Japanese see, for example, Nihon Keizai Shimbunsha, ed., *Gekitotsu: Soni tai Matsushita: bideo ni kakeru soryokusen* [Crash! Sony versus Matsushita: the all-out war wagered on video] (Tokyo, 1978); and Itami Hiroyuki, *Nihon no VTR sangyo: naze sekai o seiha dekita no ka* [Japan's VTR industry: why it was able to dominate the world] (Tokyo, 1989).

4. Marvin B, Lieberman and David B. Montgomery, "First-Mover Advantages," *Strategic Management Journal* 9 (1988): 41–58; and Michael A. Porter, *Competitive Advantage: Creating and Sustaining Superior Performance* (New York, 1985), 186–89.

5. For discussions of these cases, see Richard S. Rosenbloom and Karen J. Freeze, "Ampex Corporation and Video Innovation," in *Research on Technological Innovation, Management, and Policy,* ed. R. S. Rosenbloom (Greenwich, Conn., 1985), 2: 113–86; Franklin M. Fisher, James W. McKie, and Richard B. Mancke, *IBM and the U.S. Data Processing Industry: An Economic History* (New York, 1983); and Gary Jacobson and John Hilkirk, *Xerox: American Samurai* (New York, 1986).

6. This definition of "first movers" is used in Alfred D. Chandler, Jr., *Scale and Scope: The Dynamics of Industrial Capitalism* (Cambridge, Mass., 1990).

7. Richard S. Rosenbloom and Michael A. Cusumano, "Technological Pioneering and Competitive Advantage: The Birth of the VCR Industry," *California Management Review* 1, 4 (1987): 51–76.

8. David J. Teece, "Profiting from Technological Innovation: Implications for Integration, Collaboration, Licensing, and Public Policy," in *The Competitive Challenge,* ed. David J. Teece (Cambridge, Mass., 1987), 185–219.

9. Lieberman and Montgomery, "First-Mover Advantages": Porter, *Competitive Advantage;* and Richard N. Foster, *Innovation: The Attacker's Advantage* (New York, 1986).

10. See, for example, M. L. Katz and C. Shapiro, "Technology Adoption in the Presence of Network Externalities," *Journal of Political Economy* 94 (1986): 822–41; J. Farrell and G. Saloner, "Installed Base and Compatibility: Innovation, Product Preannouncements, and Predation," *American Economics Review* 76 (1986): 940–55; and W. Brian Arthur, "Positive Feedbacks in the Economy," *Scientific American,* Feb. 1990, 92–99.

11. See Rosenbloom and Cusumano, "Techno-logical Pioneering and Competitive Advantage," and Rosenbloom and Freeze, "Ampex Corporation and Video Innovation."

12. Useful discussions of the concept of a dominant design as well as "architectural" variations, which seem to describe VHS and Beta as refinements of the U-Matic, can be found in Kim B. Clark, "The Interaction of Design Hierarchies and Market Concepts in Technological Evolution," *Research Policy* 14 (1985): 235–51; and Rebecca M. Henderson and Kim B. Clark, "Architectural Innovation: The Reconfiguration of Existing Product Technologies and the Failure of Established Firms," *Administrative Science Quarterly* 35 (1990): 9–30.

13. Nihon Keizai Shimbunsha, ed., *Gekitotsu;* and Rosenbloom and Cusumano, "Technological Pioneering and Competitive Advantage."

14. Nihon Keizai Shimbunsha, ed., *Gekitotsu;* Nomura Management School, "VTR Sangyo noto" [VTR industry note] (Tokyo, 1984); and Richard S. Rosenbloom interviews with Nobutoshi Kihara and Masaaki Morita, Senior Managing Directors, Sony Corporation, July 1980.

15. Rosenbloom and Freeze, "Ampex Corporation and Video Innovation."

16. Katz and Shapiro, "Technology Adoption in the Presence of Network Externalities."

17. Arthur, "Positive Feedbacks in the Economy."

18. Nihon Keizai Shimbunsha, ed., *Gekitotsu,* 151–54.

19. Akio Morita, *Made in Japan* (New York, 1986).

20. Nick Lyons, *The Sony Vision* (New York, 1976).

21. Lardner, *Fast Forward,* 84; and *TV Digest,* 21 April 1975.

22. Nihon Keizai Shimbunsha, ed., *Gekitotsu,* 13–17.

23. Nayak and Ketteringham, *Breakthroughs!* 37–38.

24. Nihon Keizai Shimbunsha, ed., *Gekitotsu,* 33–34; Lardner, *Fast Forward,* 156.

25. Quoted in *TV Digest,* 16 Feb. 1976.

26. Yoichi Yokomizo, "VCR Industry and Sony" (MS Thesis, MIT, Sloan School of Management, 1986), 79–80.

27. Nihon Keizai Shimbunsha, ed., *Gekitotsu,* 59–72.

28. See Appendix A and Tables 6.3, 6.5, and 6.6; Nayak and Ketteringham, *Breakthroughs!* 42; Nomura Management School, "VTR Sangyo noto"; and "Innovations Spur Boom in VCR Sales," *The New York Times,* 11 Dec. 1984, D1.

29. Nihon Keizai Shimbunsha, ed., *Gekitotsu,* 54.

30. JVC committed to supplying Hitachi on an OEM basis although this entailed that a large portion of its production capacity of about 2,000–3,000 units per month would be diverted to that end. This portion would have been significantly smaller for Sony, which, at the time, had a production capacity of more than 7,000 units per month. See Nomura Management School, "VTR Sangyo noto"; and *TV Digest,* 21 April 1975 and 13 Dec. 1976.

31. Nayak and Ketteringham, *Breakthroughs!* 46.

32. Michael A. Cusumano interview with Susumu Gozu, Manager, Domestic Sales Dept., Video Products Division, Victor Company of Japan, July 1989.

33. Kokichi Matsuno, message to employees in taking over as JVC President in 1975, and Shizuo Takano, JVC's Video Department manager, both quoted in Nayak and Ketteringham, *Breakthroughs!* 41. Susumu Gozu, in his interview with Cusumano, gave a similar account of JVC's approach.

34. Nayak and Ketteringham, *Breakthroughs!* 32–33; also, Gozu interview.

35. Lardner, *Fast Forward,* 148–49.

36. Alan Cawson et al., *Hostile Brothers: Competition and Closure in the European Electronics Industry* (New York, 1990).

37. "VCRs: Coming on Strong," *Time,* 24 Dec. 1984, 48; "Selecting the First VCR—Some Questions to Keep in Mind," *The New York Times,* 18 Dec. 1983, H38; Tony Hoffman, "How to Buy a VCR," *Home Video,* April 1981, 48–55.

38. Nihon Keizai Shimbunsha, ed., *Gekitotsu;* Yanagida Kunio, "VHS kaihatsu dokyumento" [Documentation of VHS development], *Shukan gendo,* May 1980; Richard S. Rosenbloom interviews with Nobutoshi Kihara and Masaaki Morita, Senior Managing Directors, Masaru Ibuka, Honorary Chairman, and Akio Morita, Chairman, Sony Corporation, July 1980.

39. Rosenbloom and Cusumano, "Technological Pioneering and Competitive Advantage"; Rosenbloom and Freeze, "Ampex Corporation and Video Innovation"; and Margaret B. W. Graham, *RCA the VideoDisc: The Business of Research* (New York, 1986).

40. Nomura Management School, "VTR Sangyo noto," 4.

41. Rosenbloom and Cusumano, "Technological Pioneering and Competitive Advantage"; Yanagida Kunio, "VHS kaihatsu dokyumento"; Michael A. Cusumano interview with Gozu of JVC as well as with Tak Matsumura, Assistant Director, Video Recorder Division, Matsushita Electric, July 1989.

42. Nihon Keizai Shimbunsha, ed., *Gekitotsu,* 21–24, 54; Lardner, *Fast Forward,* 159.

43. Yokomizo, "VCR Industry and Sony," 39–40.

44. *TV Digest,* 4 April 1977.

45. Itami, *Nihon no VTR sangyo.*

46. Lardner, *Fast Forward,* 161–63; Nayak and Ketteringham, *Breakthroughs!* 47.

47. *TV Digest,* 11 July 1979.

48. Ibid., 29 Aug. 1977.

49. Ibid., 30 May, 27 June, 7 Nov. 1977.
50. Ibid., 4 April 1977.
51. Ibid., 29 Aug., 3, 31 Oct., 7 Nov. 1977.
52. Bruce C. Klopfenstein, "Forecasting the Market for Home Video Players: A Retrospective Analysis" (Ph.D. diss., Ohio State University, 1985).
53. *TV Digest,* 9 Sept., 16 Oct. 1978, 12 April 1979.
54. Klopfenstein, "Forecasting the Market for Home Video Players," 141.
55. Richard S. Rosenbloom, personal interviews at RCA, 1982.
56. *TV Digest,* 6 Oct. 1980.
57. Ibid., 8 Dec. 1980.
58. *The Wall Street Journal,* 21 April 1986, 20D.

APPENDIX A. VCR Industry Chronology, 1974–1978

Year/Month	
1974/9	Sony proposes to Matsushita and JVC that they jointly adopt the Sony VCR under development, although development was largely completed and Sony already had begun setting its manufacturing dies and making other production preparations. Sony also shows the Betamax prototype to RCA, in the hope of persuading the U.S. firm to adopt it. (RCA subsequently abandons an attempt to develop its own VCR but rejects the Betamax because of its short 1-hour recording and playing time.) Toshiba and Sanyo introduce their own VCR, the V-Code I, with 30-minute and 1-hour tapes.
/12	Sony shows the Betamax prototype to Matsushita and JVC, but still receives no commitment from them.
1975/4	Sony introduces the Betamax SL-6300 in Japan, priced at 229,800 yen (ca. $800); 1-hour recording time. JVC announces to Matsushita that it has a competing VCR under development, the VHS.
/7	Hitachi approaches Sony as a potential licensee of the Betamax, but is rebuffed as Sony prefers to wait for Matsushita and modify the Betamax for 2 hours.
/9	Matsushita introduces its own VCR model, the VX-100, with 1-hour tape. JVC also completes a VHS prototype and demonstrates this to Matsushita and later to other firms.
/12	Hitachi adopts the VHS format.
1976/1	JVC asks Sharp and Mitsubishi Electric to adopt the VHS format; they agree by fall 1976.
/2	Sony introduces the Betamax (SL-7200) in the U.S.
/3	Hitachi, acting on behalf of JVC, asks Toshiba and Sanyo to join the VHS group. Sony again approaches Matsushita and asks that it adopt the Betamax and Matsushita shows the VHS prototype to Sony for the first time.
/4	Toshiba and Sanyo introduce the V-Code II with a 2-hour tape.
/5	Matsushita introduces the VX-2000, with a 100-minute tape. JVC begins manufacturing preparations for the VHS.
/6	Sony and JVC each ask the Ministry of International
/7	Trade and Industry (MITI) to back their standards. MITI proposes that
/8	JVC adopt the Betamax, or that the two firms negotiate on a standard, adopt one or the other or a combination, but these suggestions fail to be accepted.
/10	JVC introduces the VHS for commercial sale in Japan with a 2-hour tape.
/12	Hitachi begins marketing VHS machines supplied by JVC.
1977/1	Sharp begins marketing VHS machines supplied by JVC. Matsushita publicly adopts the VHS format.
/2	Sanyo, Toshiba, and Zenith adopt the Betamax format.
/3	Sony introduces a 2-hour color version of the Betamax (SL-8100), although it is not compatible with the 1-hour Betamax. Matsushita introduces a 4-hour version of the VHS for export to RCA, Magnavox, Sylvania, GE, and Curtis.
/4	Pioneer and Aiwa adopt the Betamax format.
/8	Sanyo reaches an agreement with Sears-Roebuck to supply it with Betamax machines.
/10	The VHS group settles on a European standard, followed by export agreements to several European distributors.
/11	NEC adopts the Betamax format.
/1	Hitachi begins in-house production of the VHS
/5	Mitsubishi begins in-house production of the VHS for export

Sources: Primarily Nihon Keizai Shimbunsha, ed., *Gekitotsu: Soni tai Matsushita: bideo ni kakeru soryokusen* [Crash! Sony versus Matsushita: the all-out war wagered on video] (Tokyo, 1978) and Sony Corporation, "Table of Sony VTR History," unpublished memorandum, 16 Aug. 1977.

APPENDIX B. Sony Product Schedule, 1975–1985

Name	Date	Yen Price	Comments
SL-6300	May-75	229,800	First Betamax
SL-7300	Jul-75	285,000	
SL-6301	Feb-76	238,000	
SL-7100	Oct-76	215,000	Price-Down/Simple Operation
SL-8100	Mar-77	255,000	2-hr Recording (Both Beta I&II)
SL-8300	Mar-77	258,000	2-hr Recording Only (Beta II)
SL-8500	Oct-77	228,000	
SL-3100	Sep-78	229,000	Portable
SL-J7	Mar-79	279,000	Multi-Function/Beta-Scan/Beta III
SL-J5	Jun-79	229,000	
SL-J1	Mar-80	198,000	Portable
SL-J9	Jul-80	298,000	Stereo
SL-F1	Jul-81	165,000	Portable
SL-F11	Jul-81	278,000	Wireless Remote Control/Stereo
SL-J10	Aug-81	158,000	Price-Down
SL-J30	Jun-82	198,000	Price-Down with Stereo
SL-J20	Jun-82	137,000	
SL-F7	Sep-82	225,000	Swing Search
SL-J25	Dec-82	178,000	
SL-F3	Mar-83	145,000	
SL-B5	Mar-83	199,000	Portable
SL-HF77	Apr-83	299,000	Hi-Fi
SL-F5	Jun-83	169,000	Micon Voice
BMC-100	Jul-83	289,000	Beta-Movie
BL-F17	Oct-83	132,000	
SL-HF66	Nov-83	249,800	Hi-Fi
SL-HF55	Apr-84	198,000	Hi-Fi
SL-HFR30	May-84	137,000	BetaPlus (Expandability for Hi-Fi)
BMC-200	May-84	289,000	Beta-Movie Auto Focus
SL-HFR60	Jul-84	145,000	BetaPlus
SL-HF300	Sep-84	189,000	Hi-Fi
FL-F33	Oct-84	108,000	
SL-HF500	Nov-84	185,000	Hi-Fi
SL-HF355	Nov-84	198,000	Hi-Fi
EV-A300	Jan-85	145,000	8mm
BMC-500	Jan-85	268,000	Beta-Movie Auto Focus
SL-HF900	Feb-85	239,800	Pro/Hi-Band
CCD-V8	Mar-85	280,000	8mm Movie
SL-HFR70	May-85	135,000	Hi-Band
SL-HF505	Jun-85	168,000	Hi-Band
EV-A300	Jun-85	145,000	8mm
EV-S700	Jun-85	249,800	8mm Digital
BMC-600	Jul-85	270,000	Hi-Band/Beta-Movie/Auto-Focus
SL-HF505	Sep-85	168,000	Hi-Band
CCD-M8	Sep-85	198,000	8mm Movie
EV-C8	Sep-85	148,000	8mm Portable
CCD-V8AF	Oct-85	299,800	8mm Movie/Auto Focus

Source: Sony Corporation, cited in Yokomizo, "VCR Industry and Sony," 83.

APPENDIX C. JVC's Product Schedule, 1976–1985

Name	Date	Yen Price	Comments
HR-3300	Oct-76	256,000	First VHS; 2-hr, 2-head
HR-3600	Dec-77	279,000	Slow/Still; Wired Remote
HR-4100	Feb-78	248,000	Portable
HR-3310	Sep-78	248,000	Microphone Mixing
HR-4000	Nov-78	198,000	VHS Player
HR-4110	Jun-79	225,000	Portable, Slow, Wireless Remote
HR-3500	Jul-79	238,000	Slow Function
HR-3750	Aug-79	268,000	Multi-Function/Speed, 4-head Stereo
HR-6700	Dec-79	268,000	Multi-Speed, 6-hrs., 2-head
HR-2200	Jul-80	188,000	Portable, 2-head
HR-6500	Nov-80	215,000	4-head, Timer & Counter*
HR-7300	Sep-81	188,000	
HR-7650	Jan-82	268,000	Front-Loading, Wireless Remote
HR-2650	May-82	208,000	
HR-C3	Jul-82	153,000	Compact (VHS-C)
HR-7500	Nov-82	165,000	Random Search Function
HR-7100	Nov-82	139,800	
HR-D120	Jul-83	148,000	
HR-D225	Sep-83	195,000	
HR-D725	Nov-83	298,000	Hi-Fi
HR-D220	Nov-83	158,000	One-Touch Timer
GR-C1	Mar-84	288,000	Compact Camcorder
HR-S10	Jul-84	158,000	
HR-D130	Jul-84	138,000	Simplified Timer
HR-D150	Nov-84	129,800	
HR-D555	Dec-84	218,000	Hi-Fi/Stereo
HR-D250	May-85	149,800	
HR-D140	Jun-85	119,800	
GR-C2	Jul-85	299,000	Compact Camcorder
HR-D565	Aug-85	189,800	Hi-Fi
HR-D160	Nov-85	104,800	
HR-D755	Dec-85	239,800	Hi-Fi, Programming Remote Control

*Note: All subsequent models are 4-head.
Source: JVC, Public Relations Dept.

APPENDIX D. Matsushita Product Schedule, 1977–1985

Name	Date	Yen Price	Comments
NV-8800	Jun-77	266,000	2-hr/4-hr Recording
NV-5500	Mar-78	238,000	
NV-6600	Jul-78	279,000	Slow/Still
NV-5000	ct-78	220,000	Portable
NV-6000	Aug-79	289,000	6-hr Rec/Slow/Still/Stereo
NV-6200	Oct-79	268,000	Stereo
NV-3000	Feb-80	198,000	Portable
NV-3500	Jun-80	215,000	Multifunction (Scan/Slow/Still)
NV-3300	Nov-80	168,000	
NV-3700	Mar-81	298,000	Wireless Remote Control/Stereo
NV-3200	Jul-81	198,000	Portable
NV-1000	Nov-81	350,000	4-head/Clean Still/Reverse/Stereo
NV-700	Nov-81	229,000	4-head/Clean Still
NV-310	Dec-81	163,000	
NV-710	Feb-82	244,000	4-head
NV-100	Feb-82	178,000	Portable
NV-350	Jun-82	169,000	
NV-300	Aug-82	139,800	
NV-200	Aug-82	163,000	Portable
NV-750	Sep-82	229,800	4-head
NV-600	Feb-83	169,800	3-head
NV-150	Feb-83	189,800	Portable
NV-330	Mar-83	149,800	3-head
NV-800	May-83	289,800	Hi-Fi/4-head
NV-370	Aug-83	132,800	3-head
NV-850HD	Oct-83	239,800	Hi-Fi/4-head
NV-630	Nov-83	169,800	
NV-360	Feb-84	123,800	
NV-180	Mar-84	189,800	Portable/4-head
NV-7700	Mar-84	189,800	4-head
NV-270	Aug-84	125,000	3-head
NV-870HD	Oct-84	219,800	Hi-Fi/4-head
NV-650	Nov-84	169,800	4-head
NV-900HD	Jan-85	229,800	Hi-Fi/4-head
NV/M1	Jan-85	298,000	VHS-Movie
NV-550	Mar-85	139,800	
NV-260	May-85	125,000	
NV-880HD	Jul-85	189,800	Hi-Fi/4-head
NV-660	Sep-85	139,800	
NV-U1	Oct-85	100,000	
NV-M3	Oct-85	298,000	VHS-Movie

Source: Matsushita Electric, cited in Yokomizo, "VCR Industry and Sony," 84.

7

Managing Product Families: The Case of the Sony Walkman

SUSAN SANDERSON
MUSTAFA UZUMERI

The virtues of speed to market and product line breadth have been extolled in both popular and scholarly literature. Several scholars have noticed that many successful firms competing in fast product cycle industries (e.g., consumer electronics and computers) were those that combined model variety and high rates of technological change [30,33,34]. This is especially true for the personal portable stereo market, pioneered by Sony when it introduced the first Walkman in 1979 [23].

Many observers consider Sony to be one of the most consistent and impressive innovators of consumer and industrial products [4,5,7,20,25]. Within Sony, there has been no greater success than the Walkman. Sony has dominated the personal portable stereo market, worth $1 billion worldwide, for over a decade and has remained the leader, both technically and commercially, despite fierce competition from world-class competitors. Sony has competed against outstanding firms such as Matsushita, Toshiba, Sanyo, Sharp, and Philips—consumer electronics giants with marketing savvy, financial strength, excellent engineering skills, a strong technology base, and world-class manufacturing. In this paper, we describe how Sony's unique approach has made it possible for the firm to receive a price premium for its models (up to $20 more than competing firms) while retaining a 50 percent revenue market share for over a decade in this highly contested market.

The Walkman is an example of outstanding product family management. Sony's strategy employed a judicious mix of design projects, ranging from large team efforts that produced major new model "platforms" to minor tweaking of existing designs. Throughout, Sony followed a disciplined and creative approach to focus its subfamilies on clear design goals and target models to distinct market segments. Sony supported its design efforts with continuous innovation in features and capabilities, as well as key investments in flexible manufacturing. Taken together, these activities allowed Sony to maintain both technological and market leadership.

Previous studies have suggested that the combination of technological leadership and the systematic spinning out of models leads to success in consumer product markets [30]. Firms' product line breadth has been attributed to their ability to introduce new models quickly [3,26]. We suggest that success in fast cycle industries like consumer electronics depends both on rapid model changeover

Reprinted from *Research Policy*, Vol. 24, No. 5, Sanderson, S. and Mustafa Uzumeri, "Managing Product Families: The Case of the Sony Walkman" pp. 761–782, 1995 with kind permission from Elsevier Science, The Netherlands.

and on model longevity. We will show that Sony was consistently as fast or faster than any of its competitors in getting new models to market. Yet Sony also offered many models that enjoyed much longer market lives than competing models. We suggest that it was the *combination* of novelty and longevity that allowed Sony to fashion its decade-long dominance in this highly competitive industry. We believe that Sony's example illustrates the potential for more sophisticated product planning. If firms follow Sony's lead, they can go beyond an obsession with speed to a more subtle understanding of product diversity and timing.

GENESIS OF THE WALKMAN PRODUCT FAMILY

Although accounts of the events surrounding the Walkman's development differ in certain particulars, all versions agreed that the first Walkman model was anything but a radical technological change [1,14,16,28]. The founding member of the Walkman product family was based far more on market insights than on technological breakthroughs.

In 1978, Sony shifted responsibility for the design of tape recorders from its audio group to a group that has since made cassette decks and boom boxes [28]. At the time, the engineers in the audio group were forced to scramble to generate new tape-based products. As it happened, several engineers were working on a stereo cassette recorder based on the compact, high performance Pressman recorder that had been launched in 1977. The Pressman was a handy device designed to be used by reporters to record interviews. In trying to master the technique for installing recording and playback mechanisms in small spaces, Sony engineers decided first to develop a prototype equipped only with a playback mechanism. This prototype reproduced sound of such high quality that the engineers believed there might be a potential market for the player alone. Music lovers were already buying prerecorded music cassettes, but the tape recorders available at the time were large and only marginally portable.

Meanwhile, another research team was working in the same building on a set of lightweight headphones. This was part of an ongoing research effort to miniaturize components of all kinds. The program was driven by Sony's twin design goals of greater audio fidelity and greater portability for all its products. At the time, the lightest headphones in the world weighed about 100 grams (3.5 oz). The team had decided to produce headphones weighing about half that and had already developed a prototype model. The miniature headphone and cassette player teams were unaware of the others' work until the spring of 1979 when Masaru Ibuka (Sony's honorary chairman) dropped by and made the connection between the two projects.

The first model weighed slightly less than a pound and was basically the Pressman without recording capabilities. When Sony's engineers married the Pressman, stereo circuitry, and the small headphone, each component was already fully developed. Sony did not even remove the RECORD button from the Pressman case. Instead, the engineers put in twin headphone jacks so two people could listen at the same time and changed the RECORD button to a mute button so the couple could stop the music while talking to one another.

Production of a prototype Walkman began in Japan in 1978. Sony introduced the Walkman to the Japanese market in July 1979. Thirty thousand units of this model were produced, and the entire stock was sold within three months. The Walkman was a stunning hit and production couldn't keep up with demand. The first full production model, the TPS-L2, sold 1.5 million units in just two years.

Although Sony's Japanese competitors can usually introduce high quality imitations very quickly, it took them a year to offer a similar product. Sony has led the worldwide market for personal stereos with its worldwide market share on a unit basis hovering around 40 percent for over a decade (see Figure 7.1). On a revenue basis Sony's market share has remained around 50 percent. As shown in Table 7.1. Sony has comparable market shares in both the U.S. and Japanese markets.

Sony has dominated the worldwide market despite the competence of its competitors. General

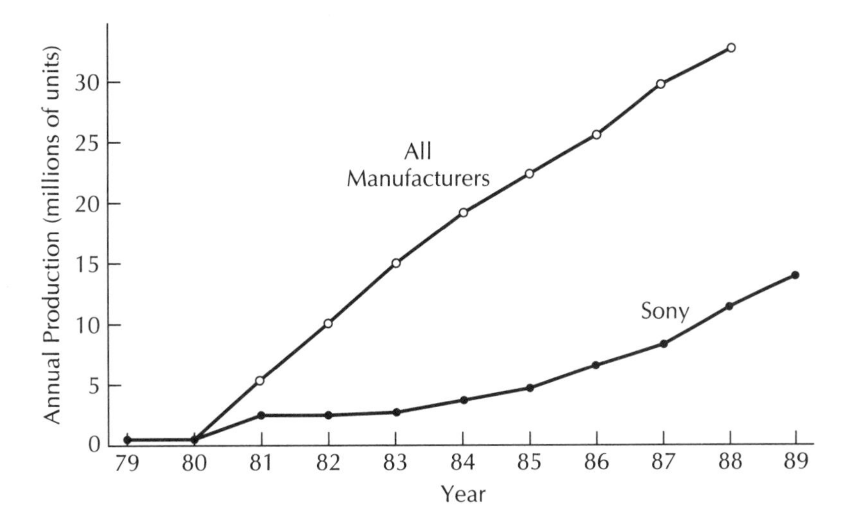

Fig. 7.1. Worldwide headphone stereo market. (*Source*: Sony estimates.)

Electric held about 10 percent of the market in the United States, but was primarily a distributor of lower-priced models made overseas. Among manufacturers, Panasonic's 6 to 8 percent market share is less than a fifth of Sony's U.S. sales. In the Japanese market, Sony's dominance appears to be challenged by Aiwa, but this is misleading, because Sony owns just over 50 percent of Aiwa. The individuals we interviewed reported that Sony and Aiwa compete aggressively with one another at the operating level. At the corporate level, however, Sony plays an active role in Aiwa's management and influences Aiwa's strategic direction [19]. Aiwa has historically been perceived as offering slightly upscale models, while Sony was considerably more aggressive in its exploration of new market segments. If Aiwa's sales are combined with Sony's, their market shares easily exceeded 50 percent in both the United States and Japan.

PATTERNS OF PRODUCT COMPETITION

Given the caliber of Sony's major competitors, how did Sony maintain its dominance in this product family for more than a decade? Moreover, Sony was widely acknowledged to have charged premium prices for its Walkman models. How did Sony maintain this practice for more than a decade? To try to answer this question, we collected and analyzed product data from the major firms to reconstruct a detailed history of the Sony Walkman product family. To build this history, we relied on three types of data.

Models, Features, and Prices. We extracted and coded model and price data from newspaper advertisements on a quarterly basis from 1980–91. We used this data to identify the models that were available in the U.S. market and the prices at which these models were offered. Information on the design, features, and specifications of each model was obtained from manufacturers' sales brochures supplied by the manufacturer. More than 550 models of personal stereos appeared on the U.S. market during the 1980s. While it was possible to get technical information on many of them, documentation for others was difficult to obtain. We had particular difficulty obtaining information for many low-priced models that were manufactured in Southeast Asia and sold under private labels. Consequently, we focused on the personal stereo models produced by the major consumer electronics firms, Sony, Panasonic, Toshiba, and Aiwa. Detailed information was obtained for 260 models sold in the United

TABLE 7.1. Personal Stereo Revenue Market Share (1989–90)

Rank	Firms	% U.S. Market Share[a]	Firms	% Japanese Market Share[b]
1	Sony	45–50	Sony	46
2	General Electric	10–12	Aiwa	20
3	Panasonic	6–8	Matsushita	8
4	Sanyo	4–6	Toshiba	7
5	Aiwa	4–6	Sanyo	6
	OTHERS	25	OTHERS	13

[a]*Source*: Conversation with Integrated Marketing Intelligence, confirmed by conversations with marketing managers at Sony.
[b]*Source*: Market Share in Japan 1989, Yano Research Institute Ltd. New York.

States between 1980 and 1991. This set also included a number of stereo-like radios. We also gathered data and published materials on models offered for sale in Japan, the home market for the five major manufacturers. This was supplemented by model descriptions from Sony's European product brochures. This data was used to compare Sony models across international markets.

Interviews. Interviews were conducted with industrial designers, product managers, and marketing and sales managers with Sony and its major competitors in the United States and in Japan. In addition, we visited Sony's Bonson factory in Japan, where much of the production for the Japanese market takes place and spoke at length with the director of manufacturing and the chief engineer.

Supplementary Published Information. We collected data from company records and newsletters, trade and industry journals, conference proceedings, and academic journals reporting engineering advances and market share data.

Sony's Product Variety

In personal portable stereos, as well as in many other product categories, the number of customers buying consumer products in the U.S. and European markets is larger than in the Japanese market. For example, 42 percent of Sony's Walkman units were sold in North America, 40 percent in Europe, 11 percent in Japan, and 7 percent in the rest of the world in 1988. To be successful in markets outside their home base, firms must have models that appeal to customers around the world. Porter has suggested that if firms have their headquarters in markets that *anticipate* the demands of markets elsewhere, they are likely to be more successful [21]. We looked at this issue by comparing personal stereo models sold in the United States and Japan.

Figure 7.2 compares the price and size of the personal portable stereo models sold by the five major Japanese manufacturers in U.S. and Japanese markets in 1989–1990. The Japanese prices were converted to comparable New York discount prices by using price data on identical models sold in both markets to estimate an effective exchange rate.

The majority of personal stereo models in the U.S. sold for $55 to $60, while Japanese models typically cost between $130 and $150.[1] Figure 7.2 shows that these price differences are generally related to the physical sizes of the models sold in each market. The Japanese models in Figure 7.2 are generally smaller and have higher sound quality than the average model sold in the United States. Nonetheless, small models display similar prices in both markets. Larger models cost less, but are not offered for sale in Japan. This suggests that the differences in average selling price are due to sale mix differences, rather than to pricing differences.

The differences in the model mix between Japan and the United States are visible in the product features (see Table 7.2). From conversations with product managers in the United States and Japan, it would appear that the explanation for the

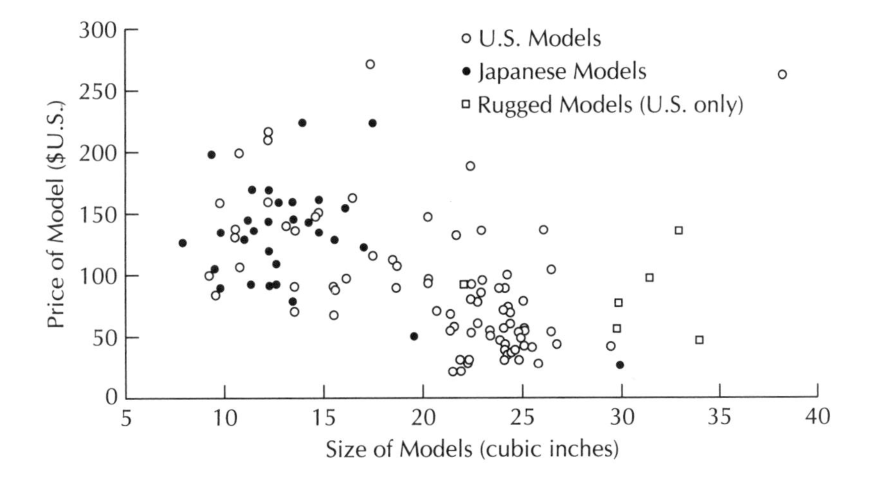

Fig. 7.2. Personal stereos in United States and Japan (1989–90).

different product mix lies with differences in consumer lifestyle. Japanese consumers tend to have more urbanized lifestyles. They live in cramped (by American standards) accommodations and commute long distances to work, often by subway or train. Not surprisingly, Japanese consumers prefer small, high-performance, rechargeable models. Japanese models tend not to have a radio tuner, as there are far fewer radio stations in Japan than in the United States. Finally, many Japanese models have remote controls that allow the tape to be controlled from a wand built into the headphone cord. This feature means little to most American joggers, but has great value to Japanese commuters packed into subway cars so tightly that they cannot move their hands. Finally, the shapes and colors of models sold in Japan by all manufacturers are quite similar to one another but distinctly different from those sold in the United States. Only the features and performance are significantly different.

By contrast, models in the U.S. market were much more diverse. American consumers are more likely to participate in active sports such as exercising, jogging, or cycling. Others spend leisure time in outdoor areas or at beaches. While Sony's rugged, water-resistant models were extremely popular in the United States, they were not even offered for sale in Japan. In the United States, most models had radio tuners and U.S. models tended to stress low price over small size.

Although we did not conduct a systematic analysis of the European market, we learned from Sony designers that European models differ from both American and Japanese models. European customers tend to demand higher quality sound and classic design styles, with less emphasis on either cost or size. In Germany, for example, all of Sony's models were designed to be sturdy and many had dark metal cases.

To satisfy these markedly different preferences, all major manufacturers made models exclusively for the U.S. market. While the models that Sony and its competitors' sold in Japan looked very similar, Sony far outstripped its competitors in bringing varied and unique models to the United

TABLE 7.2. Percentage of Models with Given Features in the United States and Japan

Feature		Japan (%)	US (%)
Remote control	Sony	58	10
	Others	77	11
AM and/or FM radio	Sony	47	61
	Others	30	93
Water resistance	Sony	0	18
	Others	0	1

TABLE 7.3. Number of Models in Product Line (1989–90)

Firm	United States	Japan
Sony	24	18
Panasonic	8[a]	15
Sanyo	8[a]	8
Aiwa	20	*n/a*
Toshiba	10[a]	*n/a*
Sharp	*n/a*	7
Kenwood	*n/a*	5

n/a = not available.
[a]data from 1989 product lineup.

States and Europe. Table 7.3 shows the number of models offered by each manufacturer in the U.S. and Japanese markets. Sony's broad product line is evident in both cases. In the U.S. market. Sony consistently offered more models than any of its competitors. Only Aiwa approaches Sony's variety and, as late as 1988, Aiwa offered only nine models in the U.S. market.

Sony's variety did not develop overnight. Each symbol in Figure 7.3 represents the median value of the prices listed for a model in the discount electronics ads.[2] The increasing breadth of Sony's product line is striking. Over the period, Sony is represented by 572 points, more than the total offerings from Aiwa, Toshiba, Sanyo, and Panasonic combined. In fact, Figure 7.3 understates Sony's model variety. An inspection of the sales catalogs from the various manufacturers shows that the model-time-price points for the other manufacturers in Figure 7.3 include many models that compete directly against one another. Sony's competitors produced mainly generic models that were aimed at the large market segments and tended to be larger and less expensive than those sold in Japan.

Sony, on the other hand, offered numerous models that were designed specifically for the U.S. market. The "Sports Walkman" was an excellent example. Characterized by its bright yellow waterproof case, the "sports" theme was used in other Sony product lines such as compact disc (CD) players, "Karaoke" units (sing-along tape players), camcorders, and walkie-talkies. Because of this diversity, many of Sony's designs, especially those sold in the United States and Europe, face little direct competion from other firms. This also helped Sony to avoid cannibalization among its own designs.

Model Longevity and Manufacturer Variety

From the model-price-time points in Figure 7.3, it is possible to estimate the market availabil-

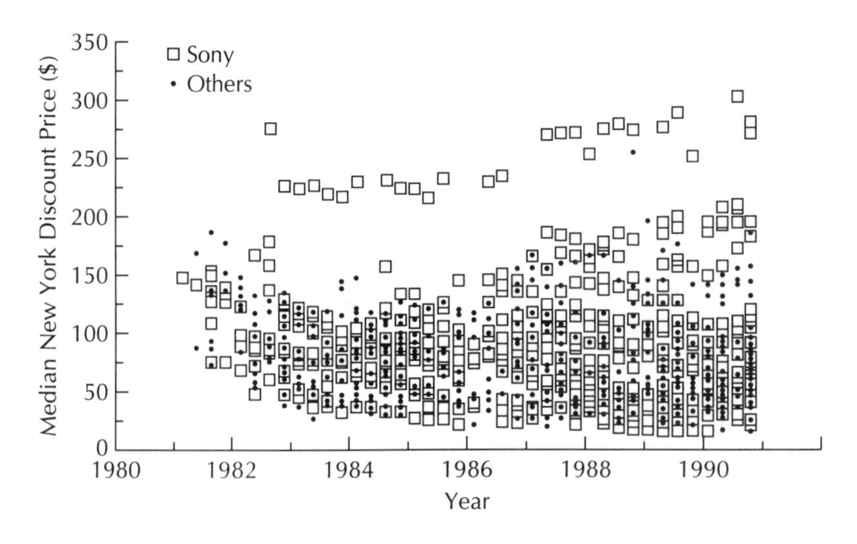

Fig. 7.3. Variety of personal stereo models in the U.S. market.

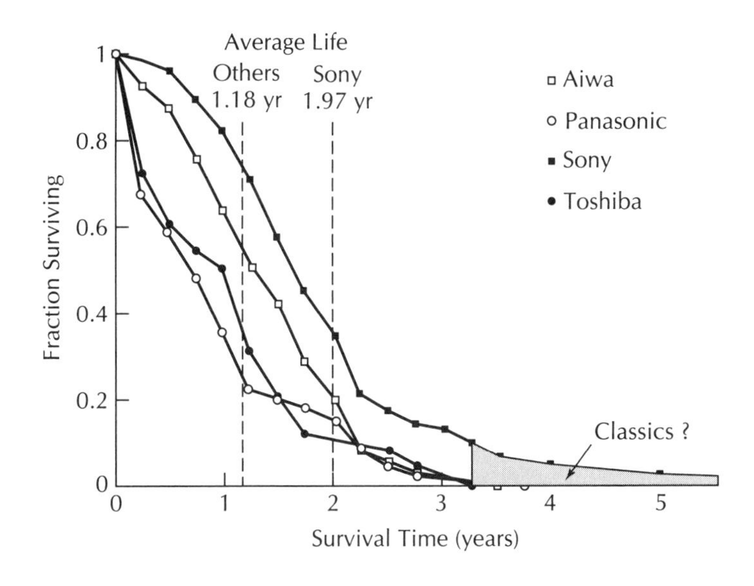

Firm	Model	Entered	Left
Aiwa	HSP02	8/82	11/84
Panasonic	RX1936	8/86	11/88
	RXSA80	11/86	5/89
	RXSA60	2/87	5/89
	RXSA73	5/88	*
Sony	TBSL2	9/80	11/93
	WMF2	8/81	8/83
	WMR2	8/81	11/84
	WM3	8/81	5/85
	WMF1	2/83	5/85
	WMF5	5/83	11/85
	WMDC2	8/84	8/86
	WMD6C	8/84	*
	WMF100	5/86	5/89
	WMF45	2/97	5/89
	WMF73	8/87	8/90
	WMAF64	5/88	*
Toshiba	KTR2	8/81	11/83
	KTS1	8/81	2/84
	KTS3	8/81	8/84

**Still on market in last period of study.*

Fig. 7.4. *Left*: Survival of personal stereo models in U.S. market (1980–1989). *Right:* Models surviving more than 2

ity of the various models. The period of availability serves as an estimate of the market life of each model. The price quotes provide additional information about the nature of that availability. If the price quotes near the end of the model life fall sharply, it may indicate that the model has actually been withdrawn. To eliminate such spurious values, model life was estimated as the period of time during which the model sold for at least 50 percent of its initial price. Figure 7.4*a* shows the survival curves for 162 models offered for sale in the United States by four major manufacturers. From 1980–88, the average market life of a Sony model in our database was 1.97 years. The average life of models from Sony's major competitors was 1.2 years. This difference is significant at the $p < 0.001$ level, based on a χ^2 test.

The data show an interesting pattern. Sony was widely considered the most innovative of the four firms and was the undisputed market share leader, yet its models survived almost twice as long as its competitors [6,9,13]. Of particular interest are the long-lived models in the upper tail of the curve in Figure 7.4*a*. We have labeled these "classics" and listen them separately in Figure 7.4*b*.[3] Sony generated more classic models than any of its competitors. Most of these were designs that were so well adapted to the needs of their target market segments that Sony felt little market pressure to redesign them. Several models enabled Sony to make an early entry into key market segments and effectively preempt them. With the WMF45 Sports Walkman, for example, Sony identified a specific consumer lifestyle and its basic solution remained unchallenged throughout the decade. In 1990, the WMD6C (a high-quality professional model) had been on the market for over six years.

The long-lived models allowed Sony to keep more models on the market with the same level of design effort that its competitors expended on models with shorter lives. Sony's competitors were apparently unable or unwilling to invest in new models for market niches that would have permitted their product lines to achieve comparable longevity and sustainable variety. They elected to compete in the undifferentiated, price-sensitive markets where the design changes tended to be minor and easy to copy. This increased the pressure for frequent redesign and led to rapid model attrition. When its models competed directly against designs of other manufacturers, Sony changed the designs as often as its competitors.

The importance of long-lived models has not been fully recognized by innovation researchers. The Sony example shows that model longevity can have a major impact on the evolutionary development of product families. Sony's longer-lived models extended its product line and multiplied the impact of any investment in new model design. If firms are to plan and effectively manage rapidly evolving product families, it is important to identify and catalog these mechanisms and effects.

SONY'S TECHNOLOGICAL EVOLUTION

Although Figure 7.3 seems chaotic, the evolution of Sony's Walkman product family has actually been very systematic and disciplined. By examining their cost and novelty, it is possible to classify the design changes that produced the pattern in Figure 7.3. The framework by Wheelwright and Clark provides a convenient starting point for this task, but requires elaboration. Design changes in personal stereos can be placed in two of Wheelwright and Clark's five categories. A small number of "platform projects" provided the foundation for "derivative projects" that elaborated and diversified the product line [32].

Discussions with product managers in the competing firms confirmed that all personal stereo makers relied heavily on platforms. In Sony's case, the design process for platforms differed substantially from the process for managing derivative designs. With platforms, there is generally a sharp break from preceding designs that we have termed a *generational* design change. Sony's generational design changes entailed substantial technical novelty, required a major design effort and often required major changes to the manufacturing process. Generational designs were not technological discontinuities, because they retained crucial links to the dominant design. Although materials, component technologies, features, and performance were altered significantly, the underlying family resemblance remained intact. Consequently, generational platforms were able to coexist within the Walkman product family and support the development of important subfamilies.

Surprisingly few generational design changes were made during the first decade of the personal stereo, and all of those were pioneered by Sony. According to Sony's published account, five significant innovations occurred in the Walkman product family. These fall into two categories: 1) those producing new electromechanical platforms; and 2) those creating key component technologies. These milestones are shown in Figure 7.5.

In 1981, Sony launched a full-scale, worldwide sales campaign for the second-generation WM2. The WM2 was the first major technological innovation in the Walkman family. It was more compact and had superior sound quality. The body of the WM2 weighed 9.9 oz., and the headphones weighed only 1 oz. At the time, it was the lightest product of its kind in the world. Despite the WM2's success, managers and engineers at Sony continued to innovate. The next generational innovation was the WM20. It was half the thickness of the WM2. To reach this size goal, miniature parts were developed and the model was designed to run on just one 1.5-volt AA battery. This meant that the mechanism had to consume half the power of the WM2. The project team developed a revolutionary flat motor, together with numerous minor improvements. The team took two years to finish its work, and Sony introduced the WM20 in October 1983.

While one team worked on the WM20, another team sought to improve sound quality. In the WM2, a belt connected the motor to the flywheel that dampened the tape drive. The second team developed a direct drive mechanism that used a friction coupling to transmit the motor's rotation directly to the flywheel. The resulting WMDD model also incorporated a servo system that precisely controlled the speed of the capstan. The system mini-

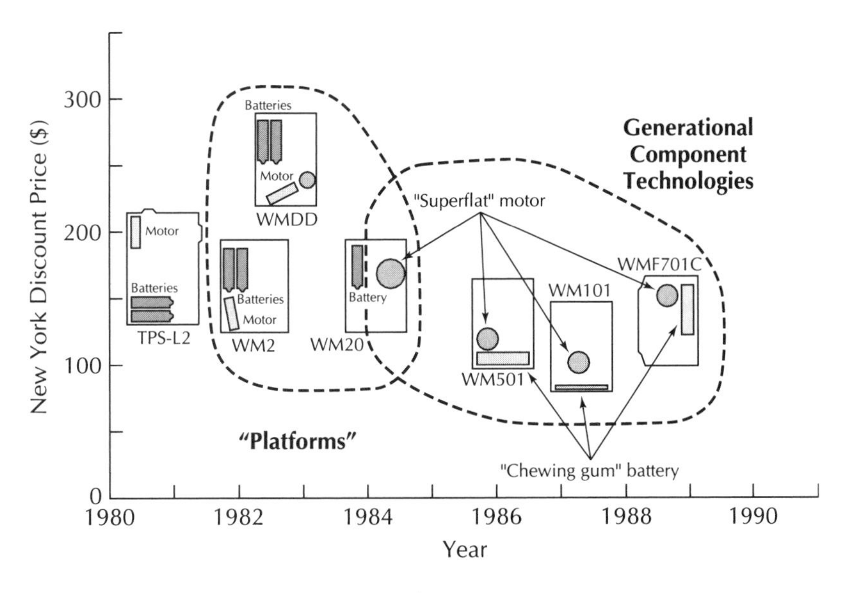

Fig. 7.5. Generational Walkman innovations.

mized the sound distortion that results from uneven tape speed and external vibration. This technology first appeared in the Walkman Professional (WMD6), introduced in February 1982. The immediate successors of the WMD6 (the WMD3 and WMD6C) were still in production in 1990.

The WM2, the WMDD, and the WM20 were the basic platforms on which all subsequent Walkman models were built. The basic mechanisms in each platform were continually refined and remained in production through the end of the decade. A fourth platform was introduced in 1989 to support the development of low-priced models. The generational changes in Walkman platforms required major advances in the battery and motor technology. Battery advances allowed further miniaturization without loss of performance. The rechargeable NiCd battery, called the "chewing gum battery," was first used in 1986 in the WM101 and WM501 models and later in the classic WM701C. Together with the superflat motor, it became a key component in all of Sony's smaller units.

Wheelwright and Clark suggest that the development of generic technologies should be characterized as R&D, not product engineering [32]. This would seem appropriate for some of Sony's innovations, but not for all. Sony relied on an R&D project to develop the miniaturized headphone that was the key to the first Walkman. However, the "superflat" motor and the "chewing gum" battery were designed by the same teams that created new platforms. Sony's process of generational design change was primarily guided by its focused view of the portable consumer electronics market. From the outset, Sony's designers pursued superior sound quality and the smallest, most portable package possible. By consistently pursuing these twin goals of audio quality and miniaturization, Sony was able to coordinate the development of platforms with the development of enabling technologies.

However, the platforms are only a small part of the story. Sony offered as many as twenty new models each year and almost 250 models during the 1980s in the U.S. market. With the platforms as a basis, most of these models were achieved by making small changes in features, packaging, and appearance Wheelwright and Clark's "derivative" project neatly describes the relationship between individual models and the platforms they were

TABLE 7.4. Sony's Technological Leadership in Personal Portable Stereos

Feature	Firm	Date	Imitated?
First "Walkman"	Sony	79	Y
AM/FM stereo radio	Sony	80	Y
Stereo recording	Sony/Aiwa	80–81	Y
FM tuner cassette	Toshiba	80–81	Y
Autoreverse	Sony	81–82	Y
FM headphone radio	Sony	81–82	Y
Dolby	Sony/Aiwa	82	Y
Shortwave tuner	Sony	83	
Remote control	Aiwa	83	Y
Separate speakers	Aiwa	83	Y
Water resistance	Sony	83–84	
Graphic equalizer	Sony	85	Y
Solar-powered	Sony	86	
Radio presets	Panasonic	86	Y
Dual cassette	Sony	86	
TV audio band	Sony	86–87	Y
Digital tuning	Panasonic	86–87	Y
Child's model	Sony	87	
Enhanced bass	Sony	88	Y

based on. At the same time, Wheelwright and Clark's derivative category is still very broad. If all nonplatform projects were labeled "derivative," 99 percent of Sony's personal stereo innovations would fall into that one category. To develop a better understanding of the dynamics of Sony's product line management, we divided nonplatform design changes into two further categories: *incremental* and *topological* innovations.

Table 7.4 lists the incremental changes that have appeared in the U.S. market. These models generally did not require retooling and carried little technical risk for the firms involved. Many features (e.g., Dolby noise reduction) were widely used in other consumer electronics products and could be implemented with a minor redesign of the electronic circuits. With a few notable exceptions, these innovations were also easy to copy. When Toshiba introduced a cassette-shaped FM tuner to fit into playback-only units, the other manufacturers added radio reception in less than a year.

Although incremental changes were less demanding than generational changes, Sony consistently led in this innovation category as well. Of greater importance is the fact that several of Sony's incremental innovations were never copied by competitors. The Sports Walkman was introduced in 1983, but remained unchallenged through the end of the decade. Sanyo, Panasonic, and manufacturers in Southeast Asia offered similar models at various times, but each manufacturer quickly backed off and dropped those models from their lines. Given the simplicity and popularity of the sports series, it is interesting that Sony's competitors did not seriously attempt to imitate it.

Despite Sony's aggressive pursuit of incremental design changes, such models only accounted for twenty to thirty of the 250 odd new models that Sony introduced in the United States during the 1980s. The remaining 85 percent of Sony's models were produced from minor rearrangements of existing features and cosmetic redesigns of the external case. Sony generated these designs much as a child would build with LEGO. We refer to this final category as *topological* design changes.

Once the platforms, components, and incremental innovations are in place, topological changes can be made with little cost or risk. Sony standardized its basic design elements and relied on flexible, automated manufacturing processes to keep the costs of these changes to a minimum. At the same time, topological changes often appear to be totally novel to the customer. Topological changes keep product lines fresh and help producers deal with the proliferation of markets, distribution channels, and market segments. Seiko Epson, for example, offers more than 1,000 watch models and introduces 200 new ones each year based on just four or five basic watch movements [35]. Carried to extremes, topological changes can lead to dysfunctional "product churning." Faced with frequent model changes from competitors, firms may accelerate their own introductions of new topological designs. Some observers have suggested that Japanese consumer electronics companies, including Sony, have fallen into this trap [12]. As noted in the previous section, however, an inspection of Sony's product line during the 1980s shows that the firm often created topological designs that were more than a meaningless reshuffling of components. A Sony sales brochure from 1986–90 is visibly more diverse than its counterparts from other firms.

During the 1980s, the Sony Walkman organized its topological evolution around three basic platforms: (1) an effort to miniaturize the Walkman started by the WM20, (2) direct-drive models targeted to audiophils; and (3) the low-cost models targeted to price-sensitive consumers. The pattern that emerges is summarized in Figure 7.6. The shaded regions in the figure show the range of model prices supported by the three platforms and can be compared to the models represented in Figure 7.3.

Figure 7.6 illustrates the difficulty that can arise in predicting the effect of a specific design change. The WM20, for example, was seen by its designers as the logical successor to the WM2 [28]. It served the same market, but was smaller and better in every way. Although this would seem to be an argument for one design to replace another, the WM20 did not replace the WM2. Instead, Sony took the WM2 platform and steadily refined its manufacturability so that it became the basis for a whole series of low-priced units. By changing its mission, Sony retained the WM2 platform through the end of the decade. The WMDD and its successors, by contrast, show what a generational design can do to widen a product line. Their superior sound quality and high prices created an entirely new product that customers were unlikely to confuse with any of Sony's other models.

Ideally, the next step in analyzing the competitive pattern of product evolution would be to examine the profitability of the various features, models, and product families. Unfortunately, this sort of historical data is generally not gathered or retained at the necessary level of disaggregation, and the firms involved were reluctant to release the data that they did have.

Instead, we conducted a "hedonic" price analysis to explore the relationship between the sales prices and design features [2,24]. The *model price at a specific point in time* was used as the dependent variable.[4] The independent variables included the model's measurable *characteristics or features*. Most features, with the exception of size, were represented by binary indicator variables. In addition, the major brand names were noted by indicator variables, and two time variables were calculated from the database of price quotes: 1) the year that the model was first introduced (*vintage*),

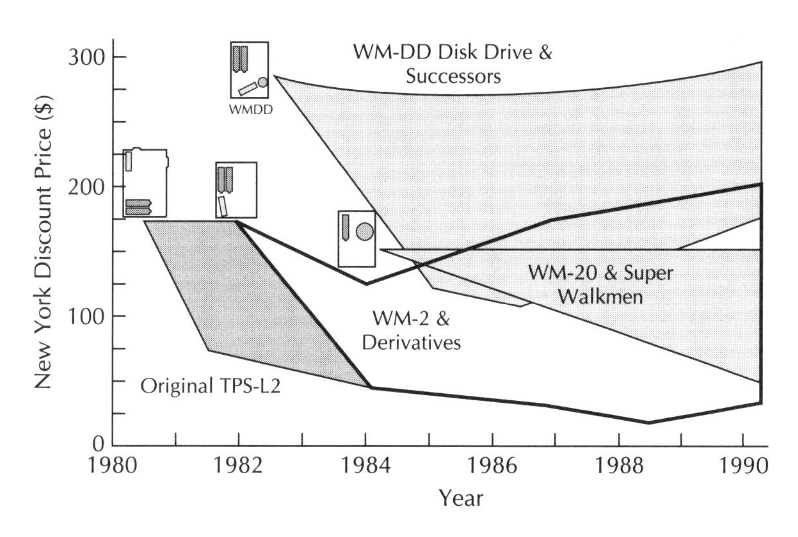

Fig. 7.6. Evolution of prices for generational platform innovations in the U.S. market.

and 2) the number of years that the model had been offered for sale when the price quote was advertised (*age*). Finally, interaction terms between the features and the two time variables (vintage and age) were added to the model. A MLS regression model was fitted to this data from 1980 to 1990.[5] The model explained several important aspects of the variance in the published discount prices.

From the resulting model, it is possible to impute the price increment that was charged for each variable in the model. Some features (e.g., digital tuners, water resistance, and FM radio) appear to have earned a price increment that did not change with either vintage or age. More common were features (e.g., graphic equilizers, Dolby™, and enhanced bass) whose imputed price increments diminished in value as time went by.

There were a few features (autoreverse, AM radio, the PANASONIC brand name,[6] and remote control) that actually seemed to become more valuable (or at least less of a liability) as vintage increased. Finally, there were features (e.g., AIWA and TOSHIBA brand names) that seemed to command high prices only when the individual model was successful enough to survive in the marketplace to a reasonable age.

The key point about this analysis is its empirical support for the view that design features drive prices in complex ways over time. This provides further support for our view that Sony's product planning decisions were critical to its success. For example, the results suggest that Sony's long-lived "sports" models enjoyed a price premium in the range of $15 to $20 over the entire study period. Given the inexpensive topological innovations required for this design change, it is hard to see how Sony could have failed to make a substantial profit with its sports models.

SONY'S MANAGEMENT OF TECHNOLOGY

A strong case can be made that Sony's pattern of innovation was well suited to achieving dominance in personal portable stereos. This section examines the organizational structures and managerial practices that Sony used to generate that pattern. Figure 7.7 shows how the elements in Sony's

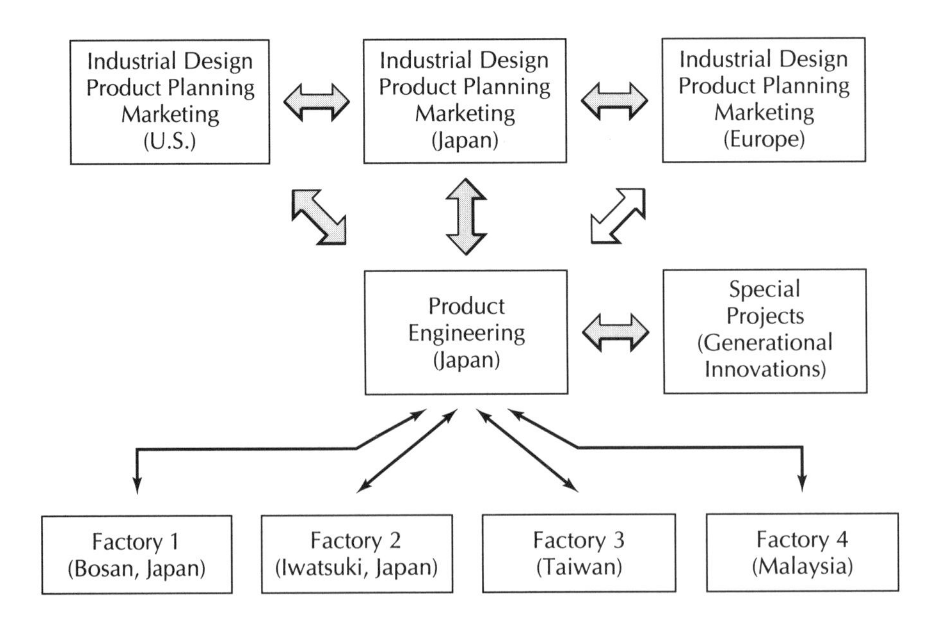

Fig. 7.7. The organization of design and manufacturing for Sony Walkman.

Walkman business network relate to one another. Marketing, product planning, and industrial design functions are located in Sony's major regional markets: the United States, Japan, and Europe. Product and manufacturing engineering is done entirely in Japan, while manufacturing occurs in Japan, Singapore, and Malaysia.

All Walkman engineering is done at Sony's audio products division in Japan. Sony's flagship manufacturing plants are in Japan. These plants make the small, highly sophisticated models that sell to Japanese consumers. Sony's factories in Taiwan and Malaysia manufacture the larger, more price-sensitive and rugged models that are shipped to the United States and the rest of the world [8].

The engineering and manufacturing facilities are connected to a worldwide network of marketing and sales offices. To exploit its strong brand image, Sony maintains close links with consumers in each market. Because each market area has its own distribution and retail infrastructure, Sony sales and marketing functions are needed to provide the necessary access and support. Marketing and sales link the Sony innovation system with the retail outlets for their products. Sony's Industrial Design Center, with branch operations in New Jersey and Europe, is the final element in the Walkman network. This organization provides Sony with a means to understand the lifestyle trends that influence the way customers use personal stereos.

When we interviewed Sony employees, we had no difficulty in piecing Figure 7.7 together, but it was almost impossible to get a simple answer from Sony as to how this system worked. We kept wondering who was in charge, who approved design decisions, and what the reporting relationships were. Each time we asked, we would get a different answer, or no answer at all. It took time and much discussion for the realization to sink in (confirmed in later discussions with Sony managers) that Sony does design in several different ways, depending mainly on the nature of the design change. Generational changes are led by engineering, with keen interest by top management. Marketing and sales lead for certain classes of topological changes. Finally, numerous incremental and topological design changes that do not fall into either category are the province of Sony's industrial design organization.

Generational Platform Teams

We reviewed published accounts of the projects that created Sony's generational innovations. We also interviewed members of Sony's American and Japanese design centers and manufacturing operations who had firsthand design experience with the Walkman. The major platforms were the products of intense engineering efforts, involving fully dedicated, multidisciplinary design teams that Sony assembled from among its best design and manufacturing engineers. Sony routinely has several of these teams working in parallel on key technologies to support its audio products.

The accounts of these projects agree that the leadership of these generational efforts rested with the senior management and with Sony's design and manufacturing engineers. Projects took a year or more to complete and were supported at the highest management levels. Teams were given clear development targets. In the case of the WM20, the project team made a wooden model to visualize the final size. The model was barely larger than the standard cassette tape the unit would have to play. However, within these parameters, engineers had considerable freedom to find a workable solution.

Sony's generational design teams typically developed the manufacturing processes to produce each platform. Sony's philosophy has been to manage design and manufacturing of product families, assuming and planning for modifications over time and making up-front investments in manufacturing systems to ease model changeover. In assembling its high-end tape transport mechanism in Japan, Sony used robotics for the assembly of precision mechanical components such as gears, flywheels, and other control buttons. With its highly automated and flexible system and judicious use of assembly and testing labor, Sony could produce a new Walkman every twenty seconds from each production line [11]. The goal of this flexible system was not only to assemble a Walkman quickly, but to rapidly change from one design to the next, even when the design changes were significant. The Sony assembly system is sufficiently flexible that Sony used it to assemble VCR mechanisms and even marketed it to other firms [11].

In pursuing its goals for innovation and variety, Sony developed a flexible assembly system call SMART (Sony Multi-Assembly Robot Technology) and the Advanced Parts Orientation System (APOS) designed specifically with flexibility and small-lot production in mind. The SMART and APOS systems were created for ease of model change. Although the multifunction machine cost almost twice as much as a comparable single-function machine, the greater flexibility possible using manufacturing equipment designed with multiple products and rapid changeover in mind offsets its initial cost. Single-purpose machines have other disadvantages. Single-function machines make complicated assembly difficult. Although firms frequently use simple machines to perform simple tasks and manual assembly for more complicated tasks, these mixed lines have certain disadvantages. If there are two or three shifts, substantial labor costs may result. In addition, when simple and advanced machines are mixed together, it is difficult to change product mixes. Simpler automated assembly lines can be developed, but they have the disadvantage in increasing material handling and loading. Finally, equipment life is longer for multifunction machines because they are more flexible and able to cope with product changes [11].

The development of Sony's multifunction assembly system required a breakthrough in parts acquisition and orientation. To solve the problem of parts acquisition for the multifunction machines, Sony developed an advanced parts-orientation system (APOS). The APOS is a flexible, automatic parts-orienting machine that uses vibration to align parts automatically on pallets. The APOS handles parts and components of various shapes and sizes for easy product changes. A part that has been correctly positioned in a cavity will remain there while misaligned parts will fall out. The APOS was designed for small and medium lot production and can handle up to six different parts on one machine.

By itself, however, Sony's manufacturing system gave Sony few direct competitive advantages. Matsushita (Panasonic), for example, operated systems with similar capabilities [10]. Nor was Sony able to escape pressure from low-cost manufacturers in Southeast Asia. Sony's low-cost units are increasingly manufactured in Taiwan and in a new (1988) facility in Malaysia. The manufacturing systems at these plants emphasize less flexible techniques for high-volume, low-cost models.

SONY'S APPROACH TO UNDERSTANDING CUSTOMER LIFESTYLE

Apart from the five or six generational changes, the vast majority of Walkman design changes were either incremental or topological in nature. The challenge for these designs lay in Sony's ability to effectively read the minds of its customers. In this respect, Sony's design changes appear to fall into two categories: 1) channel-driven changes; and 2) lifestyle-driven changes.

Sources of Design Changes

Sony uses its marketing and sales organization to keep track of channel needs in each of its major market regions. Ideas for new models come from many sources, including customer inquiry cards, contacts in distribution channels, and direct customer surveys, as well as sales force meetings with their distribution accounts. Sony is often pressed to provide models with subtle differences that appeal to particular channels. These changes are identified and championed by the marketing and sales organization. Sony sells through a variety of outlets, including department stores, mass merchants, and catalogs. A department store may want a model with more expensive features that it can use to justify higher margins. A catalog store may prefer a lower cost model that appeals to price-conscious customers.

While some of Sony's low-cost competitors try to satisfy these requirements by private branding (i.e., offering models customized for a single distributor), Sony has avoided this practice and seeks to meet channel needs with its official product line. According to product managers at Sony and several competitors, pressure for variety is also generated by the wholesale pricing practices in the American consumer electronics industry. Large retailers typically receive price discounts that are based on the total dollar value of their order, rather than on the volumes of the individual models that they buy. This provides an incentive for retailers to purchase all of their models from one or two manufacturers. With its large product line, Sony was well positioned to attract the business of these larger retailers. With its ability to rapidly make topological changes, Sony was also in a position to be responsive to their needs.

Sony's marketing organization rarely proposes a new model that requires anything more than a topological design change. For their purposes, a subtle difference in color or minor features (e.g., adding a stop watch to the digital AM/FM channel tuner) is often enough to customize a model for a customer or channel. The marketing groups and sales forces collect ideas about the needs of the regional markets and work with Sony's local industrial design groups to create versions of the Walkman that address those needs. Together, these teams feed information about special needs in their respective markets to product planning and engineering in Japan.

Lifestyle design changes are a traditional Sony strength, and industrial designers are Sony's masters of incremental and topological change. By changing the shape or color of a product's case or the arrangement of its functions, an industrial designer can make a product appeal to an entirely different customer, often at minimal cost or risk. Industrial designers were both creative and systematic in spinning out new Walkman models. According to Liz Powell, former director of Sony America's Design Center,

> Each time we embark on a project, the design process begins with an analysis of the product that's already out there that we are selling and what other people are selling and also new technologies and new possibilities that people haven't thought of yet.

Sony's Industrial Design division traces its roots to designers who are assigned to each production unit in the late 1950s and early 1960s, making it one of the pioneering efforts in the consumer electronics industry. These designers were integrated into a Design Division in 1960, and a U.S. design unit was established in 1969. The division was dissolved back into the production units during the 1970s, resurrected during the early 1980s, and given expanded responsibility in the late 1980s. Currently, the Design Center is an equal partner with engineering and marketing in the Consumer System Products and Design Group, which is the major corporate entity charged with product innovation. Figure 7.8 shows Sony's view of the role played by industrial design. The placement of the Industrial Design Center (ID Center in the diagram) at the heart of the Sony corporate structure illustrates the extraordinary emphasis that it has received.

In contrast to similar groups in its competitors, the Sony Design Center has received top-level support from its inception. As a result, it has acquired a level of credibility within Sony that exists in few other major manufacturers [7]. Industrial designers contribute an ability to visualize, as well as an understanding of mechanics, new technology, product planning, consumers, and the marketplace. In addition, Sony expects its designers to negotiate with engineers, work on improvements in production tools with manufacturing, educate subcontractors about the critical role of product design and to hold up their end of corporate commitments to reduce costs. Sony also sees designers as filling a very special role with respect to marketing. Sony believes that "marketers cannot be designers, but designers can be marketers." Sony looks for industrial designers who have "capabilities of expressing corporate policy and corporate identity through designing" [29]. To keep ideas fresh, industrial designers and design engineers are rotated from one product family to another.

The Industrial Design Center also places a strong emphasis on the quality of the design. According to Liz Powell, designers are expected to design models that adhere to Sony "look." While

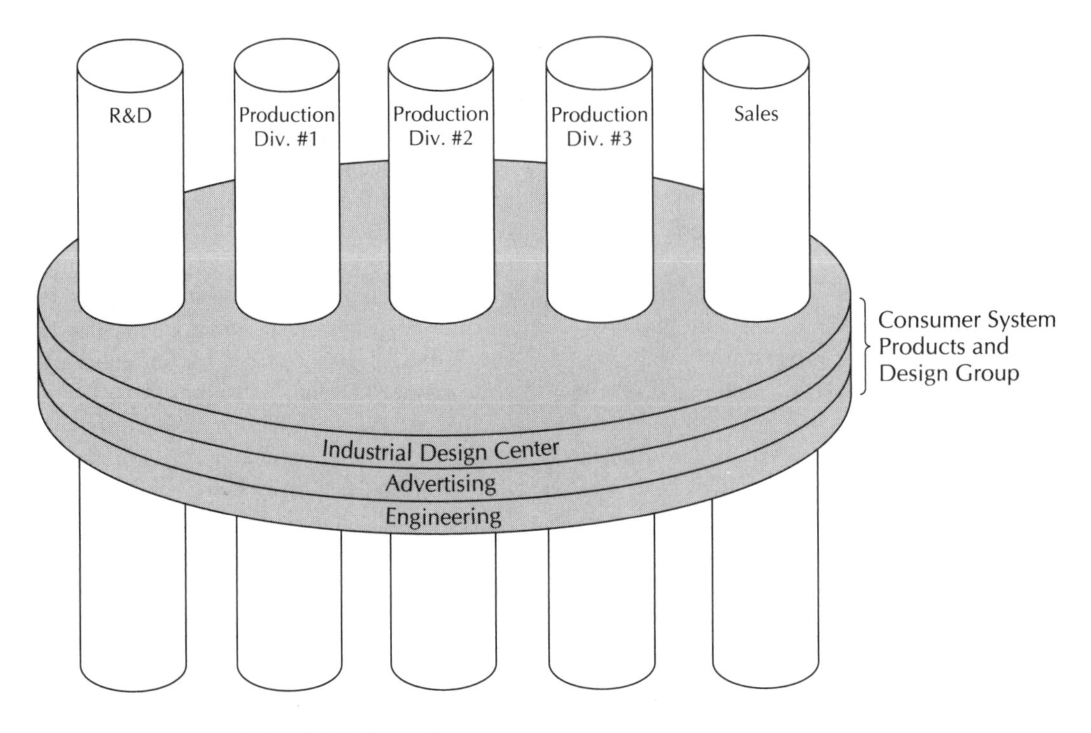

Fig. 7.8. The role of industrial design at Sony.

the guidelines are not formally written, industrial designers hold continual discussions to maintain a consensus about what that look is and is not. According to Powell, the designer must consider what the customer really wants to use. The designer must also understand what the engineer is saying about technical and cost factors. The engineers reflect the pressures from manufacturing, and it is the synthesis of aesthetics and engineering that produce the beautiful, high-quality products for which Sony is known.

From an engineering point of view, the design changes that produced the original Walkman were barely incremental. Nonetheless, it created a new product family for consumers and a worldwide market for Sony. Customers could go anywhere and listen to fine stereo music. Later Walkman models had similar impacts. The Sports Walkman allowed customers to hear stereo music in new locations (e.g., at the beach). The professional series of Walkman offered unprecedented sound quality in any location. The My First Sony gave stereo sound to children in a fascinating and (hopefully) indestructible case. These lifestyle-driven design changes were neither engineering challenges, nor were they first suggested by Sony's customers or distributors.

Designers try to come up with a product that nobody realized they wanted or needed by looking at how people live, how they spend their leisure time, what they presently do with a Walkman, and what they might like to do with a Walkman that they can't do now. The combination of topological design changes and manufacturing flexibility meant that Sony's designers could take chances that other firms might not be able to afford. Each topological change cost so little to design, that Sony only needed to sell 30,000 units worldwide to break even on a new model [1]. When designers are trying to understand complex customer lifestyle choices, even failed models are immensely valuable [15]. An example of a failed product was the

WM8000. This was a dual cassette Walkman that was motivated by the success of using two cassette drives in "Boomboxes." It survived on the market for about six months. Even so, by reducing the cost of failure, Sony could afford to fail more often and hence learn more quickly [27].

The lower risk also allowed Sony to take more chances. Sony could throw a model into an unfamiliar or marginal market segment with little fear of major loss. As Raubitschek has pointed out, inherent market uncertainty means that these marginal markets will occasionally produce a pleasant surprise [22]. By keeping down the costs of introducing new models, Sony had a built-in incentive to expand its variety. It never knew when its designers might next hit the jackpot.

Industrial Design Outposts: Understanding Foreign Markets

Sony's product lines in Japan, Europe, and the United States differed far more than did the product lines of its competitors. One of the key reasons for Sony's leadership in markets outside Japan has been its ability to interpret foreign market preferences to its design and engineering operations in Japan. Sony relied on the industrial design centers in Japan, New Jersey, and Milan to bridge the gap between the sales network's understanding of customer requirements, and the new product opportunities that were latent in Sony's technology. None of Sony's major competitors locate their industrial design staffs outside Japan.

Sony's U.S. design group developed the concept and prototype for the Sports Walkman. By putting its standard Walkman in packages that were sealed and ruggedized, Sony made personal stereos more useful for customers who wished to take them to wet or dusty environments, or who might subject them to wear and tear in the course of strenuous athletics. The "My First Sony" line was first developed by a joint planning, marketing, and industrial design team in the United States. "My First Sony" is a line of Sony products for children that was first introduced as a Walkman in 1987 and then extended to other Sony products. The team, led by Aki Amanuma, former director of design for Sony's U.S. subsidiary, came up with the idea and convinced key management at Sony headquarters to support it. Because these products were built on platforms, Sony was able to launch the products in record time. In the case of "My First Sony Walkman," it took less than one year from inception to market.

Sometimes the impetus for a new model comes from Sony America Design, and the Sony America team will prepare a prototype and present it to the engineers in Japan. At other times, the American product planners will go directly to the Japanese engineers and say that they would like an updated version of a specific model to satisfy particular market needs. On those occasions, the Japanese engineers prepare the sketches and develop prototypes that they present to the American designers and product planners. The marketing and industrial designers must agree on a new line-up of products that is presented at meetings in Tokyo in the fall and spring. The design review meetings are held both in English and Japanese and typically last a week, with a day devoted to each product family. A full day must be devoted to the Walkman, or to tape recorders, or to microcassette recorders.

Line-up meetings are opportunities to review new product ideas and to review the entire market situation including forecasts of where the industry is heading. Both Sony America and the home office make presentations describing what is happening in their respective markets. They review each model in each product family and discuss ideas for new models. Industrial designers show sketches indicating what the new models will look like. The product planners consider the possible features that the new products will have. Generally, there is general agreement before the meeting among the U.S. team, and often there has been some consultation in advance with industrial designers and engineers in Tokyo. In addition, the engineers and industrial designers in Japan may have ideas for models they believe will appeal to the American market. Finally, all of the models, both old and new, are reviewed and a negotiation takes place over which models will remain the same, which will be changed, what will be added, and what will be dropped. Timing of new model introduction, as well as price, is negotiated.

Sony's distributed approach offers several advantages and may have contributed to Sony's dominant position in markets outside Japan. Industrial designers combine their creative skills with intelligent product planning and market research to design models targeted for specific national markets. By linking industrial designers with product planners and marketing specialists and by making design mock-ups in regional design centers, Sony can tailor models more effectively than if it centered its design activities exclusively in Japan. When the industrial designer brings a mock-up of the proposed new model to Japan and discusses the design in great detail with the engineering staff, both the concept and the form are communicated directly.

ALTERNATIVE EXPLANATIONS

Commenting on earlier versions of this paper, several reviewers have wondered about alternative explanations for the Walkman success. It is not uncommon to view innovative success as the natural result of managerial leadership and effective marketing. Many observers have admired the leadership achievements of Sony, especially its founders, Akio Morita and Masaru Ibuka. Others have been impressed by Sony's marketing skills and internationally respected brand name. Why, then, should one believe that these are not sufficient to explain the Walkman success?

While we believe that Sony's strengths in these areas are extremely important, we began this study with an interest in *patterns of innovation* and the structure of the *innovation system.* As we gathered information about the competitive history of the Walkman, we came to the conclusion that the system view offers a better explanation of the observed variance in competitive behavior. In effect, we made a conscious decision to allocate more time to the study of the system and less to the study of traditional leadership and marketing issues. This decision was not taken lightly.

Senior managers played key roles in launching the first Walkman, redesigning the WM2 platform, and deciding to market "My First Sony." However, as we examined various written and interview accounts of the Walkman's birth, the stories differed significantly as to the persons involved, the rationales, and even the events [1,14,16,28]. It would seem that success has many mothers, even at Sony. We concluded that the ground was too badly trampled for us to fully unravel top leadership actions over a ten-year study period.

At the same time, it became clear that top management had little need to be involved on a day-to-day basis. In Sony's system, most innovations are topological, or at most incremental. In conversations with Sony managers, both in the United States and in Japan, we found evidence of considerable movement among the individuals responsible for the Walkman's success. Project teams had been formed and disbanded and managers had moved around. Yet, Sony followed a consistent pattern of innovation for over a decade. This suggested that the real answer lay in the structural organization of innovation.

Finally, empirical support for the system view came from the analysis of model lives and prices. We do not believe that any leader, however inspired, could personally have created Sony's decade-long string of classic designs (see Figure 7.4). Similarly, our hedonic price analysis found that the imputed price premium for the Sony brand name was modest compared to the values created by its long string of design innovations.

CONCLUSIONS

Our interest in the Walkman story began with a desire to understand why Sony was so successful, even though it lacked the traditional forms of sustainable competitive advantages generally required by management theorists. Sony's major Japanese competitors have equally fine engineering staffs, world-class manufacturing plants, international marketing organizations, broad consumer electronics product lines, strong financing, and educated, highly committed workforces. Sony's competitors have fared well against Sony in several other, closely related consumer electronics categories. In the case of the Walkman, moreover, Sony held no determining patents and was unable to defend any technological barriers to entry. While Sony's brand name undoubtedly carried weight, its major competitors (Panasonic, Toshiba, and Sanyo) were also well known and highly regarded.

As we assembled and analyzed historical data on the Walkman, we began to suspect that Sony's handling of the Walkman represented something more interesting and complex than we initially imagined. The Walkman story displays a rich complexity that we believe is characteristic of effective management of product families. If Sony's methods are indications of emerging best practice, future competitive survival in this, and similar industries, will be a daunting task. Firms will have to acknowledge and deal with the parallelism, the differences in designs and design processes and, ultimately, with the complex interactions that exist among incremental design changes.

In examining Sony's decade-long dominance, we believe there is reason to suspect the existence of an important new form of competitive advantage, namely a firm's skill at managing the evolution of its product families. In fact, Sony's founder, Akio Morita, recently made the following comment:

> Japanese business and government leaders, like many of their American counterparts, continue to believe that the United States must address its agenda of domestic economic challenges more vigorously and with more attention to enhancing the competitiveness of U.S. business over the long-term. My personal view is that as long as MBA's, lawyers, and financial wizards are valued over engineers and *product planners*, American society will be in danger of moving in the wrong direction. (emphasis added) [17].

Mr. Morita places the "product planner" on the same footing as financiers, engineers, and lawyers. We believe that our review of the history of the Walkman makes it is easier to understand his choice of words. Managers, designers, and engineers at Sony faced a challenge of managing the complex technologies, perceptions, and organizational interactions that ultimately shaped the evolution of the Walkman product family. In response, they marshaled a pattern of incremental innovation that was coherent, efficient, and extremely productive. The persistence of internal consistency with which Sony pursued the development of the Walkman constitutes a strong argument for Mr. Morita's prescription.

Sony started with a long-term plan of miniaturization and performance improvement that predated the Walkman. It aggressively invested in key component technologies to support progressively smaller, consumer electronics designs. It invested in the development of key generational platforms, which simplified the task of introducing large numbers of incremental and topological design changes. To support its large variety, Sony developed its own system of flexible manufacturing so that it could make money on small production runs.

Sony relied on its industrial design centers to understand lifestyles outside Japan and to exploit the potential for incremental and topological changes to serve customer perceptions and lifestyles. By combining industrial design with manufacturing flexibility, Sony generated designs that had large market impacts and little marginal cost. This approach ultimately allowed Sony to create a climate of design freedom that fostered risk taking by its industrial designers.

The creativity, low cost, and low risk associated with Sony's incremental and topological designs were crucial to Sony's variety-intensive strategy. By creating a model for many different customer perceptions and lifestyle, Sony preempted several important market segments. These niches were too small to attract challenges from other firms, and Sony did not have to refresh the designs of these models as rapidly as it did for more competitive segments. The long-lived models in these niches further eased Sony's design burden.

In achieving its success with the Walkman, Sony capitalized on several factors that may not be present for other products or firms. Sony started with an important first-mover advantage that it was careful not to relinquish. Sony carefully maintained its mastery of the Walkman technology. In addition, the personal stereo is particularly sensitive to lifestyle considerations, a fact that was to Sony's particular advantage. Finally, Sony was able to develop the personal stereo secure in the knowledge that international acceptance of the standard audio cassette tape guaranteed the Walkman product concept would remain viable for a considerable period of time. Despite these peculiarities, there are important elements in Sony's handling of the Walkman that deserve consideration by other manufacturing firms. In particular, Sony's management of the Walkman family highlights four specific tactics of product planning that may be applied to other manufacturers operating in an environment of incremental change.

The first element is Sony's *variety-intensive product strategy*. Sony pushed hard to create a large model variety. Sony preempted market segments that competitors have still not contested. These segments include the market associated with outdoor lifestyles (Sports Walkman), audiophiles (WMD6C and other direct drive models), and children (My First Sony). Because models in these segments required less frequent redesign, they did not call on design resources that were needed elsewhere to update models in more competitive segments.

A second tactic was Sony's *multilateral management of product design*. By locating marketing, product planning, and industrial design resources in each of its major markets, Sony placed responsibility for design changes in the hands of the people who could best perform them. Japanese engineers led the development of generational platforms. Marketers carried requests for new models from the sales channels. Industrial designers in key markets around the world led the development of models that tapped into local lifestyles. While Walkman development teams always included engineers, marketers, and designers, the locus of the leadership shifted. Engineers led when technology was the hurdle. Marketers led when appealing to specific customers was the goal. Industrial designers led when the challenge was to understand and respond to a potential lifestyle. By decentralizing its leadership, Sony was able to bring more contributors into the design process for the incremental and topological designs. This made it easier to generate the large number of models and made them more relevant.

A third element in Sony's strategy was its *judicious use of industrial designers*. By fully exploiting their talents, Sony built a bridge between the varied and changing needs of customers and the relatively stable technical capabilities embodied in its platforms. The industrial designers found creative ways to use the platforms to enter new market niches using incremental and topological design changes.

The fourth element was Sony's commitment to *minimizing its design cost*. In a world in which variety and change interact to increase designers' workloads, Sony carefully controlled the costs of new models by building all of its models around key modules and platforms. Modular design and flexible manufacturing allowed Sony to produce a wide variety of models with high quality and low cost. What is more important, by keeping down the design and manufacturing costs of new variants, Sony gave its designers the creative freedom they needed. Moreover, the greater longevity of several of the models contributed to the variety of models available to customers. Rapid model turnover was combined with greater model longevity of key models to provide greater model line breadth.

We believe that the Walkman example makes a significant contribution to our understanding of innovation. Sony's management of the Walkman product family is consistent with, but goes deeper than the normative product planning process proposed by Wheelwright and Clark [32]. While their framework contains powerful insights, it does not capture the rich interactions and dynamics that occur in and among the topological design changes. Nor does it acknowledge the impact of customer perceptions, something Sony's industrial designers do so well.

Sony's focus on the design process has not waned now that the Walkman technology has become mature. Sony introduced the compact disc in 1982. Several months later, it introduced four more products (two priced higher and two lower). In less that two years, Sony introduced eight new low-priced models. It took substantially longer for other competitors to get comparable models into the market place. As the price of CD players dropped, demand for compact disc players soared, and thirty companies eventually entered the market, but many of those are reselling Sony units or components. Despite intense competition, which eventually forced many players out of the market, Sony has remained both the variety and the market share leader.

Notes

1. Product managers at each of the major manufacturers confirmed this observation.

2. The number of models in Figure 7.3 is larger than the variety of models that each firm was actually producing, because retailers sometimes advertise units that linger in the distribution channel after the manufacturer has ceased production.

3. The estimates of model lifetimes were complicated by truncated time series data sets. The availability of individual models may be zero because the study period has ended rather than because the model has died a natural death. In this application, the problem was largely avoided by a conservative choice of study period. Figure 7.4*b* shows that only three of the classic models had not died by the end of the study period.

4. Gleaned from the *New York Times* discount electronics ads.

5. The personal stereos' characteristics were obtained from manufacturers' sales brochures, although information on some models from obscure manufacturers was gleaned from ads. The relationship was fitted only for personal stereo models whose features were fully described. Altogether, 883 price quotes for 129 different models were used to estimate sixty-one model parameters. The analysis used SAS Proc REG and the CP model selection method to select the ten best subjects from the available independent variables (using Mallows' Cp criterion [18]). The overall best-fit model had an adjusted R^2 of 0.8714 and, among the ten best models, the coefficient estimates showed good stability in both magnitude and sign.

6. Sony was the base case brand name in the regression models. When the indicator variables for AIWA, PANASONIC, TOSHIBA, and the various MINOR MGF are set to zero, the parameter coefficient estimates for the remaining variables described SONY models. The interaction terms between brand name and feature indicator variables were not considered in the model. If more data were available, that might offer an interesting line of inquiry.

BIBLIOGRAPHY

1. Interview with Aki Amanuma, director of Industrial Design Center, Sony of America, May 1990.
2. E.R. Berndt and Z. Griliches. Price Indexes for Microcomputers: An Exploratory Study. Paper presented at Conference on Research in Income and Wealth Workshop on Price Measurements and Their Uses, Washington, D.C. March 22–23, 1990.
3. J.L. Bower and T.M. Hout. "Fast-Cycle Capability for Competitive Power." *Harvard Business Review*, (1988), 66 (6), 110–118.
4. Gian Carlo Cainarca, Massimo Colombo, and Sergio Mariotti. "An Evolutionary Pattern of Innovation Diffusion: The Case of Flexible Automation." *Research Policy*, 18(1989), 59–86.
5. Nigel Cope. "Walkman's Global Stride." *Business (UK)*, Mar. 1990, 52–59.
6. Jacques Delacroix and Anand Swaminathan. "Cosmetic, Speculative and Adaptive Organizational Change in the Wine Industry: A longitudinal study." *Administrative Science Quarterly*, 36(1991), 631–661.
7. J. Fairhead. *Design for Corporate Culture*. National Economic Development Office, London, March 1988.
8. Carl Goldstein. "Sound Electronics Strategy." *Far Eastern Economic Review*, 1988, 140 (24), 120–121.
9. Ralph E. Gomory. "From the 'Ladder of Science' to the Product Development Cycle." *Harvard Business Review*, 1989, 67 (6), 99–105.
10. John Hartley. "Japan Favors Flexible Cells for Electronics Manufacture." *Assembly Automation*, 1988, 8 (1), 49–51.
11. H. Hitakwa. "Advanced Parts Orientation System Has Wide Application." *Assembly Automation*, 1988, 8 (3), 147–150.
12. Bob Johnstone. "Creative Confusion." *Far Eastern Economic Review*, 1991, 154 (51) , 42, 44.
13. Sunder Kekre and Kannan Srinivasan. "Broader Product Line: A Necessity to Achieve Success?" *Management Science*, 36(1990), 1216–1231.
14. Larry Klein. "Happy 10th Anniversary, Sony Walkman!" *Radio-Electronics*, 1989, 60, 72–73.
15. Modesto A. Maidique and Billie Jo Zirger. "The New Product Learning Cycle." *Research Policy*, 14(6), (1985), 299–313.

16. Akio Morita. *Made in Japan: Akio Morita and SONY*. E.P. Dutton, New York, 1986.
17. Akio Morita. "Partnering for Competitiveness: Role of Japanese Business." *Harvard Business Review*, 70 (3), 76–83.
18. John Neter, William Wasserman, and John Michael Kutner. *Applied Linear Regression Models*. Irwin, Homewood, Ill., 1989.
19. *"Sony Official to AIWA Unit." New York Times*, January 2, 1992, D2, column 6.
20. Shigeru Hozu. "Japanese CEOs: How They View Their Jobs and Life." *Tokyo Business Today*, 1991, 59 (12), 58–60.
21. Michael Porter. *The Competitive Advantage of Nations*. Free Press, New York, 1991.
22. Ruth Raubitschek. "A Model of Proliferation with Multiproduct Firms." *Journal of Industrial Economics*, 25(1987), 269–279.
23. Susan Sanderson and Vic Uzumeri. "Industrial Design: The Leading Edge of Product Development for World Markets." *Design Management Journal*, 1992, 3 (2), p. 28.
24. P.P. Saviotti. "An Approach to the Measurement of Technology Based on the Hedonic Price Method and Related Methods." *Technological Forecasting and Social Change*, 27 (1985), 309–334.
25. Brenton Schlender. "How Sony Keeps the Magic Going." *Fortune*, 1992, 195 (4), 76–84.
26. Theodore W. Schlie and Joel D. Goldhar. "Product Variety and Time Based Manufacturing and Business Management: Achieving Competitive Advantage Through CIM." *Manufacturing Review*, 2(1), (1989), 32–42.
27. Sam S. Sitkin. "Learning Through Failure: The Strategy of Small Loss." *Research in Organizational Behavior*, 14(1992), 231–266.
28. Sony Corporation's promotional publication. "Walkman, the First Decade," 1989.
29. Presentation materials supplied by Sony's Industrial Design Center, Bosan, Japan, 1989.
30. George Stalk, Jr., and Thomas M. Hout. *Competing Against Time*. The Free Press, New York, 1990.
31. Michael L. Tushman and Philip Anderson. "Technological Discontinuities and Organizational Environments." *Administrative Science Quarterly*, 31(1986), 439–465.
32. Steven C. Wheelwright and Kim B. Clark. "Creating Project Plans to Focus Product Development." *Harvard Business Review*, 1992, 70 (2), 70–82.
33. Christoph-Friedrich von Braun. "The Acceleration Trap. *Sloan Management Review*, 1990, 32, 49–58.
34. Christoph-Friedrich von Braun. "The Acceleration Trap in the Real World." *Sloan Management Review*, 1991, 32, 43–52.
35. Lewis Young. "Product Development in Japan: Evolution vs. revolution." *Electronic Business*, 1991, 17 (12), 75–77.

8

The Technology-Product Relationship: Early and Late Stages

RALPH GOMORY

This paper will examine, and distinguish between, two kinds of relationship between technology and product: One which is characteristic of the early stages of an industry and one which is characteristic of later stages. The first one is more or less familiar; and it is the one that has shaped most people's thinking on the subject of science, technology, and product. The second is less familiar. My terminology for these, which you will find in other articles of mine, are the "ladder" and the "cycle."

First of all, let me discuss the more familiar relation of science, technology and product. This is the one that relates to why people came to believe that scientific supremacy should mean supremacy in product markets. This first type of relationship is exemplified by the transistor. The transistor was the product of decades of fundamental scientific research which eventually reached a point of practicality and then, through a series of rapid developments, resulted in the first semiconductor chips and went on to be the start of an enormous industry. And it is that paradigm which is the more familiar one to most people who discuss this subject. This I call the ladder paradigm because the new thing descends from the realm of science—step by step—into practice and becomes the genesis of an industry. Molecular biology is, today, in that state, having emerged from a period of tremendous scientific progress, and is starting to generate a whole new industry.

The belief that this kind of scientific dominance should translate into product dominance is probably, in many cases, the residue of the Second World War. Those of us who can remember that with any clarity remember the enormous impression made by the science-led, science-developed process of the atomic bomb. And, after the war, we emerged with a picture that scientific dominance does translate into economic dominance. The processes of which that is an example—the atom bomb, the transistor, molecular biology—are scientist-led. Scientists play the dominant role both in the basic research and in the early phases of the industry because they are the only people who understand what's going on in enough detail. So, in the early stages of a new industry, the ladder paradigm predominates. Everything revolves around the new technology. There are no old plants to accommodate; there are new people, new ideas, and new facilities—you are writing on a blank slate. That this kind of activity should produce industrial dominance is an attractive idea. But we should have known better, because the United States, before the Second World War, was already the dominant in-

Reprinted from *Technology and the Wealth of Nations*, edited by Rosenberg, N., R. Landau, and D. Mowery with the permission of the publishers, Stanford University Press.

dustrial power of the world, and had been for several decades. And it was not this kind of breakthrough on which that dominance was built, but much more on the manufacturing skills and so forth which are characteristic of the second type of innovation, which I will describe.

Secondly, we should realize that our problems today are not caused by the lack of this "ladder" type of innovation because those industries which cause the negative balance of payments—semiconductor memories, consumer electronics such as TVs and VCRs, and the automobile industry—were not industries that someone else started by stunning innovations. They were industries which U.S. people started and reduced to practice, led and dominated. And it was only in the later stages of these industries that we lost control of them. So, the familiar paradigm is really a paradigm for getting things started. But it is not a paradigm for winning the longer race.

Let me turn then to a second paradigm, a second relationship of technology and science, which I call "the cyclic process." The cyclic development process is a process of repeated, continuous, incremental improvement. It's a problem of getting out a better semiconductor chip next year, based on what you already have in production; or, you are already in production with automobiles, and you are working on next year's model. It is that process of following up what you have in manufacturing with the next model, which is designed, built and prototyped, tested, redesigned for manufacturing, put into manufacturing, and then you turn around and start on the next generation. This process is characteristic of the later (not the earlier) states of an industry.

The type of industry that I am talking about is discrete manufacturing, of which automobiles and transistors are very good examples. It is this cyclic development process that determines in the long run, then, who will be dominant in this industry. It is not as glamorous as the breakthrough type of thing; but, nevertheless, the progress which it brings about is enormous. By and large, the small, innovative companies tend to be very early phases—not necessarily of an industry, but of an industry sector, or of something new. Whenever that settles down and becomes major, then it turns into the cyclic development process. And I think that one way to characterize the difficulties of the American semiconductor industry, perhaps a superficial way, but with a grain of truth, is that it consists of the companies that started it. And those companies have the characteristics that enabled them to succeed in the ladder phase, but they do not necessarily have the scale or other characteristics that would enable them to succeed in the cyclic development process.

Certainly, firms do make a transition from being small startup firms to becoming large ones. But, it is notorious that many of the firms do not; that they have the wrong management style or whatever. Now, I think to succeed today in the DRAM business requires the ability to make enormous investments in plant and in tooling, and to make those steadily because the competition will make them steadily; which means to make them in bad times as well as in good. Now, how would you develop such a capability? I am not quite sure. I think Texas Instruments, for example, has reached the scale where they seem to be able to do that. So, scale is, clearly, one indicator. The second might be to be a vertically integrated company and be a part, say, of a company that makes things that depend on semiconductors. And that when the semiconductor cycle is down, the rest of the thing may still be up. And that might give you the stability needed to make those continuing investments. But you have got to change gears from the rapid one-pass motion of the early stages, to the ability to invest, sustain, and rapidly turn over the crank. That does happen, but a lot of firms fall by the wayside.

For example, in semiconductor chips, by the process of cyclic development, they have gone from sixteen bits on a chip to one or four million bits on a chip, in some twenty years. All of this by the straightforward process of turning the crank: refinement, manufacturing, refine, manufacture. This process is the one which is not clearly visible in most people's minds, and which I want to describe in a little more detail.

First of all, the length of the cycle itself—the cycle from design to manufacture—is a very important parameter and is often mistaken for technological innovation or technological leadership. But consider, if you have one company that can bring a product from design to manufacture in three and a half years and another company that can do it in two and a half years—and suppose they both came out in 1989—the technical notions embedded in the product with the shorter cycle will be one year more advanced. This is simply because, even if both companies are using the same set of ideas, the product with the shorter cycle will embody a later technology. And it will be technically more advanced, though no invention at all has taken place and both companies are simply using the common stream of ideas.

Thus, it is difficult to overestimate the importance of getting through each turn of the cycle more quickly than a competitor. It requires only a few turns for the company with the shortest cycle time to build up a commanding lead. Even if a company starts out with an inferior product, it is possible to overtake the industry leader if it has the capacity to turn out a new line six to twelve months more quickly. In fact, our Japanese competitors believe that, in the shortest of long runs, quick development triumphs over market research every time. I once made the mistake of asking a Japanese colleague, my counterpart at that time in an electronics company, whether he had undertaken any research on how customers were likely to respond to a particular kind of ink jet for printers. Why, he politely retorted, should he study whether customers are likely to respond positively to this or that jet if his company can get out a wholly redesigned printer in a year to eighteen months? Why not simply adapt to actual buying patterns? (Why, he implied, should I be bothering with such questions in the first place?)

My conclusion is that a company that can establish and maintain a shorter cycle can develop a decisive advantage over its competitors. The speed with which the design is translated into manufacture, turned around and redone, is enormously important in determining commercial success.

I have taken the case in which technology is changing and, therefore there is something new to be exploited a year or so later. If market is the issue, of course, the short cycle also gives you quicker ability to adjust to market feedback. This kind of cyclic process is very different from the ladder process. It is not scientist-based. It is based on what is already there, the existing product and its restrictions. If you made a printer last year, and you want to put a new head in the next round of cycle, it has to fit into that printer. If you want to put a better metallurgy in the next round, it has to be one that the development team can deal with, and accept, and get done quickly. If there are only eighteen months for development, there is not necessarily time to do something over from the beginning. So this type of development is very much restrained, not by a totally new idea, but rather by what is already there, whether that is the plant, or the tools, or the engineering team, or what they understand. And if new technologies are going to be part of this, they must fit into that very special world.

If the engineers are partway through the cycle, that is, if they started eight months ago to design the next round of printers, and you approach them with a new head design, you will find that they are not interested. A typical reaction to that is to decry the development people as having the not-invented-here syndrome, to conclude that the company engineers are simply impervious to ideas from outside sources. But this is an inappropriate psychological phrase to describe a genuine, objective difficulty. The resistance of designers to new ideas cannot simply be ascribed to a certain mental inertia—an inertia that accounts for the resistance of U.S. car designers to disc brakes, radial tires, and computer-governed electronic fuel-injection systems or that accounts for how long it took consumer electronics companies to replace metal parts and casings with molded plastics. All this reflects a lack of perception of what the real problem is. To revert to the case of the new printer, the real problem is that, if the engineers are going to bring the new printer out on time, and they are already eight months down the pipeline, they cannot accept a new head. Thus, if you want to get new technology into the cyclic development process, you have a genuine problem: the new technology has to arrive very close to the beginning of the next cycle.

Further, new ideas from any source need to be very thoroughly fleshed out. Many things are hard to accept into this process because the process is bound by the necessities of schedule. It is simplistic to ignore the scheduling compulsions of the cyclic development process and to conclude that what is at issue can be reduced to resistance to outside ideas. If people are building something as part of a development and manufacturing team, what they can accept depends a great deal on what they presently have. It depends on what tools they have, whether they accept square shapes or round, whether their engineers are familiar enough with any particular approach being proposed.

A related point is that it is often also quite difficult for ideas originating in the university community to be quickly accepted into this rather closed world of industry. You can have a great idea at the university, and you can then go to the product developers and say: "Look, I can make a better valve." But even if it is a better valve, the odds are overwhelming that they cannot accept it. Putting a better valve into their equipment may mean that they have got to change 43 tools, and they may quite possibly conclude that is not worth that much disruption.

On the other hand, if the people in industry are familiar with ongoing advances in science and technology, they can go out and pick the ones that they can accept—the ones that fit. That is the phenomenon of "pull." By contrast, it is very important not to assign to universities the responsibility for "push"—for transferring ideas into industry. I do not think they can do it (I am distinguishing here between the "ladder" stages, when universities do indeed play a major role, and the later "cyclical," or product improvement stages, which is the context of the present discussion).

In this respect, one of Japan's great advantages is that much of its technical strength is in entities that are tied to industry or actually in industry, whereas America's scientific strength is heavily concentrated in the universities. As a result, even though Japan's universities have a weak research capability compared to those of the United States, and do not help as much in the idea-generation process, Japan's industries function extremely well in the process of appropriating new ideas.

The essential point is that the primary way that ideas enter the product development cycle is not outside push, but inside pull. The engineers, who are themselves the prime movers in this, go out and select from the great world of outside ideas which are visible to them at engineering society meetings or at universities, those few which fit into their constraints. If you are the owner of a new, technical idea and you go to them, the odds are overwhelming that it will not fit. Your notion will call for stamping something round, but their stamps will stamp square things; and both the time and the cost of changeover make your idea unacceptable.

So, in the cyclic development process, which is the main process of a developed industry, not at startup phase, the key pullers of technological process have to be the people there, not the people outside. This is very different from the ladder process, the early-stage process. Again, because of the importance of cyclic development time, the close cooperation between the design and manufacturing is absolutely essential.

Of course, the manufacturing skill, itself, is a tremendous factor, as are the closely related issues of quality, but I don't propose to get into that issue here because they have been exhaustively described by many other people.

I just want to mention, though, the importance in a short cycle of designing for manufacture, on the part of the design team; and I will illustrate that by an example which perhaps you have heard of, but perhaps not in as much detail. And that is the example of the IBM ProPrinter.

The IBM ProPrinter was a matrix printer, exactly the sort of a thing you would put on a table next to your personal computer. It hammers out letters by driving little wires into a ribbon, the choice of wires determining the letter that emerges. It is basically a mechanical device. IBM was not originally competitive in the production of these printers. And when the early PCs, the first PCs, came out, they were all equipped with competitors' printers—notably Epson printers. And when IBM's development team looked into how it might have a competitive machine, one thing that was immediately discovered was that the IBM designs had many more parts than the competitive designs. The number of parts in the IBM design was, I think, roughly 120 or so in the Epson-type designs. For the first time, the development team took very seriously the consequences of designing things that way. They concentrated on designing the next printer as a manufacturable printer. And they took as their goal 60 parts. The number of parts is not a bad surrogate for the complexity of assembly.

There also was a second theme, which was pure manufacturing. They put together a robotic line. They cleaned out an old plant and put in 35 robots, two-armed, highly programmable robots to assemble this thing—and spent a lot of money doing that. I, myself, was very pleased to see the two-arm robots go in there because they were the product of a project which I had started in IBM Research in 1972. So, I thought this would be very spectacular. Here is what happened.

The design goal was almost met. They emerged with 62 parts. They designed the machine to make it easy for the robots to put it together; and that meant that the parts were all put down top-to-bottom, no insertion sideways. There was also a rule—no screws and no springs—because screws and springs are also hard for a robot to assemble. They made it with 62 parts, no screws, and no springs. Now, when they assembled this printer, they found that a human being could assemble this in roughly three to four minutes. It was that easy to put together because it had been designed to be manufacturable.

Another thing that occurred, by the way, was that the development time was cut approximately in half from previous development times. So, they affected the development cycle by making it simple, and they made it so easy to manufacture that, in fact, the gain due to the investment in the robots was almost completely obviated. Because if the human being could put it together in three and a half minutes, a very expensive robot had a real problem in just funding itself.

I have difficulty separating the managerial issues of this process from the structure. Is it structure that makes you turn the cycle fast—or other things—or is it management? I do not know. The point is, one way or another, you do it. Now, in the case of the ProPrinter, someone, I do not know whether it was management or other people, took the competing thing apart, and found a very small number of parts. And someone had the wit to say, "Let's have less." Now that may seem very trivial, but it is a change, or change of view. It is a change to say, "Our job, as engineers, is to make this thing manufacturable." I don't know whether that is management or organization. I am really describing what has to be done. I think, in some cases, it may be a spectacular individual that does that. And, sometimes, it will simply be the sharp spur of necessity, knowing that the other guy is going to bring out the next round two years from today and you'd better have something there. And there may be no spectacular individual at all involved. Perhaps the "Champion" concept matters less in the world of cyclic development. I think the system itself can drive it, if you know where you are going wrong.

I bring all this out to emphasize the importance of the cycle and of design for manufacturing. This printer went on to become the best selling printer of that type in the United States. This was not due to a scientific breakthrough. It was due to proper design for manufacturing. By the way, that is a learnable thing; and, when people talk, as they often do, about the difficulties of competing with the Japanese because of various and mystical factors, that we must change the entire U.S. culture in order to compete, we have got to redo the educational system, etc.—let me point out that competitiveness was achieved, in this instance, by changing the culture, not of a country, but of 70 people; the 70 people who made up that design team. So, there is a great deal of competitiveness that is concrete, focusing on the manufacturing/development cycle, on finding ways to shorten it, on finding ways to design for manufacturing, and finding ways for the engineering team to pull ideas into the next product.

And let me say once more that, in the taxonomy that I use, the cyclic development process is, in the long run, the decisive one for many manufacturing industries.

Now, if we do look at the competition—at the Japanese competition—we find that they do all the things I am talking about. They do design for manufacturability extremely well. In fact, the design and manufacturing teams are often very much tied together. They do have very short cycle times. If you read the press, you would think that the Japanese owe a great deal of their success to the very advanced technology programs which MITI has sponsored. Now those programs are useful, but they relate to the early stages of technology, not to the rapid turnover of the cyclic development process. And in the areas which I have seen, and with the people with whom I have discussed these matters, it is the cycle, not the ladder, that has been decisive in the industries in which we are in trouble—automobiles, semiconductors, TV; that sort of thing.

Another striking contrast with the Japanese is relevant here. In most American firms the people who are responsible for development, and those who are responsible for manufacturing, tend to live in different worlds. In U.S. industry, the high prestige is in development. The manufacturing task is considered to be relatively dull, repetitive, and for the uneducated. Therefore, production considerations have not been an early part of the design, say, of a new car or computer or printer. Until very recently, the product was not designed to be manufactured. It was designed to work, to make good print, or to run fast, but it was not designed to be manufactured. When the prototype was ready, you simply handed it to a different group of people—the manufacturing people—and you said, "Okay, I made one, you make 10,000."

In Japan, curiously enough, it is the other way around: Manufacturing calls the shots. Manufacturing people are highly trained, but people also move back and forth a good deal between various functions. The distinction we tend to make between the engineers in development and those who replicate their designs does not exist in Japan.

So, let me sum up what I am saying here. Much of our thinking about innovation comes from the ladder process because it has been so visible and so spectacular, involving the emergence of a new scientific effort and new products emerging from that effort for the first time. The United States has, in fact, historically been very good at that, while it has been steadily losing in the cyclic development process which is the phase characteristic of the mature industry. Now, that does not mean that we should let go of the first process because you can lose at that, too, and the Japanese are making much more of an effort in those early phases than they did in years past. But it does mean that we also have to make the cyclic development process more visible and to understand it better.

I believe strongly that the invisibility of that process does hurt us. Many of the words that we use to describe this subject hide, by their very meaning, the fact that there is a cyclic development process.

I think the wrong mental picture hurts us a great deal. Let me be concrete. Consider the picture brought to mind by the phase, "commercialization of new technology." If you talk to most people, I think you will find there is a notion that commercialization means to take something new and make it commercial, whereas the essence of the cyclic development process is you are refining something which is already commercial. When you talk about the commercialization of scientific discovery, those words already specialize you to phase one, or the ladder process.

The United States, as a whole, seems to resonate far more to the spectacular events of ladder innovation. Take, for example, superconductivity—high-temperature superconductivity. High-temperature superconductivity, from the point of view of industrial competitiveness, is something that is way, way out there—at least, optimistically, ten years away, if ever. And yet, when that great scientific event occurred, we had meetings in Washington, with the President of the United States in attendance, special legislation in Congress, set up special committees to advise the government, etc. A tremendous focus on a ladder-type event and, in all the discussion about it, the close tie with industrial supremacy which, in fact, does not exist.

It is necessary, therefore, for us to learn the ins and outs of the cyclic process, because, very often, that is the decisive process. And, if you look at that picture, I think you will find that it does turn around what you have to think about.

Take, for example, the notion of investing in R&D. A lot of people, if you are talking to them, will say, "Gee, the Japanese are investing a lot in R&D and that means they are going to jump past us," or something like that. And, again, the picture is, you put the R&D in here and the product pops out. That is very, very different from the one you really encounter if you are a participant in a cyclic development process, because in the cyclic development process it runs much more like this. You have a product. The product is selling. That gives you a certain stream of revenue. You can take that stream of revenue and put some of it into R&D for the next round. Some of it has to be reserved for the manufacturing, some of it for profits. Now, if you are on an upward swing and your product is succeeding, you have a flow back of money to invest into R&D; and if it isn't, you don't. And, in my experience, and the experience of many other people, oddly enough, R&D is determined, more or less, as a percent of sales. It is not an independent variable.

Let me say that once more. R&D is often a fixed percent of sales. Now I exaggerate to make the point. Ten percent is a very reasonable sort of number in a high-tech industry. Now, there may be some nonlinearity in relation with production. But, as you produce more, you spend more on R&D. It is not at all clear that in this phase of the industry, the cyclic development phase, that the R&D decision is an independent decision. As it is often discussed, "We have decided to invest in R&D" is a phrase that fits, again, much better at the front end of an industry. Later, what you can afford is determined by your success in the cyclic process. And it may be that, in the correlation, which has often been remarked on, between R&D spending and industrial success, it is the industrial success which causes the R&D spending, not the other way around.

I think that high-definition TV is just one more step along the continuum of TV products, but it is being represented as something totally new, and a tremendous opening to go in. I think that it would not be very different to decide that you want to get back into TV. I think getting into HD-TV is not that different a decision; and the only way to do it would not be the sort of things that were originally happening—which is, "Let's have DARPA spend some money in R&D." This is a total misconception related to mistaking the ladder for the cycle. You are still in a cycle process, in my opinion. The only way that you can enter HD-TV is to build up a manufacturing capability, on a very large scale, and back it up with a major R&D, and start the cycle churning. It is not that kind of an event where you can come in from the outside with a new technology. It is a cycle event, not a ladder event.

So, I think we need to rework our mental picture if we are to make progress with this matter. And in the subject of cycle versus ladder, and the characteristics of the cycle, there is much to learn. Of course, the ladder process is not always quite as separate from the cyclical process as I depict. A good company is continually investing in both kinds of developments and the problems within both of them. It is how to move things along that is the problem. But if we do not realize that there is a cyclic development process with its own peculiar resistances, then things will not move. Believe me, I had a very expensive and extensive education along those lines. So, I do very much believe in doing the research. But you then have to pay attention to how it will get into product. Much research is done, and it never will get in, because it will violate the rules. It will deal with an object that is not there. It will not understand that there is an issue of timing, that there is an issue of familiarity, and so forth. You have to understand that this cyclic beast is grinding away, and this is the beast you have to affect, and that your invention is only a part, and you have to match the timing and you have to match what is there. And, otherwise, it just does not happen. It is not enough to have done it. To have done it is to put it on a shelf, along with many other inventions or innovations; and then, various companies will come along and pull from that shelf the thing that fits their cycle. Now, which company does that—who knows? I think it is very wise for companies to be assiduous in that process. If you are talking about things that a company itself develops, then it had better pay attention to what is known as technology transfer, a very nontrivial process, because the thing you are transferring to has its own life structure.

Let me close with an important caveat with regard to the transfer of technology. Success or failure in technology transfer depends critically on the characteristics of the receptor. If the receptor knows very little, he can do very little even with a simple idea, because he cannot generate the mass of detail that is typically required to put a new technology into execution. On the other hand, if he knows a great deal and is capable of generating the necessary details, then from just a few sentences or pieces of technology he will be capable of filling in all the rest. That is why it is hard to transfer technology to the Third World and very hard not to transfer it to Japan.

9

Gunfire at Sea: A Case Study of Innovation

ELTING MORISON

In the early days of the last war when armaments of all kinds were in short supply, the British, I am told, made use of a venerable field piece that had come down to them from previous generations.* The honorable past of this light artillery stretched back, in fact, to the Boer War. In the day of uncertainty after the fall of France, these guns, hitched to trucks, served as useful mobile units in the coast defense. But it was felt that the rapidity of fire could be increased. A time-motion expert was, therefore, called in to suggest ways to simplify the firing procedures. He watched one of the gun crews of five men at practice in the field for some time. Puzzled by certain aspects of the procedures, he took some slow-motion pictures of the soldiers performing the loading, aiming, and firing routines.

When he ran these pictures over once or twice, he noticed something that appeared odd to him. A moment before the firing, two members of the gun crew ceased all activity and came to attention for a three-second interval extending throughout the discharge of the gun. He summoned an old colonel of artillery, showed him the pictures, and pointed out this strange behavior. What, he asked the colonel, did it mean. The colonel, too, was puzzled. He asked to see the pictures again. "Ah," he said when the performance was over, "I have it. They are holding the horses."

This story, true or not, and I am told it is true, suggests nicely the pain with which the human being accommodates himself to changing conditions. The tendency is apparently involuntary and immediate to protect oneself against the shock of change by continuing in the presence of altered situations the familiar habits, however incongruous, of the past.

Yet, if human beings are attached to the known, to the realm of things as they are, they also, regrettably for their peace of mind, are incessantly attracted to the unknown and things as they might be. As Ecclesiastes glumly pointed out, men persist in disordering their settled ways and beliefs by seeking out many inventions.

The point is obvious. Change has always been a constant in human affairs; today, indeed, it is one of the determining characteristics of our civilization. In our relatively shapeless social organization, the shifts from station to station are fast and easy. More important for our immediate purpose, America is fundamentally an industrial society in a time of tremendous technological development. We are thus constantly presented with new devices or new forms of power that in their refinement and extension continually bombard the fixed structure

*This essay was delivered as one of three lectures at the California Institute of Technology in 1950. It has been reprinted in various truncated forms a good many times since.

of our habits of mind and behavior. Under such conditions, our salvation, or at least our peace of mind, appears to depend upon how successfully we can in the future become what has been called in an excellent phrase a completely "adaptive society."

It is interesting, in view of all this, that so little investigation, relatively, has been made of the process of change and human responses to it. Recently, psychologists, sociologists, cultural anthropologists, and economists have addressed themselves to the subject with suggestive results. But we are still far from a full understanding of the process and still further from knowing how we can set about simplifying and assisting an individual's or a group's accommodation to new machines or new ideas.

With these things in mind, I thought it might be interesting and perhaps useful to examine historically a changing situation within a society; to see if from this examination we can discover how the new machines or ideas that introduced the changing situation developed; to see who introduces them, who resists them, what points of friction or tension in the social structure are produced by the innovation, and perhaps why they are produced and what, if anything, may be done about it. For this case study the introduction of continuous-aim firing in the United States Navy has been selected. The system, first devised by an English officer in 1898, was introduced in our Navy in the years 1900 to 1902.

I have chosen to study this episode for two reasons. First, a navy is not unlike a society that has been placed under laboratory conditions. Its dimensions are severely limited; it is beautifully ordered and articulated; it is relatively isolated from random influences. For these reasons the impact of change can be clearly discerned, the resulting dislocations in the structure easily discovered and marked out. In the second place, the development of continuous-aim firing rests upon mechanical devices. It therefore presents for study a concrete, durable situation. It is not like many other innovating reagents—a Manichean heresy, or Marxism, or the views of Sigmund Freud—that can be shoved and hauled out of shape by contending forces or conflicting prejudices. At all times we know exactly what continuous-aim firing really is. It will be well now to describe, as briefly as possible, what it really is. This will involve a short investigation of certain technical matters. I will not apologize, as I have been told I ought to do, for this preoccupation with how a naval gun is fired. For one thing, all that follows is understandable only if one understands how the gun goes off. For another thing, a knowledge of the underlying physical considerations may give a kind of elegance to the succeeding investigation of social implications. And now to the gun and the gunfire.

The governing fact in gunfire at sea is that the gun is mounted on an unstable platform, a rolling ship. This constant motion obviously complicates the problem of holding a steady aim. Before 1898 this problem was solved in the following elementary fashion. A gun pointer estimated the range of the target, ordinarily in the nineties about 1600 yards. He then raised the gun barrel to give the gun the elevation to carry the shell to the target at the estimated range. This elevating process was accomplished by turning a small wheel on the gun mount that operated the elevating gears. With the gun thus fixed for range, the gun pointer peered through open sights, not unlike those on a small rifle, and waited until the roll of the ship brought the sights on the target. He then pressed the firing button that discharged the gun. There were by 1898, on some naval guns, telescope sights which naturally greatly enlarged the image of the target for the gun pointer. But these sights were rarely used by gun pointers. They were lashed securely to the gun barrel, and, recoiling with the barrel, jammed back against the unwary pointer's eye. Therefore, when used at all, they were used only to take an initial sight for purposes of estimating the range before the gun was fired.

Notice now two things about the process. First of all, the rapidity of fire was controlled by the rolling period of the ship. Pointers had to wait for the one moment in the roll when the sights were brought on the target. Notice also this: There is in every pointer what is called a "firing interval"—that is, the time lag between his impulse to fire the gun and the translation of this impulse into the act of pressing the firing button. A pointer, because of this reaction time, could not wait to fire the gun until the exact moment when the roll of the ship brought the sights onto the target; he had to will to fire a little before, while the sights were off the target. Since the firing interval was an individual matter, varying obviously from man to man, each pointer had to estimate from long practice his own interval and compensate for it accordingly.

These things, together with others we need not here investigate, conspired to make gunfire at sea relatively uncertain and ineffective. The pointer, on a moving platform, estimating range and firing interval, shooting while his sight was off the target, became in a sense an individual artist.

In 1898, many of the uncertainties were removed from the process and the position of the gun pointer radically altered by the introduction of continuous-aim firing. The major change was that which enabled the gun pointer to keep his sight and gun barrel on the target throughout the roll of the ship. This was accomplished by altering the gear ratio in the elevating gear to permit a pointer to compensate for the roll of the vessel by rapidly elevating and depressing the gun. From this change another followed. With the possibility of maintaining the gun always on the target, the desirability of improved sights became immediately apparent. The advantages of the telescope sight as opposed to the open sight were for the first time fully realized. But the existing telescope sight, it will be recalled, moved with the recoil of the gun and jammed back against the eye of the gunner. To correct this, the sight was mounted on a sleeve that permitted the gun barrel to recoil through it without moving the telescope.

These two improvements in elevating gear and sighting eliminated the major uncertainties in gunfire at sea and greatly increased the possibilities of both accurate and rapid fire.

You must take my word for it, since the time allowed is small, that this changed naval gunnery from an art to a science, and that gunnery accuracy in the British and our Navy increased, as one student said, 3000% in six years. This does not mean much except to suggest a great increase in accuracy. The following comparative figures may mean a little more. In 1899 five ships of the North Atlantic Squadron fired five minutes each at a lightship hulk at the conventional range of 1600 yards. After twenty-five minutes of banging away, two hits had been made on the sails of the elderly vessel. Six years later one naval gunner made fifteen hits in one minute at a target 75 by 25 feet at the same range—1600 yards; half of them hit in a bull's-eye 50 inches square.

Now with the instruments (the gun, elevating gear, and telescope), the method, and the results of continuous-aim firing in mind, let us turn to the subject of major interest: how was the idea, obviously so simple an idea, of continuous-aim firing developed, who introduced it into the United States Navy, and what was its reception?

The idea was the product of the fertile mind of the English officer Admiral Sir Percy Scott. He arrived at it in this way while, in 1898, he was the captain of H.M.S. *Scylla.* For the previous two or three years he had given much thought independently and almost alone in the British Navy to means of improving gunnery. One rough day, when the ship, at target practice, was pitching and rolling violently, he walked up and down the gun deck watching his gun crews. Because of the heavy weather, they were making very bad scores. Scott noticed, however, that one pointer was appreciably more accurate than the rest. He watched this man with care, and saw, after a time, that he was unconsciously working his elevating gear back and forth in a partially successful effort to compensate for the roll of the vessel. It flashed through Scott's mind at that moment that here was the sovereign remedy for the problem of inaccurate fire. What one man could do partially and unconsciously perhaps all men could be trained to do consciously and completely.

Acting on this assumption, he did three things. First, in all the guns of the *Scylla*, he changed the gear ratio in the elevating gear, previously used only to set the gun in fixed position for range, so that a gunner could easily elevate and depress the gun to follow a target throughout the roll. Second, he rerigged his telescopes so that they would not be influenced by the recoil of the gun. Third, he rigged a small target at the mouth of the gun, which was moved up and down by a crank to simulate a moving target. By following this target as it moved and firing at it with a subcaliber rifle rigged in the breech of the gun, the pointer could practice very day. Thus equipped, the ship became a training ground for gunners. Where before the good pointer was an individual artist, pointers now became trained technicians, fairly uniform in their capacity to shoot. The effect was immediately felt. Within a year the *Scylla* established records that were remarkable.

At this point I should like to stop a minute to notice several things directly related to, and involved in, the process of innovation. To begin with, the personality of the innovator. I wish there were time to say a good deal about Admiral Sir Percy Scott. He was a wonderful man. Three small bits of evidence must here suffice, however. First, he had a certain mechanical ingenuity. Second, his personal life was shot through with frustration and bitterness. There was a divorce and a quarrel with that ambitious officer Lord Charles Beresford, the sounds of which, Scott liked to recall, penetrated to the last outposts of empire. Finally, he possessed, like Swift, a savage indignation directed ordinarily at the inelastic intelligence of all constituted authority, especially the British Admiralty.

There are other points worth mention here. Notice first that Scott was not responsible for the invention of the basic instruments that made the reform in gunnery possible. This reform rested upon the gun itself, which as a rifle had been in existence on ships for at least forty years; the elevating gear, which had been, in the form Scott found it, a part of the rifled gun from the beginning; and the telescope sight, which had been on shipboard at least eight years. Scott's contribution was to bring these three elements appropriately modified into a combination that made continuous-aim firing possible for the first time. Notice also that he was allowed to bring these elements into combination by accident, by watching the unconscious action of a gun pointer endeavoring through the operation of his elevating gear to correct partially for the roll of his vessel. Scott, as we have seen, had been interested in gunnery; he had thought about ways to increase accuracy by practice and improvement of existing machinery; but able as he was, he had not been able to produce on his own initiative and by his own thinking the essential idea and modify instruments to fit his purpose. Notice here, finally, the intricate interaction of chance, the intellectual climate, and Scott's mind. Fortune (in this case, the unaware gun pointer) indeed favors the prepared mind, but even fortune and the prepared mind need a favorable environment before they can conspire to produce sudden change. No intelligence can proceed very far above the threshold of existing data or the binding combinations of existing data.

All these elements that enter into what may be called "original thinking" interest me as a teacher. Deeply rooted in the pedagogical mind often enough is a sterile infatuation with "inert ideas"; there is thus always present in the profession the tendency to be diverted from the *process* by which these ideas, or indeed any ideas, are really produced. I well remember with what contempt a class of mine which was reading Leonardo da Vinci's *Notebooks* dismissed the author because he appeared to know no more mechanics than, as one wit in the class observed, a Vermont Republican farmer of the present day. This is perhaps the expected result produced by a method of instruction that too frequently implies that the great generalizations were the result, on the one hand, of chance—an apple falling in an orchard or a teapot boiling on the hearth—or, on the other hand, of some towering intelligence proceeding in isolation inexorably toward some prefigured idea, like evolution, for example.

This process by which new concepts appear, the interaction of fortune, intellectual climate, and the prepared imaginative mind, is an interesting subject for examination offered by any case study of innovation. It was a subject as Dr. Walter Cannon pointed out, that momentarily engaged the attention of Horace Walpole, whose lissome intelligence glided over the surface of so many ideas. In reflecting upon the part played by chance in the development of new concepts, he recalled the story of the three princes of Serendip who set out to find some interesting object on a journey through their realm. They did not find the particular object of their search, but along the way they discovered many new things simply because they were looking for *something*. Walpole believed this intellectual method ought to be given a name, in honor of the founders, serendipity; and serendipity certainly exerts a considerable influence in what we call original thinking. There is an element of serendipity, for example, in Scott's chance discovery of continuous-aim firing in that he was, and had been, looking for some means to improve his target practice and stumbled upon a solution by observation that had never entered his head.

Serendipity, while recognizing the prepared mind, does tend to emphasize the role of chance in intellectual discovery. Its effect may be balanced by an anecdote that suggests the contribution of the adequately prepared mind. There has recently been much posthaste and romage in the land over the question of whether there really was a Renaissance. A scholar has recently argued in print that since the Middle Ages actually possessed many of the instruments and pieces of equipment associated with the Renaissance, the Renaissance could be said to exist as a defined period only in the mind of the historians such as Burckhardt. This view was entertainingly rebutted by the historian of art Panofsky, who pointed out that although Robert Grosseteste indeed did have a very rudimentary telescope, he used it to examine stalks of grain in a field down the street. Galileo, a Renaissance intelligence, pointed his telescope at the sky.

Here Panofsky is only saying in a provocative way that change and intellectual advance are the products of well-trained and well-stored inquisitive minds, minds that relieve us of "the terrible burden of inert ideas by throwing them into a new combination." Educators, nimble in the task of pouring the old wine of our heritage into the empty vessels that appear before them, might give thought to how to develop such independent, inquisitive minds.

But I have been off on a private venture of my own. Now to return to the story, the introduction of continuous-aim firing. In 1900 Percy Scott went out to the China Station as commanding officer of H.M.S. *Terrible*. In that ship he continued his training methods and his spectacular successes in naval gunnery. On the China Station he met up with an American junior officer, William S. Sims. Sims had little of the mechanical ingenuity of Percy Scott, but the two were drawn together by temperamental similarities that are worth noticing here. Sims had the same intolerance for what is called spit and polish and the same contempt for bureaucratic inertia as his British brother officer. He had for some years been concerned, as had Scott, with what he took to be the inefficiency of his own Navy. Just before he met Scott, for example, he had shipped out to China in the brand new pride of the fleet, the battleship *Kentucky*. After careful investigation and reflection he had informed his superiors in Washington that she was "not a battleship at all—but a crime against the white race." The spirit with which he pushed forward his efforts to reform the naval service can best be stated in his own words to a brother officer: "I am perfectly willing that those holding views differing from mine should continue to live, but with every fibre of my being I loathe indirection and shiftiness, and where it occurs in high place, and is used to save face at the expense of the vital interests of our great service (in which silly people place such a child-like trust), I want that man's blood and I will have it no matter what it costs me personally."

From Scott in 1900 Sims learned all there was to know about continuous-aim firing. He modified, with the Englishman's active assistance, the gear on his own ship and tried out the new system. After a few months' training, his experimental batteries began making remarkable records at target practice. Sure of the usefulness of his gunnery methods, Sims then turned to the task of educating the Navy at large. In thirteen great official reports he documented the case for continuous-aim firing, supporting his arguments at every turn with a mass of factual data. Over a period of two years, he reiterated three principal points: first, he continually cited the records established by Scott's ships, the *Scylla* and the *Terrible*, and supported these with the accumulating data from his own tests on an American ship; second, he described the mechanisms used and the training procedures instituted by Scott and himself to obtain these records; third, he explained that our own mechanisms were not generally adequate without modification to meet the demands placed on them by continuous-aim firing. Our elevating gear, useful to raise or lower a gun slowly to fix it in position for the proper range, did not always work easily and rapidly enough to enable a gunner to follow a target with his gun throughout the roll of the ship. Sims also explained that such few telescope sights as there were on board our ships were useless. Their cross wires were so thick or coarse they obscured the target, and the sights had been attached to the gun in such a way that the recoil system of the gun plunged the eyepiece against the eye of the gun pointer.

This was the substance not only of the first but of all the succeeding reports written on the subject of gunnery from the China Station. It will be interesting to see what response these met with in Washington. The response falls roughly into three easily identifiable stages.

First stage: At first, there was no response. Sims had directed his comments to the Bureau of Ordnance and the Bureau of Navigation; in both bureaus there was dead silence. The thing—claims and records of continuous-aim firing—was not credible. The reports were simply filed away and forgotten. Some indeed, it was later discovered to Sims's delight, were half-eaten-away by cockroaches.

Second stage: It is never pleasant for any man's best work to be left unnoticed by superiors, and it was an unpleasantness that Sims suffered extremely ill. In his later reports, beside the accumulating data he used to clinch his argument, he changed his tone. He used deliberately shocking language because, as he said, "They were furious at my first papers and stowed them away. I therefore made up my mind I would give these later papers such a form that they would be dangerous documents to leave neglected in the files." To another friend he added, "I want scalps or nothing and if I can't have 'em I won't play."

Besides altering his tone, he took another step to be sure his views would receive attention. He sent copies of his reports to other officers in the fleet. Aware as a result that Sims's gunnery claims were being circulated and talked about, the men in Washington were then stirred to action. They responded, notably through the Chief of the Bureau of Ordnance, who had general charge of the equipment used in gunnery practice, as follows: (1) our equipment was in general as good as the British; (2) since our equipment was as good, the trouble must be with the men, but the gun pointer and the training of gun pointers were the responsibility of the officers on the ships; and most significant (3) continuous-aim firing was impossible. Experiments had revealed that five men at work on the elevating gear of a six-inch gun could not produce the power necessary to compensate for a roll of five degrees in ten seconds. These experiments and calculations demonstrated beyond peradventure or doubt that Scott's system of gunfire was not possible.

This was the second stage—the attempt to meet Sims's claims by logical, rational rebuttal. Only one difficulty is discoverable in these arguments; they were wrong at important points. To begin with, while there was little difference between the standard British equipment and the standard American equipment, the instruments on Scott's two ships, the *Scylla* and the *Terrible*, were far better than the standard equipment on our ships. Second, all the men could not be trained in continuous-aim firing until equipment was improved throughout the fleet. Third, the experiments with the elevating gear had been ingeniously contrived at the Washington Navy Yard—on solid ground. It had, therefore, been possible to dispense in the Bureau of Ordnance calculation with Newton's first law of motion, which naturally operated at sea to assist the gunner in elevating or depressing a gun mounted on a moving ship. Another difficulty was of course that continuous-aim firing was in use on Scott's and some of our own ships at the time the Chief of the Bureau of Ordnance was writing that it was a mathematical impossibility. In every way I find this second stage, the apparent resort to reason, the most entertaining and instructive in our investigation of the responses to innovation.

Third stage: The rational period in the counterpoint between Sims and the Washington men was soon passed. It was followed by the third stage, that of name-calling—the *argumentum ad hominem*. Sims, of course, by the high temperature he was running and by his calculated overstatement, invited this. He was told in official endorsements on his reports that there were others quite as sincere and loyal as he and far less difficult; he was dismissed as a crackbrained egotist; he was called a deliberate falsifier of evidence.

The rising opposition and the character of the opposition were not calculated to discourage further efforts by Sims. It convinced him that he was being attacked by shifty, dishonest men who were the victims, as he said, of insufferable conceit and ignorance. He made up his mind, therefore, that he was prepared to go to any extent to obtain the "scalps" and the "blood" he was after. Accordingly, he, a lieutenant, took the extraordinary step of writing the President of the United States, Theodore Roosevelt, to inform him of the remarkable records of Scott's ships, of the inadequacy of our own gunnery routines and records, and of the refusal of the Navy Department to act. Roosevelt, who always liked to respond to such appeals when he conveniently could, brought Sims back from China late in 1902 and installed him as Inspector of Target Practice, a post the naval officer held throughout the remaining six years of the Administration. And when he left, after many spirited encounters we cannot here investigate, he was universally acclaimed as "the man who taught us how to shoot."

With this sequence of events (the chronological account of the innovation of continuous-aim firing) in mind, it is possible now to examine the evidence to see what light it may throw on our present interest: the origins of and responses to change in a society.

First, the origins. We have already analyzed briefly the origins of the idea. We have seen how Scott arrived at his notion. We must now ask ourselves, I think, why Sims so actively sought, almost alone among his brother officers, to introduce the idea into his service. It is particularly interesting here to notice again that neither Scott nor Sims invented the instruments on which the innovation rested. They did not urge their proposal, as might be expected, because of pride in the instruments of their own design. The telescope sight had first been placed on shipboard in 1892 by Bradley Fiske, an officer of great inventive capacity. In that year Fiske had even sketched out on paper the vague possibility of continuous-aim firing, but his sight was condemned by his commanding officer, Robley D. Evans, as of no use. In 1892 no one but Fiske in the Navy knew what to do with a telescope sight any more than Grosseteste had known in his time what to do with a telescope. And Fiske, instead of fighting for his telescope, turned his attention to a range finder. But six years later Sims, following the tracks of his brother officer, took over and became the engineer of the revolution. I would suggest, with some reservations, this explanation: Fiske, as an inventor, took his pleasure in great part from the design of the device. He lacked not so much the energy as the overriding sense of social necessity that would have enabled him to *force* revolutionary ideas on the service. Sims possessed this sense. In Fiske, who showed rare courage and integrity in other professional matters not intimately connected with the introduction of new weapons of his own design, we may here find the familiar plight of the engineer who often enough must watch the products of his ingenuity organized and promoted by other men. These other promotional men when they appear in the world of commerce are called entrepreneurs. In the world of ideas they are still entrepreneurs. Sims was one, a middle-aged man caught in the periphery (as a lieutenant) of the intricate webbing of a precisely organized society. Rank, the exact definition and limitation of a man's capacity at any given moment in his career, prevented Sims from discharging all his exploding energies into the purely routine channels of the peacetime Navy. At the height of his powers he was a junior officer standing watches on a ship cruising aimlessly in friendly foreign waters. The remarkable changes in systems of gunfire to which Scott introduced him gave him the opportunity to expend his energies quite legitimately against the encrusted hierarchy of his society. He was moved, it seems to me, in part by his genuine desire to improve his own profession but also in part by rebellion against tedium, against inefficiency from on high, and against the artificial limitations placed on his actions by the social structure, in his case, junior rank.

Now having briefly investigated the origins of the change, let us examine the reasons for what must be considered the weird response we have observed to this proposed change. Why this deeply rooted, aggressive, persistent hostility from Washington that was only broken up by the interference of Theodore Roosevelt? Here was a reform that greatly and demonstrably increased the fighting effectiveness of a service that maintains itself almost exclusively to fight. Why then this refusal to accept so carefully documented a case, a case proved incontestably by records and experience? Why should virtually all the rulers of a society so resolutely seek to reject a change that so markedly improved its chances for survival in any contest with competing societies? There are the obvious reasons that will occur to all of you—the source of the proposed reform was an obscure, junior officer 8000 miles away; he was, and this is a significant factor, criticizing gear and machinery designed by the very men in the bureaus to whom he was sending his criticisms. And furthermore, Sims was seeking to introduce what he claimed were improvements in a field where improvements appeared unnecessary. Superiority in war, as in other things, is a relative matter, and the Spanish-American War had been won by the old system of gunnery. Therefore, it was superior even though of the 9500 shots fired at various but close ranges, only 121 had found their mark.

These are the more obvious, and I think secondary or supporting, sources of opposition to Sims's proposed reforms. A less obvious cause appears by far the most important one. It has to do with the fact that the Navy is not only an armed force; it is a society. Men spend their whole lives in it and tend to find the definition of their whole being within it. In the forty years following the Civil War, this society had been forced to accommodate itself to a series of technological changes—the steam turbine, the electric motor, the rifled shell of great explosive power, case-hardened steel armor, and all the rest of it. These changes wrought extraordinary changes in ship design, and, therefore, in the concepts of how ships were to be used; that is, in fleet tactics, and even in naval strategy. The Navy of this period is a paradise for the historian or sociologist in search of evidence bearing on a society's responses to change.

To these numerous innovations, producing as they did a spreading disorder throughout a service with heavy commitments to formal organization, the Navy responded with grudging pain. For example, sails were continued on our first-line ships long after they ceased to serve a useful purpose mechanically, but like the holding of the horses that no longer hauled the British field pieces, they assisted officers over the imposing hurdles of change. To a man raised in sail, a sail on an armored cruiser propelled through the water at 14 knots by a steam turbine was a cheering sight to see.

This reluctance to change with changing conditions was not limited to the blunter minds and less resilient imaginations in the service. As clear and untrammeled an intelligence as Alfred Thayer Mahan, a prophetic spirit in the realm of strategy, where he was unfettered by personal attachments of any kind, was occasionally at the mercy of the past. In 1906 he opposed the construction of battleships with single-caliber main batteries—that is, the modern battleship—because, he argued, such vessels would fight only at great ranges. These ranges would create in the sailor what Mahan felicitously called "the indisposition to close." They would thus undermine the physical and moral courage of a commander. They would, in other words, destroy the doctrine and the spirit, formulated by Nelson a century before, that no captain could go very far wrong who laid his ship alongside an enemy. The fourteen-inch rifle, which could place a shell upon a possible target six miles away, had long ago annihilated the Nelsonian doctrine. Mahan, of course, knew and recognized this fact; he was, as a man raised in sail, reluctant only to accept its full meaning, which was not that men were no longer brave, but that 100 years after the battle of the Nile they had to reveal their bravery in a different way.

Now the question still is, why this blind reaction to technological change, observed in the continuation of sail or in Mahan's contentions or in the opposition to continuous-aim firing? It is wrong to assume, as it is frequently assumed by civilians, that it springs exclusively from some causeless Bourbon distemper that invades the military mind. There is a sounder and more attractive base. The opposition, where it occurs, of the soldier and the sailor to such change springs from the normal human instinct to protect oneself, and more especially, one's way of life. Military organizations are societies built around and upon the prevailing weapons systems. Intuitively and quite correctly the military man feels that a change in weapon portends a change in the arrangements of his society. Think of it this way. Since the time that the memory of man runneth not to the contrary, the naval society has been built upon the surface vessel. Daily routines, habits of mind, social organization, physical accommodations, conventions, rituals, spiritual allegiances have been conditioned by the essential fact of the ship. What then happens to your society if the ship is displaced as the principal element by such a radically different weapon as the plane? The mores and structure of the society are immediately placed in jeopardy. They may, in fact, be wholly destroyed. It was the witty cliché of the twenties that those naval officers who persisted in defending the battleship against the apparently superior claims of the carrier did so because the battleship was a more comfortable home. What, from one point of view, is a better argument? There is, as everyone knows, no place like home. Who has ever wanted to see the old place brought under the hammer by hostile forces whether they hold a mortgage or inhabit a flying machine?

This sentiment would appear to account in large part for the opposition to Sims; it was the product of an instinctive protective feeling, even if the reasons for this feeling were not overt or recognized. The years after 1902 proved how right, in their terms, the opposition was. From changes in gunnery flowed an extraordinary complex of changes: in shipboard routines, ship design, and fleet tactics. There was, too, a social change. In the days when gunnery was taken lightly, the gunnery officer was taken lightly. After 1903, he became one of the most significant and powerful members of a ship's company, and this shift of emphasis naturally was shortly reflected in promotion lists. Each one of these changes provoked a dislocation in the naval society, and with man's troubled foresight and natural indisposition to break up classic forms, the men in Washington withstood the Sims onslaught as long as they could. It is very significant that they withstood it until an agent from outside, outside and above, who was not clearly identified with the naval society, entered to force change.

This agent, the President of the United States, might reasonably and legitimately claim the credit for restoring our gunnery efficiency. But this restoration by *force majeure* was brought about at great cost to the service and men involved. Bitternesses, suspicions, wounds were made that it was impossible to conceal and were, in fact, never healed.

Now this entire episode may be summed up in five separate points:

1. The essential idea for change occurred in part by chance but in an environment that contained all the essential elements for change and to a mind prepared to recognize the possibility of change.
2. The basic elements, the gun, gear, and sight, were put in the environment by other men, men interested in designing machinery to serve different purposes or simply interested in the instruments themselves.
3. These elements were brought into successful combination by minds not interested in the instruments for themselves but in what they could do with them. These minds were, to be sure, interested in good gunnery, overtly and consciously. They may also, not so consciously, have been interested in the implied revolt that is present in the support of all change. Their

temperaments and careers indeed support this view. From gunnery, Sims went on to attack ship designs, existing fleet tactics, and methods of promotion. He lived and died, as the service said, a stormy petrel, a man always on the attack against higher authority, a rebellious spirit; a rebel, fighting in excellent causes, but a rebel still who seems increasingly to have identified himself with the act of revolt against constituted authority.

4. He and his colleagues were opposed on this occasion by men who were apparently moved by three considerations: honest disbelief in the dramatic but substantiated claims of the new process, protection of the existing devices and instruments with which they identified themselves, and maintenance of the existing society with which they were identified.
5. The deadlock between those who sought change and those who sought to retain things as they were was broken only by an appeal to superior force, a force removed from and unidentified with the mores, conventions, devices of the society. This seems to me a very important point. The naval society in 1900 broke down in its effort to accommodate itself to a new situation. The appeal to Roosevelt is documentation for Mahan's great generalization that no military service should or can undertake to reform itself. It must seek assistance from outside.

Now with these five summary points in mind, it may be possible to seek, as suggested at the outset, a few larger implications from this story. What, if anything, may it suggest about the general process by which any society attempts to meet changing conditions?

There is, to begin with, a disturbing inference half-concealed in Mahan's statement that no military organization can reform itself. Certainly civilians would agree with this. We all know now that war and the preparation for war are too important, as Clemenceau said, to be left to the generals. But as I have said before, military organizations are really societies, more rigidly structured, more highly integrated, than most communities, but still societies. What then if we make this phrase to read, "No society can reform itself"? Is the process of adaptation to change, for example, too important to be left to human beings? This is a discouraging thought, and historically there is some cause to be discouraged. Societies have not been very successful in reforming themselves, accommodating to change, without pain and conflict.

This is a subject to which we may well address ourselves. Our society especially is built, as I have said, just as surely upon a changing technology as the Navy of the nineties was built upon changing weapon systems. How then can we find the means to accept with less pain to ourselves and less damage to our social organization the dislocations in our society that are produced by innovation? I cannot, of course, give any satisfying answer to these difficult questions. But in thinking about the case study before us, an idea occurred to me that at least might warrant further investigation by men far more qualified than I.

A primary source of conflict and tension in our case study appears to lie in this great word I have used so often in the summary, the word "identification." It cannot have escaped notice that some men identified themselves with their creations—sights, gun, gear, and so forth—and thus obtained a presumed satisfaction from the thing itself, a satisfaction that prevented them from thinking too closely on either the use or the defects of the thing; that others identified themselves with a settled way of life they had inherited or accepted with minor modification and thus found their satisfaction in attempting to maintain that way of life unchanged; and that still others identified themselves as rebellious spirits, men of the insurgent cast of mind, and thus obtained a satisfaction from the act of revolt itself.

This purely personal identification with a concept, a convention, or an attitude would appear to be a powerful barrier in the way of easily acceptable change. Here is an interesting primitive example. In the years from 1864 to 1871 ten steel companies in this country began making steel by the new Bessemer process. All but one of them at the outset imported from Great Britain English workmen familiar with the process. One, the Cambria Company, did not. In the first few years those companies with British labor established an initial superiority. But by the end of the seventies, Cambria had obtained a commanding lead over all competitors. The President of Cambria, R. W. Hunt, in seeking a cause for his company's success, assigned it almost exclusively to the labor policy. "We started the converter plant without a single man who had ever seen even the outside of a Bessemer plant. We thus had willing pupils with no prejudices and no reminiscences of what they had done in the old country." The Bessemer process, like any new technique, had been constantly improved and refined in this period from 1864 to 1871. The British laborers of Cambria's competitors, secure in the performance of their own original techniques, resisted and resented all change. The Pennsylvania farm boys, untrammeled by the rituals and traditions of their craft, happily and rapidly adapted themselves to the constantly changing process. They ended by creating an unassailable competitive position for their company.

How then can we modify the dangerous effects of this word "identification"? And how much can we tamper with this identifying process? Our security—much of it, after all—comes from giving our allegiance to something greater than ourselves. These are difficult questions to which only the most tentative and provisional answers may here be proposed for consideration.

If one looks closely at this little case history, one discovers that the men involved were the victims of *severely limited* identifications. They were presumably all part of a society dedicated to the process of national defense, yet they persisted in aligning themselves with separate parts of that process—with the existing instruments of defense, with the existing customs of the society, or with the act of rebellion against the customs of the society. Of them all the insurgents had the best of it. They could, and did, say that the process of defense was improved by a gun that shot straighter and faster, and since they wanted such guns, they were unique among their fellows, patriots who sought only the larger object of improved defense. But this beguiling statement, even when coupled with the recognition that these men were right and extremely valuable and deserving of respect and admiration—this statement cannot conceal the fact that they were interested too in scalps and blood, so interested that they made their case a militant one and thus created an atmosphere in which self-respecting men could not capitulate without appearing either weak or wrong or both. So these limited identifications brought men into conflict with each other, and the conflict prevented them from arriving at a common acceptance of a change that presumably, as men interested in our total national defense, they would all find desirable.

It appears, therefore, if I am correct in my assessment, that we might spend some time and thought on the possibility of enlarging the sphere of our identifications from the part to the whole. For example, those Pennsylvania farm boys at the Cambria Steel Company were, apparently, much more interested in the manufacture of steel than in the preservation of any particular way of making steel. So I would suggest that in studying innovation, we look further into this possibility: the possibility that any group that exists for any purpose—the family, the factory, the educational institution—might begin by defining for itself its grand object and see to it that that grand object is communicated to every member of the group. Thus defined and communicated, it might serve as a unifying agent against the disruptive local allegiances of the inevitable smaller elements that compose any group. It may also serve as a means to increase the acceptability of any change that would assist in the more efficient achievement of the grand object.

There appears also a second possible way to combat the untoward influence of limited identifications. We are, I may repeat, a society based on technology in a time of prodigious technological advance, and a civilization committed irrevocably to the theory of evolution. These things mean that we believe in change; they suggest that if we are to survive in good health we must, in the phrase that I have used before, become an "adaptive society." By the word "adaptive" is meant the ability to extract the fullest possible returns from the opportunities at hand: the ability of Sir Percy Scott to select judiciously from the ideas and material presented both by the past and present and to throw them into a new combination. "Adaptive," as here used, also means the kind of resilience that will enable us to accept fully and easily the best promises of changing circumstances without losing our sense of continuity or our essential integrity.

We are not yet emotionally an adaptive society, though we try systematically to develop forces that tend to make us one. We encourage the search for new inventions; we keep the mind stimulated, bright, and free to seek out fresh means of transport, communication, and energy; yet we remain, in part, appalled by the consequences of our ingenuity, and, too frequently, try to find security through the shoring up of ancient and irrelevant conventions, the extension of purely physical safeguards, or the delivery of decisions we ourselves should make into the keeping of superior authority like the state. These solutions are not necessarily unnatural or wrong, but they historically have not been enough, and I suspect they never will be enough to give us the serenity and competence we seek.

If the preceding statements are correct, they suggest that we might give some attention to the construction of a new view of ourselves as a society which in time of great change identified with and obtained security and satisfaction from the wise and creative accommodation to change itself. Such a view rests, I think, upon a relatively greater reverence for the mere *process* of living in a society than we possess today, and a relatively smaller respect for and attachment to any special *product* of a society, a product either as finite as a bathroom fixture or as conceptual as a fixed and final definition of our Constitution or our democracy.

Historically such an identification with *process* as opposed to *product,* with adventurous selection and adaptation as opposed to simple retention and possessiveness, had been difficult to achieve collectively. The Roman of the early republic, the Italian of the late fifteenth and early sixteenth century, or the Englishman of Elizabeth's time appears to have been most successful in seizing the new opportunities while conserving as much of the heritage of the past as he found relevant and useful to his purpose.

We seem to have fallen on times similar to theirs, when many of the existing forms and schemes have lost meaning in the face of dramatically altering circumstances. Like them we may find at least part of our salvation in identifying ourselves with the adaptive process and thus share with them some of the joy, exuberance, satisfaction, and security with which they went out to meet their changing times.

I am painfully aware that in setting up my historical situation for examination I have, in a sense, artificially contrived it. I have been forced to cut away much, if not all, of the connecting tissue of historical evidence and to present you only with the bare bones and even with only a few of the bones. Thus, I am also aware, the episode has lost much of the subtlety, vitality, and attractive uncertainty of the real situation. There has, too, in the process, been inevitable distortion, but I hope the essential if exaggerated truth remains. I am also aware that I have erected elaborate hypotheses on the slender evidence provided by the single episode. My defense here is only that I have hoped to suggest possible approaches and methods of study and also possible fruitful areas of investigation in a subject that seems to me of critical importance in the life and welfare of our changing society.

10

How Established Firms Respond to Threatening Technologies

ARNOLD C. COOPER
CLAYTON G. SMITH

The emergence of a new industry based on a major product innovation (such as the electronic calculator industry in the 1960s) often poses a threat of substitution to companies with a base in a more established industry (e.g., the producers of electromechanical calculators). During the early stages of industry development, the extent to which a substitution effect will occur is rarely clear. Nevertheless, managers of firms in the established, threatened industry must decide how to respond to an innovation that has the potential to alter or destroy their companies' existing business.[1]

This article considers the decisions confronting managers of firms in threatened industries and, in particular, the alternative of entering the emerging young industry. Such a response would seem to represent a natural extension of a company's existing business, one that, at a minimum, would allow the firm to build upon its established marketing resources and skills (including brand names, customer relationships, and channels of distribution). Further, at a time when it may be unclear whether the new product will substitute for the old, a strategy of participation in the young industry also provides a hedge, an opportunity to replace lost sales if the traditional product is displaced.

Our specific focus is on the challenges and pitfalls associated with this strategic response to a technological threat.[2] The analysis is based upon a study of eight young industries and twenty-seven "threatened firms," with the firms being chosen based upon the strong competitive positions that they enjoyed in their home industries. Table 10.1 lists the industries, the firms, and the time periods during which the young industries were examined. (The young industries and the firms were studied during the period that roughly corresponded to the "introduction" and "rapid growth" stages of development.) In total, nearly 250 sources of information from the secondary literature were considered in the analysis.

Each young industry involved the commercialization of a major product innovation. The nature of each innovation was such that, while firms from the threatened industry were sometimes able to build upon their existing technical capabilities, important new capabilities were required that were not associated with the traditional product. For example, the CT scanner harnessed both x-ray and data processing technology. Thus, while entrants from the x-ray equipment industry were able to build upon existing technical capabilities, they also

Reprinted by permission of *The Academy of Management Executive*. "How Established Firms Respond to Threatening Technologies" by Cooper, A. and Smith, C., Vol. 6, No. 2 1992.

TABLE 10.1. Summary of Industries and Firms Considered

Young Industry (Time Period)*	Established Industry Firms Considered	Young Industry (Time Period)	Established Industry Firms Considered
Ball-point pens (1945–1962)	Fountain pens Eversharp Inc. Parker Pen Co. Sheaffer Pen Co.	Electronic Calculators (1962–1976)	Electromechanical calculators Litton Industries (Monroe) SCM Corp. Singer Co. (Friden) Victor Comptometer Corp.
CT scanners (1973–1979)	X-ray/nuclear medical equip. General Electric Co. G.D. Searle & Co. Technicare Corp.	Electronic Watches (1969–1980)	Mechanical watches Bulova Watch Co. K. Hattori & Co. (Seiko) Timex Corp.
Diesel-electric locomotives (1924–1953)	Steam locomotives American Locomotive Co. Baldwin Locomotive Works	Microwave Ovens (1955–1983)	Gas/electric ovens General Electric Co. Magic Chef, Inc. Roper Corp. Tappan Co.
Electric typewriters (1925–1965)	Mechanical typewriters Remington Rand Royal Typewriter Co. Smith-Corona, Inc. Underwood Company	Transistors (1948–1968)	Receiving tubes General Electric Co. RCA Raytheon Mfg. Co. Sylvania Electric Products, Inc.

*Using decision rules which were based upon yearly changes in unit sales volume and industry development histories, the young industries and the firms were studied during the time period that roughly corresponded to the "introduction" and "rapid growth" stages of development.

had to develop important new resources and skills pertaining to computer technology and product design. Indeed, several innovations were what has been termed "competence-destroying"—so fundamentally different that the technical capabilities for the traditional product were largely irrelevant for the new technology.[3] Finally, these new industries attracted not only firms from the threatened industries, but also start-up firms and companies from other industries as well.

Prior work has presented frameworks for analyzing the potential of new technologies, and the industry attributes that favor "first-movers" versus those who follow.[4] Other studies have found that where established firms enter threatening young industries, they do not pursue the new product aggressively, and that they continue to make substantial commitments to their old product even after its sales begin to decline;[5] the studies have largely attributed these patterns to the constraining effects of sunk costs and internal political difficulties. Recent research has examined the ways in which innovations may enhance or destroy existing competences, and has suggested that unrelated new technologies can be especially challenging.[6] However, little consideration has been given to the pitfalls that may be encountered by entrants from an established, threatened industry. In particular, the ways in which the historic experience of firms from a threatened industry can affect their perceptions of how to compete in the new field have received little attention.

We begin by considering the characteristics of young industries that typically develop around new technologies, characteristics that have important implications for companies that enter an emerging field. The participation strategies of the leading threatened firms that entered these new industries are then considered. As noted in Table 10.2, a strategy of participation involves decisions along a spectrum concerning the timing of entry and magnitude of commitments, as well as how the new business should be organized and the strategy it should follow. Our findings indicate that entrants from the threatened industries made these decisions in characteristic ways—and often encountered particular pitfalls in the process. Special attention is given in the discussion to these pitfalls, and to possible reasons for them. The implications of the analysis for decisions concerning a strategy of participation in a threatening young industry are examined in the final section. While there are no assured success formulas, our intention is to highlight some of the problems that can be encountered and suggest possibilities for avoiding them.

YOUNG INDUSTRIES AND TECHNOLOGICAL THREATS

Any assessment of a technological threat obviously involves an appraisal of the new technology and the extent to which it will have advantages in meeting user needs. What may be less obvious, however, is the necessity of appraising the nature of the industry that is likely to develop around a major product innovation, and how competition in the new field may differ from competition in the established, threatened industry. Our study of these eight young industries highlights the common patterns of industry development—patterns that managers of threatened firms should be sensitive to if they consider entering an emerging field that poses a threat of substitution.

In the eight industries, the first commercial introduction of the new product was made by a firm from the established industry in three cases (electric typewriters, electronic watches, and transistors); firms from outside the industry introduced the others. But in every instance, early versions of the new product were crude and expensive, and there was great uncertainty about how rapidly the market would develop. For example, the first electronic watches were unreliable, and so bulky and unattractive that they were sometimes referred to as "quarter-pounders." Consumers were also deterred from purchasing early microwave ovens by the high price—roughly $1,500, in 1950s dollars—and the fact that microwave cooking left cold spots and changed food colors. In some instances, the initial lack of complementary products (e.g., cookbooks and cookware for microwave cooking) retarded the market's development as well.

In this context, early entrants invariably had to overcome substantial technical difficulties, and they often had to remain patient when the market developed slowly. Indeed, in some instances, the new product may never gain widespread acceptance. But in all of the cases studied here, an industry did begin to emerge, and a time of rapid growth followed that promised substantial opportunities to participating firms. Beyond the prospect of significant increases in revenues, the environment created by burgeoning demand often limited competitive pressures and permitted lucrative profit margins. In addition to later entrants from the established, threatened industry, the growing promise of the new product encouraged start-up firms and

TABLE 10.2. Decisions Concerning a Strategy of Participation

Decision	From	→ To
Timing of entry	Early	Late
Magnitude of Commitments	Token	Major
Degree of organizational separation	Close linkages with established organization	Separate/independent organization
Competitive strategy for new business	Traditional ways of competing	New ways of competing

Note: The decisions outlined above range along a spectrum, such as from early to late entry, etc. Further, some decisions may change over time, with, for instance, a token early commitment followed by later commitments of greater magnitude.

established companies from other industries to enter the field as well.

Even in the "growth stage," however, these industries were characterized by high levels of uncertainty and risk. During this period, there usually was no clear success formula and firms often followed different strategies; only time proved—or disproved—the validity of the assumptions on which these strategies were based. Thus, steam locomotive producers found that their new competitor in the diesel-electric field, General Motors, disdained their traditional custom manufacturing methods and produced standard products for inventory instead. In fact, firms from the established industry and entrants from outside that industry often had different ideas about how to compete. In such cases, the traditional firms had to appraise and respond to the novel strategies that their new competitors chose to pursue.

During the time before "dominant designs" became established, there was much experimentation with different technical approaches and product designs.[7] In the transistor industry's early years, for example, it was not clear whether germanium or silicon transistors would prevail. Similarly, in the electronic watch industry, models with light-emitting diode (LED) and liquid crystal displays (LCDs) dueled for market acceptance. Industry participants had to decide which approaches "to bet on," and some companies committed to what later proved to be blind alleys.

Rapid rates of change also characterized the industries. To stay in the game, participants often had to field successive generations of the product as the state of the art advanced; in CT scanners, four generations followed one another in almost as many years. Companies also had to cope with constant changes in manufacturing methods. In transistors, one vacuum tube firm blundered when it built the most automated plant for germanium transistors, shortly before silicon transistors and newer manufacturing processes came to dominate. And while the competitive environment was usually not intense initially, cut-throat competition often quickly emerged when the period of rapid growth began to wane—in some cases with little warning.[8] In this context, participating firms needed strong R&D and financial capabilities to remain competitive, and to absorb the inevitable setbacks and risks.

> For firms from the established industry, longstanding competitive strengths did not necessarily provide an advantage in the new field. All of the industries required technical resources and skills that were not associated with the traditional product; in several cases, the traditional technical capabilities were largely irrelevant. For example, the ability of electromechanical calculator firms to produce precise mechanical parts was of little value for electronic calculators.

And surprisingly, even established marketing capabilities for the traditional product did not always provide a long-term advantage, as several vacuum tube firms discovered. While their distribution networks were important in the tube business as a means of quickly delivering replacement components, the value of this capability for transistors waned as the new product became more and more reliable.

In summary, these industries were characterized by high levels of uncertainty, new competitors that had different ideas about how to compete, alternative and unproven technical approaches and product designs, rapid rates of change and, in some cases, by the obsolescence of established technical and marketing capabilities for the traditional product. The characteristics of the young industries and the corresponding managerial implications are displayed in Table 10.3. A typical pattern of sales development for a new technology (transistors), and its impact on the traditional counterpart (receiving tubes) is also shown in Figure 10.1. Because of their dynamic nature, competitive positions were often unstable; success was often transient. Commitment to these new technologies was like entering a poker game in which the stakes kept increasing and the rules were not at all clear. Overall, these industries differed markedly from the mature, relatively stable industries that were threatened.

TABLE 10.3. Characteristics of Young Industries

Attribute	Managerial Implications
New product crude and expensive at first, Sometimes, an initial lack of complementary products.	Difficult to judge rate at which market will develop. Often must overcome substantial technical difficulties, and remain patient if market develops slowly.
Entrants often start-up firms and established firms from other industries.	Difficult to predict competitors' actions. Must appraise and respond to "different" strategies.
Alternative and unproven technical approaches and product designs.	Risks associated with "betting on" particular approaches/designs.
Rapid rates of change, especially in product and manufacturing methods.	Strong R&D and financial capabilities needed to remain competitive over time.
New technical resources and skills required. Existing technical/marketing capabilities sometimes of limited or little value.	Established competitive strengths may not provide a long-term advantage.

THE STRATEGY OF PARTICIPATION: CHALLENGES AND PITFALLS

We now consider the strategic issues confronting a firm that chooses to respond to a technological threat by entering the new field. A decisions to participate in the young industry involves determinations concerning the: (1) timing of entry; (2) magnitude of commitments; (3) degree of organizational separation between new and traditional product activities; and (4) competitive strategy for the new business. In the sections that follow, the participation strategies of the twenty-seven threatened firms are examined along these

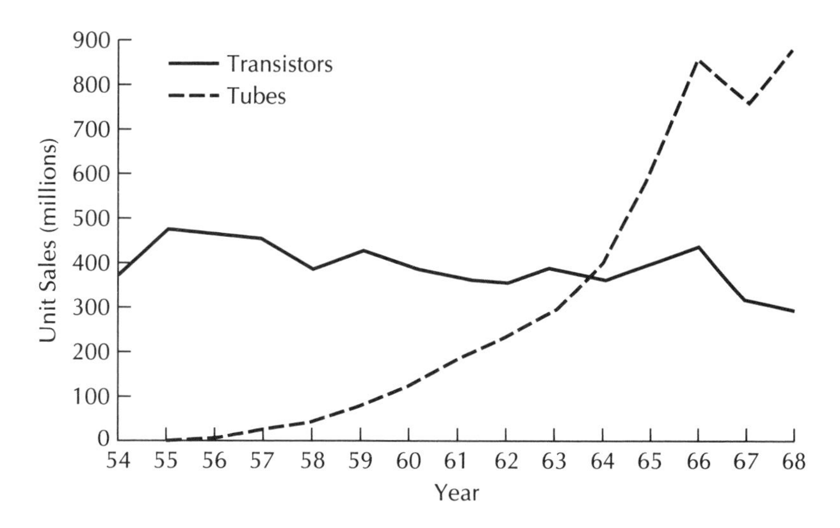

Fig. 10.1. Transistor/receiving tube unit volume comparison. (*Sources: Business Week* (various issues); Electronic Industries Association.)

dimensions. Particular attention is given to the problems that many of the firms encountered, and to possible reasons for their occurrence.

Timing of Entry

Prior research indicates that early entry by established firms into a young industry is associated with higher levels of long-term performance.[9] Balanced against this, however, are the risks of an early entry, including the risk that the new product may never achieve commercial success. When faced with a major product innovation, a firm often must contend with the challenges of assessing the potential of an unproven technology and the speed with which market acceptance will occur. Further, it must appraise the resources and skills required to compete at a time when the requirements for success are not clear. Overall, a firm must decide whether and when to enter under rather uncertain circumstances.

There is a common view that established firms, when threatened by major product innovations, lack the vision and will to commit to the new technology. (For example, Levitt's classic, "Marketing Myopia," gives example after example of firms that failed to define their business broadly, and thereby missed opportunities or subsequently failed.[10]

> In reviewing these twenty-seven leading firms, however, it was found that all entered the young industries. Indeed, twenty-one entered relatively early, before sales of the new product began to grow rapidly. Far from ignoring the new technology, these firms seemed to recognize its possibilities at an early date. But, surprisingly, eight of these twenty-one early entrants made abortive commitments, in which they entered but then withdrew before achieving much success.

For example, Eversharp and Sheaffer, leading producers of fountain pens, were two of the earliest companies to introduce ball-point pens in 1946. Unfortunately, their new pens—and those of many new entrants as well—tended to skip, blot, and even leak into pockets. One industry observer condemned the early ball-point pen as "the only pen which would make eight copies and no original." After an initial fad phase, public disenchantment set in, and the ball-point pen industry virtually disappeared. Eversharp and Sheaffer both withdrew from the market and many of the new entrants went bankrupt. In 1950, improved ball-point pens were introduced by a new firm, Papermate. Subsequently, the young industry began to develop rapidly, as shown in Figure 10.2.

Similar patterns were observed with producers of mechanical typewriters. Remington and Royal had electric typewriters as early as 1925. However, their initial offerings were bulky, noisy, and prone to break down; the two firms discontinued their efforts after three or four years. (IBM, which never made mechanical models, introduced the first commercially successful electric in 1934.) And American Locomotive introduced a number of experimental diesel-electric engines beginning in 1924. However, because of their high weight-to-power ratio and other difficulties, American subsequently ceased its development efforts. (In 1934, General Motors, which never produced steam locomotives, made the first successful introduction of a diesel-electric based upon improved diesel-engine technology developed in its laboratories.)

In several cases (e.g., Eversharp, Sheaffer, and Royal), it appeared that the firm had little in the way of an R&D orientation that might have provided experience in managing the process of introducing *and improving* major product innovations. While they recognized the new technology's potential, they seemed unable to make the further advancements that later led to commercial success. Further, in industries such as diesel-electric locomotives and electronic calculators, the new technology was so different from the traditional that the technical capabilities for the old product were largely irrelevant. At the outset, the firms may not have fully recognized the fact that their existing engineering/production capabilities would be of little value in the new field. But this fact undoubtedly

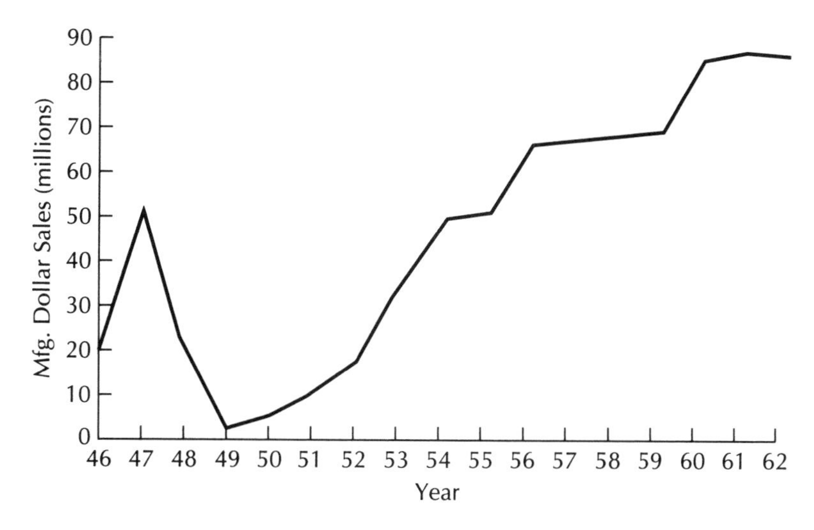

Fig. 10.2. Ball point pen sales. (*Source*: Writing Instrument Manufacturers Association.)

made it more difficult for them to overcome the technical obstacles that all of the early entrants faced.

Of the eight firms whose initial entry was abortive, two companies resumed their commercial efforts within a year. However, on average, the other six firms did not do so for more than nine years. For example, American Locomotive did not renew its efforts with diesel-electrics until 1936, two years after General Motors made the first successful introduction of the new product. Having been "burned" once, these firms seemed content to leave the pioneering to others. With one exception (General Electric in microwave ovens), the six firms did not resume their commercial efforts until after the new product's viability had been demonstrated by other firms from outside the established industry.

Magnitude of Commitments

Where corporate management chooses to enter the new field, important decisions must be made over time concerning the magnitude of commitments. The firm's initial involvement could vary widely, ranging from a token effort with only a few prototypes to major and immediate investments. The magnitude of commitments may also change over time as the young industry evolves. Prior research has argued that major early investments should often lead to greater long-term success, as the firm reaps the benefits of a stronger competitive position.[11] However, such investments may also lead to substantial commitments to the "wrong" technical approach (e.g., germanium versus silicon transistors) or to a market that develops very slowly (e.g., microwave ovens).

Of the twenty-seven companies examined, twenty-four made substantial investments over time. But while four of the twenty-four firms took an aggressive stance from the start, the investments made over time by most of the remaining twenty were uneven. It was quite common for these latter firms to make a limited initial commitment (relative to their capabilities), and permit other companies—usually from outside the established industry—to lead in improving the new product and in gaining market acceptance for it. Typically, they would then mount a more vigorous effort as the industry developed. This approach had the virtue of delaying major investments until after many of the technical and market uncertainties had faded.

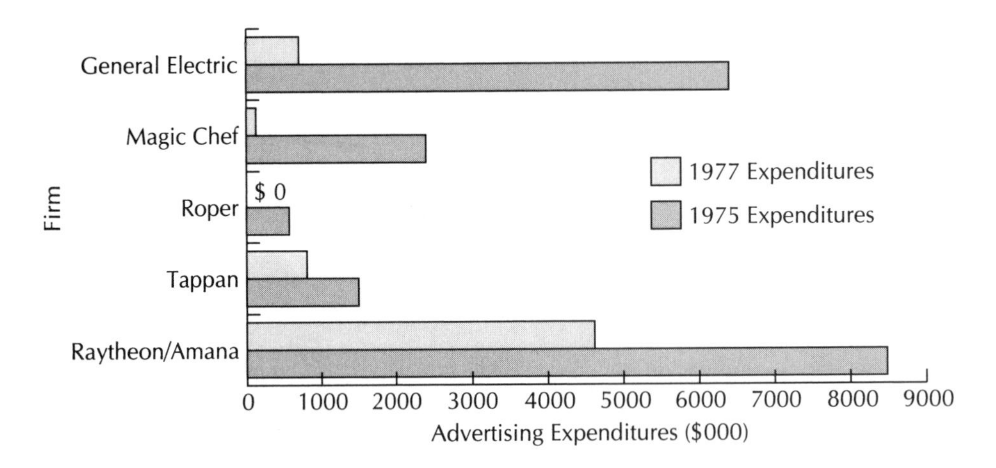

Fig. 10.3. 1975/1977 Microwave oven advertising expenditures. (*Source*: Leading National Advertisers, Inc.)

However, as it turned out, the net effect in most cases was to fall behind and to make it much more difficult to establish a viable long-term competitive position.

During the microwave oven industry's early growth stages (1972–1975), for instance, there were challenges in stimulating primary demand, in teaching consumers how to cook with the new product, and in overcoming fears about radiation. Here, the conventional oven producers made very limited initial commitments relative to entrants from outside the range industry, notably Raytheon/Amana. As an example of this, Figure 10.3 displays company advertising expenditures for 1975 and 1977; the contrast in emphasis between the conventional range producers and Amana is striking. As sales growth accelerated further in the years after 1975, the conventional oven producers began to make more significant commitments. Still, it was through efforts of new entrants such as Amana that the image of the microwave oven was changed from that of an expensive "hot-dog cooker" to a legitimate cooking device.[12]

While Timex entered electronic watch manufacturing early, the firm continued to concentrate on mechanical watches in its marketing efforts. In fact, at a trade shown in 1975, the firm's director of sales refused to be photographed holding a model from its electronic line; he said that the company's big push that year was still in traditional watches. The firm was also slow to develop internal R&D/manufacturing capabilities for semiconductor components (which were critical to the new product's performance, quality, and cost), preferring instead to rely on outside suppliers for its requirements. Only after prices for electronic watches began to tumble into the firm's low-price niche did Timex begin to make a strong commitment. However, the combination of increasing competition and a late start in mastering the technology proved to be a severe handicap; each time the firm introduced a newer watch line, other producers had already fielded superior models at lower prices. In 1980, after four years of heavy investments, Timex was still struggling to build and market electronic watches profitably.

Similarly, while Litton/Monroe and SCM entered the electronic calculator industry at an early date, their initial efforts largely took the form of marketing electronic desk-top models that were produced by Canon and Toshiba (firms that had never been active in electromechanical calculators). To a considerable degree, this was due to their belief that electronic calculators would be mainly used by scientists and engineers—small, specialized customer groups that had previously relied on large computers for computational needs. And as the division president of Monroe said later of his company's early participation:

> Our effort in electronics, I think logically, was not to create competition for these electromechanical machines. It was not to take away the established base, but to seek new business over that base.[13]

Subsequently the broader applicability of the new product in the office equipment and other segments became more apparent, and Monroe and SCM began a serious effort to develop their own skills for electronic calculators. But by this time, Canon, Toshiba, and other new entrants had gained a formidable lead in the requisite technical and manufacturing capabilities, and were rapidly becoming established in the market under their own brand names.

Finally, while American and Baldwin Locomotive entered the market for diesel-electrics, they continued to devote most of their energies to steam locomotives for several years while General Motors was gaining a foothold in the new field. (As late as 1938, American's president spoke to the Western Railway Club, saying, "For a century, as you know, steam has been the principal railroad motive power. It still is and, in my view, will continue to be."[14]) By the time they began to make a serious effort in diesel-electrics, General Motors had already gained a significant lead in product design and had developed facilities and methods for mass-producing its product. In addition, GM had also established a solid reputation and strong customer relationships with many of the major railroads in the United States.

While most of the firms examined made substantial commitments over time, these investments were made only after the potential of the new product had become apparent. Such firms seemed to harbor the expectation (initially) that the new product would not penetrate the core markets of the traditional business. In several cases, there were also concerns that the new product's early imperfections could tarnish the firm's reputation; as such, there was a reluctance to make a full commitment until the product was "proven." In virtually every case, however, these companies appeared to underestimate the ability of firms from outside the established industry to overcome important technological obstacles, to gain market acceptance for the new product, and to establish a defensible competitive position. Only after the miscalculation became apparent did these firms begin to mount a more vigorous effort.

Degree of Organization Separation

The third area to be considered concerns the degree of organizational separation between new and traditional product activities. In choosing to enter the young industry, it would be natural to consider using the established organization for the new product. Potentially, this arrangement could provide significant cost savings, facilitate the coordination of efforts with the traditional product, and permit the new business to benefit from the skills of executives from the traditional business. At the same time, however, it is important to have decision-makers who are enthusiastically committed to the new product, even though it may cannibalize sales of the traditional product.[15] Further, it may also be necessary to violate the conventional wisdom that has developed over time in the established organization.

Specific information on parent-business organizational decisions was available for seventeen of the twenty-seven firms; eleven of the seventeen appeared to have established close organizational linkages between new and traditional product activities. For example, based upon information obtained from secondary sources (e.g., annual reports, business articles, and industry studies), it appeared that Remington, Royal, Smith-Corona, and Underwood handled the production, sales, and distribution of electric typewriters through the departments that had been used for mechanical models. Similarly, Tilton's study of the transistor industry and other sources indicated that the entrants from the vacuum tube industry had placed their semiconductor operations in their tube divisions, where tube experts were the principal decision makers.[16]

However, in a number of cases where the new and traditional technologies were fundamentally different, decisions to use the established organization proved to be ill-advised. Thus, while other firms were lowering costs by building new manufacturing facilities, Victor Comptometer decided to use its existing organization and build electronic calculators at an old three-story plant in Chicago, where electromechanical machines were being made. Victor eventually made nearly everything for its electronic calculators but the semiconductor chips at this facility. Unfortunately, the net result was a line of high-priced, high-cost machines that brought little or no return. As an executive from a rival firm said later, "Victor failed to realize that its mechanical and technical expertise meant nothing in the electronics age."

More generally, when the new product was closely tied to the established organization, business strategy decisions were often constrained by concerns about the obsolescence of existing investments. This appears to have been true for RCA, where transistor activities were initially placed within the established vacuum tube division. In 1960, eight years after entering the new field, RCA recognized the problem and separated its tube and semiconductor operations. However, the two divisions were recombined three years later when the transistor unit began to experience losses. The latter decision, in addition to reintroducing the original problem, apparently undermined the morale of the semiconductor people, as their unit's strategic direction kept changing with each reorganization.[17]

These examples illustrate the fact that there were no easy answers to the question of how to organize the new business. There were tradeoffs between the synergies to be gained from having close linkages with the established organization, and the benefits of having the flexibility and drive of an independent management group. However, it appeared that the tendency among these firms was to forge close organizational linkages between activities for the new and traditional products. This decision may also have influenced the ways in which they chose to compete in the young industry, which we will now consider.

> Even where an independent unit was created for the new product, other problems sometimes emerged. At Timex for example, the decision to establish a separate organization for electronic watches in 1976 led to fierce rivalry between the mechanical and electronic groups. One old-guard manager acknowledged that his people had stood by while new managers in the electronic unit made mistakes that the company had made years earlier. Indeed, the rivalry became so intense that the electronic people accused the mechanical group of withholding essential information, and senior management had to enter the fray to lay the charge ro rest.[18] Needless to say, these organizational difficulties compounded the firm's competitive problems in the electronic watch market.

Competitive Strategy for the New Business

The final issue to be examined concerns the strategy that is developed for competing in the young industry. In making decisions concerning competitive strategy, firms from the established, threatened industry have a substantial base of experience to draw upon. Particularly when the firm's strategy in the established industry has historically been a successful, it would be natural to consider employing the same basic approach in the new field (essentially to fold the new product into the existing strategy). However, if the new technology requires, or makes possible, different product concepts or ways of competing, strategic approaches that are based on "the conventional wisdom" may be less likely to succeed.

In cases where a new product concept emerged, the traditional firms rarely pioneered the new design. For example, when Tappan and General Electric entered the infant microwave oven industry in the mid-1950s, they utilized product designs (built-in models and free-standing double ovens) that were quite similar to conventional models. However, because such designs implied the replacement of a household's existing range and, also since microwave ovens proved to be ill-suited for cooking meats and other foods, sales were very limited. Indeed, rapid market growth did not begin until after Raytheon/Amana entered the field with its countertop design in 1967—a configuration that was intended to supplement, not replace, a household's existing range. Further, while other new competitors quickly followed Raytheon's lead, conventional oven producers were slow to respond to the promising countertop innovation. (Tappan and GE did not begin self-manufacture of countertop ovens until 1973 and 1974, respectively, more than half a decade after Raytheon-Amana pioneered the countertop design.)[19]

Similarly, where new ways of competing emerged, traditional producers were usually slow to recognize the implications. In electromechanical calculators, for example, direct-sales organizations and service networks had historically been important requirements for success. In fact, the leading producers of such machines each had approximately 1500 sales and service personnel in several hundred company-owned branches. As a senior executive at Friden said at the time, "You don't have a chance in this business without this capability." However, firms from outside the established industry chose to sell their electronic desk-top calculators through office equipment dealers. Because this permitted them to deal with a few hundred dealers rather than with thousands of end-users, a sales force of less than fifty people was sufficient.

Office equipment dealers did not have the facilities for servicing electronic calculators. (Friden attempted to exploit this fact in magazine ads that were aimed at foreign producers of electronic calculators. The ads asked potential customers, "Does an imported repairman come with your imported calculator?") However, as the reliability of the new product (which had few moving parts) improved relative to electromechanical models (which had an average downtime of ten percent), the need for a strong service network waned. The new competitors also engaged in aggressive price competition, a practice that was uncommon in electromechanical calculators. As the new technology improved over time, market prices for most desk-top models fell to levels that made it difficult to cover the costs of direct-selling efforts. As a result, several firms from the established industry, such as SCM Corp., were eventually forced to develop a network of office equipment dealers for their electronic calculators.[20]

Analogous patterns were found in other industries as well. In ball-point pens, the concept of a throw-away pen was pioneered not by a fountain pen producer, but by BIC. In diesel-electric locomotives, it was General Motors that pioneered the concept of a standardized locomotive produced for inventory. And in transistors, firms from the vacuum tube industry emphasized process innovation in their business strategies, as they had in vacuum tubes. In contrast, young semiconductor firms, such as Texas Instruments, focused on product innovation—and sometimes obsoleted the designs of transistors that the vacuum tube companies were trying to produce in volume.[21]

These patterns suggest that the historic experience of threatened firms in the established industry often "colored" their perceptions of how to compete effectively in the emerging young industry. As noted previously, these firms were often among the earliest entrants in the young industries. But they rarely took the lead in developing new product concepts or ways of competing that might have allowed them to better capitalize upon the new technology's potential. Rather, they tended to use the same basic approaches that had been successful in the established industry.

Patterns of Performance

Given the difficulties that have been described, an obvious question is, "What were the overall patterns of performance for these twenty-seven firms?" To explore this issue, the firms were classified as "successful" or "unsuccessful," based upon their survival or non-survival in the given industry, market share data, and descriptive information concerning business performance obtained from secondary sources (e.g., annual reports, busi-

TABLE 10.4. The Successful Firms

Company (Young Industry)	Time of Entry	Magnitude of Commitments	Degree of Org. Separation	Competitive Strategy for New Business
Parker Pen Co. (Ball point pens)	Late entrant	Strong commitment	—	Traditional strategic approach employed (ignored throw-away pen segment)
General Electric Co. (CT scanners)	Entered at outset of growth stage	Strong commitment	High	Traditional strategic approach employed
Technicare Corp. (CT scanners)	Entered at outset of growth stage	Strong commitment	High	Traditional strategic approach employed
Smith-Corona, Inc. (Elec. typewriters)	Late entrant	Significant commitment	Low	Traditional strategic approach employed
Litton/Monroe (Electronic cals.)	Early entrant	Ltd. → Strong commitment	—	Traditional strategic approach employed (slow response to emergence of new distribution methods)
K. Hattori/Seiko (Electronic watches)	Early entrant Initial entry Abortive	Strong commitment	—	Traditional strategic approach employed
General Electric Co. (Microwave ovens)	Early entrant Initial entry Abortive	Ltd. → Strong commitment	High	Traditional strategic approach employed (slow response to countertop Innovation)

Note: These firms were all relatively successful as of the end of the study period. However, it should be noted that the fortunes of several subsequently waned when they later faced severe competitive pressures from overseas manufacturers or from still newer technologies.

ness articles, industry surveys, and stock reports). As noted earlier, these twenty-seven companies were all leading firms in the established industries that were threatened. However, despite their presumed advantages, twenty of the twenty-seven firms were unable to develop and sustain a strong competitive position in the young industries.[22]

The remaining seven firms were classified as having been successful, through the end of the study period. The firms were:

- Parker Pen (ball-point pens)
- General Electric (CT scanners)
- Technicare (CT scanners)
- Smith-Corona (electric typewriters)
- Litton/Monroe (electronic calculators)
- K. Hattori/Seiko (electronic watches)
- General Electric (microwave ovens)

For these companies, there was no single path to a strong competitive position, and they, too, encountered the problems that have been described (see Table 10.4). The entry of five of the seven firms either preceded or coincided with the "take-off" in market sales. Of the five firms, GE discontinued its microwave oven effort for several years when the slow rate of market development became apparent; K. Hattori/Seiko withdrew from electronic watches for nearly a year because of technical problems with its original offering. The two remaining firms, Parker Pen and Smith-Corona, were late entrants. Both had limited R&D and financial capabilities, but their late entry allowed them to avoid some of the technical challenges that earlier entrants faced. They also focused on niche markets (high-priced ball-point pens and portable electrics) and largely avoided the more competitive mainstream segments.

Relative to their capabilities, five of the firms made significant or strong commitments from the outset. While Litton/Monroe and General Electric (in microwave ovens) made only limited initial investments, their considerable R&D and financial capabilities enabled them to establish a strong position in the new field after they began to make a whole hearted effort. Regarding organizational decisions, three of the four firms for which data were available appeared to maintain a high degree of separation between new and traditional product activities. Finally, with respect to competitive strategy decisions, these firms rarely pioneered new product concepts or ways of competing. As it turned out, however, Parker Pen was able to ignore the throw-away pen because it never threatened its high-priced niche. And while GE and Litton were eventually compelled to respond to new product concepts and methods of distribution, they did so before their position was severely undermined.[23]

AVOIDING THE PITFALLS

The decision to enter an emerging young industry from a base in an established, threatened industry is clearly fraught with challenges and pitfalls. The "typical" participation strategy decisions of the firms examined, and the corresponding pitfalls are summarized in Table 10.5. The limited number of successful firms makes any discussion of assured success formulas problematic. Indeed, much may depend upon how a young industry develops, including the extent to which there prove to be sizable market segments that are not attacked by new competitors. But based upon the experiences of these twenty-seven firms, it is possible to highlight some of the potential problems that are associated with a strategy of participation, and to suggest possibilities for avoiding them.

In relation to the time-of-entry decision, most of the firms examined recognized the possibilities of the new technology at an early date, and entered the field before the market began to grow rapidly. There was little evidence that management "buried its head in the sand" and refused to recognize the new product's potential. However, some of the companies may not have appreciated the implications of an early commitment. These firms often lacked the resources and skills required to overcome the initial technical obstacles, and they sometimes became disenchanted when the market did not develop quickly.

The implication is that, in considering an early entry, management should carefully appraise the firm's technical capabilities relative to the challenges involved. The challenge is not simply to introduce the new product; it is to be able to make the further advancements in performance, quality, and cost that will be required for commercial success. Where the firm's capabilities are lacking, it should be understood that the inevitable technical

TABLE 10.5. Pitfalls of Typical Strategy Decisions

Typical Strategic Decision	Pitfalls Encountered
Entered early (before market began to grow rapidly).	Often lacked resources and skills required to overcome initial technical obstacles; sometimes became disenchanted when market did not develop quickly.
Limited initial commitment; mounted more substantial effort as industry developed.	Often allowed new competitors to establish formidable technical lead; became much more difficult for firm to develop viable long-term competitive position.
Established close organizational linkages between new and traditional product activities.	Decision-making attuned to competition in established industry/needs of traditional business. Problem of divided loyalties sometimes constrained firm's actions in new field.
Utilized product concepts/ways of competing that had been successful in established industry.	Wedded to "conventional wisdom." Often slow to recognize potential/implications of new product concepts and ways of competing.

obstacles will be especially difficult; developing the necessary competences will be a critical task. Further, management should consider the extent to which the firm has the patience and the will to overcome substantial technical difficulties, to work closely with early customers, and to make continuing investments even if the market develops slowly. If these qualities are lacking, a decision to enter early is likely to be regretted.

In terms of the magnitude of commitments, these firms often made limited early commitments—but then mounted a more vigorous effort as the industry developed. Perhaps this was due to the high initial levels of technical and market uncertainty, or organizational resistance to more substantial investments. However, the result in most cases was to fall behind more aggressive entrants, usually from outside the threatened industry. Thus, where a limited early commitment is being considered, management should weigh the impact of this on the firm's ability to compete later in the industry's development, possibly when other firms have established a commanding technical lead. Deferring heavy investments may be possible if the firm has very strong R&D and financial capabilities that it can bring to bear (e.g., GE in microwave ovens), or if it can find markets where competition is not established (e.g., portable electrics for Smith-Corona). But, in general, limited early commitments seem likely to impair the firm's ability to establish a viable long-term competitive position.

Concerning the degree of organizational separation, close linkages between new and traditional product activities appeared to be fairly common among these firms. Indeed, there will often be synergies to be gained from utilizing the existing organization for the new product. But in considering this approach, management should recognize that decisions concerning the new product will take place within an administrative system and culture that is attuned to competition in the established industry, and the needs of the traditional business. Management should be especially sensitive to the problem of divided loyalties that may result if the existing organization is used, and of the constraining effect that this may have upon the firm's actions in the young industry. In addition, if the decision is made to create an independent unit for the new product, senior managers should recognize that a high level of sponsorship and protection on their part may be required for this arrangement to succeed.

Finally, with respect to the strategy that is developed for competing in the new field, there seemed to be a tendency for firms to use the same basic approaches that were successful in the established industry—to fold the new product into the traditional strategy. It may be that they viewed the new product simply as a new way of meeting needs they had previously served (for example, cooking food with a microwave oven rather than a gas or electric oven). Nevertheless, such firms seemed to be slow in recognizing possibilities for employing new product concepts and ways of competing that might have allowed them to capitalize upon the new technology's potential.

The implication is that management should be willing to allow new or experimental strategies, and should carefully monitor the approaches that are pursued by other entrants. Appraising the strategies of new competitors may be especially important, since they will often possess different resources and skills, different ideas about how to compete, and little interest in the status quo. Overall, the conventional wisdom may no longer apply. Those who view an emerging young industry through the lens of their experience in an established, threatened industry may see what is familiar more clearly than what is different. A sensitivity to the problems encountered by the firms studied here may increase the chances for success.

Notes

1. The potential impact of major product innovations upon established industries was described by Schumpeter, who spoke of the "creative destruction" which "strikes not at the margins of the profits and the outputs of . . . existing firms, but at their foundations and their very lives." J.A. Schumpeter, *The Theory of Economic Development*, (Cambridge, MA: Harvard University Press, 1934).

2. Other strategic options that are oriented toward defending the traditional product against a substitution threat (for example, by attempting to improve its performance/cost, or by focusing on market segments that are less likely to adopt the new) are discussed by A.C. Cooper and D. Schendel, "Strategic Responses to Technological Threats," *Business Horizons*, 19(1), 1976, 61–69; and M. Porter, *Competitive Advantage*, (New York, NY: The Free Press, 1985).

3. M.L. Tushman and P. Anderson, "Technological Discontinuities and Organizational Environments," *Administrative Science Quarterly*, 31, 1986, 439–465.

4. See M.E. Porter, op cit., 1985; and D.J. Teece, "Profiting from Technological Innovation: Implications for Integration, Collaboration, Licensing and Public Policy." in D.J. Teece (Ed.), *The Competitive Challenge: Strategies for Industrial Innovation and Renewal* (Cambridge, MA: Ballinger, 1987).

5. R.N. Foster, *Innovation: The Attacker's Advantage* (New York: Summit Books, 1986); and A. Cooper and D. Schendel, op cit., 1976.

6. W.J. Abernathy and K.B. Clark, "Innovation: Mapping the Winds of Creative Destruction," *Research Policy*, 14, 3–22; M.L. Tushman and P. Anderson, op cit., 1986; and C.G. Smith. "Responding to Substitution Threats: A Framework for Assessment," *Journal of Engineering and Technology Management*, 7(1), 1990, 17–36.

7. W.J. Abernathy and J.M. Utterback, "Patterns of Industrial Innovation," *Technology Review*, 80(7), 1978, 41–47.

8. In CT Scanners, for example, sales fell substantially after the advent of restrictive "certificate of need" legislation. (The legislation required health care providers accepting federal reimbursements for medical procedures to obtain prior approval for all capital expenditures in excess of $100,000.)

9. C.G. Smith and A.C. Cooper, "Established Companies Diversifying Into Young Industries: A Comparison of Firms with Different Levels of Performance," *Strategic Management Journal*, 9(2), 1988, 111–121.

10. T. Levitt, "Marketing Myopia," *Harvard Business Review*, 38(4), 1960, 26–37.

11. R.E. Biggadike, *Corporate Diversification: Entry, Strategy, and Performance* (Cambridge, MA: Harvard University Press, 1979).

12. C.G. Smith, "Established Companies Diversifying into Young Industries: A Comparison of Firms With Different Levels of Performance," unpublished Ph.D. disseration, Purdue University, 1985.

13. B.A. Majumdar, "Innovations, Product Developments and Technology Transfers: An Empirical Study of Dynamic Competitive Advantage. The Case of Electronic Calculators," unpublished Ph.D. dissertation. Case Western Reserve University, 1977.

14. A Study of the Antitrust Laws, Hearings Before the Subcommittee on Antitrust and Monopoly of the Committee on the Judiciary, U.S. Senate, First Session, December 9, 1955, 3975.

15. R.N. Foster, op cit., 1986.

16. J.E. Tilton, *International Diffusion of Technology: The Case of Semiconductors* (Washington, DC: Brookings Institute, 1971).

17. W.R. Soukup, "Strategic Response to Technological Threat in the Electronics Components Industry," unpublished Ph.D. dissertation, Purdue University, 1979.

18. M. Magnet, "Timex Takes the Torture Test," *Fortune*, June 27, 1983, 112.

19. C.G. Smith, op cit., 1985.

20. Majumdar, op cit., 1977; and Creative Strategies International, *Electronic Calculators* (San Jose, CA: Creative Strategies, Inc., 1978).

21. J.E. Tilton, op cit., 1971.

22. Of these twenty firms, six withdrew from the given industry, and one surrendered its independence to become a division of another company. In each case, information obtained from secondary sources indicated that the decision was due to financial difficulties stemming from the firm's participation in the young industry. The other thirteen firms survived, but market share data and secondary source descriptions of business performance made it clear that they had been relegated to marginal or second tier positions by the end of the study period.

23. Interestingly, while GE was slow to respond to the countertop oven, it gradually learned what the customer was looking for. In 1978, the firm introduced the "Spacemaker" oven which was designed to be mounted below a kitchen cabinet. The design quickly gained acceptance among space-cramped apartment dwellers and helped GE to emerge as the industry leader in 1982.

CHAPTER THREE

ORGANIZATION ARCHITECTURES AND MANAGING INNOVATION

Coping with historical innovation cycles is an *organizational* challenge. Consider an analogy. Helping an employee diagnose a problem and find the most effective response to it is difficult. Implementing the response is more difficult, even if we know what the "best" response is. Growing employees who are capable of adapting to problems *in general* is the most difficult of all. Organizations that thrive in a world of technological change do so because they have developed an *architecture* that helps them innovate and adapt to situations no leader can fully foresee or understand. How do executives create such architectures?

Nadler and Tushman set forth a framework for thinking about the problem of adaptation and congruence, providing a clear, step-by-step method for analyzing organizational problems. We then examine the key elements of an organization's innovation architecture. Prahalad shows how to identify an organization's core competencies—both technical and nontechnical—and describes how successful firms build on those competencies to create value and define new competitive spaces. Katz explores how to manage jobs and careers, particularly those of technical professionals, who are often the bedrock of an innovative organization. O'Reilly describes and illustrates how successful companies create a culture that supports innovation and change. Pfeffer turns our attention to an organization's power structure, which governs the way firms get things done and manage the paradox of balancing innovation and efficiency. Schoonhoven and Jelinek question the old notion that innovative firms have flexible, "organic" structures while firms emphasizing efficiency rely on routine, "mechanistic" ways of organizing. The Silicon Valley companies they study combine organic and mechanistic structures, balancing these pressures through frequent reorganization that is not at all chaotic. Finally, Leonard-Barton discusses the way in which a successful architecture creates the seeds of its own destruction; a firm's very competencies become obstacles to change. She discusses how managers of development projects surmount this paradox, illustrating her ideas with a series of comparative case studies.

11

A Congruence Model for Organization Problem Solving

DAVID A. NADLER
MICHAEL L. TUSHMAN

CASE: REDESIGNING XEROX

By 1991, the Xerox Corporation had engineered one of the most successful comebacks in U.S. corporate history. Under the leadership of David Kearns, Xerox became the first major American company to overcome an onslaught by Japanese competitors. Its intense emphasis on quality had earned the company the first annual Baldridge National Quality award. A massive, systematic rethinking of its strategy was finally focusing Xerox's scattered energy and resources on a sharply defined identity as "The Document Company."

But much remained to be done. As Paul A. Allaire, Kearns' successor, surveyed the competitive landscape, he became convinced that Xerox had to become a fundamentally different company, "The New Xerox," if it was to capitalize on its new strategy. The problems inherent in the old Xerox were manifesting themselves in every aspect of the operation.

Xerox continued to operate as a traditional "functional machine." So many key decisions were bottled up in an all-powerful and overstaffed corporate office that it begged comparison with the old central committee of the Soviet regime. The complexity of the bureaucracy discouraged key operating managers from making important decisions on their own, and blurred accountability for the development and delivery of products and services. Everything about Xerox was too slow—technological innovation, the development of products to meet customer needs, and the capacity to adapt to new market conditions. People continued to focus too much on internal operations rather than customer demands.

In short, despite all its progress, Xerox continued to rumble along as an overly centralized, overly complex institution that had failed to unleash the initiative, judgment, creativity, and energy of its people. And its failure to radically change the way the company was run was costing it dearly in terms of productivity, innovation, customer service, and ultimately, financial performance.

UNDERSTANDING THE ORGANIZATION

What Paul Allaire faced in 1991 was a classic case of an organization in need of an entirely new architecture. Typically, the radical reshaping of an organization begins with strategy, as executives search for solutions to the problems created by shifting business conditions. But, as is often the case, Xerox discovered that its old organization was incapable of fully implementing the new strategy; the two just didn't fit.

Moreover, Allaire understood this wasn't a problem that would be solved by realigning the boxes on his table of organization. This was truly a case in which the company needed an entirely new architecture that would foster the development of new organizational capabilities—speed, innovation, accountability, customer focus, simplicity, self-management. If it was to succeed, Allaire knew the change would have to touch every aspect of the organization—its work, its structure, its people, and its culture.

What, then, are the most crucial issues and conflicting demands that should be uppermost in the mind of anyone setting out to redesign an organization? Two sets of questions are involved (see Table 11.1). First, the designer must consider what kind of structure will enable the organization to best manage its work to meet its strategic objectives. At the same time, the designer must take into consideration how those new structures will affect—and in turn, be influenced by—the culture, politics, and informal behavior patterns of the people who make up the organization.

TABLE 11.1. Two Design Perspectives

Strategy/Task Performance	Individual/Social/Cultural
Design supports the implementation of strategy	How will existing people fit into the design?
Design facilitates the flow of work	How will the design affect power relations among different groups?
Design permits effective managerial control	How will the design fit with people's values and beliefs?
Design creates doable, measurable jobs	How will the design affect the tone and operating style of the organization?

Both perspectives, structural and social, are valid, but considering either one in isolation from the other invariably leads to trouble. Focusing exclusively on structures and formal work processes is likely to lead to an organization that looks impressive on paper, but somehow can't get the right people doing the right work the right way. Indeed, restructuring done from that perspective is likely to create more problems than it solves. On the other hand, basing organizational design solely on social and cultural issues may well result in a working environment where people eagerly look forward to coming to work each day; the problem is that one day they may show up and find the place locked and shuttered because they were all happily working away on activities that had little to do with the organization's crucial strategic objectives.

These two perspectives reflect sharply contrasting notions about the fundamental purpose of organizations. The strategic/task perspective views organizations purely as mechanisms to perform the work required to execute strategies which, if successful, will create value and thereby benefit customers, shareholders, and society as a whole. It perceives organizations essentially as economic mechanisms created to achieve results that would be impossible for individuals working alone. The social/cultural perspective, on the other hand, sees organizations as devices for satisfying the needs, desires, and aspirations of various shareholders, both inside and outside the organization. Consequently, it views organizations as social organisms that exist for the express purpose of meeting individual needs, aiding in the exercise of power, and expressing individual or collective values.

Conceptual Models as Managerial Tools

Clearly, there is some validity to both perspectives. So it is essential for any manager engaged in redesign to always keep in mind the importance of balancing both underlying forces driving the organization. And to do that, he or she has got to have a conceptual tool for sorting out those complex interests while reconfiguring the elements of the organization. That tool is a model, or framework, that clearly identifies the important elements of the organization and describes their interrelationships. In a sense, a sound model provides a road map to the terrain of organizational behavior.

This isn't just a theoretical exercise. The selection of a model is critical; in reality, it will guide the designer's analysis and action. Here's why: Problem-solving in any organizational situation, including design, involves the collection of information, the analysis and interpretation of that information to determine specific problems and causes, and the development of appropriate responses. And the fact is that models—explicit ones we draw on flip charts or informal ones we might not even realize are tucked away in a corner of our mind—influence the kind of information we collect, guide our interpretation and analysis, and then help shape our alternative courses of action and ultimate decisions.

Over the past four decades, extensive research and the development of sophisticated theories based on real-life observation have dramatically expanded our understanding of organizational models. Today, that wealth of knowledge is available—and essential—to managers serious about solving organizational design problems.

Our intent here is to offer a general model for understanding the dynamics of organizations, and then to explore the model's implications for key decisions about organization design. The first step is to develop a framework for looking at the organization as a whole. And from that perspective, we propose a model that is based on the premise that the organization is a highly integrated system, and the components of that system must be constantly managed and sometimes restructured to ensure they maintain their proper relationship to one another.

That general model also builds on the notion of organizational architecture. It flows from the concept of an overall organizational design that provides the framework within which managers mold both the formal structure and social dimensions that shape the patterns and performance of work.

A Basic View of Organizations

When asked to "draw a picture of an organization," most managers usually respond the same way: They sketch some version of the traditional, pyramid-shaped structure that has characterized hierarchical organizations for centuries. It is a perspective based on a snapshot in time; it focuses on a stable configuration of jobs and work units as the most critical factor in an organization. Certainly, that view provides some helpful ways to conceptualize the enterprise. But in the long run, it's a seriously limited model, one that excludes such critical factors as leadership, external influences, informal behavior, and power relationships. The traditional model captures only a fraction of what's really going on within any complex organization.

Over the past two decades, there's been a growing tendency to replace the traditional static model with one that views the organization as a system. This new perspective stems from repeated observations that social organisms display many of the same characteristics as mechanical and natural systems. In particular, some theorists argue that organizations are better understood if they are thought of as dynamic and "open" social systems.[1]

Simply stated, a *system* is a set of interrelated elements—that is, a change in one element affects the others. Taking it one step further, an *open system* is one that interacts with its environment; it draws input from external sources and transforms it into some form of output. Think of a manufacturing plant as a simple example. The plant consists of different but related components—departments, job, technologies, and employees. It receives input from the environment—capital, raw materials, production orders, cost control guidelines—and it transforms them into products or output. Organizations display some of the same characteristics as any basic system. Some of the most important include the following.

Internal interdependence. Changes in one component of an organization frequently have repercussions for other parts; the pieces are interconnected. In our hypothetical manufacturing plant, a change in one element—the skill level of the employees, for example—will affect production speed, the quality of the output and the kind of supervision that's required.

Capacity for feedback. Information about the output can be used to control the system. Feedback allows organizations to correct errors and even change themselves. In our plant, if management receives information indicating that product quality is declining, it can use the information to identify factors in the system that are contributing to the problem. But human organizations are, after all, only human; the availability of feedback doesn't ensure it will be used to correct the system.

Equilibrium. When an event throws the system out of balance, its response is to try to right itself. If, for some reason, one work group in our plant were to suddenly increase its output, the system would be seriously out of balance. Its suppliers would have trouble keeping pace, while inventory would start piling up on the other end. One way or another, the system would work itself back into balance. Either the rest of the plant would somehow be changed to catch up with the accelerated pace or—and this is more likely—measures would be taken to slow down the overachievers, such as trimming staff or limiting its suppliers.

Equifinality. There is no "one best way" to structure the system—different configurations can lead to the same output. This point is particularly critical in the context of organizational design. The challenge is to identify alternative designs that are technically comparable in terms of the output they would produce, and then select the design that makes the most sense from the standpoint of social/cultural considerations.

Adaptation. For a system to survive, it must maintain a favorable balance of input and output transactions with the environment or it will run down. Our plant might have done a spectacular job of printing circuit boards until the day it shut down because the world was shifting to microchips or until we modified our skills, equipment, and technology to produce something the marketplace wanted. Very simply, any system must constantly adapt to changing environmental conditions.

Compared with the old pyramid-shaped drawing, the open systems theory certainly provides a more helpful way of thinking about organizations in more complex and dynamic terms. And yet, it's still too abstract to be of much practical use in solving day-to-day organizational problems; what we need is a model that encompasses the general concepts of the open system, but provides us with a more pragmatic problem-solving tool.

THE CONGRUENCE MODEL

We began building that model in the mid-1970s, benefiting from the earlier work of noted theorists such as Katz and Kahn and of Seiler and Lorsch, among others. At the same time, we were doing our work at Columbia University, Jay Galbraith at MIT, and Harold Leavitt at Stanford were preceeding along the same lines. From that collective thinking emerged a new way of looking at organizations, which has proven helpful to thousands of managers in scores of organizations over the years. We call this conceptual framework "the congruence model of organizational behavior."

According to this model, the components of any organization exist together in various states of balance and consistency—what we call "fit." The higher the degree of fit—"or congruence"—among the various components, the more effective the organization. This model clearly illustrates the critical role of interdependence within the system and places special emphasis on the transformation process—the means by which the organization converts input into output. So to understand how the entire organization works, it's essential to understand the input feeding into the system, the output it must produce, the elements of the transformation process, and the interaction among all of these components.[2]

Input

The organization's input includes the elements that, at any point in time, make up the "givens" with which it has to work. There are three main categories of input, each of which affects the organization in different ways.

1. *The Environment.* Every organization exists within and is influenced by a larger environment, which includes people, other organizations, social and economic forces, and legal constraints. More specifically, the environment includes markets (clients or customers), suppliers, governmental and regulatory bodies, technological and economic conditions, labor unions, competitors, financial institutions, and special interest groups. The environment affects the way organizations operate in three ways.

First, the environment makes *demands* on the organization. For example, customer requirements and preferences will play a large role in determining the quantity, price, and quality of the offerings—the products and/or services—the organization provides. Second, the environment often places *constraints* on the organization. These range from limitations imposed by scarce capital or insufficient technology to legal prohibitions rooted in government regulation, court action, or collective bargaining agreements. Third, the environment provides *opportunities*, for example, the potential for new markets resulting from technological innovation, government deregulation, or the removal of trade barriers. Any worthwhile analysis of an organization must take into account these environmental factors and the demands, constraints, and opportunities they imply.

2. *Resources.* The second element of input is the organization's resources. This includes the full range of assets to which it has access, including employees, technology, capital, and information. Resources may also include less tangible assets, such as the perception of the organization in the marketplace or a positive organizational climate. Two features are of primary interest to anyone analyzing the organization's capabilities and design alternatives: the relative quality of those resources or their value in the context of the current or future environment, and the extent to which resources can be reshaped or redeployed.

3. *History.* There is considerable evidence that the way an organization functions today is greatly influenced by events in its past. It's impossible to fully understand an organization's capacity to act now or in the future without an appreciation of the developments that shaped it over time—the strategic decisions, behavior of key leaders, response to past crises, and the evolution of values and beliefs.

Organization Strategy

Environmental conditions, organizational resources, and history cannot be changed in the short run—they are "givens" that provide the setting within which managers make strategic decisions. The term "organization strategy" describes the set of major decisions that allocate scarce resources against the demands, constraints, and opportunities of a given environment (review Table 11.2).

More specifically, strategy can be defined as explicit choices of products, markets, technology, and distinctive competence. Taking into consideration the threats and opportunities presented by the environment, the organization's strengths and weaknesses, and the pattern of performance suggested by the company's history, managers have to make decisions such as what products and services to offer to which markets, and how to distinguish their organization from others in ways that will provide sustainable competitive advantage. Then these general, long-term strategic objectives must be refined into a set of internally consistent, short-term objectives and supporting strategies.

For managers, these decisions about offerings, markets, and competitive advantage are crucial. Organizations that make the wrong strategic decisions will underperform or fail. No amount of organization design can prop up an ill-conceived strategy. But by the same token, no strategy, no matter how dazzling it looks on paper, can succeed unless it's consistent with the structural and cultural capabilities of the organization. The manager's challenge, consequently, is to design and build an organization capable of accomplishing the strategic objectives.

In practice, strategy flows from a shared vision of the organization's future—a coherent idea

TABLE 11.2. Organizational Context

Input	Environment	Resources	History	Strategy
Definition	All factors, including institutions, groups, individuals, events, and so on, that are outside the organization being analyzed but that have a potential impact on that organization	Various assets to which the organization has access, including human resources, technology, capital, information, and so on, as well as less tangible resources (recognition in the market, and so forth)	The patterns of past behavior, activity, and effectiveness of the organization that may affect current organizational functioning	The stream of decisions about how organizational resources will be configured to meet the demands, constraints, and opportunities within the context of the organization's history
Critical Features for Analysis	1. What demands does the environment make on the organization? 2. How does the environment put constraints on organizational action?	1. What is the relative quality of the different resources to which the organization has access? 2. To what extent are resources fixed rather than flexible in their configuration(s)?	1. What have been the major stages or phases of the organization's development? 2. What is the current impact of such historical factors as strategic decisions, acts of key leaders, crises, and core values and norms?	1. How has the organization defined its core mission, including the markets it serves and the products/services it provides to these markets? 2. On what basis does it compete? 3. What supporting strategies has the organization employed to achieve the core mission? 4. What specific objectives have been set for organizational output?

Strategy
Objectives/Tactics

of its aspirations, its competitive strengths, its relative position of leadership in the market, and its operating culture. Beyond the general view of the future, however, strategy is closely linked to specific, measurable objectives. The guiding aspirations are molded into a strategic intent—a set of specific goals that must be reached if the organization is to fulfill its guiding vision.

As they embark on periods of major change, successful organizations begin by molding and articulating that vision, strategy, and set of objectives, as demonstrated by Xerox's Leadership Through Quality program of the mid-1980s and the Document Company vision of the 1990s. At Ford, the vision was articulated as Ford 2000, a set of goals for market leadership and superior quality. At Corning, Inc., in the 1980s, the redesign work followed a guiding vision known as the Corning Wheel, an entirely new way of looking at the relationships involving the conglomerate's five major areas of activity. And at Sun Microsystems, a radical new architecture built around the notion of Sun and its "planets" was designed in the early 1990s; the idea, which proved immensely successful, was to leverage the competitive potential of its internal operations by reconfiguring its traditional internal divisions as a loosely-linked network of semiautonomous business involving hardware, software, sales, service, and distribution—the "planets" revolving around a central coordinating unit—the "Sun."

While strategic designs differ, their conceptual foundations are the same: Strategic design, in its fullest sense, flows directly from clarity on strategy, objectives, and vision; from a strategic intent that clearly articulates a vision and aspirations for the future course of the organization.

Output

In our organizational model, "output" is a broad term describing what the organization produces, how it performs, and how effective it is. It involves not only the organization's effectiveness at creating products and services or providing a certain level of economic return, but also refers to the performance of individuals and groups within the organization.

There are three criteria for evaluating performance at the organizational level:

Goal attainment or how successfully the organization meets the objectives specified by its strategy

Resource utilization or how well the organization uses its available resources to meet its objectives (and how successfully it meets those goals by building resources rather than "burning them up")

Adaptability or the organization's success at continuously repositioning itself to seize the opportunities and ward off the threats posed by a constantly changing environment.

Of course, the performance of groups and individuals contribute directly to overall organizational performance. But in certain situations, changes in both individual and collective attitudes and capabilities such as the satisfaction, stress, dysfunctionally poor morale, or the acquisition of important experience can all be seen as output or the end results of the transformation process.

For the manager involved in organization design, the central issue is the identification of "performance gaps." That entails a comparison of the specific objectives articular in the "strategic intent"—which we will discuss later in this chapter in more detail—with actual output. The "gaps" help spotlight those activities where output is falling short of objectives, and provide essential guidelines in figuring out where in the organization the redesign efforts need to be focused.

The Organization as a Transformation Process

The heart of the model is the transformation process, embodied in the organization, which draws on inputs from the environment, resources, and history to produce a set of outputs. The organization contains four key components: the work, the people who perform the work, the formal arrangements

TABLE 11.3. The Four Organizational Components

Component	Task	Individual	Formal Organizational Arrangements	Informal Organization
Definition	The basic and inherent work to be done by the organization and its parts	The characteristics of individuals in the organization	The various structures, processes, methods that are formally created to get individuals to perform tasks	The emerging arrangements including structures, processes, relationships
Critical Features of Each Component	Degree of uncertainty associated with the work, including such factors as interdependence and routineness Types of skills and knowledge demands the work poses Types of rewards the work inherently can provide Constraints on performance demands inherent in the work (given a strategy)	Knowledge and skills individuals have Individual needs and preferences Perceptions and expectancies Background factors Demography	Grouping of functions, structure of units Coordination and control mechanisms Job design Work environment Human resource management systems Reward systems Physical location	Leader behavior Norms, values Intragroup relations Intergroup relations Informal working arrangements Communication and influence patterns Key roles Climate Power, politics

that provide structure and direction to their work, and the informal arrangements—sometimes referred to as "culture"—that reflect their values, beliefs and patterns of behavior (See Table 11.3). For managers involved in organizational design, the issue is finding the best way to configure those components to create the output necessary to meet the strategic objectives. To do that, it's essential to understand each of the components of the organization and its relationship to the others.

1. *The Work.* We use this general term to describe the basic and inherent activity engaged in by the organization and its units in furtherance of its strategy. Because the performance of this work is one of the primary reasons for the organization's existence, any analysis from a design perspective has to start with an understanding of the nature of the tasks to be performed and patterns of work flow, as well as an assessment of the more complex characteristics of the work—the knowledge or skills it demands, the rewards it offers, and the stress or uncertainty it involves.

2. *The People.* From the standpoint of organization "fit," the key issue is identifying the important characteristics of the people responsible for the range of tasks involved in the core work. That means looking at the workforce in terms of skills, knowledge, experience, expectations, behavior patterns, and demographics.

3. *Formal Organizational Arrangements.* These are the explicit structures, processes, systems, and procedures developed to organize work and guide the activity of individuals in their performance of activities consistent with the strategy.

4. *The Informal Organization.* Coexisting alongside the formal arrangements is a set of informal, unwritten guidelines that exert a powerful influence on the behavior of groups and individuals. Also referred to as organizational culture and operating environment, the informal organization encompasses a pattern of processes, practices, and political relationships that embody the values, beliefs, and accepted behav-

ioral norms of the individuals who work there. It's not unusual for informal arrangements to actually supplant formal structures and processes that have been in place so long that they've lost their relevance to the realities of the current work environment.

THE CONCEPT OF CONGRUENCE

In our model, which views the organization as a system, the components themselves are less important than the relationships among them—and more importantly, the ways in which those relationships affect output. At any given time, each organizational component maintains some degree of congruence with each of the others. The congruence between two components is defined as the degree to which the needs, demands, goals, objectives, and/or structures of one component are consistent with the needs, demands, goals, objectives, and/or structures of another component.

Congruence, therefore, is a measure of how well pairs of components fit together. Consider, for example, two components: the task and the individual. At the simplest level, the task presents skill and knowledge demands on individuals who would perform it. At the same time, the individuals available to do the tasks have certain characteristics—including their levels of skill and knowledge. The greater the fit between the individual's characteristics and the demands of the task to be performed, the more efficient the performance is likely to be. Obviously, the fit between the individual and the task involves more than just knowledge and skill; performance will be influenced by a wide range of factors such as job fulfillment, anxiety, uncertainty, expectation of rewards, and so forth. Similarly, each congruence relationship in the model has its own specific characteristics, though research and informed theory can guide the assessment of fit in each relationship. For an overview of the critical elements of each congruence relationship, see Table 11.4.

The Congruence Hypothesis

The organization, as a whole, displays a relatively high or low degree of system congruence in the same way that each pair of components has a high or low degree of congruence. The basic hypothesis of the model is this: Other things being equal, the greater the total degree of congruence or fit among the various components, the more effective the organization.[3] In this context, effectiveness is defined as the degree to which actual organizational output matches the output specified by the strategy.

The basic dynamic of congruence sees the or-

TABLE 11.4. Definitions of Fit Among Components

Fit	Issues
Individual/Organization	How are individual needs met by the organizational arrangements? Do individuals hold clear perceptions of organizational structures? Is there a convergence of individual and organizational goals?
Individual/Task	How are individual needs met by the tasks? Do individuals have skills and abilities to meet task demands?
Individual/Informal Organization	How are individual needs met by the informal organization? How does the informal organization make use of individual resources consistent with informal goals?
Task/Organization	Are organizational arrangements adequate to meet the demands of the task? Do organizational arrangements motivate behavior that is consistent with task demands?
Task/Informal Organization	Does the informal organization structure facilitate task performance? Does it help meet the demands of the task?
Organization/Informal Organization	Are the goals, rewards, and structures of the informal organization consistent with those of the formal organization?

ganization as most effective when its pieces fit together. If we also consider strategy, this view expands to include the fit between the organization and its larger environment; an organization is most effective when its strategy is consistent with its environment (in light of organizational resources and history) and when the organizational components are congruent with the tasks necessary to implement that strategy.

One important implication of the congruence hypothesis is that diagnosing organizational problems requires describing the system, identifying problems, and analyzing fit to determine the sources of poor fit. The model also implies that various components can be configured in different ways and achieve some degree of desired output. Therefore, the question becomes one not necessarily of finding the "one best way of managing," but determining the combinations of components that will lead to the greatest degree of congruence.

Another way to think about congruence is to think of it in terms of architecture—more specifically, in this case, the architecture of technology. It involves three elements:

The Hardware. The computers, monitors, keyboards, modems, disk drives, servers, cables, printers, scanners, and other physical pieces of equipment that make up a computer system

The Software. The encoded sets of instructions that allow the hardware components to function, both individually and collectively as a system or network

The Human Interface. The people who actually select and use the software that makes the hardware perform its assigned functions.

Clearly, the effectiveness of each component is dependent on how well it meshes with the rest of the system. Neither hardware nor software is of any use whatsoever in the absence of the other. Some hardware and software are simply incompatible; together, they're as good as useless. Beyond that, there are all kinds of software that will run on all kinds of hardware—but some combinations work much more smoothly than others. And even if you've found the optimal combination of hardware and software, it doesn't do any good unless the people using it understand how to use the software and the equipment, have a clear idea of the work they're supposed to perform, and are motivated to do it in a swift and conscientious way.

The parallels with organizational architecture are clear. In many organizations, the term "hardware" has become synonymous with the technical/structural axis of the model—the work and the formal structural arrangements. "Software" is analogous to the informal operating environment; like software, you can't actually see or feel to touch it, but you can easily see its powerful impact on the way both the machine and organization operate. And in both cases, it is the skills, motivation, and guiding values and beliefs of the individuals involved who determine how productively the work gets done. In both forms of architecture—technological and organizational—it is the proper fit among all the components that determines the effectiveness of performance.

While the congruence model and the concept of architecture provide a general framework, organization designers will need other, more specific models to help define high and low congruence, particularly in the relationship between the formal organizational arrangements and the other components. In Chapter 4, we will describe a model that focuses specifically on those crucial relationships.

In summary, then, we have just described a general model for analyzing organizations (see Figure 11.1). It embodies a view of the organization as a system and, in particular, as a process that transforms input into output; that process, in turn, is driven by the relationships among the four basic components. The critical dynamic is the fit or congruence among the components. We now turn our attention to how this model can be used to analyze organizational problems.

A Process for Organizational Problem Analysis

Organizations typically face a set of constantly changing conditions. Consequently, effective man-

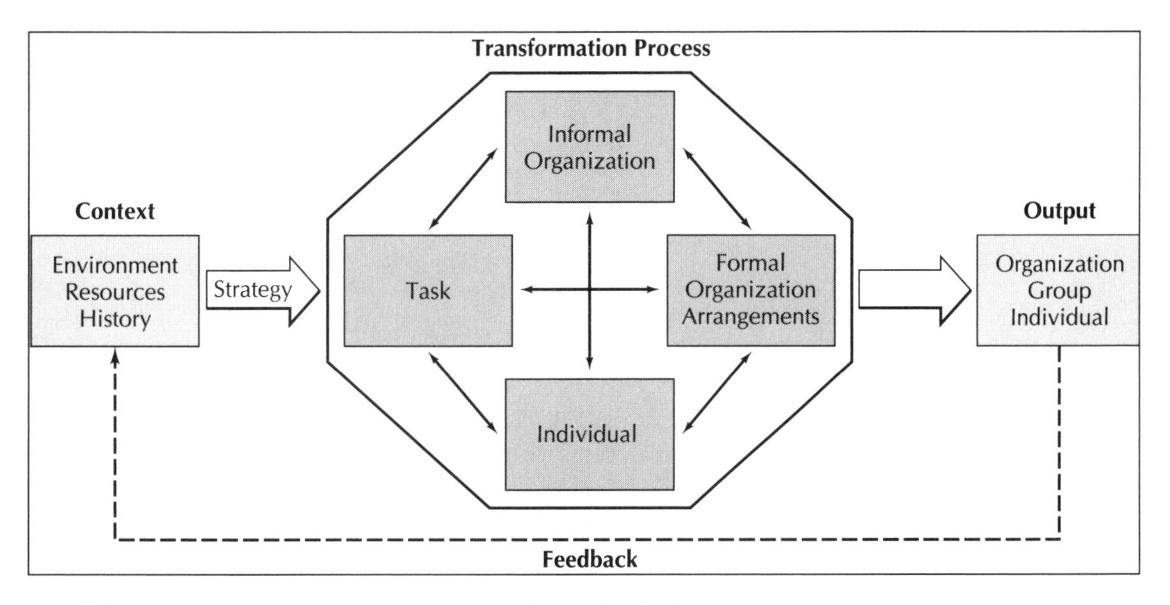

Fig. 11.1. A congruence model for diagnosing organizational behavior.

agers are continually working to identify and solve problems. That entails gathering data on performance, matching actual performance against goals, identifying the causes of problems, selecting and developing action plans, and finally, implementing and then evaluating the effectiveness of those plans. Any organization's long-term success requires ongoing problem-solving activities along these lines.

Our experience has led us to develop a general approach to using the congruence model for solving organizational problems (see Table 11.5). It involves the following steps:

1. *Identify symptoms.* In any situation, initial information may reveal symptoms of poor performance without pinpointing real problems and their causes. Still, this information is important because it focuses the search for more complete data.
2. *Specify input.* With the symptoms in mind, the next step is to collect data concerning the organization's environment, its resources, and critical aspects of its history. Input analysis also involves identifying the organization's overall strategy—its vision, supporting strategies, and objectives.
3. *Identify output.* The third step is to analyze the organization's output at the individual, group, and organizational levels. Output analysis involves defining precisely what output is required at each level to meet the overall strategic objectives, and then collecting data to measure precisely what output is actually being achieved.
4. *Identify problems.* The next step is to pinpoint specific gaps between planned and actual output, and to identify the associated problems—organizational performance, group functioning, or individual behavior, for example. Where information is available, it's often useful to identify the costs associated with the problems or with the failure to fix them. The costs might be in the form of actual costs, such as increased expenses or missed opportunities, such as lost revenue.
5. *Describe organizational components.* This is where analysis goes beyond merely identifying problems and begins focusing on causes. It begins with the collection of data concerning each of the four major components of the organization. A word of caution: Not all problems have

TABLE 11.5. Steps in Organizational Problem Analysis

Step	Explanation
1. Identify symptoms	List data indicating possible existence of problems.
2. Specify input	Identify the system. Determine nature of environment, resources, and history. Identify critical aspects of strategy.
3. Identify output	Identify data that defines the nature of output at various levels (individual, group/unit, organization)—should include desired output (from strategy) and actual output being obtained.
4. Identify problems	Identify areas where there are significant and meaningful differences between desired and actual output. To the extent possible, identify penalities, i.e., specific costs (actual and opportunity costs) associated with each problem.
5. Describe components of the organization	Describe basic nature of each of the four components with emphasis on its critical features.
6. Assess congruence (fit)	Analyze relative congruence among components (draw on more specific models as needed).
7. Generate hypotheses	Analyze to associate fit with specific problems.
8. Identify action steps	Indicate what possible actions might deal with causes of problems.

internal causes. Some result from external developments—a shift in the regulatory environment, a new competitor, a major technological breakthrough—that make existing strategies insufficient or obsolete. It's important to consider strategic issues before focusing too narrowly on organizational causes for problems; otherwise, the organization is in danger of merely doing the wrong thing more efficiently.

6. *Assess congruence (fit).* Using the data collected in step five, in the context of applicable models, the degree of congruence among the various components is assessed.
7. *Generate hypotheses about problem causes.* The next step is to look for correlations between poor congruence and problems that are affecting output. Once these problem areas have been identified, available data is used to test whether poor fit is, indeed, a key factor influencing output and a potential leverage point for forging improvement.
8. *Identify action steps.* The final stage is to identify action steps, which might range from specific changes aimed at relatively obvious problems to more extensive data collection designed to test hypotheses about complex problems. Additionally, problem solving involves predicting the consequence of various actions, choosing a course of action, implementing it, and evaluating its impact.

OPENING CASE REVISITED

Faced with the awesome task of redesigning the basic architecture of Xerox, Allaire began by appointing a team of managers from throughout the corporation to explore alternative design models used by other organizations and then present him with four recommendations for redesigning Xerox. After the "Future-techture" group completed its assignment, Allaire and his senior team refined Xerox's basic business and organization model. In essence, Xerox stood the organization on its side; instead of organization around the traditional functions of product development, production, sales, and service, it created a set of independent business units, each offering various products and services aimed at specific segments of the market. The focus was shifted radically from internal operations to customer needs in an entirely new organization architecture—both new hardware and software.

After deciding on an overall design, Allaire then appointed a second team, the Organization Transition Board (OTB), made of executives two and three levels below the corporate office, to flesh out the general plan and formulate the tactics for implementing it.

At the same time, the OTB devoted considerable time and energy to formulating a list of key dimensions required of managers in the new Xerox. These characteristics—some of them markedly different from the styles and traits exhibited by Xerox managers in the past—then became the guideline for assessing and assigning managers to top positions in the new organization.[4]

The Xerox redesign process succeeded, in large part, because the executives guiding it were thoroughly conscious of the need to make changes—and consistent changes—in every aspect of the organization. They not only restructured the formal structures and processes—the work and formal arrangements, in terms of our model—but also paid close attention to defining the skills, characteristics, and management styles essential for the managers responsible for building a new operating environment.

The Xerox experience illustrates the relationship of the congruence model to the task of redesign. Management's first job, in most instances, is to develop strategies in the context of the organization's history, resources, environment, and internal capacities. The second step is to implement strategies through the creation, shaping, and maintaining of an organization. This means constantly defining the key tasks to be performed, making sure that individuals are motivated and capable of performing those tasks, and developing formal and informal organizational arrangements consistent with the strategy, work requirements, culture, and people.

Notes

1. Katz, D. and R. Kahn. *Social Psychology of Organizations*. New York: Wiley, 1966; K. Weick. *Social Psychology of Organizing*. Reading, Mass.: Addison-Wesley, 1969; March, J. and H. Simon. *Organizations*. New York: Wiley, 1959.

2. Other similar approaches include: J. Galbraith. *Organization Design*. Reading, Mass.: Addison-Wesley, 1977; Peters, T. and R. Waterman. *In Search of Excellence*. New York: Harper & Row, 1982; Leavitt, Hal. "Applied Organizational Change in Industry." In March, J., *Handbook of Organizations*. Chicago: Rand McNally, 1965; N. Tichy. *Managing Strategic Change*. New York: Wiley, 1983.

3. Other approaches and research on the notion of fit or congruence include: Miles, R. and C. Snow. *Fit, Failure, and the Hall of Fame*. New York: Free Press, 1994; Miller, D. "The Architecture of Simplicity." *Academy of Management Review*, 18:116–38, 1993; Greshov, C. "Exploring Fit and Misfit with Multiple Contingencies." *Administrative Science Quarterly*, 34:431–53, 1989; Donaldson, L. *In Defence of Organization Theory*. Cambridge, Mass.: Cambridge University Press, 1985.

4. See also Howard, R. "The CEO as Organization Architect: An Interview with Xerox's Paul Allaire." *Harvard Business Review*, October 1992.

12

The Role of Core Competencies in the Corporation

C. K. PRAHALAD

The debate about the competitiveness of Western firms in a wide variety of industries inevitably triggers debates about technology policy and investment levels in technology. While the preoccupation with investment in R&D and the list of "critical technologies" are important inputs to that debate, the real issue for most Western firms is internal capacity for new business development. Technology leadership is but an enabler.

Therefore, the debate on technology should benefit from a top management focus—the perspective of general management and, more importantly, that of the CEO. Such a perspective would provide a very different vantage point from which to examine the role of technology in the growth of the company; specifically, the underlying rationale and logic for growth in a globally competitive environment. The CEO perspective, as opposed to that of the chief technical officer (CTO), will allow us to put technology investment in a business perspective.

I shall illustrate my ideas with examples primarily drawn from high-volume electronics. I believe these concepts are equally applicable to industries as diverse as agricultural processing, chemicals, and, at least in some cases, defense electronics. I do not claim that the concepts presented here apply universally to all industries. I have found them useful in a wide variety of industries.

In this article, I present the conceptual framework. A previous article, by Presbylowicz and Faulkner, described a process for operationalizing these ideas in the context of a specific firm.[1]

THE MANAGEMENT SCORECARD

It will be useful to start with a scorecard for Western top management during the last 35–40 years. If we consider the period 1950–80, immediately following World War II—in a wide variety of industries, from automotive, to semiconductors, to tires, to medical systems, to earth-moving equipment, and to reprographics—almost all the world leaders were Western companies. For example, in the automotive industry General Motors and Ford dominated the world. In the merchant semiconductor business Texas Instruments and Motorola were the leaders. However, if we consider the period 1980–1990 and ask: "Who is providing the intellectual leadership in these industries?" we come up with a very different list. For example, the intellectual leadership in the automotive industry is increasingly provided by companies such as Honda and Toyota, be it in terms of use of new technology, new features, new standards of quality, customer orientation and service, and price-performance relationships. The changing pattern of

Reprinted from *Research/Technology Management*, November-December 1993, pp. 40–47, with permission from the Industrial Research Institute, Washington D.C.

TABLE 12.1. Industrial Leadership Shifted in a Variety of Industries during 1975–1985

Industry	Leaders 1950–80	Challengers 1980–85
Automotive	GM Ford	Toyota Nissan Honda
Semiconductors	TI Motorola	NEC Toshiba Fujitsu Hitachi
Tires	Goodyear Firestone*	Michelin Bridgestone
Medical systems	GE Philips Siemens	Hitachi Toshiba
Consumer electronics	GE* RCA* Philips	Matsushita Sony
Photographic	Kodak	Fuji
Xerography	Xerox	Canon

*Has been acquired

industrial leadership in a wide variety of industries is illustrated in Table 12.1.

We should reflect on this change: *How did the intellectual and market leadership in so many industries shift in such a short period of time?* In view of the many advantages U.S. firms enjoyed during the period 1950–1980—such as superior technology, larger size, global distribution, reputation, and management know-how—how did the intellectual leadership slip away? If it had happened in just one industry, we could attribute the decline to a wide variety of external factors, including the role of MITI in Japan, the cost of capital, and the attitude of unions. But why and how did the U.S. lose leadership in so many industries during the 1975–1985 period?

This loss of intellectual leadership is just one part of the scorecard. Consider *internally generated growth* during the same period. Let us focus on the top 25 electronics companies during the decade of the 1980s. The giants of 1980 were IBM, GE, ITT, Philips, and Siemens. Most Japanese firms were small. Hitachi at around $12 billion was about half the size of IBM. Sony at $3 billion was one-fifth of Philips. And high-volume electronics was a "fortunate" industry during this decade. The industry overall experienced an average growth rate of 14 percent through the decade of the 1980s; but if we consider the rate of growth of various companies in this industry, we see wide variations: 12 percent for IBM, 11 for GE, 5 for Philips, 8 for Siemens. But Hitachi grew at 17 percent, Matsushita at 16, Toshiba at 15, NEC at 23. This is the annual compounded growth rate during the decade. The disparity in growth rates among firms, essentially in the same industry category, demands an explanation: Why do some firms grow at 5 percent and others at 20 percent for 10 years in the same industry?

Large Western firms have not fared well on the opportunity management (growth) dimension. Let us use a simple metric to evaluate the scorecard of a company's capacity to grow. Consider the period 1985–1991. If sales were 100 for the year 1985, the capacity-to-grow index during the period 1985–1991 can be computed as follows:

> Capacity-To-Grow Index = (sales revenue for 1991 − acquisitions during the period 1985 to 1991 − inflation + divestments during the period 1985–1991).

By eliminating growth through acquisitions and inflation, this index measures *internal capacity to grow*. Unfortunately, for most large firms, the index is not very flattering. Top managers must ask themselves the following questions: What is the opportunity that we have lost? Why? Who is responsible for it? Who should pay the price for lost opportunities?

Let us take this analysis one step forward with a paired comparison of companies: Westinghouse, Hitachi and General Electric. Westinghouse grew from $8.5 billion to $12 billion during the decade. Westinghouse primarily divested itself of major businesses and acquired a few. GE also aggressively pursued a strategy of portfolio shuffling—acquisitions and divestments. However, Hitachi grew through internal development. This pattern is not different if we consider RCA, Sony and Matsushita, or GTE and NEC.

Paired comparisons allow us to further re-examine our scorecard. How can one company

(Hitachi) grow from $12 to $50 billion, primarily through internal development, during a decade, while another company (Westinghouse), with a similar starting portfolio, grows from $8 billion to $12 billion only? The market opportunities were similar around the globe and the technological capabilities were comparable. In fact, all during the 1970s and 1980s, U.S. firms like Westinghouse led in technology in almost every field.

A reflection on this scorecard suggests that explanations must focus not on the differences in starting resource positions but on the differences in the ability of managers of firms to leverage corporate resources. We need answers to such questions as: Is our orientation to management and exploitation of technology and market opportunities appropriate? Is there a distinctly different underlying logic (as compared to the financial portfolio logic) to profitable growth? What is that logic? And what should top managers do in order to change their orientation from cost-cutting to opportunity management and growth?

RETHINKING THE SCORECARD

The real issue for the 1990s is *growth.* That is the agenda for top management—not down-sizing, not restructuring and de-cluttering of organizations. Restructuring without rethinking the role of management inevitably leads to further restructuring. Many large firms have restructured themselves more than once in the last ten years and the problems do not seem to go away. If growth and new business development are the real issues, *value creation* will be the scorecard for managers during this decade. This scorecard consists of two parts: 1) managing the *performance gap*; i.e., improving performance across a wide variety of dimensions such as quality, cost, cycle time, productivity, and profitability; 2) simultaneously, managers should focus on the *opportunity gap*, profitably deploying resources to create new markets, new businesses and a sense of broad strategic direction.

The twin aspects of value creation are illustrated in Figure 12.1. During the past decade, man-

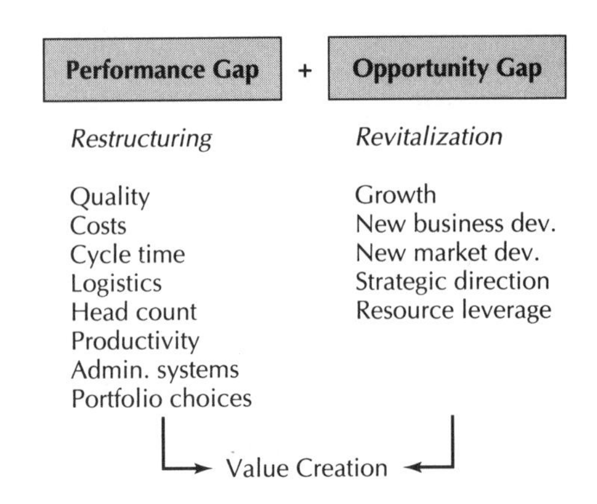

Fig. 12.1. Value creation is not just catching up with the performance—it is the active management of the opportunity gap as well.

agement attention has been primarily focused on the performance gap. It is the legitimate task of the management to fix problems of profitability, cost, quality, cycle time, logistics, and productivity. Managing the performance gap, if done well, ought to create a *large investment pool.* The question for managers, then, is how a redeploy the investment pool, created by focusing on the performance gap, in the pursuit of new opportunities for growth. To create value, concerns for operational improvement (performance gap) and strategic direction (opportunity gap), must coexist.

In this article, I focus on the opportunity gap through revitalization and growth. New business development, growth, new market development, and leveraging of corporate resources are an integral part of the value creation process. So I start with the assumption that value creation is not just catching up with the performance gap; it is the active management of the opportunity gap as well. Is there an underlying logic to opportunity management?

A NEW FRAMEWORK FOR VALUE CREATION

I suggest that the logic of opportunity gap management consists of at least four interlinked parts, as shown in Figure 12.2

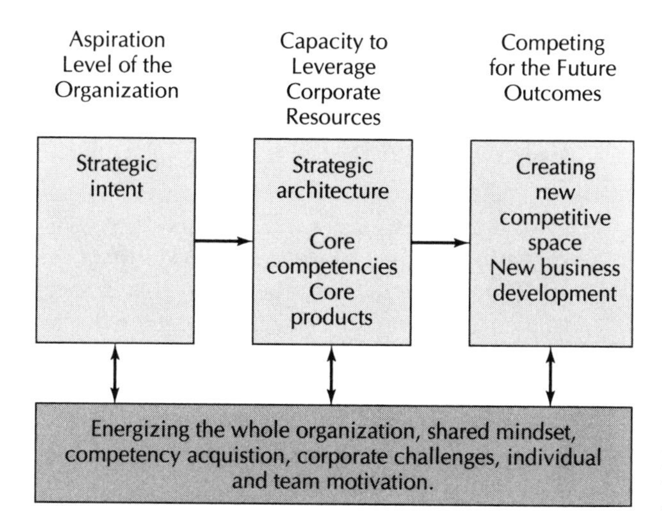

Fig. 12.2. Opportunity gap management begins with establishing an aspiration level (strategic intent) for the organization.

1. How can top managers establish an aspiration level (strategic intent) for the organization? Motivation for change results from an aspiration that all employees can identify with and feel committed to. Aspirations must represent a stretch and must by definition exceed the current resources of the company. Therefore, by design, strategic intent must cause a "misfit" between aspirations and current resources and current approaches to using resources. The aspiration must focus the energies of the organization toward innovation (changing the rules of the game) in the way the firm competes.

2. A high aspiration level (compared to the resources available) leads to the need for resource leverage. The issue for managers is: How do you create the capacity in a large organization to leverage corporate resources? The process of resource leverage is accomplished through the development of a *strategic architecture* (a way to capture the pattern of likely industry evolution), identifying *core competencies* and *core products*. Reusability of invisible assets, as well as core products, in new and imaginative configurations to create new market opportunities is at the heart of the process of leverage.

3. An internal capacity to leverage resources is a prerequisite for inventing new businesses, creating *new competitive space*. This is competing for the future and requires a framework for identifying new opportunities, focusing on functionalities rather than on current products and services, and dramatically altering the price-performance relationships in an industry.

4. This new approach is not just a technical task or a senior management task—it is a task for the whole organization. The role of top management, therefore, is essentially one of energizing the whole organization—all people, at all levels, in all functions, and in all geographies. It involves developing a shared mindset and shared goals, and developing strategies for acquiring competency. Senior managers must focus on such questions as: How do we stretch the imagination of the total employee pool? How do we challenge the organization? How do we focus on individual and team motivation?

Using this framework, I will examine one building block at a time, starting with strategic intent.[2]

AN OBSESSION WITH WINNING

Strategic intent is a way of creating an obsession with winning that encompasses the total organization (all levels and all functions). It is a shared competitive agenda, sustained over a long period of time, for global leadership. Extraordinary accomplishment is often based on a clearly articulated strategic intent.

The U.S. has experienced the power of a clear strategic intent. Consider the Apollo program: "Man on the moon by the end of the decade" was a "stretch" target. It meant global leadership and the domination of space. The goal was competitively focused; Russians were the enemy. It was very clear. While the goal was clear, managers of the project had to discover the means, and a lot of new technologies had to be developed under enormous time pressure. How do we account for the inventiveness that was characteristic of the NASA efforts during the 1960s? Why is the "spirit" of the Apollo program not replicable in firms?

Consider some specific company examples, such as NEC's goal of *C&C* (computers and communications) or Kodak's strategic intent: to remain a *world leader in imaging*, not just chemical or electronic imaging, but the creative combination of both. Once *imaging* is accepted as the strategic intent, debate inside the company on whether chemical imaging is superior to electronic or vice versa, dies down. The focus shifts to creating new hybrids—products and services that creatively combine both capabilities. In Komatsu, the strategic intent was to encircle Caterpillar. Strategic intent may be stated in different ways in different firms—from C&C to Leadership in Imaging. But in all cases, it must represent an agenda for the whole company, not just for a function—be it manufacturing, marketing, top management, technologists, or sales.

Strategic intent provides a basis for stretching the imagination of the total organization and a focus for developing "barrier-breaking" initiatives.

FRAMEWORK FOR LEVERAGE

Once we have developed a shared aspiration, we need a framework for leveraging corporate resources that is consistent with the strategic intent. We start with a strategic architecture, which is a way of developing a point of view regarding the evolution of an industry. How will the interface with customers change? What are the new technological possibilities? How are our current and future competitors positioning themselves to approach this industry? Strategic architecture is a *distillation of a wide variety of information.* It is a way of capturing major discontinuities and trends in the industry. It does not attempt to identify a specific product or business opportunity, but captures the direction and major likely milestones. It provides a framework for focused resource allocation over a long period, allows managers to maintain consistency in their efforts, and provides a logic for managing linkages across business units in a large company.

Strategic architecture can be used to identify targeted acquisitions, and alliance partners. Most important of all, it is a useful framework for effectively managing innovation. The underlying assumption is that innovation is a line job and not a staff job. Further, innovation cannot be left to skunk works or "off-line activities" such as internal entrepreneurship or internal venture teams. *Innovation is the fundamental job of a general manager.* We need to develop a framework in which innovation can be planned and managed. Strategic architecture provides one such framework for proactively managing the innovation process.

An example of strategic architecture is NEC's concept of C&C, or the convergence of computing and communications, shown in Figure 12.3. The evolution of computing driven by the need for decentralized processing as a trend, coupled with the changes in communication and component technologies, lead to the convergence called C&C.[3] Obviously, in this architecture, there are no specific product plans, but the basic milestones are obvious. Accomplishing the aspiration of C&C is a stage-managed process. For over 15 years, the broad framework of C&C represented by the strategic architecture was used as an organizing idea. The value of such a framework to provide consistency and direction to technical resource allocation is obvious.

The fact is that anyone in this industry could have drawn this picture. However, even though all

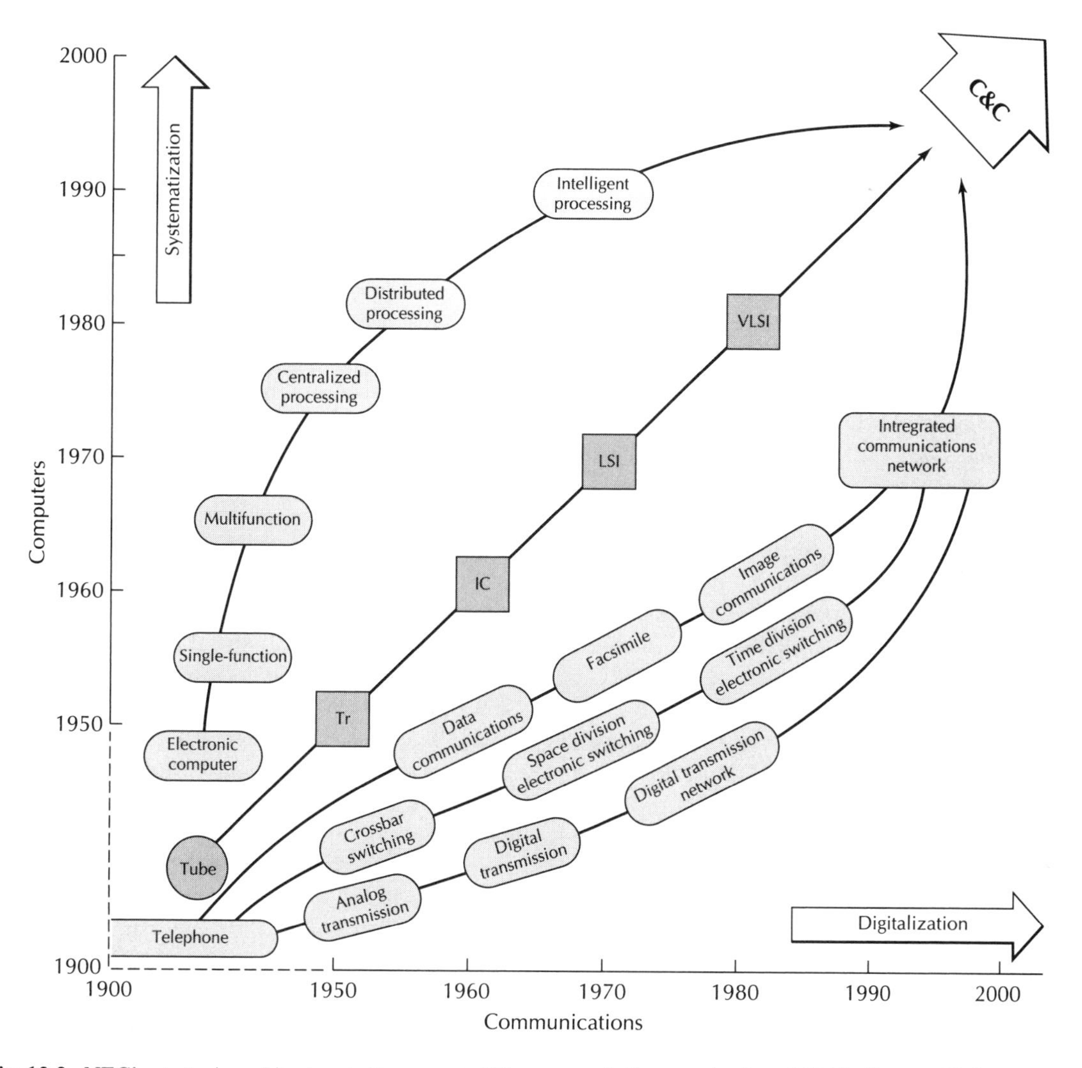

Fig. 12.3. NEC's strategic architecture is its concept of Computers & Communications (C&C). (*Source*: Koji Kobayashi, *Computers and Communications*.)

of us could visualize C&C, why didn't other companies use it as the organizing and stage-setting concept for mobilizing the efforts of the total company? Why did C&C not provide a logic for resource allocation? These are the critical questions. It is not enough for a small group of technical people to have a bold concept. There has to be widespread agreement and understanding of the concept. An architecture, such as C&C, can be easily developed by the technical community. But to get agreement among several levels of managers inside a company is an entirely different task. It is an effort that takes time and patience.

What is the benefit of this approach? Consider R&D expenditures. NEC's R&D budget, during the period 1980–1990, was considerably smaller than either IBM or AT&T. But NEC generated a 23 percent per year growth record, for over a decade, on significantly less investments in R&D. NEC supplemented its investments in R&D by a series of carefully targeted alliances. Between 1965 and 1987, it was involved in more than 130 alliances. The logic for this network of alliances can be derived from the strategic architecture. In fact, one can track all the alliances between 1965 and 1987 and position them in the overall strategic architecture of the company. The strategic architecture also provided the motivation to learn from these alliances. It is important to recognize that NEC used its alliance partners as multipliers to its internal resources. NEC used its architecture not only as an organizing framework inside the firm but as a way of communicating to the rest of the world what it was all about. NEC's advertisements embodied the C&C theme.

The NEC portfolio encompasses enormous product variety. However, there is no problem in understanding the logic behind that diverse portfolio. They are all derivatives of C&C. This consistency—in strategic intent, architecture, alliance strategies, and businesses—contributed to NEC's ability to leverage its resources. NEC, in 1990, occupied one of the top five positions in telecom, computing and semiconductors worldwide. And it was a $3 billion company in 1980!

Other firms such as Vickers (a division of TRINOVA), Sharp, Colgate-Palmolive, Kodak, and others have developed similar strategic, architectures to guide their managerial actions.

IDENTIFYING CORE COMPETENCIES

A strategic architecture allows managers to identify what core competencies we have and what we need to get. Core competencies are an important link in the process of leverage. The concept of core competencies tends to be confused with core technologies and/or capabilities. Core technologies are a component part of core competencies. Core competency results when firms learn to *harmonize multiple technologies*. For example, consider miniaturization, which has been the unique signature of Sony. Miniaturization requires core technologies, such as microprocessors, miniature power sources, power management, packaging, and manufacturing. It certainly also requires knowledge and understanding of user-friendly design and a knowledge of ergonomics. In addition, miniaturization is a result of deep sensitivity to emerging life styles. A core competency does not represent just technical capabilities in microprocessors, or packaging, or passive components; it also means understanding how to exploit life style knowledge using electronics. The point I want to stress is that it is not just technical capabilities that matter. What matters is the *creative bundling* of multiple technologies and customer knowledge and intuition, and managing them as a harmonious whole.

A core competency can be identified by applying three simple tests: 1) Is it a significant source of competitive differentiation? Does it provide a unique signature to the organization, like miniaturization for Sony, or user-friendliness at Apple? 2) Does it transcend a single business? Does it cover a range of businesses, both current and new? 3) Is it hard for competitors to imitate? It is hard for someone to visit Matsushita or Sony and come back and outline why they are good at manufacturing or miniaturization, respectively. Competence permeates the whole organization, and it represents tacit learning in an organization.

The difference between technology and competence is that technology can be stand-alone (e.g., design of very-large-scale-integration). Competence, on the other hand, is getting consistently high yields in VLSI. This transcends design capabilities. The process of converting good designs into high yields requires that multiple levels (e.g., shop floor to product development engineers) and multiple functions (e.g., application engineers and manufacturing groups) work very closely. A lot of the understanding and learning is tacit. And the recognition that competence represents tacit as well as explicit learning and is the cumulative knowledge base involving a large number of people is critical to understanding core competence. Technical capabilities, as stand-alone skills, are not the key to understanding core competencies. Competence is embedded in the whole organization.

Miniaturization at Sony, network management at AT&T, billing at the regional Bell operating companies, user friendliness at Apple, and high-volume manufacturing at Matsushita, are examples of core competencies.

Often, core competencies are confused with capabilities. Capabilities are, in some cases, prerequisites to being in a business. For example, "just-in-time" delivery is a prerequisite to be a Tier 1 supplier to the auto industry. It is the price one has to pay to get into the game. If one is a gambler, one may call it "table stakes." Capability is crucial for survival but, unlike a core competency, does not confer any specific differential advantage over other competitors in that industry.

COMPETENCE IS GOVERNANCE TOO

The key to understanding competence is that although it incorporates a technology component, it also involves the *governance process* inside the organization (the quality of relationships across functions, across business units), and *collective learning* across levels and functions) inside the company. We may conceptualize competence as follows: Competence = (Technology × Governance Process × Collective Learning).

We can examine the implications of this view with a hypothetical example. Consider a typical U.S. firm. The assumption is that if we pour a lot of money into technology the competitiveness problems will go away. Using the expression above, let us consider this hypothetical firm to be rich in technology, say 1,000 units of technology. However, let us assume that the various businesses within this corporation do not work together: Let's give them 20 units for the governance process—capability to work across business and functional unit boundaries. Let us also assume that in this firm the capability for collective learning is low. So let us give it 5 units for this dimension. Using the formula above, we now have an overall competence score of 100,000 units.

Consider another company that is not blessed with as much technology. It qualifies for not more than, say, 200 units, using the same formula we used in considering the previous firm. However, this firm has fostered the capacity to work across organizational boundaries and is fully focused on organizational learning. Let us give it 100 units for governance and 500 for collective learning, leading to a competence of 10 million. The message is clear: Investments in technology, if they are not, in tandem, accompanied by investments in *governance* and *creation* of a *learning environment* at all levels in the organization will remain under-leveraged. So, the logical point of leverage for Western firms resides in investments to improve the *quality of organization.*

Honda has been an example of this kind of thinking. Honda's multiple businesses are built on the basis of a competence in engines. If every one of those business units, be it power tillers or lawn mowers, behaved as if it were a discrete and stand-alone business, and each business unit only focused on (and was willing to pay for) the functionalities it needed, Honda's engine competence could be easily compromised. For example, in power tillers, managers may demand and be willing to pay for a light and robust engine. However, noise reduction is not a major priority for power tiller managers as they target their products for use in villages in the developing world. On the other hand, the business unit manager developing lawn mowers for sale in the United States may need not only a light and robust engine but one with low noise level as well. Business unit managers can uniquely define the functionalities that they need. But if each one of them tends to optimize their needs without maintaining a perspective on the implications of their parochial approach for the protection and development of an overall competence—in this case, engines—the skill base will be eroded. Often, a single-minded focus on SBU (strategic business unit) structure without checks and balances can destroy the very basis of nurturing and exploiting core competencies.

If the only model we use represents the company as a portfolio of businesses, then cost reduction within those businesses appears to be the primary task, sometimes followed by focus on product line extensions. When we model the corporation as a portfolio of core competencies, we tend to focus on new application opportunities. This perspective enhances the focus of management on new business development.

The portfolio of businesses at Sharp, Sony and Canon makes the case for focus on core competencies. But in order to focus on growth based on core competencies, we need to create a new set of managerial capabilities—one where sharing knowledge and components across organizational boundaries is relatively routine and painless. For example, Canon has a wide variety of businesses (end products) like copiers, laser printers, FAX, cameras, and camcorders. But those business units are fed by core products, such as lens systems, laser engines and miniature motors. These core products are supported by core competencies such as miniaturization, mechatronics and so on. Each business has an independent identity. It focuses on a specific set of customers and markets. But underlying that is a structure of shared core products and core competencies. As a result, within the firm, there is an opportunity for gaining economies of scale, an ability to provide new functionalities, and leverage.

Core products are often the physical embodiment of one or more core competencies. Compressors in Matsushita and laser printer engines in Canon are examples of core products. Canon not only uses its laser printer engine in several businesses, it also markets it outside. Firms such as Canon distinguish between their market share of end products (e.g., copiers), manufacturing share (e.g., share of market gained by providing manufactured products under private label to others), and share of core products (e.g., laser printer engines sold to others). Canon has gained a significant market share of over 85 percent worldwide in laser printer engines. It remains a small player in laser printers—the end product, worldwide. There is a market that is developing for core products. And we need to recognize that competition for core products is distinctly different from competition for end products and services.

We need to go beyond market share for end products. Consider the color television business, for example. In order to succeed, firms must have access to core products such as picture tubes, signal processing ICs, tuners, and line output transformers. If we disaggregate businesses at the core product level—be it TVs, VCRs, Camcorders, or laptop computers—we find very few Western firms that dominate worldwide. This perspective allows us to evaluate competitive outcomes differently. For example, with this perspective we can explain why Matsushita and JVC won the battle for VCRs. Their combined market share for VHS VCRs was only 24 percent. However, manufacturing share was 41 percent, format share through licensing was 80 percent, and core product share for decks was 85 percent. Eight-five percent of the world's requirements for decks were made by one company!

We underestimate the power that accrues to these companies because they dominate core products. Managers tend to underestimate the power of core product dominance. The issue is who controls critical technologies. Technological superiority without competence may represent a hollow victory.

The emerging competitive picture should force senior managers to ask themselves such questions as: How long can this erosion of core product capability in the West be sustained? Who are the custodians of the technical virtuosity of our companies? While business unit managers have no natural inclination to concern themselves with core product share or competencies, should group and sector executives transcend the concerns of the business units and play a role in protecting the basis for long-term competitiveness of our firms? Who should protect the disciplines that require multiple business units to work together?

It is important to recognize that competition today takes place on multiple planes. First, there is competition for end product markets and services; that is, price-performance competition represented by market share battles for today's market. Managers have to fight in that arena. There is also a less visible battle for dominance in core products that create the capacity to lead in the development of new functionalities. Finally, there is competition for competence—the capacity to create new businesses. The three levels of competition are shown in Figure 12.4. We need to learn to compete on all three levels.

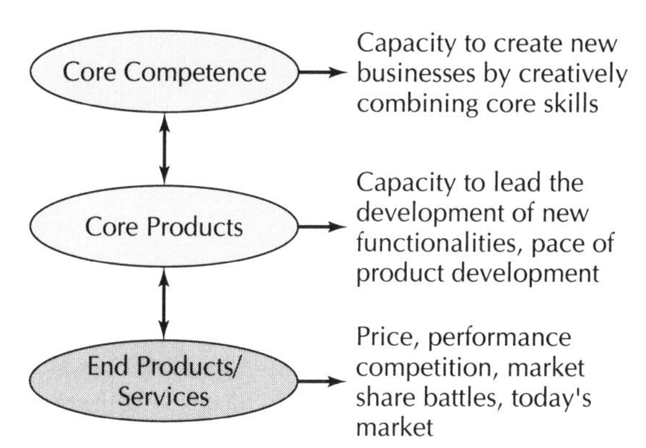

Fig. 12.4. We need to learn to compete on all three levels of competition.

CREATING NEW COMPETITIVE SPACE

How can firms create totally new products and services? Consider new businesses such as personal FAX. Global Positioning Satellites for hikers, photo CD, and camcorders. I believe the *opportunity list* for the next decade in high-volume electronics could be five pages long. The question for consideration by senior managers is: What do we have to do to capture our share of new business development?

I believe we need to develop a new mindset that is characterized by the following:

1. *Challenging existing price-performance assumptions.* Why can't we create a color FAX for $200? Why should it be $5,000? Canon in copiers, and Lexus or Honda in the luxury car segment are good examples of firms dramatically challenging the existing price-performance assumptions in the industry.

2. *Understanding the "meaning of customer-led."* In most firms, this means listening to customers and giving them what they ask. That's important, but it is also important to lead customers. As customers, many of us may not have anticipated our emerging dependence on a FAX in our home 10 years ago. Being customer-led is important, but leading customers is what competing for the future is about: understanding functionalities and needs, and creating products with a price performance that makes it attractive for people to buy.

3. *Escaping the tyranny of the "served market" orientation among managers.* A served market orientation puts too much emphasis on current businesses, and reduces the capacity to foresee new opportunities, especially opportunities that fall between two or more current business units. If we want to re-fire corporate imagination, we need to do the following:

We must de-emphasize the served market orientation and emphasize an orientation that focuses on the opportunity horizon. Managing the served market is important, but exploiting the opportunity horizon is what leads to profitable growth. We must not just defend markets, but create markets. Instead of incrementalism in price-performance there must be stretch price-performance goals; instead of simply benchmarking, currently popular in many firms, we must outpace the competition. We want to move from satisfying needs to anticipating needs; from being close to customers to leading customers; from thinking in terms of products to focusing on functionality and rapid market incursions; from focusing on core business to diversifying around core competencies. This is a very different mindset.

Managers have traditionally focused on current customers, product-markets, and corresponding business units. My research has convinced me that there is something beyond understanding business units and customers. We need to start with a *strategic intent, create a strategic architecture, understand core competencies and products*, such that there is a logic for business units, both current and new, and that leverage is based on continuous reconfiguration of these competencies.

To realize both the stretch and the leverage that this set of ideas promotes, we need to develop a set of values and beliefs that are consistent with this orientation to profitable growth. What is the unit of analysis for resource allocation? How do we manage inter-business unit linkages, inter-functional linkages? How do we create organizational capabilities, such as global-local capability or cycle time? And then, how do we think about administrative processes such as budgeting or planning?

The next challenge for senior management is: How do you connect individual employees' motivation and contribution with customers through a transparent process inside the company, where everybody understands what the shared aspirations are and how the various businesses interlink with each other, and the logic for nesting individual products and new initiatives? That, to me, is the next round of challenge.

I conclude with the following thoughts: 1) Growth is the agenda—not restructuring. 2) Dramatic growth will not take place if we focus on technology; it will take place when we focus on the organization, with technology as a part of it. 3) Dramatic growth requires a radical rethinking of current management paradigms.

REFERENCES

1. Przbylowicz, Edward P. and Faulkner, Terrence W. "Kodak Applies Strategic Intent to the Management of Technology." *Research Technology Management*, No. 1 (1993), p. 31–38.
2. Hamel, Gary and Prahalad, C. K. "Strategic Intent." *Harvard Business Review*, No. 3 (1989), pp. 63–76; Prahalad, C. K. and Hamel, Gary. "The Core Competence of the Corporation." *Harvard Business Review*, No. 3 (1990), pp. 79–91; Hamel, Gary and Prahalad, C. K. "Corporate Imagination and Expeditionary Marketing." *Harvard Business Review*, No. 4 (1991), pp. 81–92.
3. Kobayashi, Koji. *Computers and Communications: A vision of C&C*. Cambridge, The MIT Press, 1986.

13

Managing Professional Careers: The Influence of Job Longevity and Group Age

RALPH KATZ

Any serious consideration of organizational careers must eventually explore the dynamics through which the concerns, abilities, and experiences of individual employees combine and mesh with the demands and requirements of their employing work environments. How do employees' motivational needs for security, equitable rewards, and opportunities for advancement and self-development, for example, interact with the needs of organizations for ensured growth, profitability, and innovativeness? More important, how should they interact so that both prescription sets are filled satisfactorily?

Further complexity is added to this "matching" process with the realization that interactions between individuals and organizations are not temporally invariant but can shift significantly throughout workers' jobs, careers, and life cycles. As employees pass from one phase in their work lives to the next, different concerns and issues are emphasized; and the particular perspectives that result produce different behavioral and attitudinal combinations within their job settings. Over time, therefore, employees are continuously revising and adjusting their perspectives toward their organizations and their roles in them. And it is the perspective that one has formulated at a particular point in time that gives meaning and direction to one's work and professional career.

Because the effectiveness of any given group of professional employees ultimately depends on the combined actions and performances of its membership, we need to analyze more precisely the different kinds of concerns and behaviors that seem to preoccupy and characterize employees as they enter their work environments and proceed through respective jobs, project groups, and organizational career paths. Clearly, a better understanding of the changes taking place over time will help clarify the accommodation processes between organizations and individuals so that eventual motivational and performance problems can be dealt with more quickly and satisfactorily.

A MODEL OF JOB LONGEVITY

Based on some findings on how job satisfaction and performance relate to certain task dimensions of job challenge, Katz (1980) has developed a general theory for describing how employees' perspectives unfold and change as they proceed through their own discrete sequences of job situations. In particular, a three-transitional stage model of job longevity has been proposed to illustrate how certain kinds of concerns might change in importance according to the actual length of time an employee has been working in a given job position. Generally speaking, each time an employee is assigned to a new job position within an organization, either as a recent recruit or through transfer or promotion, the individual enters a relatively brief but nevertheless important "socialization" period. With increasing familiarity about his or her new job environment, however, the employee soon passes from socialization into the "innovation" stage, which, in turn, slowly shifts into a "stabilization" state as the individual gradually adapts to extensive job longevity (i.e., as the employee continues to work in the same overall job context for an extended period of time). Figure 13.1 summarizes the sequential nature of these three stages by comparing some of the different kinds of issues affecting employees as they cycle through their various job positions.[1]

Socialization

As outlined under the initial socialization stage, employees entering new job positions are concerned primarily with reality construction, building more realistic understandings of their unfamiliar social and task environments. In formulating their new perspectives, they are busily absorbed with problems of establishing and clarifying their own situational roles and identities and with learning all the attitudes and behaviors that are appropriate and expected within their new job settings. Estranged from their previous work environments and supporting relationships, newcomers must construct situational definitions that allow them to understand and interpret the myriad of experiences associated with their new organizational memberships. They need, for example, to learn the customary norms of behavior, decipher how reward systems actually operate, discover supervisory expectations, and more generally learn how to function meaningfully within their multiple group contexts (Schein, 1978). Through information communicated by their new "significant others," newcomers learn to develop perceptions of their own roles and skills that are both supported within their new surroundings and which permit them to organize their activities and interactions in a mean-

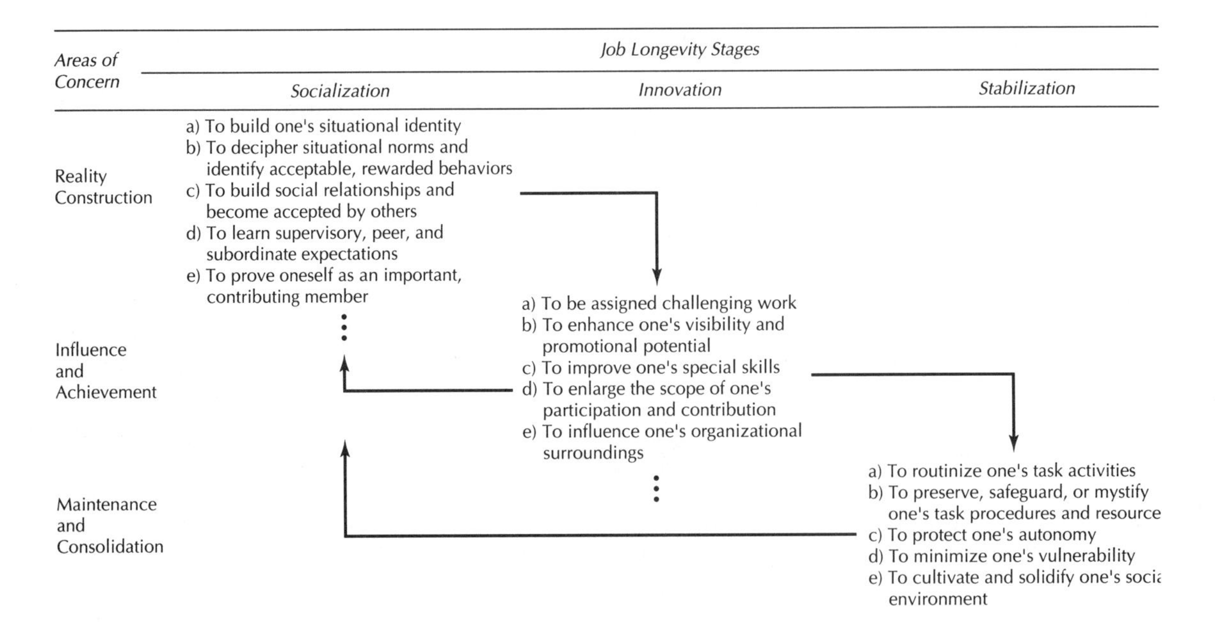

Fig. 13.1. Examples of special issues during each stage of job longevity.

ingful fashion. As pointed out by Hughes (1958) in his discussion of "reality shock," when new employees suddenly discover that their somewhat "overglorified" work-related expectations are neither realistic nor mutually shared by their boss or co-workers, they are likely to feel disenchanted and will experience considerable pressure to either redefine more compatible expectations or terminate from their work settings.

The importance of such a "breaking-in" period has long been recognized in discussions of how social processes affect recent organizational hires trying to make sense out of their newfound work experiences. What is also important to recognize is that veteran employees must also relocate or "socialize" themselves following their displacements into new job positions within their same organizations (Wheeler, 1966). Just as organizational newcomers have to define and interpret their new territorial domains, veteran employees must also restructure and reformulate perceptions regarding their new social and task realities.[2] As they assume new organizational positions and enter important new relationships, veterans must learn to integrate their new perceptions and experiences with prior organizational knowledge in order to develop fresh situational perspectives, including perceptions about their own self-images and their images of other organizational members.

Such perceptual revisions are typically necessary simply because work groups and other organizational subunits are often highly differentiated with respect to their idiosyncratic sets of norms, beliefs, perceptions, time perspectives, shared language schemes, goal orientations, and so on (Lawrence and Lorsch, 1967). As communications and interactions within an organizational subunit continue to take place or intensify, it is likely that a more common set of understandings about the subunit and its environment will develop through informational social influence. Such shared meanings and awarenesses not only provide the subunit's members with a sense of belonging and identity but will also demarcate the subunit from other organizational entities (Pfeffer, 1981). Consequently, as one shifts job positions and moves within the organization, one is likely to encounter and become part of a new set of groups with their correspondingly different belief systems and perspectives about themselves, their operations, and their operating environments. It is in this initial socialization period, therefore, that organizational employees, and newcomers in particular, learn not only the technical requirements of their new job assignments but also the interpersonal behaviors and social attitudes that are acceptable and necessary for becoming a true contributing member.

Since employees in the midst of socialization are strongly motivated to reduce ambiguity by creating out of their somewhat vague and unfamiliar surroundings, it becomes clear why a number of researchers have discovered organizational newcomers being especially concerned with psychological safety and security and with clarifying their new situational identities (Kahn et al., 1964; Hall and Nougaim, 1968). In a similar vein, Schein (1971) suggests that to become accepted and to prove one's competence represent two major problems that newcomers and veterans must face before they can function comfortably within their new job positions. It is these kinds of concerns that help to explain why Katz (1978a) discovered that during the initial months of their new job positions, employees are not completely ready to respond positively to all the challenging characteristics of their new task assignments. Instead, they appear most responsive to job features that provide a sense of personal acceptance and importance as well as a sense of proficiency through feedback and individual guidance.[3] As illustrations of just how influential such communication and feedback processes can be during socialization, the studies of professionals' careers by Lee (1992) and Katz and Tushman (1983) have shown very strong relationships between employees' communication networks and their subsequent degree of turnover and level of performance. In fact, Katz and Tushman reported that young engineers and scientists who left their organization over the 5-year interval in which the research study took place, had had more than four to five times less work-related interaction with their immediate supervisors and department heads during socialization as those engineers and scientists who still remained during this 5-year period.

How long this initial socialization period lasts probably depends on how long it takes employees to feel accepted and competent within their new work environments. Not only is the length of such a time period greatly influenced by the abilities, needs, and prior experiences of individual workers and influenced as well by the clarity and usefulness of the interpersonal interactions that take place, but it also probably differs significantly across occupations. Based on the retrospective answers of his hospital employee sample, for example, Feldman (1977) reports that on the average, accounting clerks, registered nurses, and engineering tradesmen reporting feeling accepted after one, two, and four months, respectively although they did not feel completely competent until after three, six, and eight months, respectively. Generally speaking, one might posit that the length of one's initial socialization period varies positively with the level of complexity within one's job and occupational requirements, ranging perhaps from as little as a month or two on very routine, programmed-type jobs to as much as a year or more on very skilled, unprogrammed-type jobs, as in the engineering and scientific professions. With respect to engineering, for example, it is generally recognized that a substantial socialization period is often required before engineers can fully contribute within their new organizational settings, using their particular knowledge and technical specialties. Thus, even though one might have received an excellent education in mechanical engineering principles at a university or college, one must still figure out from working and interacting with others in the setting how to be an effective mechanical engineer at Westinghouse, DuPont, or Procter and Gamble.[4]

Innovation

With time, interaction, and increasing familiarity, employees soon discover how to function appropriately in their jobs and to feel sufficiently secure in their perceptions of their workplace. Individual energies can now be devoted more toward task performance and accomplishment instead of being expended on learning the previously unfamiliar social knowledge and skills necessary to make sense out of one's work-related activities and interactions. As a result, employees become increasingly capable of acting in a more responsive, innovative, and undistracted manner.

The movement from socialization to the innovation stage of job longevity implies that employees no longer require much assistance in deciphering their new job and organizational surroundings. Having adequately constructed their own understandings of their situations during the socialization period, employees are now freer to participate within their own conceptions of organizational reality. They are now able to divert their attention from an initial emphasis on psychological safety and acceptance to concerns for achievement and influence. Thus, what becomes progressively more pertinent to employees as they proceed from socialization to the innovation stage are the opportunities to participate and grow within their job settings in a very meaningful and responsible manner.

The idea of having to achieve some reasonable level of psychological safety and security in order to be fully responsive to challenges in the work setting is very consistent with Kuhn's (1963) concept of "creative tensions." According to Kuhn, it is likely that only when conditions of *both* stability and challenge are present can the creative tensions between them generate considerable innovative behavior. Growth theorists such as Maslow (1962) and Rogers (1961) have similarly argued that the presence of psychological safety is one of the chief prerequisites for self-direction and individual responsiveness. For psychological safety to occur, however, individuals must be able to understand and attach sufficient meaning to the vast array of events, interactions, and information flows involving them throughout their workdays. Of particular importance to growth theorists is the idea that employees must be able to expect positive results to flow from their individual actions. Such a precondition implies that employees must have developed sufficient knowledge about their new job situations in order for there to be enough predictability for them to take appropriate kinds of actions.[5]

A similar point of view is taken by Staw (1977) when he argues that if employees truly expect to improve their overall job situations, they must first learn to predict their most relevant set of behavioral-outcome contingencies before they try to influence or increase their control over them. One must first construct a reasonably valid perspective about such contingencies before one can sensibly strive to manage them for increasingly more favorable outcomes. In short, there must be sufficient awareness of one's environment, sufficient acceptance and competence within one's setting, and sufficient openness to new ideas and experiences in order for employees to be fully responsive to the "richness" of their job demands.

Stabilization

As employees continue to work in their same overall job settings for a considerable length of time, without any serious disruption or displacement, they may gradually proceed from innovation to stabilization in the sense of shifting from being highly involved in and receptive to their job demands to becoming progressively unresponsive. For the most part, responsive individuals prefer to work at jobs they find stimulating and challenging and in which they can self-develop and grow. With such kinds of work-related activities and opportunities, they are likely to inject greater effort and involvement into their tasks which, in turn, will be reflected in their performances (Hackman and Oldham, 1975; Katz, 1978b). It seems reasonable to assume, however, that in time even the most challenging job assignments and responsibilities can appear less exciting and more habitual to jobholders who have successfully mastered and become increasingly accustomed to their everyday task requirements. With prolonged job longevity and stability, therefore, it is likely that employees' perceptions of their present conditions and of their future possibilities will become increasingly impoverished. They may begin essentially to question the value of what they are doing and where it may lead. If employees cannot maintain, redefine, or expand their jobs for continual challenge and growth, the substance and meaning of their work begins to deteriorate. Enthusiasm wanes, for what was once challenging and exciting may no longer hold much interest at all.

At the same time, it is also important to mention that if an individual is able to increase or even maintain has or her own sense of task challenge and excitement on a given job for an extended period of time, then instead of moving toward stabilization, the process might be the reverse (i.e., continued growth and innovation). As before, the extent to which an individual can maintain his or her responsiveness on a particular job strongly depends on the complexity of the underlying tasks as well as on the individual's own capabilities, needs, and prior experiences. With respect to individual differences, for example, Katz's (1978b) findings suggest that employees with high growth needs are able to respond to the challenging aspects of their new jobs sooner than employees with low growth needs. At the same time, however, high-order-need employees might not retain their responsiveness for as long a job period as employees with low-growth-need strength.

It should also be emphasized that in addition to job longevity, many other contextual factors can affect a person's situational perspective strongly enough to influence the level of job interest as one continues to work in a given job position over a long period of time. New technological developments, rapid growth and expansion, the sudden appearance of external threats, or strong competitive pressures could all help sustain or even enhance an individual's involvement in his or her job-related activities. On the other hand, having to work closely with a group of unresponsive peers might shorten an individual's responsive period on that particular job rather dramatically. Clearly, the reactions of individuals are not only influenced by psychological predispositions and personality characteristics but also by individuals' definitions of and interactions with their overall situational settings (Homans, 1961; Salancik and Pfeffer, 1978).

Generally speaking, however, as tasks become progressively less stimulating to employees with extended job longevity, they can either leave the setting or remain and adapt to their present job situations (Argyris, 1957). In moving from innovation to stabilization, it is suggested that employees who continue to work in their same overall job situations for long periods of time gradually succeed in adapting to such steadfast employment by becoming increasingly indifferent and unresponsive to the challenging task features of their job assignments (Katz, 1978a). In the process of adaptation, they may also redefine what they consider to be important, most likely by placing relatively less value on intrinsic kinds of work issues. The findings of Kopelman (1977) and Hall and Schneider (1973) suggest, for example, that when individuals perceive their opportunities for intrinsic-type satisfactions and challenges to be diminishing, they begin to match such developments by placing less value on such types of expectations. And as employees come to care less about the intrinsic nature of the actual work they do, the greater their relative concern for certain contextual features, such as salary, benefits, vacations, friendly co-workers, and compatible supervision.

The passage from innovation to stabilization is not meant to suggest that job satisfaction necessarily declines with long-term job longevity. On the contrary, it is likely that in the process of adaptation, employees' expectations have become adequately satisfied as they continue to perform their familiar duties in their normally acceptable fashions. If aspirations are defined as a function of the disparity between desired and expected (Kiesler, 1978), then as long as what individuals desire is reasonably greater than what they can presently expect to attain, there will be energy for change and achievement. On the other hand, when employees arrive at a stage where their chances for future growth and challenges in their jobs are perceived to be remote, then as they adapt, it is likely that existing situations will become accepted as the desired and aspirations for growth and change will have been reduced. As a result, the more employees come to accept their present circumstances, the stronger the tendency to keep the existing work environment fairly stable. Career interests and aspirations become markedly constricted, for in a sense, adapted employees simply prefer to enjoy rather than try to add to their present job accomplishments.

Uncertainty and Motivation

Underpinning the descriptive changes represented by the stabilization stage is the basic idea that over time individuals try to organize their work lives in a manner that reduces the amount of stress they must face and which is also low in uncertainty (Pfeffer, 1980; Staw, 1977). Weick (1969) also relies on this perspective when he contends that employees seek to "enact" their environments by directing their activities toward the establishment of a workable level of certainty and clarity. In general, one can argue that employees strive to bring their work activities into a state of equilibrium where they are more capable of predicting events and of avoiding potential conflicts.[6]

Given such developmental trends, it seems reasonable that with considerable job longevity, most employees have been able to build a work pattern that is familiar and comfortable, a pattern in which routine and precedent play a relatively large part. According to Weick (1969), as employees establish certain structures of interlocked behaviors and relationships, these patterns will in time become relatively stable simply because they provide certainty and predictability to these interlinked employees. It is further argued that as individuals adapt to their long-term job tenure and become progressively less responsive to their actual task demands, they will come to rely more on their established modes of conduct to complete their everyday job requirements. Most likely, adapted employees feel safe and comfortable in such stability, for its keeps them feeling secure and confident in what they do, yet requires little additional vigilance or effort. In adapting to extended job longevity, therefore, employees become increasingly content and ensconced in their customary ways of doing things, in their comfortable routines and interactions, and in their familiar sets of task demands and responsibilities.

If change or uncertainty is seen by individuals in the stabilization period as particularly disruptive, then the preservation of familiar routines and patterns of behavior is likely to be of prime concern. Given such a disposition, adapted employees are probably less receptive toward any change or toward any information that might threaten to disturb their developing sense of complacency. Rather than striving to enlarge the scope of their job demands, they may be more concerned with maintaining their comfortable work environments by protecting themselves from sources of possible interference, from activities requiring new kinds of attention, or form situations that might reveal their shortcomings. Adapted employees, for example, might seek to reduce uncertainty in their day-to-day supervisory dealings perhaps by solidifying their attractiveness through ingratiating kinds of behavior (Wortman and Linsenmeier, 1977) or perhaps by isolating themselves from such supervisory contacts (Pelz and Andrews, 1966). Or they might seek to reduce uncertainty by trying to safeguard their personal allocations of resources and rewards through the use of standardized practices and policies. Whatever the specific behaviors that eventually emerge in a given setting, it is likely that employees who have become unresponsive to the challenging features of their assigned tasks will strongly resist events threatening to introduce uncertainty into their work environments.

One of the best examples of the effects of such long-term stability can still be found in Chinoy's (1955) classic interviews of automobile factory workers. Chinoy discovered that although almost 80% of the workers had wanted to leave their present jobs at one time or another, very few could actually bring themselves to leave. Most of the workers were simply unwilling to give up the predictability and comfortableness of their presently familiar routines and cultivated relationships for the uncertainties of a new job position. Individuals do not resist change in and of itself. What individuals resist is the uncertainty provoked by change. For as long as employees are able to process information that allows them to reduce the uncertainties surrounding the changes, their apprehensions and concerns become progressively diminished.

SITUATIONAL VERSUS INDIVIDUAL CONTROL OF INTERPRETATION

In presenting this three-stage model of job longevity, I have tried to describe some of the major concerns affecting employees as they enter and adapt to their particular job positions. Of course, the extent to which any specific individual is affected by these issues depends on the particular perceptual outlook that has been developed over time through job-related activities and through role-making processes with other individuals, including supervisors, subordinates, and peers (Weick, 1969; Graen, 1976). Employees, as a result, learn to cope with their particular job and organizational environments through their interpretations of relevant work experiences as well as their expectations and hopes of the future. To varying degrees, then, situational perspectives are derivatives of both retrospective and prospective processes, in that they are built and shaped through knowledge of past events and future anticipations.

One of the more important aspects of the socialization process, however, is that the information and knowledge previously gathered by employees from their former settings are no longer sufficient or necessarily appropriate for interpreting or understanding their new organizational domains. Newcomers, for instance, have had only limited contact within their new institutional surroundings from which to construct their perceptual views. Similarly, the extent to which veterans who are assuming new job positions can rely on their past organizational experiences and perspectives to function effectively within their new work settings can also be rather limited, depending of course on their degrees of displacement.

Essentially, individuals in the midst of socialization are trying to navigate their way through new and unfamiliar territories without the aid of adequate or even accurate perceptual maps. During this initial period, therefore, they are typically more malleable and more susceptible to change (Schein, 1968). In a sense, they are working under conditions of high "situational control" in that they must depend on other individuals within their new situations to help them define and interpret the numerous activities taking place around them. The greater their unfamiliarity or displacement within their new organizational areas, the more they must rely on their situations to provide the necessary information and interactions by which they can eventually construct their own perspectives and reestablish new situational identities. And it is precisely this external need or "situational dependency" that enables these individuals to be more easily influenced during their socialization processes through social interactions (Salancik and Pfeffer, 1978; Katz, 1980).

As employees become increasingly cognizant of their overall job surroundings, however, they also become increasingly capable of relying on their own perceptions for interpreting events and executing their everyday task requirements. In moving from socialization into the innovation or stabilization stage, employees have succeeded in building a sufficiently robust situational perspective, thereby freeing themselves to operate more self-sufficiently within their familiar work settings. They are now working under conditions of less "situational" but more "individual" control, in the sense that they are now better equipped to determine for themselves the importance and meaning of the various events and information flows surrounding them. Having established their own social and task supports, their own perceptual outlooks, and their own situational identities, they become less easily changed and less easily manipulated. As pointed out by Schein (1973), when individuals no longer have to balance their situational perspectives against the views of significant others within their settings, they become less susceptible to change and situational influences. Thus, movement through the three stages of job longevity can also be characterized, as shown in Figure 13.2, by relative shifts to more individual and less situational control of interpretation.

As the locus of "control of interpretation" shifts with increasing job longevity and individuals continue to stabilize their situational definitions, other important behavioral tendencies can also materialize. In particular, strong biases can develop in

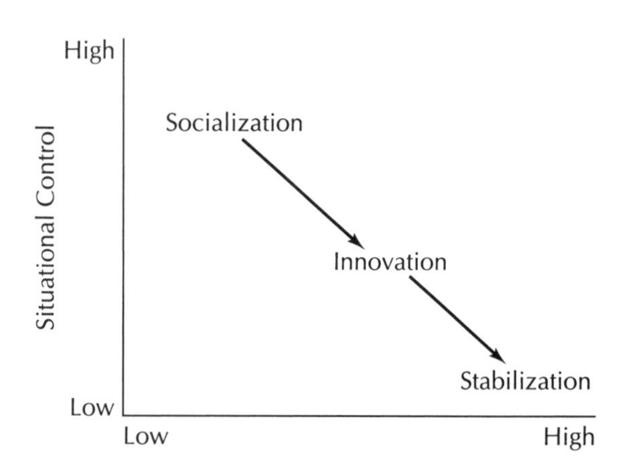

Fig. 13.2. Situational versus individual control of interpretation along the job-longevity continuum.

the way individuals select and interpret information, in their cognitive abilities to generate new options and strategies creatively, and in their willingness to innovate or implement alternative courses of action. Table 13.1 outlines in more detail some of the specific possibilities within each of these three general areas. Furthermore, it is the capacity either to prevent or overcome these kinds of tendencies that is so important to the long-term success of organizations; for, over time, each of these trends could lead to less effective performance and decision-making outcomes.

TABLE 13.1. Representative Trends Associated with Long-term Job Longevity

Problem-solving processes
Increased rigidity
Increased commitment to established practices and procedures
Increased mainlining of strategies
Information processes
Increased insulation from critical areas
Increased selective exposure
Increased selective perception
Cognitive processes
Increased reliance on own experiences and expertise
Increased narrowing of cognitive abilities
Increased homophyly

Problem-Solving Processes

It has been argued throughout this paper that as employees gradually adapt to prolonged periods of job longevity, they may become less receptive toward any change or innovation threatening to disrupt significantly their comfortable and predictable work practices and patterns of behavior. Individuals, instead, are more likely to develop reliable and effective routine responses (i.e., standard operating procedures) for dealing with their frequently encountered tasks in order to ensure predictability, coordination, and economical information processing. As a result, there may develop over time increasing rigidity in one's problem-solving activities—a kind of functional fixedness that reduces the individual's capacity for flexibility and openness to change. Responses and decisions are made in their fixed, normal patterns while novel situations requiring responses that do not fit such established molds are either ignored or forced into these molds. New or changing situations either trigger responses of old situations or trigger no responses at all. It becomes, essentially, a work world characterized by phrases such as, "business as usual," or "we've never done that before," or "we're not in that business," etc.

Furthermore, as individuals continue to work by their well-established problem-solving strategies and procedures, the more committed they may become to such existing methods. Commitment is a function of time, and the longer individuals are called upon to follow and justify their problem-solving approaches and decisions, the more ingrained they are likely to become. Drawing from his work on decision making, Allison (1971) strongly warns that increasing reliance on regularized practices and procedures can become highly resistant to change, since such functions become increasingly grounded in the norms and basic attitudes of the organizational unit and in the operating styles of its members. Bion (1961) and Argyris (1969) even suggest that it may be impossible for individuals to break out of fixed patterns of activity and interpersonal behavior without sufficiently strong outside interference or help.

With extended job tenure, then, problem-solving activities can become increasingly guided by consideration of methods and programs that have worked in the past. Moreover, in accumulating this experience and knowledge, alternative ideas and approaches were probably considered and discarded. With such refutations, however, commitments to the present courses of action can become even stronger—often to the extent that these competing alternatives are never reconsidered.[7] In fact, individuals can become overly preoccupied with the survival of their particular approaches, protecting them against fresh approaches or negative evaluations. Much of their energy becomes directed toward "mainlining their strategies," that is, making sure their specific solution approaches are selected and followed. Research by Janis and Mann (1977) and Staw (1980) has demonstrated very convincingly just how strongly committed individuals can become to their problem-solving approaches and decisions, even in the face of adverse information, especially if they feel personally responsible for such strategies. And under these kinds of conditions, professional knowledge workers are more likely to come to tough meetings, not with open-mindedness and problem-solving information, but with preconceived solutions based on filtered information.

Information Processes

One of the potential consequences of developing this kind of "status-quo" perspective with respect to problem-solving activity is that employees may also become increasingly insulated from outside sources of relevant information and important new ideas. As individuals become more protective of and committed to their current work habits, the extent to which they are willing or even feel they need to expose themselves to new or alternative ideas, solution strategies, or constructive criticisms becomes progressively less and less. Rather than becoming more vigilant about events taking place outside their immediate work settings, they become increasingly complacent about external environmental changes and new technological developments. Studies, such as those of D'Aveni (1994) and Miller

(1990), have convincingly shown just how important it is to pay attention to outside sources of information if organizations truly hope to survive under the pressures of long-term competition.

In addition to this possible decay in the amount of external contact and interaction, there may also be an increasing tendency for individuals to communicate only with those whose ideas are in accord with their current interests, needs, or existing attitudes. Such a tendency is referred to as selective exposure. Generally speaking, there is always the tendency for individuals to communicate with those who are most like themselves (Rogers and Shoemaker, 1971). With increasing adaptation to long-term job longevity and stability, however, this tendency is likely to become even stronger. Thus, selective exposure may increasingly enable these individuals to avoid information and messages that might be in conflict with their current practices and dispositions.

One should also recognize, of course, that under these kinds of circumstances, any outside contact or environmental information that does become processed by these long-tenured individuals might not be viewed in the most open and unbiased fashion. Janis and Mann (1977), for example, discuss at great length the many kinds of cognitive defenses and distortions commonly used by individuals in processing outside information in order to support, maintain, or protect certain decisional policies and strategies. Such defenses are often used to argue against any disquieting information and evidence in order to maintain self-esteem, commitment, and involvement. In particular, selective perception is the tendency to interpret information and communication messages in terms favorable to one's existing attitudes and beliefs. And it is this combination of increasing insulation, selective exposure, and selective perception that can be so powerful in keeping critical information and important new ideas and innovations from being registered.

Cognitive Processes

As individuals become more comfortable and secure in their long-tenured work environments, their desire to seek out and actively internalize new knowledge and new developments may begin to deteriorate. Not only may they become increasingly isolated from outside sources of information, but their willingness to accept or pay adequate attention to the advice and ideas of fellow experts may become less and less. Unlike the socialization period in which individuals are usually very attentive to sources of expertise and influence within their new job settings, individuals in the stabilization stage have probably become significantly less receptive to such information sources. They may prefer, instead, to rely on their own accumulated experience and wisdom and consequently are more apt to dismiss the approaches, advice, or critical comments of others. As a result, adapted employees may be especially defensive with regard to critical evaluations and feedback messages, whether they stem from sources of outside expertise or from internal supervision. Long-tenured veteran employees, for example, would not regard "constructive" performance appraisals with nearly as much enthusiasm as newcomers who might welcome such discussions in order to calibrate more accurately their organizational career paths and contributions.

It should also not be surprising that with increasing job stability one is more likely to become increasingly specialized, that is, moving from broadly defined capabilities and solution approaches to more narrowly defined interests and specialties. Without new challenges and opportunities, the diversity of skills and of ideas generated are likely to become narrower and narrower. And as individuals welcome information from fewer sources and are exposed to fewer alternative points of view, the more constricted their cognitive abilities can become. This often results in a more restricted perspective of one's situation, coupled with a more limited set of coping responses—all of which affect creativity. Such a restricted outlook can also be very detrimental to the organization's overall effectiveness, for it could lead at times to the screening out of some vitally important environmental and competitive information cues.

Homophyly refers to the degree to which interacting individuals are similar with respect to certain attributes, such as beliefs, values, education, and social status (Rogers and Shoemaker, 1971). Not only is there a strong tendency for individuals to communicate with those who are most like themselves, but it is also likely that continued interaction can lead to greater homophyly in knowledge, beliefs, and problem-solving behaviors and perceptions (Burke and Bennis, 1961; Pfeffer, 1980). The venerable proverb "birds of a feather flock together" makes a great deal of sense, but it may be just as sensible to say that "when birds flock together, they become more of a feather." Accordingly, as individuals stabilize their work settings and patterns of communication, a greater degree of homophyly is likely to have emerged between these individuals and those with whom they have been interacting over the long tenure period. And any increase in homophyly could lead in turn to further stability in the communications of the more homophilous pairs, thereby increasing their insulation from heterophilous others. Thus, it is possible for the various trends to feed on each other. For as previously discussed, although individuals may be able to coordinate and communicate with similar partners more easily and economically, such homogeneous interactions can also yield less creative and innovative outcomes (Pelz and Andrews, 1966).

Longevity and Performance

These problem-solving, informational, and cognitive tendencies, of course, can be very serious in their consequences, perhaps even fatal. Much depends, however, on the nature of the work being performed and on the extent to which such trends actually transpire. The performances of individuals working on fairly routine, simple tasks in a rather stable organizational environment, for example, may not suffer as a result of these trends, for their own knowledge, experiences, and abilities become sufficient. Maintaining or improving on one's routine behaviors is all that is required—at least for as long as there are no changes and no new developments. However, as individuals function in a more rapidly changing environment and work on more complex tasks requiring greater levels of change, creativity, and informational vigilance, the effects of these long-term longevity trends are likely to become significantly more dysfunctional.

GROUP AGE

The degree to which any of these previously described trends actually materializes for any given individual depends, of course, on the overall situational context. Individuals' perceptions and responses do not take place in a social vacuum but develop over time as they continue to interact with various aspects of their job and organizational surroundings (Crozier, 1964; Katz and Van Maanen, 1977). And in any job setting one of the most powerful factors affecting individual perspectives is the nature of the particular group or project team in which one is a functioning member (Schein, 1978, Katz and Kahn, 1978).

Ever since the well-known Western Electric Studies (Cass and Zimmer, 1975), much of our research in the social sciences has been directed toward learning just how strong group associations can be in influencing individual member behaviors, motivations, and attitudes (Asch, 1956; Shaw, 1971; Katz, 1977). From the diffusion of new innovations (Robertson, 1971) to the changing of meat consumption patterns, to less desirable but more plentiful cuts such as liver (Lewin, 1965), to the implementation of job enrichment (Hackman, 1978), group processes and effects have been extremely critical to more successful outcomes. The impact of groups on individual responses is substantial, if not pervasive, simply because groups mediate most of the stimuli to which their individual members are subjected while fulfilling their everday task and organizational requirements. Accordingly, whether individuals experiencing long-term job longevity enter the stabilization period and become subjected to the tendencies previously described may strongly depend on the particular reinforcements, pressures, and behavioral norms encountered within their immediate project or work groups (Likert, 1967; Weick, 1969).

Generally speaking, as members of a project group continue to work together over an extended period of time and gain experience with one another, their patterns of activities are likely to become more stable, with individual role assignments becoming more well-defined and resistant to change (Bales, 1955; Porter et al., 1975). Emergence of the various problem-solving, informational, and cognitive trends, therefore, may be more a function of the average length of time the group members have worked together (i.e., group age or group longevity) rather than varying according to the particular job longevity of any single individual. A project group, then, might either exacerbate or ameliorate the various trends (e.g., insulation from outside developments and expertise), just as previous studies have shown how groups can enforce or amplify certain standards and norms of individual behavior (e.g., Seashore, 1954; Stoner, 1968). Thus, it may be misleading to investigate the responses and reactions of organizational individuals as if they functioned as independent entities; rather, it may be more insightful to examine the distribution of responses as a function of different project teams, especially when project teams are characterized by relatively high levels of group stability as measured by group age.

To investigate such a possibility, Professor Tom Allen and I conducted a field study in a large chemical company to investigate the effects of group age on the group's overall project performance, where group age was measured by the average length of time the project or group members had worked together. A comparative analysis of some sixty R&D project teams within this organization revealed a very strong curvilinear relationship between group age and project performance, the lower performing project groups either having worked together for less than a year or they had been working together for at least four years.

Further analyses suggested that this curvilinear relationship may be the result of two component forces, as shown in Figure 13.3. One component term rises rapidly with mean group tenure, showing the positive effects of "team-building." Group members develop better understanding of each other's capabilities, contributions, working styles, etc. and such improvements in communication and working relationships result in higher levels of group performance. At the same time, however, a decay component term sets in, resulting in part from the previously described problem-solving, communication, and cognitive processes that become established, reinforced, and habitual as in-

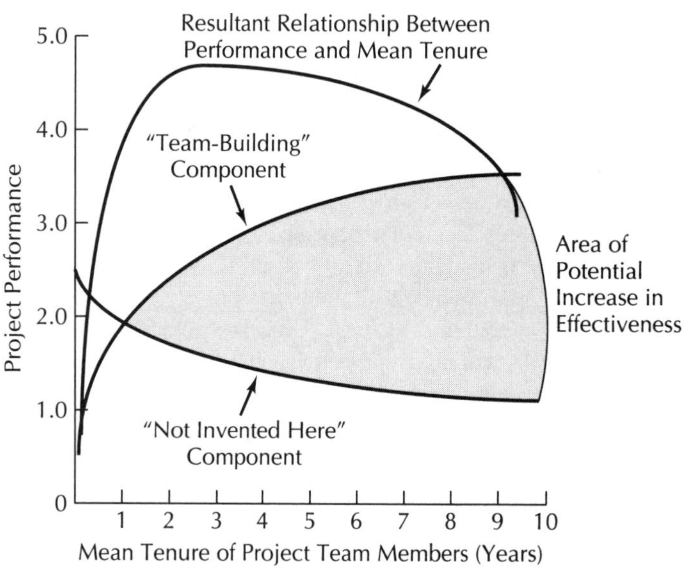

Fig. 13.3. The relationship between mean tenure project and performance analyzed into its components.

dividuals reduce uncertainty together within their stabilized group setting. Katz and Allen describe this decay component as the well-recognized Not Invented Here (N.I.H.) syndrome. Between these two component curves lies the area for potentially influencing a project's performance. Managers need to balance the advantages of team-building by also making sure that over time they take those steps necessary to counter the negative effects of N.I.H.

CONCLUSIONS

What is suggested by this discussion of job and group longevities is that employee perspectives and behaviors, and their subsequent effects on performance, might be significantly managed through staffing and career decisions. One could argue, for example, that the energizing and destabilizing function of new team members can be very important in preventing a project group from developing some of the tendencies previously described for long-tenured individuals, including insulation from key communication areas. The benefit of new team members is that they may have a relative advantage in generating fresh ideas and approaches. With their active participation, existing group members might consider more carefully ideas and alternatives they might have otherwise ignored or dismissed. In short, project newcomers can represent a novelty-enhancing condition, challenging and improving the scope of existing methods and accumulated knowledge.[8]

The longevity framework also seems to suggest that periodic job mobility or rotation might help prevent employees from moving into a stabilization stage. As long as the socialization period is positively negotiated, employees can simply cycle from one innovation period into another.[9] Put simply, movements into new positions may be necessary to keep individuals stimulated, flexible, and vigilant with respect to their work environments. Within a single job assignment, the person may eventually reach the limit to which new and exciting challenges are possible or even welcomed. At that point, a new job position may be necessary. To maintain adaptability and to keep employees responsive, what might be needed are career histories containing sequences of job positions involving new challenges and requiring new skills (Kaufman, 1974). As pointed out by Schein (1968), continued growth and development often come from adaptations to new or changing work environments requiring individuals to give up familiar and stable work patterns in favor of developing new ones.

As important as job mobility is, it is probably just as important to determine whether individuals and project groups can circumvent the effects of longevity without new assignments or rejuvenation from new project members. Rotations and promotions are not always possible, especially when there is little organizational growth. As a result, we need to learn considerably more about the effects of increasing job and group longevities. Just how deterministic are the trends? Can long-tenured individuals and project teams remain high-performing, and if so, how can it be accomplished?

In a general sense, then, we need to learn how to detect the many kinds of changes that either have taken place or are likely to take place within a group as its team membership ages. Furthermore, we need to learn if project groups can keep themselves energized and innovative over long periods of group longevity, or whether certain kinds of organizational structures and managerial practices are needed to keep a project team effective and high-performing as it ages.

In response to this issue, a more extensive study in twelve different organizations involving over 300 R&D project teams has been undertaken by Katz and Allen. Within this overall sample of project groups, approximately fifty of the groups had been working together for more than four years. More importantly, a large number of these long-tenured groups were still judged to have very high levels of group performance. The research question, of course, is how have these particular long-tenured groups managed to circumvent the NIH problem and maintain their high levels of performance. Although the data are still being processed, preliminary analyses seem to indicate that the nature of the project's supervision may be the most important factor differentiating the more effective long-tenured teams from those less effective. In particular, engineers belonging to the high-performing, long-tenured groups perceived their project supervisors to be superior in dealing with conflicts between groups and individuals, in obtaining necessary resources for project members, in setting project goals, and in monitoring the activities and progress of project members toward these goals. Furthermore, in performing these supervisory functions, the more effective project managers of long-tenured groups were not very participative in their approaches, instead, they were extremely demanding of their teams, challenging them to perform in new ways and directions. In fact, the most participative managers (as viewed by project members) were significantly less effective in managing teams with high group longevity. Our study also revealed that not all managers may be able to gain the creative performances out of long-term technical groups. Typically, the managers of the higher performing long-term groups had been with their teams less than 3 years and had come to this assignment with a strong history of prior managerial success. It was not their first managerial experience! To the contrary, most were well-respected technical managers who had "made things happen" and who had developed strong power bases and strong levels of senior managerial support within their R&D units or divisions. It was this combination of technical credibility and managerial respect and power that enabled these managers to be effective with their long-term stabilized R&D project teams.

What these and other initial findings seem to indicate is that in managing long-tenured groups, project managers who have the appropriate credibility and experience should place less emphasis on participative decision making and empowerment and place more emphasis on direction and control. As long as members of long-tenured groups are unresponsive to the challenges in their tasks, participative management will only be related to job satisfaction—not project performance.

In a broader context, we need to learn how to manage workers, professionals, and project teams as they enter and proceed through different stages of longevity. Clearly, different kinds of managerial styles and behaviors may be more appropriate at different stages of longevity. Delegative or participative management, for example, may be very effective when individuals are vigilant and highly responsive to their work demands, but such supervisory activities may prove less successful when employees are unresponsive to their job environments, as in the stabilization stage. Furthermore, as perspectives and responsiveness shift over time, the actions required of the managerial role will also vary. Managers may be effective, then, to the extent they are able to recognize and cover such changing conditions. Thus, it may be the ability to manage change—the ability to diagnose and manage between socialization and stabilization—that we need to learn so much more if we truly hope to provide careers that keep employees responsive and also keep organizations effective.

Notes

1. For a more extensive discussion of the job-longevity model, see Katz (1980). In the current presentation, the term "stabilization" is used in place of "adaptation" since individuals are in effect adapting to their job situations in all three stages, albeit in systematically different ways.

2. The extent to which a veteran employee actually undergoes socialization depends on how displaced the veteran becomes in undertaking his or her new job assignment. Generally speaking, the more displaced veterans are from their previously familiar task requirements and interpersonal associations, the more intense the socialization experience.

3. After comparing the socialization reactions of veterans and newcomers, Katz (1978a) suggests that newcomers may be especially responsive to interactional issues involving personal acceptance and "getting on board," whereas veterans may be particularly concerned with reestablishing their sense of competency in their newly acquired task assignments.

4. One of the factors contributing to the importance of this socialization period lies in the realization that engineering strategies and solutions within organizations are often not defined in very generalizeable terms but are peculiar to their specific settings (Allen, 1977; Katz and

Tushman, 1979). As a result, R&D project groups in different organizations may face similar problems yet may define their solution approaches and parameters very differently. And it is precisely because technical problems are typically expressed in such "localized" terms that engineers must learn how to contribute effectively within their new project groups.

5. It is also interesting to note that in discussing his career-anchor framework, Schein (1978) points out that career anchors seem to represent a stable concept around which individuals are able to organize experiences and direct activities. Furthermore, it appears from Schein's research that it is within this area of stability that individuals are most likely to self-develop and grow.

6. There are, of course, alternative arguments, such as in activation theory (Scott, 1966), suggesting that people do in fact seek uncertainty, novelty, or change. The argument here, however, is that as individuals adapt and become increasingly indifferent to the task challenges of their jobs, it is considerably more likely that they will strive to reduce uncertainty and maintain predictability rather than the reverse.

7. As shown by Allen's (1966) research on parallel project efforts, such reevaluations can be very important in reaching more successful outcomes.

8. As discussed by Katz and Allen (1981), the socialization process of individuals can greatly affect the extent to which newcomers may be willing to try to innovate on existing "widsoms."

9. A discussion on effectively managing the socialization process is beyond the scope of this paper. The reader is referred to the descriptive theory presented in Schein (1968), Kotter (1973), Hall (1976), Katz (1980), and Wanous (1980).

REFERENCES

Allen, T.J. "Studies of the problem-solving processes in engineering designs." *IEEE Transactions in Engineering Management*, 1966, *13,* 72–83.

Allen, T.J. *Managing the Flow of Technology.* Cambridge, Mass.: M.I.T. Press, 1977.

Allison, G.T. *Essence of Decision: Explaining the Cuban Missile Crisis.* Boston: Little, Brown, 1971.

Argyris, C. *Personality and Organization.* New York: Harper Torch Books, 1957.

Argyris, C. "The incompleteness of social psychological theory: examples from small group, cognitive consistency and attribution research." *American Psychologist*, 1969, *24*, 893–908.

Asch, S.E. "Studies of independence and conformity: a minority of one against a unanimous majority." *Psychological Monographs*, 1956, *70*.

Bales, R.F. "Adaptive and integrative changes as sources of strain in social systems." In A.P. Hare, E.F. Borgatta, and R.F. Bales, eds., *Small Groups: Studies in Social Interaction.* New York: Knopf, 1955, pp. 127–31.

Bion, W.R. *Experiences in Groups.* New York: Basic Books, 1961.

Burke, R.L., and Bennis, W.G. "Changes in perception of self and others during human relations training." *Human Relations*, 1961, *14*, 165–82.

Cass, E.L., and Zimmer, F.G. *Man and Work in Society.* New York: Van Nostrand Reinhold, 1975.

Chinoy, E. *Automobile Workers and the American Dream.* Garden City, N.Y.: Doubleday, 1955.

Crozier, M. *The Bureaucratic Phenomenon.* Chicago: University of Chicago Press, 1964.

D'Aveni, R.A. *Hypercompetition.* New York: Free Press, 1994.

Dewhirst, H., Avery, R., and Brown, E. "Satisfaction and performance in research and development tasks as related to information accessibility." *IEEE Transactions on Engineering Management*, 1978, *25*, 58–63.

Dubin, S.S. *Professional Obsolescence.* Lexington, Mass.: Lexington Books, D.C. Heath, 1972.

Feldman, D. "The role of initiation activities in socialization." *Human Relations*, 1977, *30*, 977–90.

Graen, G. "Role-making processes within complex organizations." In M.D. Dunnette, ed., *Handbook of Industrial and Organizational Psychology*, Chicago: Rand McNally, 1976.

Hackman, J.R. "The design of self managing work groups." In B. King, S. Streufert, and F. Fielder, eds., *Managerial Control and Organizational Democracy.* New York: Wiley, 1978.

Hackman, J.R., and Oldham, G.R. "Development of the job diagnostic survey." *Journal of Applied Psychology*, 1975, *60*, 159–70.

Hall, D.T. *Careers in Organizations.* Pacific Palisades, Calif.: Goodyear, 1976.

Hall, D.T., and Nougaim, K.E. "An examination of Maslow's need hierarchy in an organizational setting." *Organizational Behavior and Human Performance*, 1968, *3*, 12–35.

Hall, D.T., and Schneider, B. *Organizational Climates and Careers.* New York: Seminar Press, 1973.

Homans, C.C. *Social Behavior: Its Elementary Forms.* New York: Harcourt, Brace and World, 1961.

Hughes, E.C. *Men and Their Work.* Glencoe, Ill.: Free Press, 1958.

Janis, I.L., and Mann, L. *Decision Making.* New York: Free Press, 1977.

Kahn, R.L., Wolfe, D.M., Quinn, R.P., Snock, J.D., and Rosenthal, R.A. *Organizational Stress: Studies on Role Conflict and Ambiguity.* New York: Wiley, 1964.

Katz, D., and Kahn, R.L. *The Social Psychology of Organizations*. New York: Wiley, 1978.

Katz, R. "The influence of group conflict on leadership effectiveness." *Organizational Behavior and Human Performance*, 1977, *20*, 265–86.

Katz, R. "Job longevity as a situational factor in job satisfaction." *Administrative Science Quarterly*, 1978a, *10*, 204–23.

Katz, R., "The influence of job longevity on employee reactions to task characteristics." *Human Relations*, 1978b, *31*, 703–25.

Katz, R. "Time and work: toward an integrative perspective." In B. Staw and L.L. Cummings, eds., *Research in Organizational Behavior*, Vol. 2. Greenwich, Conn.: JAI Press, 1980, 81–127.

Katz, R., and Allen, T. "Investigating the not-invented-here syndrome." In A. Pearson, ed., *Industrial R&D Strategy and Management*, London: Basil Blackwell Press, 1981.

Katz, R., and Tushman, M. "Communication patterns, project performance and task characteristics: an empirical evaluation and integration in an R&D setting." *Organizational Behavior and Human Performance*, 1979, *23*, 139–62.

Katz, R., and Tushman, M. "A longitudinal study of the effects of boundary spanning supervision in turnover and promotion in Research and Development." *Academy of Management Journal*, 1983, *26*, 437–56.

Katz, R., and Van Maanen, J. "The loci of work satisfaction: job, interaction, and policy." *Human Relations*, 1977, *30*, 469–86.

Kaufman, H.G. *Obsolescence of Professional Career Development*. New York: AMACOM, 1974.

Kiesler, S. *Interpersonal Processes in Groups and Organizations*. Arlington Heights, Ill.: AHM Publishers, 1978.

Kopelman, R.E. "Psychological stages of careers in engineering: an expectancy theory taxonomy." *Journal of Vocational Behavior*, 1977, *10*, 270–86.

Kotter, J. "The psychological contract: managing the joining-up process." *California Management Review*, 1973, *15*, 91–99.

Kuhn, T.S. *The Structure of Scientific Revolutions*. Chicago: University of Chicago Press, 1963.

Lawrence, P.R., and Lorsch, J.W. *Organizational and Environment*. Boston: Harvard Business School, 1967.

Lee, D. "The effects of socialization on performance." Unpublished Working Paper Manuscript, Suffolk University, 1992.

Lewin, K. "Group decision and social change." in H. Proshansky and B. Seidenberg, eds., *Basic Studies in Social Psychology*. New York: Holt, Rinehart, and Winston, 1965, pp. 423–36.

Likert, R. *The Human Organization*. New York: McGraw-Hill, 1967.

Maslow, A. *Toward a Psychology of Being*. Princeton, N.J.: D. Van Nostrand, 1962.

Menzel, H. "Information needs and uses in science and technology." In C. Cuadra, ed., *Annual Review of Information Science and Technology*, New York: Wiley, 1965.

Miller, D. *The Icarus Paradox*. New York: Harper Collins, 1990.

Pélz, D.C., and Andrews, F.M. *Scientists in Organ-izations*. New York: Wiley, 1966.

Pfeffer, J. "Management as symbolic action: the creation and maintenance of organizational paradigms." In L.L. Cummings and B. Staw, eds., *Research in Organizational Behavior*, Vol. 3. Greenwich, Conn.: JAI Press, 1981.

Porter, L.W., Lawler, E.E., and Hackman, J.R. *Behavior in Organizations*. New York: McGraw-Hill, 1975.

Robertson, T.S. *Innovative Behavior and Communi-cation*. New York: Holt, Rinehart, and Winston, 1971.

Rogers, C.R. *On Becoming a Person*. Boston: Houghton-Mifflin, 1961.

Rogers, E.M., and Shoemaker, F.F. *Communication of Innovations: A Crosscultural Approach*. New York: Free Press, 1971.

Salancik, G.R., and Pfeffer, J. "A social information processing approach to job attitudes and task design." *Administrative Science Quarterly*, 1978, *23*, 224–53.

Schein, E.H. "Organizational socialization and the profession of management." *Industrial Management Review*, 1968, *9*, 1–15.

Schein, E.H. "The individual, the organization, and the career: a conceptual scheme." *Journal of Applied Behavioral Science*, 1971, *7*, 401–26.

Schein, E.H. "Personal change though interpersonal relationships." In W.G. Bennis, D.E. Berlew, E.H. Schein, and F.I. Steele, eds., *Interpersonal Dynamics: Essays and Readings on Human Interaction*. Homewood, Ill.: Dorsey Press, 1973.

Schein, E.H. *Career Dyanamics*. Reading, Mass.: Addison-Wesley, 1978.

Scott, W.E., "Activation theory and task design." *Organizational Behavior and Human Performance*, 1966, *1*, 3–30.

Seashore, S.F. "Group cohesiveness in the industrial work group." Ann Arbor, Mich.: Survey Research Center, University of Michigan, 1954.

Shaw, M.E. *Group Dynamics: The Psychology of Small Group Behavior*. New York: McGraw-Hill, 1971.

Staw, B. "Motivation in organizations: toward synthesis and redirection." In B. Staw and G.R. Salancik, eds., *New Directions in Organizational Behavior*, Chicago: St. Clair Press, 1977.

Staw, B. "Rationality and justification in organizational

life." In B. Staw and L.L. Cummings, eds., *Research in Organizational Behavior*, Vol. 2. Greenwich, Conn.: JAI Press, 1980, pp. 45–80.

Stoner, J.A. "Risky and cautious shifts in group decisions: the influence of widely held values." *Journal of Experimental Social Psychology*, 1968, *4*, 442–59.

Tushman, M., and Katz, R. "External communication and project performance: an investigation into the role of gatekeepers." *Management Science*, 1980, *26*, 1071–1085.

Wanous, J. *Organizational Entry*. Reading, Mass.: Addison-Wesley, 1980.

Weick, K.E., *The Social Psychology of Organizing*. Reading, Mass.: Addison-Wesley, 1969.

Wheeler, S. "The structure of formerly organized socialization settings." In O.G. Brim and S. Wheeler, eds., *Socialization after Childhood: Two Essays*. New York: Wiley, 1966.

Wortman, C.B., and Linsenmeier, J. "Interpersonal attraction and techniques of ingratiation in organizational settings." In B. Staw and G.R. Salancik, eds., *New Directions in Organizational Behavior*. Chicago: St. Clair Press, 1977.

14

Using Culture for Strategic Advantage: Promoting Innovation Through Social Control

CHARLES A. O'REILLY III
MICHAEL L. TUSHMAN

In previous readings you have been introduced to the congruence model as a method to help managers diagnose and solve problems. The congruence model calls attention to both *formal organizational control systems* (e.g., critical tasks, reward systems, and organizational structure) and *informal or social control systems* (e.g., motivating people and managing culture). While both types of control are important for implementing strategy, it is clear that the horizontal axis of the model, or the social control system, is the lever most managers have difficulty using. As we shall show, it is also the case that formal controls are ill-suited to promoting innovation and change. Paradoxically, the culture of the organization can act as a powerful social control system that can be a critical way managers can foster innovation. Unfortunately, managers are often not attuned to managing culture, seeing it as imprecise and difficult to manage. What is soft, turns out to be quite hard.

In this chapter we focus on organizational culture, not as an abstract concept but as a social control system that managers can use to promote innovation and change. We show how thinking about culture as social control, expressed in the norms and values shared by an organization's members, can enable managers to use the horizontal axis of the congruence model to achieve strategic advantage.

USING CULTURE TO PROMTE INNOVATION AND CHANGE

While both the academic and business press are filled with references to culture, the concept itself often seems abstract and difficult for managers to get a handle on, especially those who favor a more detached, rational approach. The stories may be interesting, but what can a manager actually do to affect the culture in his or her own unit or organization? What are the levers to manage culture? To address this critical topic, let's think not about *culture* but about a subject most managers are more comfortable with: *control* in organizations. Much of management is really about control; about getting people to do what is necessary to get the work done, preferably in a way that uses their full potential and leaves them feeling motivated and engaged. In this fundamental sense, management is really about coordinating and controlling collective action.

But where does control come from? How does it work? At its essence, *Control comes from the knowledge that someone who matters to us is paying close attention to what we are doing and will tell us how we are doing.* From this perspective, organizational control systems, whether they are financial planning systems, budgets, inventory control processes, or safety programs, are effective when those being monitored are aware that others who matter to them, such as a boss or staff department, know how they're doing and will provide rewards or punishments for compliance or noncompliance.

An illustration of this was brought home to us several years ago while we were doing research in a large manufacturing company. The company had, at great effort and expense, installed an MBO system. Employees were trained and the requisite procedures developed. But senior management was puzzled. After a year of operation, they had noticed that the system was working in only part of the organization. As we conducted our research, they asked us for any opinions about why the MBO system seemed only partially effective. The answer was as simple as it was surprising. In the part of the organization where the system was working, we heard a very consistent story from the divisional manager on down. "This is a useful system. It isn't easy and requires some effort, but it really does help us get coordinated." In the part of the organization where the system was not working, the divisional manager had a clearly stated opinion that the system was another one of those personnel department exercises. He was direct in telling his subordinates that they should complete the paperwork in case they were audited but should get on with the more important work they were doing. In the part of the firm where people who mattered, in this case divisional management, were paying attention, the MBO system was implemented and worked. When the signals from senior management were that the system was unimportant, the control system did not work. Thus, in spite of all the time and money the company had spent, the control system worked, consistent with our definition, only when people who mattered were paying attention.

While the generic definition of control we have offered applies to formal control systems, it is also useful in thinking about culture as a *social control system.* As we indicated earlier, culture is really about those norms and values shared in a unit or organization. If we care about others, for instance, we like them and we want others to like us, and if we have some agreement about what is important, then we are under their control. That is, if I care about you and we have shared agreements about what is important and how to act, then whenever we are together, we effectively "control" each other.

In organizations, we have numerous agreements about what's right or wrong and what's important and unimportant. Whatever senior managers may claim is important, most people in organizations soon figure out what is really important and how to behave. From a managerial perspective, this is what culture really is. It is a social control system that exists in all organizations and subunits within them. The relevant question is whether this social control system is supporting or hindering managers in accomplishing their critical tasks. These shared agreements and social expectations constitute a powerful and pervasive social control system within groups and organizations—a system that we will show can be more powerful and effective than some formal control systems.

For example, in the early 1980s both Roger Smith, then CEO of GM, and Don Petersen, CEO of Ford, announced quality efforts at their respective organizations. At the time, Ford's quality levels were the worst in Detroit and three times worse than the Japanese. Half a dozen years later, Ford's quality was the highest among the Big 3 U.S. automakers. This was done without reorganizing or an infusion of outsiders. Ford engaged in a cultural revolution driven by Don Petersen and his senior team. Over 80 percent of these improvements came from the human side, not from automation. During this period, GM, except for NUMMI, largely failed in its quality effort. Rather than truly focus on quality, the reliance was on technology and financial numbers. Senior management at both companies proclaimed "quality" efforts. While Ford experienced the cultural change, at General Motors, the emphasis on management by the numbers continued.

The Downside of Managing with Formal Control

Because control is at the heart of organizations, it is not surprising that both managers and academics should be preoccupied with it. Without a means to coordinate or control collective action, organizations would provide no advantage over individual efforts. When most of us think about designing control systems, we typically try to measure either outcomes or behavior or both. Our first inclination is to design measurement and reward systems based on outcomes. If we can measure things like sales per hour, margins, number of defects, or customer satisfaction, we feel as though we are in control. In those cases where assessing outcomes is difficult or impossible, we settle for measures of behavior. Did the person follow prescribed methods? Was the paperwork completed properly? Is the greeting suitably enthusiastic? When measures of both outcomes and behavior are available, we can know with precision whether the work is being done the way we prescribed. Our uncertainty and dependence on others is reduced because we can quickly identify deviations from the right way to do things. We can also be more confident that we are rewarding the high performers and punishing those who fail to measure up. The question, however, is whether this really is the most effective way for managers to operate?

Think for a minute about the types of jobs associated with extensive formal control systems, especially those that rely on monitoring both outcomes and behavior. Jobs in this category include traditional assembly-line work and many retail sales positions. But how do incumbents feel about their work? What is the typical motivation of people in these jobs? Anyone who has held such a job can answer quickly; they feel they are not trusted, overcontrolled, a cog in a machine, and, of course, they are often unmotivated. Under these circumstances, innovation often consists of efforts to beat the system. Too often, this is the situation characterizing sales people in department stores and workers on automobile assembly lines. And yet, we have the counterexamples of Nordstrom and NUMMI. How can their employees be motivated when they too are in "controlled" jobs?[1]

The difference is in the use of social control to supplant or supplement formal control. For instance, there are *no* industrial engineers at NUMMI. Does this mean there is no attention to work design and efficiency? Hardly. At NUMMI work teams assume responsibility for these activities. New workers are trained in industrial engineering techniques and the team itself does the redesign and improvements. One of the authors was walking through the NUMMI plant with a team member who proudly pointed out how he and his team had figured out that by repositioning some soda pop machines they were able to reduce the time for the just-in-time inventory run by two seconds. In a typical Detroit plant industrial engineers would do time and motion studies on the workers and instruct them in how and when to behave. Under this extensive control, the workers would, in turn, likely be unmotivated and, where possible, expend effort and ingenuity in figuring out ways to beat the system rather than improve it. At NUMMI, the culture or social control system includes norms about continuous improvement and team responsibility. Instead of feeling unmotivated, the workers feel a sense of autonomy and responsibility. Instead of the formal control system requiring possibly expensive monitoring, the team members control each other.

How can managers develop control systems for nonroutine and unpredictable jobs where it is difficult to assess reliably and accurately either outcomes or behaviors, such as R&D, consulting, and technical specialties like information systems? These attributes characterize any job in which change is frequent and unpredictable. This can easily encompass almost all service jobs where nonstandard requirements are the norm. In the face of work requirements that are becoming more complex, uncertain, and changing, control systems cannot be based on static, formal control. Rather, control must come in the form of social control systems that allow directed autonomy and rely on the judgment of employees informed by clarity about the vision and objectives of the business.

The flaws in formal control systems become apparent when the underlying assumptions are surfaced. Three such assumptions seem obvious. First, formal control relies on the idea that almost everything of importance can be anticipated in advance. This is reasonable only in a stable or slowly changing environment. The second assumption underlying formal control systems is that this type of top-down authority is legitimate and worthy of compliance. This smacks of the old scientific management ideas that people at work will comply with directives in exchange for a fair day's pay. While most people continue to accept authority in organizational hierarchies as legitimate, there has been a growing acknowledgment that people want more from work, including an opportunity to contribute and use their talents. The third assumption implicit in the use of formal control is that extrinsic rewards are largely sufficient to direct job-relevant behavior and that these can be based on the measurement of outcomes. But it doesn't take a degree in psychology to realize that this is, at best, an incomplete story. Not only do we understand that what gets measured often overwhelms what doesn't, we also know that intrinsic motivation and nonmonetary rewards can sometimes be more powerful motivators than money or promotions.

This, of course, is not an argument *not* to use formal control systems. A close examination of successful firms and managers suggests the need for both *formal* as well as *social* control systems, with the latter often being an overlooked but powerful alternative to the former. Indeed, the more organizations are downsized and made more lean, the more these firms must rely on social control to both motivate and monitor behaviors. In *What America Does Right*, Bob Waterman explores the success of Federal Express, a company that relies on extensive formal and social control systems.[2] Waterman notes that, ". . . everyone I talked to at FedEx—from couriers to middle managers to people at the top—say the thing they like most about the company is the *freedom to do things their way*. Managing this delicate balance—the balance between potentially oppressive systems and the cheerful attitudes of a clearly enthusiastic work force—turns out to be the key to FedEx's success." In his view, FedEx has it both ways—formal systems that serve their strategic ends and social controls that motivate and empower employees. The irony is that social control systems can be more powerful and less intrusive than formal controls.

In the logic of the congruence model, we want to align the horizontal axis, or social control system, in a way that is consistent with the critical tasks to be performed. Core norms are those with performance consequences for the particular unit, such as quality, customer service, teamwork, and innovation. In this chapter we will show how culture as a social control system can be a critical element in implementing strategy and promoting innovation and change. To do this, we show how culture can be used to promote or hinder organizational innovation. We then generalize this to suggest how culture and social control can be used to implement corporate and business unit strategy.

Using Culture To Promote Innovation and Change

To illustrate this perspective, consider the impact of culture on innovation. First, as we described earlier, it is obvious that organizations must be innovative to survive. At the same time, designing formal control systems to insure innovation is difficult, because by its very nature innovation involves unpredictability, risk taking, and nonstandard solutions. If there were a guaranteed way to promote innovation, such as there is to optimally schedule inventory or manage cash flow, most organizations would adopt it, and there would be little competitive advantage gained. It is precisely because of this uncertainty that a competitive advantage is possible for innovative firms like 3M, Intel, Rubbermaid, and Federal Express. The intriguing issue raised in examples such as these is *How do they keep doing it?* How is it that some firms sustain a competitive advantage based on innovation over long periods?

An important piece of the puzzle is in how they use culture or social control. Think about those norms, which if they were widely shared and strongly held in your organization would help promote innovation. Imagine you could introduce whatever norms you liked into your organization. What are the shared expectations about attitudes and behavior that would result in higher levels of innovation? This is the challenge we have investigated in our research for the past several years.

To answer this, we must first recognize that innovation is an outcome. We know innovation has occurred only because something has changed. Our definition of innovation is "the successful introduction into an applied situation of means and ends that are new to the situation." Note that this definition emphasizes that a new way of doing something has occurred. It does not mean that something unique has happened, only that an idea that is new to the particular situation has been implemented. It includes ideas that may be old hat to some but are new to the specific context. This typically involves the invention, discovery, modification, or copying of an idea, technology, or process that is new to the organization, followed by the development and implementation of this approach. This means there are two component processes that underlie all innovation: (1) *creativity* or the generation of a new idea; and (2) *implementation* or the actual introduction of the change. Innovation occurs *only* when both components are present. Thus, to promote innovation in organizations requires that managers both stimulate new ideas and put these ideas into practice. Simply being creative is not a guarantee of innovation. By focusing on these two components, we now have a more tractable question: What are the norms that can promote creativity and implementation?

NORMS TO PROMOTE INNOVATION

To understand how managers and organizations enhance innovation, we began six years ago to systematically ask participants in executive programs on strategic innovation and change to suggest norms that helped promote innovation in their organizations. We challenged participants to identify those expectations that, if they were widely shared and strongly held in their business, would significantly enhance both creativity and implementation. Over two thousand managers from Asia, Europe, Africa, and the United States have participated in these discussions. Managers from industries as diverse as mining, financial services, health care, heavy manufacturing, consumer products, and high technology have responded. The consistency across these disparate geographies and industries is remarkable. Consider the data.

Table 14.1 reports a set of norms associated with innovation as identified by managers from five very different samples; a South African natural resources company, a European pharmaceutical company, a U.S. financial services institution, an international group of R&D managers, and a Japanese beer company. Although these suggestions came from widely differing national cultures and industries, there is great consistency in how these managers think about promoting innovation. All recognize the fundamental importance of challenging the status quo, designing systems that tolerate some failure, encouraging risk taking, providing resources, and open communication. A complete reporting of the data we have collected would show even more consensus. Repeatedly, managers identified a common set of norms needed to promote creativity and implementation.

After hearing similar suggestions from many sources, we decided to consolidate and validate these findings. To do this, we conducted a study involving over 200 managers from 29 groups in high technology firms in the Silicon Valley. Table 14.2 summarizes the results of that study.[3] First, the idea that innovation is reflected in norms that promote the two component processes of creativity and implementation was confirmed. Further, those groups that had comparatively strong norms for both were rated as the most innovative. Simply promoting creativity or focusing on being quick to implement did not guarantee innovation. Both are

TABLE 14.1. Norms for Innovation, Five Firms

South African Natural Resources Company	European Pharmaceutical Company	U.S. Financial Services Firm	International R&D Managers	Japanese Beer Company
Mistakes are OK	Rewards	Accept failure	Freedom to fail	Cooperation
Recognition	Accept failure	Free to try things	Risk taking	Mistakes OK
Rewards	Careful	Time	Fast	Openness
Mutual respect	Learn	Support	Prudent	Flexibility
Open communication	Calculated	Resources	Cheap	Clear direction
Freedom to experiment	Clear objectives	Clear goals	Rewards	Ideas are valued
Expect change	Share information	Celebrate success	Involvement	Rewards for innovation
Challenge the status quo	Teamwork	Remove barriers to change	Tolerate dissent	Resources
Equal partners	Commitment from the top	Set the example	Listen	
	Empowerment		Positive attitude	
			Resources	

needed. To illustrate in more detail, consider the four main norms shown in Table 14.2 that were strongly associated with team-based innovation.

Stimulating Creativity

The results of the study suggested that there are two main ingredients to stimulating creativity: (1) support for risk taking and change; and (2) a tolerance of mistakes. In discussions, managers invariably identified these elements as critical for innovation. When pushed about how to apply these ideas, some very practical advice emerged. First, think about how employees in any organizational setting might learn that innovation is important. Said differently, why should members of a group or organization expend effort to come up with ways to do things differently, whether it is to serve customers better, deliver a higher quality product, or develop a new way of producing goods and services? Change is uncomfortable, so why bother?

TABLE 14.2. Norms That Promote Innovation and Change

Norms That Promote Creativity
Support for risk taking and change
Rewards and recognition for innovation
Positive attitudes and role models for change by management
People are expected to challenge the status quo
Tolerance of mistakes
Mistakes are accepted as a normal part of the job
People are given the freedom to make changes
"It's better to be safe than sorry" isn't an accepted practice
Norms That Promote Implementation
Effective group functioning
Teamwork is emphasized
People share common goals
Information is shared openly
Speed of action
Decisions are made quickly
Flexibility and adaptability are emphasized
Sufficient autonomy is given to insure implementation

Support for Risk Taking and Change

Clearly, one important way of signaling that something is important is by rewarding it. Most managers we spoke to, however, were not convinced that monetary rewards were very effective at promoting creativity. They believed that recognition from management, colleagues, and others was more powerful. Thus, at companies like 3M, Dow, and the French petrochemical company, Elf Acquitane, managers rely on nonmonetary rewards like recognition to signal that innovation is rewarded. At BOC, the U.K. gases company, winners of innovation awards receive team trophies, individual plaques, and trips to London for meetings with senior management. The recognition that innovation often requires extensive teamwork means that individual-based awards, or rewards that rely primarily on cash awards, has moved many firms toward more symbolic awards. Bryan Sanderson, the CEO of BP Chemicals, which has recently revamped its reward program for innovation, has noted that "Rapid cost reduction and focusing is a miserable management job, but conceptually it is easy. But getting the next phase of profit improvement is difficult and increasingly it will have to come from more innovative thinking which increases performance and productivity."[4] This requires the careful design of reward and recognition systems to promote the norms that encourage innovation.

In addition to suggesting the need to recognize creativity, managers were also quick to note that to stimulate creativity, you had to be prepared to encourage risk taking and accept failures. Indeed, the issue of designing mechanisms for these seemed more important than a formal monetary reward system, no matter how cleverly designed. At DuPont, for instance, one manager recalled how a group in his department had tried an innovative approach to a problem that had failed miserably. Rather than ignore the failure, he decided to have a recognition celebration and label the failure as a "good try." At Hershey Foods, one manager instituted the "Exalted Order of the Extended Neck" for employees who bucked the system and exhibited entrepreneurship. At Nordstrom and Federal Express, the heroes often include those who tried to satisfy customers and failed as well as those who succeeded. For instance, at Nordstrom each customer complaint is put in a binder with the actions taken to remedy it. These are then judged at regional meetings for the best response. The winner can win prizes up to $1,000. Each complaint is seen as an opportunity to improve and people are rewarded for this.[5] In one Canadian company, a monthly award is given for the best idea that didn't work: a toilet plunger with a duck head.

At Rubbermaid, which introduced 365 new, not simply improved, products during 1993, management continually emphasizes that it's OK to take risks. A former head of the company used to make it a point to tell his subordinates how early in his career he started a product line that flopped. The important lesson was that when the firm closed the business, the then-chairman came down and told him and his team that the failure would not be held against them and that their futures were still bright. In each of these incidents, managers are trying to signal that creativity and trying is valued, either through rewards and recognition of success or by recognizing good faith efforts that failed.

As suggested by the congruence model, aligning culture sometimes involves modifying the formal reward system. More often, however, it requires that managers simply use foresight and imagination in providing small rewards and informal recognition for attempts at being creative. To be effective at this also requires that managers genuinely understand the values and needs of their employees. What motivates Ph.D. chemists in Switzerland is likely to be different than for retail clerks in San Francisco or managers in a Japanese manufacturing firm. Understanding what motivates a group of employees insures that appropriate types of rewards can be given. Once the rewards are adapted to the particular context and type of employees, there is good evidence that rewards and recognition can stimulate creativity.

For example, national cultural norms in Switzerland include expectations of public modesty. These preclude ostentatious public recognition ceremonies like those used at Mary Kay Cosmetics where successful "beauty consultants" (never saleswomen) are crowned as beauty queens in front of hundreds of their cheering friends and colleagues. However, this does not mean that recognition is less effective in Switzerland. It simply means that the specific rewards and settings are suitable for the Swiss national sense of modesty, for instance, more private recognition. Thus, while Mary Kay uses an array of pins, trophies, and pink Cadillacs, which are valued by her employees, organizations like Apple and 3M use elevation to technical societies as recognition for their scientists. Recognition works. The managerial challenge is to use it in a manner consistent with the underlying values of the employees.

Another lever for encouraging support for creativity and risk taking that emerged from the study is the importance played by a positive attitude toward change shown by management. People in organizations are continually watching and listening to their bosses and senior managers. The test is whether management is consistent in modeling and supporting risk taking. It is no surprise that creativity is more likely when managers demonstrate in word and deed how people throughout the organization should behave. When Fred Smith, CEO of Federal Express, tells employees to be mavericks, he also tells them how he proposed the original idea for the company in a college term paper and received a "C" with the admonition that it would never work. The message is that the success of Federal Express comes from taking chances, even when so-called experts don't agree. One FedEx worker is quoted by Levering and Moskowitz as saying, "I don't have any fear that if I try something that doesn't work, there will be repercussions. There have been a few things that didn't work. We just didn't do them the next day."[6] At Alagasco, a natural gas distributor in Alabama, the president passed out a variety of business cards to his employees encouraging them to try things differently. He told them each card was like the 'get out of jail free' card in Monopoly. If you try something and it fails, turn in the card and you're forgiven. Jack Welch constantly acknowledges that he and senior managers make mistakes. Herb Kelleher routinely engages in playful behavior, like wearing a dress and feathered boa, to get employees to loosen up and take risks. Even John Rollwagen at Cray Research has appeared at company meetings dressed as a duck. Robert Goodale, president of Finevest Foods, has two pictures of himself on his office wall. One shows him shaking hands with former President Reagan; in the other he's wearing a hula skirt. "The real me is the one in the hula skirt," he says.[7]

Underneath this seemingly silly behavior is a serious message: it's OK to think out of the box. Inspiring employees to be creative requires that managers convey a consistent message that challenging the status quo is expected. It isn't acceptable to continue to do things the way they've always been done. For instance, at Odetics, a small aerospace company, management has an explicit policy of "structured spontaneity," which refers to a policy of deliberately not institutionalizing things. Management believes that it is important to convey the message that it isn't acceptable to rest on past successes by developing tried and true procedures. Procter & Gamble drives this message home using its performance appraisal process. When subordinates set their annual goals with their managers, they are expected to show how they will change their job during the company year. At Rubbermaid, responsibility for each one of the 4,700 products is assigned to a team. Each team's compensation is tied in part to new product introductions, including a corporate goal of 33 percent of corporate revenues from products introduced in the past five years. Each division may have an active list of hundreds of new products and revisions. The message in each of these examples is that change is an expected part of the goals of everyone.

Tolerance of Mistakes

Another major ingredient in the development of organizational creativity is developing a tolerance for mistakes, both personally and organizationally. This is easy to say and hard to do. All of us have heard bosses say that "it's OK to make mistakes." We've also seen careers that have been damaged when people took this exhortation at face value. Most of us have a realistic cynicism about statements like this. How do managers get past this resistance? Part of the answer is in great clarity about what types of risk taking and mistakes are OK. At Johnson & Johnson, a common statement is R. W. Johnson's maxim that "Failure is our most important product." J&J managers communicate what constitutes a "reasonable" mistake. That is, mistakes are permitted if they are based on analysis, foster learning, and are modest in impact. At IBM, mistakes are also accepted if there is learning and the same mistake is not made again. The important lesson is that each manager needs to clearly communicate what is meant by "acceptable" mistakes. Then, as illustrated above, there must be continual support for those who try and do not succeed. As a Delta Airlines superviser observed, "One thing that helps us to admit that we have made a mistake is that people who are above us don't hold it against us. In this company, if you make a mistake and you recognize it and something positive comes out of that, that's a positive for the company."[8]

Together with the design of fault-tolerant systems, people must also be given the freedom to try things and fail. If management at 3M and other innovative organizations routinely punished employees who tried new things and failed, innovation would soon cease. Part of this is managerial style. Do superiors send signals that they trust their subordinates? Are they willing to give people the latitude to try new things? Are jobs designed in ways that encourage incumbents to learn new things, or is the emphasis on strict adherence to fixed routines? At 3M, William McKnight, the spiritual father of the company whose legacy is still alive in the company, laid down the principle that, "The mistakes that people will make are of much less importance than the mistake that management makes if it tells them exactly how to do a job." This leads to another oft-repeated saying within 3M, "The captain bites his tongue until it bleeds." This refers to the requirement that sponsors of projects often have to mightily restrain their inclination to evaluate and say "no." Indeed, the eleventh commandment at 3M is "never kill a new product idea."

The extreme version of this philosophy can be found at W. L. Gore and Associates, where people have neither titles nor formal job descriptions. Newcomers are expected to find ways they can make a contribution. Initiative is required. Organizations like these create strong norms that trying is valued and being cautious is not acceptable. Under these circumstances, creativity or new ways of doing things cannot help but be increased as employees see innovation modeled by managers, watch co-workers get rewarded, hear it talked about by their peers, and observe that good-faith mistakes are not only tolerated, but may even be approved of. As a salesman at Nordstrom said, "You will never be faulted for erring on the side of the customer." Jack Welch is clear when he says "The trick is not to punish those who fall short. If they improve, you reward them—even if they haven't reached the goal."[9] Under these conditions (support for risk taking, positive attitudes toward change from management, and fault-tolerant reward systems), it is easy to see how norms that promote creativity can develop. It is also easy to see how little creativity would occur in settings where management emphasized not making mistakes and always punished those who made them.

Promoting Implementation

The second set of norms that are critical in enhancing innovation is those that promote implementation or execution. Once creative ideas and approaches are identified, the question is "which ones will be acted upon?" The former CEO of Philips once said, "Around here we talk a lot, but everything stays the same." Although Philips was often a pioneer in developing new technologies, it frequently failed to capture value in the market. At that time, Philips' culture was characterized by norms of consensus, negotiation, and politeness, not norms that promote action. Contrast this with the "Just do it" culture at Nike. What, then, are the norms that ensure action? Our study suggests two important types: (1) norms that emphasize effective teamwork and group functioning, and (2) those that emphasize speed and urgency. Consider why these are important.

Effective Group Functioning

In complex organizations innovation almost always involves getting people to alter the way they operate. As we noted in Chapter 3, when tasks are characterized by sequential or reciprocal interdependence, any change in work flow requires adjustment by those doing the work. To introduce a new product, process, or technology means that the old way of doing things must be disrupted. Implementation of the required changes is enhanced when groups operate effectively. Our research suggests that in order for groups to work well together, three factors are important: (1) teamwork, (2) common goals, and (3) the open sharing of information.

The first element, teamwork, is enhanced when members like and respect each other, understand others' perspectives and operating styles, can resolve disagreements, and can communicate effectively. This is based partly on skills and abilities that can be learned (e.g., problem solving) and partly on shared expectations within the group (e.g., norms about mutual respect and how to resolve differences). For example, Hewitt Associates and J. P. Morgan are firms that believe that their competitive advantage rests partly on their ability to act as a team to meet their clients' needs. At Goldman, Sachs, where teamwork has long been a core value, a primary consideration for promotion is the ability to be a team player. "If you can't sublimate your ego or work with others, you have a problem," says a top partner. One of the guiding principles at Goldman is, "We stress teamwork in everything we do."[10]

More recently, however, as Goldman has expanded rapidly and emphasized more proprietary trading, this emphasis on teamwork has come under some stress with more individualistic rewards for trading profits undercutting teamwork norms. This has led to a questioning of whether teamwork is important for the future. Says Stephen Friedman, the recently retired CEO, "Every major company I have seen that has suffered by failing to adapt has been too rigid in worshiping the dysfunctional aspects of its prior culture." However, the new CEO, Jon Corzine, sees teamwork as a critical success factor for Goldman and is working hard to reinforce this belief. Ironically, one of Goldman's major competitors, Bankers Trust, the quintessential trading bank, has recently begun moving toward relationship banking and away from a reliance on proprietary trading. Its previous culture reflected the strategy and critical tasks needed for trading: individualistic, innovative, and a concern with transactions rather than with clients. This new strategy has required senior management to make changes in the structure of the firm (e.g., merging its equities sales force with the equities derivatives sales group) and the reward and measurement systems, and to begin emphasizing teamwork and understanding the client's needs as a critical cultural ingredient. While the structural and systems changes can be made quickly, the cultural changes will clearly take longer. As Thomas Theobald, former senior manager at Citibank and chairman of Continental Bank, observed, "This is multiyear if not decade long stuff to move a cultural positioning. You can't flip-flop in a year or two." Yet the balance between the norms required by a trading culture and those emphasizing relationship banking is a difficult one to achieve, as shown by the destructive conflict that split Shearson-Lehman during the 1980s. The challenge is for managers like Charles Sanford at Bankers Trust and Jon Corzine at Goldman to do inculcate both, a formidable task.

At Odetics, a software engineer said, "Everyone works very hard and everyone is very team-oriented. They do their own part, but they support everyone else. If we're working on a deadline or a program, as soon as one person gets their work done, they will go and help the next person to get theirs done."[11] At one of P&Gs successful self-managing plants team members are responsible for training and qualifying other team members. Bob Waterman, after studying innovation at Rubbermaid, claims that ". . . teams are the heart and soul of Rubbermaid" and "They're almost like tiny companies unto themselves. Team participants live and breathe their products." The explicit expectation in these organizations is that people will be helpful to each other rather than competitive. Norms that encourage effective group functioning make the implementation of new ideas quicker and easier. In this sense, these norms facilitate the coordination and adjustment necessary for organizational innovation and change.

A second factor enhancing implementation is the development of shared values. These help people understand both what they are trying to do as a group and also provide predictability and efficiency in how to interact. A common vision, whether it is for the corporation or work group, can be a powerful way to coordinate people and involve them emotionally. At Motorola, for instance, people often say "I am a Motorolan," not "I work at Motorola." Having a shared vision means that employees understand what the fundamental goals and values of the organization are and can act more quickly because of this. In organizations and groups where no such commonality exists, the implementation of new approaches must proceed more slowly as members negotiate disparate perspectives and beliefs.

A third element in effective group functioning is the open sharing of information, including good news and bad. The reason for this is clear: if groups are to function effectively, they must deal with the way the world is, not the way they would like it to be. This is one key success factor in the success at Springfield ReManufacturing. The CEO, Jack Stack, believes that unless all employees know and understand the financial firm's data, they won't really understand how their jobs contribute. Therefore, all employees are educated in how to read a financial statement and interpret accounting data. One line worker said, "They are training us on financial statements right now and they are hard to understand. They show us the sales and profits, what they ship, they go through every item. Every week you sit down with your supervisor and he gives you the numbers. You can see how your work affects the statement. . . ." If you are not working up to standard, it's going to show up on that paper."[12] The wide dissemination of information not only enhances understanding and motivation, it also signals to all employees that they are trusted and there is a reciprocal obligation between the firm and the worker. Jim Morgan, the CEO of Applied Materials, a very successful manufacturer of semiconductor manufacturing equipment, has an aphorism that he recites to promote information sharing: "Good news is no news. Bad news is good news. No news is bad news."

Speed of Action

In addition to the promotion of effective group functioning, another set of norms that promotes the implementation of innovation and change in organizations is related to speed and urgency. Three important elements comprise this factor; norms that decisions should be made quickly, expectations that promote flexibility and adaptability, and a sense of personal autonomy that encourages action. Each of these supports getting things done and is distinct from the norms that enhance creativity. These norms are about action.

The first of these three relates to the speed with which decisions are taken. Think about how quickly or slowly decisions are made in different organizations. The pace of work can vary considerably even within an organization. Why should this be? In some places, the norms may include expectations that speed is important, the pace of work is rapid, and people are expected to work long hours. In other situations, the pace is leisurely and working hard is not valued. (One of the oldest terms in industrial sociology, dating to the Hawthorne studies at Western Electric, is that of a "rate buster" or someone who does not comply with the group productivity norm.) Where do these differences come from? They are clearly socially constructed expectations, often sufficiently widely shared so that everyone knows what is expected. At Microsoft, for instance, the corporate espresso bars are open twelve hours a day, and people are there at all hours of the day and night. There is also a sense of insecurity that drives people to excel. Jay Blumenthal, director of program management, said, "One of the things that characterizes Microsoft is insecurity. People never believe that they've got the best product. It's always the fear that somebody is going to come up with a better product."[13] At Intel, Andy Grove refers to this attitude as "competitive paranoia," and notes that "The pace of work these days isn't easy to live with, but welcome to the nineties." Compare these to Steve Simmons, a government worker in San Francisco, who was quoted as saying, "I can't think of a day when I haven't done nothing. I do nothing three fourths of the time, not including sleeping."

At Goldman, Sachs, sixty- to 80-hour weeks are the norm, and the expectation is that people are prepared to sacrifice their personal lives to ensure the work gets done. In this environment, there is often a sense that doing things quickly is important. It is not OK to delay things. At Federal Express, for instance, there is a policy that every employee or customer query is answered that day, even if the answer is only that the issue is being worked on. At other firms, norms sometimes convey an expectation that inaction is the preferred alternative and initiative is not expected or desired. For example, at Preston Trucking a worker (now called an "associate") described how under the old system employees never did anything without authorization, including something as simple as changing a light bulb. Officers of the company developed the plans and managers instructed workers what to do. People were expected to come to work and do what they were told. Since the early 1990s, there has been a cultural revolution emphasizing initiative and innovation. Norms now emphasize the expectation that workers will be flexible and use their initiative. In 1991, the 5,000 plus associates generated over 7,500 suggestions for improvement. For its part, management now shares performance and financial information, actively encourages innovation, and emphasizes trust and teamwork. Meanwhile, since the deregulation of the trucking industry in 1980, over 75 percent of Preston's direct competitors have failed.[14]

The second element in promoting implementation through social control emphasizes flexibility and adaptability. These are norms, for instance, which encourage and legitimate groups and individuals to take initiative and adapt to changing conditions. At Avis, the employee-owned rental car company, there is an emphasis on EPGs (employee participation groups). These allow groups of employee–owners to develop locally responsive actions to help the company. In New York, for example, they developed a job-swap program in which employees at the local airports swapped jobs for short periods with others at the corporate headquarters to learn more about each other's work. In Hawaii, the local EPG designed their own Avis outfits to fit the tropical environment (all in Avis red). At BE&K, a large construction company, there is a deliberate policy of not having rigid job descriptions to encourage responsibility. At Chaparral Steel, employees are given encouragement and opportunities to learn every job in their department. From the employee's perspective this provides growth and challenge. From the firm's viewpoint, it enhances flexibility.

As H-P has transformed itself, one of the new values that now gets emphasized with the H-P ways is flexibility and the need for change. In these and other firms the underlying value being emphasized is flexibility. The intent is to create expectations within the organization's workforce that adapting to changing circumstances, taking initiative and responsibility, and modifying the way work gets done are expected parts of the job. Bob Waterman quotes Mike Lauderdale, a FedEx employee working in the billing department where attention to detail is important, as saying "I never look at things as being a rule. I look at them as a guideline. I always have the freedom to take the guidelines and adjust them to bet suit me." The clear shared expectation at Federal Express is that people are expected to use their good judgment and adapt to the circumstances. At Nordstrom, to deliver outstanding customer service, the informal rule is "Never say no" to a customer. It is expected that salespeople will sell outside their department. So when a customer buys a suit and would like a pair of shoes, the salesperson accompanies the customer to the shoe department and continues to serve the customer.

A final set of norms that helps to promote speed are those associated with a sense of autonomy. These create expectations that the individual has the latitude and obligation to take action, not wait for approval. A core value at Johnson & Johnson is the notion of decentralization that emphasizes the autonomy of the group companies. Managers are expected to run their own companies without interference from the corporations. So, for example, no strategic planning is done in the headquarters at New Brunswick. Companies do their own. Early on, H-P adopted as a core part of its philosophy the idea that managers should give subordinates a well-defined objective, allowing as much freedom as possible in working toward that objective, and seeing that their contributions were recognized throughout the company. 3M has similar values as an important part of its culture, including norms within the auditing group to adopt a "how can I help you" attitude and not a "how can I stop you."

At a more micro-level, many firms emphasize

autonomy for their employees in the way they minimize rules and design jobs. At Chaparral Steel, one employee said "The main thing that makes me happy about getting up in the morning and coming to work is the fact that I like what I do and have freedom." This is echoed by a credit collection manager at Dayton-Hudson, "I think the thing that stands out about this company is the culture and value they place on challenging individuals and letting them run their business." At TDIndustries, an employee-owned plumbing and air-conditioning company, senior management emphasizes "freedom" as a core value. This translates into continual efforts to free up employees so they can do their jobs in the way they best see fit. Gary Brackett, a service technician, said "Around here you're basically your own boss. I do big jobs for my boss. He sends me out, I go look at the job, I bid on it, and I put it in. Half the time he doesn't even see the job or know what's involved unless I tell him."[15] A NUMMI worker claimed that the best thing about the Toyota system was that he rarely saw his supervisor. The expectation was that the workers themselves would get the job done the best way they saw fit. A Nordstrom salesman said, "Where else could I get paid so well and have so much autonomy? . . . No one tells me what to do, and I feel I can go as far as my dedication will take me. I feel like an entrepreneur." This is in a setting with strong conformity around norms such as dress, positive attitudes, and treating customers in a particular way. Yet, the feeling is one of freedom.

SUMMARY

It would be a mistake to pretend that inculcating norms such as those we have described above was easy or can be done quickly. To develop shared expectations among employees around risk taking, tolerance of mistakes, teamwork, and speed, takes time. It requires absolute consistency by managers at all levels and constant repetition of the message. This is not because people are stupid or don't listen. It is quite the opposite. Most people are perceptive and have good memories. They learn quickly how things really work in organizations. They understand how easy it is to talk about the need for something and how hard it is to really change it. Some companies, however, seem to be able to use social control to promote innovation over long periods.

A CASE IN POINT: 3M

In our view, the evidence for the importance of culture as a social control system for promoting innovation and change in organizations is convincing. First, it is difficult to conceive of formal control systems that will reliably promote innovation. Second, innovation is too important to be left uncontrolled or unmanaged. If you accept the notion that innovation is an outcome of two complementary processes, *creativity* and *implementation*, then it is easy to see how norms that promote these can be used as crucial levers to enhance innovation. The opposite is even more striking. Imagine an organization whose culture was characterized by the following norms: no rewards or support for innovation; people were routinely punished for taking risks that failed; individuals did not believe in teamwork; information was not shared; there was no sense of urgency; and people were proud of their past accomplishments and highly conservative and resistant to change. In a competitive world, most people would see this as a recipe for disaster. Yet, these are the norms that accompany social and structural inertia in older, successful firms.

The importance of social control for fostering innovation in organizations and providing competitive advantage is illustrated most convincingly at 3M, the $15 billion dollar company with over 60,000 products that routinely gets over 25 percent of its revenues from products that are less than five years old. As one reporter who studied 3M noted, "The cultural rules work—and go a long way toward explaining why an old-line manufacturing company, whose base products are sandpaper and tape, has become a master at innovation." How can a company filled with "loyal lifers" be so innovative? When asked what types of people would fit in at 3M, former CEO Alan Jacobsen responded, "There are three things that I think are real obstacles to personal success in this company. The first is having a big ego because this is a team. The same for the person who is political. . . . The other thing is somebody who is turfy: the job or their department is their own turf. To make a business like ours work, you have to be able to cross boundaries."[16]

A careful look at the 3M culture and the systems that support it shows how consistently they reinforce norms that promote creativity and implementation. For example, there are explicit norms that promote bootlegging or the pursuit of unauthorized projects, the so-called 15 percent rule that provides technologists with time to pursue any idea they want. The budgeting system is also designed to provide extra resources to support this, including multiple ways in which funding for a new product idea can be obtained—even if the idea has been rejected by other sources. This is further supported by measurement and reward systems that highlight the proportion of divisional revenues that come from products that are less than five years old. The target used to be 25 percent but has recently been increased to 30 percent.

The formal and informal reward systems also emphasize creativity and implementation. There is an awareness among 3Mers that to have a successful career, you need to be involved in a successful new product development, both because successful champions get to run their own project and because senior managers are frequently those who were champions themselves. Several years ago, the company developed an explicit career track, called the "Venture Career Path," to formally acknowledge the importance of pursuing innovative product ideas. Informal, cross-functional groups are encouraged to form at an early stage. The idea is for new product ideas to continually surface, with those that are successful becoming first project teams, then separate business units, then profit centers and finally even divisions, a "grow and divide" philosophy that ensures that individuals and teams see the fruit of their ideas. An individual or team with an idea should stay with it, even if for strategic reasons the idea is transferred to another unit. After a new product's annual sales exceed $1 million, the title of the team leader is changed to that of "product champion."

Other norms emphasize risk taking and teamwork. Failure is not a career-modifying event. It is expected that many new product ideas will not work out. Those involved simply return to their original units and begin again. As a former CEO observed, "Before a program succeeds it has to be crushed at least once. If it isn't, the people possessed by the research into which they've poured body and soul in an effort to find a way will never be able to reach a truly correct answer." This fosters leaders who act as defiant role models and are highly visible. Autonomy is fostered in the design of jobs and groups. Deliberate attempts are made to kill bureaucratic obstacles. So, for example, proposals for new product ideas are restricted to five pages. Managers are encouraged not to nitpick and to let people run with ideas. Recognition is deliberate and extensive. Numerous efforts are made to continually share information across units, whether in the form of formal information systems with over 100,000 technical reports on-line, or informal means such as monthly meetings among technologists. The entire system, formal and informal, is one that survives through continual innovation driven by a culture designed to foster creativity and implementation. The current CEO, L. D. DeSimone, claims that "the soul of 3M is innovation."

The importance of culture in the long-term success of firms like 3M was highlighted in a recent study by two Harvard Business School professors, John Kotter and James Heskett.[17] They began with a simple hypothesis: if culture was really important, then firms with strong cultures should be successful over long periods. To explore this, they collected data from over 200 U.S. firms in twenty-two different industries. They then calculated measures of economic performance for roughly a ten-year period, including ROI and shareholder return. To their surprise, an initial analysis found no evidence for the link between culture and performance. Strong culture firms were no more successful than those with weak cultures. After reflecting on these disappointing results, they recognized that a strong culture might not be strategically appropriate, in which case you wouldn't expect any link to performance. They then reanalyzed the data focusing on firms with strong, strategically appropriate cultures. Again, to their dismay, they found no strong association between culture and performance. They then had an insight. As we have suggested, strong cultures may, over time, become a barrier to reorientation and lead to failure. The authors then concentrated on firms in their sample that had strong, strategically appropriate cultures that also incorporated innovation and change, and found strong links between culture and organizational performance. They concluded that ". . . our research shows that even contextually or strategically appropriate cultures will not promote excellent performance over long periods unless they contain norms and values that help firms adapt to a changing environment."

LINKING STRATEGY, INNOVATION, AND CULTURE

To understand the importance of the strategy-innovation-culture fit, recall the logic of the congruence model. The first step required of a manager is to be clear about strategy, objectives, and vision. Every manager needs to answer fundamental questions such as: "How will we compete?; From the customer's perspective, what is our competitive advantage?" Once clarity about these issues has been achieved, the next step is to understand the critical tasks that must be accomplished to successfully execute the strategy. This is not an abstract exercise. Literally describe what are the specific, concrete tasks that must be accomplished. Then, think about the culture or norms needed to accomplish the critical tasks.

Intel, the $8 billion semiconductor firm, offers another example of the benefits of linking culture to strategy. Intel's strategy has long been to be a first-mover or innovator. Whereas a firm like National Semiconductor has historically chosen to compete as a low-cost producer, Intel has always been explicit about being first to market with the latest technology, reaping high margins, and then leaving the product class as competitors move in and margins fall. Andy Grove, the CEO, has always been clear about what norms are necessary to successfully pursue this strategy. In his view, "speed is what matters most. It's the only way to keep our advantage."[18]

With speed as an underlying value, what norms need to be widely shared and strongly felt to ensure fast innovation? One norm for which Intel is noted is what is referred to as "constructive confrontation," an expectation that all employees will challenge not only their colleagues, but also their superiors. The reason for this is quite strategic. In an era of shortened product life-cycles and rising costs, Intel's competitive advantage is directly linked to how quickly new products can be developed. Grove reasons that in this environment, Intel doesn't have time to be polite. It's imperative that issues and disagreements are surfaced and dealt with immediately. One practice at Intel that supports this is the so-called "ARs" or action requirements. Any employee can demand an AR of any executive on any issue. The executive is then required to make a decision.

To ensure that employees understand the importance of constructive confrontation, managers not only model the behavior, but a specific training course is given to all employees in how to be confrontational. This makes Intel an aggressive, combative place to work, which does not appeal to all people. For instance, a tongue-in-cheek response in a company magazine to a new employee who wanted to know the limits of constructive confrontation was that "When your co-worker criticizes your work, that's constructive confrontation. If he criticizes your work and calls you an idiot, that may or may not be constructive confrontation (remember, you may, in fact, be an idiot). If he criticizes your work, calls you an idiot, and tosses your work out the window . . . he should sign up for situational management." The norm of constructive confrontation helps encourage bottom-up initiatives and generates an obsessive concern with continuous improvement, both attributes crucial for innovation.

Grove also understands the need to modify the

culture to meet changes in the environment. "A corporation is a living organism and it has to continue to shed its skin." Ten years ago, Intel was a semiconductor memory company. Today it is a microprocessor company with a strong thrust toward selling products to end users. This has required an increased need for norms that promote customer orientation. The degree to which Intel has been successful at creating this change is evident in the current put-down used with technologists who still have not gotten the message, "C'mon, stop being a chip-head." Grove understands that "success can trap you" and can see the need to transform the organization again. He understands that part of this transformation involves modifying the formal control systems, including structure, measurements, and rewards. Perhaps more important, he also understands the importance of changing the social control system to promote the new norms that help execute the changed strategy.

A failure to heed this lesson can be seen in the problems encountered by a large integrated electronics company a decade ago. This firm was at the technological frontier. It made everything from satellites to microwaves to semiconductors. A core part of its engineering culture was never to let a product go until it was absolutely right. This makes good sense when you are in the business of building satellites. However, its success was largely dependent on government funding, with over 50 percent of its annual revenues coming from federal sources. The CEO, a smart man, anticipated that the level of government funding would not continue. He believed that it was strategically important that the company develop products that could be sold in nongovernmental markets. Therefore, he commissioned a study to identify technologies within the company that could be developed into private-sector products.

The result was the identification of a small computer the firm had developed for military applications. The consensus was that this technology could be modified and used to enter the then growing word processor market. The CEO established a word processing division, transferred some of his best managers, and ran into serious problems. When we visited the company, the product roll-out was two years late. Ultimately, it wrote off a $40 million investment. Why? With the managers and engineers, the company also transplanted its culture; a culture that caused continuous development delays until it missed the market. Determined to enter the word processing market, it bought a word processing company only to discover that it didn't have the culture needed to successfully compete in this market. Today, this company is no longer in the word processing market. It had the resources, technology, and even the strategic intent, but not the culture.

Whether it is retailing, steel, pharmaceuticals, financial services, or telecommunications, there are similar examples illustrating how firms have been forced to alter previously successful strategies. As shown here, however, this is not enough. Larry Bossidy, the CEO of Allied-Signal and veteran of GE, states, "The competitive difference is not in deciding what to do, but in how to do it." T. J. Rodgers, CEO of hard-driven Cypress Semiconductor and author of the book *No Excuses Management*, echoes this view when he says, "Most companies don't fail for a lack of talent or strategic vision. They fail for lack of execution—the routine blocking and tackling that great companies consistently do well and always strive to do better." Organizations often fail when existing cultures become inward or fail to support the implementation of new strategies. In our view, culture is a key to the successful execution of revolutionary change.

NOTES

1. O'Reilly, C. and Chatman, J. (1996). "Culture as social control: Corporations, cults, and commitment. In B. Staw and L. Cummings (eds.), *Research in Organizational Behavior*, vol. 18, in press.

2. Waterman, R. (1994). *What America Does Right: Learning From Companies That Put People First.* New York: Norton.

3. Caldwell, D. and O'Reilly, C. (1995). "Promoting team-based innovation: The use of normative influence." Paper presented at the 54th Annual Meetings of the Academy of Management, Vancouver.

4. Holder, V. (1994). Rewards for bright ideas. *Financial Times*, (32), 543.

5. Spector, R. and McCarthy, P. (1995). *The Nordstrom Way*. New York: John Wiley.

6. Waterman, op. cit.

7. Levering, R. and Moskowitz, M. (1993). *The 100 Best Companies to Work for in America*. New York: Currency Doubleday.

8. Levering and Moskowitz, op. cit.

9. Tichy, N. and Sherman, S. (1993). *Control Your Destiny or Someone Else Will*. New York: Currency Doubleday.

10. Raghavan, A. (1994). "Goldman faces pinch as partners leave, clubby culture wanes." *Wall Street Journal,* Dec. 16.

11. Levering and Moskowitz, op. cit.

12. Stack, J. (1992). *The Great Game of Business*. New York: Currency Doubleday

13. Schlender, B. (1995). What Bill Gates really wants. *Fortune*, 31, 34–63.

14. Levering and Moskowitz, op. cit.

15. Levering and Moskowitz, op. cit.

16. O'Reilly, B. (1994). J&J is on a roll. *Fortune*. Dec. 26, 1994, 130, 178–192.

17. Kotter, J. and Heskett, J. (1992). *Corporate Culture and Performance*. New York: Free Press.

18. Sherman, S. (1993). Andy Grove: How Intel makes spending pay off. *Fortune*, 127, 56–61.

15

Understanding Power in Organizations

JEFFREY PFEFFER

Norton Long, a political scientist, wrote, "People will readily admit that governments are organizations. The converse—that organizations are governments—is equally true but rarely considered."[1] But organizations, particularly large ones, are like governments in that they are fundamentally political entities. To understand them, one needs to understand organizational politics, just as to understand governments, one needs to understand governmental politics.

Ours is an era in which people tend to shy away from this task. As I browse through bookstores, I am struck by the incursion of "New Age" thinking, even in the business sections. New Age can be defined, I suppose, in many ways, but what strikes me about it are two elements: (1) a self-absorption and self-focus, which looks toward the individual in isolation; and (2) a belief that conflict is largely the result of misunderstanding, and if people only had more communication, more tolerance, and more patience, many (or all) social problems would disappear. These themes appear in books on topics ranging from making marriages work to making organizations work. A focus on individual self-actualization is useful, but a focus on sheer self-reliance is not likely to encourage one to try to get things done with and through other people—to be a manager or a leader. "Excellence can be achieved in a solitary field without the need to exercise leadership."[2] In this sense, John Gardner's (former secretary of HEW and the founder of Common Cause) concerns about community are part and parcel of a set of concerns about organizations and getting things accomplished in them.[3] One can be quite content, quite happy, quite fulfilled as an organizational hermit, but one's influence is limited and the potential to accomplish great things, which requires interdependent action, is almost extinguished.

If we are suspicious of the politics of large organizations, we may conclude that smaller organizations are a better alternative. There is, in fact, evidence that the average size of establishments in the United States is decreasing. This is not just because we have become more of a service economy and less of a manufacturing economy; even within manufacturing, the average size of establishments and firms is shrinking. The largest corporations have shed thousands, indeed hundreds of thousands of employees—not only middle managers, but also production workers, staff of all kinds, and employees who performed tasks that are now contracted out. Managers and employees who were stymied by the struggles over power and influence that emerge from interdependence and differences in point of view have moved to a world of smaller,

Reprinted with permission of Harvard Business School Press. From *Managing with Power: Politics and Influences in Organizations* by Jeffrey Pfeffer. Boston: 1993. In *California Management Review,* Winter 1992.

simpler organizations, with less internal interdependence and less internal diversity, which are, as a consequence, less political. Of course, such structural changes only increase interdependence among organizations, even as they decrease interdependence and conflict within these organizations.

I see in this movement a parallel to what I have seen in the management of our human resources. Many corporations today solve their personnel problems by getting rid of the personnel. The rationale seems to be that if we can't effectively manage and motivate employees, then let's turn the task over to another organization. We can use leased employees or contract workers, or workers from temporary help agencies, and let those organizations solve our problems of turnover, compensation, selection, and training.

It is an appealing solution, consistent with the emphasis on the individual, which has always been strong in U.S. culture, and which has grown in recent years. How can we trust large organizations when they have broken compacts of long-term employment? Better to seek security and certainty within oneself, in one's own competencies and abilities, and in the control of one's own activities.

There is, however, one problem with this approach to dealing with organizational power and influence. It is not clear that by ignoring the social realities of power and influence we can make them go away, or that by trying to build simpler, less interdependent social structures we succeed in building organizations that are more effective or that have greater survival value. Although it is certainly true that large organizations sometimes disappear,[4] it is also true that smaller organizations disappear at a much higher rate and have much worse survival properties. By trying to ignore issues of power and influence in organizations, we lose our chance to understand these critical social processes and to train managers to cope with them.

By pretending that power and influence don't exist, or at least shouldn't exist, we contribute to what I and some others (such as John Gardner) see as the major problem facing many corporations today, particularly in the United States—the almost trained or produced incapacity of anyone except the highest-level managers to take action and get things accomplished. As I teach in corporate executive programs, and as I compare experiences with colleagues who do likewise, I hear the same story over and over again. In these programs ideas are presented to fairly senior executives, who then work in groups on the implications of these ideas for their firms. There is real strength in the experience and knowledge of these executives, and they often come up with insightful recommendations and ideas for improving their organizations. Perhaps they discover the wide differences in effectiveness that exist in different units and share suggestions about how to improve performance. Perhaps they come to understand more comprehensively the markets and technologies of their organizations, and develop strategies for both internally oriented and externally oriented changes to enhance effectiveness. It really doesn't matter, because the most frequently heard comment at such sessions is, "My boss should be here." And when they go back to their offices, after the stimulation of the week, few managers have either the ability or the determination to engineer the changes they discussed with such insight.

I recall talking to a store manager for a large supermarket chain with a significant share of the northern California grocery market. He managed a store that did in excess of $20 million in sales annually, which by the standards of the average organization makes him a manager with quite a bit of responsibility—or so one would think. In this organization, however, as in many others, the responsibilities of middle-level managers are strictly limited. A question arose as to whether the store should participate in putting its name on a monument sign for the shopping center in which the store was located. The cost was about $8,000 (slightly less than four hour's sales in that store). An analysis was done, showing how many additional shoppers would need to be attracted to pay back this small investment, and what percentage this was of the traffic count passing by the center. The store manager wanted the sign. But, of course, he could not spend even this much money without the approval of his superiors. It was the president of the northern California division who decided, after a long meeting, that the expenditure was not necessary.

There are many lessons that one might learn from this example. It could be seen as the result of a plague of excessive centralization, or as an instance of a human resource management policy that certainly was more "top down" than "bottom up." But what was particularly interesting was the response of the manager—who, by the way, is held accountable for this store's profits even as he is given almost no discretion to do anything about them. When I asked him about the decision, he said, "Well, I guess that's why the folks at headquarters get the big money; they must know something we don't." Was he going to push for his idea, his very modest proposal? Of course no, he said. One gets along by just biding one's time, going along with whatever directives come down from the upper management.

I have seen this situation repeated in various forms over and over again. I talk to senior executives who claim their organizations take no initiative, and to high-level managers who say they can't or won't engage in efforts to change the corporations they work for, even when they know such changes are important, if not essential, to the success and survival of these organizations. There are politics involved in innovation and change. And unless and until we are willing to come to terms with organizational power and influence, and admit that the skills of getting things done are as important as the skills of figuring out what to do, our organizations will fall further and further behind. The problems is, in most cases, not an absence of insight or organizational intelligence. Instead the problem is one of passivity, a phenomenon that John Gardner analyzed in the following way:

> In this country—and in most other democracies—power has such a bad name that many good people persuade themselves they want nothing to do with it. The ethical and spiritual apprehensions are understandable. But one cannot abjure power. Power, as we are now speaking of it . . . is simply the capacity to bring about certain intended consequences in the behavior of others. . . . In our democratic society we make grants of power to people for specified purposes. If for ideological or temperamental reasons they refuse to exercise the power granted, we must turn to others. . . . To say a leader is preoccupied with power is like saying that a tennis player is preoccupied with making shots his opponent cannot return. Of course leaders are preoccupied with power! The significant questions are: What means do they use to gain it? How do they exercise it? To what ends do they exercise it?[5]

If leadership involves skill at developing and exercising power and influence as well as the will to do so, then perhaps one of the causes of the so-called leadership crisis in organizations in the United States is just this attempt to sidestep issues of power. This diagnosis is consistent with the arguments made by Warren Bennis and his colleagues, who have studied leaders and written on leadership. For instance, Bennis and Nanus noted that one of the major problems facing organizations today is not that too many people exercise too much power, but rather the opposite:

> These days power is conspicuous by its absence. Powerlessness in the face of crisis. Powerlessness in the face of complexity. . . . power has been sabotaged. . . . institutions have been rigid, slothful, or mercurial.[6]

They go on to comment on the importance of power as a concept for understanding leadership and as a tool that allows organizations to function productively and effectively:

> However, there is something missing . . . POWER, the basic energy to initiate and sustain action translating intention into reality, the quality without which leaders cannot lead. . . . power is at once the most necessary and the most distrusted element exigent to human progress. . . . power is the basic energy needed to initiate and sustain action or, to put it another way, the capacity to translate intention into reality and sustain it.[7]

Such observations about power are not merely the province of theorists. Political leaders, too, confirm that the willingness to build and wield power is a prerequisite for success in public life. In this consideration of power and leadership, Richard Nixon offered some observations that are consistent with the theme of this article:

> Power is the opportunity to build, to create, to nudge history in a different direction. There are few satisfactions to match it for those who care about such things. But it is not happiness. Those who seek happiness will not acquire power and would not use it well if they did acquire it.
>
> A whimsical observer once commented that those who love laws and sausages should not watch either being made.

By the same token, we honor leaders for what they achieve, but we often prefer to close our eyes to the way they achieve it. . . .

> In the real world, politics is compromise and democracy is politics. Anyone who would be a statesman has to be a successful politician first. Also, a leader has to deal with people and nations as they are, not as they should be. As a result, the qualities required for leadership are not necessarily those that we would want our children to emulate—unless we wanted them to be leaders.

In evaluating a leader, the key question about his behavioral traits is not whether they are attractive or unattractive, but whether they are useful.[8]

OUR AMBIVALENCE ABOUT POWER

That we are ambivalent about power is undeniable. Rosabeth Kanter, noting that power was critical for effective managerial behavior, nevertheless wrote, "Power is America's last dirty word. It is easier to talk about money—and much easier to talk about sex—than it is to talk about power."[9] Gandz and Murray did a survey of 428 managers whose responses nicely illustrate the ambivalence about power in organizations.[10] Some items from their survey, along with the percentage of respondents reporting strong or moderate agreement, are reproduced in Table 15.1. The concepts of power and organizational politics are related; most authors, myself included, define organizational politics as the exercise or use of power, with power being defined as a potential force. Note that more than 90% of the respondents said that the experience of workplace politics is common in most organizations, 89% said that successful executives must be good politicians, and 76% said that the higher one progresses in an organization, the more political things become. Yet 55% of these same respondents said that politics were detrimental to efficiency, and almost half said that top management should try to get rid of politics within organizations. It is as if we know that power and politics exist, and we even grudgingly admit that they are necessary to individual success, but we nevertheless don't like them.

TABLE 15.1. Managers' Feelings about Workplace Politics

Statement	Percentage Expressing Strong or Moderate Agreement
The existence of workplace politics is common to most organizations	93.2
Successful executives must be good politicians	89.0
The higher you go in organizations, the more political the climate becomes	76.2
Powerful executives don't act politically	15.7
You have to be political to get ahead in organizations	69.8
Top management should try to get rid of politics within the organization	48.6
Politics help organizations function effectively	42.1
Organizations free of politics are happier than those where there are a lot of politics	59.1
Politics in organizations are detrimental to efficiency	55.1

Source: Jeffrey Gandz and Victor V. Murray, "The Experience of Workplace Politics," *Academy of Management Journal*, 23 (1980).

This ambivalence toward, if not outright disdain for, the development and use of power in organizations stems from more than one source. First, there is the issue of ends and means—we often don't like to consider the methods that are necessary to get things accomplished, as one of the earlier quotes from Richard Nixon suggests. We are also ambivalent about ends and means because the same strategies and processes that may produce outcomes we desire can also be used to produce results that we consider undesirable. Second, some fundamental lessons we learn in school really hinder our appreciation of power and influence. Finally, in a related point, the perspective from which we judge organizational decisions often does not do justice to the realities of the social world.

Ends and Means

On Saturday, September 25, 1976, an elaborate testimonial dinner was held in San Francisco for a man whose only public office was as a commissioner on the San Francisco Housing Authority board. The guest list was impressive—the mayor, George Moscone; Lieutenant Governor Mervyn Dymally, at that time the highest-ranking Afro-American in elected politics; District Attorney Joe Freitas; Democratic Assemblyman Willie Brown, probably the most powerful and feared individual in California politics; Republican State Senator Milton Marks; San Francisco Supervisor Robert Mendelsohn; the city editor of the morning newspaper; prominent attorneys—in short, both Democrats and Republicans, a veritable who's who of the northern California political establishment. The man they were there to honor had recently met personally with the president's wife, Rosalynn Carter. Yet when the world heard more of this guest of honor, some two years later, it was to be with shock and horror at what happened in a jungle in Guyana. The person being honored that night in September 1976—who had worked his way into the circles of power in San Francisco using some of the very same strategies and tactics described in this article—was none other than Jim Jones.[11]

There is no doubt that power and influence can be acquired and exercised for evil purposes. Of course, most medicines can kill if taken in the wrong amount, thousands die each year in automobile accidents, and nuclear power can either provide energy or mass destruction. We do not abandon chemicals, cars, or even atomic power because of the dangers associated with them; instead we consider danger an incentive to get training and information that will help us to use these forces productively. Yet few people are willing to approach the potential risks and advantages of power with the same pragmatism. People prefer to avoid discussions of power, apparently on the assumption that "If we don't think about it, it won't exist." I take a different view. John Jacobs, now a political editor for the *San Francisco Examiner*, co-authored a book on Jim Jones and gave me a copy of it in 1985. His view, and mine, was that tragedies such as Jonestown could be prevented, not by ignoring the processes of power and influence, but rather by being so well schooled in them that one could recognize their use and take countermeasures, if necessary—and by developing a well-honed set of moral values.

The means to any end are merely mechanisms for accomplishing something. The something can be grand, grotesque, or, for most of us, I suspect, somewhere in between. The end may not always justify the means, but neither should it automatically be used to discredit the means. Power and political processes in organizations can be used to accomplish great things. They are not always used in this fashion, but that does not mean we should reject them out of hand. It is interesting that when we use power ourselves, we see it as a good force and wish we had more. When others use it against us, particularly when it is used to thwart our goals or ambitions, we see it as an evil. A more sophisticated and realistic view would see it for what it is—an important social process that is often required to get things accomplished in interdependent systems.

Most of us consider Abraham Lincoln to have been a great president. We tend to idealize his accomplishments: he preserved the Union, ended slavery, and delivered the memorable Gettysburg Address. It is easy to forget that he was also a politician and a pragmatist—for instance, the Emancipation Proclamation freed the slaves in the Confederacy, but not in border states that remained within the Union, whose support he needed. Lincoln also took a number of actions that far overstepped his Constitutional powers. Indeed, Andrew Johnson was impeached for continuing many of the actions that Lincoln had begun. Lincoln once explained how he justified breaking the laws he had sworn to uphold:

> My oath to preserve the Constitution imposed on me the duty of preserving by every indispensable means that government, that nation, of which the Constitution was the organic law. Was it possible to lose the nation and yet preserve the Constitution? . . . I felt that measures, otherwise unconstitutional, might become lawful by becoming indispensable to the preservation . . . of the nation.[12]

LESSONS TO BE UNLEARNED

Our ambivalence about power also comes from lessons we learn in school. The first lesson is that life is a matter of individual effort, ability, and achievement. After all, in school, if you have mastered the intricacies of cost accounting, or calculus, or electrical engineering, and the people sitting on either side of you haven't, their failure will not affect your performance—unless, that is, you had intended to copy from their papers. In the classroom setting, interdependence is minimized. It is you versus the material, and as long as you have mastered the material, you have achieved what is expected. Cooperation may even be considered cheating.

Such is not the case in organizations. If you know your organization's strategy but your colleagues do not, you will have difficulty accomplishing anything. The private knowledge and private skill that are so useful in the classroom are insufficient in organizations. Individual success in organizations is quite frequently a matter of working with and through other people, and organizational success is often a function of how successfully individuals can coordinate their activities. Most situations in organizations resemble football more than golf, which is why companies often scan resumes to find not only evidence of individual achievement but also signs that the person is skilled at working as part of a team. In achieving success in organizations, "power transforms individual interests into coordinated activities that accomplish valuable ends."[13]

The second lesson we learn in school, which may be even more difficult to unlearn, is that there are right and wrong answers. We are taught how to solve problems, and for each problem, that there is a right answer, or at least one approach that is more correct than another. The right answer is, of course, what the instructor says it is, or what is in the back of the book, or what is hidden away in the instructor's manual. Life appears as a series of "eureka" problems, so-called because once you are shown the correct approach or answer, it is immediately self-evident that the answer is, in fact, correct.

This emphasis on the potential of intellectual analysis to provide the right answer—the truth—is often, although not invariably, misplaced. Commenting on his education in politics, Henry Kissinger wrote, "Before I served as a consultant to Kennedy, I had believed, like most academics, that the process of decision-making was largely intellectual and all one had to do was to walk into the President's office and convince him of the correctness of one's view. This perspective I soon realized is as dangerously immature as it is widely held."[14] Kissinger noted that the easy decisions, the ones with right and wrong answers that can be readily discerned by analysis, never reached the president, but rather were resolved at lower levels.

In the world in which we all live, things are seldom clearcut or obvious. Not only do we lack a book or an instructor to provide quick feedback on the quality of our approach, but the problems we face often have multiple dimensions—which yield multiple methods of evaluation. The consequences of our decisions are often known only long after the fact, and even then with some ambiguity.

AN ALTERNATIVE PERSPECTIVE ON DECISION MAKING

Let me offer an alternative way of thinking about the decision-making process. There are three important things to remember about decisions. First, a decision by itself changes nothing. You can decide to launch a new product, hire a job candidate, build a new plant, change your performance evaluation system, and so forth, but the decision will not put itself into effect. As a prosaic personal example, recall how many times you or your friends "decided" to quit smoking, to get more exercise, to relax more, to eat healthier foods, or to lose weight. Such resolutions often fizzle before producing any results. Thus, in addition to knowledge of decision science, we need to know something about "implementation science."

Second, at the moment a decision is made, we cannot possibly know whether it is good or bad. Decision quality, when measured by results, can only be known as the consequences of the decision become known. We must wait for the decision to be implemented and for its consequences to become clear.

The third, and perhaps most important, observation is that we almost invariably spend more time living with the consequences of our decisions than we do in making them. It may be an organizational decision, such as whether to acquire a company, change the compensation system, fight a union-organizing campaign; or a personal decision, such as where to go to school, which job to choose, what subject to major in, or whom to marry. In either case, it is likely that the effects of the decision will be with us longer than it took us to make the decision, regardless of how much time and effort we invested. Indeed, this simple point has led several social psychologists to describe people as rationalizing (as contrasted with rational) animals.[15] The match between our attitudes and our behavior, for instance, often derives from our adjusting our attitudes after the fact to conform to our past actions and their consequences.[16]

If decisions by themselves change nothing; if, at the time a decision is made, we cannot know its consequences; and if we spend, in any event, more time living with our decisions than we do in making them, then it seems evident that the emphasis in much management training and practice has been misplaced. Rather than spending inordinant amounts of time and effort in the decision-making process, it would seem at least as useful to spend time implementing decisions and dealing with their ramifications. In this sense, good managers are not only good analytic decision makers; more important, they are skilled in managing the consequences of their decisions. "Few successful leaders spend much time fretting about decisions once they are past. . . . The only way he can give adequate attention to the decisions he has to make tomorrow is to put those of yesterday firmly behind him."[17]

There are numerous examples that illustrate this point. Consider, for instance, the acquisition of Fairchild Semiconductor by Schlumberger, an oil service company.[18] The theory behind the merger was potentially sound—to apply Fairchild's skills in electronics to the oil service business. Schlumberger wanted, for example, to develop more sophisticated exploration devices and to add electronics to oil servicing and drilling equipment. Unfortunately, the merger produced none of the expected synergies:

> When Schlumberger tried to manage Fairchild the same way it had managed its other business units, it created many difficulties. . . . resources were not made available to R&D with the consequence of losing technical edge which Fairchild once had. Creative . . . technical people left the organization and the company was unable to put technical teams together to pursue new technological advancement.[19]

A study of 31 acquisitions found that "problems will eventually emerge after acquisitions that could not have been anticipated. . . . both synergy and problems must be actively managed."[20] Moreover, firms that see acquisitions as a quick way of capturing some financial benefits are often insensitive to the amount of time and effort that is required to implement the merger and to produce superior performance after it occurs. Emphasis on the choice of a merger partner and the terms of the deal can divert focus away from the importance of the activities that occur once the merger is completed.

Or, consider the decision to launch a new product. Whether that decision produces profits or losses is often not simply a matter of the choices made at the time of the launch. It also depends on the implementation of those choices, as well as on subsequent decisions such as redesigning the product, changing the channels of distribution, adjusting prices, and so forth. Yet what we often observe in organizations is that once a decision is made, more effort is expended in assigning credit or blame than in working to improve the results of the decision.

I can think of no example that illustrates my argument as clearly as the story of how Honda entered the American market, first with motorcycles, and later, of course, with automobiles and lawn mowers. Honda established an American subsidiary in 1959, and between 1960 and 1965, Honda's sales in the United States went from $500,000 to $77 million. By 1966, Honda's share of the U.S. motorcycle market was 63%,[21] starting from zero just seven years before. Honda's share was almost six times that of its closest competitors, Yamaha and Suzuki, and Harley-Davidson's share had fallen to 4%. Pascale showed that this extraordinary success was largely the result of "miscalculation, serendipity, and organizational learning," not of the rational process of planning and foresight often emphasized in our efforts to be successful.[22]

Sochiro Honda himself was more interested in racing and engine design than in building a business, but his partner, Takeo Fujisawa, managed to convince him to turn his talent to designing a safe, inexpensive motorcycle to be driven with one hand and used for package delivery in Japan. The motorcycle was an immediate success in Japan. How and why did Honda decide to enter the export market and sell to the United States? Kihachiro Kawashima, eventually president of American Honda, reported to Pascale:

> In truth, we had no strategy other than the idea of seeing if we could sell something in the United States. It was a new frontier . . . and it fit the "success against all odds" culture that Mr. Honda had cultivated. I reported my impressions . . . including the seat-of-the-pants target of trying, over several years, to attain a 10 percent share of U.S. imports. . . . We did not discuss profits or deadlines for breakeven.[23]

Money was authorized for the venture, but the Ministry of Finance approved a currency allocation of only $250,000, of which less than half was in cash and the rest in parts and motorcycle inventory. The initial attempt to sell motorcycles in Los Angeles was disastrous. Distances in the United States are much greater than in Japan, and the motor cycles were driven farther and faster than their design permitted. Engine failures were common, particularly on the larger bikes.

The company had initially focused its sales efforts on the larger, 250cc and 350cc bikes, and had not even tried to sell the 50cc Supercub, believing it was too small to have any market acceptance:

> We used the Honda 50s . . . to ride around Los Angles on errands. They attracted a lot of attention. One day we had a call from a Sears buyer. . . . we took note of Sears' interest. But we still hesitated to push the 50cc bikes out of fear they might harm our image in a heavily macho market. But when the larger bikes started breaking, we had no choice. We let the 50cc bikes move. And surprisingly, the retailers who wanted to sell them weren't motorcycle dealers, they were sporting goods stores.[24]

Honda's "you meet the nicest people on a Honda" advertising campaign was designed as a class project by a student at UCLA, and was at first resisted by Honda. Honda's distribution strategy—sporting goods and bicycle shops rather than motorcycle dealers—was made *for* them, not *by* them. And its success with the smaller motorbike was almost totally unanticipated. It occurred through a combination of circumstances: the use of the motorbike by Honda employees, who couldn't afford anything fancier; the positive response from people who saw the bike; and the failure of Honda's larger bikes in the American market.

Honda did not use decision analysis and strategic planning. In fact, it is difficult to see that Honda made any decisions at all, at least in terms of developing alternatives and weighing options against an assessment of goals and the state of the market. Honda succeeded by being flexible, by learning and adapting, and by working to have decisions turn out right, once those decisions had been made. Having arrived with the wrong product for a market they did not understand, Honda spent little time trying to find a scapegoat for the company's predicament; rather, Honda personnel worked vigorously to change the situation to their benefit, being creative as well as opportunistic in the process.

The point is that decisions in the world of organizations are not like decisions made in school. There, once you have written down an answer and turned in the test, the game is over. This is not the case in organizational life. The important actions may not be the original choices, but rather what happens subsequently, and what actions are taken to make things work out. This is a significant point because it means that we need to be somewhat less concerned about the quality of the decision at the time we make it (which, after all, we can't really know anyway), and more concerned with adapting our new decisions and actions to the information we learn as events unfold. Just as Honda emerged as a leader in many American markets more by accident and trial-and-error learning than by design, it is critical that organizational members develop the fortitude to continue when confronted by adversity, and the insight about how to turn situations around. The most important skill may be managing the consequences of decisions. And, in organizations in which it is often difficult to take any action, the critical ability may be the capacity to have things implemented.

WAYS OF GETTING THINGS DONE

Why is implementation difficult in so many organizations, and why does it appear that the ability to get decisions implemented is becoming increasingly rare? One way of thinking about this issue, and of examining the role of power and influence in the implementation process, is to consider some possible ways of getting things done.

One way of getting things to happen is through hierarchical authority. Many people think power is merely the exercise of formal authority, but it is considerably more than that, as we will see. Everyone who works in an organization has seen the exercise of hierarchical authority. Those at higher levels have the power to hire and fire, to measure and reward behavior, and to provide direction to those who are under their aegis. Hierarchical direction is usually seen as legitimate, because the variation in formal authority comes to be taken for granted as a part of organizational life. Thus the phrase, "the boss wants . . ." or "the president wants . . ." is seldom questioned or challenged. Who can forget Marine Lieutenant Colonel Oliver North testifying, during the Iran-contra hearings, about his willingness to stand on his head in a corner if that was what his commander-in-chief wanted, or maintaining that he never once disobeyed the orders of his superiors?

There are three problems with hierarchy as a way of getting things done. First, and perhaps not so important, is that it is badly out of fashion. In an era of rising education and the democratization of all decision processes, in an era in which participative management is advocated in numerous places,[25] and particularly in a country in which incidents such as the Vietnam War and Watergate have led many people to mistrust the institutions of authority, implementation by order or command is problematic. Readers who are parents need only reflect on the difference in parental authority between the current period and the 1950s to see what I mean. How many times have you been able to get your children to do something simply on the basis of your authority as a parent?

A second, more serious problem with authority derives from the fact that virtually all of us work in positions in which, in order to accomplish our job and objectives, we need the cooperation of others who do not fall within our direct chain of command. We depend, in other words, on people outside our purview of authority, whom we could not command, reward, or punish even if we wanted to. Perhaps, as a line manager in a product division, we need the cooperation of people in human resources for hiring, people in finance for evaluating new product opportunities, people in distribution and sales for getting the product sold and delivered, and people in market research for determining product features and marketing and pricing strategy. Even the authority of a chief executive is not absolute, since there are groups outside the focal organization that control the ability to get things done. To sell overseas airline routes to other domestic airlines requires the cooperation of the Transportation and Justice Departments, as well as the acquiescence of foreign governments. To market a drug or medical device requires the approval of the Food and Drug Administration; to export products overseas, one may need both financing and export licenses. The hierarchical authority of all executives and administrators is limited, and for most of us, it is quite limited compared to the scope of what we need in order to do our jobs effectively.

There is a third problem with implementation accomplished solely or primarily through hierarchical authority: what happens if the person at the apex of the pyramid, the one whose orders are being followed, is incorrect? When authority is vested in a single individual, the organization can face grave difficulties if that person's insight or leadership begins to fail. This was precisely what happened at E.F. Hutton, where Robert Fomon, the chief executive officer, ruled the firm through a rigid hierarchy of centralized power:

> Fomon's strength as a leader was also his weakness. As he put his stamp on the firm, he did so more as monarch than as a chief executive. . . . Fomon surrounded himself with . . . cronies and yes men who would become the managers and directors of E.F. Hutton and who would insulate him from the real world.[26]

Because Fomon was such a successful builder of his own hierarchical authority, no one in the firm challenged him to see the new realities that Hutton, and every other securities firm, faced in the 1980s.[27] Consequently, when the brokerage industry changed, Hutton did not, and it eventually ceased to exist as an independent entity.

Another way of getting things done is by developing a strongly shared vision or organizational culture. If people share a common set of goals, a common perspective on what to do and how to accomplish it, and a common vocabulary that allows them to coordinate their behavior, then command and hierarchical authority are of much less importance. People will be able to work cooperatively without waiting for orders from the upper levels of the company. Managing through a shared vision and with a strong organizational culture has been a very popular prescription for organizations.[28] A number of articles and books tell how to build commitment and shared vision and how to socialize individuals, particularly at the time of entry, so that they share a language, values, and premises about what needs to be done and how to do it.[29]

Without denying the efficacy and importance of vision and culture, it is important to recognize that implementation accomplished through them can have problems. First, building a shared conception of the world takes time and effort. There are instances when the organization is in crisis or confronts situations in which there is simply not sufficient time to develop shared premises about how to respond. For this very reason, the military services rely not only on techniques that build loyalty and esprit de corps,[30] but also on a hierarchical chain of command and a tradition of obeying orders.

Second, there is the problem of how, in a strong culture, new ideas that are inconsistent with that culture can penetrate. A strong culture really constitutes an organizational paradigm, which prescribes how to look at things, what are appropriate methods and techniques for solving problems, and what are the important issues and problems.[31] In fields of science, a well-developed paradigm provides guidance as to what needs to be taught and in what order, how to do research, what are appropriate methodologies, what are the most pressing research questions, and how to train new students.[32] A well-developed paradigm, or a strong culture, is overturned only with great difficulty, even if it fails to account for data or to lead to new discoveries.[33] In a similar fashion, an organizational paradigm provides a way of thinking about and investigating the world, which reduces uncertainty and provides for effective collective action, but which also overlooks or ignores some lines of inquiry. It is easy for a strong culture to produce groupthink, a pressure to conform to the dominant view.[34] A vision focuses attention, but in that focus, things are often left out.

An organization that had difficulties, as well as great success, because of its strong, almost evangelical culture is Apple Computer. Apple was founded and initially largely populated by counterculture computer hackers, whose vision was a computer-based form of power to the people—one computer for each person. IBM had maintained its market share through its close relations with centralized data processing departments. IBM was the safe choice—the saying was, no one ever got fired for buying IBM. The Apple II was successful by making an end run around the corporate data processing manager and selling directly to the end-user, but "by the end of '82 it was beginning to seem like a good idea to have a single corporate strategy for personal computers, and the obvious person to coordinate that strategy was the data processing manager."[35] Moreover, computers were increasingly being tied into networks; issues of data sharing and compatibility were critical in organizations that planned to buy personal computers by the thousands. Companies wanted a set of computers that could run common software, to save on software purchasing as well as training and programming expenses. Its initial vision of "one person–one machine," made it difficult for Apple to see the need for compatibility, and as a consequence:

> The Apple II wouldn't run software for the IBM PC; the PC wouldn't run software for Lisa, Lisa wouldn't run software for the Apple II; and none of them would run software for the Macintosh. . . . Thanks largely to Steve [Jobs], Apple had an entire family of computers none of which talked to one another.[36]

Apple's strong culture and common vision also helped cause the failure of the Apple III as a new product. The vision was not only of "one person–one machine," but also of a machine that anyone could design, modify, and improve. Operating systems stood between the user and the machine, and so the Apple culture denigrated operating systems: The problem with an operating system, from the hobbyist point of view, was that it made it more difficult to reach down inside the computer and show off your skills; it formed a barrier between the user and the machine. Personal computers meant power to the people, and operating systems took some of that power away. . . . It wasn't a design issue; it was a threat to the inalienable rights of a free people.[37]

Apple III had an operating system known as SOS for Sophisticated Operating System, which was actually quite similar to the system Microsoft had developed for IBM's personal computer—MS DOS (Microsoft Disk Operating System), except it was even better in some respects. Yet Apple was too wary of operating systems to try to make its system *the* standard, or even *a* standard, in personal computing. As a result the company lost out on a number of important commercial opportunities. The very zeal and fervor that made working for Apple like a religious crusade and produced extraordinary levels of commitment from the work force made it difficult for the company to be either cognizant of or responsive to shifts in the marketplace for personal computers.

There is a third process of implementation in organizations—namely, the use of power and influence. With power and influence the emphasis is on method rather than structure. It is possible to wield power and influence without necessarily having or using formal authority. Nor is it necessary to rely on a strong organizational culture and the homogeneity that this often implies. Of course, the process of implementation through power and influence is not without problems of its own. What is important is to see power and influence as one of a set of ways of getting things done—not the only way, but an important way.

From the preceding discussion we can see that implementation is becoming more difficult because: (1) changing social norms and greater interdependence within organizations have made traditional, formal authority less effective than it once was, and (2) developing a common vision is increasingly difficult in organizations composed of heterogeneous members—heterogeneous in terms of race and ethnicity, gender, and even language and culture. At the same time, our ambivalence about power, and the fact that training in its use is far from widespread, mean that members of organizations are often unable to supplement their formal authority with the "unofficial" processes of power and influence. As a result their organizations suffer, and promising projects fail to get off the ground. This is why learning how to manage with power is so important.

THE MANAGEMENT PROCESS: A POWER PERSPECTIVE

From the perspective of power and influence, the process of implementation involves a set of steps, which are outlined below.

- Decide what your goals are, what you are trying to accomplish.
- Diagnose patterns of dependence and interdependence; what individuals are influential and important in your achieving your goal?
- What are their points of view likely to be? How will they feel about what you are trying to do?
- What are their power bases? Which of them is more influential in the decision?
- What are your bases of power and influence? What bases of influence can you develop, to gain more control over the situation?
- Which of the various strategies and tactics for exercising power seem most appropriate and are likely to be effective, given the situation you confront?
- Based on the above, choose a course of action to get something done.

The first step is to decide on your goals. It is, for instance, easier to drive from Albany, New York, to Austin, Texas, if you know your destination than if you just get in your car in Albany and drive randomly. Although this point is apparently obvious, it is something that is often overlooked in a business context. How many times have you attended meetings or conferences or talked to someone on the telephone without a clear idea of what you were trying to accomplish? Our calendars are filled with appointments, and other interactions occur unexpectedly in the course of our day. If we don't have some clear goals, and if we don't know what our primary objectives are, it is not very likely that we are going to achieve them. One of the themes Tom Peters developed early in his writing was the importance of consistency in purpose. Having the calendars, knowing the language, what gets measured, and what gets talked about—all focus on what the organization is trying to achieve.[38] It is the same with individuals; to the extent that each interaction, in each meeting, in each conference, is oriented toward the same objective, the achievement of that objective is more likely.

Once you have a goal in mind, it is necessary to diagnose who is important in getting your goal accomplished. You must determine the patterns of dependence and interdependence among these people and find out how they are likely to feel about what you are trying to do. As part of this diagnosis, you also need to know how events are likely to unfold, and to estimate the role of power and influence in the process. In getting things accomplished, it is critical to have a sense of the game being played, the players, and what their positions are. One can get badly injured playing football in a basketball uniform, or not knowing the offense from the defense. I have seen, all too often, otherwise intelligent and successful managers have problems because they did not recognize the political nature of the situation, or because they were blindsided by someone whose position and strength they had not anticipated.

Once you have a clear vision of the game, it is important to ascertain the power bases of the other players, as well as your own potential and actual sources of power. In this way you can determine your relative strength, along with the strength of other players. Understanding the sources of power is critical in diagnosing what is going to happen in an organization, as well as in preparing yourself to take action.

Finally, you will want to consider carefully the

various strategies, or, to use a less grand term, the tactics that are available to you, as well as those that may be used by others involved in the process. These tactics help in using power and influence effectively, and can also help in countering the use of power by others.

Power is defined here as the potential ability to influence behavior, to change the course of events, to overcome resistance, and to get people to do things that they would not otherwise do.[39] Politics and influence are the processes, the actions, the behaviors through which this potential power is utilized and realized.

WHAT DOES IT MEAN, TO MANAGE WITH POWER?

First, it means recognizing that in almost every organization, there are varying interests. This suggests that one of the first things we need to do is to diagnose the political landscape and figure out what the relevant interests are, and what important political subdivisions characterize the organization. It is essential that we do not assume that everyone necessarily is going to be our friend, or agree with us, or even that preferences are uniformly distributed. There are clusters of interests within organizations, and we need to understand where these are and to whom they belong.

Next, it means figuring out what point of view these various individuals and subunits have on issues of concern to us. It also means understanding why they have the perspective that they do. It is all too easy to assume that those with a different perspective are somehow not as smart as we are, not as informed, not as perceptive. If that is our belief, we are likely to do several things, each of which is disastrous. First, we may act contemptuously toward those who disagree with us—after all, if they aren't as competent or as insightful as we are, why should we take them seriously? It is rarely difficult to get along with those who resemble us in character and opinions. The real secret of success in organizations is the ability to get those who differ from us, and whom we don't necessarily like, to do what needs to be done. Second, if we think people are misinformed, we are likely try to "inform" them, or to try to convince them with facts and analysis. Sometimes this will work, but often it will not, for their disagreement may not be based on a lack of information; it may, instead, arise from a different perspective on what our information means. Diagnosing the point of view of interest groups as well as the basis for their positions will assist us in negotiating with them in predicting their response to various initiatives.

Third, managing with power means understanding that to get things done, you need power—more power than those whose opposition you must overcome—and thus it is imperative to understand where power comes from and how these sources of power can be developed. We are sometimes reluctant to think very purposefully or strategically about acquiring and using power. We are prone to believe that if we do our best, work hard, be nice, and so forth, things will work out for the best. I don't mean to imply that one should not, in general, work hard, try to make good decisions, and be nice, but that these and similar platitudes are often not very useful in helping us get things accomplished in our organizations. We need to understand power and try to get it. We must be willing to do things to build our sources of power, or else we will be less effective than we might wish to be.

Fourth, managing with power means understanding the strategies and tactics through which power is developed and used in organizations, including the importance of timing, the use of structure, the social psychology of commitment and other forms of interpersonal influence. If nothing else, such an understanding will help us become astute observers of the behavior of others. The more we understand power and its manifestations, the better will be our clinical skills. More fundamentally, we need to understand strategies and tactics of using power so that we can consider the range of approaches available to us, and use what is likely to be effective. Again, as in the case of building sources of power, we often try not to think about these things, and we avoid being strategic or purposeful about employing our power. This is a mistake. Although we may have various qualms, there will be others who do not. Knowledge without power is of remarkably little use. And power without the skill to employ it effectively is likely to be wasted.

Managing with power means more than knowing the ideas discussed in this article. It means being, like Henry Ford, willing to do something with that knowledge. It requires political savvy to get things done, and the willingness to force the issue. For years in the United States, there had been demonstrations and protests, court decisions and legislative proposals attempting to end the widespread discrimination against minority Americans in employment, housing, and public accommodations. The passage of civil rights legislation was a top priority for President Kennedy, but although he had charisma, he lacked the knowledge of political tactics, and possibly the will to use some of the more forceful ones, to get his legislation passed. In the hands of someone who knew power and influence inside out, in spite of the opposition of Southern congressmen and senators, the legislation would be passed quickly.

In March 1965, the United States was wracked by violent reactions to civil rights marches in the South. People were killed and injured as segregationists attacked demonstrators, with little or no intervention by the local law enforcement agencies. There were demonstrators across from the White House holding a vigil as Lyndon Johnson left to address a joint session of Congress. This was the same Lyndon Johnson who, in 1948, had opposed federal antilynching legislation, arguing that it was a matter properly left to the states. This was the same Lyndon Johnson who, as a young congressional secretary and then congressman, had talked conservative to conservatives, liberal to liberals, and was said by many to have stood for nothing. This was the same Lyndon Johnson who in eight years in the House of Representatives had introduced not one piece of significant legislation and had done almost nothing to speak out on issues of national importance. This was the same Lyndon Johnson who, instead, had used some of his efforts while in the House to enrich himself by influencing colleagues at the Federal Communications Commission to help him obtain a radio station in Austin, Texas, and then to change the operating license so the station would become immensely profitable and valuable. This was the same Lyndon Johnson who, in 1968, having misled the American people, would decide not to run for reelection because of his association with both the Vietnam War and a fundamental distrust of the presidency. On that night Johnson was to make vigorous use of his power and his political skill to help the civil rights movement:

> With almost the first words of his speech, the audience . . . knew that Lyndon Johnson intended to take the cause of civil rights further than it had ever gone before. . . . He would submit a new civil rights bill . . . and it would be far stronger than the bills of the past. . . . "their cause must be our cause, too," Lyndon Johnson said. "Because it is not just Negroes, but really it is all of us, who must overcome the crippling legacy of bigotry and injustice. . . . And we shall overcome."[40]

As he left the chamber after making his speech, Johnson sought out the 76-year-old chairman of the House Judiciary Committee, Emmanuel Celler:

> "Manny," he said, "I want you to start hearings tonight."
>
> "Mr. President," Cellar protested, "I can't push that committee or it might get out of hand. I am scheduling hearings for next week.:
>
> . . . Johnson's eyes narrowed, and his face turned harder. His right hand was still shaking Celler's, but the left hand was up, and a finger was out, pointing, jabbing.
>
> "Start them *this* week, Manny," he said. "And hold night sessions, too."[41]

Getting things done requires power. The problem is that we would prefer to see the world as a kind of grand morality play, with the good guys, and the bad ones easily identified. Obtaining power is not always an attractive process, nor is its use. And it somehow disturbs our sense of symmetry that a man who was as sleazy, to use a term of my students, as Lyndon Johnson was in some respects, was also the individual who almost single-handedly passed more civil rights legislation in less time with greater effect than anyone else in U.S. history. We are troubled by the issue of means and ends. We are perplexed by the fact that "bad" people sometimes do great and wonderful things, and that "good" people sometimes do "bad" things, or often, nothing at all. Every day, managers in public and private organizations acquire and use power to get things done. Some of these things may be, in retrospect, mistakes, although often that depends heavily on your point of view. Any reader who always does the correct thing that pleases everyone should immediately contact me—we will get very wealthy together. Mistakes and opposition are inevitable. What is not inevitable is passivity, not trying, not seeking to accomplish things.

In many domains of activity we have become so obsessed with not upsetting anybody, and with not making mistakes, that we settle for doing nothing. Rather than rebuild San Francisco's highways, possibly in the wrong place, maybe even in the wrong way, we do nothing, and the city erodes economically without adequate transportation. Rather than possibly being wrong about a new product, such as the personal computer, we study it and analyze it, and lose market opportunities. Analysis and forethought are, obviously, fine. What is not so fine is paralysis or inaction, which arise because we have little skill in overcoming the opposition that inevitably accompanies change, and little interest in doing so.

Theodore Roosevelt, making a speech at the Sorbonne in 1910, perhaps said it best:

> It is not the critic who counts; not the man who points out how the strong man stumbles, or where the doer of deeds could have done them better. The credit belongs to the man who is actually in the arena, whose face is marred by dust and sweat and blood; who strives valiantly; who errs, and comes short again and again; because there is not effort without error and shortcoming; but who does actually strive to do the deeds; who knows the great enthusiasms, the great devotions; who spends himself in a worthy cause, who at the best knows in the end the triumphs of high achievement and who at the worst, if he fails, at least fails while daring greatly, so that his place shall never be with those cold and timid souls who know neither victory or defeat.[42]

It is easy and often comfortable to feel powerless—to say, "I don't know what to do, I don't have the power to get it done, and besides, I can't really stomach the struggle that may be involved." It is easy, and now quite common, to say, when confronted with some mistakes in your organization, "It's not really my responsibility, I can't do anything about it anyway, and if the company wants to do that, well that's why the senior executives get the big money—it's their responsibility." Such a response excuses us from trying to do things; in not trying to overcome opposition, we will make fewer enemies and are less likely to embarrass ourselves. It is, however, a prescription for both organizational and personal failure. This is why power and influence are not the organization's last dirty secret, but the secret of success for both individuals and their organizations. Innovation and change in almost any arena requires the skill to develop power, and the willingness to employ it to get things accomplished. Or, in the words of a local radio newscaster, "If you don't like the news, go out and make some of your own."

REFERENCES

1. Norton E. Long, "The Administrative Organization as a Political System," in S. Mailick and E.H. Van Ness, eds., *Concepts and Issues in Administrative Behavior* (Englewood Cliffs, NJ: Prentice-Hall, 1962), p. 110.
2. Richard M. Nixon, *Leaders* (New York, NY: Warner Books, 1982), p. 5.
3. John W. Gardner, *On Leadership* (New York, NY: Free Press, 1990).
4. Michael T. Hannan and John Freeman, *Organizational Ecology* (Cambridge, MA: Harvard University Press, 1989).
5. Gardner, *op. cit.*, pp. 55–57.
6. Warren Bennis and Burt Nanus, *Leaders: The Strategies for Taking Charge* (New York, NY: Harper and Row, 1985), p. 6.
7. Ibid., pp. 15–17.
8. Nixon, *op. cit.*, p. 324.
9. Rosabeth Moss Kanter, "Power Failure in Manage-

ment Circuits," *Harvard Business Review, 57* (July/August 1979): 65.

10. Jeffrey Gandz and Victor V. Murray, "The Experience of Workplace Politics," *Academy of Management Journal, 23* (1980):237–251.
11. Tim Reiterman with John Jacobs, *Raven: The Untold Story of the Rev. Jim Jones and His People* (New York, NY: E.P. Dutton, 1982), pp. 305–307.
12. Nixon, *op cit.*, p. 326.
13. Abraham Zaleznick and Manfred F.R. Kets de Vries, *Power and the Corporate Mind* (Boston, MA: Houghton Mifflin, 1975), p. 109.
14. Henry Kissinger, *The White House Years* (Boston, MA: Little Brown, 1979), p. 39.
15. Elliot Aronson, *The Social Animal* (San Francisco, CA: W.H. Freeman, 1972), chapter 4; Barry M. Staw, "Rationality and Justification in Organizational Life," *Research in Organizational Behavior*, B.M. Staw and L.L. Cummings, eds. (Greenwich, CT: JAI Press, 1980), vol. 2, pp. 45–80; Gerald R. Salancik, "Commitment and the Control of Organizational Behavior and Belief," *New Directions in Organizational Behavior*, Barry M. Staw and Gerald R. Salancik, eds. (Chicago, IL: St. Clair Press, 1977), pp. 1–54.
16. Leon Festinger, *A Theory of Cognitive Dissonance* (Stanford, CA: Stanford University Press, 1957).
17. Nixon, *op. cit.*, p. 329.
18. Alok K. Chakrabarti, "Organizational Factors in Post-Acquisition Performance," *IEEE Transactions on Engineering Management, 37* (1990): pp. 259–268.
19. Ibid., p. 259.
20. Ibid., p. 266.
21. D. Purkayastha, "Note on the Motorcycle Industry—1975," #578-210, Harvard Business School, 1981.
22. Richard T. Pascale, "Perspectives on Strategy: The Real Story Behind Honda's Success," *California Management Review, 26* (1984): 51.
23. Ibid., p. 54.
24. Ibid., p. 55.
25. William A. Pasmore, *Designing Effective Organizations: The Sociological Systems Perspective* (New York, NY: John Wiley, 1988); David L. Bradford and Allan R. Cohen, *Managing for Excellence* (New York, NY: John Wiley, 1984).
26. Mark Stevens, *Sudden Death: The Rise and Fall of E.F. Hutton* (New York, NY: Penguin, 1989), p. 98.
27. Ibid., p. 121.
28. Thomas J. Peters and Robert H. Waterman, Jr., *In Search of Excellence* (New York, NY: Harper and Row, 1982); Terrence Deal and Allan A. Kennedy, *Corporate Cultures* (Reading, MA: Addison-Wesley, 1982); Stanley Davis, *Managing Corporate Culture* (Cambridge, MA: Ballinger, 1984).
29. Richard T. Pascale, "The Paradox of 'Corporate Culture': Reconciling Ourselves to Socialization," *California Management Review, 26* (1985): 26–41; Charles O'Reilly, "Corporations, Culture, and Commitment: Motivation and Social Control in Organizations," *California Management Review, 31* (1989):9–25.
30. Sanford M. Dornbusch, "The Military Academy as an Assimilating Institution," *Social Forces 33* (1955): 316–321.
31. Richard Harvey Brown, "Bureaucracy as Praxis: Toward a Political Phenomenology of Formal Organizations," *Administrative Science Quarterly, 23* (1978): 365–382.
32. Janice Lodahl and Gerald Gordon, "The Structure of Scientific Fields and the Functioning of University Graduate Departments," *American Sociological Review, 37* (1972): 57–72.
33. Thomas S. Kuhn, *The Structure of Scientific Revolutions*, 2d ed. (Chicago, IL: University of Chicago Press, 1970).
34. Irving L. Janis, *Victims of Groupthink* (Boston, MA: Houghton Mifflin, 1972).
35. Frank Rose, *West of Eden: The End of Innocence at Apple Computer* (New York, NY: Viking Penguin, 1989), p. 81.
36. Ibid., p. 85.
37. Ibid., p. 97.
38. Thomas J. Peters, "Symbols, Patterns, and Settings: An Optimistic Case for Getting Things Done," *Organizational Dynamics, 7* (1978): 3–23.
39. Jeffrey Pfeffer, *Power in Organizations* (Marshfield, MA: Pitman Publishing, 1981); Kanter, *op. ct.*,; Richard M. Emerson, "Power-Dependence Relations," *American Sociological Review 27* (1962): 31–41.
40. Robert A. Caro, *Means of Ascent: The Years of Lyndon Johnson* (New York, NY: Alfred A. Knopf, 1990), pp. xix–xx.
41. Ibid., p. xxi.
42. Nixon, *op. cit.*, p. 345.

16

Dynamic Tension in Innovative, High Technology Firms: Managing Rapid Technological Change Through Organizational Structure

CLAUDIA BIRD SCHOONHOVEN
MARIANN JELINEK

People look at organizational change as, "Somebody did something wrong." That's nuts, because there's absolutely no reason why an organization that you created two years ago has any relevance to the organization that you need two years from now. The beauty of this business is that the technology will always change (and thus) the organization, the interfaces, the customer interfaces, and the vendor interfaces are always going to change, because of the technology.

—LES VADASZ, SENIOR VICE PRESIDENT, INTEL

Eternal innovation seems the price of survival in contemporary economic competition, and this is especially true for high technology firms. Survival depends on maintaining a steady stream of innovative microelectronic products manufactured in high volume by frequently changing state-of-the-art technologies. In this chapter we ask how organizations that must continuously innovate over time can manage the dominant competitive issue of innovation. Our focus is not merely on the creation of a single high technology innovation; rather we are concerned with how companies create a *stream of innovations* that are commercially successful. Innovations that have repeated marketplace success are our concern.

We examine how companies organize for innovation, drawing from a longitudinal study of five highly innovative U.S. electronic firms that manufacture and compete in the semiconductor components, systems, and computer marketplaces. We interviewed more than 100 high technology managers and engineers and discovered that they do not view the organizing process in dichotomous, simplistic, either/or terms. Instead we found complex organizational processes in these companies, which did not conform to the simple organizing models that pervade contemporary management textbooks. Today's textbooks still describe organizations as either organic, and thus appropriately structured for innovation and change, or mechanistic, and thus appropriately structured for steady-state technologies and environmental stability. One best-selling yet typical management text notes that "a particular

Reprinted from Von Glinow, M., and S. Mohram (Eds.), *Managing Complexity in High Technology Organizations*, Oxford University Press, 1990, pp. 90–118.

organization . . . should be structured depending upon whether it must be (1) relatively efficient and productive or (2) adaptive and flexible" (Donnelly, Gibson, and Ivancevich, 1984, p. 196). The dominant prescription, therefore, is that organizations should be structured for either efficiency or flexibility. However, what if conditions require both *flexibility and efficiency*? Neither of these simplistic recipes for organizing comes close to adequately describing the subtleties and nuances required to organize for a continuous stream of state-of-the-art innovations that also enjoy success in the marketplace.

It is clear from our research that commercial success is a function of an organization's innovative ability as well as its ability to efficiently produce what it has created. The old flexible-versus-efficient formula will not work because these companies must do four things successfully and simultaneously, not merely one. First, they must innovate high technology products. Second, they must innovate their manufacturing processes, because success depends heavily on more and more sophisticated manufacturing techniques. Third, they must efficiently manufacture in high volume a continuously changing set of products and manufacturing techniques—neither remains stable for long—so steady-state manufacturing is not possible, and yet manufacturing efficiency is incredibly important here, just as it is in other non-high technology industries. And, fourth, their innovative products must have high market cachet: it is the market test that ultimately determines success for these companies. This imposes a dynamic tension on the organizations to be simultaneously clearly structured for efficiency while adapting to organizational modifications required by changes in their technical and market conditions. The companies we have studied pass this test, time after time, product after product. The old organizing formulas do not work because they underestimate the complex set of activities with which high technology companies must simultaneously contend to be successful, and thus for which they must be organized simultaneously.

The old formulas will not work for a second reason. Microelectronic and contemporary electronic firms compete in environments that require both innovation and efficiency. Thus, external as well as internal conditions impose the dynamic tension. We agree with Lawrence and Dyer's assumption that a key environmental characteristic for these companies is high rates of technological change (1983). Where they and other contemporary theorists err, however, is in their failure to analyze the nature of the technological change. Both product and manufacturing process innovations characterize the semiconductor, systems, and computer industries. Whenever process change is required, new efficiencies must be gained quickly despite uncertainties associated with applying the new manufacturing methods. The "end game" of manufacturing, even in the highly dynamic semiconductor industry, nonetheless dictates a strong need for efficiency. Manufacturing costs must be as low as possible to compete in these industries. Similarly, if these companies are to efficiently develop new products, clearly an organic, amorphous structure will not allow them to create new product designs in the least wasteful manner possible. Efficiency is defined as performing or functioning in the least wasteful manner possible—an essential characteristic of contemporary electronic competition.

The companies included in our research are Intel, National Semiconductor, Texas Instruments, Hewlett-Packard, and Motorola Semiconductor—all highly successful in creating rapid innovations over time. We used a longitudinal research design and tracked the companies from 1981 to 1988, to record changes and shifts in the companies' products, markets, technologies, and organizations. We interviewed over 100 innovation-knowledgeable people in these firms, starting with their founders and chief executive officers, down through vice presidents, division general managers, and R & D managers, to design and manufacturing engineers. We sought the perspectives of those involved in innovation and its management from the top to the bottom of these companies. The reader will recognize some of the leaders of innovation we interviewed—Charlie Sporck, David Packard, John Young, Gordon Moore, Les Vadasz, and Jerry

Junkins. Others equally crucial to our research have been less publicly visible: Ted Hoff, the inventor of the microprocessor, Ed Boleky, a fabrication manager, and Steve Pease, an applications engineer. It is through their words that we illustrate our major findings on how contemporary high technology firms organize for a stream of commercially successful innovations over time.

OLD RECIPES AND NEW REALITIES

How does Hewlett-Packard coordinate and effectively direct the work of 84,000 employees to produce a constant stream of new, complex products? How does National Semiconductor keep track of the roughly 3500 different types of complex microelectronic products it offers at any given time—products that account for less than half of its overall revenues? How do any of these companies manage the innumerable and shifting activities of thousands of people, products, and technologies that are their world? For a company to continuously produce commercially successful innovations, year after year, suggests that an explicit management of the innovation process takes place. It also suggests that a mechanism exists to capture the attention of employees in a systematic way.

The companies we have studied are large, by any standard, and very complex organizationally. Their employees create, develop, and manufacture complicated microelectronic devices that govern satellites, robots, and computers of all sizes. How do these companies manage to pull off such an incredibly complicated set of simultaneous activities whose results form the heart of the information technology essential to the success and well-being of every industrial nation in the years ahead?

It has been clear to us from the beginning of this research that no single innovative idea is enough. Any single idea, be it for a new product or a new technology, is overwhelmingly insufficient. Good, even great ideas must be marshaled, financially supported, in tune with the marketplace, and efficiently manufactured. At virtually every step of the way, these processes involve complex activities requiring a host of contributors whose participation must be effectively timed, coordinated, monitored, and redirected as the need arises. In short, these activities must be organized and managed effectively. Ad hoc chance events will not do as an organizing philosophy. Explicit organization is the key to capturing the attention of thousands of employees who routinely create complicated end products. The pace of technical development is so fast that entire existing product lines can be obsolete within two years. As Intel's chairman, Gordon Moore, observed, "Every two years almost our entire product line turns over. We can almost go out of business in two years if we don't do it right."

To cope with the torrid pace of technology development, we found four key aspects of structure in these companies that interact to coordinate the many complex activities required in constant innovation: (1) a dynamic tension between (2) formal structures and reporting relationships, (3) quasiformal structures, and (4) informal structures of the organization. We will deal with each in turn; however, throughout this chapter the underlying emphasis remains on innovation, technical excellence, manufacturing efficiency, and marketplace success, and thus predictability amid near-constant change. This seemingly contradictory reality is central to success in the high technology markets of today and, we believe, will be increasingly important in many industries in the future.

We first reviewed the existing literature as a guide for our research. Although there are literally hundreds of studies relevant to the organization and management of innovation,[1] most advice to managers is derived from a groundbreaking study by British researchers Tom Burns and G. M. Stalker (1961). Basically, the best advice from the past was that innovation is best served by organizing according to principles of "organic management." Organic management is thought to enable firms to deal successfully with unpredictable technical and market contingencies, and these abound in the companies studied. In the original study of the British electronics industry, the companies that were high performers and innovative were organically managed to deal with the changing technical and environmental conditions they faced. Since high rates of technological innovation and environmental change have been well documented in the semiconductor, systems, and computer industries (Webbink, 1977; Jelinek, 1979; Schoonhoven, 1985), organic systems appeared to be the structure of choice for the companies we were about to study.

What specifically are these "organic structures" and management systems? The authors described organic systems as "adapted to unstable conditions, when . . . requirements for action arise which cannot be broken down and distributed among specialists . . . within a clearly defined hierarchy. Jobs lose much formal definition [and] . . . have to be redefined continually. Interaction runs laterally as much as vertically. Communication between people of different ranks tends to resemble lateral consultation rather than vertical command" (Burns and Stalker, 1961, pp. 5–6).

This definition makes intuitive sense, and yet a closer look reveals that it is unsatisfactory. When their ideas are analyzed more carefully, four separate dimensions of organic structure are implied: (1) ambiguous reporting relationships, with an unclear hierarchy; (2) unclear job responsibilities; (3) decision making is consultative and based on task expertise rather than being centralized in the management hierarchy; and (4) communication patterns are lateral as well as vertical (Schoonhoven and Eisenhardt, 1985). Burns and Stalker illustrated ambiguous reporting relationships with unclear hierarchy by their experience with the president of an electronics firm. When he began drawing the management hierarchy, his sketch "petered out in unresolved dilemmas. . . . There was, therefore, a deliberate avoidance of clearly defined functions and of lines of responsibility at the top level of management which . . . carried down through the rest of the management structure" (1961, pp. 83–85).

The second element of organic structure described by Burns and Stalker is that job responsibilities are indeterminant or unspecified. Managers and engineers in the electronic firms studied were unable to clearly state their job responsibilities. The authors reported that a product engineer said "nobody is very clear about his title or status or even his function." A foreman said, "Of course, nobody knows what his job is in here" (1961, p. 93).

Because these findings have remained unchallenged, we were led to expect ambiguous organizational structures, unclear reporting relationships, and indeterminant job responsibilities, given the competitive conditions facing semiconductor and computer firms. In short, we should have expected to find what contemporary writers have called unstructured, loosely coupled, amorphous adhocracies (Toffler, 1970; Weick, 1976; Mintzberg, 1979).

THE CONTEMPORARY STRUCTURE OF INNOVATION

Clear Organizations and Jobs

Our microelectronic firms certainly faced unpredictable, unstable conditions with complex problems. However, the problem is that the first two dimensions of an organic structure are worse than useless for providing the precise and definitive coordination, tight controls, high efficiency,and tough decision making needed in the fiercely competitive global marketplace. In contrast, contemporary semiconductor and computer firms, imbedded as they are in highly turbulent and changeful technical environments, display well-articulated structures, definite reporting relationships, and clear job responsibilities.

These firms are far from unstructured adhocracies. Quite the contrary, we found universally explicit reporting structures and clear hierarchies in which executives, managers, and engineers know who their bosses are, who their bosses' bosses are, who their reporting subordinates are, and who their organizational peers are at equivalent hierarchical levels in their organizations. There was simply no evidence of amorphous reporting, unclear hierarchy, or not knowing who is supposed to do what, as Burns and Stalker reported finding. All of the semiconductor and computer firms we studied had explicit organizational charts, readily available, and formalized from the top to the bottom of their companies.

In a few of our 100-plus interviews, managers were cautious in sharing these data with us, because organization charts are regarded as serious competitive information that reveal how the company organizes its work. Organization charts also typically reveal valued key employees whom some companies prefer to remain anonymous, obscured from their competitors' knowledge. In these cases, when a manager was not comfortable showing the explicit chart to us, he instead sketched one in our presence. None "petered out into unresolved dilemmas." None had unclear reporting relationships. The difference between our findings and those of Burns and Stalker is striking.

In addition to having clear structures, the innovative and successful companies we have studied also have clear job responsibilities. This, too, contrasts with the earlier organic management research, as well as with contemporary theories about adhocracies. We found that employees' titles and jobs are clear; they easily stated their titles and readily described their responsibilities. These are also printed on their business cards and listed on the organization charts, which were shared with us, providing more evidence of clarity. Thus independent of title, there is high predictability of who has responsibility for what throughout these companies. There was no evidence in any of our interviews of Burns and Stalker's famous quote, "nobody knows what his job is around here." In contrast, we found that these are high predictability organizations with clear job responsibilities and reporting relationships.

To illustrate how clearly structured these organizations are, Edward Boleky, a fourth-level manager at Intel, described his title, job responsibilities, and place in the organization, while sketching his organization on a chalkboard during the interview.

> I'm the California Site Fabrication Manager, responsible for three of Intel's wafer fabrication areas located in California. I'm one of four people reporting to Gerry Parker [corporate vice president for technology development]. My organization right now is relatively simple [writing on the board] . . . so I have three fab managers reporting directly to me, and I have a fellow who is running a task force . . . that involves a 32-bit microcomputer, the 8386. The fellow that's heading that reports to me, also.

Boleky continued by describing the remaining levels of reporting relationships in his organization, for which he was directly responsible. He drew out the positions that reported to the four positions that reported to him. One might predict clear structuring in a manufacturing subunit like Boleky's, but what about the structure of technology development, the heart of the innovative activity at Intel? Surely uncertain research activities would be more ambiguous and less clearly organized.

To understand how companies like Intel achieve a smooth and expeditious transfer of responsibility from development into high-volume production, we interviewed several managers and engineers in technology development. One of these was Kim Kokkonen, project manager in Static Random Access Memory (SRAM) Development. Kokkonen supervised a relatively small group of engineers who were developing an advanced semiconductor process technology, complementary metal oxide semiconductor (CMOS) process, and the first two SRAM products to be produced using the new CMOS technology. The semiconductor process and the new products would be transferred into manufacturing—actually into Boleky's fabs, described earlier.

When Kokkonen was asked to explain the transfer process, he immediately picked up paper and pencil and began to sketch an organization chart and a time line to express relationships over time. The organization Kokkonen drew started with the chief operating officer, Andrew Grove, followed by a direct report, Gene Flath, followed by Gerry Parker, the vice president and director of technology development, down to the groups that reported to Gerry Parker, including Kokkonen's in Static RAM Development. This was the exact same set of relationships Gerry Parker had drawn for us in a prior interview. There was high agreement on what the development organization was, who reported to whom, and what the separate positions and departments were. In Kokkonen's own department, SRAM Technology Development, there were thirty-five engineers and six program managers. Having drawn the entire organization, Kokkonen used the sketch to describe responsibilities of the various functions:

> Okay, this is me down here [pointing to the organization chart], and this [other] group is working on an advanced CMOS process. There's another group that's very similar to mine that's working on a different definition of a product that attacks different market segments. There is yet another group doing something similar. There's a group which does nothing but circuit designs, which take advantage of technologies which have already been developed and designs new circuits using those technologies. Finally, there's a basic module development group that develops some of the basic capabilities that we need to have: equipment development—things that aren't product or market related at all. An example would be lithography.

After describing his own organization, Kokkonen continued to explain how new developments are transferred into manufacturing, again going to paper and pencil. Here he drew a three-year timeline which he used to describe technology development at Intel and all of the various groups that come into play along the development cycle. In doing so, he described a clear set of relationships between eight separate development and manufacturing groups, each of which was bounded by time horizons and milestones along the path of development. In addition to his own SRAM development project group, the eight groups with which he and his engineers interfaced were new technology engineering (NTE), fab sustaining engineering, process reliability engineering, product reliability engineering, assembly engineering, product engineering, technology development test engineering, and technology exploitation group (TXG).

What is remarkable about these eight separate organizations is that Kokkonen, the project manager of six engineers, had crystal clear recall of these "interfaces," to use his term. Kokkonen was clear on what each of these groups was called, when each came into play along the development path, what each group was responsible for—in short, the entire picture. There was no fuzziness here, nor any failure to understand who is responsible for what or ambiguity regarding who these groups are. Kokkonen succinctly described their responsibilities, the key interfaces, and the transition points:

> My technology in general has about the most complex set of relationships to other subunits because we have a lot of internal ties. We also tie into all of the manufacturing arms of the company, in trying to transfer things. So the . . . things that we tie into: first the fab environment. There's a particular group set up in fab called "New Technology Engineering" (NTE) whose *defined job* is to spend the last year of development with us and the year after transfer of a [new manufacturing] process in cleaning up bugs and in just getting it ready for a full manufacturing environment.

As it developed, NTE was only the beginning of this complicated set of relationships. As Kokkonen continued, the entire diagram he had drawn came to life, with a heavy emphasis on reliability for the process as well as for the new products. As Kokkonen moved through the complete description of the life cycle of innovation at Intel, the extensive size of the Intel development and manufacturing activities became apparent, as did the complexities of dealing with the multiple groups. The project manager had fluently described the large number of people and groups with which he and his engineers had relationships, and the logic of the interfaces was perfectly clear to the listener. As to how he had developed this clear overview, Kokkonen remarked, "I think it's forced on you. The first time I didn't know all of this, but to get to this transfer point [into manufacturing], you have to deal with all of these people. . . . You just learn."

Similarly, in each of the other interviews at our five companies, no executive, project manager, or engineer was unable to clearly describe his title, job responsibilities, or the organization's structure of reporting relationships and the myriad of subunits within the structure.

The clear structure of these firms is deliberate, not a bureaucratic mistake. John Young, president and chief executive officer of Hewlett-Packard, provided some insight into this unexpected finding. He succinctly explained why there are clear reporting relationships at Hewlett-Packard: "I just think it helps 84,000 people, particularly when you're having a lot of changes, to see what you've done, what we're trying to accomplish, to make it as clear as you can what are the central thrusts of the organization."

We believe this is an important finding regarding the contemporary, innovating organization. As both a product of prior successes and a requisite for future success, these companies are highly complex organizations. They all produce thousands of products in multiple divisions, using state-of-the-art equipment and processes. The pace of technology development required to remain among the most successful innovators is murderous and will continue to be so. As a consequence, employees must have high predictability within their internal organizational environments. Clear structures and job responsibilities provide this predictability.

The lesson we derive from these data is the importance of a managed structure. With explicit reporting relationships, there is little of the predicted ambiguity of organic systems, and none of the wasteful uncertainty about to whom to turn. Structure is actively used to guide employees in the firms, for delimiting responsibilities, identifying connections between positions and people, and insuring that attention is actively allocated to appropriate tasks. In the midst of frequent product and manufacturing process changes and external competitive uncertainty, structure within these companies is clear. It provides high predictability for the behavior for organizations and subunits and rapid identification of needed resources and decision-making authority.

How, then, do these companies organize for repeated innovation? Very explicitly. These firms are characterized by highly effective, very explicit organizations, with well-understood and clearly articulated relationships. In retrospect, it could scarcely be otherwise. The activities undertaken by these firms are complicated, and cannot be undertaken except by large, sophisticated arrays of multidisciplined talent, marshaling substantial resources of equipment, dollars, and information. Wafer fabrication lines frequently cost millions of dollars, and very large-scale integrated circuit fabrication processes have hundreds of complicated steps. Without meticulous coordination and control, these processes simply will not be successful. Ad hoc organizations or loose, organic structures often recommended for innovation simply *will not do*, for the long run. Even if ad hoc arrangements might once have worked, their repeated success would be impossible, given the complexity of relationships, technologies, and product changes. Innovation in these firms is too important to leave to chance, and too complex to handle without explicit coordination. These companies need reliable, repeatable methods to bring good ideas from concept to product to market.

There is a puzzlement, however. How can we account for the high clarity of formal structure when Burns and Stalker and others have argued that a fluid, ambiguous, ad hoc structure is adapted to high rates of external change? If they do not have fluid, ambiguous structures, how do these highly innovating organizations adapt?

FREQUENT REORGANIZATION AND DYNAMIC TENSION

Two major differences are found when our data are compared with the earlier work of Burns and Stalker and more contemporary theorists. First, these innovating organizations *reorganize frequently*, and as a consequence, responsibilities and reporting relationships change to meet modified external and technical circumstances. This finding illustrates that one way in which high technology organizations adapt to technological and environmental change is by reorganizing, changing the formal components of the organization, rather than by so-called organic structures in which position responsibilities and reporting relationships are continuously ambiguous. This finding is also consistent with earlier research on reorganization in high technology industries (Schoonhoven, 1980). Thus, although clarity of formal reporting relationships and of position responsibilities is a consistent pattern in the successfully innovative companies we have studied, these companies are far from structurally inert. We have discovered that these organizations adapt to change by reorganizing their formal structures.

The second difference in our findings from that of others is that a *dynamic tension* is inherent in semiconductor and computer firms faced with changing markets, technologies, and manufacturing processes. On the one hand, continuity, control, and integration between functions and departments are required for the orderly process of new technology development and reliable, high-quality manufacturing. On the other hand, changing environments and technologies require organizational adaptability and flexibility. We refer to this as maintaining dynamic tension: the ability to be flexible through reorganizations as well as sufficiently systematic to be efficient producers. Organic structures do not include this dynamic tension, we have found—tension between clear structures and frequent reorganizations—but instead focus only on constant, free-form flexibility through amorphous organizational forms. Organic structures ignore the continuity and predictability of action which a clear structure provides for the organization and for its external constituencies.

It is important to recognize that these companies do not move from one adhocracy to the next, as some futurists have envisioned or as some have suggested innovative, high technology firms do (Toffler, 1970; Mintzberg, 1979). Adhocracies have been described as the "structure du jour"—the structure of whatever works for today's problems. However, if managers in an adhocracy perceived their problems as changed tomorrow, the structure du jour would be changed. Adhocracies also carry a highly temporal, spurious connotation, in which "you had to be there to appreciate it." Others in a large organization may not know how your subunit is structured if they have not checked recently. Lack of shared understandings and predictability of behavior is theorized to characterize adhocracies in a fast-changing, turbulent environment.

Rather than being organic adhocracies, the highly innovative companies we studied changed their structures when the problems for which the current structure was designed had changed; they shifted from one existing, clear structure designed for specific problems to another clear structure, maintaining a dynamic tension among organizational elements.

MANAGING STRUCTURAL CHANGE

We found that changes in formal structure are one of the principal tools by which the organization's executives can continue to keep their company innovative but attuned to the current and anticipated future problems to face the company. Reorganizing the company signals an adaptation to a changed environment and technical circumstances for which the current organization is no longer appropriate, rather than a mistake in the original organizational structure as outsiders who do not understand the dynamic nature of the business often assume. Les Vadasz of Intel Corporation remarked, "People look at organizational changes as, 'Somebody did something wrong.' And that's nuts, because there's absolutely no reason why an organization that you created two years ago has any relevance to the organization that you need two years from now."

Vadasz has the distinction of wearing Intel Badge #3, signifying that he was Intel employee number three, after its two founders, Gordon Moore and Robert Noyce. Currently an executive vice president and a member of the Intel board of directors, Vadasz has witnessed how Intel's environment and technology have indeed been modified over the years: "And things *have* changed. The beauty of this business is that the technology will always change, the organizations, the organizational interfaces, the customer interfaces, the vendor interfaces. That's always going to change, because of the technology. To assume that your organization makes sense today, just because it made sense five years ago is really incorrect."

The prevalence and frequency of reorganization was indeed interesting to observe in these companies. We first began our interviews in the summer of 1981, and we completed the most recent interviews in 1988. Over this eight-year period, all of the companies made relatively a large number of changes to their formal organizations, both at the corporate level and at the division level. Obviously higher-level changes are more visible from the outside and may represent more significant reorganizations. Many more changes at operating levels occur with fewer external signs, and these modify the structure in very important ways. The experience of just one of our study companies, Hewlett-Packard, will suffice to illustrate the frequency and pervasiveness of structural change in these firms.

Hewlett-Packard

An explicit example of frequency of reorganization can be seen by tracing the organizational changes at Hewlett-Packard over the period of our work. In 1982, within six months of the beginning of our research, Hewlett-Packard began to make some externally visible organizational changes. John Young, H-P's chief executive, explained to us that these were the first of a set of organizational modifications designed to address the changing technical and market conditions in their computer business.

In 1982 H-P had "a really big problem, which, in 1985, we are still in the final throes of executing," said Young. The problem was to change their entire computer line around to a new architecture and to create a new family of computers. The long-term technical issues were still very much in the development stage, but H-P nonetheless had innovative computer products to offer its customers. As a consequence, in December 1982 H-P reorganized the computer group itself into more of a centralized marketing activity as well as a product division. Young explained his thinking on the reorganization in 1982:

> I really didn't want to disturb the product side of things until we had that really bolted down [technically]. Now [in 1985] that we have made more [technical] progress, I've made another change in the organization, and I think we feel we're in a position to deal with that. Again, that's a difference in that particular product, and now we've engineered things. We have one whole group that's doing nothing but engineering this one computer system: 800 professionals in it. This is quite a different [organizational] arrangement than we've ever had before.

In July 1984 H-P again reorganized as its markets continued to converge. For example, their traditional industrial computer users who needed test instruments, also needed to link the instruments to personal computers. H-P's structure, composed of relatively autonomous divisions, had made it difficult to take advantage of the converging markets. One sales force sold instruments and a separate sales force sold computers, even to the same customers. The result, according to one observer, was that "You had within H-P an instrument company and a computer company treating each other at arm's length while trying to do battle with IBM" (*Business Week*, 1984).

In 1984, within eighteen months of the creation of the personal computer group, H-P introduced three new products: a lap-sized computer named the Portable and two new printers called ThinkJet and LaserJet. In October 1984 John Young remarked that "creating the personal computer group was an extremely good way of getting a focus on marketing. It was a way of communicating to everyone that this marketing was okay; that it's okay to eat quiche" (Saporito, 1984). It was also a way of signaling to H-P insiders as well as to outsiders that the organization was formally addressing a hitherto untapped activity.

In mid-1984, the new position of "chief operating officer" was created to centralize authority and relationships among the divisions somewhat. Dean Morton was appointed, and a corporate marketing division was also created at that time, merging the computer and instrument sales forces to sell what is now referred to as an "integrated solution," not merely separate machines. At this time, while the management and organization of the overall computer division was modified, the personal computer group's organization was not changed.

The creation of a corporate marketing division was a major departure from Hewlett-Packard's usual methods of organizing. But the products and the markets were changing, and so was its structure. Long renowned for its engineering elegance and as a seller to an appreciative, technically trained customer base, H-P's employees had worked hard during the following two years to develop an equally elegant marketing capacity. Its vice president of the personal computer group was quoted as saying, "I keep telling my engineers that they now have five minutes to make a sale, not five hours like we used to. We have to focus on apparent user benefits." Dick Alberding, in 1984 the newly appointed corporate marketing head, spoke explicitly of the Hewlett-Packard shift from simply great technical products to market share. "The personal computer industry is a market-share business. If you own 10 percent of the business, it's important to know what the incremental costs are to get more." Responding to implicit criticisms of H-P's marketing expertise at the time, John Young publicly stated, "I don't think HP is bad at marketing. It's just that we needed to accelerate the marketing to complement the engineering."

In 1985 Young described the causes of changes in organization structure made in the late summer of that year. His comments are worth lengthy quotation, because they lay out so clearly the connection between marketplace changes, technical shifts, and how restructuring helped deal with these. Both formal structures and what we call "quasi-structure" are apparent in Young's thinking.

> We see a lot of changes in the marketplace, over a period of years, I guess, both on the product as well as the customer side. The customer for our kinds of products used to be technically sophisticated people, buying things for them to work on, on their own. As we both changed the kinds of business to more information systems content, and even the technical side of that, customers have been growing less and less willing to do it themselves.
>
> Let's say the customer has been looking for more complete answers to their business or technical problems. This is what put pressure on the factory side, to conceive and deliver and support and document solutions that were very sophisticated in nature. Of course, all of this means more teamwork, more things that have to work together, more interaction between pieces of the organization, and less and less departmentalization, such is possible in traditional, division organizations.
>
> So, we still believe in small work groups, and we call those divisions at Hewlett-Packard, but we simply have to find ways of getting them to work more cooperatively, and to make sure that the system disciplines that are so essential are put in place.
>
> So, basically what we did on the factory side, is we organized things more on market centers, as opposed to product centers. On the field side, we dissolved those product linkages and turned them around into more of a customer team organization. That's been going along for a year now, with a lot of work in reestablishing those informal communications back and forth.

Explicit in Young's comments is a focus on structure as a management tool and on the importance of the interaction between formal organization and the quasi-formal activities of teams. Equally visible, however, is a shift from the prior focus on technical elegance as the selling point to *solving customers' problems*. The problems have changed, and thus the structure was reorganized.

Expecting Structural Change

Employees come to regard structural change as a part of life in companies like H-P, National, and Texas Instruments. They expect future changes, as well. Some of these anticipated changes are finer tunings of structure at lower levels. For example, Douglas Spreng, the division general manager of H-P's computer systems division, shared a recent formal organizational chart with us. This chart reflected the changes in structure that John Young describes in the preceding paragraphs. Spreng offered the chart with a telling observation about structural change at H-P: "Do you want a copy of the corporate organization chart? In fact I've got a couple of slides I can give you that show the relationships too. They're probably in transition; six to nine months from now, they will probably be changed."

Spreng's comment was a straightforward observation on anticipated future modifications in H-P's structure, offered neither flippantly nor cynically. Instead, he was simply reflecting on likely consequences of changes in the relevant technologies that affected current businesses—and important marketplace realities. For example, Spreng speculated on what the near future changes in the structure might be, relevant to the businesses he was most familiar with:

> My guess is what we'll finally end up seeing is three businesses separately managed. But we're still sorting that out, because the last year has been rather complex, and it's not as clearly definable as the other two sets of changes in the organization (which have been made). We're taking a little longer time, because we do not want to take false steps. It's really important I think for us, if we make a step, that we make it firmly, and we go from there. [We won't] . . . do any herky-jerky . . . as some of the things in the past [have been].

This thoughtful division general manager was among that set of participants whose formerly highly autonomous operations were seeing their autonomy somewhat curtailed by the reorganization. Nonetheless he felt the changes were positive and in the needed directions. Spreng commented on these events:

> We're reorganizing the structure of the company, I think, in a much more solid way. The company is changing tremendously from the old days of what our vice president calls the futile-abilities of the instrument divisions to a vastly integrated, highly leveraged company. At the same time we are growing at a very good clip, which is quite a challenge, but it is absolutely necessary for us to achieve that.
>
> We don't want to become a centralized company like IBM—we're not going to become that. But we're moving toward a more coordinated, strategically oriented kind of a management balance between the entrepreneurship of the divisional units and the strategic overview, the guiding hands if you will. This computer business, which has basically overtaken the company, has dramatically changed the company at the same time.

Managing the Process of Structural Change

The process by which these companies reorganize is substantially participative. After clarifying changes in strategic directions that the company will take, the members of the top management team (the management committee in H-P's case) sketch out the major skeletal changes in the structure. However, it is at the group and division levels that middle- and lower-tier executives determine structural changes in their own organizations.

Details of the reorganized structure *must* be worked out by lower-level managers closer to the impacted groups. Unless the company's top executives want to take forever to design the optimal structure, some interim approximations must be made. It is generally not possible for the top management team to be omniscient when reorganization occurs. As a consequence, they place their best bets on what they believe the structure should be, implement it, and then observe the results. Frequently there are some elements of the new structure whose changes could not be anticipated in advance. For example, Doug Spreng remarked on the consequences of some changes that took place at Hewlett-Packard in 1985:

> You change things around . . . and then complexity of the interactions becomes awesome. You think that you've got this situation bounded, territories understood and so forth, and then you find out: "No, these three other organizations over there are intimately affected by what you just did here. You had no idea." Yet, you know unless you want to take forever in reorganizing you're going to have to make some of these decisions on the front end and hope that you haven't screwed something else up. Or hope that you can fix it.

The necessity for fine-tuning the reorganized structure is anticipated. These necessary structural adjustments provide an important opportunity for lower-level managers to impact the structure in ways more finely tuned to the actual problems at lower levels of the organization.

When the reorganization results in problems, the lower-level managers participate to refine the structure for better working relationships. Some of these structural changes hint at the important relationship between the formal structure and informal interaction. For example, Doug Spreng described why he and a related division general manager were going to have lunch to discuss which of their divisions two laboratories should join in the newly reorganized company:

> There's a lot of concessions going on . . . when you get down to some of the details. Like the guy across the hall, here [another division general manager]. I believe that at least one and maybe two of his lab groups ought to be within an organization that I've got. . . . I've already approached the subject and asked him not to make any changes in his organization until we get it resolved, because I believe that we've got it wrong right now, and I don't want to keep it wrong. It's easy for us to do that because we both trust each other, and we know we're not trying to build an empire. We're just trying to do the right thing for the company and get these people to where they can make their best contributions. . . . We're going to get together over lunch and talk about it some more. We're creating a document that shows what the recommendations are and why. I think the best thing about H-P is that we can work these things out.

Self-Designing Organizations

The preceding narrative about changes in H-P's formal structure over the four-year period from 1982 to 1986 amply illustrates the kinds of changes we saw taking place at each of the companies we studied. For each firm, the changes in structure were customized to meet the modifications managers perceived in their company's environmental and technological situations. As a consequence, the timing of when each company reorganized varied. The aspect of structure modified by any given reorganization also varied by company. It is important, nonetheless, that the changes in structure were customized to deal with each of the company's localized task environments, its markets, and its technologies.

Companies whose managers engage in this kind of behavior are said to produce "self-designing organizations" (Weick, 1977). These companies can be self-designing because of two factors. First, the people working within these organizations are highly self-critical regarding how they are operating and whether the current structure is facilitating goal accomplishment fast enough. The managers and executives we interviewed provide strong evidence that it is a norm—a shared agreement regarding appropriate and desirable behavior—to evaluate the quality and utility of the organizational status quo. The need for constructive self-criticism is an important shared belief. If an element of the organization is not functioning adequately, alternatives are considered. The current organization is revised if necessary, or major elements of the current structure are changed if more modest organizational changes are deemed inadequate. The key is that these executives and managers openly evaluate the utility of the status quo of the structure. There is strong agreement among the managers and engineers we interviewed that organizational self-criticism is appropriate.

Our research suggests that for organizations to be self-designing over time, widely shared agreements regarding this activity are essential. In our firms, such agreements are part of the organizations' cultures and visible in the managers' operating behaviors. It is expected and widely found from top to bottom, down to the level of engineers. Founders and top-level leaders are very important in helping to shape expectations and norms, of course. We found that members of the top management teams of these companies expressed a willingness to reorganize if it seemed appropriate. More important, this expressed readiness was acted on: real changes in structure took place with some frequency; the changes were well publicized throughout the companies. All of this produced a high tolerance for change among the people at lower levels in the organization with whom we spoke.

The second reason these companies managed structural change so effectively was their awareness of market, scientific, and technology conditions outside their organizations. The external environment was given scrupulous attention. Effective organizational change must target and adapt to major environmental and technological shifts. Structure is a deliberate artifact intended to suit an explicit purpose, not mere random change. Random structural changes will have unpredictable effects on organizational performance—an extremely dangerous situation in the intensely competitive environment these firms face. Consequently, awareness of the context to be adapted to must be blended with a willingness to change in addition to skill at structural arrangements.

Managers in these companies systematically try to understand their businesses even better. They do so by measuring nearly everything internally, monitoring their competition, tracking technology and product shifts, and seeking to understand impending competitive changes. These are quite deliberately self-conscious organizations, with information shared and used in the process of structural modification.

For example, it is simply routine that the companies rigorously measure manufacturing "yields," the percent of usable components ("chips") produced on a single wafer of silicon. It is pro forma that manufacturing cycle times are continuously recorded and analyzed and that changes in manufacturing equipment, organizations, and people are instituted to drive the yields up and cycle times further and further down. The equipment changes frequently; new manufacturing processes are introduced frequently; existing technologies are improved through process development; and new products are introduced on new and continuously improving technologies. It is not sufficient to be right once; it is a continuous battle under difficult technical circumstances.

Such extraordinary attention is driven by broad awareness of competitive hazards. "We've got to beat the Japanese through speed of development and consistently high quality" was one phrase we encountered frequently in our interviewing. The awareness of the external market extends deeply within the internal levels of these companies. For example, the chief executive of National Semiconductor, Charlie Sporck, spoke of a newly graduated engineer, a circuit designer, who had joined National just the year before. "His views of what the [semiconductor] business is all about are far from perfect. But he knows where his product stands relative to everybody else's. Very clearly. He knows it better than I do."

The changes in organizational structure previously described also illustrate the *dynamic tension* principle of organizing in each of these companies. Each time a reorganization took place, clear new reporting relationships were worked out, clear responsibilities were assigned to individuals and groups, and these responsibilities were refined to clarify any problems or ambiguities. These companies can move rapidly to introduce new products or technologies in pursuit of corporate goals because people in the companies know what other people are doing and how they can count on others to facilitate their own local subgoals.

If these companies need to reorganize or modify their structures, their managers are fairly clear as to why. Something is not working optimally. The managers of these companies are highly self-critical and analytic about organizational arrangements. As conditions change, the highly innovative commercially successful companies we have studied reorganize to adapt to the new and anticipated future conditions.

QUASI-FORMAL STRUCTURE

Teams, Task Forces, and Dotted-Line Organizations

Clarity of and changes in the formal structure are important ways in which these companies have organized for innovation; however, formal elements of structure are only a part of the story. Organization theory seems to recognize only two dimensions of organizational structure: formal structure captured in the organization chart as subunits, positions, and reporting relationships; and informal structure, the unsanctioned patterns of interaction devised around social and task requirements the formal organization has failed to take into account. We believe it is time to recognize a third, intermediate level of structure that prevails in the companies we studied.

The intermediate level of structure we have observed is labeled "quasi-formal structure." This intermediate level of structure is not formally characterized on the organization charts. Nor is it "informal," in the usually covert, unsanctioned, purely social, and even adversarial sense of this term (Blau, 1955; Connor, 1980; Gouldner, 1954; Leavitt, 1973; Roethlisberger and Dickson, 1939; Selznick, 1949). The term generally means "used in combination with"; something is "quasi" because it "resembles but is not the thing in question."

In these organizations, quasi-formal structure consists of an extensive use of committees, task forces, teams, and dotted-line relationships. The quasi-formal structures we have observed in these companies resemble formal structure because they are formally sanctioned by the organization and explicitly recognized as legitimate. They are typically not designated in form, specified in a formally defined charter, or delimited rigidly in terms of membership. Nor do they take the place of formal authority and the usual responsibilities. Instead, quasi-formal structure is used *in conjunction with* the formal structure but does not replace it.

Quasi-formal structure should not be confused with the "informal structure" of the organization because it is *not unsanctioned*, illegitimate, covert, or in any way outlawed or disapproved of. On the contrary, it is explicitly recognized, condoned, and encouraged by management at all levels. Because managers themselves participate in creating this level of the companies' structures, this is almost a formalized element of these organizations; quasi-structure is the accepted mode for the entire range of employee problem-solving groups, from engineers and technicians to high-level executives.

One way in which quasi-formal elements are different from formal structure is that they change with an even greater frequency than the formal structure, as problems are resolved or new problems appear. Another difference is that these groups are relevance-based and problem-focused and typically transcend whatever formal structural boundaries that might otherwise impede problem solving. Thus, quasi-structure serves as an important organizational facilitator, enabling explicit formal structures to work, amid the urgent need to solve frequently changing technical and competitive problems. Quasi-structure augments the formal structure and bridges the gap between the skeleton of the organization and the informal mechanisms for doing things.

Quasi-structure also saves companies from simply becoming adhocracies, paradoxically because it makes change and adaptation to problems much easier. The organization need not move to a "structure du jour," and continuous, formal reorganization need not be a constant problem, with structure losing its clarity or people constantly working in a high readjustment mode as a consequence of continuous change. Quasi-structure augments the formal structure: it demonstrates the utility of forming problem-focused groups across internal boundaries and thus helps create a norm of helpful interaction and relationships across boundaries, which makes the formal boundaries and their limitations less constraining.

In each of the companies studied we observed an extensive use of committees, teams, task forces, and dotted-line relationships that augmented their clear formal structures. These were used for one-time or continuously problematic circumstances that required off-line, additional technical and managerial time. The genesis of these units of quasi-structure varies. Sometimes they arise spontaneously, forming around common problems with people in other functions. Sometimes "management" at some level explicitly requests that a task force or committee be convened to deal with a specific problem. A concrete example will illustrate the speed with which elements of quasi-structure are created.

Richard Walleigh, a manufacturing engineering manager in the computer systems division at H-P, described an example of a spontaneously formed task force around the concept of total quality control (TQC). First, someone has to perceive the need for a joint focus on a problem and take the initiative. In this case Walleigh himself was the initiator, with the following observation:

"Here's a great idea, we ought to do this." Then once you have the light bulb turned on, the next thing you do in H-P is get a group of people together and you start talking about it. It always requires a group of people to do practically anything. There's so few things that you can do in your own department without influencing other departments, that it seems everything we do is in groups.

So . . . you get them charged up about it, get them working on it in their departments through an informal mechanism. You just put together a team of the relevant players and say this is a great idea, talk about it, and we then say, how can we implement it here? There everyone starts working on [implementation]. The team sort of meets, works on the project until the project's done, and it gets developed. . . . It's our style here and it's almost necessary. We do lots of things in teams, some formal and some informal.

Walleigh is not alone at H-P in his enthusiasm for teams. His CEO, John Young, fully understands and approves of the kind of teamwork Walleigh's narrative describes. Young corroborated Rick's account with his own examples:

We have a million task forces around, particularly a lot of them on technical and marketing kinds of things, that really are very complex subsets of the organization—things that never show on this chart. It also happens in our administrative things: all the way from the compensation task force, to you name it, to . . . just finding knowledgeable and respected people who want the chance to participate in this problem is probably the most frequent pattern we use for the kind of coordination and cooperative activity.

Not Matrix Management

It is important to define quasi-structure in terms of what it is not, as well as what it actually is. Quasi-structure is not "matrix management." In the 1960s an organizational form called the matrix structure evolved within U.S. aerospace firms, often under contract to the U.S. Department of Defense and other federal agencies. In matrix management, the corporation or its subunits are reconfigured so that employees within functional departments have two bosses: a functional boss and a second project boss. This is called a dual-authority structure because both bosses have the traditional rights of supervisors to assign tasks, evaluate work, and influence the rewards and promotions their matrix subordinates receive.

Matrix structures were regarded as a genuine structural innovation, since they were well adapted to the vicissitudes of congressional funding, long-term projects with overlapping time lines, and shifting organizational and personnel needs of the aerospace industry. Matrix forms were a clear departure from rigidly hierarchical, vertical organizations typically seen in most large organizations until that time. They are configured on paper as a grid with dotted-line relationships, in which an engineer, for example, would "report out" of his or her functional engineering group to a project organization as well. Thus, "dotted-line" relationships have become a shorthand for one's matrix boss in addition to one's functional boss.

Occasionally the managers and executives we interviewed spoke of matrix relationships. Invariably further investigation revealed that neither the entire corporation nor its divisions were structured as a pure matrix form. Rather project groups were created for technology or product development, which typically drew engineers from several organizations. Other relationships initially described as matrix were later revealed to be of the dotted-line form, rather than a pure matrix form of subordinates with multiple bosses in a dual-authority structure.

More often than not, a dotted-line relationship in these companies designated a "for information only" relationship, in which it was considered important for two positions on different levels, existing diagonally across organizational lines, to communicate with regularity. In such a case, a dotted line was used to characterize the relationship. In another case a project supervisor explained that he was "matrixed" to several other groups and divisions; this, too, was revealed to be an information-sharing relationship characterized by committees or meetings to report on and solve problems with other disciplines and departments. A structure of dual-reporting relationships was clearly not the meaning when the term "matrix" was encountered in our research; rather, a number of organizational units were involved in sharing information, in the typical case.

More to the point, quasi-structures are both more formal in their recognized status as problem-solving groups and less formal in their authority relationships. These are colleague-based, problem-focused entities that exist for the duration of the problem. They are also far more pervasive, more deliberate, and more explicitly used than "informal" relationships or "cultural" linkages.

The matrix concept, pure or otherwise, was familiar enough that many of the managers were rather clear that they did not have a matrix structure. Indeed, some believed that a formal matrix structure reduces the organization's clarity and reduces its effectiveness. Peter Rosenbladt, an R & D manager of about 120 people at H-P observed the following:

> Matrix management, wherever it's been tried [at H-P], it's been very distasteful, because it is always resulting in undue complexity, unclear objectives, so we like to do it informally wherever we can. We have never successfully implemented a matrix of management that really does work, because it immediately means that everyone has at least two bosses, and that doesn't go well with things like who determines how well you did, what the priorities are, and those kinds of things.

Managing Quasi-Structure: Costs as Well as Benefits

Though pervasive, quasi-structure is not without its costs. At Intel, for example, it is used extensively. Les Vadasz described the prevalence of both task forces and dotted-line relationships: "We have a lot of multiple reporting relationships, dotted-line reporting relationships and shifts in organizations, especially if you look at project organizations, as the project reaches different stages. People from multiple organizations participate. . . ."

The use of multiple reporting relationships on Intel's technical projects efficiently allocates scarce human talent to needed technical projects. However, some complications are associated with dotted-line relationships, according to Vadasz. Consistent with other research, dotted-line and team-based relationships in technical projects and other organizational applications produce what he described as "some interim ambiguity." As Vadasz explained, "There is a difference between knowing that you have ambiguity and knowing that "Hey, I want to work out of it." If you just ignore it, then you get chaos. What you have to struggle for is clear organization lines, knowing fully well intellectually that you will not be able to accomplish it perfectly, because you want to do what's right for getting the project done. But if you always struggle to get the cleanest possible line [of organization], you have less chaos." In short, in a rapidly changing technical environment, constant vigilance is the price of organizational clarity for a firm striving to prosper and survive, even though it may have a clear formal structure.

Another cost of maintaining innovation in a large corporation is that considerable time must be spent integrating across organizational boundaries. This is another aspect of quasi-structure: it simply takes time, even if that time is invisible at first glance. While Hewlett-Packard executives consciously strive to keep its organizational units small with divisions of approximately 1000 participants, it is nonetheless clear that coordination across some 50 such divisions is not a trivial task. Task forces, dotted-line relationships, and teams are omnipresent within H-P, as indeed they are in all the firms we studied. These quasi-formal elements of structure are an absolutely essential mechanism of organizing for innovation. Task forces and dotted-line relationships are part of the solution.

At each company people were cognizant of the essential part that committees and task forces played in problem-solving and coordination, and yet they were simultaneously critical of the time required for effective task force participation. At each company, managers were aware of the need to eliminate any elements of organization no longer essential.

However, elimination was not always easily accomplished. Les Vadasz at Intel felt that quasi-structure had to be tended, just as any aspect of one's garden might be. Less useful elements must be cut back and pruned. "Do we have too many dotted-line relationships? I would say 'Yes, we do, because it is not easy to kill something.' People get used to the fact that they are there, they have a committee of some sort. . . . Although the committee has done its job and the problem is resolved, some committees nonetheless evolve into standing committees. . . . And we have to be able to prune them."

Vadasz indicated that Intel was cognizant of the need to eliminate unessential functions, committees that survive beyond their functional utility, and meetings that simply take up time. A key observation, however, is that when quasi-structural elements become ineffective, thinking gets "stale."

> We are trying more consciously these days [to address the longevity of teams and committees]. For example, we have a very extensive council system. About a year ago, one of the best council systems we've had from the beginning was our die production work. Because of so many factories all around the world, we have to make sure the better processes are common enough in all the factories so that we have the flexibility to move things around. Well, those committees over the years got a little bit stale. So we made a major change and pruned it dramatically. There were more people who felt that it was a breath of fresh air and so it wasn't a big issue . . . to prune it.

The latent value of task forces, committees, and councils is that they present opportunities for lower-level engineers and managers to begin developing their leadership and group skills—both essential components for future managerial success in an innovative company. This value is not without its costs, however. Vadasz continued, "But I think the biggest problem [with task forces] is that somebody has to do the supervising. Someone has to guide them. You don't just create something and let them go. If you were to hire a manager and let the manager go without supervision, you generally end up with undesirable results."

We have concluded that the quasi-formal parts of structure—the committees, the task forces, the project groups—require as much guidance and monitoring as any other part of the organization. Managing quasi-structure is, for all its necessary openness and participation, a demanding responsibility. Despite the fact that lower-level managers "chair" such committees, the burden on management for guidance is nonetheless heavy. They are "absolutely not a freebie" according to Intel's Vadasz, because they require the participants' time as well as some additional managerial time. Time is scarce in the big innovation technology marathon these companies are running. In balance, however, the value of councils, task forces, and committees is in essential component if these companies are to accomplish the required interaction that must take place across disciplines, departments, and divisions to achieve the remarkable multidisciplinary technologies and products these companies produce year after year. Councils, task forces, and committees are an effective tool in the management of innovation. Though not without costs, these organizational arrangements deliver tremendous benefits in the management of innovation if they are monitored and attended, just like any other aspect of the organization that you expect to bear fruit.

COLLEGIAL INTERACTIONS IN A TEAM ENVIRONMENT

One element of classical "organic management" for which we did find substantial evidence was collegial interaction. In place of strongly hierarchy-based patterns of interaction found in more traditional organizations, our firms embody very strong norms and practices of nonhierarchical, problem-, expertise-, and interest-based interaction. The extensive use of committees and teams that cross organizational boundaries and are composed of multilevel people clearly facilitates a nonhierarchically based pattern of interaction. Norms that support collegial interaction patterns and expertise-based authority clearly facilitate a nonhierarchically based pattern of interaction.

Fundamentally in innovative high technology companies, multitudes of novel problems exist for which no established solutions are readily apparent. The sensible approach is to rely on each and every resource available. Thoughtful reflection on the problem and substantive contributions, rather than hierarchy or status, are more relevant in the highly competitive environment these companies face. Fast decision making is essential to effective competition, so standing on form and position merely impedes problem solving.

John Young at Hewlett-Packard described this phenomenon and the part it plays in maintaining an innovative organization rather well: "To make the structure as clear as you can is a long way from saying there are no ambiguities. There is a very big need for a collegial environment . . . to make it work . . . I think it's far more the style and the freedom of activity and association and communication back and forth that makes an innovative environment."

Consider that these companies deliberately hire the very best people they can. People who have the most recent technical training are highly valued: recent graduates from engineering and business schools, as well as experienced engineers, executives, managers, in addition to those they can hire or develop internally for support functions. Even production workers and equipment engineers need to be of very high quality. In an industry of class 1 clean rooms, which counts contaminants to integrated circuit production in parts per million or less, the users and maintainers of highly sophisticated electronic production equipment must be very knowledgeable. All of these employees are the highest quality that can be found.

Now, the cynical reader will remark, "Well, surely there are some laggards, lower quality, just plain zeros in each of these companies." Undoubtedly there are. What distinguishes these companies, however, is the widely shared belief that we must use whatever human resources we have to bear on a given problem. The norm that permeates these companies is that anyone who seems to have good ideas or who is reflective will be relied on if their experience, expertise, or intellectual abilities appear to be relevant to the problem at hand.

Selection from among the thousands of people available in each of these companies is facilitated by the companies' structures. There are smaller work groups whenever possible, and the overall structure of each of the companies is clear. One can determine without great difficulty where additional help might be found, if the people in your current organization do not have the requisite knowledge or help resolve the problems. People become known to those around them, and there are continuous opportunities to solve problems.

There is a pervasive norm that people in these companies call a "team mentality." The team mentality involves the shared perception that "we are in this together, and it will likely take more than just myself to solve a given problem." Indeed, if the team mentality were stated as a norm to guide the behavior of new recruits to these companies, it would resemble this statement: "Given the complex technical solutions that we are seeking in addition to the complicated and uncertain market conditions we face, it is likely that more than just a single individual will be required to solve a given problem."

We asked a division general manager at Hewlett-Packard if the highly competitive computer market their company was facing might change the willingness of managers to cooperate with one another internally. Was the pressure on H-P as a company going to make it tougher for individual division general managers to survive and for their careers to prosper, and thus increase internal competition among such managers for future promotions—and perhaps reduce their willingness to collaborate? In contrast to what might be expected, the manager responded as follows:

> Oh, no, you have to be *more teamwork-oriented*. Teamwork is the only way we can preserve our thrust into the markets as a company. The divisional structure will [continue to] exist, but yet [the reorganization] allows us to work at a much higher plane of cooperation as an organizational entity. So teamwork is . . . our secret. We're finding measurement of teamwork coming up more and more in evaluations and in ranking sessions of managers.

> You just don't go off and do your own thing in this company anymore.

Informal, collegial interaction was widely practiced at the technical levels of the organizations we studied, as well as among the managerial ranks. At National Semiconductor, for example.

> Sometimes [innovation] happens in the men's room. One guy's talking to another guy, and another guy's standing, eavesdropping on the conversation, scribbling on a napkin. If you dropped down right now to the cafeteria, you would see it going on. Most of [the innovation motivation] comes out of, let's say, frustration or irritation with the way something currently works. And a guy says, "You know, I really wish I had a fill-in-the-blank. What I had didn't do fill-in-the-blank." And the other guy says, "Well, you know, a few years ago I had a problem like that and what I did was x, y, and z." And then the chemistry starts, and the result is a product that solves the problem.

In the restrooms, in the cafeterias, and in the hallways—informal colleague-based interaction patterns permeate these companies.

CORPORATE SIZE: HOW BIG IS TOO BIG?

We cannot leave a discussion of structure without turning to the issue of organizational size. Large numbers of people pose significant problems for coordination in a fast-changing marketplace and with fast-changing technology. So, too, do broadly scattered operations, broad product lines, and global competition. Each of the study companies operates worldwide, in an increasingly global market and technical environment. All have revenues of over $1 billion, and their number of employees range from just under 20,000 at Intel to Hewlet-Packard's 82,000 and Motorola's more than 90,000 employees worldwide. These are large corporations by any measure. However, they have not always been large. Indeed, many of the founders and top executives remember when the companies were smaller and more manageable, as recently as ten years ago. Although Texas Instruments, Motorola, and Hewlett-Packard were founded more than thirty years ago, they experienced their most explosive growth in the 1970s, just as the younger companies, Intel and National, were also experiencing rapid growth.

The founders and their early executives unanimously recalled how much simpler it was to communicate in a small company. It was relatively easy to retain a close coupling between the various organizational functions. David Packard described how he and Bill Hewlett operated in the early days at H-P, after World War II: "When the company was small, [a close coupling] was almost automatic, because we were all involved in everything. In fact, when we were small, I'd spend a lot of time in the laboratory, and I'd spend a lot of time in the factory, and I spent a lot of time in the field. And so you did all these things yourself." Packard's experiences nearly duplicate the description Pat Haggerty gave of his early days at Texas Instruments: "It was here that the setting of objectives and being small and tightly coupled . . . was so important. We put good people at the thing, and they were very tightly coupled."

During World War II, the war boom gave Packard and his partner, Bill Hewlett, the opportunity to see the impact of boom–bust economies on organizations through their observations of the aerospace industry. Neither wanted to hire people only to lay them off. They preferred to build loyalty through a long-term commitment to the people who were hired—even at the near-certain risk of slower growth and a smaller organization. Packard remarked, "I'd been seeing a lot of activity in the aerospace industry where companies would go out and get a big contract and hire all the engineers. When that contract was over, they'd have to fire them, and they'd go and work somewhere else. So, I decided that we would try to build our business on a long-term basis and not go out and get a lot of contracts. We had a maximum of about 200 people during the war, so we were a very small business at that time."

These observations on the negative impact of growth through short-term contracts laid the foundation for Hewlett-Packard's renowned human resource policies—to build loyalty through a long-term commitment to the people who are hired. Thus, although H-P stayed small during the war, its post-war success in building high- and low-frequency measuring devices eventually necessitated hiring more people. With growth, Packard realized that "the first hurdle to get over is to recognize that you can't do all these things yourself. Then the name of the game is to try and get people to do things the way you'd do them, if you were doing it. So you try and develop people who would have the same ideas, the same approach, and the same philosophy, and so we developed a very close relationship."

As H-P grew, the two founders and their close associates developed some clear ideas on how they would prefer to continue to manage the business. Mr. Packard continued:

> As we got bigger we decided that the way to manage this business was to break it down in small enough units so that people could have expertise in a particular field. So the first thing we did, we built our engineering laboratories into two sections: one on low-frequency instruments and one on high-frequency instruments. That meant then that these people not only kept up with all of technology, but they knew what the market needed and it was their business to stay ahead of the game in their fields.

Thus, from the time when H-P began to grow, the governing principle for organizing was to break the company down into small enough units so that people could specialize on a more finite piece of the business and thus develop a high level of expertise in it. This appears to be a fundamentally important organizing principle: in some form, all of the study companies seek to maintain a "small feel," although they differ substantially on how this is done. We conclude that, in complex businesses such as these, despite the increased time and attention required, it just makes good sense to "keep it small," within the cognitive limits of the human mind to grasp and manipulate effectively.

Packard's successor, John Young, was among many we interviewed who paid deliberate attention to organizational size. He described the importance of explicitly organizing people, of explicitly keeping the organizational units within smaller, more manageable units, and of building the requisite linkages between divisions and groups in a large, innovative organization.

> We are a six billion dollar company. That's one business [and not a conglomerate]. We have 52 operating divisions, not because it's easy to have 52 operating divisions, but because we believe in those small work groups. That means about a thousand people per work group. It's more on the model of a small business, . . . because of the interaction.
>
> I think it is extremely important for people to see who you need to cooperate with and why cooperating is a growing, valued characteristic. . . . The only excuse for having a big organization, it seems to me, is one where you generally have those linkages, that can build on each other. But linkage is a big problem. And it takes a lot of time. So, we have had to articulate [why it's worth time coordinating], and the organization form is one way to help do that.

Because they have learned how to manage large size by grouping people into smaller units, the founders and executives of these companies do not appear to be overly concerned with becoming "too big" as corporations. We asked Gordon Moore, chairman of the board and a founder of Intel, if he saw any upper size limit, an optimal number of people beyond which Intel was not likely to grow as a corporation? With a broad smile, Dr. Moore observed, "We're in a business where you grow or you die. We'll grow as long as we can. We may start to worry when we get as big as IBM. . . ."

Size, then, is not the issue. Instead, it is coordination, cooperation, information sharing, maintaining a problem-solving focus, regrouping people into smaller work units, and maintaining collegial relationships through the quasi-structure of multiple teams and task forces.

CONCLUSIONS

In our judgment the key to the continued innovation success rates these companies enjoy is to be found in their careful management of structure. This includes much thoughtful reexamination of existing formal structures, in light of changing technology and market needs. Sustained enthusiasm for constant improvement in the organization's structure through reorganization, as necessary, provides a consistent signal to employees that change is neither undesirable nor to be resisted. Instead, structure is to be managed and adapted to changed conditions.

This, in turn, requires a willingness to invest valuable time in structure per se. Quasi-structural elements like crossdisciplinary, multilevel committees, task forces, and teams require fostering, attention, and support. Thus, quasi-formal relationships to coordinate and exchange information are legitimated in these firms by widespread investment of sanctioned time. This includes senior management time, given to participating in these as needed and supporting quasi-structure on behalf of the organization at large. Senior management commitment to the essential importance of quasi-structure is clearly not lip service. The interpersonal skills essential for working within organizations with extensive quasi-structural elements, like teamwork and cooperative ability, are increasingly used to assess managers' behavior. That Hewlett-Packard now includes cooperation and team skills as variables by which its managers are evaluated is among the most fundamental indicators of how important these issues are to the survival of the companies we have studied.

These firms show a consistent pattern of structure—formal, quasi-formal, and informal—that actively facilities both innovative ideas and task-relevant cooperation. While they have thousands of employees, these companies clearly organize for innovation, they reorganize for innovation, and they retain the small company flavor by carefully fostering patterns of interaction usually associated with small companies. These behaviors are clearly encouraged by structure, but they are also an expected part of the cultures of these companies. Informal interaction and coordination were found everywhere: in the hallways, cafeterias, and even in the men's and women's rooms. The dynamic tension among all these elements is an important component of the management of innovation.

Above all, it appears that effective, explicit, and deliberate management of the innovation processes is the key to success in high technology companies, a point increasingly recognized by managers themselves, if not as widely as might be hoped in the organization theory literature. Contemporary innovation, particularly in research-intensive companies, is a very elaborate process, drawing on a wide array of technical specialties. Yet technology itself is by no means enough; adept organizing for innovation is essential to continued success over time. Clear organizational structures, frequent reorganizations, and an extensive use of quasi-structure contribute significantly to the long-term innovative abilities of the high technology companies we have studied.

Note

1. See, for example, three extensive reviews of the literature:

Tornatzky, Louis, et al. (1983). *The Process of Technological Innovation: Reviewing the Literature*. National Science Foundation, May 1983.

Van de Ven, Andrew H. (1986). "Central problems in the management of innovation." *Management Science* (May 1986).

Van de Ven, Andrew H. (1988). "Processes of innovation and organizational change." Technical Report of the Strategic Management Research Center, University of Minnesota, February 1988.

In addition to these review articles, a sample of other works illustrates the extensive literature on the management of innovation.

Hatch, M. J. (1987). "Physical barriers, task characteristics, and interaction activity in research and development firms," *Administrative Science Quarterly*, 32 no. (1987): 387–399.

Jennings, D. F., and D. H. Sexton (1985). "Managing innovation in established firms: Issues, problems and the impact on economic growth and employment." *Proceedings*, 1985, Conference on Industrial Science and Technological Innovation. Sponsored by National Science Foundation. The Center for Research and Development. State University of New York at Albany.

Lundstedt, S.B., and E. W. Colglazier, Jr. (1982). *Managing Innovation*. New York: Pergamon Press.

Mintzberg, Henry, and A. McHugh (1985). "Strategy formulation in an adhocracy," *Administrative Science Quarterly*, 30 (1985): 160–197.

Tushman, M. L., and W. L. Moore (1982). *Readings in the Management of Innovation*. Boston: Pitman.

Tushman, M. L., and P. Anderson (1986). "Technological discontinuities and organizational environments." *Administrative Science Quarterly*, 31 (1986): 439–465.

17

Core Capabilities and Core Rigidities: A Paradox in Managing New Product Development

DOROTHY LEONARD-BARTON

INTRODUCTION

Debate about the nature and strategic importance of firms' distinctive capabilities has been heightened by the recent assertion that Japanese firms understand, nurture and exploit their core competencies better than their U.S.-based competitors (Prahalad and Hamel, 1990). This paper explores the interaction of such capabilities with a critical strategic activity: the development of new products and processes. In responding to environmental and market changes, development projects become the focal point for tension between innovation and the *status quo*—microcosms of the paradoxical organizational struggle to maintain, yet renew or replace core capabilities.

In this paper, I first examine the history of core capabilities, briefly review relevant literature, and describe a field-based study providing illustrative data. The paper then turns to a deeper description of the nature of core capabilities and detailed evidence about their symbiotic relationship with development projects. However, evidence from the field suggests the need to enhance emerging theory by examining the way that capabilities inhibit as well as enable development, and these arguments are next presented. The paper concludes with a discussion of the project/capabilities interaction as a paradox faced by project managers, observed management tactics, and the potential of product/process development projects to stimulate change.

THE HISTORY OF CORE CAPABILITIES

Capabilities are considered *core* if they differentiate a company strategically. The concept is not new. Various authors have called them distinctive competences (Snow and Hrebiniak, 1980; Hitt and Ireland, 1985), core or organizational competencies (Prahalad and Hamel, 1990; Hayes Wheelwright and Clark, 1988), firm-specific competence (Pavitt, 1991), resource deployments (Hofer and Schende, 1978), and invisible assets (Itami, with Roehl, 1987). Their strategic significance has been discussed for decades, stimulated by such research as Rumelt's (1974) discovery that of nine diversification strategies, the two that were built on an existing skill or resource base in the firm were associated with the highest performance. Mitchell's (1989) observation that industry-specific capabilities increased the likelihood a firm could exploit a new technology within that industry, has confirmed the early work. Therefore some

Reprinted from *Strategic Management Journal*, Leonard-Barton, D. "Core Capabilities and Core Rigiditites: A Paradox in Managing New Product Development," Vol. 13, 1992. Reprinted by permission of John Wiley & Sons, Ltd.

authors suggest that effective competition is based less on strategic leaps than on incremental innovation that exploits carefully developed capabilities (Hayes, 1985; Quinn, 1980).

On the other hand, institutionalized capabilities may lead to 'incumbent inertia' (Lieberman and Montgomery, 1988) in the face of environmental changes. Technological discontinuities can enhance or destroy existing competencies within an industry (Tushman and Anderson, 1986). Such shifts in the external environment resonate within the organization, so that even 'seemingly minor' innovations can undermine the usefulness of deeply embedded knowledge (Henderson and Clark, 1990). In fact, all innovation necessarily requires some degree of 'creative destruction' (Schumpeter, 1942).

Thus at any given point in a corporation's history, core capabilities are evolving, and corporate survival depends upon successfully managing that evolution. New product and process development projects are obvious, visible arenas for conflict between the need for innovation and retention of important capabilities. Managers of such projects face a paradox: core capabilities *simultaneously* enhance and inhibit development.[1] Development projects reveal friction between technology strategy and current corporate practices; they also spearhead potential new strategic directions (Burgelman, 1991). However, most studies of industrial innovation focus on the new product project as a self-contained unit of analysis, and address such issues as project staffing or structure (Souder, 1987; Leonard-Barton, 1988a; Clark and Fujimoto, 1991. Chapter 9).[2] Therefore there is little research-based knowledge on managing the interface between the project and the organization, and the interaction between development and capabilities in particular. Observing core capabilities through the lens of the project places under a magnifying glass one aspect of the 'part-whole' problem of innovation management, which Van de Ven singles out as '[p]erhaps the most significant structural problem in managing complex organizations today. . . ' (1986:598).

Recent field research on 20 new product and process development projects provided an opportunity to explore and conceptually model the relationship between development practices and a firm's core capabilities. As described in the Appendix, four extensive case studies in each of five companies (Ford, Chaparral Steel, Hewlett Packard, and two anonymous companies, Electronics and Chemicals) were conducted by joint teams of academics and practitioners.[3] (Table 17.1). Before describing the interactions observed in the field, I first define core capabilities.

Dimensions of Core Capabilities

Writers often assume that descriptors of core capabilities such as 'unique,' 'distinctive,' 'difficult to imitate,' or 'superior to competition' render the term self-explanatory, especially if reference is also made to 'resource deployment' or 'skills.' A few authors include activities such as 'collective learning' and explain how competence is and is not cultivated (Prahalad and Hamel, 1990). Teece, Pisano and Shuen provide one of the clearest definitions: 'a set of differentiated skills, complementary assets, and routines that provide the basis for a firm's competitive capacities and sustainable advantage in a particular business' (1990:28).

In this article, I adopt a knowledge-based view of the firm and define a core capability as the knowledge set that distinguishes and provides a competitive advantage. There are four dimensions to this knowledge set. Its content is embodied in (1) employee *knowledge and skills* and embedded in (2) *technical systems*. The processes of knowledge creation and control are guided by (3) *managerial systems*. The fourth dimension is (4) the *values and norms* associated with the various types of embodied and embedded knowledge and with the processes of knowledge creation and control. In managerial literature, this fourth dimension is usually separated from the others or ignored.[4] However, understanding it is crucial to managing both new product/process development and core capabilities.

The first dimension, knowledge and skills embodied in people, is the one most often associated with core capabilities (Teece *et al.*, 1990) and the

TABLE 17.1. Description of Projects Studied

Company	Product/Process Description
Ford Motor Company	FX15 Compressor for automobile air conditioning systems
	EN53 New full-sized car built on carryover platform
	MN12 All new car platform including a novel supercharged engine
	FN9 Luxury automobile built on carryover platform with major suspension system modifications
Chaparral Steel	Horizontal Caster New caster used to produce higher grades steel
	Pulpit Controls Furnace control mechanism upgrade from analog to digital
	Microtuff 10 New special bar quality alloy steel
	Arc Saw Electric arc saw for squaring ends of steel beams
Hewlett-Packard Company	Deskjet Low cost personal computer and office printer using new technology
	Hornet Low cost spectrum analyzer
	HP 150 Terminal/PC linked to high-end computer
	Logic Analyzer Digital logic analyzer
Chemicals	Special use camera
	Large format printer for converting digital input to continuous images
	New polymer used in film
	21st century 'factory of the future'
Electronics	New RISC/UNIX workstation
	Local area network linking multiple computer networks
	Software architecture for desktop publishing
	High-density storage disk drive

one most obviously relevant to new product development. This knowledge/skills dimension encompasses both firm-specific techniques and scientific understanding. The second, knowledge embedded in technical systems, results from years of accumulating, codifying and structuring the tacit knowledge in peoples' heads. Such physical production or information systems represent compilations of knowledge, usually derived from multiple individual sources; therefore the whole technical system is greater than the sum of its parts. This knowledge constitutes both information (e.g. a data base of product tests conducted over decades) and procedures (e.g. proprietary design rules). The third dimension, managerial systems, represents formal and informal ways of creating knowledge (e.g. through sabbaticals, apprenticeship programs or networks with partners) and of controlling knowledge (e.g. incentive systems and reporting structures).

Infused through these three dimensions is the fourth: the value assigned within the company to the content and structure of knowledge (e.g. chemical engineering vs. marketing expertise; 'open-systems' software vs. proprietary systems), means of collecting knowledge (e.g. formal degrees v. experience) and controlling knowledge (e.g. individual empowerment vs. management hierarchies). Even physical systems embody values. For instance, organizations that have a strong tradition of individual vs. centralized control over information prefer an architecture (software and hardware) that allows much autonomy at each network node. Such 'debatable, overt, espoused values' (Schein, 1984:4) are one 'manifestation' of the corporate culture (Schein, 1986:7).[5]

Core capabilities are 'institutionalized' (Zucker, 1977). That is, they are part of the organization's taken-for-granted reality, which is an accretion of decisions made over time and events in

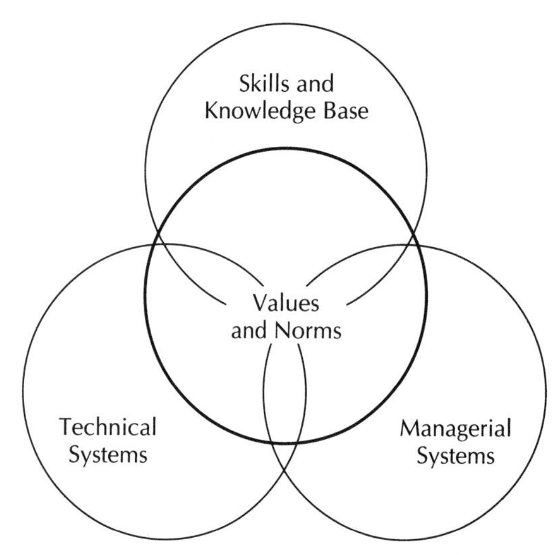

Fig. 17.1. The four dimensions of a core capability.

corporate history (Kimberly, 1987; Tucker, Singh and Meinhard, 1990; Pettigrew, 1979). The technology embodied in technical systems and skills usually traces its roots back to the firm's first products. Managerial systems evolve over time in response to employees' evolving interpretation of their organizational roles (Giddens, 1984) and to the need to reward particular actions. Values bear the 'imprint' of company founders and early leaders (Kimberly, 1987). All four dimensions of core capabilities reflect accumulated behaviors and beliefs based on early corporate successes. One advantage of core capabilities lies in this unique heritage, which is not easily imitated by would-be competitors.

Thus a core capability is an interrelated, interdependent knowledge system. See Figure 17.1. The four dimensions may be represented in very different proportions in various capabilities. For instance, the information and procedures embedded in technical systems such as computer programs are relatively more important to credit card companies than to engineering consulting firms, since these latter firms likely rely more on the knowledge base embodied in individual employees (the skills dimension).[6]

Interaction of Development Projects and Core Capabilities: Managing the Paradox

The interaction between development projects and capabilities lasts over a period of months or years and differs according to how completely aligned are the values, skills, managerial and technical systems required by the project with those currently prevalent in the firm. (See Figure 17.2). Companies in the study described above identified a selected, highly traditional and strongly held capability and then one project at each extreme of alignment: highly congruent vs. not at all (Table 17.2). Degree of congruence does not necessarily reflect project size, or technical or market novelty. Chaparral's horizontal caster and Ford's new luxury car, for instance, were neither incremental enhancements nor small undertakings. Nor did incongruent projects necessarily involve 'radical' innovations by market or technological measures. Electronic's new workstation used readily available, 'state-of-the-shelf' components. Rather, unaligned projects were nontraditional for the organization along several dimensions of the selected core capability.

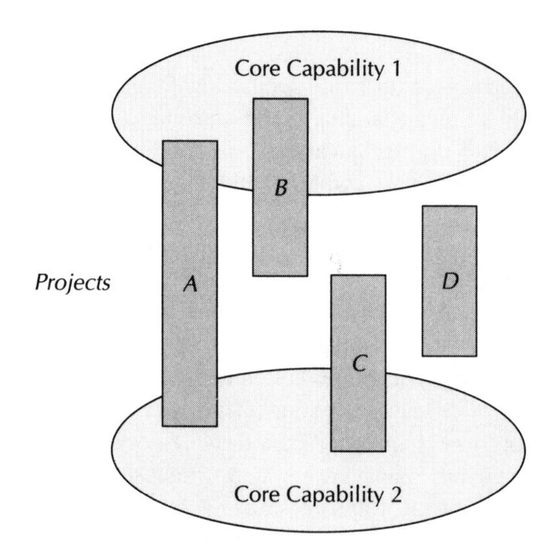

Fig. 17.2. Possible alignments of new product and process development projects with current core capabilities at a point in time.

TABLE 17.2. Relationship of Selected Projects with a Very Traditional Core Capability in Each Company Studied

		Degree of Alignment	
Company Name	Traditional Core Capability	Very High	Very Low
Ford Motor Co.	Total vehicle architecture	Luxury car built on carryover platform (FN9)	Compressor for air conditioner system (FX15)
Chaparral Steel	Science of casting molds	Horizontal caster	Electric arc saw
Hewlett Packard	Measurement technology	Low cost spectrum analyzer	150 Terminal/ personal computer
Chemicals	Silver halide technology	New polymer for film	Factory of the future
Electronics	Networking	Local area network link	Stand-alone workstation

For instance, Chemicals' project developing a new polymer used in film drew heavily on traditional values, skills and systems. In this company, film designers represent the top five percent of all engineers. All projects associated with film are high status, and highly proprietary technical systems have evolved to produce it. In contrast, the printer project was nontraditional. The key technical systems, for instance, were hardware rather than chemical or polymer and required mechanical engineering and software skills. Similarly, whereas the spectrum analyzer project at Hewlett Packard built on traditional capabilities in designing measurement equipment, the 150 terminal as a personal computer departed from conventional strengths. The 150 was originally conceived as a terminal for the HP3000, an industrial computer already on the market and as a terminal, was closely aligned with traditional capabilities. The attempt to transform the 150 into a personal computer was not very successful because different technical and marketing capabilities were required. Moreover, the greater system complexity represented by a stand-alone computer (e.g. the need for disk drives) required very untraditional cross-divisional cooperation.

Similar observations could be made about the other projects featured in Table 17.2. Chaparral's horizontal caster pushed the traditional science of molds to new heights, whereas the arc saw required capabilities that turned out to be unavailable. The local area networks project at Electronics grew directly out of networking expertise, whereas the new RISC/UNIX workstation challenged dominant and proprietary software/hardware architecture. At Ford, the three car projects derived to varying degrees from traditional strengths—especially the new luxury car. However, the air-conditioner compressor had never been built in-house before. Since all new product development departs somewhat from current capabilities, project misalignment is a matter of degree. However, as discussed later, it is also a matter of kind. That is, the type as well as the number of capability dimensions challenged by a new project determines the intensity of the interaction and the project's potential to stimulate change.

THE UP SIDE: CAPABILITIES ENHANCE DEVELOPMENT

In all projects studied, deep stores of knowledge embodied in people and embedded in technical systems were accessed; all projects were aided by managerial systems that created and controlled knowledge flows, and by prevalent values and norms. That is, whether the projects were aligned or not with the prominent core capability identified by the company, *some* dimensions of that capability favored the project. However, the closer the alignment of project and core knowledge set, the stronger the enabling influence.

In order to understand the dynamic interaction of project with capabilities, it is helpful to tease apart the dimensions of capabilities and put each dimension separately under the microscope. However, we must remember that these dimensions are interrelated; each is supported by the other three. Values in particular permeate the other dimensions of a core capability.

Skills/Knowledge Dimension

Excellence in the Dominant Discipline

One of the most necessary elements in a core capability is excellence in the technical and professional skills and knowledge base underlying major products. The professional elite in these companies earn their status by demonstrating remarkable skills. They expect to 'achieve the impossible'—and it is often asked of them. Thus managers of development projects that draw upon core capabilities have rich resources. In numerous cases, seemingly intractable technical problems were solved through engineering excellence. For instance, although engineers working on the thin film media project at Electronics had little or no prior experience with this particular form of storage technology, (because the company had always used ferrite-based media) they were able to *invent* their way out of difficulties. Before this project was over, the geographically dispersed team had invented new media, new heads to read the data off the thin film media, as well as the software and hardware to run a customized assembly and test line for the new storage device.

Pervasive Technical Literacy

Besides attracting a cadre of superbly qualified people to work in the dominant discipline, time-honored core capabilities create a reservoir of complementary skills and interests outside the projects, composed of technically skilled people who help shape new products with skilled criticism. In the Electronics Software Applications project, the developers enlisted employees through computer networks to field test emerging products. After trying out the software sent them electronically, employees submitted all reactions to a computerized 'Notes' file. This internal field testing thus took advantage of both willing, technically able employees and also a computer system set up for easy world-wide networking. Similarly, Electronics Workstation developers recruited an internal 'wrecking crew' to evaluate their new product. Employees who found the most 'bugs' in the prototype workstations were rewarded by getting to keep them. At Chemicals, developers tested the special purpose camera by loading down an engineer going on a weekend trip with film, so that he could try out various features for them. In these companies, internal testing is so commonplace that it is taken for granted as a logical step in new product/process creation. However, it represents a significant advantage over competitors trying to enter the same market without access to such technically sophisticated personnel. Internal 'field testers' not only typify users but can translate their reactions into technical enhancements; such swift feedback helps development teams hit market windows.

The Technical Systems Dimension

Just as pervasive technical literacy among employees can constitute a corporate resource, so do the systems, procedures and tools that are artifacts left behind by talented individuals, embodying many of their skills in a readily accessible form. Project members tap into this embedded knowledge, which can provide an advantage over competitors in timing, accuracy or amount of available detail. At Ford Motor Company, the capability to model reliability testing derives in part from proprietary software tools that simulate extremely complex interactions. In the full-sized car project, models simulating noise in the car body allowed engineers to identify nonobvious root causes, some originating from interaction among physically separated components. For instance, a noise apparently located in the floor panel could be traced instead to the acoustical interaction of sound waves reverberating between roof and floor. Such simulations cut development time as well as costs. They both build on and enhance the engineers' skills.

The Management Systems Dimension

Managerial systems constitute part of a core capability when they incorporate unusual blends of skills, and/or foster beneficial behaviors not observed in competitive firms. Incentive systems encouraging innovative activities are critical components of some core capabilities, as are unusual educational systems. In Chaparral Steel, all employees are shareholders. This rewards system interacts with development projects in that employees feel that every project is an effort to improve a process they own. 'I feel like this company partly belongs to me,' explains a millwright. Consequently, even operators and maintenance personnel are tenacious innovation champions. The furnace controls upgrade (incorporating a switch from analog to digital) was initiated by a maintenance person, who persevered against opposition from his nominal superiors. Chaparral Steel also has a unique apprenticeship program for the entire production staff, involving both classroom education and on-the-job training. Classes are taught by mill foremen on a rotating basis. The combination of mill-specific information and general education (including such unusual offerings as interpersonal skills for furnace operators) would be difficult to imitate, if only because of the diversity of abilities required of these foremen. They know what to teach from having experienced problems on the floor, and they must live on the factory floor with what they have taught. This managerial system, tightly integrating technical theory and practice, is reflected in every development project undertaken in the company (Leonard-Barton, 1991).

Values Dimension

The values assigned to knowledge creation and content, constantly reinforced by corporate leaders and embedded in management practices, affect all the development projects in a line of business. Two subdimensions of values are especially critical: the degree to which project members are empowered and the status assigned various disciplines on the project team.

Empowerment of Project Members

Empowerment is the belief in the potential of every individual to contribute meaningfully to the task at hand and the relinquishment by organizational authority figures to that individual of responsibility for that contribution. In HP, 'Electronics,' and Chaparral, the assumption is that empowered employees will create multiple potential futures for the corporation and these options will be selected and exercised as needed. The future of the corporation thus rests on the ability of such individuals to create new businesses by championing new products and processes. Since strategy in these companies is 'pattern in action' or 'emergent' rather than 'deliberate' (Mintzberg, 1990), empowerment is an especially important element of their core capabilities, and project members initiating new capabilities were exhilarated by the challenges they had created. The Hewlett Packard printer and the Electronics storage teams actually felt that they had turned the course of their mammoth corporate ship a critical degree or two.

High Status for the Dominant Discipline

A business generally recognized for certain core capabilities attracts, holds, and motivates talented people who value the knowledge base underlying that capability and join up for the challenges, the camaraderie with competent peers, the status associated with the skills of the dominant discipline or function. Each company displays a cultural bias towards the technical base in which the corporation has its historical roots. For Chemicals, that base is chemistry and chemical engineering;

for Hewlett Packard and Electronics, it is electronics/computer engineering and operating systems software. A history of high status for the dominant discipline enables the corporation and the projects to attract the very top talent. Top chemical engineers can aspire to become the professional elites constituting the five percent of engineers who design premier film products at Chemicals. At Hewlett Packard and Electronics, design engineers are the professional elite.

A natural outgrowth of the prominence of a particular knowledge base is its influence over the development process. In many firms, a reinforcing cycle of values and managerial systems lends power and authority to the design engineer. That is, design engineers have high status because the new products that are directly evaluated by the market originate in design engineering; in contrast, the expertise of manufacturing engineers is expended on projects less directly tied to the bottom line and more difficult to evaluate. The established, well-paid career path for product designers attracts top engineering talent, who tend to perform well. The success (of failure) of new products is attributed almost entirely to these strong performers, whose high visibility and status constantly reinforce the dominance of their discipline.

As the above discussion suggests, projects derive enormous support from core capabilities. In fact, such capabilities continually spawn new products and processes because so much creative power is focused on identifying new opportunities to apply the accumulated knowledge base. However, these same capabilities can also prove dysfunctional for product and process development.

THE DOWN SIDE: CORE RIGIDITIES INHIBIT DEVELOPMENT

Even in projects that eventually succeed, problems often surface as product launch approaches. In response to gaps between product specifications and market information, or problems in manufacture, project managers face unpalatable choices. They can cycle back to prior phases in the design process (Leonard-Barton, 1988a), revisiting previous decisions higher up the design hierarchy (Clark, 1985), but almost certainly at the cost of schedule slippage. Or they may ship an inadequate product. Some such problems are idiosyncratic to the particular project, unlikely to occur again in the same form and hence not easily predicted. Others, however, occur repeatedly in multiple projects. These recurring shortfalls in the process are often traceable to the gap between current environmental requirements and a corporation's core capabilities. Values, skills, managerial systems, and technical systems that served the company well in the past and may still be wholly appropriate for some projects or parts of projects, are experienced by others as core rigidities—inappropriate sets of knowledge. Core rigidities are the flip side of core capabilities. They are not neutral; these deeply embedded knowledge sets actively create problems. While core rigidities are more problematic for projects that are deliberately designed to create new, nontraditional capabilities, rigidities can affect all projects—even those that are reasonably congruent with current core capabilities.

Skills and Knowledge Dimension

Less Strength in Nondominant Disciplines

Any corporation's resources are limited. Emphasizing one discipline heavily naturally makes the company somewhat less attractive for top people in a nondominant one. A skilled marketing person knows that she will represent a minority discipline in an engineering-driven firm. Similarly, engineers graduating from top U.S. schools generally regard manufacturing in fabrication industries less attractive than engineering design, (see Hayes *et al.*, 1988) not only because of noncompetitive salaries, but because of a lower level of expertise among potential colleagues.

In each of the nonaligned and hence more difficult projects (Table 17.2), specific nontraditional types of knowledge were missing. Chaparral Steel's electric arc saw project required understanding electromagnetic fields for a variety of alloys—a very different knowledge set than the usual metallurgical expertise required in casting. The Hewlett Packard 150 project suffered from a lack of knowledge about personal computer design and manufacture. The company has a long history of successful instrument development based on 'next-bench' design, meaning the engineering designers based their decisions on the needs and skills of their colleagues on the bench next to them. However, such engineers are not representative of personal computer users. Therefore traditional sources of information and design feedback were not applicable for the 150 project. Similarly, the new workstation project of Electronics met with less than optimal market acceptance because the traditional focus on producing a 'hot box,' i.e. excellent hardware, resulted in correspondingly less attention to developing software applications. The knowledge relevant to traditional hardware development flows through well-worn channels, but much less knowledge exists about creating application software. Therefore, the first few working prototypes of the UNIX/RISC workstation were shipped to customers rather than to third-party software developers. While this practice had worked well to stimulate interest in the company's well-established lines of hardware, for which much software is available, it was less appropriate for the new hardware, which could not be used and evaluated without software.

Technical Systems Dimension

Physical systems can embody rigidities also, since the skills and processes captured in software or hardware become easily outdated. New product designers do not always know how many such systems they are affecting. For example, in the RISC/UNIX workstation project at Electronics, the new software base posed an extreme challenge to manufacturing because hundreds of diagnostic and test systems in the factory were based on the corporate proprietary software. The impact of this incompatibility had been underestimated, given the very tight 8 month product delivery targets.

Management Systems Dimension

Management systems can grow just as intractable as physical ones—perhaps more so, because one cannot just plug in a new career path when a new project requires strong leadership in a hithertofore underutilized role. Highly skilled people are understandably reluctant to apply their abilities to project tasks that are undervalued, lest that negative assessment of the importance of the task contaminate perceptions of their personal abilities. In several companies, the project manager's role is not a strong one—partly because there is no associated career path. The road to the top lies through individual technical contribution. Thus a hardware engineer in one project considered his contribution as an engineering manager to be much more important than his simultaneous role as project manager, which he said was 'not my real job.' His perception of the relative unimportance of project leadership not only weakened the power of the role in that specific project but reinforced the view held by some that problem-solving in project management requires less intelligence than technical problem-solving.

Values Dimension

Core rigidities hampered innovation in the development projects especially along the values dimension. Of course, certain generic types of corporate cultures encourage innovation more than others (Burns and Stalker, 1961; Chakravarthy, 1982). While not disagreeing with that observation, the point here is a different one: the very same values, norms and attitudes that support a core capability and thus enable development can also constrain it.

Empowerment as Entitlement

A potential down side to empowerment observed is that individuals construe their empowerment as a psychological contract with the corporation, and yet the boundaries of their responsibility and freedom are not always clear. Because they undertake heroic tasks for the corporation, they expect rewards, recognition and freedom to act. When the contract goes sour, either because they exceed the boundaries of personal freedom that the corporation can tolerate, or their project is technically successful but fails in other ways, or their ideas are rejected, or their self-sacrifice results in too little recognition, they experience the contract as abrogated and often leave the company—sometimes with a deep sense of betrayal.

Engineers in projects that fall towards the 'incongruity' end of the spectrum speak of 'betting their [corporate identification] badges,' on the outcome, and of having 'their backs to the cliff' as ways of expressing their sense of personal risk. One engineering project manager describes 'going into the tunnel,' meaning the development period, from which the team emerges only when the job is done. 'You either do it or you don't . . . You don't have any other life.' Such intrapreneurs seem to enjoy the stress—as long as their psychological contract with the company remains intact. In this case the manager believed her contract included enormous freedom from corporate interference with her management style. When corporate management imposed certain restrictions, she perceived her contract as abrogated, and left the company just 2 months before product launch, depriving the project of continuity in the vision she had articulated for an entire stream of products.

Empowerment as a value and practice greatly aids in projects, therefore, until it conflicts with the greater corporate good. Because development requires enormous initiative and yet great discipline in fulfilling corporate missions, the management challenge is to channel empowered individual energy towards corporate aims—without destroying creativity or losing good people.

Lower Status for Non-dominant Disciplines

When new product development requires developing or drawing upon technical skills traditionally less well respected in the company, history can have an inhibiting effect. Even if multiple subcultures exist, with differing levels of maturity, the older and historically more important ones, as noted above, tend to be more prestigious. For instance, at Chemicals, the culture values the chemical engineers and related scientists as somehow 'more advanced' than mechanical engineers and manufacturing engineers. Therefore, projects involving polymers or film are perceived as more prestigious than equipment projects. The other companies displayed similar, very clear perceptions about what disciplines and what kinds of projects are high status. The lower status of nondominant disciplines was manifested in pervasive but subtle negatively reinforcing cycles that constrained their potential for contributions to new product development and therefore limited the cross-functional integration so necessary to innovation (Pavitt, 1991). Four of these unacknowledged but critical manifestations are: who travels to whom, self-fulfilling expectations, unequal credibility and wrong language.[7]

One seemingly minor yet important indication of status affecting product/process development is that lower status individuals usually travel to the physical location of the higher. Manufacturing engineers were far more likely to go to the engineering design sites than vice versa, whether for one-day visits, or temporary or permanent postings. Not only does such one-way travel reinforce manufacturing's lower status, but it slows critical learning by design engineers, reinforcing their isolation from the factory floor. The exception to the rule, when design engineers traveled to the manufacturing site, aided cross-functional coordination by fostering more effective personal relationships. Such trips also educated the design engineers about some of the rationale behind design for manufacture (Whitney, 1988). A design engineer in one project returned to alter designs after seeing 'what [manufacturing] is up against' when he visited the factory floor.

Expectations about the status of people and roles can be dangerously self-fulfilling. As dozens of controlled experiments manipulating unconscious interpersonal expectations have demonstrated, biases can have a 'pygmalion effect': person A's expectations about the behavior of person B affect B's actual performance—for better or worse (Rosenthal and Rubin, 1978). In the engineering-driven companies studied, the expectation that marketing could not aid product definition was ensured fulfillment by expectations of low quality input, which undermined marketers' confidence. In the Electronics Local Area Network project, the marketing people discovered early on that users would want certain very important features in the LAN. However, they lacked the experience to evaluate that information and self-confidence to push for inclusion of the features. Not until that same information was gathered directly from customers by two experienced consulting engineers who presented it strongly was it acted upon. Precious time was lost as the schedule was slipped four months to incorporate the 'new' customer information. Similarly, in the Hewlett Packard printer project, marketing personnel conducted studies in shopping malls to discover potential customers' reactions to prototypes. When marketing reported need for 21 important changes, the product designers enacted only five. In the next mall studies, the design engineers went along. Hearing from the future customers' own lips the same information rejected before, the product developers returned to the bench and made the other 16 changes. The point is certainly not that marketing always has better information than engineering. Rather history has conferred higher expectations and greater credibility upon the dominant function, whereas other disciplines start at a disadvantage in the development process.

Even if nondominant disciplines are granted a hearing in team meetings, their input may be discounted if not presented in the language favored by the dominant function. Customer service representatives in the Electronics LAN project were unable to convince engineering to design the computer boards for field repair as opposed to replacing the whole system in the field with a new box and conducting repairs back at the service center, because they were unable to present their argument in cost-based figures. Engineering assumed that an argument not presented as compelling financial data was useless.

Thus, nondominant roles and disciplines on the development team are kept in their place through a self-reinforcing cycle of norms, attitudes and skill sets. In an engineering-dominated company, the cycle for marketing and manufacturing is: low status on the development team, reinforced by the appointment of either young, less experienced members or else one experienced person, whose time is splintered across far too many teams. Since little money is invested in these roles, little contribution is expected from the people holding them. Such individuals act without confidence, and so do not influence product design much—thus reinforcing their low status on the team.

THE INTERACTION OF PRODUCT/PROCESS DEVELOPMENT PROJECTS WITH CORE RIGIDITIES

The severity of the paradox faced by project managers because of the dual nature of core capabilities depends upon both (1) the number and (2) the types of dimensions comprising a core rigidity. The more dimensions represented, the greater the misalignment potentially experienced between project and capability. For example, the Arc Saw project at Chaparral Steel was misaligned with the core metallurgical capability mostly along two dimensions: technical systems (not originally designed to accommodate an arc saw), and more importantly, the skills and knowledge-base dimension. In contrast, the Factory-of-the-Future project at Chemicals challenged all four dimensions of the traditional core capability. Not only were current proprietary technical systems inadequate, but existing managerial systems did not provide any way to develop the cross-functional skills needed. Moreover, the values placed on potential knowledge creation and control varied wildly among the several sponsoring groups, rendering a common vision unattainable.

The four dimensions vary in ease of change. From technical to managerial systems, skills and then values, the dimensions are increasingly less tangible, less visible and less explicitly codified. The technical systems dimension is relatively easy to alter for many reasons, among them the probability that such systems are local to particular departments. Managerial systems usually have greater organizational scope (Leonard-Barton, 1988b), i.e. reach across more subunits than technical systems, requiring acceptance by more people. The skills and knowledge content dimension is even less amenable to change because skills are built over time and many remain tacit, i.e. uncodified and in employees' heads (see von Hippel, 1990). However, the value embodied in a core capability is the dimension least susceptible to change; values are most closely bound to culture, and culture is hard to alter in the short term (Zucker, 1977), if it can be changed at all (Barney, 1986).

Effects of the Paradox on Projects

Over time, some core capabilities are replaced because their dysfunctional side has begun to inhibit too many projects. However, that substitution or renewal will not occur within the lifetime of a single project. Therefore, project managers cannot wait for time to resolve the paradox they face (Quinn and Cameron, 1988). In the projects observed in this study, managers handled the paradox in one of four ways: (1) abandonment; (2) recidivism, i.e. return to core capabilities; (3) reorientation; and (4) isolation. The arc saw and factory-of-the-future projects were abandoned, as the managers found no way to resolve the problems. The HP150 personal computer exemplifies recidivism. The end product was strongly derivative of traditional HP capabilities in that it resembled a terminal and was more successful as one than as a personal computer. The special-use camera project was reoriented. Started in the film division, the stronghold of the firm's most traditional core capability, the project languished. Relocated to the equipment division, where the traditional corporate capability was less strongly ensconced, and other capabilities were valued, the project was well accepted. The tactic of isolation, employed in several projects to varying degrees, has often been invoked in the case of new ventures (Burgelman, 1983). Both the workstation project at Electronics and the HP Deskjet project were separated physically and psychologically from the rest of the corporation, the former without upper management's blessing. These project managers encouraged their teams by promoting the group as hardy pioneers fighting corporate rigidities.

Effects of the Paradox on Core Capabilities

Although capabilities are not usually dramatically altered by a single project, projects do pave the way for organizational change by highlighting core rigidities and introducing new capabilities. Of the companies studied, Chaparral Steel made the most consistent use of development projects as agents of renewal and organization-wide learning. Through activities such as benchmarking against best-in-the-world capabilities, Chaparral managers use projects as occasions for challenging current knowledge and for modeling alternative new capabilities. For instance, personnel from vice presidents to operators spent months in Japan learning about horizontal casting and in the case of the new specialty alloy, the company convened its own academic conference in order to push the bounds of current capabilities.

In other companies, negative cycles reinforcing the lower status of manufacturing or marketing were broken—to the benefit of both project and corporation. In the workstation project at Electronics, the manufacturing engineers on the project team eventually demonstrated so much knowledge that design engineers who had barely listened to 20 percent of their comments at the start of the project, gave a fair hearing to 80 percent, thereby allowing manufacturing to influence design. In the deskjet printer project at Hewlett Packard, managers recognized that inequality between design and manufacturing always created unnecessary delays. The Vancouver division thus sought to raise the status of manufacturing engineering skills by creating a manufacturing engineering group within R&D and then, once it was well established, moving it to manufacturing. A rotation plan between manufacturing and R&D was developed to help neutralize the traditional status differences; engineers who left research to work in manufacturing or vice versa were guaranteed a 'return ticket.' These changes interrupted the negative reinforcing cycle, signalling a change in status for manufacturing and attracting better talent to the position. This same project introduced HP to wholly unfamiliar market research techniques such as getting customer reactions to prototypes in shopping malls.

As these examples indicate, even within their 1—8-year lifetime, the projects studied served as small departures from tradition in organizations providing a 'foundation in experience' to inspire eventual large changes (Kanter, 1983). Such changes can be *precipitated* by the introduction of new capabilities along any of the four dimensions. However, for a capability to become *core*, all four dimensions must be addressed. A core capability is an interconnected set of knowledge collections—a tightly coupled system. This concept is akin to Pfeffer's definition of a paradigm, which he cautions is not just a view of the world but 'embodies procedures for inquiring about the world and categories into which these observations are collected. Thus', he warns, 'paradigms have within them an internal consistency that makes evolutionary change of adaptation nearly impossible' (1982:228). While he is thinking of the whole organization, the caution might apply as well to core capabilities. Thus, new technical systems provide no inimitable advantage if not accompanied by new skills. New skills atrophy or flee the corporation if the technical systems are inadequate, and/or if the managerial systems such as training are incompatible. New values will not take toot if associated behaviors are not rewarded. Therefore, when the development process encounters rigidities, projects can be managed consciously as the 'generative' actions characteristic of learning organizations (Senge, 1990) only if the multidimensional nature of core capabilities is fully appreciated.

CONCLUSION

This paper proposes a new focus of inquiry about technological innovation, enlarging the boundaries of 'middle range' project management theory to include interactions with development of capabilities, and hence with strategy. Because core capabilities are a collection of knowledge sets, they are distributed and are being constantly enhanced from multiple sources. However, at the same time that they enable innovation, they hinder it. Therefore in their interaction with the development process, they cannot be managed as a single good (or bad) entity.[8] They are not easy to change because they include a pervasive dimension of values, and as Weick (1979:151) points out, 'managers unwittingly collude' to avoid actions that challenge accepted modes of behavior.

Yet technology-based organizations have no choice but to challenge their current paradigms. The swift-moving environment in which they function makes it critical that the 'old fit be consciously disturbed . . . ' (Chakravarthy, 1982:42). Itami points out that 'The time to search out and develop a new core resource is when the current core is working well,' (1987:54)—a point that is echoed by Foster (1982). Development projects provide opportunities for creating the 'requisite variety' for innovation (Van de Ven, 1986:600; Kanter, 1986). As micro-level social systems, they create conflict with the macro system and hence a managerial paradox. Quinn and Cameron argue that recognizing and managing paradox is a powerful lever for change: 'Having multiple frameworks available . . . is probably the single most powerful attribute of self-renewing . . . organizations' (1988:302).

Thus project managers who constructively 'discredit' (Weick, 1979) the systems, skills or values traditionally revered by companies may cause a complete redefinition of core capabilities or initiate new ones. They can consciously manage projects for continuous organizational renewal. As numerous authors have noted, (Clark and Fujimoto, 1991; Hayes *et al.*, 1988; Pavitt, 1991) the need for this kind of emphasis on organizational learning over immediate output alone is a critical element of competition.

ACKNOWLEDGMENTS

The author is grateful to colleagues Kim Clark, Richard Hackman and Steven Wheelwright as well as members of the research team and two anonymous reviewers for comments on earlier drafts of this paper, to the Division of Research at

Harvard Business School for financial support, and to the companies that served as research sites.

A full report on the research on which this paper is based will be available in Kent Bowen, Kim Clark, Chuck Holoway and Steven Wheelwright, *Vision and Capability: High Performance Product Development in the 1990s*, Oxford University Press, New York.

Notes

1. According to Quinn and Cameron,'(t)he key characteristic in paradox is the simultaneous presence of contradictory, even mutually exclusive elements' (1988:2).

2. Exceptions are historical cases about a developing technical innovation in an industry (see for example, Rosenbloom and Cusumano, 1987).

3. Other members of the data-collection team on which I served are: Kent Bowen, Douglas Braithwaite, William Hanson, Gil Preuss and Michael Titelbaum. They contributed to the development of the ideas presented herein through discussion and reactions to early drafts of this paper.

4. Barney (1986) is a partial exception in that it poses organizational culture as a competitive advantage.

5. Schein distinguishes between these surface values and 'preconscious' and 'invisible' 'basic assumptions' about the nature of reality (1984:4).

6. Each core capability draws upon only *some* of a company's skill and knowledge base, systems and values. Not only do some skills, systems and norms lie outside the domain of a particular core capability, but some may lie outside *all* core capabilities, as neither unique nor distinctly advantageous. For instance, although every company has personnel and pay systems, they may not constitute an important dimension of any core capability.

7. Such cycles, or 'vicious circles' as psychiatry has labeled them, resemble the examples of self-fulfilling prophecies cited by Weick (1979: 159–164).

8. This observation is akin to Gidden's argument that structure is always both constraining and enabling (1984:25).

REFERENCES

Barney, J. B. 'Organizational culture: Can it be a source of sustained competitive advantage?' *Academy of Management Review*, 11(3), 1986, pp. 656–665.

Burgelman, R. 'A process model of internal corporate venturing in the diversified major firms', *Administrative Science Quarterly*, 28, 1983, pp. 223–244.

Burgelman, R. 'Intraorganizational ecology of strategy making and organizational adaptation: Theory and field research', *Organization Science* 2(3), 1991, pp. 239–262.

Burns, T. and G. M. Stalker. *The Management of Innovation*, Tavistock, London, 1961.

Chakravarthy, B. S. 'Adaptation: A promising metaphor for strategic management', *Academy of Management Review*, 7(1), 1982, pp. 35–44.

Clark, K. 'The interaction of design hierarchies and market concepts in technological evolution' *Research Policy*, 14, 1985, pp. 235–251.

Clark, K. and T. Fujimoto. *Product Development Performance*, Harvard Business School Press, Boston, MA 1991.

Foster, R. 'A call for vision in managing technology,' *Business Week*, May 24, 1982, pp. 24–33.

Giddens, A. *The Constitution of Society: Outline of the Theory of Structuration*. Polity Press, Cambridge, UK, 1984.

Hayes, R. H. 'Strategic planning—forward in reverse?', *Harvard Business Review*, November-December 1985, pp. 111–119 (Reprint # 85607).

Hayes, R. H., S. C. Wheelwright and K. B. Clark. *Dynamic Manufacturing: Creating the Learning Organization*, Free Press, New York, 1988.

Henderson, R. and K. B. Clark. 'Architectural innovation: The reconfiguration of existing product technologies and the failure of established firms', *Administrative Science Quarterly*, 35, 1990, pp. 9–30.

Hitt, M. and R. D. Ireland. 'Corporate distinctive competence, strategy, industry and performance', *Strategic Management Journal*, 6, 1985, pp. 273–293.

Hofer, C. W. and D. Schendel. *Strategy Formulation: Analytical Concepts*. West Publishing, St. Paul, MN, 1978.

Huber, G. and D. J. Power. 'Retrospective reports of strategic-level managers: Guidelines for increasing their accuracy', *Strategic Management Journal*, 6(2), 1985, pp. 171–180.

Itami, H. with T. Roehl. *Mobilizing Invisible Assets*, Harvard University Press, Cambridge, MA, 1987.

Kanter, R. M. *The Change Masters*. Simon and Schuster, New York, 1983.

Kanter, R. M. 'When a thousand flowers bloom: Structural, collective and social conditions for innovation in organizations', Harvard Business School Working Paper # 87–018, 1986.

Kimberly, J. R. 'The study of organization: Toward a biographical perspective'. In J. W. Lorsch (ed.), *Handbook of Organizational Behavior*, Prentice-Hall, Englewood Cliffs, NJ, 1987, pp. 223–237.

Leonard-Barton, D. 'Implementation as mutual adaptation of technology and organization', *Research Policy*, 17, 1988a, pp. 251–267.

Leonard-Barton, D. 'Implementation characteristics in organizational innovations', *Communication Research*, 15(5), October 1988b, pp. 603–631.

Leonard-Barton, D. 'The factory as a learning laboratory', Harvard Business School Working Paper # 92–023, 1991.

Lieberman, M. and D. B. Montgomery. 'First-mover advantages', *Strategic Management Journal*, 9, Summer 1988, pp. 41–58.

Mintzberg, H. 'Strategy formation: Schools of thought'. In J. W. Fredrickson (ed.), *Perspectives on Strategic Management*, Harper & Row, New York, 1990.

Mitchell, W. 'Whether and when? Probability and timing of incumbents' entry into emerging industrial subfields', *Administrative Science Quarterly*, 34, 1989, pp. 208–230.

Pavitt, K. 'Key characteristics of the large innovating firm', *British Journal of Management*, 2, 1991, pp. 41–50.

Pettigrew, A. 'On studying organizational cultures', *Administrative Science Quarterly*, 24, 1979, pp. 570–581.

Pfeffer, J. *Organizations and Organization Theory*, Ballinger Publishing, Cambridge, MA, 1982.

Prahalad, C. K. and G. Hamel. 'The core competence of the corporation', *Harvard Business Review*, 68(3), 1990, pp. 79–91, (Reprint # 90311).

Quinn, J. B. *Strategies for Change: Logical Incrementalism*, Richard D. Irwin, Homewood, IL, 1980.

Quinn, R. and K. Cameron. 'Organizational paradox and transformation'. In R. Quinn and K. Cameron (eds), *Paradox and Transformation*, Cambridge, MA, Ballinger Publishing, 1988.

Rosenbloom, R. and M. Cusumano. 'Technological pioneering and competitive advantage: The birth of the VCR industry', *California Management Review*, 29(4), 1987, pp. 51–76.

Rosenthal, R. and D. Rubin. 'Interpersonal expectancy effects: The first 345 studies', *The Behavioral and Brain Sciences*, 3, 1978, pp. 377–415.

Rumelt, R. P. *Strategy, Structure and Economic Performance*, Harvard Business School Classics, Harvard Business School Press, Boston, MA, 1974 and 1986.

Schein, E. 'Coming to a new awareness of organizational culture', *Sloan Management Review*, Winter, 1984, pp. 3–16.

Schein, E. *Organizational Culture and Leadership*, Jossey-Bass, San Francisco, CA, 1986.

Schumpeter, J. *Capitalism, Socialism, and Democracy*, Harper, New York, 1942.

Senge, P. 'The leader's new work: Building a learning organization', *Sloan Management Review*, 32(1), 1990, pp. 7–23. (Reprint # 3211).

Snow, C. C. and L. G. Hrebiniak. 'Strategy, distinctive competence, and organizational performance', *Administrative Science Quarterly*, 25, 1980, pp. 317–335.

Souder, W. E. *Managing New Product Innovations*, Lexington Books, Lexington, MA, 1987.

Teece, D. J., G. Pisano and A. Shuen. 'Firm capabilities, resources and the concept of strategy', Consortium on Competitiveness and Cooperation Working Paper # 90–9, University of California at Berkeley, Center for Research in Management, Berkeley, CA, 1990.

Tucker, D., J. Singh and A. Meinhard. 'Founding characteristics, imprinting and organizational change'. In J. V. Singh (ed.), *Organizational Evolution: New Directions*, Sage Publications, Newbury Park, CA, 1990.

Tushman, M. L. and P. Anderson. 'Technological discontinuities and organizational environments', *Administrative Science Quarterly*, 31, 1986, pp. 439–465.

Van de Ven, A. 'Central problems in management of innovations', *Management Science*, 32(5), 1986, pp. 590–607.

von Hippel, E. 'The impact of 'Sticky Data' on innovation and problem-solving', Sloan Management School, Working Paper # 3147-90-BPS, 1990.

Weick, K. E. 'Theory construction as disciplined imagination', *Academy of Management Review*, 14(4), 1989, pp. 516–531.

Weick, K. E. *The Social Psychology of Organizing*, Random House, New York, 1979.

Whitney, D. 'Manufacturing by design', *Harvard Business Review*, 66(4), 1988, pp. 83–91.

Zucker, L. G. 'The role of institutionalization in cultural persistence', *American Sociological Review*, 42, 1977, pp. 726–743.

APPENDIX: METHODOLOGY

Structure of Research Teams

Four universities (Harvard, M.I.T., Stanford and Purdue) participated in the 'Manufacturing Visions' project. Each research team was composed of at least one engineering and one management professor plus one or two designated company employees. The research was organized into a matrix, with each research team having primary responsibility for one company and also one or more specific research 'themes' across sites and companies. Some themes were identified in the research protocol; others (such as the capabilities/project interaction) emerged from initial data analysis. In data collection and analysis, the internal company and outside researchers served as important checks on each other—the company insiders on the generalizability of company observations from four cases and the academics on the generalizabiltiy of findings across companies.

Data-gathering

Using a common research protocol, the teams developed case histories by interviewing development team members, including representatives from all functional groups who had taken active part and project staff members. These in-person interviews, conducted at multiple sites across the U.S., each lasted 1–3 hours. Interviewers toured the manufacturing plants and design laboratories and conducted follow-up interview sessions as necessary to ensure comparable information across all cases. The data-gathering procedures thus adhered to those advocated by Huber and Power (1985) to increase reliability of retrospective accounts (e.g. interviews conducted in tandem, motivated informants selected from different organizational levels, all responses probed extensively). In addition, the interviewers' disparate backgrounds guarded against the dominance of one research bias, and much archival evidence was collected. I personally interviewed in 3 of the 5 companies.

Data Analysis

Notes compiled by each team were exchanged across a computer network and joint sessions were held every several months to discuss and analyze data. Company-specific and theme-specific reports were circulated, first among team members and then among all research teams to check on accuracy. Team members 'tested' the data against their own notes and observations and reacted by refuting, confirming or refining it. There were four within-team iterations and an additional three iterations with the larger research group. Thus observations were subjected to numerous sets of 'thought trials' (Weick, 1989).

Each team also presented interim reports to the host companies. These presentations offered the opportunity to check data for accuracy, obtain reactions to preliminary conclusions, fill in missing data and determine that observations drawn from a limited number of projects were in fact representative of common practice in the company. The examples of traditional core capabilities presented in Table 17.2 were provided by the companies as consensus judgments, usually involving others besides the company team members. While the 20 projects vary in the degree of success attributed to them by the companies, only two were clear failures. The others all succeeded in some ways (e.g. met a demanding schedule) but fell short in others (e.g. held market leadership for only a brief period).

CHAPTER FOUR

TECHNOLOGY AND BUSINESS STRATEGY

Well-designed, highly capable organizations may nonetheless fail to surmount the challenges of cyclical, technological change if they do not pursue strategies that fit the moving competitive landscape. History is replete with examples of firms that pioneered innovations, only to watch rivals capture most of the profits. An even greater number of firms are blind-sided by innovations whose strategic importance they dismissed. In a shifting environment, perhaps the only constant is that a strategy effective at one point in a cycle will not work as an industry continues to evolve. Strategies must adapt as conditions change.

Burgelman and Rosenbloom capture this evolutionary idea by taking a process perspective on technology strategy. They choose not to focus on articulating the "right" strategy for any set of circumstances. Instead, they concentrate on describing how successful executives enact strategies that build up and build on consistent capabilities, while adapting as the organization learns through experience. Teece turns the spotlight to the problem of strategic scope: when should a firm exploit an innovation by itself, when should it seek strategic partners to realize the value of an innovation, and when are the most benefits created by licensing the innovation? His framework explains when innovators are expected to capture the most value from a new technology, and when following is a superior strategy. Strategies are realized by the innovation project portfolio a firm selects. Mitchell and Hamilton respond to the frustration managers experience when traditional financial analysis is used to select a set of R&D projects. They introduce the notion that the value of a project depends on the options it creates, and describe alternative tools rooted in option pricing theory that may help managers think more clearly about the real value of a particular undertaking.

18

Technology Strategy: An Evolutionary Process Perspective

ROBERT A. BURGELMAN
RICHARD S. ROSENBLOOM

I. INTRODUCTION

This chapter concerns itself with the strategic management of technology. In our view, technology is a resource that is as pervasively important in the organization as are financial and human resources. Its management is a basic business function. Viewing technology as a functional capability implies the need to develop a technology strategy, analogous to financial and human resource strategies. For technology, such a strategy is defined by a set of interrelated decisions encompassing, among others, technology choice, level of technology competence, level of funding for technology development, timing of technology introduction in new products/services, and organization for technology application and development (e.g., Maidique and Patch, 1978). Technology strategy conceived in this fashion is much broader than R&D strategy (Mitchell, 1985).

Technology, itself, is defined here as the ensemble of theoretical and practical knowledge, knowhow, skills and artifacts that are used by the firm to develop, produce and deliver its products and services. Technology can be embodied in people, materials, facilities, procedures, and in physical processes. Key elements of technology may be implicit, existing only in embedded form (e.g., trade secrets based on knowhow). "Craftsmanship" and "experience" usually have a large tacit component, so that important parts of technology may not be expressed or codified in manuals, routines and procedures, recipes, rules of thumb or other explicit articulations (Burgelman et al., 1988). In the broadest sense, a firm's "technology" encompasses the entire set of technologies employed in the sequence of activities that constitute its value chain (Porter, 1985). This set can be decomposed for analysis into constituent elements according to type of technology, role in the value chain, or contribution to end product (or service) categories.

In the so-called "hi-tech" firms, technology is usually a major force driving the firm's competitive strategy. But all firms have in recent years become more aware of the critical role of technology in strategic decisions, and of the need to integrate technology strategy into the strategic management process. As a consequence, technology increasingly is recognized as an important element of business definition and competitive strategy. Abell (1980) identifies technology as one of three prin-

Reprinted from *Research on Technological Innovation, Management and Ploicy*, by Rosenbloom, R. and R. Burgelman, Vol. 4, 1989, by permission of JAI Press.

cipal dimensions of business definition. As he notes, "technology adds a dynamic character to the task of business definition, as one technology may more or less rapidly displace another one over time." Porter (1983) observes that technology is among the most prominent factors that determine the rules of competition. Friar and Horwitch (1985) explain the growing prominence of technology as the result of historical forces: disenchantment with strategic planning, success of hi-tech firms in emerging industries, the surge of Japanese competition, recognition of the competitive significance of manufacturing, and the emergence of an academic interest in technology management.

The present chapter elaborates the view of technology as a functional capability, and develops an evolutionary process framework for discussing technology strategy. The perspective is that of the general manager responsible for the overall business strategy rather than that of the functional manager. The analysis, however, may help the functional-level technology manager think about his or her approach to the job. The framework focuses on technology strategy, but emphasizes its link with the overall business strategy.

The remainder of this chapter is organized as follows. The next section discusses an evolutionary process perspective on strategy-making and technology strategy. The following section examines the forces shaping technology strategy and provides examples of some companies' technology strategies. This is followed by a discussion of the main substantive dimensions of technology strategy and of the ways in which it is enacted. A brief final section summarizes the main themes of the paper.

II. TECHNOLOGY STRATEGY: AN EVOLUTIONARY PROCESS PERSPECTIVE

Paul Adler's chapter in this volume provides an overview of the growing literature on technology strategy and suggests the need for developing conceptual frameworks that would allow for cumulative knowledge to be generated. An evolutionary process perspective may be useful for this purpose. This perspective is emerging in economics (Nelson and Winter, 1982), organization theory (e.g., Weick, 1979; Aldrich, 1979; Hannan and Freeman, 1984) and strategic management (e.g., Burgelman, 1984; 1988b). It focuses on variation, selection, retention mechanisms for explaining dynamic behavior over time.

Gould (1987) has warned against the fallacy of unwarranted analogy in applying concepts from biological evolution to processes of cultural evolution. Taking into account Gould's admonition, the framework of evolutionary theory has nevertheless been fruitfully applied in the cultural evolutionary perspective that has emerged in recent years (e.g., Campbell, 1979; Weick, 1979; Boyd and Richerson, 1985). Gould's (1987) own interpretation of the establishment of QWERTY (David, 1985) as the dominant, if inferior, approach to laying out keys on typewriter keyboards in terms of the "panda thumb principle," shows the power of evolutionary reasoning in identifying and elucidating interesting phenomena concerning technological evolution.

Cultural evolutionary theory recognizes the importance of history, irreversibilities, invariance and inertia in explaining the behavior of social systems. But it also considers the effects of individual and social learning processes. A cultural evolutionary perspective may be useful for integrating extant literatures on technology. The study of technological development, for instance, contains many elements that seem compatible with the variation-selection-retention structure of evolutionary theory (e.g., Rosenberg, 1979; Krantzberg and Kelly, 1978; Abernathy, 1978; Clark, 1985). In fact, several other chapters in this volume adopt an evolutionary perspective. Metcalfe and Gibbons present an application of the evolutionary perspective at the level of populations of technologies. While Van de Ven and Garud criticize applications of a Darwinian view of evolution to the emergence of new industries around new technologies, their analysis emphasizes the social context of innovation and appears consistent with a cultural evolutionary perspective.

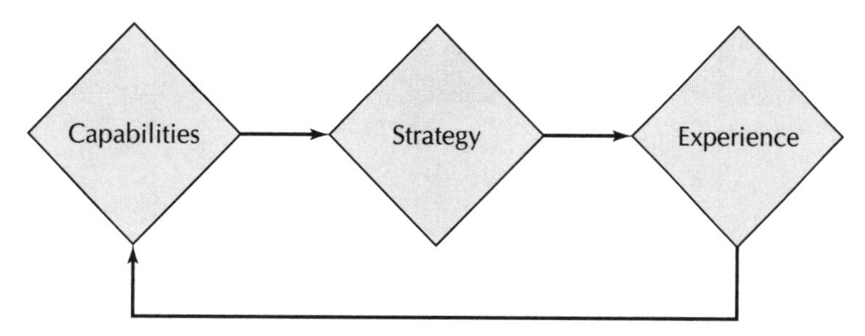

Fig. 18.1. An organizational learning framework of strategy-making.

A. Capabilities-Based Perspective

Embedded in the cultural evolutionary perspective is a view of strategy-making as a social learning process. In this view, strategy is inherently a function of the quantity and quality of organizational capabilities. Organizational capabilities are the source of opportunities which are discovered, selected, and retained in the strategy-making processes. Experience with ("performing") a strategy is expected to have feedback effects on the set of organizational capabilities. The general structure of the capabilities-based perspective (Burgelman, 1984; 1988a) is presented in Figure 18.1.

It is useful to note several key ideas underlying this framework. One is that successful firms develop distinctive strategies in the course of their development and that the direction of this development cannot be completely determined at the outset. A second key idea is that increasing the firm's capabilities is a mechanism for stimulating strategic development. This is consistent with Itami's (1987) discussion of the accumulation of "invisible assets" as a key factor in an organization's development. It is also consistent with Maidique and Zirger's (1985) study of the product learning cycle. A third idea is that "unlearning" is an important aspect of organizational learning (Imai, Nonaka and Takeuchi, 1985; Levitt and March, 1988).

Within this perspective, "performance" is viewed in terms of experience with actually performing the different tasks involved in carrying out a strategy. This view of performance is akin to the use of the term in craftsmanship, the arts, and sports. Studying the details of actually performing a strategy may shed light on exactly how skills are accumulated and how organizational learning and unlearning, in their various forms actually come about.

III. FORCES SHAPING TECHNOLOGY STRATEGY

The factors shaping technology strategy comprise a number of varied forces, mediating the influence of capabilities and experience. In this section, we explore a simple framework—depicted in Figure 18.2—which augments the theoretical context for the multilevel analysis of technology proposed by Rosenbloom (1978). This concept differentiates the factors according to locus—within or outside the firm—and functions—generative or integrative—in the strategy process. We hope that this classification might help to focus research and enable managers to examine their firm's strategies within a theoretical context.

In summary, the idea expressed in Figure 18.2 is that technology strategy emerges from organizational capabilities, shaped by the generative forces of the firm's strategic behavior and evolution of the technological environment, and by the integrative mechanisms of the firm's organizational context and the environment of the industry in which it operates. Each of these mechanisms is briefly addressed in what follows.

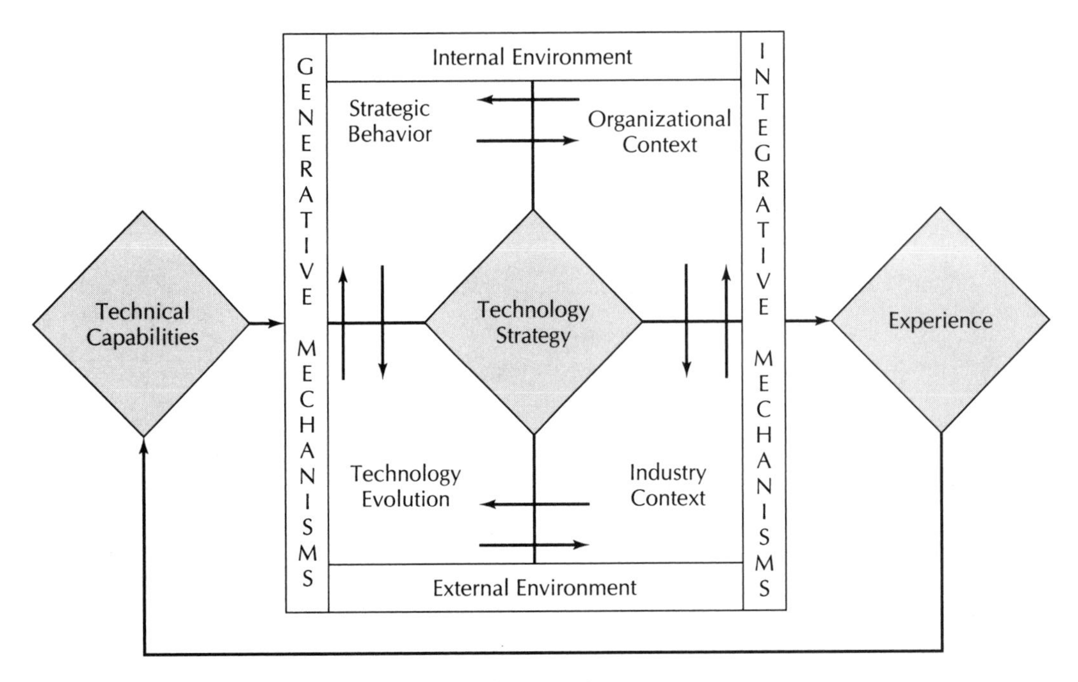

Fig. 18.2. An evolutionary process framework for technology strategy.

1. Internal Forces

Generative mechanisms: strategic behavior. Strategy-making with respect to technology is a subset of the broader strategy-making of the firm. From an evolutionary perspective, a firm's concept of strategy impounds the social learning about the distinctive competences on which past success is based (Selznick, 1957; Burgelman, 1988b). McKelvey and Aldrich (1983) view distinctive competence, "comps," as "the combined workplace (technological) and organizational knowledge and skills . . . that together are most salient in determining the ability of an organization to survive" (p. 112). Nelson and Winter (1982), in a similar vein, use the concept of "routines," which they consider to play a role similar to that of genes in biological evolution.

The analogy between genes and distinctive competence (capabilities) raises the issue of where strategic change and new technical capabilities, ultimately, come from. It seems plausible to expect that the "comps" and "routines" that have evolved in the course of achieving organizational survival will be naturally applied over and over again in the course of the firm's strategic action. To a large extent strategic behavior is, indeed, *induced* by the prevailing concept of strategy of the firm and results, through new product and process development efforts falling within its scope, in the enhancement and augmentation of *existing* technical capabilities. The induced strategic process is likely to manifest a degree of inertia relative to the cumulative changes in the external environment (Hannan and Freeman, 1984; Burgelman, 1988b). Cooper and Schendel (1978), for instance, found that established firms, when confronted with the threat of radically new technologies, were likely to increase their efforts to improve the existing technology rather than switch to the new technology even after the latter had passed the threshold of viability.

But firms will usually also exhibit some amount of autonomous strategic behavior, aimed at getting the firm into new areas of business (e.g., Penrose, 1968; Burgelman, 1983). These initiatives often are rooted in technology development efforts.

In the course of their work, for example, technical people may serendipitously discover results that provide the basis for redirection or replacement of major technologies of the firm. The existence of a corporate R&D capability often provides a substratum for the emergence of such new technical possibilities (Rosenbloom and Kantrow, 1982; Burgelman and Sayles, 1986).

Participants engaging in autonomous strategic behavior serve to explore the boundaries of a firm's capabilities and corresponding opportunities sets (Burgelman, 1983). As Itami (1983) has observed, "in reality, many firms do not have . . . complete knowledge and discover the full potential of their ability only after the fact" (p. 15). The degree to which autonomous strategic behavior can be exploited, however, depends critically on the organizational context.

Integrative mechanisms: organizational context. Internal context encompasses administrative (Bower, 1970) and cultural (Ouchi, 1980) factors which affect participants' expectations about which types of strategic behavior are likely to be supported in the organization. From an evolutionary point of view, organizational context serves as an internal selection mechanism (Burgelman, 1988) affecting the strategic management capacity of the firm, that is, (1) the firm's ability to exploit opportunities associated with its current strategy (induced process), (2) its ability to take advantage of opportunities that emerge spontaneously outside the scope of the current strategy (autonomous process), and (3) its ability to balance the different modes of strategic action associated with (1) and (2) over time (Burgelman and Sayles, 1986).

The evolutionary process perspective draws attention to the fact that organizational context takes shape over time, and reflects the dominant culture (values) of the firm. The dominant culture as it relates to technology may be different depending on whether the firm's distinctive competences are rooted in science (e.g., pharmaceutical firms), engineering (e.g., semiconductor firms) or manufacturing (e.g., Japanese firms); whether the product development process has been driven by technology push, need pull or a more balanced approach; whether operations are viewed as strategic or not; and so on.

2. External Forces

Generative mechanisms: technology evolution. A firm's technology strategy is rooted in the evolution of its technical capabilities. However, the dynamics of these capabilities, and hence the technology strategy, are not completely endogenous. A firm's technical capabilities are affected in significant ways by the evolution of the broader areas of technology of which they are part. Different aspects of technological evolution have been discussed in several studies: (1) technology regimes and their development along a particular trajectory (e.g., Twiss, 1982; Foster, 1986); (2) the emergence of dominant designs, design hierarchies, industry standards given a technological regime (Abernathy, 1978; Clark, 1985; and Metcalfe and Gibbons in this volume); (3) the interplay between product and process technology development within design configurations under a particular technological regime (Abernathy, 1978); (4) the emergence of new technological regimes and their trajectories (S-curves) (Foster, 1986); (5) the capability (competence) enhancing or destroying nature of new technological regimes (Astley, 1985; Tushman and Anderson, 1986); (6) dematurity (Abernathy, Clark and Kantrow, 1983); and (7) the locus of origin of new technological regimes (von Hippel, 1988). These studies highlight some of the major evolutionary forces associated with technological development which, as noted earlier, affect and are affected by the strategic behavior within firms.

Integrative mechanisms: industry context. One important aspect of industry context is the competitive market place. Competitive strategy and the quest for competitive advantage take into account five major forces (Porter, 1980); (1) rivalry among existing firms, (2) bargaining power of buyers, (3) bargaining power of suppliers, (4) threat of new entrants, and (5) threat of substitute products or services. The interplay of these five forces is expected to determine the appropriate content of competitive strategy. Technological change affects each of the five forces (Porter, 1983) and technology strategy may serve as a potentially powerful tool for pursuing the generic competitive strategies (overall cost leadership, overall differentiation, focused cost leadership, focused differentiation).

Contextual factors affect the choice of leadership versus followership (Porter, 1983). Examining this choice further, Teece (1986) argues that the fundamental building blocks of a decision framework include: (1) the appropriability regime associated with a technological innovation, (2) complementary assets needed to commercialize a new technology, and (3) the dominant design paradigm. The interplay of these factors affects the likely distribution of profits generated by a technological innovation among the different parties involved as well as the strategic choices concerning the optimal boundaries of the innovating firm's capabilities set.

The chapter by Metcalfe and Gibbons in this volume shows how the economic aspects of the industry context can be operationalized from an evolutionary perspective. Industry context selects among technological alternatives and the most efficient technological alternative need not necessarily become dominant. Still another important aspect of industry context is the social embeddedness of economic transactions. The chapter by Van de Ven and Garud in this volume examines the broader institutional forces associated with the industry context.

3. Applying the Framework: Some Examples

Before moving on to discuss the implications of the evolutionary perspective for the substance and deployment of technology strategy, it seems appropriate to examine how the framework may help gain insight in a firm's technology strategy. Several examples can serve this purpose.

• Crown Cork and Seal has been able to do very well over a 30-year period as a relatively small player in a mature industry. Technology strategy seems to have contributed significantly to Crown's success. When taking over the company in 1957, CEO John Connelly recognized the existence of Crown's strong skills in metal formation (capabilities) and built on these to specialize in "hard to hold" applications for tin cans. He developed strong relations with the steel companies and convinced them to quickly adapt a major external technological innovation (*technology evolution*)—the two-piece can—initiated by an aluminum company for use with steel. He did not want Crown to have an R&D department, but developed strong links between a highly competent technical sales force and an applications-oriented engineering group to be able to provide complete technical solutions for customers' "filling needs"—a dominant value at CC&S (*organizational context*). Over the years, CC&S continued to stick to what it could do best (*strategic behavior*) while its competitors were directing their attention to diversification and gradually lost interest in the metal can industry (*industry context*). CC&S has continued to refine the skill set that made it the only remaining independent metal can company of the four original major players (*experience*).

• Marks and Spencer (M & S), a British retailer with a worldwide reputation for quality, and Bank One Corporation, a Midwest banking group that consistently ranks among the most profitable U.S. banking operations, are two companies in the services sector that have used technology effectively to gain competitive advantage.

The success of M&S is based on a consistent strategy founded on an unswerving commitment to giving the customer "good value for money." The genesis of its technology strategy was the transformation, in 1936, of a small textile testing department into a "Merchandise Development Department," designed to work closely with vendors to bring about improvements in quality. Thirty years later, the M&S technical staffs, then numbering more than two hundred persons working on food technology as well as textiles and home goods, allowed M&S, quite literally, to control the cost structure of its suppliers. The development of the technical capability itself was driven by the strong value of excellent supplier relationships held and continuously reinforced by top management.

In 1958, the new CEO of City National Bank of Columbus Ohio (CNB), John G. McCoy, persuaded his Board to invest 3% of profits each year to support a "research and development" activity. Over the next two decades, CNB, which became the lead bank of Banc One Corporation, developed capabilities that made it a national leader in the application of electronic information-processing technologies to retail banking. It was the first bank to install automatic teller machines and a pioneer in the development of bank credit cards, point-of-sale transaction processing, and home banking. While not all of its innovative ventures succeeded, each contributed to the cumulative development of a deep and powerful technical capability that remains a distinctive element of the bank's highly successful competitive strategy.

The three companies cited above are notable for the consistency of their strategic behavior over several decades. The following example illustrates the problems that can arise in a time of changing technology and industry context when a fundamental change in strategy is not matched by corresponding adaptation of the organizational context.

• The National Cash Register Company (NCR) built a dominant position in world-wide markets for cash registers and accounting machines on the basis of superior technology and an outstanding salesforce created by the legendary John H. Patterson. By 1911, NCR had a 95% share in cash register sales. Scale economies in manufacturing, sales, and service presented formidable barriers to entry to its markets (*industry context*), preserving its dominance for another 60 years. Highly developed skills in the design and fabrication of complex low-cost machines (*capabilities*) not only supported the strategy, they also shaped the culture of management, centered on Dayton where a vast complex housed engineering and fabrication for the traditional product line (*organizational context*). In the 1950s, management began to build new capabilities in electronics (*a revolution, not just evolution in technology*) and entered the emerging market for Electronic Data Processing (EDP). A new strategic concept tried to position traditional products (registers and accounting machines) as "data-entry" terminals in EDP systems (*changing strategic behavior*). But a salesforce designed to sell stand-alone products of moderate unit cost proved ineffective in selling high-priced "total systems." At the same time, the microelectronics revolution destroyed the barriers inherent in NCR's scale and experience in fabricating mechanical equipment. A swarm of new entrants found receptive customers for their new electronic registers (*changing industry context*). As market share tumbled and red-ink washed over the P&L in 1972, the chief executive was forced out. His successor was an experienced senior NCR manager who had built his career entirely outside of Dayton. He moved swiftly to transform the ranks of top management, decentralize manufacturing (reducing employment in Dayton by 85%), and restructure the salesforce along new lines. The medicine was bitter, but it worked; within 2 years, NCR had regained leadership in its main markets, and was more profitable in the late 1970s than it had been at any point in the 1960s.

IV. TECHNOLOGY STRATEGY; SUBSTANCE AND ENACTMENT

A strategic view of technology has many facets, but the substance of a firm's technological strategy can be grasped in terms of a few fundamental themes. In essence, a business uses technology to create enhanced value in its products or services and to gain sustainable advantage in relation to its rivals. Two important dimensions of strategy, then, are the way it positions the business competitively and in relation to the value chain. But the implications of a particular posture in these respects can be various, depending upon choices in two other dimensions, those defining the scope and depth of technological resources. In this section we first consider, briefly, each of these four dimensions of the substance of technology strategy. We then examine the notion of performance, the ways that strategy is realized in practice. The section concludes with conjectures about the significance of comprehensiveness and integration.

1. Substance of Technology Strategy

Competitive positioning. Technology strategy, viewed in competitive terms, is an instrument of a more comprehensive corporate strategy, as we have noted earlier. As part of the broader strategy, a business adopts a particular *competitive stance* toward technology, defining the role that technology should play in establishing generic competitive advantage: How *process* technology, *product* technology and technical *support* capabilities are used to achieve cost leadership or differentiation (Porter, 1985).

From an evolutionary perspective, however, a firm's competitive advantage is more likely to arise from the *unique* aspects of its strategy than from characteristics it shares with others. Companies that have been successful over long periods of time, such as Crown Cork and Seal, and Marks and Spencer, have developed capabilities that are quite distinct from those of their competitors and not easily replicable. The strategies of such companies cannot easily be classified simply in terms of differentiation *or* cost leadership; they combine both. The technical skills and capabilities that they have assembled and built into their organizations over the span of many years, and which they relentlessly hone and augment, are not available in the market and would take a long time to be replicated. The ability to maintain uniqueness that is salient in the market place, however, implies continuous alertness to what competitors are doing and should not be confused with insulation and an inward-looking orientation.

Technology, from a competitive point of view, can be used in a defensive role, sustaining achieved advantage in product differentiation or cost, or, offensively, as an instrument of expansion, to create new advantage in established lines of business or to open the door to new products and markets. The implications of "technological leadership" have been explored in earlier writings on technology and strategy (e.g., Ansoff and Stewart, 1967; Maidique and Patch, 1979). As noted earlier, Porter (1981) relates this concept more directly to a broader framework of competitive advantage, identifying conditions under which leadership is likely to be rewarded in the market place. Teece (1986) extends the analysis by identifying the importance of appropriability regimes and control of specialized assets. These discussions define leadership in terms of the timing (relative to rivals) of commercial use of new technology. A broader strategic definition would identify "leadership" in terms of relative advantage in the command of a body of technology. This sort of leadership results from commitment to a "pioneering" role in the development of a technology (Rosenbloom and Cusumano, 1987), as opposed to a more passive "monitoring" role. Technological leaders thus have the capability to be first-movers, but may elect not to do so.

Looking at technological leadership from an evolutionary perspective draws attention to the importance of *accumulation* of capabilities as a competitive tool (Itami, 1987). Given the importance of history, irreversibilities, and inertia, leadership is not something that can easily be bought in the market or quickly "plugged" into the organization. Rather, it involves painstaking, patient and persistent building of technical capabilities in all areas that have been determined as having strategic importance for the future development of the firm, even though it may seem cheaper ("efficient") to rely on outsiders for supplying the capability. Thinking strategically about technology means raising the question of how a particular technical capability may affect a firm's future degrees of freedom and the control over its fate. Concretely, this involves identifying and tracking key technical parameters, considering the impact on speed and flexibility of product and process development as technologies move through their life cycles. This, in turn, requires distinguishing carefully between technologies that are common to all players in the industry and have little impact on competitive advantage and those that are proprietary and likely to have a major impact on competitive advantage. It also depends on paying attention continuously to new technologies that are beginning to manifest their potential for competitive advantage and those that are as yet only beginning to emerge (A.D. Little, 1981; Booz-Allen and Hamilton, 1981; Burgelman et al., 1988).

Technology and the value chain. A second dimension of the substance of technology strategy defines a value chain stance toward the use of technology: that is, how technology is to be used to add value in process, products or technical service delivered to customers or users. A firm that lacks important capabilities in a given constituent technology (by choice, or otherwise) may elect to embrace it through supply relationships or strategic collaborations (see the chapter by Pisano and Teece in this volume). Hence the value chain stance also answers the question of how value-creating technology is sourced—in-house or outside the boundaries of the firm.

From an evolutionary process perspective, a key issue is how technology could affect the rate of *improvement* in performing the tasks involved in each of the stages of the value chain. Continuous concern with improvement in all aspects of the value creation and delivery process will guard the firm against quirky moves that endanger the accumulated learning in the organization. Similarly, viewing the issue of sourcing from this perspective highlights the importance of managing interdependencies with external providers of capabilities. One requirement of this is a continuous concern for gaining as much *learning* as possible from the relationship in terms of capabilities and skills rather than a concern solely with price. To the extent that a firm engages in alliances, it seems necessary to establish the requisite capabilities for managing the relationships. As noted earlier, Marks and Spencer's strong technical staff made it possible to have unique and valuable supplier relationships.

Scope of technology strategy. A third dimension of the substance of technology strategy defines the scope of technologies actively attended to by the firm. In one respect this goes back to the notion of the value chain and the concern with technologies in its different stages. Firms may concern themselves to a different degree with all of these. No firm, however, can hope to operate on the frontiers of all technologies relevant to its operations. Thus, it seems more useful to limit the scope of technology strategy to the set of technologies considered by the firm to have a material impact on its competitive advantage. This set of technologies can be called the *core technology*; other technologies, then, are *peripheral*. Of course, in a dynamic world, peripheral technologies today may become core technologies tomorrow and vice versa. This dimension of technology strategy is especially important in relation to the threat of new entrants in the firm's industry. All else equal, firms covering more areas of technology in their core would seem to be less *vulnerable* to attacks from new entrants attempting to gain position through producing and delivering new types of technology-based customer value. Of course, resource constraints will put a limit on how many technologies the firm can opt to develop internally.

Defining the scope of the technology strategy determines which ones within the entire range of technologies employed by the firm are placed within the core. These are the areas in which the firm needs to assess its distinctive technological competences and to decide whether to be a technological leader or follower, first-mover or second-to-market, and whether to develop requisite capabilities in-house or through vendors and allies. Such assessments and decisions identify an array of targets for technology development. The irreversibility of investments in technology makes the choice of these targets an especially salient dimension of strategy. For any given technical field, moreover, both risk and return will vary between product and process development and also according to the degree of novelty inherent in the development. Targeted technology development may range from minor improvements in a mature process to the employment of an emerging technology in the first new product in a new market (Rosenbloom, 1985).

The scope of a firm's technology strategy may be determined to a significant extent by its scale and business focus. Businesses built around large, complex systems like aircraft, automobiles, or telecommunication switches demand the ability to apply and integrate numerous distinct types of expertise creating *economies* of scope and/or scale, and *synergies*. General Electric, for instance, was reportedly able to bring to bear high powered mathematical analysis, used in several divisions concerned with military research on submarine warfare, to the development of computerized tomography (CT) products in its medical equipment division (Rutenberg, 1986 p. 28–9.). Other fields may actually contain diseconomies of scale, giving rise to the popularity of "skunkworks" and discussion of the "mythical man-month" (Brooks, 1975). The chapter by Prahalad, Doz, and Angelmar in this volume examines how the emergence of a new technology raises issues concerning the scope of the innovation that the firm should pursue and how this, in turn, may impact the delineation of the set of core technologies of the firm and the boundaries between the business units that it comprises.

Depth of technology strategy. The fourth dimension of the substance of technology strategy concerns the depth of its prowess within the core technology. This prowess can be expressed in terms of the number of technological options that the firm has available. The depth of a firm's technological strategy is determined to a significant extent by the intensity of its resource expenditures. The variation among manufacturing firms is pronounced: many firms do not spend on R&D; a few commit as much as 10% of revenues to it. While inter-industry differences can explain a large share of this variation, substantial differences in R&D intensity still remain between rivals.

From an evolutionary perspective, depth of technology is likely to be correlated with the firm's capacity to anticipate technological developments in particular areas early on. It provides the basis for acting in a *timely* way. As noted earlier, however, being able to take advantage of this lead in knowledge in terms of developing new products or services will to a large extent depend on the organizational context. Firms that can organize themselves for attaining greater technological depth may also benefit from increased *flexibility* to respond to new demands from customers/users. Imai et al. (1985) for instance, describe how Japanese firms' use of multiple layers of contractors and subcontractors in an external network allows extreme forms of specialization in particular skills which, in turn, provides both flexibility and speed in response and the potential for cost savings, since the highly specialized subcontractors operate on an experience curve even at the level of prototypes.

2. PERFORMANCE

The enactment of a strategy is embodied in the performance of the tasks that make it real. For technology, the main processes concern its acquisition, development, and support.

Acquisition. Since the sources of technology are inherently varied, so must be the mechanisms employed to make it accessible within the firm. But it is useful to make a simple distinction between inventive activity within the firm and acquisitive functions which import technology originating outside the firm.

Relatively few firms—primarily the largest ones in the R & D intensive industries—find it useful to support the kind of science-based R & D that can lead to important new technologies. But those firms, in aggregate, shape the technological directions of the economy, so their strategies are especially salient. Each firm's strategy finds partial expression in the way it funds, structures, and directs the R & D activities whose mission is to create new pathways for technology. A current example of the importance of this function is the emergence of high-temperature superconductivity, based on a discovery made in an IBM research laboratory.

Every firm finds that it must structure ways to acquire certain technologies from others. The choices made in carrying out those tasks can tell us a great deal about the underlying technology strategy. To what extent does the firm rely on on-going alliances, as opposed to discrete transactions for the acquisition of technology (Hamilton, 1986)? Is the acquisition structured to create the capability for future advances to be made internally, or will it merely reinforce dependence? For example, when Japanese electronics firms acquired technology from abroad during the 1950s and 1960s most of them structured it in ways that made it possible for them to become the leaders in pushing the frontiers of those same technologies in the 1970s and 80s.

Development. Strategy is also acted out in the functions which develop technology as products and processes. The character of product development activities embodies important aspects of the scope and depth of technology strategy and its competitive role. An understanding of what the strategy is can be gained from considering the level of resources committed, the way they are deployed, and how they are directed. How does the organization strike the delicate balance between letting technology drive development and using the marketplace to drive technology? The mammoth personal computer industry was founded when the young engineers of Apple sought to exploit the potential inherent in the microprocessor, at a time when corporate managers in Hewlett-Packard, IBM, and others were disdainful of the commercial opportunity. A decade later, however, it is clear that market needs are now the primary force shaping efforts to advance the constituent technologies. In a number of industries—autos and semiconductors would be two examples—competitiveness rests substantially on design and manufacturing process capabilities. There strategy is embodied in the links between product and process engineering, in the resources committed to capital equipment, in the concern for quality, and similar aspects of this domain.

Support. The function commonly termed "field service" creates the interface between the firm's technical function and the users of its products or services. Experience in use provides important feedback to enhance the capabilities which generate the technology (Rosenberg, 1982, Chapter 8). Airline operations, for example, are an essential source of information about jet engine technology. The technology strategy of a firm, then, finds important expression in the way it carries out this important link to users. Two-ways flows of information are relevant: expert knowledge from product developers can enhance the effectiveness of field operations, while feedback from the field informs future development.

3. Two Conjectures

Implicit in the foregoing discussion are two normative conjectures about technology strategy. The first is that the substance of technology should be comprehensive. That is, strategy, as it is enacted through the various tasks of acquisition, development, and support, should address all four substantive dimensions and do so in ways that are consistent across the dimensions. Second, we suggest that strategy should be integrated with operations. That is, that each of the tasks should be informed by the positions taken on the four substantive dimensions, in ways that create consistency across the various tasks. The matrix illustrated in Figure 18.3 is intended to suggest a framework for mapping the interactions among these factors and assessing the degree of comprehensiveness and integration in a specific situation.

V. CONCLUSION

This chapter argues that an evolutionary perspective provides a useful framework for thinking about the nature of technology strategy and about its role in the broader competitive strategy of a firm. The essence of this perspective is that strategy is built on capabilities and tempered by experience. These three main constructs—capabilities, strategy, and experience—are tightly interwoven in reality. Capabilities give strategy its force; strategy enacted creates experience that modifies capabilities.

Thus, we argue, it is useful to examine the technological strategy of a firm through the lenses of the underlying capabilities and the resultant modes of performance. Central to this idea is the notion that the reality of a strategy lies in its enactment, not in those pronouncements that appear to assert it. Through these lenses, we suggest, one can readily discern the scope and depth of the strategy and the way that it positions the firm in relation to rivals in the marketplace and in relation to others in the value chain. In other words, the substance of strategy can be found in its performance in the various modes by which technology is acquired and deployed—sourcing, development, and support. The ways in which these tasks are actually performed, and the ways in which their per-

MODES OF EXPERIENCE / SUBSTANCE	EXTERNAL TECHNOLOGY SOURCING	INVENTIVE ACTIVITY	PRODUCT DESIGN DEVELOPMENT	PROCESS DESIGN DEVELOPMENT	TECHNICAL SUPPORT SERVICE
RIVALRY STANCE					
VALUE CHAIN STANCE					
SCOPE					
DEPTH					

Fig. 18.3. Technology strategy: comprehensiveness and integration.

formance contributes, cumulatively, to capabilities convey the real substance of strategy.

A second central idea in this paper is that the on-going interactions of capabilities-strategy-experience occur within a matrix of generative and integrative mechanisms that shape strategy. These mechanisms (sketched in Figure 18.2) are both internal and external to the firm. Anecdotal evidence suggests that successful firms operate within some sort of harmonious equilibrium of these forces. Major change in one, as in the emergence of a technological discontinuity, ordinarily must be matched by adaptation in the others.

Which leads to the final conjecture of this paper, namely, that it is advantageous to attain a state in which technology strategy is both comprehensive and integrated. By comprehensive we mean that it embodies consistent answers to the issues posed by all four substantive dimensions. By integrated we mean that each of the various modes of performance is informed by the strategy.

These ideas suggest several areas where further research could be fruitfully undertaken. Longitudinal field research at the level of individual organizations could shed light on the behavioral demands of how an organization develops distinctive technological capabilities, and how these are enhanced, augmented, refocused or dissipated through internal ventures, acquisitions, alignments and spin-offs. Of particular interest are questions about how the organization builds a business strategy on its evolving capabilities and skills and how, in turn, the business strategy and its specific enactment may facilitate or impede the further development of technological capabilities and skills. The relationships between organizational growth and capabilities development would also appear to be an interesting target for further research. Comparative longitudinal research at the level of industries may shed light on why technology strategies vary among the firms they contain and may document further the forces affecting different firms' modes of adaptation.

ACKNOWLEDGMENTS

The authors gratefully acknowledge support provided by the Strategic Management Program, Graduate School of Business, Stanford University, and the Division of Research of the Harvard Business School.

REFERENCES

Abell, D., 1980. *Defining the Business*. Englewood Cliffs, NJ: Prentice-Hall.

Abernathy, W. J., 1978. *The Productivity Dilemma: Roadblock to Innovation in the Automobile Industry*. Baltimore: Johns Hopkins University Press.

Abernathy, W. J., and A. M. Kantrow, 1983. *Industrial Renaissance*. New York: Basic Books.

Aldrich, H. E., 1979. *Organizations and Environments*. Englewood Cliffs, NJ: Prentice-Hall.

Ansoff, H. I., and J. M. Stewart, 1967. ``Strategies for a Technology-Based Business," *Harvard Business Review*, November–December.

Astley, W. G., 1985. "The Two Ecologies: Population and Community Perspectives on Organizational Evolution," *Administrative Science Quarterly*, 30, 224–241.

Booz-Allen and Hamilton, 1981. "The Strategic Management of Technology," *Booz-Allen & Hamilton: Outlook*, Fall–Winter, 1981.

Bower, J. L., 1970. *Managing the Resource Allocation Process*. Boston, MA: Division of Research, Graduate School of Business Administration, Harvard University.

Boyd, R., and R. Richerson, 1985. *Culture and the Evolutionary Process*. Chicago: University of Chicago Press.

Brooks, F. P., 1975. *The Mythical Man-Month*. Reading, MA: Addison-Wesley.

Burgelman, R. A., 1983. "Corporate Entrepreneurship and Strategic Management: Insights from a Process Study," *Management Science*, 29, 1349–1364.

Burgelman, R. A., 1986. "Strategy-Making and Evolutionary Theory: Towards a Capabilities-Based Perspective," in Tsuchiya, M. (ed.), *Technological Innovation and Business Strategy*, Tokyo: Nihon Keizai Shinbunsha.

Burgelman, R. A., 1988a. "Strategy-Making as a Social Learning Process: The Case of Internal Corporate Venturing," *Interfaces*, 18, 3, (May–June) 74–85.

Burgelman, R. A., 1988b. "Intraorganizational Ecology of Strategy-Making and Organizational Adaptation," Research Paper, Graduate School of Business, Stanford, CA: Stanford University, August.

Burgelman, R. A., and L. R. Sayles, 1986. *Inside Corporation Innovation*. New York: Free Press.

Burgelman, R. A., T. J. Kosnik, and M. Van den Poel, 1988. "Toward an Innovative Capabilities Audit Framework," in R. A. Burgelman and M. A. Maidique (eds.), *Strategic Management of Technology and Innovation*. Homewood, IL: Irwin.

Campbell, D. T., 1969. "Variation and Selective Retention in Sociocultural Evolution," *General Systems*, 14, 69–85.

Clark, K. B., 1985. The Interaction of Design Hierarchies and Market Concepts in Technological Evolution," *Research Policy*, 14, 235–251.

Cooper, A. C., and D. Schendel, 1976. "Strategic Responses to Technological Threats," *Business Horizons*, (February) 61–63.

David, P. A., 1985. "Clio and the Economics of QWERTY," American *Economic Review*, 75, 2, (May) 332–337.

Foster, R. N., 1986. *Innovation: The Attacker's Advantage*. New York: Summit.

Friar, J., and M. Horwitch, 1985. "The Emergence of Technology Strategy: A New Dimension of Strategic Management," in *Technology in Society*, 7, 2/3, 143–178.

Fusfeld, A., 1978. "How to Put Technology into Corporate Planning," *Technology Review*, (May).

Gould, S. J., 1987. "The Panda's Thumb of Technology," *Natural History*, (January) 14–23.

Hamilton, W. F., 1985. "Corporate Strategies for Managing Emerging Technologies," in *Technology in Society*, 7, 2/3, 197–212.

Hannan, H. T., and J. H. Freeman, 1984. "Structural Inertia and Organizational Change," *American Sociological Review*, 43, 149–164.

Imai, K., I. Nonaka, and H. Takeuchi, 1985. "Managing the New Product Development Process: How Japanese Learn and Unlearn," in K. B. Clark, R. H. Hayes, and C. Lorenz, *The Uneasy Alliance: Managing the Productivity-Technology Dilemma*. Boston, MA: Harvard Business School Press.

Itami, H., 1983. "The Case for Unbalanced Growth of the Firm," Research Paper Series #681, Graduate School of Business, Stanford, CA: Stanford University.

Itami, H., 1987. *Mobilizing Invisible Assets*. Cambridge, MA: Harvard University Press.

Kelly, P., and M. Kranzberg, eds., 1978. *Technological Innovation: A Critical Review of Current Knowledge*. San Francisco: San Francisco Press.

Levitt, B., and J. G. March, 1988. "Organizational Learning," *Annual Review of Sociology*, 14.

Little, A. D., 1981. "The Strategic Management of Technology," *European Management Forum*.

Maidique, M. A., and P. Patch, 1978. "Corporate Strategy and Technological Policy," *Harvard Business School*, Case #9-679-033, rev. 3/80.

Maidique, M. A., and B. J. Zirger, 1985. "New Products

Learning Cycle," *Research Policy*, (December) 1–40.

McKelvey, B., and H. E. Aldrich, 1983. "Populations, Organizations, and Applied Organizational Science," *Administrative Science Quarterly*, 28, 101–128.

Mitchell, G. R., 1986. "New Approaches for the Strategic Management of Technology," in M. Horwitch (ed.), *Technology in Society*, 7, 2/3, 132–144.

Nelson, R. R., and S. G. Winter, 1982. *An Evolutionary Theory of Economic Change*. Cambridge, MA: Harvard University Press.

Ouchi, W., 1980. "Markets, Bureaucracies and Clans," *Administrative Science Quarterly*, 25, 129–141.

Penrose, E. T., 1980. *The Theory of the Growth of the Firm*. White Plains, NY: M. E. Sharpe.

Porter, M. E., 1980. *Competitive Strategy*. New York: Free Press.

Porter, M. E. 1983. "The Technological Dimension of Competitive Strategy," in R. S. Rosenbloom (ed.), *Research on Technological Innovation, Management and Policy*, 1, 1–33.

Porter, M. E., 1985. *Competitive Strategy*. New York: Free Press.

Prestowitz, C. V., Jr., 1988. *Trading Places*. New York: Basic Books.

Rosenberg, N., 1982. *Inside the Black Box*. Cambridge: Cambridge University Press.

Rosenbloom, R. S., 1978. "Technological Innovation in Firms and Industries: An Assessment of the State of the Art," in P. Kelly and M. Kranzberg (eds.), *Technological Innovation: A Critical Review of Current Knowledge*. San Francisco: San Francisco Press.

Rosenbloom, R. S., 1985. "Managing Technology for the Longer Term: A Managerial Perspective," in K. B. Clark, R. H. Hayes, and C. Lorenz, (eds.), *The Uneasy Alliance: Managing the Productivity-Technology Dilemma*. Boston, MA: Harvard Business School Press.

Rosenbloom, R. S., and A. M. Kantrow, 1982. "The Nurturing of Corporate Research," *Harvard Business Review*, (January–February), 115–123.

Rosenbloom, R. S., and M. A. Cusumano, 1987. "Technological Pioneering: The Birth of the VCR Industry," *California Management Review*, XXIX, 4, (Summer), 51–76.

Rutenberg, David, 1986. "*Umbrella Pricing*," Working Paper, Queens University.

Selznick, P., 1957. *Leadership in Administration*. New York: Harper and Row.

Teece, D. J., 1986. "Profiting from Technological Innovation: Implications for Integration, Collaboration, Licensing and Public Policy," *Research Policy*, 15, 285–305.

Tushman, M. L., and P. Anderson, 1986. "Technological and Organizational Environments," *Administrative Science Quarterly*, 31, 439–465.

Twiss, B., 1980. *Managing Technological Innovation*. London: Longman.

von Hippel, E., 1988. *The Sources of Innovation*. New York: Oxford University Press.

Weick, K., 1979. *The Social Psychology of Organizing*. Reading, MA: Addison-Wesley.

19

Capturing Value from Technological Innovation: Integration, Strategic Partnering, and Licensing Decisions

DAVID J. TEECE

Why, and under what circumstances, is the recognized technological progressiveness of a nation not sufficient to capture the benefits stemming from its capabilities in science and technology? This chapter examines why firms and nations can lose ground in the commercialization of advanced technologies at a time when they are the principal sources for major technological innovations of industrial significance; the capacity for scientific and technological innovation may be the last rather than the first advantage that a mature economy loses as it enters its declining phase.

The framework developed here helps identify the factors that determine who wins from innovation: The firm that is first to market, follower firms, or firms that have related capabilities that the innovator needs. The follower firms may or may not be imitators in the narrow sense of the term, although they sometimes are. The framework helps to explain the share of the profits from innovation accruing to the innovating firms and nations compared to its followers and suppliers.

THE PHENOMENON

A classic example of the phenomenon considered in this chapter is the computerized axial tomographic (CAT) scanner developed by the U.K. firm Electrical Musical Industries (EMI) Ltd.[1] By the early 1970s, EMI was in a variety of product lines, including phonograph records, movies, and advanced electronics. EMI had developed high-resolution televisions in the 1930s, pioneered airborne radar during World War II, and developed the United Kingdom's first all solid-state computers in 1952.

In the late 1960s, the pattern recognition research of Godfrey N. Hounsfield, an EMI senior research engineer, resulted in his being able to display a scan of a pig's brain. Subsequent clinical work established that computerized axial tomography was viable for generating cross-sectional "views" of the human body, the greatest advance in radiology since the discovery of x rays in 1895.

Although EMI was initially successful with its CAT scanner, within 6 years of its introduction into the United States in 1973, the company had lost market leadership and by the eighth year had dropped out of the CAT scanner business. Other companies successfully dominated the market, though they were late entrants, and are still profiting in the business today.

A further example is that of the Royal Crown Companies, Inc., a small beverage company that was the first to introduce cola in a can and the first to introduce diet cola. Both Coca-Cola and Pepsi-Cola followed almost immediately and deprived Royal Crown of any significant advantage from its innovation. Bowmar Instrument Corporation, which introduced the pocket calculator, was not able to withstand competition from Texas Instruments, Hewlett-Packard, and others, and went out of business. Xerox Corporation failed to succeed with its entry into the office computer business, even though Apple Computer, Inc., succeeded with the MacIntosh, which contained many of Xerox's key product ideas, such as the mouse and icons. The story of the DeHavilland Comet has some of the same features. The Comet I jet was introduced into the commercial airline business two years or so before Boeing introduced the 707, but DeHavilland failed to capitalize on its substantial early advantage. MITS introduced the first personal computer, the Altair, experienced a burst of sales, then slid quietly into oblivion.

If there are innovators who lose, there must be followers (imitators) who win. A classic example is IBM Corporation with its PC, a great success from the time it was introduced in 1981. Neither the architecture nor the components of the IBM PC were considered advanced when introduced; nor was the way the technology was packaged a significant departure from the then-current practice. Yet the IBM PC was fabulously successful and established MS-DOS as the leading operating system for 16-bit PCs. By the end of 1984, IBM had shipped more than 500,000 PCs and may have irreversibly eclipsed Apple in the PC industry.

Figure 19.1 presents a simplified taxonomy—with examples—of the possible outcomes from innovation. Quadrant 1 represents positive outcomes for the innovator. A first-to-market advantage is translated into a sustained competitive advantage that either creates a new earnings stream or enhances an existing one. Quadrant 4 and its corollary quadrant 2 are the focus of this paper.

PROFITING FROM INNOVATION: BASIC BUILDING BLOCKS

To develop a coherent framework within which to explain the distribution of outcomes illustrated in Figure 19.1, three fundamental building blocks must be put in place: the appropriability regime, the dominant design paradigm, and complementary assets.

Regimes of Appropriability

A regime of appropriability refers to the environmental factors, excluding firm and market structure, that govern an innovator's ability to cap-

	INNOVATOR	IMITATOR-FOLLOWER
WIN	1 Pilkington (Float Glass) G.D. Searle (NutraSweet) Dupont (Teflon)	2 IBM (PC) Matsushita (VHS video recorders) Seiko (quartz watch)
LOSE	4 RC Cola (diet cola) EMI (scanner) Bowmar (calculator) Xerox ("Star") DeHavilland (Comet)	3 Kodak (instant photography) Northrup (F20) DEC (PC)

Fig. 19.1. Taxonomy of outcomes from the innovation process.

ture the profits generated by an innovation. The most important dimensions of such a regime are the nature of the technology and the efficacy of legal mechanisms of protection.

It has long been known that patents do not work in practice as they do in theory. Rarely, if ever, do patents confer perfect appropriability, although they do afford considerable protection on new chemical products and rather simple mechanical inventions. Many patents can be "invented around" at modest costs. They are especially ineffective at protecting process innovations. Often patents provide little protection, because the legal requirements for upholding their validity or for proving their infringement are high.

In some industries, particularly where the innovation is embedded in processes, trade secrets are a viable alternative to patents. Protection of trade secrets is possible, however, only if a firm can put its product before the public and still keep the underlying technology secret. Usually only chemical formulas and industrial-commercial processes (for example, cosmetics and recipes) can be protected as trade secrets after they are placed on the market.

The degree to which knowledge is tacit or codified also affects ease of imitation. Codified knowledge is easier to transmit and receive and is therefore more exposed to industrial espionage and the like. Tacit knowledge by definition is not articulated, and transfer is hard unless those who possess the know-how in question can demonstrate it to others (Teece, 1981). Survey research indicates that methods of appropriability vary markedly across industries, and probably within industries as well (Levin et al., 1984).

The property rights environment within which a firm operates can thus be classified according to the nature of the technology and the efficacy of the legal system to assign and protect intellectual property. Though a gross simplification, a dichotomy can be drawn between environments in which the appropriability regime is "tight" (technology is relatively easy to protect) and "loose" (technology is almost impossible to protect). Examples of the former include the formula for Coca-Cola syrup; an example of the latter is the Simplex algorithm in linear programming.

The Dominant Design Paradigm

Two stages are commonly recognized in the evolutionary development of a given branch of a science: the pre-paradigmatic stage when there is no single, generally accepted conceptual treatment of the phenomenon in a field of study, and the paradigmatic stage, which begins when a body of theory appears to have passed the canons of scientific acceptability. The emergence of a dominant paradigm signals scientific maturity and the acceptance of agreed-upon "standards" by which what has been referred to as "normal" scientific research can proceed. These "standards" remain in force unless the paradigm is overturned. Revolutionary science is what overturns normal science, as when the Copernican theories of astronomy overturned those of Ptolemy in the seventeenth century.

Abernathy and Utterback (1978), Dosi (1982), and Utterback (in this volume) provide treatment of the technological evolution of an industry in ways that parallel Kuhnian notions of scientific evolution (Kuhn, 1970). In the early stages of industrial development, product designs are fluid, manufacturing processes are loosely and adaptively organized, and generalized capital is used in production. Competition among firms manifests itself in competition among designs, which are markedly different from each other. This might be called the pre-paradigmatic stage of an industry.

After considerable trial and error in the marketplace, one design or a narrow class of designs begins to emerge as the most promising. Such design must be able to meet a set of user needs in a relatively complete fashion. The Model T Ford, the IBM System/360, and the Douglas DC-3 are examples of dominant designs in the automobile, computer, and aircraft industries, respectively.

Once a dominant design emerges, competition shifts to price and away from design. Competitive success then shifts to a new set of variables. Scale and learning become much more important, and specialized capital is deployed as incumbents seek to lower unit costs through exploiting economies of scale and learning. Reduced uncertainty over product design provides an opportunity to amortize specialized long-lived investments.

Innovation is not necessarily halted once the dominant design emerges; as Clarke (1985) points out, it can occur at a lower level in the design hierarchy. For instance, a "v" cylinder configuration emerged in automobile engine blocks during the 1930s with the Ford V-8 engine. Niches were quickly found for it. Moreover, once the product design stabilizes, there is likely to be a surge of process innovation as producers attempt to lower production costs for the new product (see Figure 19.2).

The Abernathy-Utterback framework does not characterize all industries. It seems better suited to mass markets, in which consumer tastes are relatively homogeneous, than to small niche markets where the absence of scale and learning economies attaches a much lower penalty to multiple designs. For these niche markets, generalized equipment will be used in production.

The emergence of a dominant design is a watershed that holds great significance for the distribution of profits between innovator and follower. The innovator may have been responsible for the fundamental scientific breakthroughs as well as the basic design of the new product. However, if imitation is relatively easy, imitators may enter the fray, modifying the product in important ways, yet relying on the fundamental designs pioneered by the innovator. When the game of musical chairs stops and a dominant design emerges, the innovator might well end up in a disadvantageous position relative to a follower. Hence, when imitation is coupled with design modification before the emergence of a dominant design, followers have a good chance that their modified product will be anointed as the industry standard, often to the great disadvantage of the innovator.

Complementary Assets

Let the unit of analysis be an innovation. An innovation consists of technical knowledge about how to do something better than the existing state of the art. Assume that the know-how in question is partly codified and partly tacit. For such know-how to generate profits, it must be sold or used in the market.

In almost all cases, the successful commercialization of an innovation requires that the know-how in question be used in conjunction with other capabilities or assets. Services such as marketing, competitive manufacturing, and after-sales support are almost always needed. These services are often obtained from complementary assets that are specialized. For example, the commercialization of a

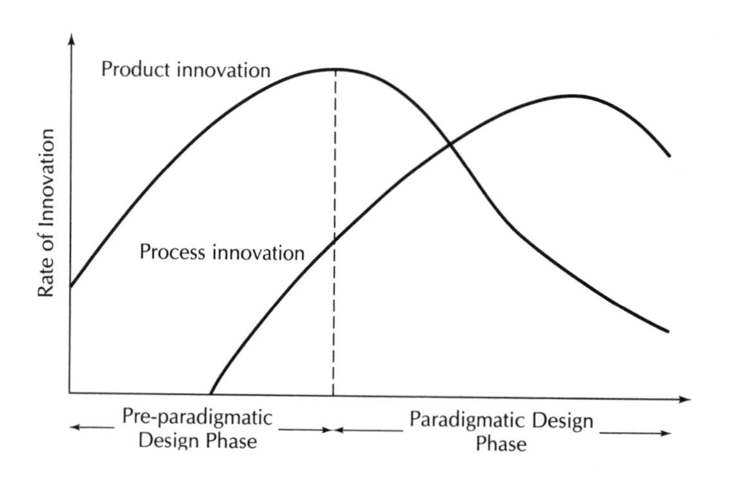

Fig. 19.2. Innovation over the product/industry life cycle.

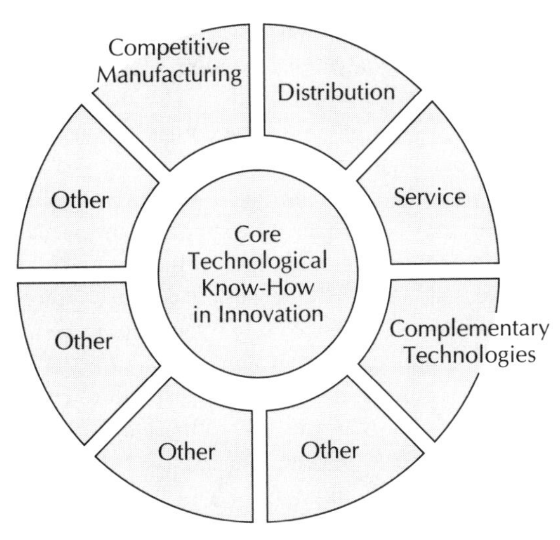

Fig. 19.3. Complementary assets needed to commercialize an innovation.

new drug is likely to require the dissemination of information over a specialized information channel. In some cases, as when the innovation is systemic, the complementary assets may be other parts of a system. For instance, computer hardware typically requires specialized software, both for the operating system and for applications. Even when an innovation is autonomous, as with plug-compatible components, certain complementary capabilities or assets will be needed for successful commercialization. Figure 19.3 summarizes this relationship schematically.

An important distinction is whether the assets required for least-cost production and distribution are specialized to the innovation. Figure 19.4 illustrates differences between complementary assets that are generic, specialized, and cospecialized.

Generic assets are general-purpose assets that need not be tailored to the innovation in question. Specialized assets are those where there is unilateral dependence between the innovation and the complementary asset. Cospecialized assets are those for which there is a bilateral dependence. For instance, specialized repair facilities were needed to support the introduction of the rotary engine by Mazda. These assets are cospecialized because of the mutual dependence of the innovation on the repair facility. Containerization similarly required the deployment of some cospecialized assets in ocean shipping and terminals. However, the dependence of trucking on containerized shipping was less than that of containerized shipping on trucking, as trucks can convert from containers to flatbeds at low cost. An example of a generic asset is the manufacturing facilities needed to make running shoes. Generalized equipment can be employed in the main, exceptions being the molds for the soles.

IMPLICATIONS FOR PROFITABILITY

These three concepts can now be related in a way that sheds light on the imitation process and the distribution of profits between innovator and follower. We begin by examining tight appropriability regimes.

Tight Appropriability Regimes

In those few instances when the innovator has an ironclad patent or copyright protection, or when

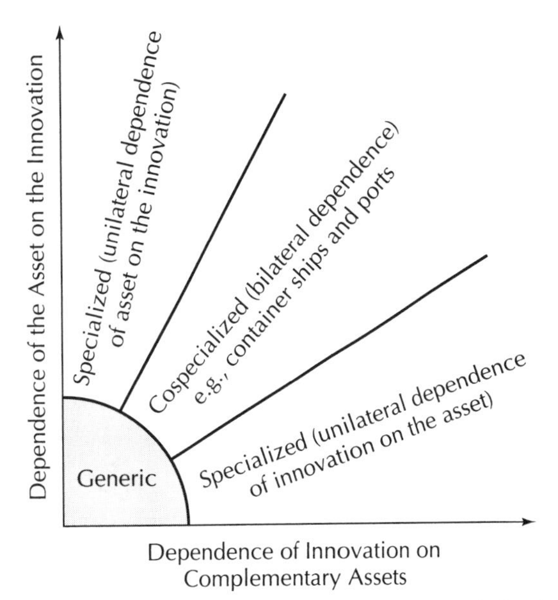

Fig. 19.4. Complementary assets: generic, specialized, and cospecialized.

the nature of the product is such that trade secrets effectively deny imitators access to the relevant knowledge, the innovator is almost assured that the innovation can be translated into market value for some period of time. Even if the innovator does not possess the desirable endowment of complementary costs, ironclad protection of intellectual property gives the innovator the time to acquire these assets. If these assets are generic, a contractual relationship may well suffice, and the innovator may simply license the technology. Specialized R&D firms are viable in such an environment. Universal Oil Products, an R&D firm that developed refining processes for the petroleum industry, is a case in point. If, however, the complementary assets are specialized or cospecialized, contractual relationships are exposed to hazards, because one or both parties will have to commit capital to certain irreversible investments, which will be valueless if the relationship between innovator and licensee breaks down. Accordingly, the innovator may find it prudent to expand by acquiring or developing specialized and cospecialized assets. Fortunately, the factors that render imitation difficult will enable the innovator to build or acquire those complementary assets without competing with imitators for their control.

Competition from imitators is muted in this type of regime, which sometimes characterizes the petrochemical industry. In this industry, the protection offered by patents is fairly easily enforced. One factor assisting the licensee in this regard is that most petrochemical processes are designed around a variety of catalysts that can be kept proprietary. An agreement not to analyze the catalyst can be extracted from licensees, affording extra protection. However, even if such requirements are violated by licensees, the innovator is still well positioned, as the most important properties of a catalyst are related to its physical structure, and the process for generating this structure cannot be deduced from structural analysis alone. Every chemical-reaction technology a company acquires is thus accompanied by an ongoing dependence on the innovating company for the catalyst appropriate to the plant design. Failure to comply with the licensing contract can thus result in a cutoff in the supply of the catalyst and possibly closure of the facility.

Similarly, if the innovator comes to market in the pre-paradigmatic phase with a sound product concept but the wrong design, a tight appropriability regime will afford the innovator the time needed to perform the trials needed to get the design right. As discussed earlier, the best initial design concepts often turn out to be hopelessly wrong, but if the innovator is protected by an impenetrable thicket of patents, or has technology that is difficult to copy, then the market may well afford the innovator the necessary time to develop the right design before being eclipsed by imitators.

Tight appropriability is the exception rather than the rule. Therefore, innovators must turn to business strategy if they are to keep imitators at bay. The nature of the competitive process will depend on whether the industry is in the paradigmatic or pre-paradigmatic phase.

Pre-paradigmatic Phase

In the pre-paradigmatic phase, the innovator must be careful to let the basic design "float" until there is sufficient evidence that the design is likely to become the industry standard. In some industries there may be little opportunity for product modification. In microelectronics, for example, designs become locked in when the circuitry is chosen. Product modification is limited to debugging and modifying software. An innovator must begin the design process anew if the product does not fit the market well. In some respects, however, the selection of designs is dictated by the need to meet compatibility standards so that new hardware can be used with existing applications software. In one sense, therefore, the design issue for the microprocessor industry today is relatively straightforward: deliver greater power and speed while meeting the computer industry standards of the existing software base. However, from time to time windows of opportunity emerge for the introduction of entirely new families of microprocessors that will define a new industry and software standard. In these instances, basic design parameters are less well defined and can be permitted to "float" until market acceptance is apparent.

The early history of the automobile industry exemplifies the importance of selecting the right design in the pre-paradigmatic stages. None of the early producers of steam-powered cars survived the early shakeout when the internal combustion engine in a closed-body automobile emerged as the dominant design. The steamer, nevertheless, had numerous early virtues, such as reliability, which the cars with internal combustion engines could not deliver.

The British fiasco with the Comet I is also instructive. De Havilland had picked an early design that had both technical and commercial flaws. By moving into production, significant irreversibilities and loss of reputation hobbled de Havilland to such a degree that it was unable to convert to the Boeing design that subsequently emerged as dominant. It was not even able to occupy second place, which went instead to Douglas.

As a general principle, it appears that innovators in loose appropriability regimes need to be intimately coupled to the market so that user needs can affect designs. When multiple parallel and sequential prototyping is feasible, it has clear advantages. Generally such an approach is prohibitively costly. When development costs for a large commercial aircraft exceed a billion dollars, variations on a theme are all that is possible.

Hence, assessing the probability that an innovator will enter the paradigmatic phase possessing the dominant design is problematic. The probabilities increase, the lower the relative cost of prototyping and the more tightly coupled the firm is to the market. The latter is a function of organizational design and can be influenced by managerial choices. The former is embedded in the technology and cannot be influenced, except in minor ways, by managerial decisions. Consequently, in industries with large costs for development and prototyping—hence significant irreversibilities—and where innovation of the product concept is easy, the probability that the innovator will emerge as a winner at the end of the pre-paradigmatic stage is low.

PARADIGMATIC STAGE

In the pre-paradigmatic phase, complementary assets do not loom large. Rivalry focuses on trying to identify the design that will be dominant. Production volumes are low, and there is little to be gained in deploying specialized assets, as scale economies are unavailable and price is not a principal competitive factor. As the leading design or designs begin to be revealed by the market, however, volumes increase and opportunities for economies of scale induce firms to begin gearing up for mass production by acquiring specialized tooling and equipment and possibly specialized distribution as well. Since these investments impose significant irreversibilities, producers are likely to proceed with caution. Islands of specialized capital will begin to appear in an industry that otherwise features a sea of general-purpose manufacturing equipment.

But as the terms of competition begin to change, and prices become increasingly important, access to complementary assets becomes critical. Since the core technology is easy to imitate, by assumption, commercial success swings upon the terms and conditions affecting access to the required complementary assets.

It is at this point that specialized and cospecialized assets become critically important. Generalized equipment and skills, almost by definition, are always available in an industry, and even if they are unavailable, they do not entail significant irreversibilities. Accordingly, firms have easy access to this type of capital, and, even if the relevant assets are not available in sufficient quantity, they can easily be put in place as this involves few risks. Specialized assets, on the other hand, imply significant irreversibilities and cannot be easily acquired by contract, as the risks are significant for the party making the dedicated investment. The firms that control the cospecialized assets, such as distribution channels and specialized manufacturing capacity, are clearly in an advantageous position relative to an innovator. Indeed, in the rare instances in which incumbent firms possess an airtight monopoly over specialized assets, and the innovator is in a regime of loose appropriability, all of the profits from the innovation could conceivably accrue to those firms, who should be able to get the upper hand.

Even when the innovator does not face competitors or potential competitors who control key assets, the innovator may still be disadvantaged. For instance, the technology embedded in cardiac pacemakers was easy to imitate, so competitive outcomes quickly came to be determined by who had easiest access to the complementary assets—in this case, specialized marketing. A similar situation has recently arisen in the United States with respect to personal computers. As an industry participant recently observed:

> There are a huge number of computer manufacturers, companies that make peripherals (e.g., printers, hard disk drives, floppy disk drives), and software companies. They are all trying to get marketing distributors because they cannot afford to call on all of the U.S. companies directly. They need to go through retail distribution channels, such as Businessland, in order to reach the marketplace. The problem today, however, is that many of these companies are not able to get shelf space and thus are having a very difficult time marketing their products. The point of distribution is where the profit and the power are in the marketplace today. [Norman, 1986, p. 438]

CHANNEL STRATEGY ISSUES

The preceding analysis indicates how access to complementary assets, such as manufacturing and distribution, on competitive terms is critical if the innovator is to avoid handing over most of the profits to imitators, or to the owners of the complementary assets that are specialized or cospecialized to the innovation. It is now necessary to delve deeper into the control structure that the innovator ideally will establish over these critical assets.

There are many possible channels that could be employed. At one extreme the innovator could integrate into all of the necessary complementary assets, but complete integration is likely to be unnecessary and also prohibitively expensive. It is important to recognize that the variety of assets and competences needed is likely to be quite large, even for only modestly complex technologies. To produce a personal computer, for instance, a company needs access to expertise in semiconductor, display, disk drive, networking, and keyboard technologies, among others. No company can keep pace in all of these areas by itself. At the other extreme, the innovator could try to gain access to these assets through straightforward contractual relationships (for example, component supply contracts, fabrication contracts, service contracts). In many instances such contracts may suffice, although they sometimes expose the innovator to hazards and dependencies that might otherwise be avoided. Between the fully integrated and full contractual extremes are many intermediate forms and channels. An analysis of the properties of the two extreme forms is presented below. A brief synopsis of mixed modes then follows.

Contractual Modes

The advantages of a contractual solution—whereby the innovator signs a contract, such as a license, with independent suppliers, manufacturers, or distributors—are obvious. The innovator will not have to make the up-front capital expenditures needed to build or buy the assets in question. This reduces risks as well as cash requirements.

Contracting rather than integrating is likely to be the optimal strategy when the innovator's appropriability regime is tight and the complementary assets are available in competitive supply (that is, there is adequate capacity and a choice of sources).

Both conditions apply in the petrochemical industry, for instance, so an innovator does not need to be integrated to be successful. Consider, first, the appropriability regime. As discussed earlier, the protection offered by patents is fairly easily enforced, particularly for process technology, in the petrochemical industry. Given the advantageous feedstock prices available to hydrocarbon-rich petrochemical exporters, and the appropriability regime characteristic of this industry, there is neither incentive nor advantage in owning the complementary assets (production facilities), as they are not typically highly specialized to the innovation. Union Carbide appears to realize this and has recently adjusted its strategy accordingly. Essentially, Carbide is placing its existing technology into a new subsidiary, Engineering and Hydrocarbons Service. The company is engaging in licensing and offers engineering, construction, and management services to customers who want to take their feedstocks and integrate them forward into petrochemicals. But Carbide itself appears to be backing away from an integration strategy.

Chemical and petrochemical product innovations are not as easily protected as process technology is, which should raise new challenges to innovating firms in developed nations as they attempt to shift out of commodity petrochemicals. There are already numerous examples of new products that made it to the marketplace, filled a customer need, but never generated competitive returns to the innovator because of imitation. For example, in the 1960s the Dow Chemical Company decided to start manufacturing rigid polyurethane foam. It was quickly imitated, however, by many small firms that had lower costs.[2] The absence of low-cost manufacturing capability left Dow vulnerable.

Contractual relationships can bring added credibility to the innovator, especially if the innovator is relatively unknown when the contractual partner is established and viable. Indeed, arms-length contracting that embodies more than a simple buy-sell agreement is becoming so common, and is so multifaceted, that the term *strategic partnering* has been devised to describe it. Even large companies such as IBM are now engaging in it. For IBM, partnering buys access to new technologies, enabling a company to "learn things we couldn't have learned without many years of trial and error."[3] IBM's arrangement with Microsoft Corporation for the use of MS-DOS operating system software on the IBM PC facilitated the timely introduction of IBM's personal computer into the market.

Smaller, less integrated companies are often eager to sign on with established companies because of the name recognition and reputation spillovers. For instance, Cipher Data Products, Inc., contracted with IBM to develop a low-priced version of IBM's 3480 half-inch streaming cartridge drive, which is likely to become the industry standard. As Cipher management points out, "one of the biggest advantages to dealing with IBM is that, once you've created a product that meets the high quality standards necessary to sell into the IBM world, you can sell into any arena."[4] Similarly, IBM's contract with Microsoft "meant instant credibility" to Microsoft (McKenna, 1985, p. 94).

It is most important to recognize, however, that strategic (contractual) partnering, which is currently fashionable, holds certain hazards, particularly for the innovator, when the innovator is trying to use contracts to acquire specialized capabilities. First, it may be difficult to induce suppliers to make costly irreversible commitments that depend for their success on the success of the innovation. To expect suppliers, manufacturers, and distributors to do so is to invite them to take risks along with the innovator. The problem this poses for the innovator is similar to the problems associated with attracting venture capital. The innovator must persuade its prospective partner that the risk is a good one. The situation is open to opportunistic abuses on both sides. The innovator has incentives to overstate the value of the innovation, while the supplier has incentives to "run with the technology" should the innovation be a success.

Instances of irreversible capital commitments by both parties nevertheless exist. Apple's Laserwriter—a laser printer that allows PC users to produce near-typeset-quality text and art department graphics—is a case in point. Apple persuaded Canon, Inc., to participate in the development of the Laserwriter by providing subsystems from its copiers—but only after Apple contracted to pay for a certain number of copier engines and cases. In short, Apple accepted a good deal of the financial risk to induce Canon to assist in the development and production of the Laserwriter. The arrangement appears to have been prudent, yet there were clearly hazards for both sides. It is difficult to write, execute, and enforce complex development contracts, particularly when the design of the new product is still "floating." Apple was exposed to the risk that its coinnovator Canon would fail to deliver, and Canon was exposed to the risk that the Apple design and marketing effort would not succeed. Still, Apple's alternatives may have been limited, inasmuch as it did not command the requisite technology to "go it alone."

In short, the current euphoria over strategic partnering may be partially misplaced. The advantages are being stressed (for example, McKenna, 1985) without a balanced presentation of costs and risks. Briefly, there is the risk that the partner will not perform according to the innovator's perception of what the contract requires; there is the added danger that the partner may imitate the innovator's technology and attempt to compete with the innovator. This latter possibility is particularly acute if the provider of the complementary asset is uniquely situated with respect to the complementary asset in question and has the capacity to imitate the technology, which the innovator is unable to protect. The innovator will then find that it has created a competitor who is better positioned than the innovator to take advantage of the market opportunity at hand. *Business Week* has expressed concerns along these lines in its discussion of the "hollow corporation."[5]

It is important to bear in mind, however, that contractual or partnering strategies in certain cases are ideal. If the innovator's technology is well protected, and if what the partner has to provide is a "generic" capacity available from many potential partners, then the innovator will be able to maintain the upper hand while avoiding the costs of duplicating downstream capacity. Even if the partner fails to perform, adequate alternatives exist (by assumption, the partner's capacities are commonly available) so the innovator's efforts to successfully commercialize the technology ought to proceed profitably.

Integration Modes

Integration, which by definition involves ownership, is distinguished from pure contractual modes in that it typically facilitates incentive alignment and tighter organizational control (Williamson, 1985). An innovator who owns rather than rents the complementary assets needed to commercialize is in a position to capture spillover benefits stemming from increased demand for the complementary assets caused by the innovation.

Indeed, an innovator might be in the position, at least before the innovation is announced, to buy up capacity in the complementary assets, possibly to great subsequent advantage. If future markets exist, though generally speaking they do not, taking forward positions in the complementary assets may suffice to capture much of the spillover.

Even after the innovation is announced, the innovator might still be able to build or buy complementary capacities at competitive prices if the innovation has ironclad legal protection (that is, if the innovation is in a tight appropriability regime). However, if the innovation is not tightly protected and once out is easy to imitate, then securing control of complementary capacities is likely to be the key success factor, particularly if those capacities are in fixed supply—so-called bottlenecks. Distribution and specialized manufacturing competences often become bottlenecks.

As a practical matter, however, an innovator may not have the time to acquire or build the complementary assets that ideally would be desirable. This is particularly true when imitation is easy, so that timing becomes critical. Additionally, the innovator may not have the financial resources to proceed. The implications of timing and cash constraints are summarized in Figure 19.5.

Accordingly, in loose appropriability regimes innovators need to rank complementary assets according to their importance. If the complementary assets are critical, ownership is warranted, although if the firm is cash constrained a minority position may well be a sensible approach.

When imitation is easy, strategic moves to build or buy specialized complementary assets must occur with due reference to the moves of competitors. There is no point in attempting to build a specialized asset, for instance, if one's imitators can do it faster and cheaper.

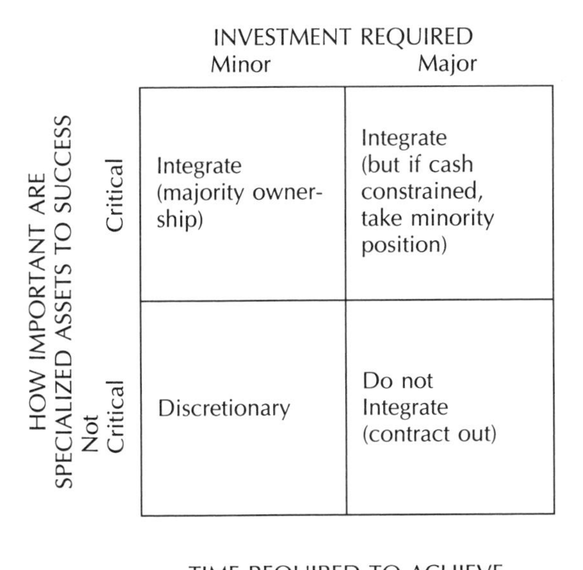

TIME REQUIRED TO ACHIEVE FAVORABLE COMPETITIVE POSITION

INVESTMENT REQUIRED	Long	Short
Minor	OK if Timing Not Critical	Full Steam Ahead
Major	Forget It	OK If Cost Position Tolerable

Fig. 19.5. Specialized complementary assets and loose appropriability: integration calculus.

It should be self-evident that if the innovator is already a large enterprise with control over many of the relevant complementary assets, integration is not likely to be the issue it might otherwise be, as the innovating firm will already control many of the relevant specialized and cospecialized assets. However, in industries experiencing rapid technological change, it is unusual that a single company has the full range of expertise needed to bring advanced products to market in a timely and cost-effective way. Hence, the integration issue is of concern to both large and small firms.

Integration Versus Contract Strategies: An Analytic Summary

Figure 19.6 summarizes some of the relevant considerations in the form of a decision flow chart. It indicates that a profit-seeking innovator faced with weak protection of intellectual property and the need to access specialized complementary assets or capabilities is forced to expand through integration to prevail over imitators. Put differently, innovators who develop new products that possess poor protection of intellectual property but require specialized complementary capacities are more likely to parlay their technology into a commercial advantage rather than see it prevail in the hands of imitators.

Figure 19.6 makes it apparent that difficult strategic decisions arise when the appropriability regime is loose and when specialized assets are critical to profitable commercialization. These situations are common and require that a thorough assessment of competitors be part of the innovator's strategic assessment of opportunities and threats. Figure 19.7 carries this discussion a step further and considers only situations where commercialization requires certain specialized capabilities. It shows the appropriate strategies for the innovators and predicts the expected outcomes for the various players.

Three classes of players are of interest: innovators, imitators, and the owners of cospecialized assets (for example, distributors). All three can potentially benefit or lose from the innovation process. The latter can potentially benefit from the additional business that the innovation may direct in the asset owner's direction. Should the asset turn out to be a bottleneck with respect to commercializing the innovation, the owner of the bottleneck facilities is obviously in a position to extract profits from the innovator or the imitators.

The vertical axis in Figure 19.7 measures how those who possess the technology (the innovator or possibly the imitators) are positioned with respect to those firms that possess required specialized assets.

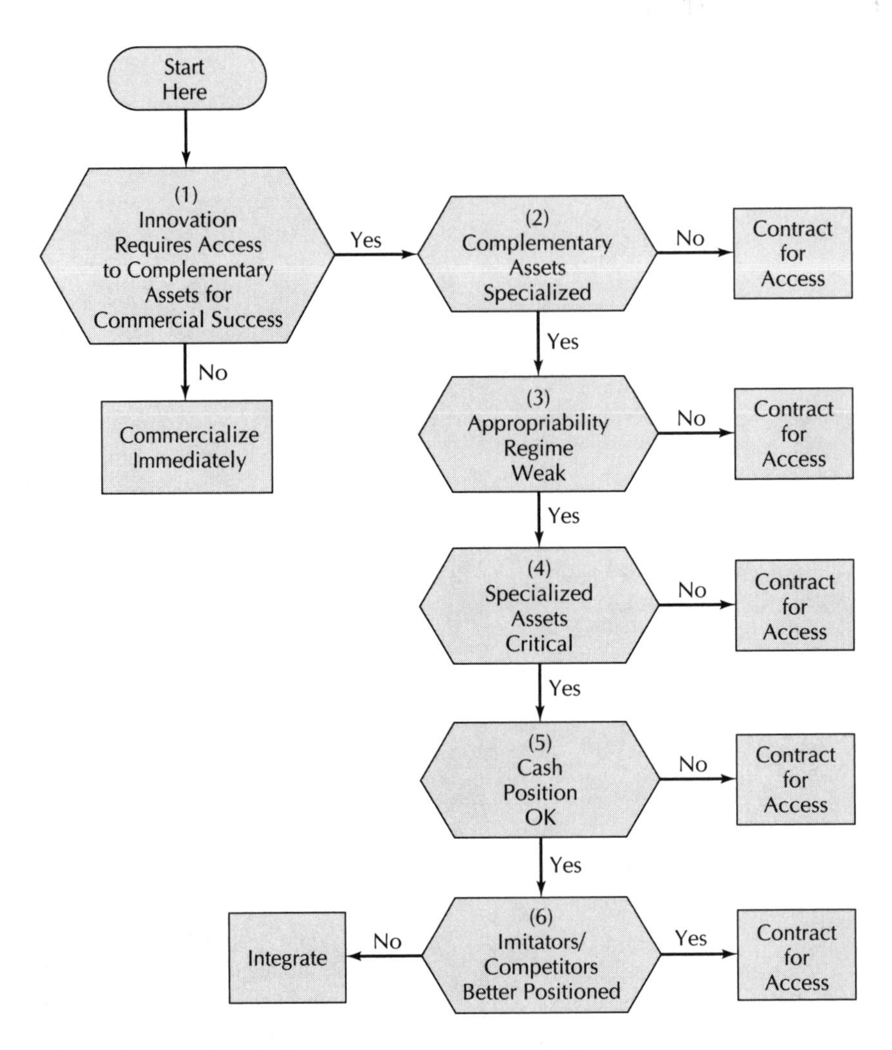

Fig. 19.6. Flow chart for integration versus contract design.

The horizontal axis measures the "tightness" of the appropriability regime, tight regimes being evidenced by ironclad legal protection coupled with technology that is difficult to copy; loose regimes offer little in the way of legal protection, and the essence of the technology, once released, is transparent to the imitator. Loose regimes are further subdivided according to how the innovator and imitators are positioned in relation to each other. This is likely to be a function of factors such as lead time and prior positioning in the requisite complementary assets.

Figure 19.7 makes it apparent that even when firms pursue the optimal strategy, other industry participants may take the jackpot. This possibility is unlikely when the intellectual property in question is tightly protected. The only serious threat to the innovator is where a specialized complementary asset is "locked up," a possibility recognized in cell 4. This can rarely be done without the cooperation of government. But it frequently occurs, as when a foreign government closes access to a foreign market, forcing the innovators to license to foreign firms, but with the government effectively cartelizing the potential licensees. With weak intellectual property protection, however, it is clear that the innovator will often lose out to imitators or asset holders, even when the innovator is pursuing the appropriate strategy (cell 6). Clearly, incorrect strategies can compound problems. For instance, if innovators integrate when they should contract, a heavy commitment of resources will be incurred for little if any strategic benefit, thereby exposing the innovator to even greater losses than would otherwise be the case. On the other hand, if

		Weak Legal/Technical Appropriability	
	Strong Logical/Technical Appropriability	Innovator Excellently Positioned Versus Imitators with Respect to Commissioning Complementary Assets	Innovator Poorly Positioned Versus Imitators with Respect to Commissioning Complementary Assets
Innovators and imitators advantageously positioned vis à vis independent owners of complementary assets	(1) Contract Innovator will win	(2) Contract Innovator should win	(3) Contract Innovator of imitator will win; asset owners will not benefit
Innovators and imitators disadvantageously positioned vis à vis independent owners of complementary assets	(4) Contract if can do so on competitive terms: Integrate if necessary Innovator should win; may have to share profits with asset holders	(5) Integrate Innovator should win	(6) Contract (to limit exposure) Innovator will probably lose to imitators and/or asset holders

Market Power of Innovators/Imitators Versus Owners of Complementary Assets

Degree of Protection for Intellectual Property

Key:

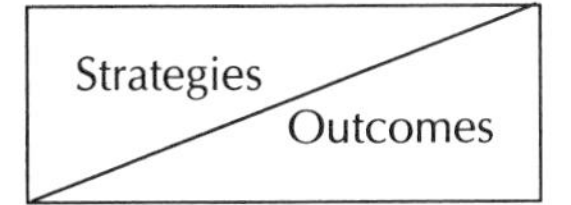

Fig. 19.7. Optimal contract and integration strategies and outcomes for innovators: specialized asset case.

an innovator tries to contract for the supply of a critical capability when it should build the capability itself, it may well find it has nurtured an imitator better able to serve the market.

Mixed Modes

The real world rarely provides extreme or pure cases. Decisions to integrate or license involve trade-offs, compromises, and mixed approaches. It is not surprising, therefore, that the real world is characterized by mixed modes of organization, involving judicious blends of integration and contracting. Sometimes mixed modes represent transitional phases. For instance, because of the convergence of computer and telecommunication technology, firms in each industry are discovering that they often lack the technical capabilities needed in the other. Since the technological interdependence of the two requires collaboration among those who design different parts of the system, intense cross-boundary coordination and information flows are needed. For separate enterprises, agreement must be reached on complex protocol issues among parties who see their interests differently. Contractual difficulties can be anticipated, as the selection of common technical protocols among the parties will often be followed by transaction-specific investments in hardware and software. There is little doubt that this was the motivation behind IBM's 1983 purchase of 15 percent of the Rolm Corporation, manufacturer of business communications systems. This position was expanded to 100 percent in 1984. IBM's stake in Intel Corporation, which began with a 12 percent purchase in 1982, is most probably not a transitional phase leading to 100 percent purchase, because both companies realized that the two corporate cultures are not compatible, and IBM may not be as impressed with Intel's technology as it once was.

The CAT Scanner and the IBM PC: Insights from the Framework

EMI's failure to reap significant returns from the CAT scanner can be explained in large measure by reference to the concepts developed above. The scanner that EMI developed was of a technical sophistication much higher than would normally be found in a hospital, requiring a high level of training support and servicing. EMI did not possess these capabilities, could not easily contract for them, and was slow to realize their importance. It most probably could have formed a partnership with a company like Siemens to gain access to the requisite capabilities. Its failure to do so was a strategic error compounded by the limited protection afforded by the law for the intellectual property embodied in the scanner. Although subsequent court decisions have upheld some of EMI's patent claims, once the product was in the market it could be reverse engineered and its essential features copied.

Two competitors, General Electric and Technicare, already possessed the complementary capabilities that the scanner required, and they were also technologically capable. In addition, both were experienced marketers of medical equipment and had reputations for quality, reliability, and service. GE and Technicare were thus able to commit their R&D resources to developing a competitive scanner and improving on the EMI scanner where they could while they rushed to market. GE began taking orders in 1976 and soon after made inroads on EMI's lead. In 1977 concern for rising health care costs caused the Carter administration to introduce "certificate of need" regulation, which required approval by the Department of Health, Education, and Welfare for expenditures on big ticket items like CAT scanners. This severely cut the size of the available market.

By 1978 EMI had lost the leadership in market share to Technicare, who was in turn quickly overtaken by GE. In October 1979 Godfrey Hounsfield of EMI shared the Nobel Prize for invention of the CAT scanner. Despite this honor, and the public recognition of EMI's role in bringing this medical breakthrough to the world, the collapse of its scanner business forced EMI in the same year into the arms of a rescuer, Thorn Electrical Industries, Ltd. GE subsequently acquired what was EMI's scanner business from Thorn.[6] Though royalties continued to flow to EMI, the company had failed to capture the largest part of the profits generated by the innovation it had pioneered and successfully commercialized.

If EMI illustrates how a company with outstanding technology and an excellent product can fail to profit from innovation while the imitators succeed, the story of the IBM PC indicates how a new product representing only a modest technological advance can yield remarkable returns to the developer.

The IBM PC, introduced in 1981, succeeded despite the fact that its architecture was ordinary and its components standard. Philip Estridge's design team in Boca Raton, Florida, decided to use existing technology rather than the state of the art to produce a solid, reliable microcomputer. With a 1-year mandate to develop a PC, Estridge's team could do little else.

However, the IBM PC did use what at the time was a new 16-bit microprocessor (the Intel 8088) and a new disk operating system (DOS) adapted for IBM by Microsoft. Other than the microprocessor and the operating system, the IBM PC incorporated existing microcomputer "standards" and used off-the-shelf parts from outside vendors. IBM did write its own basic input/output system (BIOS), which is embedded in a read-only memory chip, but this was a relatively straightforward programming exercise.

The key to the PC's success was not the technology. It was the set of complementary assets that IBM either had or quickly assembled around the PC. To expand the market for PCs, there was a clear need for an expandable, flexible microcomputer system with extensive applications of software. IBM could have based its PC system on its own patented hardware and copyrighted software. Such an approach would cause complementary products to be cospecialized, forcing IBM to develop peripherals and a comprehensive library of software in a short time. Instead, IBM adopted what might be called an induced contractual approach. By adopting an open system architecture, as Apple had done, and by making technical information about the operating system publicly available, IBM induced a spectacular output of software by third-party suppliers. IBM estimated that by mid-1983, at least 3,000 hardware and software products were available for the PC.[7] Put differently, IBM pulled together the complementary assets, particularly software, required for success and did not even use contracts, let alone integration. This was despite the fact that the software developers were creating assets that were in part cospecialized to the IBM PC, at least in the first instance.

Several special considerations made this approach a reasonable risk for the software writers. A critical element was IBM's name and commitment to the project. The reputation behind the letters *I, B, M* is perhaps the greatest cospecialized asset the company possesses. The name implied that the product would be marketed and serviced in the IBM tradition. It guaranteed that PC-DOS would become an industry standard, so that the software business would not be solely dependent on IBM, because emulators were sure to enter. It guaranteed access to retail distribution outlets on competitive terms. The consequence was that IBM was able to take a product that was at best a modest technological accomplishment and turn it into a fabulous commercial success. The case demonstrates the role of complementary assets in determining outcomes.

Though the success of the IBM PC is ongoing, the appearance of machines compatible with the IBM PC (IBM compatibles and "clones") has somewhat attenuated PC market growth for IBM. The emergence and rapid acceptance of the IBM PC established a market-based software standard. Given IBM's reputation and the quality of the product, the emergence of a market standard was predictable, as was increased price competition as competitors focused on cost reduction and performance enhancement. The fact that IBM is no longer the overwhelmingly dominant PC manufacturer—possibly because of its price umbrella and modest rate of performance improvement—does not diminish the lesson of the IBM PC program with regard to capturing returns from innovation. Despite competition from compatibles and clones, IBM's return on investment must surely have been attractive.

IMPLICATIONS FOR R&D STRATEGY, INDUSTRY STRUCTURE, AND TRADE POLICY

Allocating R&D Resources

The analysis so far assumes that the firm has developed an innovation for which a market exists. It indicates the strategies that the firm must follow to maximize its share of industry profits relative to imitators and other competitors. There is no guarantee of success even if optimal strategies are followed.

The innovating firm can improve its total return to R&D, however, by adjusting its R&D investment portfolio to maximize the probability that the technological discoveries that emerge will be easy to protect with existing property law or will require for commercialization cospecialized assets already within the firm's repertoire of capabilities. Put differently, if an innovating firm does not target its R&D resources toward new products and processes that it can commercialize advantageously relative to potential imitators or followers, then it is unlikely to profit from its investment in R&D. In this sense, a firm's history—and the assets it already has in place—ought to condition its R&D investment decisions. Clearly, an innovating firm with considerable assets already in place is free to strike out in new directions, so long as it is aware of the kinds of capabilities required to commercialize the innovation. It is therefore clear that the R&D investment decision cannot be divorced from the strategic analysis of markets and industries, and the firm's position in them.

Small Firm Versus Large Firm Comparisons

Business commentators frequently remark that many small entrepreneurial firms that generate new, commercially valuable technology fail at the same time that a large multinational firm, even with a less meritorious record in innovation, will survive and prosper. One explanation of this phenomenon is now clear. Large firms are more likely to possess the relevant specialized and cospecialized assets at the time a new product is introduced. They can therefore do a better job of using their technology, however meager, to maximum advantage. Small domestic firms are less likely to have the relevant specialized and cospecialized assets within their boundaries. They must therefore incur the expense of trying either to build the necessary assets or to develop coalitions with competitors or with owners of the assets.

Regimes of Appropriability and Industry Structure

In industries where legal methods of protection are effective, or where new products are just hard to copy, the strategic necessity for innovating firms to obtain cospecialized assets would appear to be less compelling than in industries where legal protection is weak. In cases where legal protection is weak or nonexistent, the control of cospecialized assets will be needed for long-run survival.

In this regard, it is instructive to examine the U.S. drug industry (Temin, 1979). In the 1940s, the U.S. Patent Office began to grant patents on certain natural substances that involved difficult extraction procedures. Thus, in 1948 Merck received a patent on streptomycin, which is a natural substance. However, it was not the extraction process but the drug itself that received the patent. Hence, patents were important to the drug industry, but they did not prevent imitation as, in some cases, just changing one molecule would enable a company to come up with a similar substance not violating the patent (Temin, 1979, p. 436). Had patents been more inclusive—and this is not to suggest that they should be—licensing would have been an effective mechanism for Merck to profit from its innovation. The emergence of close substitutes for patented drugs, coupled with FDA regulation that had the effect of reducing the elasticity of demand for drugs, placed high rewards on a strategy of product differentiation. This strategy required extensive marketing, including a sales force that could directly contact doctors, who were the purchasers of drugs through their ability to create prescriptions.[8] The result was exclusive production (that is, the earlier industry practice of licensing was dropped) and forward integration into marketing (the relevant cospecialized asset).

Generally, if legal protection of the innovators' profits is secure, innovating firms can select their boundaries according to their ability to identify user needs and respond to those needs through research and development. The weaker the legal methods of protection, the greater the incentive to obtain the relevant cospecialized assets. Hence, as industries in which legal protection is weak begin to mature, integration into innovation-specific cospecialized assets will occur. Often this will take the form of backward, forward, and lateral integration. (Conglomerate integration is not part of this phenomenon.) For example, IBM's purchase of Rolm can be seen as a response to the impact of technological change on the identity of the cospecialized assets relevant to IBM's future growth.

Industry Maturity, New Entry, and History

As technologically progressive industries mature, and a greater proportion of the relevant cospecialized assets are brought under the corporate umbrellas of incumbents, new entry becomes more difficult. Moreover, when it does occur it is more likely to include the early formation of coalitions. Incumbents will own the cospecialized assets, and new entrants will find it necessary to forge links with them. Here lies the explanation for the sudden surge in strategic partnering now occurring internationally, and particularly in the computer and telecommunications industry. Note that this change should not be interpreted in anticompetitive terms. Given existing industry structure, coalitions ought to be seen not as attempts to stifle competition, but as mechanisms for lowering entity requirements for innovators.

In industries in which a technological change has occurred and required deployment of specialized or cospecialized assets, a configuration of firm boundaries that no longer have compelling efficiencies may well have arisen. Considerations that once dictated integration may no longer hold, yet there may not be strong forces leading to divestiture. Hence existing firm boundaries in some industries—especially those where the technological trajectory and attendant specialized asset requirements have changed—may be fragile. In short, history is important in understanding the structure of the modern business enterprise. Existing firm boundaries cannot always be assumed to have an obvious rationale in relation to today's requirements.

The Importance of Manufacturing to International Competitiveness

Practically all forms of technological know-how must be embedded in goods and services to yield value to the consumer. An important policy issue for the innovating nation is whether the identity of the firms and nations performing this function is important.

In a world of tight appropriability and zero transactions cost—the world of neoclassical theory—it is a matter of indifference whether an innovating firm has an in-house manufacturing capability, domestic or foreign. The firm can engage in arms-length contracting (patent licensing, know-how licensing, coproduction, and so on) for the sale of the output of the activity in which it has a comparative advantage (in this case R&D) and will maximize returns by specializing in what it does best.

However, in a regime of loose appropriability, and especially where the requisite manufacturing assets are specialized to the innovation, which is often the case, participation in manufacturing may be necessary if an innovator is to appropriate the rents from its innovation. Hence, if an innovator's manufacturing costs are higher than those of an imitator, the innovator may well be put in the position of ceding the largest share of profits to the imitator.

In a loose appropriability regime, low-cost imitator-manufacturers could capture all of the profits from innovation. In a loose appropriability regime where specialized manufacturing capabilities are necessary to produce new products, an innovator with a manufacturing disadvantage may find that an early advantage at the research and development stage has no commercial value. This is a potentially crippling situation unless the innovator is assisted by governmental processes. For example, one reason why U.S. manufacturers did not capture the greatest part of the profits from the development of color TV, for which RCA was primarily responsible, is that RCA and its U.S. licensees were not competitive at manufacturing. In this context, concern that the decline of manufacturing threatens the entire economy appears to be well founded.

A related implication is that as the technology gap closes, the basis of competition in an industry will shift to the cospecialized assets. This appears to be what is happening in microprocessors. Intel is no longer out ahead technologically. As Gordon Moore, CEO of Intel points out, "Take the top 10 [semiconductor] companies in the world . . . and it is hard to tell at any time who is ahead of whom. . . . It is clear that we have to be pretty damn close to the Japanese from a manufacturing standpoint to compete."[9] It is not just that strength in one area is necessary to compensate for weakness in another. As technology becomes more public and less proprietary through easier imitation, strength in manufacturing and other areas is necessary to benefit from whatever technological advantages an innovator may possess.

Put differently, the notion that the United States can adopt a "designer role" in international commerce while letting independent firms in countries such as Japan, Korea, Taiwan, or Mexico do the manufacturing is unlikely to be a successful strategy for the long run. This is because profits will accrue primarily to the low-cost manufacturers (by providing a larger sales base over which they can exploit their special skills). Where imitation is easy, and even where it is not, it is difficult to do business in the market for know-how (Teece, 1981). In particular, there are difficulties in pricing an intangible asset whose true performance features are difficult to predict.

The trend in international business toward what Miles and Snow (1986) call dynamic networks—characterized by vertical disintegration and contracting—therefore ought to be viewed with concern. Dynamic networks, or hollow corporations, may reflect innovative organizational forms not so much as the disassembly of the modern corporation because of deterioration in manufacturing and other activities that complement technological innovation. Dynamic networks may therefore signal not so much the rejuvenation of American enterprise as its piecemeal demise.

How Trade and Investment Barriers Affect Innovators' Profits

In regimes of loose appropriability, governments can move to shift the distribution of the gains from innovation away from foreign innovators and toward domestic firms by denying innovators ownership of specialized assets. The foreign firm, by assumption an innovator, will be left with the option of selling its intangible assets on the market for know-how if both trade and investment are foreclosed by government policy. This option may appear better than the alternative (no remuneration at all from the market in question). Licensing may then appear profitable, but only because access to the complementary assets is blocked by government.

Thus, when an innovating firm generating profits needs access to complementary assets abroad, host governments, by limiting access, can sometimes milk the innovators for a share of the profits, particularly that portion that originates from sales in the host country. However, the ability of host governments to do so depends on the importance of the host country's assets to the innovator. If the cost and infrastructure characteristics of the host country are such that it is the world's lowest cost manufacturing site, and if domestic industry is competitive, then by acting as a monopsonist the government of the host country ought to be able to adjust the terms of access to the complementary assets to appropriate a greater share of the profits generated by the innovation.[10] If, on the other hand, the host country offers no unique complementary assets except access to its own market, restrictive practices by the government will only redistribute profits with respect to domestic rather than worldwide sales.

Implications for the International Distribution of the Benefits from Innovation

Thus, it is clear that innovators who do not have access to the relevant specialized and cospecialized assets may end up ceding profits to imitators and other competitors, or to the owners of the specialized or cospecialized assets. Even when the innovator possesses the specialized assets, they may be located abroad. Foreign factors of production are thus likely to benefit from research and development occurring across borders. There is little doubt, for instance, that the inability of many U.S. multinationals to sustain competitive manufacturing in the United States results in declining returns to U.S. labor. Stockholders and top management probably do as well if not better when a multinational gains access to cospecialized assets in the firm's foreign subsidiaries. However, if there is unemployment in the factors of production supporting these assets, then the foreign factors of production will benefit from innovation originating beyond national borders. This shows how important it is that innovating nations maintain competence and competitiveness in the assets—especially manufacturing—that complement technological innovation. It also shows how important it is that innovating nations enhance the protection afforded worldwide to intellectual property.

It must be recognized, however, that there are inherent limits to the legal protection of intellectual property and that business and national strategies are therefore likely to be critical in determining how the gains from innovation are shared worldwide. By making the correct strategic decision, innovating firms can move to protect the interests of stockholders. But to ensure that domestic rather than foreign cospecialized assets capture the largest share of the externalities spilling over to complementary assets, the supporting infrastructure for those complementary assets must not be allowed to decay. In short, if a nation has prowess at innovation, then in the absence of ironclad protection for intellectual property, it must maintain well-developed complementary assets if it is to capture the spillover benefits from innovation.

CONCLUSION

This chapter has attempted to synthesize from recent research in industrial organization and strategic management a framework within which to analyze the distribution of the profits from innovation. The framework indicates that the boundaries of the firm are an important strategic variable for innovating firms. The ownership of complementary assets, particularly when they are specialized or cospecialized, helps establish who wins and who loses from innovation. Imitators can often outperform innovators if they are better positioned with respect to critical complementary assets. Hence, public policy aimed at promoting innovation must focus not only on R&D but also on complementary assets as well as the underlying infrastructure. If government decides to stimulate innovation, it is important to eliminate barriers to the development of complementary assets that are specialized or cospecialized to innovation. To fail to do so will cause a large portion of the profits from innovation to flow to imitators and other competitors. If these firms lie beyond one's national borders, there are obvious implications for the international distribution of income.

When applied to world markets, results similar to those obtained from the "new trade theory" are suggested by the framework. In particular, tariffs and other restrictions on trade can in some cases injure innovating firms while simultaneously benefiting protected firms when they are imitators. However, the propositions suggested by the framework vary according to appropriability regimes, suggesting that economywide conclusions will be elusive. The policy conclusions for commodity petrochemicals, for instance, are likely to differ from those for semiconductors.

The approach also suggests that the product life cycle model of international trade will play itself out differently in different industries and markets, in part according to appropriability regimes and the nature of the assets needed to convert a technological success into a commercial one. Whatever its limitations, the approach establishes that it is not so much the structure of markets as the structure of firms, particularly the scope of their boundaries, coupled with national policies on the development of complementary assets, that determines the distribution of the profits among innovators and imitator-followers.

ACKNOWLEDGMENTS

I wish to thank Raphael Amit, Harvey Brooks, Chris Chapin, Therese Flaherty, Richard Gilbert, Bruce Guile, Heather Haveman, Mel Horwitch, David Hulbert, Carl Jacobson, Michael Porter, Gary Pisano, Richard Rumelt, Richard Nelson, Raymond Vernon, and Sidney Winter for helpful discussions relating to the subject matter of this paper. I gratefully acknowledge the financial support of the National Science Foundation under grant no. SRS-8410556 to the Center for Research in Management, University of California, Berkeley. Versions of this paper were presented at a National Academy of Engineering symposium titled "World Technologies and National Sovereignty," February 1986; at a conference on innovation at the University of Venice, March 1986; and at seminars at the Massachusetts Institute of Technology and Harvard, Yale, and Stanford universities. Helpful comments received at these conferences and seminars are gratefully acknowledged.

Notes

1. The EMI story is summarized in Michael Martin, *Managing Technological Innovation and Entrepreneurship* (Reston, Va.: Reston Publishing Company, 1984).

2. Executive Vice President, Union Carbide, Robert D. Kennedy, quoted in *Chemical Week*.

3. Comment attributed to Peter Olson III, IBM's director of business development, as reported in "The Strategy Behind IBM's Strategic Alliances," *Electronic Business*, October 1, 1985, p. 126.

4. Comment attributed to Norman Farquhar, Cipher's vice president for strategic development, as reported in *Electronic Business*, October 1, 1985, p. 128.

5. See *Business Week*, March 3, 1986, pp. 57–59. *Business Week* uses the term "hollow corporation" to describe a firm that lacks in-house manufacturing capability.

6. See "GE Gobbles a Rival in CT Scanners," *Business Week*, May 19, 1980.

7. F. Gens and C. Christiansen, "Could 1,000,000 IBM PC Users Be Wrong," *Byte*, November 1983, p. 88.

8. In the period before FDA regulation, all drugs other than narcotics were available without prescriptions. Since the user could purchase drugs directly, sales were price sensitive. Once prescriptions were required, this price sensitivity collapsed; doctors not only do not have to pay for the drugs, but in most cases they are unaware of the prices of the drugs they are prescribing.

9. "Institutionalizing the Revolution," *Forbes*, June 16, 1986, p. 35.

10. If the host country market structure is monopolistic, private actors might be able to achieve the same benefit. Government can force collusion of domestic enterprises to their mutual benefit.

REFERENCES

Abernathy, W. J., and J. M. Utterback. 1978. Patterns of industrial innovation. Technology Review 80:7 (June/July):40–47.

Clarke, K. B. 1985. The interaction of design hierarchies and market concepts in technological evolution. Research Policy 14:235–251.

Dosi, G. 1982. Technological paradigms and technological trajectories. Research Policy 11(3):147–162.

Kuhn, T. 1970. The Structure of Scientific Revolutions, 2nd ed. Chicago: University of Chicago Press.

Levin, R., A. Klevorick, N. Nelson, and S. Winter. 1984. Survey research on R&D appropriability and technological opportunity. Yale University. Unpublished manuscript.

McKenna, R. 1985. Market positioning in high technology. California Management Review XXVII:3 (Spring):82–108.

Miles, R. E., and C. C. Snow. 1986. Network organizations: New concepts for new forms. California Management Review XXVIII:3(Spring):62–73.

Norman, D. A. 1986. Impact of entrepreneurship and innovations on the distribution of personal computers. Pp. 437–439 in R. Landau and N. Rosenberg, eds., The Positive Sum Strategy. Washington, D.C.: National Academy Press.

Teece, D. J. 1981. The market for know how and the efficient international transfer of technology. Annals of the American Academy of Political and Social Science 458(November):81–96.

Temin, P. 1979. Technology, regulation, and market structure in the modern pharmaceutical industry. The Bell Journal of Economics 10(2):429–446.

Williamson, O. E. 1985. Economic Institutions of Capitalism. New York: The Free Press.

20

Managing R&D as a Strategic Option

GRAHAM R. MITCHELL
WILLIAM F. HAMILTON

There is widespread concern that the long-term competitiveness of U.S. industry is adversely affected by the short-range financial perspectives of U.S. managemen (*1*). Many of the frequently-cited indicators of this decline, such as falling market share, low productivity, poor quality, and lack of innovation, are directly linked to the underlying technical strength of U.S. corporations. Nowhere is the concern with short-range financial perspectives more frequently an issue than in the discussion of technical strategy. "Companies must abandon short-term perspectives attuned to easily quantifiable results, especially those based on the corporate shibboleth, Return on Investment. ROI eschews the risks necessary for most long-term projects—the very ones that hold out promise of major advances and entree into new market areas" (*2*). Or again, "It is inconceivable that today's successful software and computer peripheral producers—or tomorrow's biotechnology companies—started with a careful analysis of ROIs on their research projects" (*3*).

This perspective is well understood within the industrial research community, as directors of research frequently find themselves in the difficult position of arguing for research programs which they perceive to have wide-ranging and long-term benefit to the corporation, without at the same time being able to make a very robust case for ROI (*4*). If challenged to justify the programs, the research director may be drawn into suggesting future market or sales opportunities, cost reductions, or other benefits to operations. If the line management is fully supportive, there is seldom a major issue. However if there is little support from operations, perhaps because of the novelty of the proposal or the wish to avoid the turmoil and disruption it could create for the existing business, there is at least an implicit, and sometimes a very explicit, challenge to the credibility of the R&D community as compared with operating management, over the projection of future sales, markets and operating costs.

Beyond this, the research director may engender very reasonable concerns within the financial community if he appears to be violating good business practice by passing over or rejecting programs which provide shorter term, more certain benefits in favor of those having more protracted and speculative returns. In short, they may be concerned that his enthusiasm for the technology has left the research director insufficiently sensitive to some of the implications of discounted cash flow and ROI. However the use of such criteria is often deeply troubling to the research community. This is not so much because of the difficulty of generating "hard" numbers for returns, but because of a

Reprinted from *Research/Technology Management*, May/June 1988, with permission from Industrial Research Institute, Washington, D.C.

widespread sense that most investment analysis discriminates unreasonably against longer term and more risky programs, many of which may ultimately bring the largest benefits to the corporation.

In addition, the ROI analysis often fails to deal explicitly with the implications of *not* pursuing the research program. The research director is often aware that failure to *position* the corporation in key technical areas may have the effect of foreclosing numerous downstream options which could turn out to be very important to the corporation in the future.

While many in the technical community learn to live with this situation, there is also widespread recognition that the system is not serving the corporation well at either a conceptual level or during implementation. Often a deep sense exists within the R&D management that whatever the merits of ROI and similar capital budgeting frameworks in other business areas, their application to a number of very important R&D situations sets up apparently unassailable but nonetheless inappropriate selection criteria. These bias the system toward the short term, both at an individual company level and, more important, for U.S. industry in aggregate. Research directors are thus placed in the unenviable position of recognizing the increasing strategic importance of worldwide technological developments to the long-term growth and survival of their corporations, while at the same time being forced to argue their case, with "rules of evidence" which deemphasize or even exclude these longer term considerations.

FUNDING APPROACHES TO R&D

The situation just described often arises because management funding decisions for R&D are usually guided by one or another of two prevailing viewpoints:

1. *R&D As a Necessary Cost of Business.* In effect, R&D is supported as an overhead expense. This is most clearly appropriate for early-stage or exploratory research efforts and for developing or maintaining technical expertise in areas judged to be essential to future competitive advantage. However, in practice there are real limits in the amount of funding which senior management feels comfortable in treating this way.

2. *R&D As an Investment.* Underlying the treatment of R&D programs as business investments is the basic notion that scarce investment funds should be allocated to alternative company opportunities according to consistent and explicit financial criteria, whether these lie in more traditional investment domains (e.g., capital budgeting for production or distribution facilities) or in R&D. This rationale is clearly most appropriate for those technical development and engineering programs with sufficiently well-understood market and financial implications to permit meaningful quantification and analysis of important decision parameters.

The underlying problem faced by many research directors in the discussion of the overall research program portfolio is that with only two available funding models, all R&D which falls outside the limits management feels comfortable in treating as a necessary cost of business must be justified on the basis of ROI or similar capital budgeting analysis. This often requires the "force fit" of investment models to R&D situations, in which uncertainty is often directly equated with risk, and future possibilities are significantly discounted (*4*).

PUTTING R&D INTO STRATEGIC PERSPECTIVE

An essential step in dealing with the overuse of such investment criteria is to recognize that technical programs are aimed at a wide range of strategic objectives. A simple framework is shown in Figure 20.1.

Most of the technical work within large corporations is clearly directed toward a well-understood *business investment.* The technical activity involved is usually development and engineering, and the technical community is usually comfortable with the notion that the financial approach most often suited to its evaluation is an ROI or another capital budgeting framework. At the other end of the spectrum, much of the exploratory or fundamental/basic work in industry is clearly aimed toward *knowledge building*. For this work, where the business impact is often poorly defined and wide ranging, the more appropriate financial approach is that of considering R&D as a cost of doing business.

However, an important segment of the technical activity, often covering applied research, exploratory development, and sometimes feasibility demonstration, is concerned with the transition between these two broad strategic objectives. The concern is with reducing technical uncertainties and building a strong technical *position*, to the point where the corporation feels confident it can turn its technical strength into a profitable investment.

It is here where most difficulty is experienced with the two prevailing funding models. On one hand, the expenditures are often too large for management to feel comfortable treating them as an overhead or cost of doing business. On the other hand, the potential impact of the programs is often still sufficiently uncertain as to preclude meaningful ROI measurements. This is illustrated in Figure 20.1 as "strategic positioning" programs are close to neither axis.

An important first step in dealing with R&D for strategic positioning is to recognize that these expenditures are not so much directed toward an investment as they are toward the creation of an *option*. By this it is meant that the corporation is committing relatively modest R&D expenditures now to provide the opportunity to make a profitable investment at some later date.

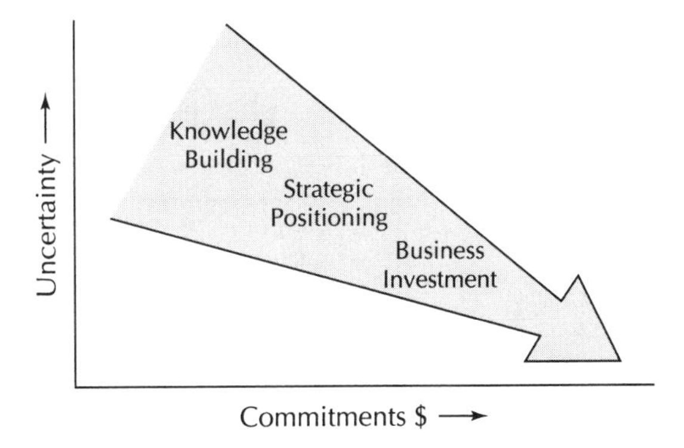

Fig. 20.1. Progression of strategic R&D programs.

The second step is to recognize that at least one class of options has been analyzed in some detail, that is, stock options, and that there are parallels with R&D options which suggest important insights that overcome some of the difficulties caused by the force fit of ROI (*5*).

R&D AS A STRATEGIC OPTION

R&D programs directed toward strategic positioning are in many ways parallel to the American call option which will permit the owner to purchase stock at a specified price (exercise price) at any time prior to an agreed-upon expiration date. The value of a stock option varies with stock price, as shown in Figure 20.2; the value of the option being at least "B" for a stock price S1. In establishing the parallel between the R&D option and the stock option:

- The price of the call option is analogous to the cost of the R&D program.
- The exercise price is analogous to the cost of the future investment needed by the company to capitalize on the R&D program (PV at some future time T) when the investment is made.
- The value of the stock for the call option is analogous in the R&D case to the returns the company will receive from the investment (PV of expected returns at a time T).

The value of a stock option has been developed formally as a function of most of the parameters involved in the options transactions (*6*). From the perspective of the industrial research community, there are two significant observations.

1. The value of an option varies in ways that are counterintuitive. That is, the value of the option moves in the opposite direction to the value of an investment with respect to volatility (uncertainty), and time. These findings apply generally to

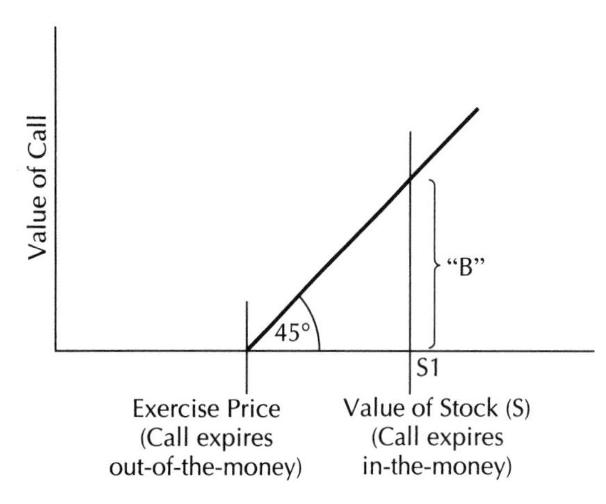

Fig. 20.2. Value of a call at expiration.

options as the relationships arise primarily as a result of limited downside risk, and are not overly sensitive to the specific assumptions of the stock options model.

2. The intuitive viewpoint often taken by the research director when justifying technical programs directed to strategic positioning, logic that often seems counterintuitive to the financial community, more closely parallels the relationships developed from the analysis of options than from the rules for business investment and ROI.

Figure 20.3 illustrates the differences between options and investments and suggests some of the direct parallels between R&D options and stock options.

Downside Risk

The downside risk for an investment, whether in the stock market or directly in the business, is that the complete investment may be lost. By contrast, the downside risk for a stock option is that the option will expire "out of the money," and that the option will not be exercised. The loss is thus limited to the price of the option, whatever the value of the stock.

FACTOR	INVESTMENT	CALL OPTION	R&D OPTION
1. Downside Risk	Risk substantial. May lose complete investment.	• Call expires out-of-the-money. • Do not exercise option, and thus lose the cost of the option.	• Company does not make major investment—lose the cost of the R&D project. (May have gained valuable insights for future R&D and other opportunities.)
2. Uncertainty (Volatility) Increases	VALUE DECLINES As uncertainty of outcome increases, NPV is discounted due to risk aversion.	VALUE INCREASES Volatility increases the Value of option upside potential without increasing downside risk.	VALUE INCREASES Wide array of speculative or partially defined applications (as with more basic innovations) increases the upside potential without increasing downside risk. (Cost of R&D.)
3. Time Increases	VALUE DECLINES Longer payback discounts value.	VALUE INCREASES Increased probability of exceeding a given exercise price.	VALUE INCREASES Option to make investments (as yet not completely defined) over a prolonged period is much more attractive than a short range or limited window of application

Fig. 20.3. Investments, options and R&D options.

The equivalent situation of an R&D option occurs when the corporation, for whatever reason, does not make the follow-up investment necessary to capitalize on the R&D program (exercise the R&D option). The equivalent loss is the cost of the R&D program, which in general will be much smaller than the follow-up investment. In practice, this represents the maximum possible loss, as the results of the R&D program, if not used directly, often provide significant insights into subsequent investments.

VOLATILITY

As volatility or general uncertainty associated with an investment increases, the value of the investment will be discounted as a result of risk aversion. Often no business investment will be made if the level of uncertainty falls above the range with which management feels comfortable.

Volatility has the reverse impact for a call option, as the downside risk is limited to the cost of the option. If the volatility of the stock price is zero, the option is worthless; increased volatility in the stock price increases the chance that it may exceed the exercise price before expiration (without increasing the downside risk).

The R&D option parallels the call option in that R&D programs which address high-impact opportunities, with a modest or low probability of success, do not imply higher risk. This distinction between options and investments is very important for R&D. Figure 20.4 illustrates two investment opportunities in which the *mean* value of expected returns is the same but the range of possible outcomes is very different. "X" might represent a relatively well-understood cost reduction program. "Y" represents the wide range of possible outcomes, including a 35 percent chance of losing money from, for example, a product still to be shown to be technically feasible in a market which has yet to be established.

Other factors being equivalent, alternative "X," having the lowest variance in expectations, would be preferred as an investment. If, on the other hand, they represent present views of two future investment opportunities, both of which require similar relatively low R&D expenditures before the decision to invest is made, the preferred R&D option may well be "Y." It is assumed that the R&D program will significantly reduce the major uncertainties associated with both programs and that in the event that "Y" is predicted by the R&D program to make a loss, no business investment would actually be made. (In much the same way that an investor would not purchase stock from a call option which expired out-of-the-money.)

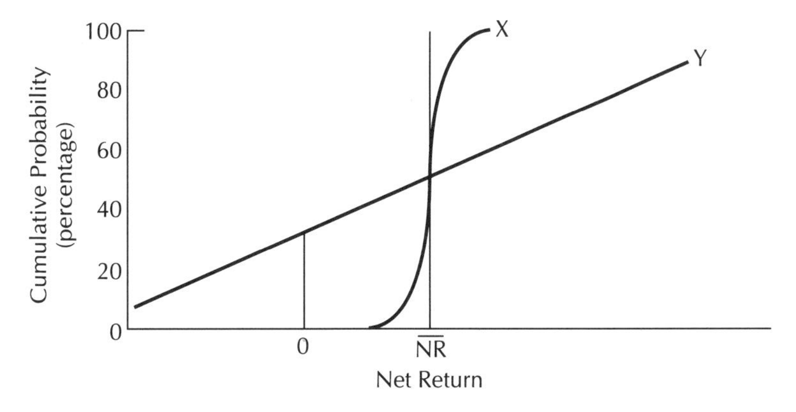

Fig. 20.4. In these two investment opportunities the mean value of expected returns is the same but the range of possible outcomes is very different. (X) Preferred if investment made on the basis of present probability estimate (since Y has 35 percent probability of significant loss). (Y) Preferred if option is taken and investment made at future date when uncertainty is resolved. Investment will not be made if $NR < 0$. Maximum loss $=$ R&D program.

Since the downside risk is limited in this way (i.e., the distribution of possible returns truncated below zero), the upside or high potential benefits, albeit uncertain, are not offset by possible losses and are thus important elements in the selection of alternative programs. If viewed from this options perspective, the priority of earlier phase, far-reaching and more basic R&D is often improved over that produced by ROI analysis.

TIME

In the investment model, the value of returns is discounted as a direct result of the time value of money. For the call option, time has the reverse effect. Increasing the time for which the option may be exercised increases the probability that stock price may exceed the exercise price during that period, and thus increases the value of the option.

The parallel situation for the R&D option is that R&D programs which offer the corporation flexibility in the timing of the subsequent investment or financial commitment, and particularly those providing the opportunity to make a series of investments over a period of time (even though the outcome of any single one may be uncertain), should be preferred to those projecting a short-range limited window of application.

POSITIVE IMPACT OF R&D ON OPTIONS

In extending the parallel, however, R&D options have one very important advantage over stock options. The purchase of a stock option has no direct effect on the exercise price or the future price of the stock, whereas the major purpose of the R&D option is to influence the future investment favorably, either by lowering costs or by increasing returns.

Figure 20.5 shows the R&D option in a framework similar to that used for the stock option. It is presumed that the corporation expects to make future investments at Cost C (analogous to the exercise price), which will yield a Return R (analogous to the value of the stock when purchased). The investment will be viable for R $>$ C, and the value of the investment to the corporation is shown as "B." However, successful R&D programs may result in a reduction of costs of the potential investment from C to C1. For example, R&D may provide the in-house skills, expertise, and sometimes proprietary designs and processes which are required for commercialization. Or, alternatively, when some of the new technology is acquired from external sources, the results of internal R&D programs often lower the cost of entry by providing

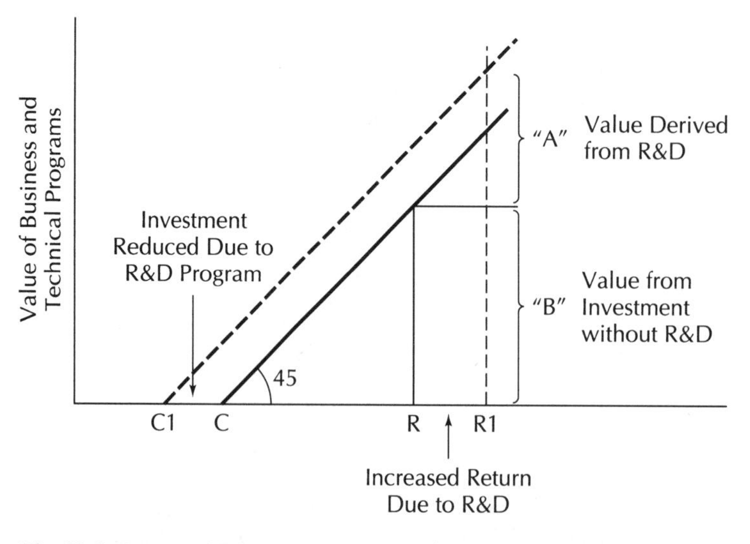

Fig. 20.5. Impact of R&D option on future investments.

Strategic Objective	Knowledge Building	Strategic Positioning	Business Investment
Technical Activity	Fundamental/ basic/exploratory research "awareness"	Focused applied research, exploratory development	Development and engineering
Financial Approach	Cost allocation	Options valuation	NPV-DCF ROI
Responsibility for Evaluation and Resource Allocation	R&D	CEO R&D/Business	Business
Market Approach	N/A	Excluded 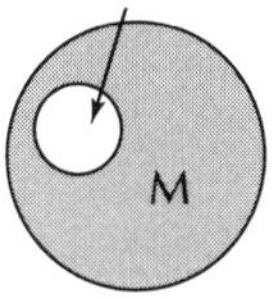 M Total market less specific exclusions	Excluded 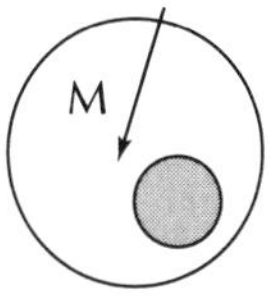 M Specific targeted product or service

Fig. 20.6. Each strategic objective requires a distinct alignment of decision-making responsibility and a different market perspective.

improved bargaining positions through patterns and know-how.

R&D programs may also produce improved returns, increasing them from R to R1. This typically results through new and improved products and services, as well as process, product and operating cost reductions. The combined result is that R&D programs have the potential to produce a benefit "A," which increases the overall value of the downstream investment.

The separate viewpoints and *a priori* positions of the business and technical communities on R&D priorities are often determined by the relative size of "A" and "B." When both "A" and "B" are large, there is usually little disagreement on the desirability of the program. However, the research director will almost certainly generate controversy if "A" is large and "B" is small. Probably the most difficult case for the R&D director to argue arises when "B" is negative and R&D programs make previously nonviable investments attractive, as when: $C > R$ and $C1 < R1$.

OVERALL FRAMEWORK FOR R&D

The broad managerial implications of characterizing R&D by strategic objective are thus not only that each objective is best addressed with a separate financial approach, but that each requires a distinct alignment of decision-making responsibility, and a different market perspective. The overall framework is summarized in Figure 20.6.

Knowledge building programs are best treated as a cost of business. The major decision-making responsibility for programs most appropriately lies with the technical community. Potential markets, to the extent they can be defined at this early stage, often provide little clear guidance on the choice of technical programs.

The central concern of this paper is *strategic positioning*, where R&D is best treated financially as an option. Decision-making responsibility is best shared between the technical and business communities, where the key role requirement for the technologist is to determine the impact of the R&D program. The choice of strategic positioning targets has long-term strategic impact and is thus of critical concern to the CEO. A broad view of potential markets must be adopted to examine strategic implications. The need to include high-impact possibilities, even if relatively uncertain, means that the potential market may often include segments and product lines beyond those in which the corporation is currently active.

By contrast, for *business investment*, ROI is often the preferred financial treatment. Decision-making is usually controlled by financial and business considerations. Markets are defined more narrowly than for strategic positioning, and returns are often estimated from sales forecasts for specific products or services to which the corporation is committed.

MANAGEMENT IMPLICATIONS

In light of the preceding discussion, three important steps should be considered by the research community:

1. *Identify Strategic Objectives.* The most direct and simplest response is to review and categorize the company-wide portfolio of technical programs to reflect the nature and purpose of R&D activities, as illustrated in Figure 20.1. It is particularly important to characterize those programs directed to strategic positioning. In many cases, explicitly recognizing that these programs are designed to produce downstream options, and are not direct investments, may contribute significantly to allaying management's discomfort. Presenting the program to business management in the different but more accurate and familiar financial framework of options often leads to an "Aha!" and a situation in which the intuitive perceptions of the business and R&D communities are more closely aligned.

2. *Review the Impact of Strategic Options.* Having roughly characterized R&D programs directed toward strategic positioning, the next step is to review their potential impact from the options perspective. This will often improve their priority in the R&D portfolio. Given the wide range of potential benefits arising from industrial R&D, the real evaluation process must be heavily judgmental rather than analytical. The goal is to form a broad estimate of "A," as shown in Figure 20.5, but the critical distinction for strategic positioning programs is to recognize that as the downside risk is limited, extended and uncertain, downstream benefits may be among the most important criteria in selecting alternative *options*. These criteria would often be heavily discounted if the programs were considered as direct investments in the business and analyzed by ROI.

For example, the situation facing manufacturers of telecommunications equipment has parallels in many other industry segments which are driven by fast-moving technology. It is recognized that relentless progress in electronics will have a powerful and continuing influence on design, performance, reliability, and cost within many market segments of the industry throughout the foreseeable future. Thus, research programs directed toward the understanding and development of microelectronics could be critical, not simply for the immediate next generation of products, but for many others yet to be fully specified. The program might be expected to provide earlier market entry and improved access to lower cost components and state-of-the-art design, from either internal or external sources, not just for the products to which the corporation is already committed, but for many which have yet to be fully defined over the next decade.

In general, there is likely to be considerable uncertainty associated with these projections of future sales since the size and timing of the future markets and the required product features will be influenced by many factors, such as the actions of competitors, which are outside the exclusive control of the corporation. If treated as an investment and analyzed using ROI, the potential benefits from these uncertain markets will be significantly discounted. However, it is the existence of these multiple downstream possibilities beyond the first application which leads to the potential long-term strategic importance of the technical area and enhances the value of the program as a strategic option.

While this kind of impact measurement almost certainly overstates the value to the corporation arising from its technical programs, the discussion brings into the open a great deal which in practice might otherwise be suppressed or significantly discounted if reviewed from an ROI perspective. The options framework legitimizes the R&D manager's concern with longer term potential and focuses attention on the danger of *not* developing positions in key technologies.

3. *Identify Strategic Positioning Targets.* The most fundamental and far-reaching step is to address explicitly the selection of strategic positioning targets to guide the development of future R&D programs. This is, or should be, an integral part of the corporation's strategic planning process.

Clearly, this challenge requires the superposition of two different kinds of information originating from two different sources. The first deals with *where* it is the corporation wishes to go. This is usually stated in terms of future market aspirations and business strategic objectives, and handled with the familiar array of strategic planning and marketing planning techniques. The second kind of information addresses the *impact* technology will have in creating competitive advantage in these markets. It is often stated in terms of projections of technical advances, forecasts or scenarios, particularly where these have the potential of being used to differentiate the corporation's performance. As the business and marketing perspectives are refined toward implementation, they will often form the basis of the estimate of "B" in Figure 20.5; detailed analysis of the technical perspective leads to estimates for "A."

Frequently, the major technical challenge in strategic positioning is to select one of the rapidly moving areas of science and technology from several exploratory alternatives and build a base from which successful future business investments will be made (7). The difficulty is that this commitment will have to be made before "hard" numbers are available for the expected returns, and that the decision to pursue one area implies that many others will not be supported to the level needed for full commercialization.

For example, Figure 20.7 illustrates that within the generic area of software, rapid progress is occurring in artificial intelligence. A number of

AREA OF TECHNOLOGY	KNOWLEDGE BUILDING (EXPLORATORY TOPICS)	STRATEGIC POSITIONING (RESEARCH FOCUS)	BUSINESS INVESTMENT (POTENTIAL APPLICATIONS)
SOFTWARE ARTIFICIAL INTELLIGENCE	PATTERN RECOGNITION SPEECH RECOGNITION SPEECH SYNTHESIS EXPERT SYSTEMS SELF IMPROVING SYSTEMS ROBOTICS USER MODELING	EXPERT SYSTEMS KNOWLEDGE-BASED SYSTEMS LOGICAL LANGUAGES LISP/PROLOG USER FRIENDLY INTERFACES	OPERATING SYSTEM IMPROVEMENTS TELECOMMUNICATIONS SWITCHING MAINTENANCE ELECTRONIC SYSTEM DESIGN NEW SERVICES

Fig. 20.7. Options positioning in software.

topics, including speech synthesis, speech recognition, visual pattern recognition, self-learning systems, and expert systems, are all experiencing rapid technical progress. All are potentially relevant to telecommunications, and all could be included in a modest exploratory and knowledge-building effort. However, significantly higher levels of effort must be devoted to any one topic if it is ever to be developed to the stage where it may have commercial payoff. In the example, expert systems has been chosen. The potential applications include several near-term opportunities in operations, design and marketing, with the promise of many more, but as yet poorly defined, possibilities.

However, the decision to commit resources and to move the work beyond the exploratory phase must be made before there is sufficient data to justify a major business investment.

IN CONCLUSION

The increasing recognition of technology as a major source of global competitive advantage has led to increased pressure on technical resources within U.S. industry. At one end of the spectrum of technical activity, the expansion in new areas of research worldwide is creating pressure for increased exploratory research within the industrial setting. At the other end of the spectrum, shorter product life-cycles and intense foreign competition are producing an increased demand for nearer-term development and engineering activities, often in an attempt to speed up the development-implementation cycle.

However, long-term survival and growth may depend more critically on strategically positioning the corporation with adequate technical strength and access to state-of-the-art technology in a limited number of the most critical technical areas. In many cases, the decisions to build technical strength in a given area, and to commit resources, must be taken before adequate returns can be established using conventional capital budgeting approaches. The recognition that these expenditures are directed to the creation of options for the corporation for downstream investments can provide a perspective which offsets to some extent the present institutional bias toward the short term.

REFERENCES

1. Robert H. Hayes and William J. Abernathy, "Managing Our Way to Economic Decline," *Harvard Business Review*, July–August 1980.
2. Leonard S. Reich, "The Making of American Industrial Research: Science and Business at G.E. and Bell, 1876–1926," Cambridge University Press, 1985.
3. Roland W. Schmitt, "Successful Corporate R&D," *Harvard Business Review*, May–June 1985.
4. Lowell W. Steele, "Selling Technology to Your Chief Executive," *Research Management*, Volume XXX No. 1, January–February 1987.
5. W. Carl Kester, "Today's Options for Tomorrow's Growth," *Harvard Business Review*, March–April 1984.
6. Robert A. Jarrow and Andrew Rudd, "Option Pricing," Myron S. Scholes, Advisory Editor, Dow Jones-Irwin 1983.
7. Graham R. Mitchell, "New Approaches to the Strategic Management of Technology," *Technology in Society*, Volume 7, Number 2/3 1985.

CHAPTER FIVE

MANAGING FUNCTIONAL COMPETENCIES

Technological change is complex and fast moving. That makes it less likely that a single generalist or group of generalists can guide a firm through cyclical patterns of change. Complex organizations manage complex innovations by pulling together highly competent groups of functional specialists, each of whom brings important, if incomplete expertise to bear on the problem. This section focuses on the special challenges innovation poses to three functional specialties: Research and Development, Marketing, and Operations. The next section deals with the problem of coordinating and integrating a group of specialists.

When technologies are complex, an organization's technical knowledge base is usually fragmented among a number of highly specialized professionals. Managing communication and cross-fertilizing ideas is a signal challenge for R&D managers, complicated when pockets of knowhow are globally dispersed. De Meyer dissects how fourteen major multinationals cope with this problem. Sakakibara and Westney add another dimension to this discussion through studying how Japanese firms cope with the pressures of globalization. They highlight a number of key differences between the way Japanese and American companies manage R&D laboratories and the technical professionals who work in them. Then Brown argues that the real mission of corporate R&D is not simply to create new products and processes, but to help companies continually reinvent themselves. Next we turn to marketing. Hamel and Prahalad contend that being marketdriven means more than linking R&D to customer needs. Instead, expeditionary marketing takes the company to a dominant position in markets that don't exist until an innovator creates them. Herstatt and von Hippel discuss how to work with "lead users," building cutting-edge designs that leverage customer knowhow. Shifting the spotlight to operations, Adler investigates how world-class manufacturing firms manage the efficiency/innovation paradox through flexible automation, highlighting the challenges posed by new production technologies. Kano then examines American attempts to implement total quality management. He finds surprisingly few differences between U.S. and Japanese quality techniques, but emphasizes that American innovative genius needs to be tempered with the systematic approach to quality that is characteristically Japanese.

21

Tech Talk: How Managers Are Stimulating Global R&D Communication

ARNOUD DE MEYER

Today, no single industrial power can pretend, as the United States could in the 1950s and 1960s, to have a quasi monopoly on high-tech development. Massachusetts and California are not the only places technology is created. Certain universities and companies in Europe, Japan, and Korea have also become drivers of technological development. Transnational companies are thus thinking about how to manage technology globally.[1]

Companies have developed several strategies: acquiring technology, either by buying laboratories or negotiating licenses; forming strategic alliances or other collaborative efforts with companies, research institutes, or governments; and creating in-house international research and development capabilities. In this paper I focus exclusively on the last strategy—developing an international network of laboratories.

Traditionally, one of the most important productivity problems in R&D is stimulating communication among researchers. How can one improve the flow of technology?[2] This problem becomes more difficult when laboratories are located far from each other. Steele observed that in many corporations coordination and communication become rather weak outside a one-day return traveling distance, and that in cases where communication and coordination do work, the corporations have long-standing multinational experience.[3]

This paper reports on the practices of fourteen large multinational companies in order to get some insight into how they manage this communication problem. It also proposes some hypotheses for how the transnational or multinational company can improve the management of its international R&D operations.[4]

THE ROLE OF COMMUNICATION IN R&D

An essential element of an R&D engineer's work is the gathering, diffusion, and creative processing of information. Many studies about R&D have shown that the productivity of an R&D engineer depends to a large extent on his or her ability to tap into an appropriate network of information flows.[5]

Allen's seminal work has repeatedly supported the notion that individual, face-to-face contact is the backbone of an efficiently operating information network.[6] To exchange information, engineers and technologists have to talk to each

Reprinted from "Tech Talk: How Managers Are Stimulating Global R&D Communicaton" by De Meyer, A., *Sloan Management Review*, (Spring1991), pp. 49–58, by permission of the publisher.

other—in each other's presence. Engineers do read. But they limit their reading mainly to textbooks, to refresh their memories, or trade journals, to figure out what the competition is doing or to see what novel components and approaches suppliers are offering. Several studies, with different technologies and in different countries, have verified this core finding.

This conclusion is quite easy to accept. To be able to understand and digest information about new technologies, specific market conditions, or competitive developments, the engineer needs more than raw information or data. He or she also needs the context of the information. Knowing about the existence of a new micro-electronic component is often insufficient. The engineer also needs to know the primary application for which the component was developed. By acting upon the new information or rephrasing it, the engineer can better understand what the raw data means. Understanding is also facilitated when the receiver can control the speed with which information is offered.

We have all experienced this: learning something new, we want to repeat to the information source what we have understood in order to verify that we understood it properly. An audience that has listened to the delivery of a written speech has lower comprehension than an audience that has been involved in building up the argument with the speaker. By controlling the speed of delivery, or by providing feedback so the observant speaker can adapt the speed of delivery, the audience receives the information more efficiently.

Why is individual, *face-to-face* communication so important? One of the interesting findings of studies in the 1960s and 1970s was that telephone communication patterns, which are individual but not face-to-face, are strongly related to the pattern of face-to-face communication. Other than calls for simple exchanges of data, one only calls the people one knows well and sees fairly often. Thus, the telephone complements but does not substitute for face-to-face communication.

Individual, face-to-face communication seems to be essential to improve and maintain the productivity of an R&D organization. But geographically decentralized R&D operations make efficient communication patterns difficult. The tendency is to form efficient communication clusters in each country, with loose couplings between clusters. This may be sufficient for development laboratories that merely adapt products, developed elsewhere, to local market conditions. As Ghoshal and Bartlett have demonstrated in their empirical study of nine large transnational companies, innovation in local subsidiaries is positively associated with *intra*subsidiary communication, though neutral with respect to *inter*subsidiary communication.[7] But both adoption and adaption by the subsidiary of innovations developed elsewhere, and diffusion of innovations created by the subsidiary to other parts of the company, are positively associated with both intra- and intersubsidiary communication. In other words, when the communication flows have to go in two or more directions, or when truly international developments are started, companies must develop mechanisms to replace or support individual, face-to-face communication.[8]

THE STUDY

My associates and I gathered data through interviews with fourteen large multinationally operating companies on their organization of international R&D operations. The number of interviews per company depended on the extent of its R&D network. In some cases we only interviewed the R&D manager. In other cases we interviewed several laboratory managers and eventual users of the research or development results (such as product or production managers). The interviews ranged from several hours to several days. Though we used a checklist of items to be discussed, in most interviews we kept the questions very open and adapted the interview format to the specific technology and market presence of the company.[9]

The fourteen companies do not form a representative sample of a particular industry or geographical region. Table 21.1 provides a description of the companies. Nine have a European head-

TABLE 21.1. The Sample

Company Activity	Number of Laboratories	Total Employees	Lab Sizes (by number of employees)
Automotive	2 (+ 1 Test ground)	3,500	2,100 and 1,400
Automotive	3	7,800	70–7,500
Electronics	6	3,100	3–2,500
Electronics	8 (2 More planned)	300	15–70
Food	1 Central research lab	450	450
	18 Local development centers	Not disclosed	20–250
Petrochemicals	3 (1 Closed after study)	400	200
Petrochemicals	2 Central research labs	3,000	2,300 and 700
	60 Business labs	2,000	10–Several hundred
Petrochemicals	8 Group central labs	4,560	150–1,600
	7 Operating labs	2,280	60–1,680
Pharmaceuticals	3	750	30–500
Pharmaceutical	9	Not available	Not available
Specialty chemicals	1 Primary research lab	660	660
	13 Development labs	550	40–100
Specialty chemicals	3	1,200	Europe: 130 Japan: 70 U.S.: 1,000
Telecommunication	13	1,300	40–250

quarters, three are North American, but with a long presence in Europe, and two are Japanese. Ten of the companies have many years of experience carrying out R&D on an international scale. The number of laboratories per company ranges from two to nineteen, and each company has several hundred if not thousands of professionals working in R&D. Judging from financial performance over the last five years and evaluations in the international business press, each company was at the time of the interviews considered successful or very successful. This does not guarantee, however, the quality and success of their R&D performance. In fact, in more than one company we were allowed to do research because managers were concerned about the performance of their international R&D and hoped to get a benchmark evaluation out of the interviews.

The activities of the laboratories ranged widely, from basic research to applied development (we discarded from the analysis simple process engineering outlets). We found enormous variety in the definition of R&D. What one company classified as research, another company (even in the same industry) would have labeled applied development. Consequently, it does not seem useful to provide a table with laboratories by technical activity.

SOLVING THE COMMUNICATION PROBLEM

We identified several elements that we expected to figure prominently in the companies' communication efforts: organizational structure, "boundary-spanning" individuals, and communications technology. The literature on R&D communication has often stressed that communication patterns are highly influenced by project structure, organizational structure, and by change over time in the project structure.[10] And since open, person-to-person communication is often a result of people having worked together,[11] reallocating personnel to different project teams can create links between teams. But altering organizational structure, the technical assignment, or project groups can also destroy communication. In addition, the style of project management can have an important influence on the creation of links between project teams. Consequently, one could expect multinational companies to take organizational measures and use career planning to manage communication patterns.

Second, a constant element in the research on communication patterns is the important role played by "boundary-spanning" individuals or "gatekeepers."[12] Information often comes from outside into a group by a two-step process. Particular individuals seem to have the capability of monitoring what is going on in the outside environment and translating that external information into messages comprehensible to the group to which they belong. They are different from the integrators or liaisons described in the literature on integration and differentiation. Indeed, gatekeepers do not integrate tasks and actions but only improve the flow of information. A boundary-spanning person, an international gatekeeper, might be able to manage the flow of information between international laboratories.

Third, recent technological innovations such as fax machines, teleconferencing, and videoconferencing have considerably improved the tool kit for complementing individual, face-to-face communication. As with telephone calls, one wonders to what extent these technological means can support or even replace face-to-face communication. The images in videoconferencing and the pictures, graphs, and tables available with fax machines, electronic mail, and computer conferencing provide a quantum improvement over the simple telephone call.

In our study we saw a bit of everything. The companies were all strongly aware of the need to improve communications and often admitted that breakdown of the communication lines was the biggest recurring problem in their organization. We sorted the solutions they were pursuing into six broad categories:

- efforts to increase socialization in order to enhance communication and information exchange;
- implementation of rules and procedures in order to increase formal communication;
- creation of boundary-spanning roles—assigning individuals to facilitate communication flows;
- creation of a centralized office responsible for managing communication;
- development of a network organization; and
- replacement of face-to-face communication by electronic systems.

None of the companies applied the methods with equal emphasis, and none limited itself to only one method. Below I describe each method in more detail.

Socialization Efforts

Though every company recognized that organizational culture can improve R&D productivity, some companies strongly emphasized socialization procedures to stimulate a positive culture. We heard the following comments:

> We need a cultural change so that people do not consider information a source of power. The amount of information you share with others should enhance your position in the company.
>
> We are like a family; we share what we have. And it should stay that way.
>
> We share our information in informal meetings. It is important that in such a meeting everybody is equal. Status or hierarchy should not play a role in evaluating the value of a piece of information. Only the factual content is important.

Companies attempted to create a "family" atmosphere in four ways. One of the most important tools was the use of *temporary assignments* to other laboratories.[13] How long should an assignment last to have a meaningful impact? No strict rule emerged from our research, except that the companies that used transfers as a major policy element expected the engineer or scientist to move into the new country. One manager said, "It has to be more than a visit where the engineer operates from a hotel. He has to get over the difficulties of settling in and increase his loyalty to the company family, which has helped him to overcome these difficulties."

Reading between the lines, we can see an effort to turn the company into a protective environment that nurtures the engineer in a quasi-hostile foreign country. Such an approach can only work, of course, if the company is organized to support the employee looking for help.

A second element of socialization was *constant traveling*. In one particular company in 1988, the R&D department alone registered more than eight thousand business trips. Travel costs ranged from 3 percent to 7 percent of the total R&D budget. In some cases there was a conscious attempt to turn visits by representatives of central headquarters into more than a work-related event. One manager said, "I try to make sure that each time I or one of my people visit one of the decentralized research companies, they have an informal get-together organized." Social events help create informal contacts, which are often the start of a healthy exchange on the task to be executed.

Some companies used *rules and procedures* to reinforce company culture. In one case, a company had two equally strong laboratories in the United Kingdom and West Germany. Training of German and British engineers is sufficiently different to create a tremendous potential for cultural clashes. But the company developed strong quality standards and procedures for execution and documentation of work; the engineers' pride in these standards helped them to identify with the company instead of with their previous training. Such a sense of pride and belonging was, in more than one company, linked to the feeling that the company had achieved technological leadership in a particular field.

A fourth method of improving company culture was the use of *training programs*. Language training was often strongly recommended. The difficulty here was to find a training program that increased identification with the company and did not negatively affect loyalty. If programs are too general, employees may use them to increase their value in the labor market.

None of the companies that emphasized organizational culture created a standard type of laboratory. On the contrary, they actively pursued diversity in research management. National characteristics and the historical background of the center and its strategic mission combined to create strong differences in the way the research laboratories operated. One manager summarized what companies hoped to achieve by improving socialization: "We want to be a multicultural [family] with common goals and values."

Rules and Procedures

Many companies used rules and procedures to enhance formal communication. These efforts usually took one of two forms: emphasizing careful reporting and documentation, or developing a planning process that stimulated communication between laboratories.

During the study it became clear to us that, in comparison with single-laboratory companies we had studied in the past, these multilaboratory companies paid much more attention to the process of writing and distributing reports.[14] Originally, we thought this emphasis was a consequence of company size or government requirements, as in the case of pharmaceutical companies, which must meet certain government requirements during the approval stage. However, if this were the case, researchers would experience the reporting structure as an administrative burden. In several companies the reporting structure was really designed to minimize the burden and maximize the diffusion of acquired know-how throughout R&D laboratories. Some companies had strict rules about report designs aimed at increasing readability and accessibility. In other companies, information sheets were organized in treelike structures that allowed the individual laboratory manager or engineer to quickly scan the new results and problems. Some companies required a quick turnaround of highlight reports. And in some companies the job description of the laboratory manager included reading reports from other laboratories and distributing them to the appropriate engineers. All these were efforts to use the reporting structure to enhance communication.

In those cases where development was strongly routinized, or in industries that had particular routine test procedures, such as pharmaceuticals, food, agro-chemical products, and mechanical engineering, the reporting structure made extensive use of electronic databases. These companies had often thought carefully about the design of the databases to give everyone in the company access (eventually with different levels of security) and to improve accessibility for inexperienced users. In laboratories or companies where the activities were more research oriented, less standardized, or more complex, most reporting was still not standardized on a database. This is probably temporary. The available commercial databases are perhaps not yet powerful enough to present unstructured information in a standardized and efficient form.

Companies also used the planning process itself to stimulate communication. A number of rituals and aspects of the planning process had more to do with rejuvenating the communication network and opening up new channels than with arriving at a plan.[15] Typical examples were scientific conferences that started the planning process, presentations made to other laboratories, and broad discussion of intermediate planning reports throughout the company. In one company, planning can be summarized as follows: the planning cycle starts with a number of regional minimeetings in preparation for the biannual conference. Representatives bring the topics from these discussions to the main conference, which has plenary sessions as well as small group meetings. Next, the business units or operating groups hold three- to four-day meetings to discuss major thrusts of the programs. Projects are not yet identified in detail at this stage. The purpose of the meetings is to obtain a portfolio of broad scientific and technical areas in which the company wants to invest and to guarantee that a balance exists between long-term and short-term research objectives, product differentiation, and cost reduction programs. The central research group participates actively in these discussions to encourage a long-term perspective, inject ideas, and identify implications for the business of emerging technologies. Several steps into the planning process, participants obtain a sort of research guide and discuss it in the laboratories. This research guide describes the technical and commercial objectives and targets of the different research programs and projects. On the basis of these discussions, the researchers provide program sheets detailing time planning and costs and discuss them with the business managers and fund providers. Once a final program is defined, the central research staff, in many cases the top R&D manager, visits each of the laboratories to review the detailed research plans and resource allocation. This whole process takes about one and a half years.

Although simplified, this is basically a description of a traditional, if elaborate, planning process. However, its length and complexity indicate that something more than a simple plan with targets and resource allocations is at stake. Planning has become a mechanism to force hundreds of researchers to talk to each other about their results, their know-how, and their technological forecasts. Later, it forces researchers to discuss these issues with the business groups, to anchor them in a realistic view of the market. Throughout, the central R&D staff plays an active role in fostering communication and diffusing the results of the planning exercise.

Though some companies were very successful in using the reporting structure or the planning process to foster communication, it became clear that rules and procedures cannot substitute for or even trigger informal communication. They probably can be a powerful mechanism for maintaining communication flows, complementary to other mechanisms that we describe here, but not a substitute for them. One of the companies, which admitted that interlaboratory communication was below expectations, had tried to use its planning procedures to create some kind of an exchange, but had failed miserably to do so. It was in the process of completely revamping the planning procedures. However, the effectiveness of rules and procedures to stimulate communication probably depends on their being built upon an already existing informal communication network.

Boundary-Spanning Roles

Many researchers have discussed the preeminent role of gatekeepers in technology transfer. We were not able to test whether a role like an international R&D gatekeeper exists—a person who has good contacts with other laboratories and can translate work from other laboratories into the jargon of his or her own laboratory. However, some companies had defined central staff's job as facilitating information flow. Central staff members had different (and sometimes inappropriate) job titles such as sponsor, liaison, or technology coordinator.

In one company, for example, the role of the sponsor was described as follows: "The goal of the system is that every sponsor would encourage and support worldwide communication of developments and results in his [or her] product group. This is mainly done through the organization of a number of technological conferences." This sponsor was placed high in the company and was responsible for organizing the communication process. In other companies, the sponsors or technological coordinators were lower in the hierarchy and had to travel around constantly to follow up on the evolution of the technology. They had to actively trigger the contacts between different individuals and groups in the laboratories. Obviously, the success of such a person was dependent on personality. At the minimum, a sponsor would need a combination of technological credibility and social listening and integrating skills.

One of the companies described this type of person as an "ambassador." Although the company had not appointed ambassadors officially, the research director commented that the success of a remote laboratory in attracting the attention of the central (and much bigger) laboratory had a lot to do with its ability to find a representative at headquarters. He said, "There is a great need for intense communication....The solution is, at least partially, to include in the remote groups people from the central site. People who know the important people in the central offices and laboratory, who know the flow of information in the different departments. These people should act as ambassadors."

In every organization we visited, we observed the use of individuals to transfer technology. In particular, it seemed to be the preferred way to construct or reconstruct a communication network when there was a major structural change in an R&D network. In those cases where such roles received a greater emphasis, or where they seemed to be the cornerstone of the communication process, the companies experienced a breakdown in communication when key individuals left the company and when new laboratories were created.

Organizational Mechanisms

We distinguished two types of organizational mechanisms in use: a *central coordination* staff, with explicit communication responsibilities, and a *network organization*. In our research they seemed to be somewhat mutually exclusive: companies tended to adhere to either one or the other. Theoretically, we see no reason why this should be so.

In seven of the fourteen companies there was a small but important central staff that was not affiliated with a particular laboratory and whose explicit task was to coordinate R&D activities and stimulate communication. Of course, all fourteen companies had a central R&D manager and office, coordinating all the laboratory activities. But in some cases this central staff was an explicit component of the oldest or biggest laboratory and in some cases it was a very small administrative staff responsible mainly for coordination of planning and budgeting activities. In the seven companies we refer to here, the central staff did more than light coordination. A group of people, perhaps small, with no explicit research task, got involved in coaching, guiding, and monitoring the laboratories. These people explicitly managed the communication process. In some cases there was a clear overlap with the previously mentioned boundary-spanning individuals. But there was a difference: the central staff had responsibility for communication as an organizational unit. These organizations looked a bit like spider webs: the central staff, sitting in the middle, monitored all movements of information between the different nodes. Weak internodal communication could exist because the main force of the communication network was in the middle. The mechanisms were fairly simple: having meetings, traveling, bringing together engineers who had problems with engineers who had answers, managing the technical reporting process, disseminating reports and results, and monitoring the company database.

In contrast, six companies had an almost negligible central staff and believed strongly in the creation of a network. The R&D directors wanted to develop networks between the different laboratories at all hierarchical levels. Each person was supposed to be a node in an extensive network of equals, relaying messages to each other and making contact with any source of information by going through two or three nodes.[16] In such a network there was no preferred route for the message, while in the centralized organization the central staff was the preferred information exchange. None of the six companies that explicitly pursued such a system had been able to get to such an ideal state.

These six companies fell into two categories: companies with a limited number of laboratories (two or three) and companies in the communication and electronics industry. It would be interesting to check whether network communications were favored by those managers who had actually been involved in electronic network design. But the fact that organizations with a limited number of laboratories favored such networks is not surprising. Organizational networks in R&D are not easy to create. Successful creation seems to require the fulfillment of three conditions, all of which are easier to implement in smaller, less complicated organizations.

First, the central staff has to accept, not only at a rational level, but also at an emotional level, that information will flow freely in the organization. Given the secrecy that is often needed in R&D, it is hard for the central office to accept that each engineer can decide what can be shared, or that there will be no secrets within the company. Not all companies are willing to shoulder the increased risk of information leaks that come with such a system.

Second, individual engineers have to be prepared to share their contacts with others. It takes time for the individual engineer to create and maintain a network, and this network is perceived to be a source of individual competitive advantage within the company. A successful network depends on the transformation of this protective attitude into one in which *sharing* is perceived as a source of competitive advantage.

Third, the company has to be prepared to support the network with the appropriate tools. At a minimum, these tools are a career planning and job rotation system, extensive meetings at different hierarchical levels, interlaboratory travel, and an electronic communication system.

Only one company was attempting both to create a network and to stimulate the network with a central R&D staff. Since the firm had only started promoting the network two years before the interviews were conducted, it was too early to judge the potential for success.

Electronic Communication

Each of the companies was experimenting in some way with electronic means of communication. Though not all had gone beyond the stage of pilot experiments, each had used computer networks, electronic mail, and computer conferencing in some way. Needless to say, fax machines were standard equipment six months after their introduction in the market. Nearly all the companies had electronic mail systems or worldwide engineering databases, if appropriate. In at least four companies, serious experiments with videoconferencing were underway. In only one company had videoconferencing become a normal way of communicating between R&D sites.

The evaluation of these electronic means was mixed. Nobody doubted that electronic communication systems could complement other channels for communication. But the mere fact that videoconferencing had not been introduced at the same sweeping pace as fax machines indicated that there was much reluctance about new forms of communication media.

Hauptmann has categorized information shared during development projects as either innovative or coordinative information.[17] Innovative information is helpful in solving technical problems. It is information on the experimental and analytical efforts researchers make to develop the product. Coordinative information includes tasks and time schedules, expected output, and the day-to-day coordination of activities. Most firms in our study used computerized communication systems to send coordinative information—schedules, results of experiments, lists of publications, and so on. They found this the most effective use of electronic media. However, we observed that the capacity to share innovative, problem-solving information was dependent on the nature of the R&D work, including its analyzability and complexity. Analyzability reflects the degree to which there exist standard procedures and methods to identify, describe, and solve the problems posed by the technology's development. The higher the analyzability and the lower the complexity of the technology, as in the case of electronic assembly or simple chemical experiments, the more effective the computer supported communication systems seemed to be. With increasing complexity and decreasing analyzability, the problem-solving content of the information exchange decreased, while the coordinative content of the information could still be communicated.

The companies that were experimenting with videoconferencing hoped that these real-time picture systems could replace some direct personal contacts. But firms did not seem confident that they could, even in their most sophisticated forms, be more than a temporary replacement for face-to-face contact.

The senior product development manager of the company with the most sophisticated electronic communication system in the sample said:

> Videoconferencing, integrated CAD/CAM databases, electronic mail, and intensive jet travel all contribute to lowering the communication barriers. All things considered, however, the most effective communication, especially in the beginning of a project, is a handshake across a table to build mutual trust and confidence. Then and only then can the electronics be really effective.

We sensed the same feeling in all the managers who had some experience with electronic communication systems. These systems can make a valuable contribution once a certain level of confidence between the partners has been established. The "handshake" is an important preceding condition for their effective use.

Weick has made a similar observation.[18] He argues that in the electronic world, information and representations can lose their meaning because the set of assumptions that give them life are lost. He sees a risk that representations of events cease forming an ordered cosmos and become chaotic. Electronic data excludes sensory information, feelings, intuition, and context. One engineer with considerable experience on videoconferencing systems summarized his own relation to the system:

> [Although it is a great system], I have two difficulties with it. I still cannot express emotions on a videoconferencing system. It seems so silly to become angry, to joke, to deviate from the subject and to talk about your family, to complain about your boss, all those things you need to do to get to know each other. And I am never sure that my colleagues at the other end will not tape me, to use my own words against me. I know it is silly, because I am not scared of being taped on the phone, but videoconferencing meetings still create many more formal commitments than a simple phone call.

Weick also argues that the people who manage the data cannot process it accurately. When people are forced to make judgments based on cryptic data, they cannot resolve their confusion by comparing different versions of the event registered in different media. When comparison is not possible, people try to clear up their confusion by asking for more data. More data of the same kind clarifies noth-

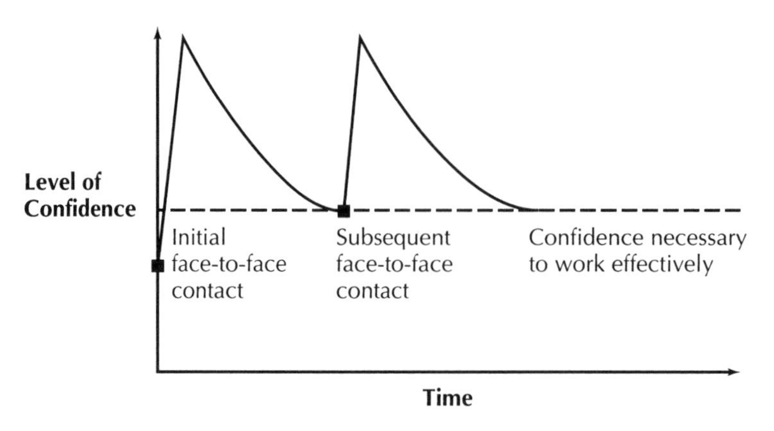

Fig. 21.1. Strategy for maintaining confidence.

ing. More and more human-processing capacity is then used to keep track of unconnected details.

We have arrived at two conclusions. First, electronically transmitted data cannot be the sole source of information. This seems obvious, but the relative convenience of electronic data communications as opposed to travel, direct contact, or even telephone calls between laboratories in different time zones gives people reason to fool themselves and to be satisfied with it.

Second, if electronic data is to be effective, there must be some level of confidence that the two parties are really communicating. This confidence comes from sensory information, feelings, and context—all of which require personal interaction. We observed that even with the best electronic communication systems, confidence between the team members of a worldwide development project seemed to decay over time. We have sometimes used the expression "the half-life effect of electronic communications" to describe, like the decay of nuclear radiation, this decreasing confidence.[19] Thus, periodic face-to-face contact seems necessary to maintain confidence at a level high enough to promote effective team work (see Figure 21.1). When one starts a new cross-laboratory project or program, one should bring the team members together to establish that initial confidence. But once everybody returns to their respective laboratory benches and starts communicating through terminals, faxes, and videoconferences, confidence will decay. One must then organize another face-to-face

TABLE 21.2. Number of Companies Deploying Mechanisms

	Chemicals and Pharmaceuticals (7 companies)	Automotive (2 companies)	Electronics and Telecommunication (4 companies)	Food (1 company)	Total
Socialization	3	2	3	1	9
Rules and procedures	3	—	2	—	5
Boundary-spanning roles	3	1	2	1	7
Centralized coordination	3	1	2	1	7
Networks	3	1	2	—	6
Electronic communication	2	1	2	—	5

meeting. The frequency of confidence-maintaining meetings can be lower with intensive use of electronic media than without them, but it will not drop below a minimum level. However, we have not been able to determine how one knows that the confidence level has decayed enough to merit another face-to-face meeting.

A PORTFOLIO OF MECHANISMS

Our categorization of communication mechanisms is, of course, somewhat artificial. The companies did not think in these categories; they consciously or unconsciously made efforts in nearly all categories. But none of the companies made strong efforts in all the categories either. Table 21.2 gives an overview of the utilization of the different categories. Socialization was the most intensively used mechanism; rules and procedures and electronic communication were used least. Electronic communication has probably not been fully deployed yet.

Table 21.3 shows that companies usually had a portfolio of mechanisms. An electronic company invested in the development of five of the six approaches. Only one company had focused its communication improvement efforts on one single activity (rules and procedures). Furthermore, there seemed to be a few patterns that we have formulated here as hypotheses:

- Building up a communication network is harder in an international environment than in a single laboratory, but just as important. Companies use portfolios of mechanisms to create and maintain communication networks.
- Integrating a new laboratory into an existing communication network, or drastically restructuring a network, requires identifying individuals to act as network builders.
- Procedural mechanisms such as cleverly designed reporting structures, or planning processes that stimulate communication, cannot create a network. However, they may help to maintain the network or adapt it slightly.
- Electronic communication cannot replace individual, face-to-face contact, but it can help to postpone the decay of confidence. It has a longer half-life effect than either written or telephone communications.
- Organizational design may influence communication patterns. We found two approaches: a spider web pattern with a central staff coordinating the flow of communication, and a network with no dominant nodes. Networks are attractive, but the difficulties in creating them should not be underestimated. They require a serious investment of effort by both the company and the individual engineers.

TABLE 21.3. Mechanisms Deployed by Companies

Number of Mechanisms	Number of Companies
5	1
4	1
3	8
2	3
1	1

CONCLUSION

Though global technology development may remain of marginal interest for most companies, those that want to compete globally must recognize the necessity of tapping sources of technology around the world. International R&D is one way to do this. If a company chooses to create an international network of laboratories, it will have to pay considerable attention to the creation and maintenance of an effective communication network. This is not a new problem in R&D management, but the problem is more complex now because of the geographical distance between laboratories. Future research should test our hypotheses and track the results of the programs we studied.

This paper was written while on sabbatical leave at the Keio University, Graduate School of Business Administration, Yokohama, Japan. I would like to thank my colleagues there for their support in providing access to Japanese companies and insight into the Japanese business environment.

REFERENCES

1. C.A. Bartlett and S. Ghoshal, *Managing across Borders* (Boston: Harvard Business School Press, 1989).
2. T.J. Allen, *Managing the Flow of Technology* (Cambridge, Massachusetts: MIT Press, 1977).
3. L.W. Steele, *Innovation in Big Business* (New York: Elsevier North Holland Publishing Company, 1975).
4. In this paper I do not discuss at length why companies internationalize their R&D operations. See: A. De Meyer, "Technology Strategy and International R&D Operations" (Fontainebleau, France: INSEAD, Working Paper No. 89/62, 1989).
5. See Allen (1977); and A. De Meyer, "The Flow of Technical Information in an R&D Department," *Research Policy* 14 (1985): 315–328.
6. Allen (1977).
7. S. Ghoshal and C.A. Bartlett, "Creation, Adoption, and Diffusion of Innovations by Subsidiaries of Multinational Companies," *Journal of International Business Studies* 19 (1988): 365–388.
8. A. De Meyer and A. Mizushima, "Global R&D Management," *R&D Management* 19 (1989): 135–146.
9. Two cases have been published after reworking into pedagogical cases. See: W.H. Davidson and J. de la Torre, *Managing the Global Corporation* (New York: McGraw Hill, 1989).
10. J.L. Utterback, "Innovation in Industry and the Diffusion of Technology," *Science* 183 (1975): 620–626.
11. M. Kanno, "Effects in Communication between Labs and Plants of the Transfer of R&D Personnel," (Cambridge, Massachusetts: MIT Sloan School of Management, Master's thesis, 1968).
12. Allen (1977); and M.L. Tushman, "Special Boundary Roles in the Innovation Process," *Administrative Science Quarterly* 22 (1977): 587–605.
13. See J. Van Maanen and E.H. Schein, "Toward a Theory of Organizational Socialization," in *Research in Organizational Behavior,* ed. B. Shaw (Greenwich, Connecticut: JAI Press, 1979); and A. Edstrom and J.R. Galbraith, "Transfer of Managers as a Coordination and Control Strategy in Multinational Organizations," *Administrative Science Quarterly* 22 (1977): 248–263.
14. See, for example, A. De Meyer, "Management of Technology in Traditional Industries: A Pilot Study in Ten Belgian Companies," *R&D Management* 13 (1983): 15–22; and R. Moenaert, J. Barbe, D. Deschoolmeester, and A. De Meyer, "Turnaround Strategies for Strategic Business Units with Aging Technology," in *The Strategic Management of Technological Innovation,* eds. R. Loveridge and M. Pitt (Chichester, England: Wiley, 1990).
15. A.P. De Geus, "Planning as Learning," *Harvard Business Review,* March-April 1988, pp. 70–74.
16. For networks in R&D, see: H. Hakansson and J. Laage-Hellman, "Developing a Network R&D Strategy," *Journal for Product Innovation Management* 4 (1984): 224–237; De Meyer (1989); and S. Ghoshal and N. Nohria, *Requisite Complexity: Organising Headquarters-Subsidiary Relations in MNCs* (Fontainebleau, France: INSEAD, Working Paper No. 90/74, 1990).
17. O. Hauptmann, "Influence of Task Type on the Relationship between Communication and Performance: The Case of Software Development," *R&D Management* 16 (1986): 127–139.
18. K.E. Weick, "Cosmos versus Chaos: Sense and Nonsense in Electronic Contexts," *Organisational Dynamics* 13 (1984): 51–64.
19. De Meyer and Mizushima (1989).

22

Japan's Management of Global Innovation: Technology Management Crossing Borders

KIYONORI SAKAKIBARA
D. ELEANOR WESTNEY

In the industries where Japan has become a major global competitor (such as computers, semiconductors, automobiles, consumer electronics), Japanese companies have developed the reputation for rapid commercialization of new product ideas and for effective and efficient incremental innovations in existing products (Rosenberg and Steinmueller 1988; Dertouzos, Lester, and Solow 1989, pp. 48–49; Stalk and Hout 1990). They have done so with relatively low investments in basic research, and with a very high degree of geographic concentration of their research and development organizations in Japan. Indeed, the success of these companies in designing in their home country laboratories products that meet the needs of customers in many different national markets has been greatly envied by U.S. firms.

These same Japanese firms, however, are now under increasing pressures to internationalize their research and development organizations and to increase their basic research activities. Japanese research managers are beginning to formulate new technology strategies to deal with these pressures and to assess the implications for institutionalized patterns of technology management.

1 NEW DIMENSIONS OF TECHNOLOGY STRATEGY

In the latter half of the 1980s, Japanese firms have increasingly confronted demands that they put more of their technology development activities overseas. Some of these pressures come from the governments and business communities within the countries which are Japan's major markets, the United States and Europe in particular, where policy makers are increasingly critical of Japanese firms' low level of local value-added, not only in manufacturing but also in product development and research (Ishikawa 1990). Western policy makers and businessmen also contend that Japan is not "pulling its weight" in investment in basic research to increase the global stock of scientific and technical knowledge (*The Economist,* 1989).

However, much of the pressure for dispersing R&D geographically is self-generated: Japanese firms want to become "true" international companies, on the model of leading Western multinationals like IBM. In addition, Japanese managers anticipate a growing shortage of scientists and engineers within Japan itself, as the aging of the

Reprinted from *Technology and the Wealth of Nations*, edited by Rosenberg, N., R. Landau, and D. Mowery, with the permission of the publishers, Stanford University Press.

Japanese population lowers the numbers of university graduates and as they must increasingly compete for those graduates with the financial services sector (which is hiring more and more scientific and technical graduates) and with foreign firms establishing R&D facilities in Japan. A survey of 177 leading Japanese firms in 1988 (*Nihon Keizai Shimbun*, September 13, 1988) found that over 80 percent of the respondents were either actively working to establish R&D bases abroad or interested in doing so.

Japanese companies are by no means alone in seeing the internationalization of their technology development capacity as an important strategic challenge. A 1985 Booz Allen study of technology management found widespread agreement within the sixteen U.S., European, and Japanese multinationals they surveyed on the perception that

> New technologies and the specialized talent that produces them will continue to develop locally in 'pockets of innovation' around the world. Nurturing those technologies, uprooting them, and cross-fertilizing them for commercialization and global distribution will continue to be major challenges in technology management. (Perrino and Tipping 1989, p. 13)

Tapping geographically dispersed "pockets of excellence"—which include government laboratories, cooperative R&D projects, and universities—can be undertaken in a variety of ways: technology scanning, cross-licensing, strategic alliances, and joint ventures. However, managers are increasingly realizing that even these can be most effectively supported by a local technology development capacity: that is, by a wholly owned critical mass of credible researchers who are part of the multinational corporation but who can function as insiders within the national technology system. More than five years ago, the Booz Allen study found a consensus that

> What we have called the 'global network' model of technology management is clearly the 'wave of the future' when it comes to competing globally. This model consists of a network of technology core groups in each major market—the United States, Japan, and Europe-managed in a coordinated way for maximum impact. (Perrino and Tipping 1989, p. 13)

In a growing number of firms, this perception is leading to increasing efforts to internationalize R&D organization, and, in those U.S. and European multinationals that already have local product development centers, to enhance and integrate these dispersed facilities (Perrino and Tipping 1989, Herbert 1989, Hakanson and Zander 1986, De Meyer and Mizushima 1989).

Within Japan, the drive to expand basic research is rooted in internal pressures that are as strong as those behind the push for internationalization, if not stronger: Japanese managers increasingly emphasize the need to generate new technology within the company, as external sources become scarcer and harder to tap (MITI 1989). In contrast to the dominant patterns in the United States and Europe, industrial companies, not government or the universities, have assumed the primary role in expanding Japan's basic research activities (Sakakibara 1988). Between 1980 and 1985, over forty of the companies listed on the Tokyo Stock Exchange have established new research facilities, many of which are oriented to basic research rather than the more traditional product-oriented R&D. In a dramatic break from tradition, these have been built away from existing corporate research and manufacturing sites, to emphasize the autonomy of the new labs.

For some companies, the drive to expand basic research is integrally linked to their internationalization strategies. Otsuka Pharmaceuticals Co. Ltd., for example, has set up research facilities in Maryland and Seattle in the United States and in Frankfurt, West Germany; their mandates cover basic research as well as clinical development. NEC has established a basic research facility in Princeton, New Jersey. Ricoh Co. Ltd. recently established a center for research in artificial intelligence in California's "Silicon Valley."

Other firms are pursuing the two agendas simultaneously but independently, expanding basic research activities at home and setting up R&D facilities overseas with less ambitious mandates. Hitachi, for example, has set its top priority on establishing the basic research facility in Japan that it set up in 1985, but it is also working on plans to build research facilities in the United States and the United Kingdom.

2 INTERNATIONAL TECHNOLOGY STRATEGY OF JAPANESE FIRMS

Theories of the multinational corporation (MNC) developed in the 1960s and early 1970s, based primarily on the experience of U.S. firms, made the internationalization of production the defining criterion of the MNC. Japanese business scholars, reflecting their own national experience, have tended to focus more broadly on the internationalization of the firm, using a three-stage model (see, for example, Saito and Itami 1986). In the first stage, marketing and distribution organizations are set up offshore but other functions, including manufacturing, remain concentrated at home. The second stage centers on the establishment of production facilities in the firm's major markets abroad. And the third—the globalization of corporate management—involves the internationalization of core corporate functions such as finance and R&D. In terms of this model, Japan's leading industrial companies have moved beyond the first stage of internationalization. They are well advanced into the second stage ("classic" foreign direct investment in production facilities abroad), and they are now moving toward the third stage (Saito and Itami 1986).

The internationalization of the technology development function of large Japanese firms has been somewhat more complex, involving five stages. The first stage, technology scanning, is associated with the first stage of internationalization, in which Japanese companies manufacture products in Japan for sale abroad. In this stage the company focuses on developing organizational systems to collect scientific and technical information and product information for use in the product development organization back in Japan. Some companies have relied heavily on sending individual "scouts" on specific technology-gathering missions; others established separate offices in the United States and Europe that were explicitly charged with technology scanning. The companies relied primarily on their own nationals in staffing these offices.

The second stage involves the creation of an organizational system to support the transfer of technology to production facilities overseas. Most production transplants have set up a technology department, following a standard Japanese pattern, in which each major factory is supported by a technology department or laboratory capable of process technology development and some incremental product improvement. In some companies these technology departments have been capable of minor modifications of product technologies to suit local markets, although new product development remained concentrated in Japan. These offshore technology departments were usually heavily staffed by Japanese, although of course local engineers were also recruited (Trevor 1988, pp. 143–44).

The formal establishment of R&D laboratories marks the beginning of the third stage. However, for many Japanese companies, the overseas laboratory, despite being called an R&D center, has done very little actual research or product development. Instead it has been a base for performing a range of other activities: technical cooperation with suppliers, support for technology transfer into production facilities, cross-licensing support, and the supervision of contract research. The overseas research laboratories of several of the leading Japanese pharmaceutical companies exemplify this stage. Instead of carrying out research directly, they contract out research to independent laboratories and specialized drug testing companies that supervise the clinical trials necessary to satisfy local regulatory requirements. In addition, they monitor technological trends and evaluate emerging technologies and new products.

In the fourth stage, overseas research laboratories embark on new product development, which becomes their central mission. These laboratories epitomize what is generally defined as the internationalization of R&D. The fifth stage extends the strategic mandate to encompass basic research, where the laboratory participates in an advanced, global division of technology development within the company.

Figure 22.1 summarizes the previous discussion and provides a way for mapping various companies' technology strategies, using two variables: the five-stages of internationalization, and the geographic dispersion of the technology function (measured by the number of regions in which the company has developed a technology capability). The Y axis represents the stages of development of the strategic motivation behind the internationalization of technology strategy, and the X axis indicates the degree of geographical dispersion of technology-generating activities. The upper left corner maps a case in which all R&D for the company, including basic research, is carried out within one country; the lower right corner represents a case in which the activities associated with technology scanning are distributed among many countries. The upper right corner, where research activities are geographically dispersed and varied, represents the "global network" of technology development described in the Booz Allen study cited above (Perrino and Tipping 1989, p. 13).

Roughly speaking, most Japanese firms are in the process of shifting from Stage II to Stage III on the Y axis and are moving gradually out on the X axis. In the computer industry, for example, NEC has advanced farthest on the Y axis, with the establishment of its basic research facility in New Jersey in the spring of 1988. Hitachi is advancing into Stage IV and is moving farther than NEC along the X axis, with its plans to set up full-scale research facilities in both the United States and the United Kingdom. Fujitsu has relied primarily on its equity partnership with Amdahl to penetrate the North American technology system, and on its nonequity strategic alliance with ICL in the United Kingdom to penetrate the European system. To date, it has not made public any plans to support these activities through a wholly owned basic R&D presence in either region. Toshiba and Mitsubishi, in comparison, have yet to propose publicly any clear internationalization strategy for technology development in their computer businesses.

Otsuka Pharmaceutical Company is another of the handful of Japanese companies who have approached Stage V on the Y axis. Otsuka has built an international system for pharmaceutical product development, including basic science, that incorporates research institutes in Maryland (near the National Institutes of Health) and Seattle in the United States and in Frankfurt, West Germany.

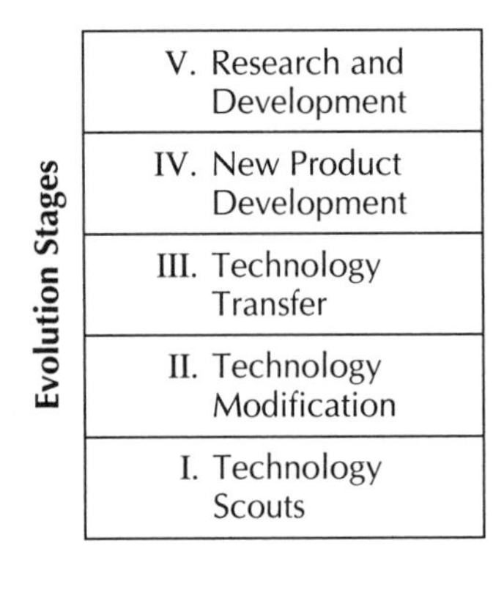

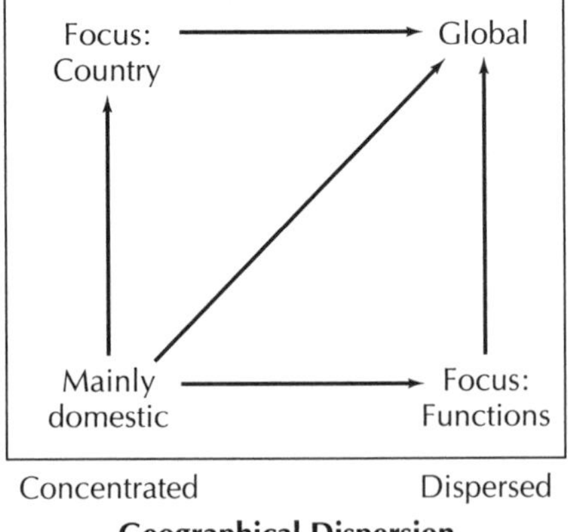

Fig. 22.1. The dimensions of international technology strategy.

The model of internationalization portrayed in Figure 22.1 differs somewhat from the model proposed by Robert Ronstadt on the basis of his research into the internationalization of technology development in U.S. multinationals, which has to date been the dominant typology in studies of the internationalization of R&D. Ronstadt identified four types of overseas technology facilities in the U.S. multinational corporation: the TTU (Technology Transfer Unit), which supported the transfer of product and production technologies from the parent to a manufacturing subsidiary; the ITU (Indigenous Technology Unit), with a capacity for new product development for the local market; the GTU (Global Technology Unit), which had the capacity to develop new products for worldwide markets; and the CTU (Corporate Technology Unit), which carries out basic research to develop advanced technology that will be applied elsewhere in the multinational (Ronstadt 1978). Ronstadt was careful to say that his typology was not a model of evolutionary stages: some of the companies he studied showed a gradual progression over time from the TTU to the GTU or CTU, while others did not.

The typology of the internationalization of technology development in Japanese firms has some similarities with Ronstadt's model. Stage II corresponds to Ronstadt's TTU, and Stage V to his GTU. However, the Japanese internationalization trajectory has been strongly influenced by two factors: the much longer reliance on export-based strategies for penetrating international markets, and Japan's lengthy experience as a technology follower. In consequence, the earliest stage of technology internationalization for Japanese firms was the development of technology-gathering outposts in the highly industrialized nations. The long period of success in gathering technology for use in their home countries' laboratories has not only given Japanese firms the capacity to develop products at home for world markets; it has become something of an impediment in their ability to develop local product development capacity. Even when top management develops a strategic commitment to develop a local product development capacity, the patterns institutionalized in Stages I and II lead the home country R&D organization to treat the developing research center as a "listening post" whose function is to report on local developments in technology and to host visiting technology scanners. In consequence, the facility finds it difficult to hire and keep good local technical people. This difficulty reinforces the belief of the home country organization that the local technical organization can never match the product development capacity in Japan.

Moving on to Stage IV and V therefore requires a very strong commitment on the part of top-level R&D managers as well as top-level general management. It often follows the development of a strong local production and marketing organization that is capable of breaking the long-established loop of gathering technical and local market information and transferring it back to Japan for embodiment in new products or product modifications, which are then transferred again to the local production organization. The lag in product development and improvement time inherent in this loop has been a major factor in stimulating Japanese and local managers to a commitment to develop a local product development capacity.

3 INTRAFIRM COORDINATION OF DISPERSED R&D

The establishment of R&D centers abroad expands the range of technological expertise on which the firm can draw effectively. However, establishing the relationships among the dispersed centers to ensure synergies in technology development strategies raises problems of internal coordination.

Figure 22.2 presents a range of configurations for intra-function coordination that develop in response to an international technology strategy. The circles represent country R&D centers (the numbers within the circles indicate different countries). The rectangles are groups of countries. In general, coordination becomes more difficult as the number of countries increases; the figure presents only a simplified set of models.

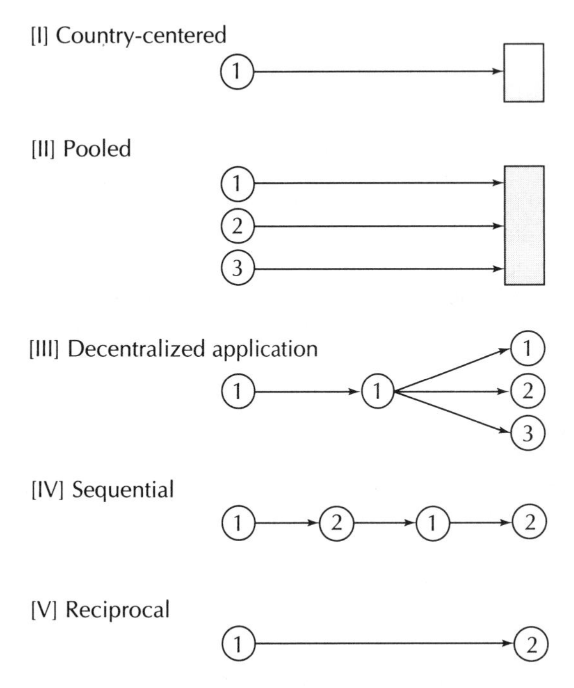

Fig. 22.2. Configurations of R&D activity flow. (Note: Circles represent countries. Each number represents a different country. Rectangles represent plural countries.)

The first model is a "country-centered" approach, which concentrates all R&D activities in one country. Strictly speaking, it is not part of an international technology strategy, even though R&D is undertaken on a global scale for multiple countries. This model makes for the easiest type of intrafirm coordination, and it preserves economies of scope and scale in R&D. Many Japanese companies still pursue this approach.

The second model is "pooled," when R&D activities are conducted at several overseas bases, and half of the research is initiated by each base, making for simultaneous, parallel R&D within the company. In this mode, some firms clearly mandate a division of labor, so that each research base has a distinctive mandate (either by product or by project segment). Others permit some duplication of R&D among their overseas bases, but usually in the same way that they permit project duplication within their home country R&D organization— to select the most promising outcome for the corporation's technology pool.

This approach is relatively simple, and close horizontal coordination across the R&D bases is not a complex problem. However, it puts a heavy load on management control systems to prevent unnecessary duplication of R&D investment. In the extreme case, it might give rise to a multidomestic strategy in which each R&D base develops a complete set of products for the national market in which it is located.

Otsuka Pharmaceutical is one example of a firm which has adopted this approach. Each overseas R&D base conducts its own basic and advanced research in the pursuit of original new pharmaceutical products. Interdependence with the home research organization is not expected; the products developed at each lab become part of the worldwide product line of the firm.

The third model is "decentralized application," in which the firm concentrates roughly half its R&D activities in Japan (particularly basic and advanced product development), and distributes the remaining half in offshore R&D centers, which focus on applied product development. The centralized part of R&D emphasizes the expansion of the basic technology portfolio of the firm; direct contact with local markets and associated local product development are pursued offshore. The actual ratio of centralized to decentralized R&D is a matter of strategic choice. If the centralized part grows too large, the R&D pattern approaches the first model described above ("country-centered"); if the offshore R&D centers come to dominate, it approaches the second model ("pooled").

This third model leads to increased complexity in managing the interdependence of the home country and local research centers. Nevertheless, many Japanese companies are taking this route. For instance, many IC makers put their custom IC development facilities abroad, where they can be closer to the customer. Several of the pharmaceutical companies test their new drugs in laboratories in Europe and the United States, close to local markets and local regulatory authorities.

The fourth model is the "sequential" strategy,

in which dispersed R&D centers share their results on a continuous basis. A typical example is the joint development of software by Xerox in the United States and Fuji-Xerox in Japan. Since 1986, these companies have built up a satellite telecommunications network. At the end of each day, Fuji-Xerox engineers in Japan electronically send their files to their U.S. counterparts. The work then continues in the United States, and at the end of the U.S. working day, the process is reversed. The goal, obviously, is to minimize development time by mobilizing development expertise in both countries, and the most important advantage of this approach is speed of development. It requires the project organization and the technology to be highly standardized across locations, and perhaps works best with routine development work, such as debugging in software development.

The fifth and final model is the "reciprocal" approach, which also features a two-way exchange in the R&D process, but which is distinguished from the sequential model by a division of labor across sites. This is ideal for mobilizing complementary expertise, but it is the most difficult in terms of coordination. A good example is the joint development of a laptop computer, the DG One, by Data General in the United States and its subsidiary in Japan, Nippon Data General. The Japanese side was in charge of hardware and the U.S. side developed the software. The project was conceived and refined through interaction between the two sites, and there was a frequent two-way flow of information throughout the development process.

There are very few actual examples of the "sequential" and "reciprocal" modes, and even fewer successful cases. For example, while Data General's laptop computer featured many noteworthy technical accomplishments, the product itself was not a market success. Other firms were able to move quickly to match its distinctive features, and were quicker to produce incremental innovations to reduce its cost and improve its features. The geographic separation between the two parts of the product development project in Data General may have inhibited those subsequent incremental innovations. Nevertheless, the rapid improvement of international data communications networks will continue to ease the technical problems of cross-border communication and may well make these modes more possible and profitable in the future.

Of these five strategies, all but the first require some degree of coordination and interaction across borders within the technology function, and even the first requires cross-border interactions between the technology function and offshore production. One of the key challenges facing the Japanese firm in internationalizing technology development is the extent to which it adapts to other societies the organizational systems which have been so successful within Japan for managing product and process innovation and for linking product development and production. In their overseas production facilities that undertake considerable local value adding (beyond simple assembly), Japanese manufacturers have made greater efforts than their U.S. or European counterparts to transfer and adapt their home country production organization. As they move into higher value adding activities in technology development, efforts to do the same with the organization of product development and technology transfer are likely, in order to maintain their competitive advantages in design for manufacturability and quality, incremental innovation, and short design cycles. The challenges in adaptation are very great, especially in North America, because of the differences in technology management systems, to which we now turn.

4 TECHNOLOGY MANAGEMENT SYSTEMS IN JAPAN AND NORTH AMERICA

The most fundamental difference between Japanese and U.S. technology management is the locus of responsibility for the career of the technical employee: in North America, that responsibility rests primarily with the individual; in Japan, it rests with the company. This difference underlies the patterns of technology transfer, incremental innovation, and human resource development in large Japanese corporations.[1]

5 RECRUITMENT

The difference in systems begins with recruitment. Leading Japanese firms still hire their technical employees from among new university graduates; those hired to work on new product development are likely to have M.S. degrees. Even today, hiring researchers with experience in other companies is rare, and limited to cases where the firm is diversifying into new technology fields. In this process Japanese companies depend heavily on close, long-term relationships with universities and with key professors within those universities. The result is an annual intake of researchers of similar age, experience, academic background, and career orientation. This similarity is reinforced in the "freshman training" given to new recruits. Japanese companies in general maintain a well-organized program for new hires; the first one or two months is usually conducted in a special training facility. The next two years are closely supervised on-the-job training, with designated "mentors" within the research organization.

In most U.S. technology-intensive firms, the technical organization is made up of people who vary widely in age, job experience, academic background, and career orientation. Few U.S. firms use formal entry-level training programs to develop more homogeneous capabilities and orientations. Instead, college graduates are hired into specific research teams and undergo whatever training they receive with the help of team members.

On the other hand, both Japanese and U.S. firms offer considerable mid-career training. In our study of technical organization in the computer industry (Westney and Sakakibara 1985), for example, we found no great differences between the Japanese and U.S. companies in the opportunities for mid-career technical training courses. The motives behind enrollment in these courses, however, were quite different, as Table 22.1 shows.

In both countries, updating existing skills was an important motivation. But many more Japanese said that they took the program because they were assigned to the course by their company. Many U.S. researchers, on the other hand, cited such motives as "to add new skills," "to improve chances of promotion," and "to improve chances of assignment to more interesting activities." They hoped to use such courses for personal growth or career advancement. Their responses reflect the strong interest in—and responsibility for—designing their own careers. U.S. companies are expected to provide opportunities for education and to provide financial and managerial support for continuing training. But the responsibility for taking advantage of those opportunities rests with the individual. Supervisors can make suggestions to individuals; they can rarely assign them to courses, as is routinely the case in Japan.

TABLE 22.1. Comparison of Motives for Taking Courses

Question: "How important was each of the following motives for taking the additional scientific, engineering, or math courses since receipt of last degree?"

	Japan ($n = 51$)	U.S. ($n = 56$)	t
Motives	Mean (S.D.)	Mean (S.D.)	
To update existing skills	4.25 (1.01)	4.00 (1.21)	1.18
To add new skills	3.71 (0.97)	4.56 (0.68)	−5.09*
To improve chances of promotion	2.02 (1.09)	2.77 (1.26)	−3.23*
To improve chances of assignment to more interesting activities	2.30 (1.29)	3.18 (1.40)	−3.30*
Assigned to course by company	3.32 (1.44)	1.32 (0.89)	8.31*

Notes: (1) Mean score of 5-point Likert scale from 1 = unimportant to 5 = very important. (2) *$p < 0.01$.

TABLE 22.2. Perceptions of Appraisal Process

Question: "The following questions concern your views about the appraisal of engineers in your organization. Please indicate how much you agree or disagree with the following statements."

Statements	Percent Indicating "Agreement"	
	Japan (n = 98)	U.S. (n = 109)
a. The performance of engineers is evaluated by the end results of their efforts rather than by the amount of effort itself.	71.4%	55.6%
b. Data collection on the performance of each engineer is highly mechanized, i.e., scoring systems exist for rating specific types of behavior.	7.1%	18.9%
c. The evaluation criteria for each engineer are individualized according to special circumstances, job and organizational situations.	64.3%	50.5%
d. In this organization judgmental or subjective appraisal by the engineer's superior is emphasized.	90.8%	59.8%
e. A regular formal face-to-face assessment interview is emphasized in the appraisal of engineers.	28.6%	70.4%
f. The performance of an engineer is evaluated over a period of five to ten years so that his potential capabilities can be taken into account.	48.0%	10.6%
g. In this organization encouragement and rewards usually outweigh criticism or negative sanctions.	34.7%	57.0%
h. In this organization people get financial rewards in proportion to the excellence of their job performance.	15.3%	55.1%

6 EVALUATIONS AND REWARDS

How technical employees are evaluated and rewarded differs dramatically between Japan and the United States, as indicated by the data in Table 22.2.

In Japan, evaluations are usually based on daily observations by managers. Some companies have elaborate formal systems that specify evaluation criteria and formats, but in practice evaluations are informal and are carried out with minimal direct feedback to the employee. The annual face-to-face evaluation meetings that play so important a role in U.S. research organizations are conspicuous by their absence in Japan. The opportunity that such interviews give employees to emphasize their accomplishments and for both parties to exchange information and receive feedback are replaced in the Japanese system by an annual written self-assessment that employees are required to give to their managers. But the employee receives no formal feedback during the process.

This kind of evaluation system works well if there is a strong, trust-based relationship between researchers and managers. Without such a relationship, frustration levels can rise, because there is no system to allow researchers to vent their dissatisfaction. Despite the prevailing image of dedicated, company-oriented researchers in Japan, the levels of frustration are growing, especially among younger and more ambitious researchers.

Frustration levels are growing over the reward system in Japanese technical organization. Even though most Japanese companies have adopted formally a more merit-based pay system, managers are reluctant to differentiate significantly among researchers in either base pay or bonuses. And they have no freedom to differentiate across functions:

base pay in research laboratories is set by the same criteria as in manufacturing or marketing within that company. In consequence, job performance has relatively little impact on salary and bonus for the individual researcher (Table 22.2, item h). On the other hand, the Japanese system does evaluate researchers over a long time period (Table 22.2, item f), and enables the company to assess and utilize the talents of each member of the technical organization.

One of the most significant rewards and incentives for Japanese researchers, therefore, is assignment to interesting and potentially important projects. But such assignments are firmly controlled by research managers in Japan, whereas in the United States researchers can seek out and volunteer for specific projects. Some U.S. companies have gone so far as to adopt an "internal market" whereby project positions are posted within the company and individual researchers encouraged to volunteer.

The greater voluntarism in the U.S. companies also leads to the possibility that researchers will leave team assignments during the project. Asked whether there was a chance of leaving their current project teams during the course of the research, 77.7 percent of the Americans said "of course" or "probably, although I might hesitate a bit." On the other hand, 71.4 percent of the Japanese respondents replied either "of course not" or "yes, but only under special conditions," meaning that in principle such departures were rare.

7 CAREER PATTERNS

Most Japanese researchers expect to move eventually into line management positions, even those whose interests and abilities lie primarily in technology and research. In interviews, we have found that Japanese researchers are more oriented than their U.S. counterparts to acquiring general managerial skills (as opposed to research management skills). In large part this is due to a widespread perception that the rewards and opportunities are much greater in the management track, and in part it is attributable to highly homogeneous expectations about the direction of careers.

A typical career path for a Japanese researcher who joins a corporate research facility is to remain there for the first seven to ten years as a researcher, usually on new product development, and then to move to a product division to work on incremental product improvement for a few years. The next step, after a few years in the divisional laboratory, is a promotion into line management in the product division. The move from the corporate to the divisional lab is usually an integral part of the technology transfer process: the researcher moves with a project on which he or she has been working, and personally follows the project through manufacturing to the market. In the divisional lab, the next few years are often spent in incremental product improvements on that same product, or a related product family. In the United States, in contrast, outlining a "typical" career for a researcher is an almost impossible task. But most researchers spend most of their working lives in the technical organization; relatively few move out into line management positions in production or general management, and even fewer aspire to do so.

8 IMPLICATIONS OF CHANGING TECHNOLOGY STRATEGIES

Clearly the Japanese system of technical management has worked extremely well in the past. The standardization of careers and the relatively low input from technical people in shaping those careers has meant that the company can move technical employees from product development into close interaction with production and assign them the responsibility for ongoing incremental improvements, regardless of individuals' personal interests and preferences. Resistance from employees is rare, because the pattern is so strongly institutionalized and taken for granted. The rewards are long term: they are enhanced prospects for advancing into the ranks of upper management. Only a very small number of the most outstanding senior researchers have a long-term career in the research organization. In consequence the corporate research organization is constantly renewed by the entry of new graduates and the movement outward of more experienced researchers.

On the other hand, the system faces challenges from the changing technology strategies of Japanese firms. Internationalization raises questions about whether the system can be transplanted and adapted to other societies. The move to basic research raises the question of whether a system so well-suited to close linkages with production and to continuous improvement of products can accommodate the different individual and managerial orientations required by a major commitment to basic research. One of the key strengths of the system is the homogeneity of technical careers. Japanese firms are beginning to struggle with a difficult dilemma: can they change their existing systems slightly to accommodate the new demands of internationalization and basic research, or must they add new units whose patterns are in sharp contrast to those in the parent technical organization? And if they choose the latter strategy, can they keep the newer patterns from eroding the patterns that have been so successful in the past?

The answers to these questions will only emerge over the next decade. Managers and scholars alike are becoming increasingly aware of the scope of the problems; the solutions still lie in the future.

Note

This discussion is based heavily on our study of the careers and organization of engineers in product development in three Japanese and three U.S. computer companies. A questionnaire was distributed to 98 Japanese and 109 U.S. engineers and interviews were conducted with individual engineers and research managers in those companies. The full research report is available as a working paper from the M.I.T.-Japan Science and Technology Program under the title of "A Comparative Study of the Training, Careers, and Organization of Engineers in the Computer Industry in Japan and the United States" by D. Eleanor Westney and Kiyonori Sakakibara.

REFERENCES

De Meyer, Arnoud and Atsuo Mizushima. 1989. Global R&D Management. *R&D Management*, vol. 19, no. 2.

Dertouzos, Michael L., Richard K. Lester, and Robert M. Solow. 1989. *Made in America: Regaining the Productive Edge*. Cambridge, MA: M.I.T. Press.

Hakanson, Lars, and Udo Zander. 1986. *Managing International Research and Development.* Stockholm: Sveriges Mekanforbund.

Herbert, Evan. 1989. Japanese R&D in the United States. *Research-Technology Management*, vol. 32, no. 6, pp. 11–20.

Ishikawa, Kenjiro. 1990. *Japan and the Challenge of Europe 1992*. London: Pinter Publishers.

MITI (Tsusho Sangyosho Sangyo Seisaku Kyoku). 1989. *NichiBei no Kigyo Kodo Hikaku* (A Comparison of Japanese and U.S. Enterprise Behavior). Tokyo: Nihon Noritsu Kyokai.

National Research Council. 1989. *Learning the R&D System: University Research in Japan and the United States.* Washington, D.C.: National Academy Press.

Perrino, Albert C., and James W. Tipping. 1989. Global Management of Technology. *Research-Technology Management*, May/June, pp. 12–19.

Ronstadt, Robert. 1977. *Research and Development Abroad by U.S. Multinationals.* New York: Praeger.

Rosenberg, Nathan, and W. Edward Steinmueller. 1988. Why Are Americans Such Poor Imitators? *American Economic Review*, vol. 78, no. 2 (May), pp. 229–34.

Saito, Masaru and Hiroyuki Itami. 1986. *Gijutsu Kaihatsu no Kokusai Senryaku* (International Strategies for Technology Development). Tokyo: Toyo Keizai Shimposha.

Sakakibara, Kiyonori. 1988. Increasing Basic Research in Japan: Corporate Activity Alone Is Not Enough. Working Paper No. 8802, Graduate School of Commerce, Hitotsubashi University.

Stalk, George, Jr., and Thomas M. Hout, 1990. Competing Against Time: How Time-Based Competition Is Reshaping Global Markets. New York: The Free Press.

Trevor, Malcolm. 1988. *Toshiba's New British Company*. London: Policy Studies Institute.

Westney, D. Eleanor, and Kiyonori Sakakibara. 1985. A Comparative Study of the Training, Careers, and Organization of Engineers in the Computer Industry in Japan and the United States. M.I.T. Working Paper.

23

Research That Reinvents the Corporation

JOHN SEELY BROWN

The most important invention that will come out of the corporate research lab in the future will be the corporation itself. As companies try to keep pace with rapid changes in technology and cope with increasingly unstable business environments, the research department has to do more than simply innovate new products. It must design the new technological and organizational "architectures" that make possible a continuously innovating company. Put another way, corporate research must reinvent innovation.

At the Xerox Palo Alto Research Center (PARC), we've learned this lesson, at times, the hard way. Xerox created PARC in 1970 to pursue advanced research in computer science, electronics, and materials science. Over the next decade, PARC researchers were responsible for some of the basic innovations of the personal-computer revolution—only to see other companies commercialize these innovations more quickly than Xerox. (See the insert "PARC: Seedbed of the Computer Revolution.") In the process, Xerox gained a reputation for "fumbling the future," and PARC for doing brilliant research but in isolation from the company's business.

That view is one-sided because it ignores the way that PARC innovations *have* paid off over the past 20 years. Still, it raises fundamental questions that many companies besides Xerox have been struggling with in recent years: What is the role of corporate research in a business environment characterized by tougher competition and nonstop technological change? And how can large companies better assimilate the latest innovations and quickly incorporate them in new products?

One popular answer to these questions is to shift the focus of the research department away from radical breakthroughs toward incremental innovation, away from basic research toward applied research. At PARC, we have chosen a different approach, one that cuts across both of these categories and combines the most useful features of each. We call it pioneering research.

Like the best applied research, pioneering research is closely connected to the company's most pressing business problems. But like the best basic research, it seeks to redefine these problems fundamentally in order to come up with fresh—and sometimes radical—solutions. Our emphasis on pioneering research has led us to redefine what we mean by technology, by innovation, and indeed by research itself. (See the insert "Letter to a Young Researcher.") Here are some of the new principles that we have identified:

1. *Research on new work practices is as important as research on new products.*

Corporate research is traditionally viewed as the source of new technologies and products. At

Reprinted by permission of *Harvard Business Review*. "Research That Reinvents the Corporation" by Brown, J. S., (January/February 1991).

PARC, we believe it is equally important for research to invent new prototypes of organizational practice. This means going beyond the typical view of technology as an artifact—hardware and software—to explore its potential for creating new and more effective ways of working, what we call studying "technology in use." Such activities are essential for companies to exploit successfully the next great breakthrough in information technology: "ubiquitous computing," or the incorporation of information technology in a broad range of everyday objects.

2. *Innovation is everywhere; the problem is learning from it.*

When corporate research begins to focus on a company's practice as well as its products, another principle quickly becomes clear: innovation isn't the privileged activity of the research department. It goes on at all levels of a company—wherever employees confront problems, deal with unforeseen contingencies, or work their way around breakdowns in normal procedures. The problem is, few companies know how to learn from this local innovation and how to use it to improve their overall effectiveness. At PARC, we are studying this process of local innovation with employees on the front lines of Xerox's business and developing technologies to harvest its lessons for the company as a whole. By doing so, we hope to turn company size, so often seen as an obstacle to innovation, into an advantage—a rich seedbed of fresh insights about technology and new work practices.

3. *Research can't just produce innovation; it must "coproduce" it.*

Before a company can learn from the innovation in its midst, it must rethink the process by which innovation is transmitted throughout the organization. Research must "coproduce" new technologies and work practices by developing with partners throughout the organization a shared understanding of why these innovations are important. On the one hand, that means challenging the outmoded background assumptions that so often distort the way people see new technologies, new market opportunities, and the entire business. On the other, it requires creating new ways to communicate the significance of radical innovations. Essentially, corporate research must prototype new mental models of the organization and its business.

4. *The research department's ultimate innovation partner is the customer.*

Prototyping technology in use, harvesting local innovation, coproducing new mental models of the organization—all these activities that we are pursuing inside Xerox are directly applicable to our customers as well. In fact, our future competitive advantage will depend not just on selling information-technology products *to* customers. It will depend on coproducing these products *with* customers—customizing technology and work practices to meet their current and future needs. One role of corporate research in this activity is to invent methods and tools to help customers identify their "latent" needs and improve their own capacity for continuous innovation.

At PARC, we've only begun to explore the implications of these new principles. Our activities in each of these areas are little more than interesting experiments. Still, we have defined a promising and exciting new direction. Without giving up our strong focus on state-of-the-art information technologies, we are also studying the human and organizational barriers to innovation. And using the entire Xerox organization as our laboratory, we are experimenting with new techniques for helping people grasp the revolutionary potential of new technologies and work practices.

The result: important contributions to Xerox's core products but also a distinctive approach to innovation with implications far beyond our company. Our business happens to be technology, but any company—no matter what the business—must eventually grapple with the issues we've been addressing. The successful company of the future must understand how people really work and how technology can help them work more effectively. It must know how to create an environment for continual innovation on the part of *all* employees. It must rethink traditional business assumptions and tap needs that customers don't even know they have yet. It must use research to reinvent the corporation.

TECHNOLOGY GETS OUT OF THE WAY

At the foundation of our new approach to research is a particular vision of technology. As the cost of computing power continues to plummet, two things become possible. First, more and more electronic technology will be incorporated in everyday office devices. Second, increased computing power will allow users to tailor the technology to meet their specific needs.

Both these trends lead to a paradoxical result. When information technology is everywhere and can be customized to match more closely the work to be done, the technology itself will become invisible. The next great breakthrough of the information age will be the disappearance of discrete information-technology products. Technology is finally becoming powerful enough to get out of the way.

Consider the photocopier. Ever since Chester Carlson first invented xerography some 50 years ago, the technology of photocopiers has been more or less the same. In a process somewhat similar to photography, a light-lens projects an image of the page onto a photoreceptor. The image is then developed with a dry toner to produce the copy. But information technology is transforming the copier with implications as radical as those accompanying the invention of xerography itself.

Today our copiers are complex computing and communications devices. Inside Xerox's high-end machines are some 30 microprocessors linked together by local area networks. They continually monitor the operations of the machine and make adjustments to compensate for wear and tear, thus increasing reliability and ensuring consistent, high copy quality. Information systems inside our copiers also make the machines easier to use by constantly providing users with information linked to the specific task they are performing. (See the insert "How Xerox Redesigned Its Copiers.") These innovations were crucial to Xerox's success in meeting Japanese competition and regaining market share during the past decade.

But these changes are only the beginning. Once copiers become computing devices, they also become sensors that collect information about their own performance that can be used to improve service and product design. For example, Xerox recently introduced a new standard feature on our high-end copiers known as "remote interactive communication" or RIC. RIC is an expert system inside the copier that monitors the information technology controlling the machine and, using some artificial-intelligence techniques, predicts when the machine will next break down. Once RIC predicts a breakdown will occur, it automatically places a call to a branch office and downloads its prediction, along with its reasoning. A computer at the branch office does some further analysis and schedules a repair person to visit the site *before* the expected time of failure.

For the customer, RIC means never having to see the machine fail. For Xerox, it means not only providing better service but also having a new way to "listen" to our customer. As RIC collects information on the performance of our copiers—in real-world business environments, year in and year out—we will eventually be able to use that information to guide how we design future generations of copiers.

RIC is one example of how information technology invisible to the user is transforming the copier. But the ultimate conclusion of this technological transformation is the disappearance of the copier as a stand-alone device. Recently, Xerox introduced its most versatile office machine ever—a product that replaces traditional light-lens copying techniques with "digital copying," where documents are electronically scanned to create an image stored in a computer, then printed out whenever needed. In the future, digital copiers will allow the user to scan a document at one site and print it out somewhere else—much like a fax. And once it scans a document, a copier will be able to store, edit, or enhance the document—like a computer file—before printing it. When this happens, the traditional distinction between the copier and other office devices like computers, printers, and fax machines will disappear—leaving a flexible, multifunctional device able to serve a variety of user needs.

Xerox

Letter to a Young Researcher

When we hire someone at PARC, there is one qualification we consider more important than technical expertise or intellectual brilliance: intuition. A well-honed intuition and the ability to trust it are essential tools for doing the kind of research we do here.

Our approach to research is "radical" in the sense conveyed by the word's original Greek meaning: "to the root." At PARC, we attempt to pose and answer basic questions that can lead to fundamental breakthroughs. Our competitive edge depends on our ability to invent radically new approaches to computing and its uses and then bring these rapidly to market.

This is different from what goes on at most corporate research centers, where the focus is on improving current technology and advancing the status quo. If you take a job somewhere else, when you embark on a project you will probably have a pretty good idea of how and when your work will pay off. The problems you address will be well defined. You will help to improve computer technology state of the art by going one step farther along a well-plotted path.

If you come to work here, there will be no plotted path. The problems you work on will be ones you help to invent. When you embark on a project, you will have to be prepared to go in directions you couldn't have predicted at the outset. You will be challenged to take risks and to give up cherished methods or beliefs in order to find new approaches. You will encounter periods of deep uncertainty and frustration when it will seem that your efforts are leading nowhere.

That's why following your instinct is so important. Only by having deep intuitions, being able to trust them, and knowing how to run with them will you be able to keep your bearings and guide yourself through uncharted territory. The ability to do research that gets to the root is what separates merely good researchers from world-class ones. The former are reacting to a predictable future; the latter are enacting a qualitatively new one.

Another characteristic we look for in our research staff is a commitment to solving real problems in the real world. Our focus is on technology *in use*, and people here are passionate about seeing their ideas embedded in products that shape the way people work, think, interact, and create.

At Xerox, both corporate executives and research scientists are strongly committed to making research pay off. Over the last few years, new channels of dialogue have opened between research and other parts of the company. In particular, corporate strategy and research shape and inform each other. PARC's strategic role will undoubtedly be further strengthened by the emergence of digital copying and the company's new focus on documents of all kinds, whether in digital or paper form. The fusion of two previously separate Xerox businesses—information systems and copying—means that the company will be able to capitalize on PARC's expertise in ways it has been unable to do in the past.

This is an exciting time to be embarking on a career in systems research. New tools and technologies make it possible to deliver large amounts of computing power to users, and this increase in power opens up possibilities for using computation in new ways.

If you come to work here, you will sacrifice the security of the safe approach in which you can count on arriving at a predictable goal. But you will have an opportunity to express your personal research "voice" and to help create a future that would not have existed without you.

Sincerely,

John Seely Brown, Corporate Vice President
Frank Squires, Vice President, Research Operations

What is happening to the copier will eventually happen to all office devices. As computing power becomes ubiquitous—incorporated not only in copiers but also in filing cabinets, desktops, white boards, even electronic "post-it" notes—it will become more and more invisible, a taken-for-granted part of any work environment, much as books, reports, or other documents are today. What's more, increased computing power will make possible new uses of information technology that are far more flexible than current systems. In effect, technology will become so flexible that users will be able to customize it ever more precisely to meet their particular needs—a process that might be termed "mass customization."

We are already beginning to see this development in software design. Increased computing power is making possible new approaches to writing software such as "object-oriented programming" (developed at PARC in the 1970s). This technique makes it easier for users to perform customizing tasks that previously required a trained programmer and allows them to adapt and redesign information systems as their needs change. From a purely technical perspective, object-oriented programming may be less efficient than traditional programming techniques. But the flexibility it makes possible is far more suited to the needs of constantly evolving organizations.

Indeed, at some point in the not-too-distant future—certainly within the next decade—information technology will become a kind of generic entity, almost like clay. And the "product" will not exist until it enters a specific situation, where vendor and customer will mold it to the work practices of the customer organization. When that happens, information technology as a distinct category of products will become invisible. It will dissolve into the work itself. And companies like ours might sell not products but rather the expertise to help users define their needs and create the products best suited to them. Our product will be our customers' learning.

HARVESTING LOCAL INNOVATION

The trend toward ubiquitous computing and mass customization is made possible by technology. The emphasis, however, is not on the technology itself but on the work practices it supports. In the future, organizations won't have to shape how they work to fit the narrow confines of an inflexible technology. Rather, they can begin to design information systems to support the way people really work.

That's why some of the most important research at PARC in the past decade has been done by anthropologists. PARC anthropologists have studied occupations and work practices throughout the company—clerks in an accounts-payable office who issue checks to suppliers, technical representatives who repair copying machines, designers who develop new products, even novice users of Xerox's copiers. This research has produced fundamental insights into the nature of innovation, organizational learning, and good product design.

We got involved in the anthropology of work for a good business reason. We figured that before we went ahead and applied technology to work, we had better have a clear understanding of exactly how people do their jobs. Most people assume—we did too, at first—that the formal procedures defining a job or the explicit structure of an organizational chart accurately describe what employees do, especially in highly routinized occupations. But when PARC anthropologist Lucy Suchman began studying Xerox accounting clerks in 1979, she uncovered an unexpected and intriguing contradiction.

When Suchman asked the clerks how they did their jobs, their descriptions corresponded more or less to the formal procedures of the job manual. But when she observed them at work, she discovered that the clerks weren't really following those procedures at all. Instead, they relied on a rich variety of informal practices that weren't in any manual but turned out to be crucial to getting the work done. In fact, the clerks were constantly improvising, inventing new methods to deal with unexpected difficulties and to solve immediate problems. Without being aware of it, they were far more innovative and creative than anybody who heard them describe their "routine" jobs ever would have thought.

Suchman concluded that formal office procedures have almost nothing to do with how people do their jobs. People use procedures to understand the goals of a particular job—for example, what kind of information a particular file has to contain in order for a bill to be paid—not to identify the steps to take in order to get from here to there. But in order to reach that goal—actually collecting and verifying the information and making sure the bill is paid—people constantly invent new work practices to cope with the unforeseen contingencies of the moment. These informal activities remain mostly invisible since they do not fall within the normal, specified procedures that employees are expected to follow or managers expect to see. But these "workarounds" enable an all-important flexibility that allows organizations to cope with the unexpected, as well as to profit from experience and to change.

If local innovation is as important and pervasive as we suspect, then big companies have the potential to be remarkably innovative—*if* they can somehow capture this innovation and learn from it. Unfortunately, it's the rare company that understands the importance of informal improvisation—let alone respects it as a legitimate business activity. In most cases, ideas generated by employees in the course of their work are lost to the organization as a whole. An individual might use them to make his or her job easier and perhaps even share them informally with a small group of colleagues. But such informal insights about work rarely spread beyond the local work group. And because most information systems are based on the formal procedures of work, not the informal practices crucial to getting it done, they often tend to make things worse rather than better. As a result, this important source of organizational learning is either ignored or suppressed.

At PARC, we are trying to design new uses of technology that leverage the incremental innovation coming from within the entire company. We want to create work environments where people can legitimately improvise, and where those improvisations can be captured and made part of the organization's collective knowledge base.

One way is to provide people with easy-to-use programming tools so they can customize the information systems and computer applications that they work with. To take a small example, my assistant is continually discovering new ways to im-

How Xerox Redesigned Its Copiers

In the early 1980s, Xerox's copier business faced a big problem. Service calls were increasing, and more and more customers were reporting that our newest copiers were "unreliable." The complaints couldn't have come at a worse time. We had been late to recognize market opportunities for low- and mid-range copiers, and Japanese competitors like Canon were cutting into our market share. Now Xerox's reputation for quality was at stake.

After interviewing some customers, we discovered that unreliability was not the real problem. Our copiers weren't breaking down more frequently than before; in fact, many of the service calls were unnecessary. But customers were finding the copiers increasingly difficult to use. Because they couldn't get their work done, they perceived the machines as unreliable.

The source of the problem was our copier design. Traditionally, Xerox technology designers—like most engineers—have strived to make machines "idiot proof." The idea was to foresee in advance all the possible things that could go wrong, then either design them out of the system or provide detailed instructions of what to do should they occur.

But as we kept adding new functions, we had to add more and more information, usually stored on flip cards attached to the machine. The copiers became so complex that it was harder for the new user to figure out how to do any particular task. To learn a new operation meant a time-consuming search through the flip cards. And whenever something went wrong—a paper jam, say, or a problem with the toner—the machines would flash a cryptic code number, which would require more flipping through the cards to find the corresponding explanation.

In many instances, users would encounter some obstacle, not be able to find out how to resolve it, and simply abandon the machine in mid-procedure. The next user to come along, unaware of the previous problem, would assume the machine was broken and call a repair person.

We had to make radical changes in copier design, but it was difficult to sell that message within the company. The idea that there might be serious usability problems with our machines met with resistance in the Xerox development organization that designs our copiers. After all, they had tested their designs against all the traditional "human factors" criteria. There was a tendency to assume that any problems with the machines must be the users' fault.

When researchers from PARC began to study the problem, we discovered that the human-factors tests used by the development group didn't accurately reflect how people actually used the machines. So, a PARC anthropologist set up a video camera overlooking one of our new copiers in use at PARC, then had pairs of researchers (including some leading computer scientists) use the machine to do their own copying. The result was dramatic footage of some very smart people, anything but idiots, becoming increasingly frustrated and angry as they tried and failed to figure out how to get the machine to do what they wanted it to do.

The videos proved crucial in convincing the doubters that the company had a serious problem. Even more important, they helped us define what the real problem was. The videos demonstrated that when people use technology like a copier, they construct interpretations of it. In effect, they have a conversation with the machine much as two people have a conversation with each other. But our traditional idiot-proof design provided few cues to help the user interpret what was going on.

We proposed an alternative approach to design. Instead of trying to eliminate "trouble," we acknowledged that it was inevitable. So the copier's design should help users "manage" trouble—just as people manage and recover from misunderstandings during a conversation. This meant keeping the machine as transparent as possible by making it easy for the user to find out what is going on and to discover immediately what to do when something goes wrong.

Xerox's most recent copier families—the 10 and 50 series—reflect this new design principle. Gone are the flip cards of earlier machines. Instead, we include enough computing power in the machines to provide customized instructions on the display panel linked to particular procedures or functions. The information the user receives is immediately put in the context of the task he or she is trying to perform. The new design also incorporates ideas from PARC's research on graphical user interfaces for computers. When something goes wrong, the display panel immediately shows a picture of the machine that visually indicates where the problem is and how to resolve it.

The results of these changes have been dramatic. Where it once took 28 minutes on average to clear a paper jam, it takes 20 seconds with a new design. And because such breakdowns are easier to fix, customers are more tolerant of them when they occur.

prove the work systems in our office. She has more ideas for perfecting, say, our electronic calendar system than any researcher does. After all, she uses it every day and frequently bumps up against its limitations. So instead of designing a new and better calendar system, we created a programming language known as CUSP (for "customized user-system program") that allows users to modify the system themselves.

We've taken another small step in this direction at EuroPARC, our European research lab in Cambridge, England. Researchers there have invented an even more advanced software system known as "Buttons"—bits of computer code structured and packaged so that even people without a lot of training in computers can modify them. With Buttons, secretaries, clerks, technicians, and others can create their own software applications, send them to colleagues throughout the corporation over our electronic mail network, and adapt any Buttons they receive from others to their own needs. Through the use of such tools, we are translating local innovation into software that can be easily disseminated and used by all.

New technologies can also serve as powerful aids for organizational learning. For example, in 1984 Xerox's service organization asked us to research ways to improve the effectiveness of their training programs. Training the company's 14,500 service technicians who repair copying machines is extremely costly and time-consuming. What's more, the time it takes to train the service work force on a new technology is key to how fast the company can launch new products.

The service organization was hoping we could make traditional classroom training happen faster, perhaps by creating some kind of expert system. But based on our evolving theory of work and innovation, we decided to take another approach. We sent out a former service technician, who had since gone on to do graduate work in anthropology, to find out how reps actually do their jobs—not what they or their managers *say* they do but what they really do and how they learn the skills that they actually use. He took the company training program, actually worked on repair jobs in the field, and interviewed tech-reps about their jobs. He concluded that the reps learn the most not from formal training courses but out in the field—by working on real problems and discussing them informally with colleagues. Indeed, the stories tech-reps tell each other—around the coffee pot, in the lunchroom, or while working together on a particularly difficult problem—are crucial to continuous learning.

In a sense, these stories are the real "expert systems" used by tech-reps on the job. They are a storehouse of past problems and diagnoses, a template for constructing a theory about the current problem, and the basis for making an educated stab at a solution. By creating such stories and constantly refining them through conversation with each other, tech-reps are creating a powerful "organizational memory" that is a valuable resource for the company.

As a result of this research, we are rethinking the design of tech-rep training—and the tech-rep job itself—in terms of lifelong learning. How might a company support and leverage the storytelling that is crucial to building the expertise not only of individual tech-reps but also of the entire tech-rep community? And is there any way to link that expertise to other groups in the company who would benefit from it—for example, the designers who are creating the future generations of our systems?

One possibility is to create advanced multimedia information systems that would make it easier for reps and other employees to plug in to this collective social mind. Such a system might allow the reps to pass around annotated videoclips of useful stories, much like scientists distribute their scientific papers, to sites all over the world. By commenting on each other's experiences, reps could refine and disseminate new knowledge. This distributed collective memory, containing all the informal expertise and lore of the occupation, could help tech-reps—and the company—improve their capacity to learn from successes and failures.

COPRODUCING INNOVATION

Our approach to the issue of tech-rep training is a good example of what we mean by "pioneering" research. We started with a real business problem, recognized by everyone, then reframed the problem to come up with solutions that no one had considered before. But this raises another challenge of pioneering research: How to communicate fresh insights about familiar problems so that others can grasp their significance?

The traditional approach to communicating new innovations—a process that usually goes by the name of "technology transfer"—is to treat it as a simple problem of transferring information. Research has to pour new knowledge into people's heads like water from a pitcher into a glass. That kind of communication might work for incremental innovations. But when it comes to pioneering research that fundamentally redefines a technology, product, work process, or business problem, this approach doesn't work.

It's never enough to just *tell* people about some new insight. Rather, you have to get them to experience it in a way that evokes its power and possibility. Instead of pouring knowledge into people's heads, you need to help them grind a new set of eyeglasses so they can see the world in a new way. That involves challenging the implicit assumptions that have shaped the way people in an organization have historically looked at things. It also requires creating new communication techniques that actually get people to experience the implications of a new innovation.

To get an idea of this process, consider the strategic implications of an innovation such as digital copying for a company like Xerox. Xerox owes its existence to a particular technology—light-lens xerography. That tradition has shaped how the company conceives of products, markets, and customer needs, often in ways that are not so easy to identify. But digital copying renders many of those assumptions obsolete. Therefore, making these assumptions explicit and analyzing their limitations is an essential strategic task.

Until recently, most people at Xerox thought of information technology mainly as a way to make traditional copiers cheaper and better. They didn't realize that digital copying would transform the business with broad implications not just for copiers but also for office information systems in general. Working with the Xerox corporate strategy office, we've tried to find a way to open up the corporate imagination—to get people to move beyond the standard ways they thought about copiers.

One approach we took a couple of years ago was to create a video for top management that we called the "unfinished document." In the video, researchers at PARC who knew the technology extremely well discussed the potential of digital copying to transform people's work. But they didn't just talk about it; they actually acted it out in skits. They created mock-ups of the technology and then simulated how it might affect different work activities. They attempted to portray not just the technology but also the technology "in use."

We thought of the unfinished document as a "conceptual envisioning experiment"—an attempt to imagine how a technology might be used before we started building it. We showed the video to some top corporate officers to get their intuitional juices flowing. The document was "unfinished" in the sense that the whole point of the exercise was to get the viewers to complete the video by suggesting their own ideas for how they might use the new technology and what these new uses might mean for the business. In the process, they weren't just learning about a new technology; they were creating a new mental model of the business.

Senior management is an important partner for research, but our experiments at coproduction aren't limited to the top. We are also involved in initiatives to get managers far down in the organization to reflect on the obstacles blocking innovation in the Xerox culture. For example, one project takes as its starting point the familiar fact that the best innovations are often the product of "renegades" on the periphery of the company. PARC researchers are part of a company group that is trying to understand why this is so often the case. We are studying some of the company's most adventuresome product-development programs to learn how the larger Xerox organization can sometimes obstruct a new product or work process. By learning how the corporation rejects certain ideas, we hope to uncover those features of the corporate culture that need to change.

Such efforts are the beginning of what we hope will become an ongoing dialogue in the company about Xerox's organizational practice. By challenging the background assumptions that traditionally stifle innovation, we hope to create an environment where the creativity of talented people can flourish and "pull" new ideas into the business.

INNOVATING WITH THE CUSTOMER

Finally, research's ultimate partner in coproduction is the customer. The logical end point of all the activities I have described is for corporate research to move outside the company and work with customers to coproduce the technology and work systems they will need in the future.

It is important to distinguish this activity from conventional market research. Most market research assumes either that a particular product already exists or that customers already know what they need. At PARC, we are focusing on systems that do not yet exist and on needs that are not yet clearly defined. We want to help customers become aware of their latent needs, then customize systems to meet them. Put another way, we are trying to prototype a need or use before we prototype a system.

One step in this direction is an initiative of Xerox's Corporate Research Group (of which PARC is a part) known as the Express project. Express is an experiment in product-delivery management designed to commercialize PARC technologies more rapidly by directly involving customers in the innovation process. The project brings together in a single organization based at PARC a small team of Xerox researchers, engineers, and marketers with employees from one of our customers—Syntex, a Palo Alto-based pharmaceutical company.

Syntex's more than 1,000 researchers do R&D on new drugs up for approval by the Food and Drug Administration. The Express team is exploring ways to use core technologies developed at PARC to help the pharmaceutical company manage the more than 300,000 "case report" forms it collects each year. (The forms report on tests of new drugs on human volunteers.) Syntex employees have spent time at PARC learning our technologies-in-progress. Similarly, the Xerox members of the team have intensively studied Syntex's work processes—much as PARC anthropologists have studied work inside our own company.

Once the project team defined the pharmaceutical company's key business needs and the PARC technologies that could be used to meet them, programmers from both companies worked together to create some prototypes. One new system, for example, is known as the Forms Receptionist. It combines technologies for document interchange and translation, document recognition, and intelligent scanning to scan, sort, file, and distribute Syntex's case reports. For Syntex, the new system solves an important business problem. For Xerox, it is the prototype of a product that we eventually hope to offer to the entire pharmaceutical industry.

We are also treating Express as a case study in coproduction, worth studying in its own right.

The Express team has videotaped all the interactions between Xerox and Syntex employees and developed a computerized index to guide it through this visual database. And a second research team is doing an indepth study of the entire Xerox-Syntex collaboration. By studying the project, we hope to learn valuable lessons about coproduction.

For example, one of the most interesting lessons we've learned from the Express project so far is just how long it takes to create a shared understanding among the members of such product teams—a common language, sense of purpose, and definition of goals. This is similar to the experience of many interfunctional teams that end up reproducing inside the team the same conflicting perspectives the teams were designed to overcome in the first place. We believe the persistence of such misunderstandings may be a serious drag on product development.

Thus a critical task for the future is to explore how information technology might be used to accelerate the creation of mutual understandings within work groups. The end point of this process would be to build what might be called an "envisioning laboratory"—a powerful computer environment where Xerox customers would have access to advanced programming tools for quickly modeling and envisioning the consequences of new systems. Working with Xerox's development and marketing organizations, customers could try out new system configurations, reflect on the appropriateness of the systems for their business, and progressively refine and tailor them to match their business needs. Such an environment would be a new kind of technological medium. Its purpose would be to create evocative simulations of new systems and new products before actually building them.

The envisioning laboratory does not yet exist. Still, it is not so farfetched to imagine a point in the near future where major corporations will have research centers with the technological capability of, say, a multimedia computer-animation studio like Lucasfilm. Using state-of-the-art animation techniques, such a laboratory could create elaborate simulations of new products and use them to explore the implications of those products on a customer's work organization. Prototypes that today take years to create could be roughed out in a matter of weeks or days.

When this happens, phrases like "continuous innovation" and the "customer-driven" company will take on new meaning. And the transformation of corporate research—and the corporation as a whole—will be complete.

24

Marketing and Discontinuous Innovation: The Probe and Learn Process

GARY S. LYNN
JOSEPH G. MORONE
ALBERT S. PAULSON

The disappointing performance of U.S. firms during the 1980s in technology-intensive, global markets (such as consumer electronics, office and factory automation, and semiconductor memories) has been widely attributed to a failure to continuously and incrementally improve products and processes. In *The Breakthrough Illusion*, Florida and Kenney wrote that "The United States makes the breakthroughs, while other countries, especially Japan, provide the follow-through" on which competitive advantage is built.[1] Gomory made a similar point, contrasting "revolutionary" innovations with "another, wholly different, less dramatic, and rather grueling process of innovation, which is far more critical to commercializing technology profitably.... Its hallmark is incremental improvement, not breakthrough. It requires turning products over again and again, getting the new model out, starting work on an even newer one. This may all sound dull, but the achievements are exhilarating."[2]

In *The Machine that Changed the World*, the most influential work on the subject of the 1980s, Womack, Jones, and Roos measured the competitive effects of this lack of attention to continuous incremental improvement through a benchmarking study of the global automobile industry.[3] Other studies reinforce this message: compared to their Japanese competitors, U.S. firms lagged in cost, quality, and speed; and in large measure, the problem stemmed from a relative weakness in and lack of attention to the process of continuous incremental improvement.[4]

While it is only just becoming apparent in the academic literature,[5] widespread anecdotal evidence suggests that this decade of concern about incremental improvement has had a dramatic effect on the management practices of American firms. To be sure, many firms have struggled to move beyond the rhetoric of this process revolution, but there is little question that large numbers of firms in virtually every industry have been trying to simplify their business processes, develop and monitor metrics of performance, benchmark themselves against world-class competitors, and continuously improve. The Malcolm Baldrige Award winners have been the most visible examples, but others abound. Perhaps the most striking comes from the very industry that was so devastatingly criticized in *The Machine that Changed the World*. By adopting the lean production and short-cycle-time development practices pioneered by its Japanese competitors, Chrysler has dramatically reversed its fortunes. The war is by no means won, but acceptance of the importance of continuous improvement is widespread, and the methodologies for achieving it have been well documented. Success is uneven and incomplete, but more often because of resistance to organizational change than because of a lack of knowledge about the need for, and the methodologies of, continuous improvement.

DISCONTINUOUS INNOVATION

While continuous improvement is critical to competitiveness, and while the failure of American firms to follow up breakthroughs with such continuous improvements accounts for at least some of U.S. industry's past decline in technology-intensive markets, there is no denying the importance of the breakthroughs themselves—the more discontinuous innovations in process and product technology that lead to new businesses and product lines. A recent study of U.S. firms that have succeeded in head-to-head competition against Japanese firms in electronics-related markets, for example, found that the U.S. successes were built on a combination of discontinuous innovations and incremental improvements.

> These businesses built and renewed, and continue to build and renew, their competitive advantage through radical and generational innovations. They sustained that advantage over time through incremental product line improvements and extensions—but it is on the basis of the riskier, failure-laden, expensive and time-consuming efforts to pioneer new businesses and new generations of technology that their competitive advantage was and still is established.[6]

By the same token, the familiar argument that Japan's success in the 1980s derived from the ability to copy and incrementally improve on U.S.-based innovations is only partially correct. Sony has been more of a pioneer in consumer electronics than a continuously improving follower[7]; Toshiba established its leadership position in laptop computers by leapfrogging the dominant players in the desktop computer market[8]; and it would be difficult to argue that Honda has been a laggard in engine technology.[9]

In contrast, the recent history of some of the most widely respected and historically innovative U.S.-based firms is replete with examples of incremental improvements to existing product lines, but almost devoid of examples of successful new business and product line creation. IBM has been a master of continuous incremental improvement in its mainframe computers, but has never managed to duplicate its success in the many new businesses it has pursued. Xerox failed to take advantage of the extraordinary opportunity in personal computing created by its Palo Alto Research Center. RCA, for all the depth and breadth of its technological capabilities, somehow failed to capitalize on the wave of new business opportunities in consumer electronics. General Motors has been a follower in the implementation of new automotive technology even though it spends more on R&D than any other firm in the world. And GE failed in every one of its attempts in the early 1980s to grow new high-tech businesses from factory automation to computer-aided design to ceramic components.[10]

None of this is meant to suggest that continuous improvements are unimportant. But a careful reading of recent industrial history leads to the conclusion that in competitive, technology-intensive global markets, advantage is built and renewed through the more discontinuous form of innovation—through the creation of entirely new families of products and businesses. Continuous, incremental product line extensions and improvements are essential for maintaining leadership, but only after it has first been established through the more discontinuous form of innovation.

MANAGERIAL DECISION-MAKING AND DISCONTINUOUS INNOVATION

The management of discontinuous or radical innovation poses a unique set of challenges for management. It is a long (often more than a decade) and investment-intensive (often more than $100 million) process, marked by setbacks and unpleasant surprises, and with no guarantee of success. Its most persistent feature is high uncertainty. Typically, the technology is evolving, the market is ill-defined, and the infrastructure for delivering the still-developing technology to the as-yet-undefined market is nonexistent. The picture is further muddied by questions of timing—time required to develop the technology, time required for the market to emerge, time required for competing technologies to develop—and by exogenous factors such as government regulations over which the firm has little or no control. Most significantly, these uncertainties interact. The form the developing technology should take depends on how the developing market responds to early versions of the technology; yet paradoxically, how the market responds depends on the form the technology takes.

This long, uncertain, failure-laden process must be managed in the context of a U.S. financial environment that is still oriented toward short-term results. The benefits, if any, of concentrating on incremental improvements are quickly reflected in the bottom line. Pursuit of more discontinuous innovations has the opposite effect. In the short term, it hurts the bottom line. In the long term, it may or may not help (depending on how the uncertainties are resolved). So while there are clear financial incentives for pursuing incremental innovation, there appear to be equally strong disincentives against pursuing the next generation of products and businesses. All the near-term consequences of ignoring new business creation are positive; the negative consequences become apparent only over the course of a decade, and by then, today's manager may well have moved on to another position.

An important tradition of scholarship has explored the subject of discontinuous innovation from an industry-level perspective. The primary questions have centered around the impact of such innovations on industry structure,[11] and the role such innovations play in the evolution and renewal of industries.[12] Much less attention has been paid to the subject from a more micromanagerial perspective.[13] What are and what should be the actual practices employed in the management of discontinuous innovations? For example, how should managers go about developing an understanding of the markets for this class of innovations? How should performance be measured in these highly uncertain activities? How should investments be justified? How should opportunities be evaluated and screened? What are the appropriate organizational forms for developing discontinuous innovations? Who are the appropriate kinds of managers? A large and rapidly growing literature on new product development answers these questions for the more incremental forms of innovation. But discontinuous innovation is very different in character, and what may be sound practice for the development of incremental improvements may be inapplicable—or worse, detrimental—to the development of discontinuous innovations. Utterback succinctly makes the point:

> Following advice to be market driven in pursuing innovation, delighting one's customers through continuous improvement of products, and seeking out lead users may be powerful concepts for success or may pave the road to failure, depending on circumstances. These are good ... lessons to follow in promoting evolutionary change in well-understood product lines. But when applied to a discontinuity, they may lead a strong firm into a dangerous trap. Similarly, ideas such as lean manufacturing ... and mass customization ... may be thought of as a way to build core competence and to be highly successful in differentiating well-known products. But these concepts may lead to a dead end when radical change is in the wind.[14]

Many conventional new product marketing approaches fall short if applied to discontinuous innovations because of the pervasive uncertainties present.[15] The familiar admonition to be customer driven is of little value when it is not at all clear who the customer is—when the market has never experienced the features created by the new technology. Likewise, analytic methods for evaluating new product opportunities (e.g., discounted cash flow and market diffusion analyses) appear to be much more appropriate for incremental than for discontinuous innovation. The projections of future performance on which these analyses rest are predicated on assumptions about the future size and growth rate of the market, the pace and degree of success of the technology development, progress made by competitors and competing technologies, and the impact of exogenous variables like government regulations—all of which are highly uncertain in the discontinuous domain.

APPROACH

This article describes the first wave of findings in a family of empirical studies about the management practices associated with such discontinuous innovation.[16] We focus here on perhaps the most difficult and poorly understood task facing developers of potentially discontinuous innovations—the problem of developing an understanding of the market for these kinds of innovations. Despite the many uncertainties that defy the conventional techniques of analysis, the history of industrial innovation is filled with examples of successful discontinuous innovations. Xerography, the VCR, personal computers, engineering plastics, and optical fibers are among the classic examples, and there are many others. All had to overcome the triple-headed uncertainties about the market, the technology, and the timing. Their success begs an obvious question: How was an understanding of the markets for these discontinuous innovations developed? If the conventional techniques are unreliable in the face of the web of uncertainties, how did the innovating firms in these many instances of successful discontinuous innovation develop understanding of markets that did not yet exist, for products that had not yet been produced? Was it management by seat of the pants and trial and error, or was there an underlying method?

To investigate these questions, we developed a diverse set of in-depth, historical, case studies of successful discontinuous innovations. Our analysis was complicated by the inherently uncertain and stochastic nature of the discontinuous innovation process. A classic approach to the study of the management of innovation is to make pairwise comparisons between successful and unsuccessful projects. But with discontinuous innovations, the many and interrelated uncertainties make it difficult to adequately link management practice to project outcome—because of the uncertainties present, good management practice can still lead to unsuccessful projects.[17]

For this reason, we chose a fundamentally different sampling approach. Rather than comparing roughly similar pairs of successes and failures, we compared cases of success that were as different as possible. By limiting the sample to cases of success, we avoid the problem of misleading conclusions based on failure projects. By maximizing the differences among the cases, we control for idiosyncratic influences in each case, and provide at least some basis for generalization. If we uncover similarities in management practices across these successful projects despite the differences, we have reason to believe that a relationship exists between the practices and outcomes.

We examined four cases of successes—optical fibers (by Corning),[18] CT scanners (by GE),[19] cellular phones (by Motorola),[20] and NutraSweet (by Searle, now Monsanto).[21] These companies and products differ from each other in a number of respects including relative size, degree of diversification, type of products sold, market orientation, and type of production and technological environments.[22]

In this research, we completed a total of seventy-eight personal interviews with people both inside and outside the innovating companies. Individuals interviewed ranged from the original innovators who conceived the ideas to their CEOs.[23] What has emerged so far from our exploratory studies is the outline of a fundamentally different approach to new product development and marketing. There appears to be a considerable, if often implicit and intuitive, method to the process of developing an understanding of the markets for discontinuous innovations.

FINDINGS

Use of Conventional Market Research Techniques

Conventional market research techniques—such as concept testing, customer surveys, conjoint analysis, focus groups, and demographic segmentation—were employed in each of the four cases. With hindsight, it is clear that some of the information generated this way proved useful. But much of it was misleading and, most important for our purposes here, almost none had a significant impact on the development of these innovations. In no instance was it the critical factor in the decision-making associated with the development of these opportunities.

In 1967, for example, in the earliest stages of Corning's optical fiber program, Robert Maurer, leader of the R&D effort, interviewed potential customers. At the time, which was well before the deregulation of the American telephone system, no potential customer was more important than AT&T. In AT&T Bell Labs' view, the market for optical fibers would "occur sometime in the 21st century." Maurer met a similar reaction at ITT. As he describes:

> New York City had an awful time with their switching system. And you couldn't ever get a dial tone. These guys [ITT] were in New York and they were hopping mad. They said, "Why are you worried about something like that [optical fibers] when we can't call across town. This [communication congestion] is the kind of stuff people in communications should be working on—not 21st century technology."

As it turned out, the first major application for low-loss optical fibers was indeed telecommunications—but in the early 1980s, rather than in the twenty-first century. Fortunately for Corning, it ignored the "voice" of its "lead customers" and continued to develop optical fibers. MCI, a company that did not even exist when Maurer conducted his interviews, proved to be the pioneering customer.

A decade after Maurer's customer survey, and five years before its market breakthrough, Corning hired a well-known consulting company to assess different optical fiber applications; develop cost, pricing, and performance targets; and decide which markets to tackle first. The consultants employed a straightforward, conventional form of analysis, matching key attributes and benefits of optical fibers with potential market applications. They concluded that Corning should focus on a "patsy market,"[24] one with high information-carrying capacity needs, but where cost was not a major stumbling block. The most likely candidate at the time appeared to be computer local area networks (LAN). Once again, the conventional techniques pointed the company away from what proved to be the most significant market opportunity—long distance telephone lines.

In the early days of Computer Axial Tomography (CT), GE had a similar experience with lead customers. In 1973, as the CT market was just emerging and before GE launched its development effort, it invited eleven of the country's leading radiologists to the Bahamas for a seven-day focus group to determine possible applications for CT. The experts identified three areas—breast, heart, and head imaging—that seemed to offer some potential, but their overall assessment was that CT would only develop into a "small niche opportunity."

Two years later, with its development effort well underway, GE interviewed local radiologists in an attempt to discern which and how many hospitals were likely to buy CT systems. At the time, GE's CT technology was far from fully developed, but it was pursuing a concept with the potential for significant performance improvement. Tom Lambert, who was responsible for marketing the GE CT venture, recalls the radiologists' reaction:

> I'd say, "This [CT] machine is purported to do this and this." Their first question would be, "How much resolution does it have?" And when I told them it had 1/10th of what they were using now, that was the end of the story. It took me several months to figure out what the problem was. Their

> view was, "I've been taught this way in medical school and this is how you do it. It's always been done that way, always will be done that way. It works fine!" They recognized their problems as being adequately addressed by available technology, so they didn't see a need for a new technology and wondered why I was wasting their time.

Motorola employed conventional marketing tools during its development of cellular telephones. In the mid-1970s, it used conjoint modeling to try to segment potential customers. Again, with hindsight, it is clear that the results were misleading. The analysis identified several market segments—for example, construction contractors, people without a car, and people with several cars—as having particularly promising potential. We now know that the actual size of the segments that were identified pale in comparison with other segments that eventually emerged, such as salespeople and the general mass market.

Motorola also used demographic and psychographic segmentation, dividing the market into segments based on customer characteristics. A marketing group within Motorola combined a mail survey of several hundred thousand potential users in fifteen cities with focus group and census data to determine the thirty demographic segments most likely to be interested in the company's portable cellular phone. The thirty-first most important group turned out to be salespeople. Fortunately for Motorola, like Corning and GE, it did not base its decision-making on the results of these analyses.

Searle used concept testing—another conventional marketing technique—in an attempt to determine the actual product forms NutraSweet should take and the population segments that would be most interested in those forms. One product that was concept tested was a NutraSweet-sweetened cereal. The tests were conducted by both General Foods and Quaker Oats. Customers were given advertisements describing the benefits of aspartame (the generic name for NutraSweet), and then asked to taste cereal sweetened with sugar (the participants were told that the food or beverage was sweetened with aspartame). The concept tests indicated that consumers loved the product. The test results "were off the chart they were so good." Unfortunately, because of technical problems, neither cereal maker was able to deliver a commercially feasible aspartame-sweetened cereal to the market.

In sum, although all four companies employed conventional marketing techniques in the development of these discontinuous innovations, the techniques proved to be of limited utility, were often ignored, and in hindsight, were sometimes strikingly inaccurate. Walt Robb, who ran GE Medical Systems throughout the development of CT, explains:

> As far as I'm concerned, this [the history of CT] is simply an indictment of marketing in that it was not able to appreciate the value of this product. And it did not get any help from the customer, who didn't realize, until they really saw the clinical evidence and technical papers that started coming out, just how important this was going to be. Here [with discontinuous innovation] is where some of you people in academia should get out in the real world and see how loosey goosey all of this is—how imprecise these estimates have to be, how difficult it is to be very accurate. You can't forecast all the events that are going to influence the market size, like new technological breakthroughs, or Certificate of Need [i.e., new government regulations] or percent growth of GNP or a competitor having a new development, or at least announcing one.

THE PROBE AND LEARN PROCESS

If conventional market analysis did not guide the development of these discontinuous innovations, what did? These companies developed their products by probing potential markets with early versions of the products, learning from the probes, and probing again. In effect, they ran a series of market experiments—introducing prototypes into a variety of market segments. The initial product was not the culmination of the development process but rather the first step, and the first step in the development process was in and of itself less important than the learning and the subsequent, better-informed steps that followed. The approach at work in these cases might best be described as probing and learning, and it is surprisingly similar to the process described by Rosenbloom and Cusumano in their study of the VCR, and to the more general works on decision-making under uncertainty by Braybrooke and Lindblom, March and Olsen, Mintzberg et al., Quinn, Simon, and others.[25]

Probing

The first step in the probe and learn process is, in effect, to experiment—to introduce an early version of the product to a plausible initial market. For example, GE's first step into the CT scanning business in the mid-1970s was with a breast scanner, even though the market seemed to be moving toward whole body scanners. The breast scanner would be half the size of a whole-body scanner and would pose a less severe technical challenge at lower developmental costs. Ted Pashler, one of GE's development engineers, explains:

> Whether or not the breast machine would be a success was a minor point. We were committed to the fan beam [a new scanning technology], and knew it [the market for CT] would develop. We needed a technology vehicle to enable us to proceed and develop a system. We could learn what the limitations were even if the final application would turn out to be different. At no time did we look at the torso [whole body] market because of the long scan times. To do a torso, you needed scan times in the 10 to 12 second range [which exceeded what GE believed to be technically feasible in such an early phase of the development process].

The breast scanner failed in market trials. Images were poor and the scanner could not differentiate well between healthy breast tissue and a tumor. But by then, GE was applying what it had learned in the development of the breast scanner to subsequent versions of its CT system.

Motorola brought its first handheld portable cellular telephone to sample markets in Chicago, New York, and Washington in 1973, ten years before the first licensed, fully commercial systems were sold. The phones were too bulky and heavy to generate much interest, but they did serve their purpose: they demonstrated to the Federal Communications Commission that "someone other than AT&T" could produce a workable cellular phone system, and they taught the company more about the market and the needs of potential customers. Marty Cooper, leader of Motorola's cellular telephone development effort at the time and a key figure in the history of several of Motorola's successful discontinuous innovations, recalls that the first generation cellular phones

> weighed about two pounds, which was obviously too heavy. You could talk on the phone for a maximum of five minutes without your hand getting tired. We learned from people, such as congressmen who tested the unit, that the portables had to be smaller and lighter. From then, continuing through to today, we continue to reduce their size and weight.

The early stages in Corning's development of optical fibers and Searle's development of NutraSweet reveal a similar pattern. Soon after Corning stunned the scientific community in the fall of 1970 with the announcement that it had succeeded in producing low-loss optical fibers (an achievement that represented literally a ninety-eight order of magnitude improvement over the then available high-loss optical fibers), it began to explore the market for the technology in telecommunications. It quickly focused on the companies that were attempting to develop picturephones, the harbinger of today's emerging videophones. As Charles Lucy, who was responsible for the marketing of Corning's optical fiber project, remembers: "Once we reached the desired attenuation [i.e., level of light loss], and could sell kilometer lengths of a few dollars, the picturephone companies became interested in playing with it." We know with hindsight that the picturephone market—like the CT breast scanner market—failed to materialize.

Searle, meanwhile, launched several early probes of the amazingly sweet chemical compound called aspartame that one of its research scientists had accidentally stumbled on while conducting research on possible treatments for ulcers. Its efforts to apply the sweetener to breakfast cereals foundered because of technical difficulties. Working with General Foods, it learned that when sugar was removed from cereal, the cereal lost much of its bulk. Two other early probes—use of aspartame in carbonated beverages and as a "spoon for spoon" substitute for table-top sugar—failed because of difficulties with the U.S. Food and Drug Administration.

Learning

Probing with immature versions of the product only makes sense if it serves as a vehicle for learning about the technology, and whether and how it can be scaled up, about the market, and which applications and market segments are most receptive to the various product features, and about the influence of exogenous factors, such as changes in government regulations and the need for regulatory approvals.

Even though GE's CT breast scanner failed miserably in the marketplace, it did demonstrate the technical feasibility of the new technological approach (i.e., the so-called "fan-beam" system) that GE believed would lead to an order of magnitude improvement in scanning speed and greatly enhanced image resolution. GE followed its breast scanner with a fan-beam based whole-body scanner. This too failed in the marketplace, but again, GE's product development engineers gained valuable experience. Among other things, this first full body scanner demonstrated that the market was indeed interested in a fundamentally faster, higher resolution CT system. When GE announced its intention to market such a product, it was flooded with orders. (In this sense, the product announcement was itself a probe.) In June 1975, when it launched its effort to develop the whole body scanner, GE had projected that for 1976, it would sell a total of five body and breast scanners. Five months later, when it announced that it would be introducing this fundamentally better whole body scanner in 1976, it received sixty-five orders (at roughly $500,000 per order)! Unfortunately, GE also learned from the experience that most of those orders were from radiologists who would insist on a whole body scanner that could also scan heads. GE's system, when it was first introduced in spring of 1976, could not do so. It also suffered from some unanticipated technical difficulties, all of which meant that most of the orders GE had received in late 1975, when it announced the product, were canceled.

Motorola's product and marketing managers learned from its initial probe into the portable cellular telephone market that its product was too bulky and heavy. Size and weight became the driving performance parameters for the cellular phone venture for the next twenty years. Motorola's experience with cellular phones was strikingly similar to its earlier experiences with portable radios and pagers (discontinuous innovations in their own right at the times they were introduced). Bill Weisz, former CEO of Motorola and now vice chairman of the board, recalls:

> The first VHF [i.e., first pager] was too big, too heavy, but it was another step in the right direction. All the versions up to the PageBoy [i.e., the first truly successful pager, introduced a decade after the first VHF] were steps along the way. Each sold a bit more, each did a bit better, each taught us more about the marketplace.[26]

What Corning's Charles Lucy learned from the initial optical fiber probe was mostly discouraging. Once the picturephone market evaporated, he turned to long distance telephone lines as the next plausible application. Lucy expected the telephone companies to "stand in line for it." But the reaction from AT&T, which controlled 80 percent of the domestic phone lines, was that there would not be a market for the fibers for another thirty years—and that by then, AT&T would have developed its own.[27] The companies that were in the business of cabling telephone lines had a similar reaction. Eventually, Lucy and his colleagues concluded from this initial experience that Corning would have to, in effect, develop the market for its new product. They therefore began establishing development agreements with foreign cable companies, like Siemens, Pirelli,

BICC, and Furukawa. These companies agreed to develop the components and cabling technology needed to make the optical fibers operational, to assist Corning in developing the market in their respective countries, and to pay Corning an annual fee. In return, Corning would grant them licenses to produce and sell the fibers in their home country. Corning's hope was that by using joint development partners, they would help stimulate demand for optical fibers. But the company had to relinquish the rights to produce optical fibers in that country if and when the demand did emerge.

Searle's reaction to its experience with the initial NutraSweet probes was entirely different from Corning's, Motorola's and GE's. In all four cases, the initial experience was mostly if not entirely negative. The breast scanner was a complete market failure; the picturephone market never materialized and the telephone and cabling companies were uninterested; the early cellular phones were too big and bulky; and NutraSweet, after the initially promising concept tests, ran into a string of technical and regulatory difficulties. GE, Motorola, and Corning responded by pressing ahead with revised approaches. But, unlike the other three, Searle's reaction was to try to extract itself from the venture. John Searle, CEO of Searle during the early stages of the project, doubted the company's ability to surmount the technical and marketing challenges associated with NutraSweet. Nor was he convinced that the potential was worth the cost. Perhaps most importantly, he did not see how it fit Searle, which was a pharmaceutical company. By the early 1970s, he had concluded that it would make more sense to sell the rights to aspartame and use the royalties to fund drug-related R&D projects. In the end, he chose not to actively try to dispose of the rights, in large measure because of the enthusiasm of the project team. But a few years later, his successor, Donald Rumsfeld, did attempt to do so.

Iteration

Probing and learning is an iterative process. The firms enter an initial market with an early version of the product, learn from the experience, modify the product and marketing approach based on what they learned, and then try again. Development of a discontinuous innovation becomes a process of successive approximation, probing and learning again and again, each time striving to take a step closer to a winning combination of product and market.[28] Thus, when GE learned that its first whole body scanner—which followed the unsuccessful breast scanner—was badly flawed, it reacted with a "long period of problem solving" to overcome the deficiencies. But even the improved version of the body scanner, which GE introduced in late 1976, proved to be a disappointment in the marketplace. Rather than leapfrog the competition, the product was merely comparable to then existing CT systems in the all-important dimension of image quality; meanwhile, the leading competitors were announcing a new generation of products with better image quality than GE, but at comparable prices. At about the same time, the market for CT scanners suddenly collapsed, as a result of a change in federal government policy regarding health care cost reimbursements. The disappointing market reaction to GE's full body system triggered yet another step in the development process, which led to a substantially improved detector system. This in turn led to a new product, introduced in 1978, that was so successful and so clearly superior to the competition's that, despite the sharply contracting market, GE's market share soared from 20 percent to over 60 percent.

In 1975, Motorola followed its initial 1973 probe with a smaller, lighter, second-generation portable cellular phone. Field tests were conducted with both external users and Motorola employees. The internal tests proved to be especially useful in helping them develop a better understanding of the market requirements and potential. As Jim Caile, head of marketing for Motorola's cellular phone venture, noted:

> In 1977, I knew it would be a business. We had a test system in Schaumburg [Illinois—Motorola's corporate offices]. I was carrying my portable and was walking across our campus to the "tower" [a distant office building] to attend a meeting. My

> phone rang; it was my secretary informing me that the meeting had been canceled. I didn't have to walk all the way there. At that moment I knew that this was the greatest thing since sliced bread. This was as close to a sure thing as there is.

In 1979, the company developed its third prototype; and in 1981, its fourth, which was designed to be manufactured in mass quantities. Each generation represented a step forward in performance, based on lessons learned in the previous step. Each step also created greater market understanding and acceptance of a product idea that had initially been met by widespread skepticism both inside and outside the company. After more than a decade of development and several product and market iterations, the FCC finally granted the first commercial license for a cellular phone system in December 1983. Motorola was well prepared. Within a year, it had introduced a complete line of cellular products and had firmly established its leadership position.

A similar pattern emerges in the other two cases. For more than a decade after Corning's initially disappointing forays into the picturephone and long distance telephone markets, the company continued to explore possible applications for optical fibers. In the mid-1970s, it probed the cable TV market, which by 1977 was composed of approximately 3,600 cable television (CATV) systems serving 12 million subscribers. With hindsight, we know that Corning was about 20 years ahead of demand for this application of fibers. Corning also pursued military applications. The company won several modest contracts ($300,000 to $500,000) for development work of optical fibers, but was unable to land any orders for more than a few kilometers of fiber. In late 1978, the company identified long-haul trunking (i.e., long distance telephone lines) as its number one priority opportunity. The high information-carrying capacity of optical fibers seemed particularly well suited to this application, but the phone companies were still reluctant to incorporate optical fibers into their long-haul systems. So, for a while, Corning began to explore the short-haul, interoffice market. This too proved disappointing, which led Corning back to the long-haul market.

All the while, Corning's development team was learning how to scaleup and improve the process for making fibers. This was a matter of explicit philosophy. Dave Duke, who led the optical fiber effort at the time and who is now Corning's vice chairman explains:

> We were doing this in 1976 and 1977. We said we had to have a curve that would get us from two dollars a meter then, to ten cents a meter some time in the future. And then we would start talking about how we were going to do that ... . Just as important, maybe even more important, is what I call the other learning curve, which is performance against time. And in our case, it was things like attenuation and bandwidth. We said attenuation had to go from 5 db/km to .5 over some period of time. And bandwidth had to go from 200 megahertz to 400 megahertz, up to several gigahertz ... We didn't know exactly how to get there, but we knew we had to be 10 or 15 years into the future.[29]

The combination of the constant probing of markets with the constant improvement of the technology (in cost and performance) eventually paid off. In late 1982, shortly after the divestiture of AT&T and after more than ten years and $100 million dollars worth of development, Corning was approached by MCI about what proved to be the first significant order for optical fibers. Within a year, the long distance telephone market for Corning's fibers was exploding. Sales grew from 200,000 kilometers and $10 million in 1982 to 1.6 million kilometers and $220 million in 1986.

Searle, meanwhile, failed in its efforts to sell the rights to NutraSweet. The company was "stuck" with the product and so decided to aggressively probe a variety of product variants and markets. These included using NutraSweet as an ingredient in carbonated soft drinks, powdered drink mixes, chewing gum, candy, and dairy products (including ice cream), and using NutraSweet as a tabletop sweetener in a "spoon-for-spoon" substitute for sugar, a tablet, and a packet. Although Searle's expectations were highest for the carbonated beverage and spoon-for-spoon applications, both were initially denied approval by the FDA. The FDA did approve, in mid-1981, the use of NutraSweet in packets (equivalent to less than two teaspoons of sugar), tablets (for use in hot beverages like coffee and tea), chewing gum, and dry bases for beverages such as coffee, tea, gelatins, puddings, and dairy product substitutes (ice cream and imitation whipped toppings). Interestingly, even when Searle launched its sugar substitute packets ("Equal") in 1982—the first successful NutraSweet application—the market reacted very differently than what the company had expected. Marcel Durot, the manager of Equal, and his colleagues had presumed that Equal would sell as a substitute for saccharin. They learned however, that the market for Equal was not saccharin users, but rather dissatisfied sugar users. Although saccharin had a bitter aftertaste, saccharin users had become used to, and in fact preferred, the aftertaste. Based on this market experience, Searle's product development and marketing team changed Equal's marketing and positioning so that the product could directly address this new and entirely unanticipated target segment.

Shortly thereafter, sales of NutraSweet took off, growing to $74 million in 1982 and $336 million in 1983. For NutraSweet—like optical fibers, CT, and cellular phones—the path to success had been littered with false starts, frequent setbacks, unexpected and mostly unpleasant developments, and when things went well, partial successes at best. Success emerged out of this long and difficult process in large measure because each step brought with it new knowledge and experience that could be incorporated into the next step. Each new probe brought with it new learning. Figures 24.1*A* through 1*D* provide graphic representations of the probing and learning process for each of the cases. The contrast to the orderly world of conventional new product development is striking and is perhaps best summarized by Bobby Bowen, a key figure in GE's development of CT and now vice president and general manager of advanced technology, GE Medical Systems:

> Several cases have been written about the history of CT, but they don't describe anything that I recognize. They tend to project what ought to have been rather than was. There is a tendency to assume that a lot more occurred by planning than what actually occurred .... In fact, one thing tended to follow from the next. There were a lot of curves on the road that we hadn't anticipated. We took things as they came. A lot of people think of product development as involving a lot of planning, but I think the key is learning and an organization's ability to learn.[30]

DISCONTINUOUS PRODUCT DEVELOPMENT VS. CONVENTIONAL PRODUCT DEVELOPMENT

Precisely how does this probe and learn process differ from the conventional new product development process? Over the past twenty-five years, the conventional new product development paradigm has come to be viewed as a staged or multiphase process. Although the specific names used to describe each phase may vary, fundamentally an innovation passes through six basic steps: idea generation, screening and evaluation, selection, development, testing, and commercial launch (see Figure 24.2). In idea generation, companies try to formulate numerous possible concepts that offer potential. In screening and evaluation, the ideas generated are checked against development costs, market potential, and company strengths and strategies. In selection, a smaller subset of the ideas generated is culled for further development. In development, the company tries to fulfill in delivering the idea(s) selected. In testing, the prototype is checked for technical and market compliance. Finally, in commercial launch, substantial resources are marshaled to plan and implement full-scale introduction.

This multiphase process is analysis driven. In the early phases, techniques such as Delphi analysis, concept tests, focus groups, conjoint analysis, and quality function deployment (QFD) are used to aid companies in answering such questions as: what market to enter (people), what product to offer (product), at what price to sell the product (price), what promotional campaign to use to sell the prod-

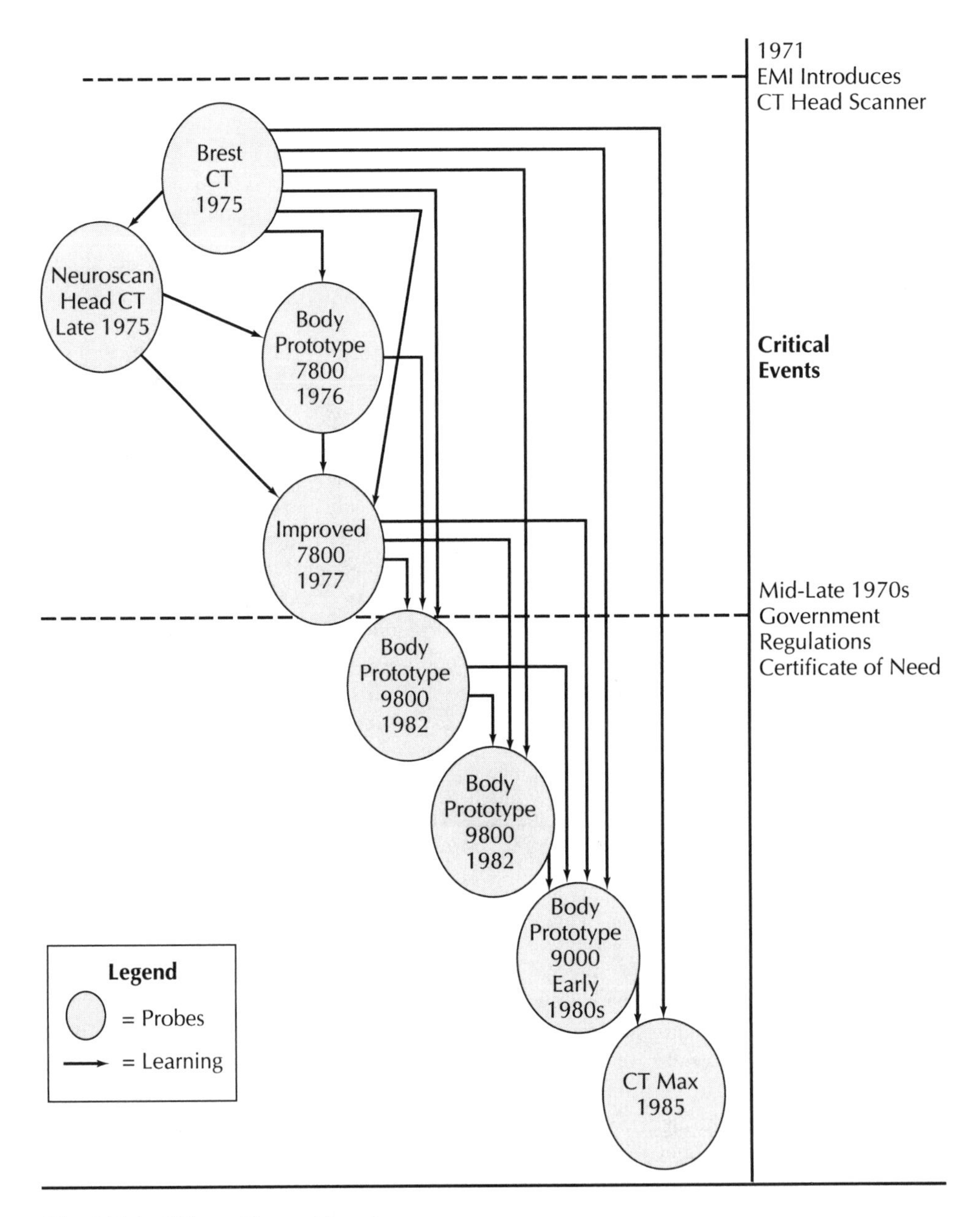

Fig. 24.1A. GE's probing and learning process.

uct (promotion), and what is the best way to move the product from producer to consumer (place). In answering these questions, marketing professionals seem to have adopted a "get it right the first time" (or very close to right) mentality. Hitting the "target" or "mark," and hitting it as soon as possible, has evolved into one of the primary objectives of marketing. The emphasis is on "target marketing," "rifle marketing," "demographic and psychographic segmentation," "market positioning," and

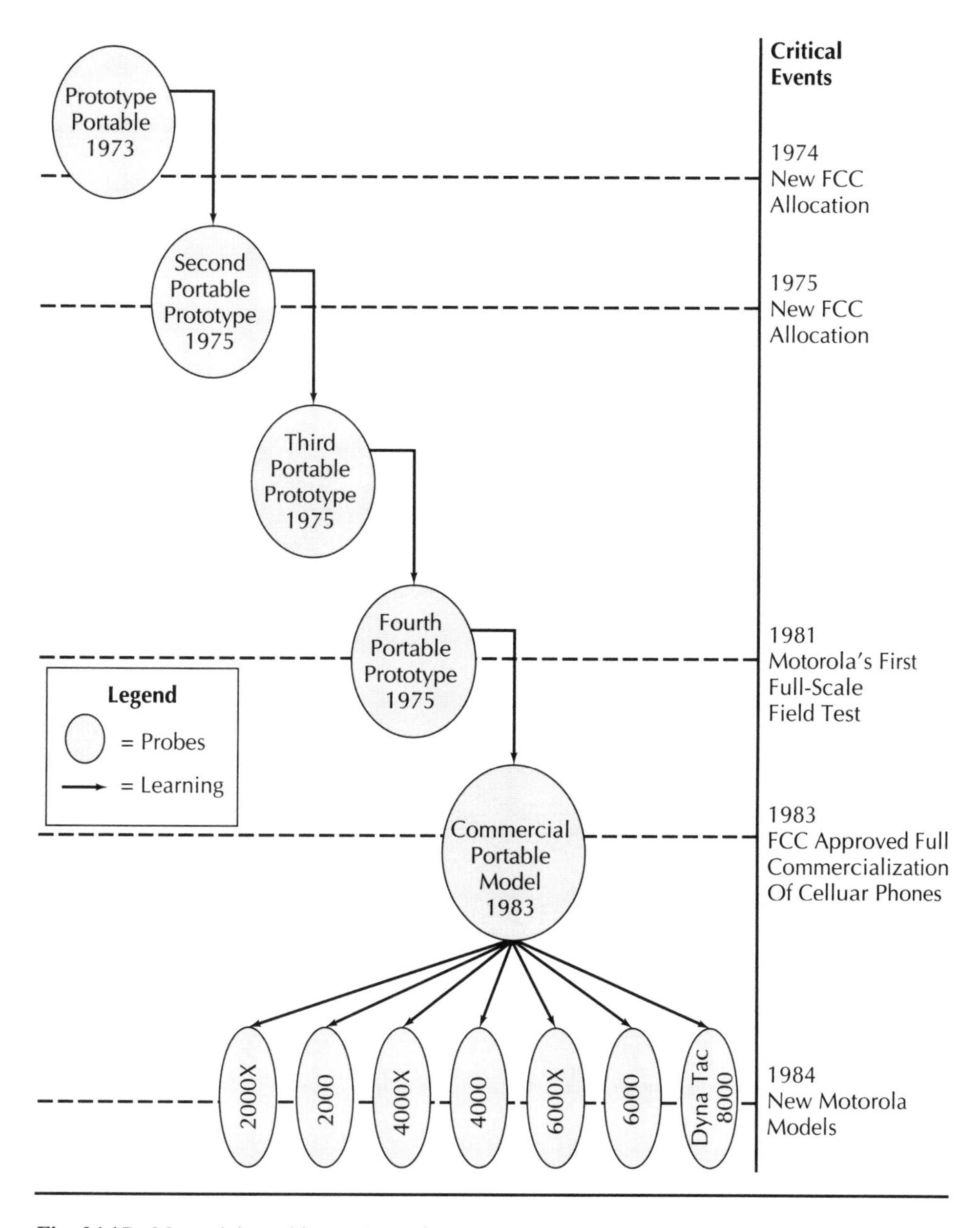

Fig. 24.1*B*. Motorola's probing and learning process.

"selective distribution." As Kotler explains, "The heart of modern strategic marketing can be described as "STP" marketing—namely, segmenting, targeting, and positioning."[31]

The discontinuous new product development process is also composed of multiple phases, but in other respects is fundamentally different from the conventional approach. Most importantly, the discontinuous process places much less emphasis on analysis and much more on probing and learning

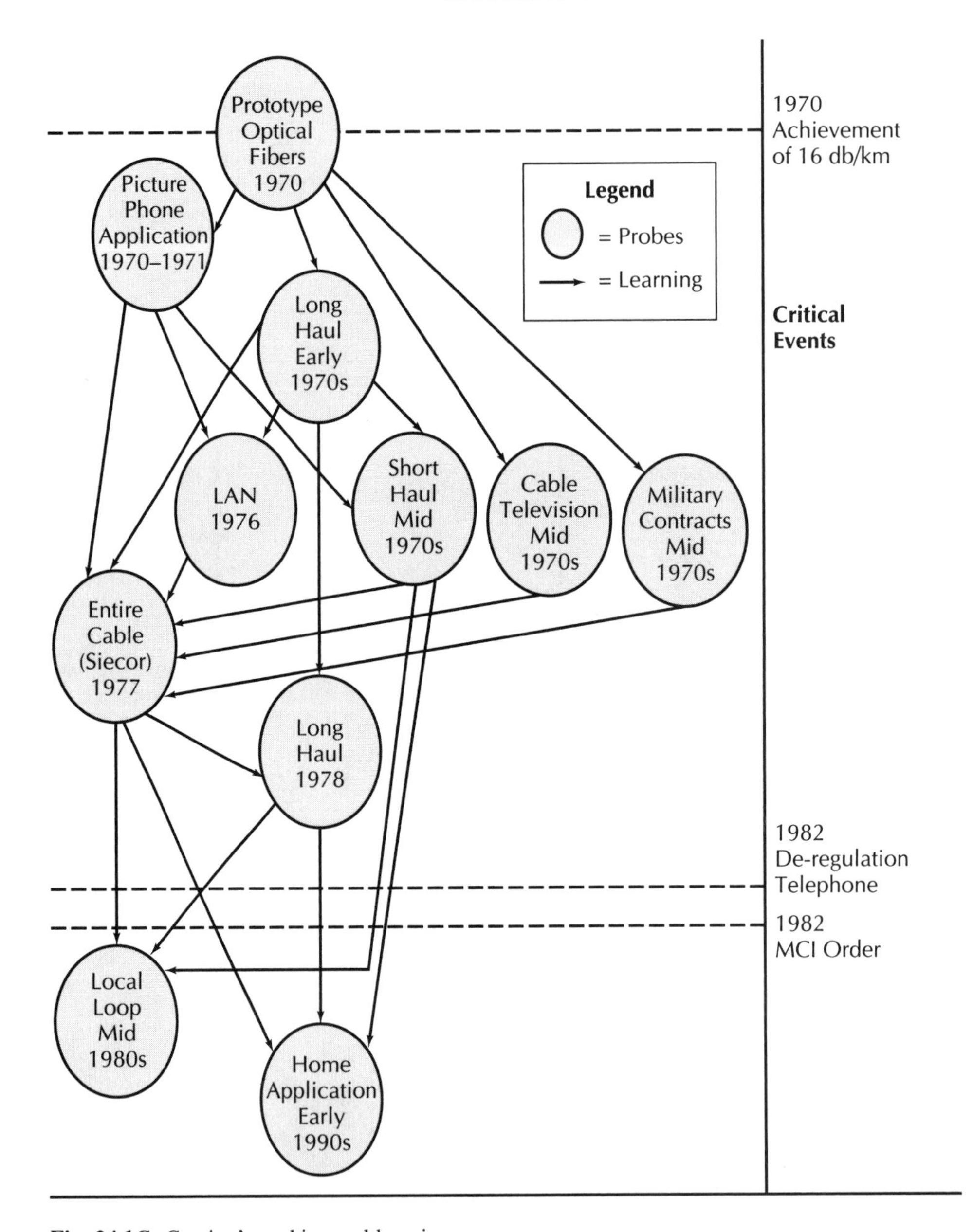

Fig. 24.1*C*. Corning's probing and learning process.

from the experience gained through each subsequent probe. The logic is far more experimental than it is analytic. This difference results from much higher levels of uncertainty associated with the discontinuous process. As we have discussed, and as our cases make abundantly clear, the market is ill-defined and evolving, the technology is ill-defined and evolving, and the two interact. As a result, and particularly in its early stages, it is virtually impossible to predict what product will even-

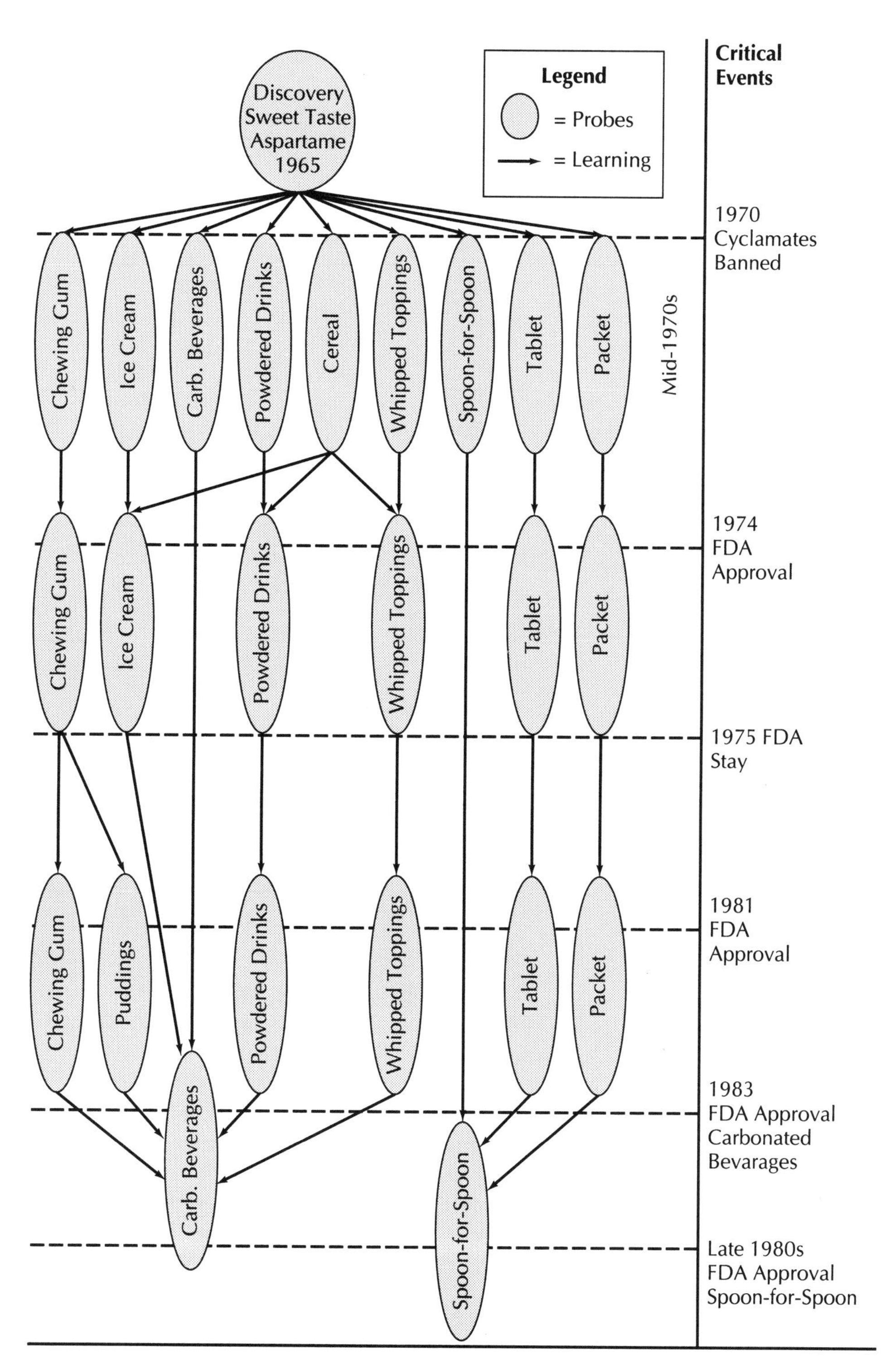

Fig. 24.1*D*. Searle's probing and learning process.

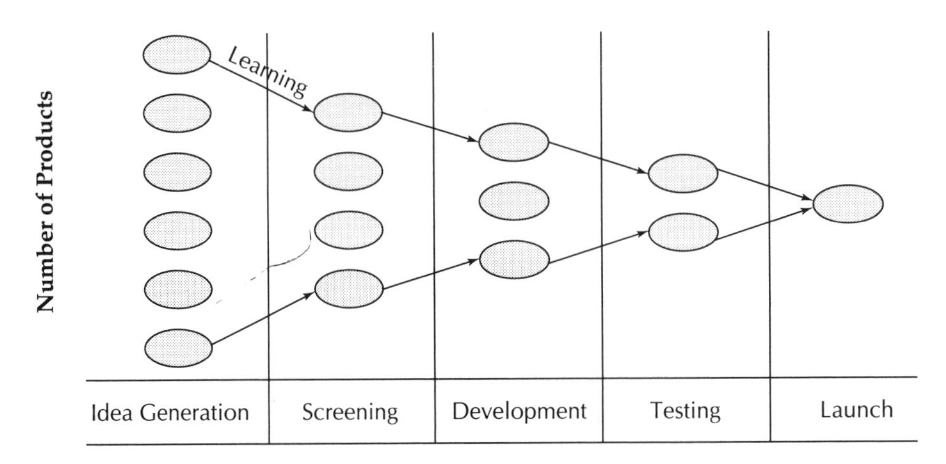

Fig. 24.2. The conventional new products.

tually be offered, at what price, to whom, when, and where. Indeed, because the process is so long and dynamic, the market and technology for the radical innovation may look entirely different at the end of the process than it did at the beginning. The product will have changed, the market will have changed, the competitive and regulatory environment will have changed, and all these factors will have interacted and influenced each other.

The net effect is that even the vocabulary used to describe the conventional process is inappropriate for discontinuous innovation. The concepts of "The Target," "The Finish Line," and even "The Launch" are misleading in the discontinuous realm. When marketing misses the "target," the marketing professional is thought to have failed. Yet missing the target is precisely what must be expected in the process of probing and learning. The ultimate target is not even known in the early stages of the probing process. Indeed, one way to view the process is as a vehicle for identifying the target.

Similarly, the idea of the "launch"—the moment of product commercialization that is commonly associated with conventional innovation—is also inappropriate in the discontinuous realm. There is no single "launch" in the discontinuous process. There are successive launches over many years. Each launch leads to a modification of the target. The finish line, in other words, keeps moving!

The discontinuous innovation process is thus a process of successive approximations that has to be managed not through analysis, but by experimental or quasi-experimental design. The key questions become not what is the right product, but rather, what steps can we take that will generate maximum information about the product and market and how do we incorporate that information into our product development. The logic is not to strive to get "it" right, but to strive to maximize learning.

PROBE AND LEARN VS. CONVENTIONAL MARKET TESTING

How does the probe and learn process differ from conventional market testing? The probe and learn process is used as a vehicle for gaining insight into what target markets to pursue, which technologies to use, and what features and benefits to incorporate. It is deployed during the front end of a new product effort and generates information to help direct or redirect a firm's development efforts. Market testing, on the other hand, occurs late in the new product development process, after a product has undergone all the necessary pretesting—just prior to full-scale introduction. As

Klompmaker et al. explain, "You should run a test market only after extensive pretesting has shown that your new product will be a winner."[32] Typically, it is applied to inexpensive consumer nondurables such as toothpaste, toilet tissue, pet foods, cigarettes, ball-point pens, and cereal[33] and is used in the development of advertising, roll-out and pricing strategies, and refinement of sales forecasts. It also serves as a final screen in the development process. Products that fare badly in a market test can be canceled. In striking contrast, "probes" that fare badly, like the early GE CT scanners, serve as vehicles for generating information for subsequent iterations.[34]

PROBE AND LEARN AND THE TECHNOLOGY CYCLE

It is illuminating to consider the probe and learn process from the perspective of the body of scholarship on technology cycles and dominant designs.[35] These four cases illustrate the patterns of innovation in the earliest stage in the technology cycle—the *pre*dominant design phase. The long and painful iterative process described in these cases is strikingly similar to what Anderson and Tushman characterize as the period of technological ferment.[36] And in each case, a new period of rapid market acceptance and growth is ushered in by the introduction of what in hindsight proved to be a dominant design. Corning, for example, probed and learned for over a decade, with very limited sales. But its optical fiber business, and the market in general, exploded with the introduction in 1983 of Corning's single-mode fibers for MCI. Precisely the same pattern occurred in the CT case, with a painful period of thrashing about, followed by rapid growth with the introduction of GE's 8800.

This leads to a closely related point. We studied how probing and learning occurs within the context of individual firms—how one firm learns from its own probes. Probing and learning also transpires across firms. The history of the personal computer and computer workstation markets are examples. Wozniak and Jobs's Apple I "computer" (a printed circuit board consisting of a microprocessor, 18 kilobytes of memory, and a number of "off-the-shelf" integrated circuits) was a refinement of earlier probes made by Altair and Call Computer.[37] Similarly, Sun Microsystems was able to use off-the-shelf components, including a Motorola CISC processor, primarily because of what its engineers had learned from earlier probes by Apollo and LISP.[38] Sun learned from the market reactions to their competitors' products and championed the open architecture concept, a technology that enabled the company to overtake Apollo, the market pioneer.

Such cross-firm probing and learning is also perfectly consistent with the dominant design literature. Sun's workstation and Apple's PC proved to be the dominant designs of the time. The point worth making is that the predominant design pattern of probing and learning that so often occurs across firms is essentially the same pattern that we have documented in our cases. The only difference is that our cases explored innovation practices within established companies that had the resources to probe and learn internally (as well as to learn externally from their competitors' probes).

THE STRATEGIC CONTEXT FOR PROBE AND LEARN

In three of our four cases, this experimental probe and learn process occurred within a well-articulated strategic context. CT, Cellular Telephones, and Optical Fibers were not just new product opportunities—they were new product opportunities that were strategically central to their businesses. Development of cellular telephones began soon after Motorola had come to view itself as having a strategic focus in mobile and portable communications—not in communications per se, or in radio-based communications, but in communications with "people and machines on the move." The cellular telephone venture followed a series of earlier discontinuous innovations in mobile and portable communications—first mobile radios, then portable radios, then pagers—and emerged as an opportunity just as the champions of these earlier efforts had risen to leadership positions in Motorola's communications business.

Likewise, Corning viewed (and still views) itself as a glass and ceramics-based company. With some notable exceptions in the late-1960s and 1970s, it has throughout its history focused on opportunities in which its glass and ceramics technology capabilities could be uniquely applied—from silicones and fiberglass, to light bulb and TV bulb housings, to Pyrex and Pyroceram (i.e., Corningware), to cellular ceramic substrates for catalytic converters and high-performance glass for liquid crystal displays. Optical fibers were thus one in a string of opportunities growing out of Corning's base technology. Moreover, optical fibers emerged as an opportunity at a time when the company was in desperate need of new sources of growth, which of course heightened the strategic importance of the venture. And so, even though Corning was in serious financial difficulties in the mid-1970s and even though there was considerable market skepticism and middle management resistance to optical fibers, the CEO of Corning, Amo Houghton, supported the costly, slow, painful development. Like cellular telephones, Corning's venture took more than a decade and $100 million to bring to fruition. In the words of one of the key figures in the history of optical fibers:

> [Houghton] was after us all the time to keep pushing. He wasn't worried about whether it was going to be a big deal. The only question was when. He wanted us to make it faster, but frankly, if it didn't happen until the end of the century for natural market reasons, he would understand that—as long as when it came, we'd be the leader.[39]

We see a similar sense of strategic centrality in the CT case. For the ten years preceding the CT venture, GE's Medical Systems business (GEMS) had struggled unsuccessfully to diversify from its traditional X-ray equipment business to a broad array of medical equipment businesses. It had given up on that effort and narrowed its focus back onto its core X-ray business just as the CT opportunity was emerging. Initially, GEMS declined to get involved, but then it became apparent that CT was not only growing explosively, but directly cannibalizing the core X-ray business. Here was an emerging technology that threatened the core of the business. Failure in this new market would be tantamount to failure of the business. Each adjustment to each new setback, each crash effort to solve each new crisis, was thus propelled by a sense of strategic urgency.

The one exception is Searle and NutraSweet. It is difficult to see how this was strategically central in any respect other than that Searle needed new growth opportunities. This was not a pharmaceutical product, even though it did require FDA approval. And because NutraSweet was tangential to the company's core business, senior management lost patience with it and tried to sell it on several occasions. In the end, Searle persisted not out of a sense of strategic urgency, but out of the failure to find a single buyer! The story of NutraSweet is much closer to a case of pure opportunism than strategic necessity.

The important point here is that the probe and learn process occurs within a strategic context—within an envelope of experiences, capabilities and competitive pressures that have a critical impact on the shape and outcome of the process. The probe and learn model suggests that the way to determine if and how to pursue a new business opportunity is to pursue it—to take a step into the market with an early version of the product, gain experience about both the technology and the market, and then modify the product and approach to the market based on that experience. New product and business development becomes a serial, iterative process, with each subsequent step building on the experiences—both positive and negative—gained from the previous step. Unfortunately, this is an extraordinarily time- and resource-consuming way to reduce uncertainty. No firm can afford to explore even a fraction of the available opportunities in this fashion. Somehow, management has to distinguish between opportunities that are worth probing about, and persisting with, and those that are not. The strategic context for the probe and learn process shapes this decision-making.

For example, to most observers in the mid-1970s, optical fibers appeared to be a remote, highly uncertain possibility. But for Corning, while the business sometimes seemed far off, it was of such direct strategic importance that even if it might not be successful for decades, CEO Houghton believed it was imperative that Corning lead the way. And the fact that it built on past experiences of Corning had a substantial impact on the way that Corning went about pursuing the opportunity. It built on an unsuccessful venture in glass lasers, a business of long standing in fused silica, and a great deal of knowledge in optical glass, which led Corning to conclude very early in the process that it would need to find a fundamentally different approach to achieving low-loss fibers than that being pursued by virtually all the other contenders.

Likewise, the "mobile and portable" religion that had emerged in Motorola heavily shaped how the cellular telephone opportunity was viewed. This was their kind of business. There was a great deal of skepticism during the 1970s about the cellular opportunity—particularly about portable cellular phones. In the words of former CEO John Mitchell, "Nobody believed us, especially the competition. They didn't believe the world was waiting for a portable. They believed the world was waiting for a mobile phone, because that's what AT&T was developing." But because of Motorola's project and marketing managers' earlier experience with portable radios and pagers, they intuitively knew cellular phones would sell.

The strategic context plays an especially important role in fueling persistence in the pursuit of these opportunities. It took Motorola fifteen years and $150 million to bring cellular telephones to market. It took Corning nearly as long and required nearly as much investment to take optical fibers to market. And while CT took considerably less time, it was a remarkable roller-coaster ride, with one setback following another. It is tempting to attribute the persistence in the face of such overwhelming difficulties, of daunting market and technical uncertainties and an ever-rising tide of internal resistance, to the vision of extraordinary individuals. But a more accurate interpretation is that senior management persisted because these opportunities made strategic sense. They fit the strategic focus of these businesses. What to the outsider appears to be an act of extraordinary risk taking or uncommon vision appears to those operating within the strategic context as good strategic sense. For Walt Robb, there was no choice but to persist with the CT effort, despite the long string of setbacks. If the controversial fan-beam approach to CT had not been successful in the end, Robb said, "we would have shifted to the fourth generation, the contending approach being pursued with some success by the competition. You just never give in. You just *have* to make it one way or another."

The implication of these experiences is that unless the opportunity for a discontinuous innovation is strategically central to the business, management will fail to muster the staying power to persist and learn from the years of twists and turns and unpleasant surprises. Unless the opportunity is strategically central, the inevitable setbacks will be interpreted as justification for disengagement rather than as springboards for new efforts. For a pharmaceutical company seeking to diversify into medical equipment in the mid-1970s, for example, the CT mammography failure would probably have been seen as just that—a failure. In contrast, not only did GE's management feel it had no choice but to continue in its pursuit of CT, it also learned important lessons from the mammography failure about both the technology (it was feasible) and the market (GEMS was aiming at the wrong target). Persistence was fueled by strategic imperative.

The NutraSweet case provides an interesting perspective on this point. As the inevitable setbacks were encountered, senior management began to question the wisdom of the venture. It had good potential, but it was not relevant to the core of the business. And thus two successive CEOs attempted to extract themselves from the venture—and even after Searle was acquired by Monsanto in the early 1980s, Monsanto also tried to sell NutraSweet for the same reason. What began as an interesting opportunity fueled by an ill-defined pursuit of new opportunities, continued in large measure because the company could not find a buyer for it.

CONCLUSION AND MANAGERIAL IMPLICATIONS

The purpose of this article was to investigate the front-end or opportunity definition phase of discontinuous new product development. Our goal was to explore how companies develop an understanding of the markets for discontinuous innovations. Our approach was to analyze a diverse set of successful cases. What we uncovered was a process of probing and learning. Rather than analyzing the market and selecting the best alternative, they develop their products through successive approximations—introducing an early version of the product to an initial market, learning from the experience, modifying the product and approach to the market based on that learning, and trying again, this time with somewhat better information and understanding and somewhat lower uncertainty.

It should also be evident from these cases that probing and learning is anything but a process of blind trial and error—it is a process of experimental design and exploration that must take place within a context of strategic relevance to the innovating firm.

This analysis raises more questions than it puts to rest. In particular, it suggests the need for development of an entirely different new product development methodology—one based on the logic of experimentation. For discontinuous innovation, the emphasis in marketing must shift from the application of well-known analytic techniques to the development of measures that ensure proper quasi-experimental design and adequate control. Proper design can be achieved by first observing a site or customer before probing, introducing the probe, and then observing the effect on the site/customer.[40] Adequate control comes from selecting sites where it is possible to specify the who, when, and what: "who" will use the discontinuous innovation—so adequate training can be provided; "when" will the innovation be used—so company personnel can be on hand to observe the experiment; and "what" will the innovation be used for—so that an effective application and protocol can be specified.[41]

The pertinent questions for further research become:

- How should companies select their probes?
- How can they minimize the cost of probing?
- How can companies maximize learning within and between probes?
- How can companies compress the time of the process, and is it possible to compress it without making it hopelessly expensive (such as by using concurrent engineering)?
- Are there times when probing outside the company's strategic context makes sense?
- What organizational and managerial mechanisms stimulate (and what mechanisms inhibit) learning?

Answers to these questions all require further analysis. This project is just the first phase in a longer-term family of studies. For now, one clear conclusion is evident: the process of understanding markets for discontinuous innovation is fundamentally different—far more experimental, far less analytic—than the conventional process.

Notes

We would like to thank the following people for their input on an earlier draft of this manuscript: Mel Adams and William Souder at the University of Alabama in Huntsville, Barry Bayus at the University of North Carolina at Chapel Hill, Arthur Shapiro at Stevens Institute of Technology, Steven Schnaars at Baruch College, and Robert Thomas at Georgetown University.

1. R. Florida and M. Kenney, *The Breakthrough Illusion* (New York: Basic Books, 1990), p. 8.

2. Ralph Gomory, "From the 'Ladder of Science' to the Product Development Cycle," *Harvard Business Review* (Nov./Dec. 1989), p. 100.

3. J. Womack, D. Jones, and D. Roos, *The Machine That Changed the World* (New York: Rawson Associates, 1990).

4. For example, see M. Dertouzos, R. Lester, and R. Solow, *Made in America: Regaining the Productive Edge* (Cambridge, MA: MIT Press, 1989); D. Kash, *Perpetual Innovation: The New World of Competition* (New York: Basic Books, 1989).

5. For example, see Noriaki Kano, "A Perspective on Quality Activities in American Firms," *California Management Review*, 35/3 (spring 1993): 12–31.

6. Joseph Morone, *Winning in High-Tech Markets* (Cambridge, MA: Harvard Business School Press, 1993), p. 217.

7. R. S. Rosenbloom and M. A. Cusumano, "Technical Pioneering and Competitive Advantage: The Birth of the VCR Industry," *California Management Review*, 29/4 (summer 1987): 51–76.

8. P. Abetti, "Toshiba Laptop," Working Paper, Rensselaer School of Management, Troy, NY, 1993.

9. Henry Mintzberg and James Brian Quinn, *The Strategy Process*, 2nd edition (Englewood Cliffs, NJ: Prentice-Hall, 1991), pp. 284–304.

10. Peter Petre, "How GE Bobbled the Factory of the Future," *Fortune*, Nov. 11, 1985, p. 52.

11. See W. Abernathy and K. B. Clark, "Innovation: Mapping the Winds of Creative Destruction," *Research Policy*, 4 (1985): 3–22; P. Anderson and M. Tushman, "Technological Discontinuities and Dominant Designs: A Cyclical Model of Technological Change," *Administrative Science Quarterly*, 35 (1990): 604–633.

12. See, for example, W. J. Abernathy and J. M. Utterback, "Patterns of Industrial Innovation," in *Readings in the Management of Innovations*, 2nd edition (New York: Ballinger Publishing Co., 1988); R. N. Foster, "Linking R&D to Strategy," in R. A. Burgelman and M. A. Maidique, eds., *Strategic Management of Technology and Innovation* (Homewood, IL: Richard D. Irwin, Inc., 1988), pp. 161–172; J. Utterback and W. Abernathy, "A Dynamic Model of Process and Product Innovation," *Omega*, 33 (1975): 639–655.

13. A notable exception is Rosenbloom and Cusumano, op. cit. See also Joseph Morone, *Winning in High-Tech Markets* (Cambridge, MA: Harvard Business School Press, 1993).

14. James Utterback, *Mastering the Dynamics of Innovation* (Cambridge, MA: Harvard Business School Press, 1995), p. 211.

15. Gary Lynn, "Understanding Products and Markets for Radical Innovations," Ph.D. dissertation, Rensselaer Polytechnic Institute, Troy, NY, 1993, pp. 15–62.

16. The studies are being conducted at the Lally School of Management and Technology in collaboration with the Industrial Research Institute and are now being funded by the Sloan Foundation.

17. The reverse is much less likely to be true. Given the low probability of success in the best of circumstances, the likelihood of successful outcomes on badly managed discontinuous projects is remote.

18. Optical fibers is a technology that allows information, in the form of light, to be sent long distances through cable. Optical fibers have many uses with their primary application in telecommunications. An optical fiber is thinner than fishing line, but has significantly more capacity than copper wire. Standard copper cables can carry 16 megabytes of information (voice, data, or visual image); optical fibers can handle over eighteen times that amount. One pound of fiber cable can carry as much information traffic as ten tons of copper cable. Also, because optical fibers do not rely on electrical current, they will not short, are immune to electromagnetic interference, and are more secure and less likely to be tapped than copper wire.

In 1966, the British Post Office (operators of the British telephone system) asked Corning to develop optical fibers for use in the telephone company's picturephone venture. In the fall of 1970, Corning stunned the technical community with its announcement that it had succeeded in making suitable optical fibers. But this proved to be only the beginning of the development of the technology. It was not until 1982—after more than a decade and $100 million of development—that Corning landed the first significant sale of optical fibers. The divestiture of AT&T earlier that year played a key role, as Corning's first major customer was MCI, a new entrant into the long distance telephone business. MCI placed an order for $90 million of Corning's optical fibers. Sales took off rapidly from there and, by 1986, Corning's annual fiber sales contributed an estimated $75 million to Corning's operating income.

19. Computer Axial Tomography (CT) is an X-ray based technology that generates cross-sectional images of the anatomy, as opposed to the two-dimensional images of conventional X-rays. The concept of CT is to send X-ray beams through a patient taking thousands of pictures of the area to be examined. The final CT cross-sectional "picture" allows doctors to discern tumor location, size, and other characteristics that previously could have only been obtained through exploratory surgery.

EMI, a British company, pioneered CT in the late 1960s and early 1970s. In January 1975, GE launched a crash effort to develop a CT system that would leapfrog EMI and others by using an entirely different technological approach, called "fan beam." After four years and many setbacks, GE introduced its fan-beam system. By 1978, GE captured 60 percent of the CT market, and dominated the CT market thereafter. By 1982, worldwide orders grew to $400 million.

20. Conventional mobile telephone systems available in the 1960s used a single high-powered transmitter to connect users within areas as large as 50 miles. With a cellular telephone system, a concept developed by AT&T, a service area could be broken into much smaller "cells," a few miles apart. This was accomplished by using lower power transmitters and sophisticated computers that would switch subscribers to another channel when they entered or left cell. The previous channel would then be free to be used by another subscriber. The result of a cellular network was that many more users could be served within a given geographic area with a fixed number of channels.

In the 1970s, AT&T proposed to the FCC a mobile cellular telephone. Motorola pursued a leapfrog strategy and developed a portable cellular telephone. The venture was very costly for the company. It took Motorola over $150 million and more than a decade in development before the business took off. In 1984, the first full year that the phones were nationally available and eleven years after the company field-tested its first portable cellular phone, Motorola's cellular telephone sales reached $200 million (although the business lost $50 million that year). By 1990, the company was well established as the leader in the field; cellular sales had reached $2 billion, and profits exceeded $300 million.

21. NutraSweet tastes remarkably similar to regular sugar—without an aftertaste. But it is approximately 200 times sweeter. As a result, it can sweeten with fewer calories. A teaspoon of sugar contains 16 calories, an equivalent amount of NutraSweet contains only 0.1 calorie.

NutraSweet was discovered by Searle accidentally. Just before Christmas of 1965, James Schlatter, a research chemist working at Searle on an ulcer treatment, licked his fingers and noticed an amazingly sweet taste. He discovered that L-aspartyl-L-Phenylalanine methyl ester, also known as aspartame (or NutraSweet), was extraordinarily sweet. Searle scientists worked on the NutraSweet project for sixteen years before it was able to be sold in this country. The first NutraSweet product was approved by the FDA in 1982. By the end of 1983, Searle's NutraSweet sales had reached $336 million.

22. Of the companies sampled, GE is the largest, Motorola is an intermediate-sized company, and Searle and Corning are much smaller companies. GE is the most diversified, Searle and Motorola are relatively undiversified, and Corning is intermediate. Regarding the nature of their product offering and market orientation: GE Medical Systems is a manufacturer of extremely expensive, large industrial equipment selling directly to businesses; Corning is primarily a components manufacturer selling to OEMs; Searle is a chemical-processing pharmaceutical company selling both to OEMs (e.g., NutraSweet as an ingredient to General Foods for use in Kool-Aid) as well as to the general public through distributors (e.g., NutraSweet in packets and powder); and Motorola Communications is a manufacturer of relatively small, inexpensive systems selling through distributors to consumers. And finally, the production environments of Motorola Communications and GE Medical are primarily assembly, whereas Corning and Searle are primarily materials-processing operations.

23. Unless otherwise noted, the information about these cases was generated from personal interviews with the following individuals: For the Searle case: Tom Bonnie, Max Downham, Marcel Durot, Betsy Hill, Barry Homler, Robert Mazur, Annette Mullendore (formerly Ripper), John Mullendore, James Schlatter, Dan Searle, Robert Shapiro, and Phil Worley—and also, Joseph McCann, author of *Sweet Success*, and Bob Ganger, director of development at General Foods. For the Motorola case: Robert Auchinleck, Jim Caile, Burnum Casterline, Marty Cooper, George Fisher, Jack Germain, Don Linder, Jim Michalak, John Mitchell, Edward Staiano, and William Weiz. For the Corning case: William Armistead, Forest Behm, Dave Charlton, Allen Dawson, David Duke, Richard Dulude, William Dumbaugh, Amory Houghton, Jr., John Hutchins, Theodore Kozlowski, Charles Lucy, Thomas MacAvoy, Robert Maurer, William Prindle, and Gail Smith. For the GE case: Morry Blumenfeld, Bobby Bowen, Arthur Chen, Lonnie Edelheit, Arthur Glenn, John Henkes, Tom Lambert, Peter (Ted) Pashler, Rowland (Red) Redington, Charles Reed, Walt Robb, Ron Schilling, Laddy Stahl, John Trani, and George Wise. This article would not have been possible without these individuals' generous assistance, for which we are most grateful.

24. Interview with Charles Lucy (Feb. 22, 1993).

25. See Rosenbloom and Cusumano, op. cit.; D. Braybrooke and C. E. Lindblom, *A Strategy of Decision* (New York: Free Press, 1970); J. G. March and J. P. Olsen, *Ambiguity and Choice in Organizations* (Bergen, Norway: Universitetsforlaget, 1976); H. Mintzberg et al., "The Structure of Unstructured Decision Processes," *Administrative Science Quarterly*, 21 (June 1985): 193–194; J. B. Quinn, *Strategies for Change* (Homewood, IL: Irwin, 1980); and H. A. Simon, *Administrative Behavior* (New York: Free Press, 1976).

26. Morone, op. cit., p. 72.

27. Ira Magaziner and Mark Patinkin, *The Silent War* (New York: Vintage Books, 1989), p. 275.

28. The patterns at work in these cases are similar in some aspects, yet different in others, with respect to what Abernathy and Utterback refer to as the dominant design. The patterns are similar in that once the new technology was developed and perfected, product sales grew exponentially. However, these patterns differ because discontinuous innovations displace the current dominant designs—e.g., GE's fan beam displaced EMI's pencil beam scanner, and Corning's single-mode fiber displaced its multimode fiber.

29. Morone, op. cit., pp. 151–152.

30. Ibid., p. 61.

31. See Philip Kotler, *Marketing Management* (Englewood Cliffs, NJ: Prentice-Hall, 1991), p. 262; and Philip Kotler and Gary Armstrong, *Principles of Marketing*, 5th edition (Englewood Cliffs, NJ: Prentice-Hall, 1991), pp. 221, 225.

32. Jay Klompmaker, G. David Hughes, and Russell Haley, "Test Marketing in New Product Development," *Harvard Business Review* (May/June 1976), p. 134.

33. See, for example, J. H. Parfitt and B. J. K. Collins, "Use of Consumer Panels for Brand Share Prediction," *Journal of Marketing Research*, 5 (May 1968): 131–145; Robert Blattberg and John Golanty, "TRACKER, An Early Test-Market Forecasting and Diagnostic Model for New Product Planning," *Journal of Marketing Research*, 15 (May 1978): 192–202; Lewis Pringle, R. Dale Wilson, and Edward Brody, "NEWS: A Decision-Oriented Model for New Product Analysis and Forecasting," *Marketing Science*, 1 (winter 1982): 1–29; Chakravarthi Narasimhan and Subrata Sen, "New Product Models for Test Market Data," *Journal of Marketing*, 47 (winter 1983): 11–24; William G. Zikmund, *Exploring Marketing Research* (New York: Dryden Press 1994), p. 326; Donald Tull and Del Hawkins, *Marketing Research: Measurement and Method*, 6th ed. (New York: Macmillan Publishing 1990), p. 257.

34. The same argument can be made about the differences between probing and learning and alpha and beta testing for industrial products. To the extent that the beta tests are used as a vehicle for getting feedback about a product in the final stages of development, it is more like the conventional market testing in consumer products than it is like the probing and learning process described here. To the extent that alpha and beta testing occurs at the very early stages in the development process—before the product is in its final form, then it is equivalent to what we are calling probing and learning.

35. The first phase of the technology cycle is the "period of technological ferment," the "era of ferment," or the "preparadigmatic design phase." Many product variations vie for acceptance and competitors fight aggressively over new versus existing technologies. The second phase begins when a dominant design emerges. The dominant design is the single product embodiment that establishes a standard in its class. Once it emerges, the industry undergoes a period of explosive growth. Well-known examples of dominant designs include: Ford's Model T, JVC's VHS video recorders, and IBM's PC. The dominant design phase leads to the third phase of the technology cycle, which is a period of incremental change. The focus in this phase shifts from product to process innovations. See Abernathy and Utterback, op. cit.; Abernathy, *The Productivity Dilemma* (Baltimore: Johns Hopkins University Press, 1978); Abernathy and Clark, op. cit.; Anderson and Tushman, op. cit.; D. Sahal, *Patterns of Technological Innovation* (Reading, MA: Addison-Wesley 1981); David J. Teece, "Profiting from Technological Innovation: Implications for Integration, Collaboration, Licensing, and Public Policy," in Robert Burgelman, Modesto Maidique, and Steven Wheelwright, eds., *Strategic Management of Technology and Innovation* (Chicago: Irwin, 1996), p. 235; Michael Tushman and Lori Rosenkopf, "Organizational Determinants of Technological Change: Toward a Sociology of Technological Evolution," in Robert Burgelman, Modesto Maidique, and Steven Wheelwright, eds., *Strategic Management of Technology and Innovation* (Chicago: Irwin, 1996), p. 191; and Utterback and Abernathy, op. cit.

Interestingly, several past studies such as Teece, op. cit., have shown that dominant designs are more applicable to mass consumer markets than industrial markets. But examples from our cases here also seem to suggest the concept of dominant designs are also relevant to industrial markets as well. Examples include GE's fanbeam system and Corning's single-mode fiber.

36. Anderson and Tushman, op. cit.

37. Jeffrey S. Young, *Steve Jobs: The Journey Is the Reward* (Glenview, IL: Scott, Foresman and Company, 1988).

38. See Krista McQuade and Benjamin Gomes-Casseres, "Mips Computer Systems," in Robert Burgelman, Modesto Maidique, and Steven Wheelwright, eds., *Strategic Management of Technology and Innovation* (Chicago: Irwin 1996), pp. 259–274.

39. Morone, op. cit., p. 150.

40. Donald Campbell and Julian Stanley, *Experimental and Quasi-Experimental Designs for Research* (Chicago: Rand McNally, 1963), pp. 7–12.

41. Ibid., pp. 34–36.

25

Developing New Product Concepts via the Lead User Method: A Case Study in a "Low-Tech" Field

Cornelius Herstatt
Eric von Hippel

INTRODUCTION

In a recent study of the market research preferences and practices of Swiss machinery manufacturers, Herstatt found that the firms viewed joint product development with users to be the most effective way to accurately understand user needs. At the same time, he found that the firms seldom employed this form of "market research," because they regarded it as being very complex, costly and difficult to implement.

Herstatt found no convincing reason *why* this form of market research would be inherently more complex or expensive than other, less-preferred methods, and so decided to conduct a "lead user" market research case study as a form of anecdotal research into the matter. In this article, we report on the procedures he used and the outcomes he obtained. As the reader will see, the method, which involves joint user–manufacturer development of new product concepts, was successfully applied in this case study. Further, anecdotal information provided by the firm participating in the case indicates that concept development with lead users was twice as fast as and half the cost of methods previously used.

WHAT IS "LEAD USER" MARKET RESEARCH?

Traditional market research methods are designed to sample the needs of a relatively large group of users, analyze the data obtained and then present the findings to product developers. In many fields, however, the richest understanding of needed new products and services is held by just a few users. Von Hippel [4,5] developed a method that exploits this fact by prescribing that firms interested in identifying needs for new products and services begin by identifying a small sample of "lead users." These especially sophisticated users are drawn into a process of *joint* development of new product or service concepts with manufacturer personnel. Then, the likely commercial appeal of the concepts developed with lead users is tested against a population of more ordinary users.

"Lead users" of a novel or enhanced product,

Reprinted by permission of the publisher from *Journal of Product Innovation Management*, Vol. 9, pp. 213–221. Copyright © 1992 by Elsevier Science Inc.

process or service have been defined by von Hippel as those who display *both* of two characteristics with respect to it:

1. They face needs that will be general in a marketplace—but face them months or years before the bulk of that marketplace encounters them.
2. They expect to benefit significantly by obtaining a solution to those needs.

Thus, a manufacturing firm with a current strong need for a process innovation that many manufacturers will need in 2 years' time would fit the definition of lead user with respect to that process.

Each of the two lead user characteristics specified above provides an independent and valuable contribution to the type of new product need and solution data lead users possess. The first is valuable because, as empirical studies in problem-solving have shown, users who have real-world experience with a need are in the best position to provide accurate data regarding it. When new product needs are evolving rapidly, as in many high-technology product categories, only users at the "front of the trend" will presently have the real-world experience that manufacturers must analyze if they are to understand accurately the needs that the bulk of the market will soon face.

The utility of the second lead user characteristic is that users who expect high benefit from a solution to a need can provide the richest need and solution data to inquiring market researchers. This is because, as has been shown by studies of industrial product and process innovations [2], the greater the benefit a given user expects to obtain from a needed novel product or process, the greater will be his or her investment in obtaining a solution.

In sum, lead users are users whose present strong needs will become general in a marketplace months or years in the future. Because lead users are familiar with conditions that lie in the future for most others, the lead user market research method can help manufacturers acquire need and solution information that will be useful in the development of "next generation" concepts for new products and services.

FOUR STEPS IN A LEAD USER STUDY

A lead user market research study involves four major steps, which are described in detail in articles by von Hippel [4] and Urban and von Hippel [3]. In brief summary, these are as follows. Step one involves specifying the characteristics lead users will have in the product/market segment of interest. That is, one must identify the trend(s) on which they lead the market, and also must specify indicators that show that they expect relatively high benefit from obtaining a solution to their trend-related needs. (One frequently useful proxy for expectations of high benefit is evidence of product development or product modification by users. As noted earlier, user investment in innovation, and user expectations of related benefit, have been found to be correlated.)

The second step is to identify a sample of lead users who meet both of the lead user criteria established in step one. Such a group will be both at the leading edge of the trend being studied and will display correlates of high expected benefit from solutions to related needs. The third step is to bring the sample of lead users together with company engineering and marketing personnel to engage in group problem-solving sessions. The outcome of these sessions is one or more "lead user" product or service concepts judged by session participants to be both responsive to lead user needs and responsive to manufacturer concerns regarding producibility, etc.

Finally, as the needs of today's lead users are not necessarily the same as the needs of the users who will make up a major share of tomorrow's predicted market, the fourth and final step in the lead user market research method is to test whether concepts found valuable by lead users also will be valued by the more typical users in the target market.

THE CASE STUDY

The company participating in this study was Hilti AG, a leading European manufacturer of components, equipment and materials used in construction. Hilti has major production facilities in Europe, the United States and Japan, and sells worldwide.

The product line we elected to concentrate on in our lead user case study was "pipe hangers"—a relatively "low-tech" type of fastening system often used in commercial and industrial buildings.

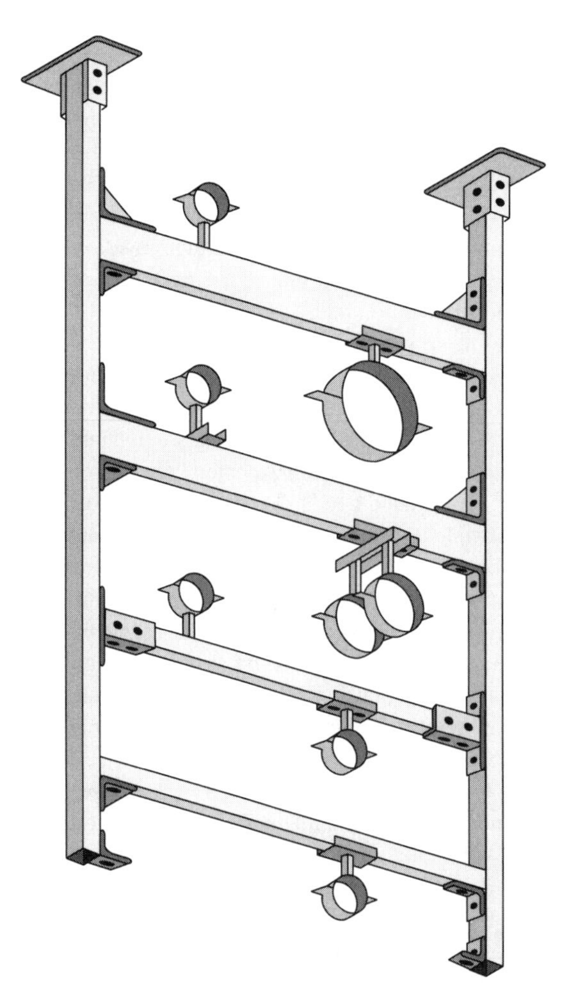

Fig. 25.1. A conventional pipe hanger configured to support several pipes.

Pipe hangers are assemblages of steel supports and pipe clamps and other hardware components used to securely fasten pipes to the walls and/or ceilings of buildings. Sometimes pipe hangers can be quite simple and support only a single pipe. Frequently, however, they are relatively complicated structures that simultaneously support and align a number of pipes of different sizes and types (Figure 25.1).

In the paragraphs that follow, we will describe how each of the four steps in a lead user study was carried out in this case.

STEP 1: SPECIFICATION OF LEAD USER INDICATORS

Recall that lead users of a product, process or service are defined as those who display two characteristics with respect to it: they have needs that are advanced with respect to an important marketplace trend(s) and they expect to benefit significantly by obtaining a solution to those needs. To identify lead users of pipe-hanging hardware, a first step was to identify important trends and users with relatively high benefit expectations related to these.

Identification of Trends

Identification of important trends in the evolution of user needs in pipe-hanging hardware began with a survey of experts. A brief analysis of the target market showed that people with expert knowledge in the relevant field would be found among "layout engineers," the specialists in charge of planning complex pipe networks in commercial and industrial buildings (layout engineers also are key decision makers with respect to determining which components will be bought and used for the pipe networks they design).

Expert advisors for this study were found in construction departments of technical universities, professional engineering organizations and municipal departments responsible for approving the design of pipe networks. Some of these were already known to the Hilti R&D department; others were identified via recommendations. Ultimately, the panel of experts who provided information for this study consisted of eight leading layout engineers in Switzerland, Germany and Austria; two researchers from the construction departments of the Swiss Federal Institute of Technology and the University of Darmstadt; one engineer from a professional organization in Bonn; and one engineer each from the municipal building departments in Bern and Berlin.

The trends identified as most important by the experts surveyed regarding pipe-hanger systems were as follows:

> Trend 1: There is an increasing need for pipe-hanger systems that are extremely easy to put together—so easy that instruction booklets will not be needed. Such systems should have significantly fewer components than at present. They should adapt to a wide range of application conditions, and should be based on a simple, consistent construction principle.
>
> Reason for trend: Education levels among installers are going down in many countries.
>
> Trend 2: There is a need for rapidly actuated, positive, interlocking fasteners to connect pipe hanger elements together securely and to attach the completed hangers securely to building walls and ceilings.
>
> Reason for trend: Safety standards in many countries are getting more stringent. Some of the multiple screws and bolts now used to assemble hangers (see Figure 1) may be inadvertently overlooked by installers—with consequent risk of field failure.
>
> Trend 3: There is a need for pipe hangers made from lighter, noncorrodible materials. Pipe hangers should therefore increasingly be made of plastics rather than of the steel elements that are used almost exclusively today.
>
> Reason for trend: Pipe-hanging systems made of steel are heavy and therefore difficult and dangerous to hang under some field conditions. In addition, steel is subject to corrosion and failure in wet environments or environments where chemicals are present.

Solutions that offered improvements with respect to these (somewhat overlapping) trends were expected to result in significant benefits for the users of pipe hangers. The skills required of installers would be reduced; fewer components would have to be stocked by users; the speed and safety of installation would be greatly increased; and the risk of field failures would be reduced.

Identification of High-Benefit Expectations

Expectations of innovation-related benefit on the part of users can be identified by survey, and this approach has been successfully applied elsewhere [3,4]. However, as mentioned earlier, innovation-related *activity* by users also can serve as a proxy for expectations of benefit, and this is the approach used here.

Users showing innovation activity were identified by conducting telephone interviews with a sample of 74 interviewees. Because, as will be described in the next section, the same sample was screened to simultaneously identify users having *both* lead user characteristics (ahead with respect to identified trends and having high expected benefit), we will defer a detailed discussion of methods and findings with respect to user innovation activity until we describe how step 2 was carried out in this study.

Here, we simply note that users engaged in innovating were determined by questions such as "Do you/did you ever build and install pipe-hanger hardware of your own design? Do you/did you ever modify commercially available pipe-hanger hardware to better suit your needs?" We also note that a high fraction of users interviewed (36%) were in fact found to display this characteristic.

STEP 2: IDENTIFICATION OF LEAD USERS

Once the trends and the user benefit characteristics were specified that would be used to identify lead users, the next step was to identify a lead user sample. This was begun by identifying, in cooperation with Hilti, a random sample of firms that buy and use pipe hangers. This sample was then screened to identify a subset of lead users within it.

The firms that install pipe-hanging systems are specialists in installing pipe networks in commercial and industrial buildings—for example, industrial plumbing firms. Installation of pipe hangers is a subtask in the larger task of pipe installation. The tradesmen who actually install pipe hangers comprised the group in which lead users would be identified. (Installers of pipe hangers have only a moderate-level technical education. In the countries from which our user sample was drawn—Switzerland, Germany and Austria—these installers complete 8 years of general schooling, and then take a 2- or 3-year vocational training program in their particular trade. Finally, they pass a municipal examination and receive a license to practice.)

Hilti has a number of geographically based sales divisions with close and frequent customer contacts. The German, Austrian and Swiss sales divisions (selected because of their geographical accessibility) were asked to provide the names of firms they thought were buyers of pipe-hanger systems made either by Hilti or its competitors. In this request no mention was made of either customer innovativeness or customer size. The three sales divisions eventually responded with the names of 120 firms they thought met the criteria.

Next, attempts were made to contact all 120 user firms for a telephone survey. Ultimately, 74 of these were in fact successfully contacted and judged suitable for and willing to undertake more detailed interviews (20 of the 120 were excluded because they could not be reached after five telephone calls; 16 were excluded because they were found to be not currently using the product type at issue and ten were not included simply because they were not willing to participate in an interview).

In the instance of the 74 firms who were willing to participate in a telephone interview, interviewers sought to identify the most expert person on the products under investigation. To do this, the first contact at each firm was asked: "Whom do you regard the most expert person on pipe-hanger systems in your company, and can we talk to that person?" The interviewers were referred to expert "fitters"—employees who actually install pipe-hanging systems in the field—in 64 of the 74 instances. In the remaining ten cases they were referred to direct supervisors of fitters, all of whom had moved into supervisory positions only after extensive experience in the field.

Interviews were next conducted with all 74 individuals. The interviews were aimed at identifying a subset of users in the total user sample who had both of the two lead user characteristics (being ahead on the trends identified by the experts and expecting high benefit from innovations along these dimensions).

The proxy used for "ahead on identified trends" was simply: (1) did the interviewees agree that advances along the trends that had been specified by the expert panel were in fact needed and important and (2) could the interviewees describe at least some technically interesting ideas regarding these trends? As we noted in our discussion of step 1 in a lead user study, the proxy used for "user innovation benefit expectations" was: had the users developed or modified pipe hangers in ways that they felt represented improvements with respect to the identified trends?

As a result of the interviews just described, a significant number of lead users of pipe-hanging hardware [22] were identified. Table 1 summarizes the findings on this matter and, as a matter of interest, compares these with data drawn from the Urban and von Hippel [3] study of PC-CAD (PC-computer-aided design) users. In both studies, there was a high overlap where users displayed the two lead user characteristics.

It is interesting to note that, as is shown in Table 25.1, 27 (36%) of our random sample of users of pipe-hanging systems had designed, built and installed hangers of their own devising in one or more cases. This compares very favorably with the 25% of innovating users found in the technically sophisticated field of PC-CAD.

STEP 3: LEAD USER PRODUCT CONCEPT DEVELOPMENT

A group of 22 lead users of pipe hangers had now been identified. The next task was to determine whether some of these lead users could be joined with expert Hilti personnel to produce novel

TABLE 25.1. Percent of Sample Found to Have Lead User Characteristics in Two Studies

	Sample of Pipe-hanger Users	Sample of PC-CAD Users[a]
Users at front of selected trend(s):	30% (22)	28% (38)
Users who built own prototype products:	36% (27)	25% (34)

[a]Data source: Urban and von Hippel [3].

product concepts that would be judged by Hilti marketing researchers and by routine users to be the basis of valuable commercial products, and that would be judged to be practicably manufacturable by Hilti engineers.

Selection of Lead User Concept Group

Recall that, in the method step described just above, a group of 22 lead users had been identified among a total user sample of 74 users. Two more tests were next applied to this sample to identify those few lead users who seemed to be most appropriate to invite to join with Hilti engineers and other experts in a 3-day concept generation workshop. These additional tests were intended to select the users most likely to be effective in such a workshop, and consisted simply of the judgment of the person who had interviewed the user on two matters: Did the interviewer judge that the user could describe his experiences and ideas clearly? Did the user seem to have a strong personal interest in the development of improved pipe-hanger systems? Fourteen of the 22 lead users met these additional tests and were invited to join the workshop.

Twelve of the 14 lead users contacted—ten pipe fitters plus two supervisors of fitters—agreed to join the product concept development workshop. Interestingly, the two that did not were users who had patented their own pipe-hanger system designs. These two were not willing to present their ideas in a workshop, most probably because they were concerned about the diffusion of their proprietary-technical know-how.

All users who joined the workshop formally agreed that any inventions or ideas developed during the sessions would be the property of Hilti. As compensation, every participant was offered a small honorarium. Interestingly, most of the participants did not accept this; they felt sufficiently rewarded by simply attending and contributing to the planned workshop.

Three-Day Product Concept Generation Workshop

The goal of the product concept generation workshop sponsored by Hilti was to develop the conceptual basis for a novel pipe-hanger system with characteristics identified in the technical trend analysis described earlier. To most effectively meet this goal and to efficiently transfer the workshop findings to Hilti, the lead users at the workshop were joined by two of the expert layout engineers who had participated in the trend analysis segment of our study. Invitees from Hilti consisted of the marketing manager, the product manager and three engineers who worked on the design of pipe-fastening systems.

The workshop was carried out over a 3-day period, and was organized as follows:

Day 1. The entire group conducted a review of important trends and problems in pipe-hanging systems. Next, five relatively independent problem areas were defined by the group, and a subgroup was established to work on each. The five subgroup topics were (1) methods of attaching pipe hangers to ceilings or walls; (2) design of support elements extending between the wall attachment and pipe clamp itself; (3) design of the pipe clamps; (4) design of the methods of attaching various system components to each other in the field and (5) methods of conveniently adjusting length of supporting members at the field site. Membership in the subgroups was at the option of workshop participant, and shifts in membership were made from time to time to avoid the possible danger of premature fixation on individual problem-solving ideas championed by individual users. Each of the subgroups was assisted by technicians from Hilti or external layout engineers.

Day 2. The five subgroups worked on their problem areas in the morning, and in the afternoon all took a break from the specific problems at hand and participated in some general problem-solving and creativity exercises such as role-playing and team-building exercises. The purpose of these was both to lessen pressure on participants and to make them more comfortable with each other. After a short while, the workshop was in fact characterized by very strong group cohesion and intensive, cordial interaction.

Day 3. The subgroup ideas were presented to the entire group for evaluation and suggestions. As an aid to this evaluation effort, each of the subgroup ideas was evaluated on the three criteria of originality (how revolutionary and novel is the solution from a technical point of view?), feasibility (how quickly can the solution be realized employing currently available technology?) and comprehensiveness of solution (does the idea represent a single solution or does it resolve several user problems simultaneously?). Next, membership in the subgroups was changed, work on the most promising concepts was continued and informal engineering drawings were produced by participants. Finally, these were critiqued and modified by the entire group and merged into one joint concept.

Results of Product Concept Generation Workshop

At the conclusion of the workshop, the single pipe-hanger system design was selected by the total group as incorporating the best of all the elements discussed in the subgroups, and this was the system recommended to Hilti.

After the workshop, the technical and economic feasibility of the new product concept proposed by the lead users was evaluated further by Hilti personnel. At the conclusion of this work, it was decided that the lead users had indeed developed a very valuable new pipe-hanger system. In the judgment of company experts it was well in advance of the offerings of competitors. Hilti, based in Lichtenstein, is a leading European manufacturer of components, equipment and materials used in construction. The firm's products range from fastening systems to drilling and cutting equipment to specialty chemicals. It has major production facilities in Europe, the United States and Japan, and sells worldwide. In 1990 the worldwide sales of Hilti were approximately 2 billion Swiss francs.

STEP 4: TESTING WHETHER LEAD USER CONCEPTS APPEAL TO ORDINARY USERS

The fourth and final step in the lead user market research method involves testing whether typical users in a marketplace find the product or service concept developed by lead users to be attractive. Because Hilti's internal evaluation showed the potential commercial value of the lead user concept to be very high, they were not willing to present it to a random sample of ordinary users for evaluation but instead decided to simply test the lead user product concept on a sample of 12 "routine" users.

The companies selected for this "routine user" sample were drawn from the sample of 74 interviewed companies. The selection criteria were that the telephone interview data showed them *not* to be lead users, and also that they must have had a long, close relationship with Hilti. (The latter requirement was added because the company wished to have confidence that these users would be willing to honor a request to keep the details of the new system secret.) The interviewees selected were buyers as well as users. They had the dominant role in the purchasing decisions of their own companies with respect to pipe hangers.

The 12 user–evaluators were asked to review the proposed pipe-hanger system in detail, noting particular strengths and weaknesses. Their response was very positive. Ten of the 12 preferred the lead user product concept over existing, commercially available solutions. All except one of the ten expressed willingness to buy such a pipe-hanger system when it became available, and estimated that they would be willing to pay a 20% higher price for it relative to existing systems.

COMPARISON OF LEAD USER METHOD WITH METHOD ORDINARILY USED BY HILTI

The case study was, as reported above, very successful. Interestingly, Hilti personnel informally judged that the lead user method, beginning with a technological trend identification and ending with a novel product concept, was significantly faster and cheaper than the more conventional marketing research methods they normally used. Unfortunately, data needed to test this judgment carefully did not exist in the firm. However, it was possible to compare the time and costs expended in this first lead user study by Hilti with the time and costs expended on a project that they had recently conducted, and judged to be of very similar scope and complexity (Table 25.2).

The process Hilti conventionally used took a total (elapsed time) of 16 months from start to final agreement on the specifications of the product to be developed, and cost $100,000. The work began with marketing personnel collecting and evaluating data on needs and problems from customers (5 months; $56,000); then marketing explained to engineering what it had found, and these two groups jointly developed tentative product specifications (2 months; $5,000). Next, engineering went off on its own to develop technical approaches to meeting the agreed-upon specifications (4 months; $23,000). Then, engineering got together with marketing to evaluate and adjust these (3 months; $10,000). Finally, both engineering and marketing wrote up a formal product specification and submitted it to management for formal approval (2 months; $5,000).

TABLE 25.2. Anecdotal Time and Cost Comparison Between Two Product Concept Generation Efforts at Hilti

Concept Generation Method Employed	Time and Cost Expenditure for Concept Generation, Evaluation and Acceptance by Hilti	
Lead user method	9	$51,000
Conventional method	16	$100,000

In contrast, the lead user method took a total (elapsed time) of 9 months and cost $51,000 from the start of work to final agreement on the specifications of the product to be developed. In this instance, the major steps were all conducted by a project group headed by the manager of the pipe-hanger product line. The group membership consisted of two development engineers and two market specialists. One of the latter was responsible for pipe hangers specifically, and one was a market research methods expert from Hilti's central market research group. The steps carried out by this group (and described in detail earlier in the article) were survey of experts (2 months; $9,000); telephone survey (2 weeks; $8,000); Lead User Workshop (3 days; $24,000); internal evaluation of lead user concept (3 months; $4,000); concept test on routine user group (2 months; $4,000); writing of formal product specification submission to management for formal approval (2 months; $2,400).

In sum, the lead user method consumed only 56% of the time used for the project put forward by Hilti as comparable. In our estimation and that of Hilti personnel, the reason for the time saving appeared to lie mainly in the systematic, parallel involvement of engineers, marketing people and highly qualified users, as opposed to the serial involvement of these groups used in the earlier method. Because of this, time-consuming feedback loops or reconsiderations, often produced by misinterpretations or information-filtering in the serial method, were avoided.

The cost of the lead user process also was found to be significantly lower than market research methods previously used by Hilti (approximately 50%). An informal evaluation of the reasons for this, conducted with Hilti personnel, suggests that the cost saving had two principal causes. First, the costs for customer surveys were smaller in the lead user method (in the lead user project, only 12 selected users were involved in joint, face-to-face discussions; in the conventional project, approximately 130 interviews with a randomly selected group of users, each involving face-to-face visits by manufacturer personnel, were carried out in three different countries). Second, the solutions provided by the lead user group required less work on the part of Hilti technical departments than did the ideas provided by marketing researchers in the conventional method (in the lead user project, people from Hilti's technical departments had direct user contact and had been involved in concept development from the start; they therefore had richer data regarding user needs in the lead user project than they did in the conventional project).

DISCUSSION

In this case study, the lead user method worked well in a relatively "low-tech" product category whose users were not characterized by advanced technical training. A significant fraction of all users sampled was found to have lead user characteristics. A group selected from among these proved very effective in working with company personnel on new product concept development. They did in fact develop a new system judged to be very valuable by both the manufacturer and a group of nonlead users. Also and importantly, study participants found participation to be both useful and enjoyable. Bailetti and Guild [1] report on a study that explicitly measured the responses of design engineers to visits with lead users. They also found that participants judged this experience to be very valuable.

An additional, unanticipated result of the lead user method was an observed improvement of teamwork within Hilti, manifested in a significant improvement in the level of cooperation between the technical and marketing groups in the company. One reason for this was apparently that the teamwork built into the lead user method had a carryover effect. Also, as product and performance requirements of innovative users were immediately translated into language meaningful to *both* engineers and marketing people, a shared language was created that made further cooperation easier.

Although the lead user method worked well in this case study, the reader should note that it is still a very new method. Details of method application will appropriately differ from study to study—and we are all still learning.

REFERENCES

1. Bailetti, Antonio J. and Guild, Paul D. Designers' impressions of direct contact between product designers and champions of innovation. *Journal of Product Innovation Management* 8(2):91–103 (June 1991).
2. Mansfield, Edwin. *Industrial Research and Technological Innovation: An Econometric Analysis.* New York, NY: W.W. Norton & Company, 1968.
3. Urban, G. and von Hippel, E. Lead user analyses for the development of new industrial products. *Management Science* 34(5):569–582 (May 1988).
4. von Hippel, E. Lead users: A source of novel product concepts. *Management Science* 32(7):791–805 (July 1986).
5. von Hippel, Eric. *The Sources of Innovation.* New York, NY: Oxford University Press, 1988.

RELATED READINGS

Biegel, U. *Kooperation zwischen Anwender und Hersteller im Forschungs- und Entwicklungsbereich.* Frankfurt/New York/Paris: 1987.

Foxall, G. User initiated product innovations. *Industrial Marketing Management* 18(2):95–104 (May 1989).

Gemunden, H. G. *Investitionsguetermarketing, Interaktionsbeziehungen zwischen Hersteller und Verwender Innovativer Investitionsgueter.* Tübingen, 1981.

Geschka, H. Erkenntnisse der Innovationsforschung—Konsequenzen für die Praxis. *VD-Berichte* 724. Neue Produkte-Anstosse, Wege, Realisierte Strategien. Dusseldorf 1989.

Shaw, B. The role of the interaction between the user and the manufacturer in medical equipment innovation. *R&D Management* 15(4):283–92 (October 1985).

Voss, C. The role of users in the development of applications software. *Journal of Product Innovation Management* 2(2):113–21 (June 1985).

26

Managing Flexible Automation

PAUL S. ADLER

How can industry capitalize on the manufacturing technology developments of the last decades? These innovations have created a range of programmable automation systems that seem to allow for a much greater flexibility in manufacturing; but what challenges does a shift towards greater flexibility generate for management? In particular, what human resource management policies can support these more flexible operating models? This article suggests some guidelines to tackling these difficult issues.

The challenge posed by the new, flexible technologies is primarily that posed by their higher "knowledge intensity." The relative importance of knowledge compared to that of labor or capital resources increases for two reasons. First, there is the increased importance of individual and organizational learning in systems that can not only manufacture a larger range of pre-specified products but can also adapt more easily to new product designs. Second, knowledge intensity is also increased by the programmable nature of the new technologies—which is at the origin of their flexibility—since this adds yet another dimension (software) to the range of types of knowledge (electrical, electronic, materials, mechanical, etc.) objectified in machine systems.

The management of the development and the utilization of systems so "dense" in knowledge poses new challenges. But while the management of knowledge has become the central task of firms wanting to survive in a world of rapidly evolving technological possibilities, knowledge management is an activity for which we have few models.[1]

The difficulty is that knowledge is an asset with particular properties. It is even more peculiar than the standard public good which doesn't get used up by being used, because with most forms of knowledge, the more you use, the more you have. The management of the production and distribution of this asset therefore imposes special challenges for management. Neither decentralization nor centralized planning, nor even any astute combination of these two modes of organization is entirely satisfactory. The emerging management literature is rich in insights into new forms of organization that are appropriate to the challenge of knowledge management.

The central argument of this article is thus that developing more enlightened forms of knowledge management is perhaps the key to unlocking flexible automation's potential.

TOWARDS FLEXIBLE COMPUTER INTEGRATED MANUFACTURING

While the range of recent technological developments in the manufacturing area is staggering it is breadth—encompassing new processes, new

This is a revised version of a report to the Organization for Economic Cooperation and Development. It has benefitted from comments by M. Graham, S. C. Wheelwright, and two *California Management Review* reviewers.

materials, and new products—the central tendency has been the convergence of three strands:

- *design automation:* computer-assisted drafting, design, and engineering;
- *manufacturing automation:* computer-controlled processes in the fabrication/assembly industries (particularly in extensions of machine tool Numerical Control); automatic materials handling; automatic storage and retrieval systems; and
- *administrative (or control) automation:* computerized accounting, inventory control systems, and shop-floor tracking systems.

A recent report by the U.S. Office of Technology Assessment described the principal elements of these technologies.[2] The most compelling aspect is their convergence towards what is now commonly called Computer Integrated Manufacturing (CIM):

- The linking of design and manufacturing is the most established initiative in this area, with numerically controlled machine tools permitting the designer to automatically generate tapes for machine control. Direct Numerical Control now permits the designer to download the program directly to the machine. The communication is, moreover, not only one-way: criteria of producibility can be built into design databases alerting the designer to the constraints of the available manufacturing technologies before, rather than after, the design is sent to Manufacturing.
- Links between design and administration permit the establishment of Bills of Materials and process plans directly from the design database.
- Links between all three elements—design, manufacturing, and administration, such as we find in Flexible Manufacturing Systems—permit the direct control of inventory and materials flow integrated with tightly coupled DNC machine systems.

The value of the new automated technologies can be assessed on several dimensions:

- *Cost:* dramatic improvements can be made not only in direct manufacturing costs but also, and perhaps more importantly, indirect personnel costs, materials costs, inventory carrying costs, and space costs.
- *Quality:* not only can defects be reduced by a greater degree of conformance to specification, but, more fundamentally, new performance characteristics become possible with the enhanced process capabilities.
- *Time:* throughput time for a given product can be reduced, but even more important can be the impact of reduced setup times and shorter change-over times on the ability to switch between products and the impact of CAD/CAM integration on the new product development cycle time.

For any given industry, a more refined analysis would also include other dimensions, such as Service. But in the general discussions of cross-industry effects, it is the third dimension, Time, that has attracted the most attention. Indeed, the new technologies seem to open up the possibility of a dramatic reorientation of industry away from the long runs of standardized commodities that seem to have been at the heart of the post-WWII prosperity.[3]

From this perspective, the key promise of the new technology would seem to be in shorter design cycles (through CAD), shorter turnaround-times for prototype testing (through CAD/CAM links), and faster manufacturing startup (through integrated manufacturing capabilities like FMS). These new parameters would allow for more rapid product turnover and for a broader range of products to be economically produced in the same facility—what Goldhar and Jelinek term "economies of scope."[4] The cost analyses of Boothroyd[5] and Hutchinson and Holland[6] confirm this potential.

AN "ERA OF FLEXIBILITY"?

From a managerial perspective, a key question is whether greater flexibility will be the hallmark of the new automation's actual implementation. Will we see a major change in the variety and turnover of products? This is less obvious than it may seem, if only because managers might find the other dimensions of automation's advantages more important than flexibility. Programmable automation could shift management priorities towards greater flexibility because it can dramatically reduce the cost and quality penalties of improvements in the time dimension, but there will no doubt remain some trade-offs to be made between these three terms, and business strategy will need to assess which benefits offer the greatest competitive leverage.

The prospects for a major change in the degree of manufacturing flexibility are further conditioned by the managerial context: one of the scarcest resources in the firm is management attention, and flexibility consumes a lot of it. Managing the complexity of these new systems and managing the learning required to realize their flexibility potential require not only a new level of expertise in manufacturing, but also a greater focus at general management levels on the manufacturing function. Will these be forthcoming?

The reluctance of U.S. managers to exploit opportunities which would disrupt established *modus operandi* has been described in vivid terms by Hayes and Abernathy,[7] Reich,[8] and Buffa.[9] Their basic argument can be summarized in the convergence of the following factors:

- financial considerations have dominated corporate strategy in the U.S. more than elsewhere;
- this has given precedence to the objective of a well-diversified portfolio so as to minimize risk across a set of given opportunities, rather than to the more entrepreneurial objective of creating new opportunities;
- the low organizational status of production engineering as compared to design engineering has both expressed and reinforced the absence of manufacturing as an active element of corporate strategy;
- the predominant philosophies of organizational design and management have served to encourage over-specialization and lack of integration; and
- competition has tended, as if by tacit agreement, to focus on nonmanufacturing dimensions—distribution, packaging, advertising, etc.

This diagnosis has become widely accepted, albeit with important nuances between authors. Analysis of the origins of these proximate causes has, however, been scanty. The most plausible explanation would appear to be the "fat and happy" hypothesis: the position of world dominance enjoyed by U.S. industry in the 1950s and 1960s left U.S. producers without the challenge necessary to keep the entrepreneurial spirit sharpened.

> While the current economic turbulence may subside, the likelihood that any new stability in markets will be at a somewhat higher level of flexibility is pushing management to explore flexibility with a new aggressiveness.

In this context, certain ideas have reinforced and contributed to U.S. managers' apparent conservatism. Amongst them is the long-held belief that in general higher levels of automation are by nature less flexible. This generalization can be found expressed in the work of Woodward[10] and was at the origin of the association implicit in the "product-process matrix" developed by Hayes and Wheelwright.[11] (See Figure 26.1.) The axes of this two-dimensional matrix trace the life-cycle of products (ranging from one-of-a-kind to higher volume and finally to standardized commodity) and of processes (from jumbled flow of the jobshop to batch processing to assembly line and finally to continuous flow refinery). In this perspective, the more "mature" processes are also typically more automated. The competitively viable positions are generally on the diagonal, other positions being

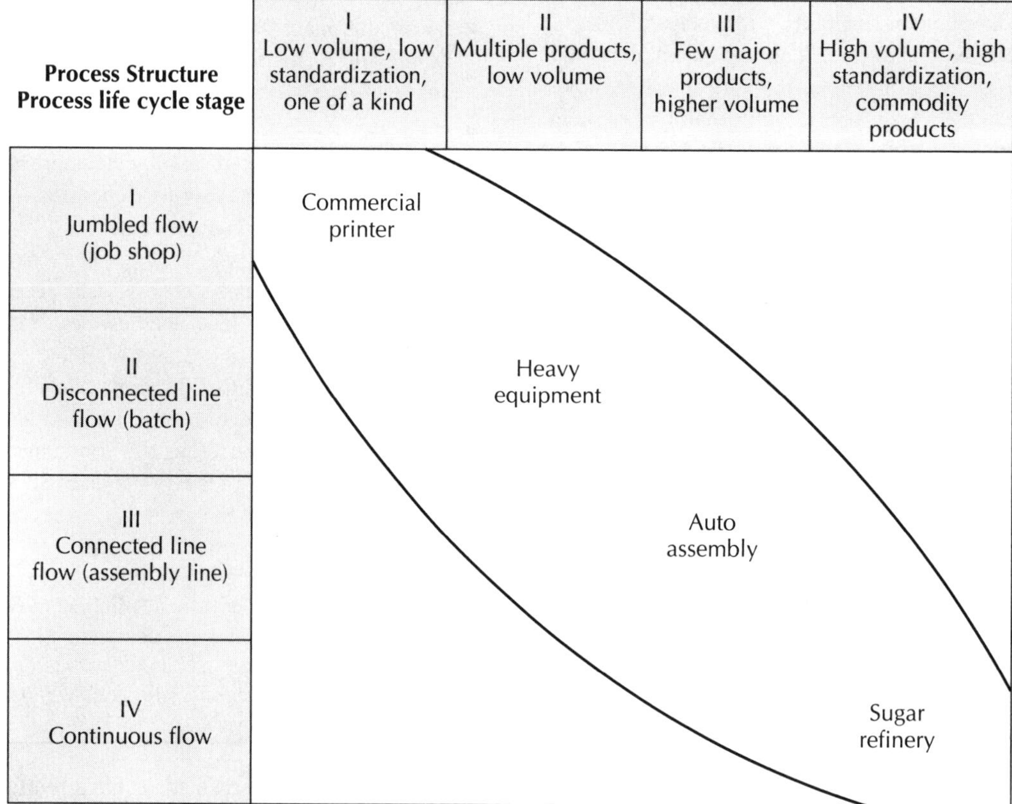

Fig. 26.1. Hayes-Wheelwright "product-process matrix." (*Source:* R. H. Hayes, S. C. Wheelwright, *Restoring Our Competitive Edge*. New York: John Wiley and Sons, Inc., 1984.)

typically inefficient mismatches of product and process: one-of-a-kind products, for example, should optimally be produced in jobshops, not on assembly lines.

Flexible automation has challenged the managerial wisdom of the preceeding decades that was captured in this proposition. It now appears clear that the matrix's normative implication is valid only at a given level of technology. In a more dynamic perspective, we need to consider the implications of recent automation tendencies which make it possible to envisage the production of less standardized products in a quasi-continuous process. Analyses like Boothroyd's[12] suggest that the new matrix's diagonal should be flattened (or at least bowed out downward), as in Figure 26.2. Such is the new argument of Hayes and Wheelwright[13] and Ferdows.[14]

The implications of such a revision are profound. It undermines a deeply and widely felt intuition of a corollary between efficiency and rigidity. The proposition that there is a fundamental tension between innovation and efficiency—and consequently a dilemma for management in the choice between them—is one of the themes of Abernathy's landmark study of the auto industry[15] and reflects widely-held beliefs culled from the last few decades' experience.

These challenges to conventional managerial philosophy are reinforced by the recent associated critique of the more mechanical applications of product (and process) life-cycle theory. The old idea according to which products and processes went through a life-cycle that took them along the respective axes of the product/process matrix has been challenged by the recent experience of many industries, starting with the auto industry.

When, during the 1970s, the U.S. auto industry came under serious competitive pressure from the Japanese and the import ratio began climbing, a debate emerged as to the nature of this challenge and the appropriate public policy response.[16] Life-cycle theories would suggest that this was a natural—and unavoidable—evolution based on the inevitable standardization of the product, and that this evolution naturally gave the Japanese lower labor costs greater competitive significance. The policy implication is that the U.S. should get out of the auto industry. But closer examination by Abernathy and others suggested that the product design, far from becoming more standardized, was undergoing a profound transformation. The Japanese were not beating the U.S. producers primarily by virtue of their lower wage rates, but by a combination of new product and new process designs.

In numerous older and supposedly more "mature" industries, international competition and new technologies have recreated the kind of uncertainty regarding basic product and process parameters that we normally associate with the infancy of an industry. The auto industry served as a compelling example for the analysis of this process of "de-maturation" presented by Abernathy, Clark, and Kantrow.[17]

A new dimension of flexibility is opened by this double-ended attack on old assumptions: the ability to pursue a path of constant product and process renovation at speeds which may vary sharply from one period to the next—flexibility, in other words, not merely within the parameters of a given product generation and process type, but in the ability to pursue painstakingly detailed manufacturing cost-reduction programs and marketing refinements at the same time as the company maintains its ability to make the next breakthrough in basic product technology or in marketing concepts.

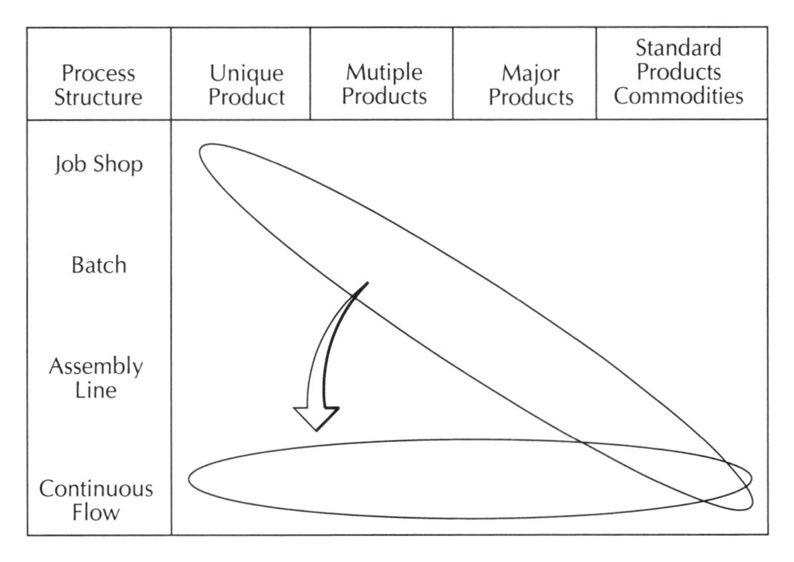

Fig. 26.2. The effect of automation on the product-process matrix.

(Abernathy, Clark, and Kantrow provide a framework—the "transilience matrix" for mapping the degree and type of change.)

These new developments in manufacturing thinking represent true gales of creative destruction demolishing established mental models. But whether they correspond to such dramatic shifts in external reality is still debatable. Indeed, to the extent that current interest in manufacturing flexibility represents a reaction to the particularly unstable conditions that many markets have experienced in recent years, it might contain elements of overreaction. The debate on long waves in economic growth continues,[18] but so far little evidence has emerged that would encourage us to believe that *all* of the turbulance of the last decade is to be considered a permanent feature of the economic landscape. If, as seems plausible, the next decade sees the emergence of new, somewhat more stable configurations of demand and of supply, some flexibility efforts may prove to have been over-reactions.

The best evidence suggests that the current economic turbulance may subside, but that both survival in the interim and the likelihood that any new stability in markets will be at a somewhat higher level of flexibility are pushing management to explore the flexibility of new manufacturing capabilities with a new aggressiveness. What, then, are the implications of such an effort for management practices?

IN SEARCH OF FLEXIBILITY

Pursuit of competence in more flexible manufacturing requires rethinking a considerable range of managerial practices.

STRATEGY

The achievement of the more advanced forms of production flexibility calls for a new status for manufacturing strategy, as outlined by Wheelwright and Hayes.[19] They outline four "stages" of manufacturing's strategic role: internally neutral, externally neutral, internally supportive, externally supportive.

> A key challenge in capitalizing on the new technological potential for flexibility is to adapt organizational cultures to maximize spontaneous cooperation.

It is the most advanced stage, where manufacturing becomes a real competitive asset, that appears to be the prerequisite for capitalizing on new opportunities for flexibility: "Stage 4 implies a deep shift in manufacturing's role, in its self-image, and in the view held of it by managers in other functions. It is, at last, regarded as an equal partner and is therefore expected to play a major role in strengthening a company's market position. Equally important, it helps the rest of the organization see the world in a new way."[20]

In relation to the specific issues posed by new, more-flexible automation, the contrast between the more and the less advanced strategic approaches is profound. To capitalize on rapidly evolving opportunities requires a "dynamic" perspective on process technology, rather than the traditional "static" perspective. In these more proactive manufacturing strategies:

- the responsibility for new technology is broadly defined to include not merely the specialists and lower-level managers, but also suppliers, workers, and senior managers;
- the process of technological change cuts across functional boundaries and is continuous over time, rather than consisting of discrete, episodic, and functionally specialized projects;
- the objective of change is the enhancement of a broadly-defined set of capabilities, rather than cost-reduction for a specific product;
- the criteria for evaluating projects include the long-term and nonfinancial benefits, rather than being limited to the traditional two or three year payback analysis; and
- the competitive contribution of automation is seen as deriving from a continual, incremental

process of improvement, rather than occasional big steps.

The principal lesson of Wheelwright and Hayes is that the dynamic enhancement of process capabilities requires a strategic vision in each function. In other words, strategy, far from being restricted to the firm's external orientation, and far from being the sole prerogative of the general manager, is the necessary foundation of excellence in all the functions, and as such the *sine qua non* of realizing the potential of advanced automation.

It is perhaps in their commitment to this kind of strategic thinking for manufacturing that Japanese companies can be most advantageously compared to U.S. companies. The Japanese excellence in creating and maintaining taut production systems (such as Just-in-Time inventory control) is perhaps not their most powerful weapon in international competition. It merely reflects the priorities that emerged at one point in time in their ongoing efforts in the manufacturing area. Today, surveys indicate that the emphasis in Japan is on new product development. The Wheelwright-Hayes argument alerts us to the possibility that U.S. fascination with the tactic—for example, Just-in-Time—may obscure the strategic lesson, namely, that continual improvement in manufacturing is critical.

Evaluating Projects

The advantages of the new technologies pose a major challenge to the traditional approaches to the justification of capital investments.

A first difficulty lies in defining flexibility. In the Appendix, I have summarized the principal approaches. No consensus has emerged yet on the most appropriate definition. The second difficulty lies in quantifying the "value of flexibility" so as to integrate it into the standard financial methodology. Here too numerous approaches are still in competition.[21]

The biggest challenge, however, is perhaps more profound. The Manufacturing Studies Board recently surveyed the CAD/CAM scene in the U.S. and concluded that "the usual financial measures, such as return on investment, were inadequate for assessing the results of integration. . . . The best measures, these companies say, are responsiveness, productivity, quality, lead time, design excellence, flexibility, and work-in-process inventory."[22]

Such eclectic approaches challenge the deeply ingrained habits of accounting. Kaplan[23] and Gold[24] have presented particularly incisive critiques of current practice in the U.S. in this regard. Their critiques focus on the excessively narrow range of benefits (and costs) included in the evaluation.

Hayes and Garvin went somewhat further in their critique, focusing on the myopia generated by the discounting approach—at least as it is commonly implemented.[25] Their key argument was that discounting practices usually give insufficient weight to the costs of *not pursuing* an investment opportunity. Hodder and Riggs recently clarified the types of misuses to which discounting has been subjected—in particular, improper treatment of inflation, over-adjustment for risk, and failure to include the effect of possible management action to mitigate risks.[26] Their conclusion is that Net Present Value can be beneficially used as a framework in which we can combine in a reasoned manner the more analytical and the more "entrepreneurial-intuitive" elements, both of which are needed for a sound investment decision.

There appears to be a consensus that evaluation of far-reaching technological changes should be conducted on an explicitly strategic foundation. It remains an open question as to whether the representations of strategy afforded by the "universal solvent" of finance are adequate to support decision making in the world of difficult-to-aggregate criteria relevant to manufacturing technology investment decisions.

Managing Change

The third set of challenges posed by the new technological potential for flexibility relates to the ability of management to identify the types of organizational changes required to pursue the realization of that potential.

A recent survey by McKinsey and Co. of CAD/CAM efforts highlights the problem.[27] Their discussions with 18 major users in the U.S. led them to identify three types of CAD/CAM configurations. In the first type, the CAD system was being used as an "electronic pencil" devoted to drafting. The second and more advanced group was composed of those using CAD as a tool for the whole design function and using this opportunity to rethink the operation of that function: restructuring jobs to combine design, checking and detailing, for example. Some two-thirds of their sample had progressed in developing this type of application. The third and most advanced configuration was one that "takes the slash out of CAD/CAM" to forge direct links with the manufacturing database. Only a minority of firms surveyed had even begun down this path (and the sample was chosen to reflect best, not average, practice).

The hardware is similar across these different configurations; the software is available, although often not entirely satisfactory. Moreover, the McKinsey report estimated that the third stage would triple the benefits of CAD/CAM as measured in reduced design cycle time. But what is as yet lacking for most firms, even those actively pursuing CAD/CAM, is the managerial commitment to realizing that potential.

Our recent research[28] on the implementation difficulties of CAD/CAM integration in the electronics and aerospace industries has generated a framework for identifying these managerial challenges. Following Pava,[29] we find that the level or type of learning required to realize the benefits of a new technology varies systematically with the magnitude of the technological step being taken. Simple technical changes are implemented effectively at the cost of some retraining and the development of some new *skills,* for example, software capabilities in manufacturing engineering and process technology understanding in design engineering. More sophisticated changes require changes of *procedure* (for example, new manufacturability review procedures) and of *structure* (for example, centralizing an Advanced Manufacturing Engineering group). For the major technological changes of the kind well exemplified by the new, more flexible automation, changes are necessary in *strategy* (in particular, the development of functional technology strategies) and in *culture* (especially in the development of a more egalitarian relationship between the functions).

The final challenge, culture, is particularly critical. To make flexibility work in practice requires a cultural foundation linking new roles and new expectations. While the corporate culture fad will probably soon be supplanted by some other, Ouchi has presented a cogent argument as to why the underlying reality it refers to is an abiding one.[30] Forms of organization which rely on socialization and a common set of norms as the principal mechanism for control, can be shown to be more effective than either market or bureaucratic systems when "performance ambiguity" is high.

Culture thus becomes increasingly important, since both the new level and the continuing acceleration of technological change increase performance ambiguity: the new level of technology demands a higher commitment to learning rather than routine execution, and, with technology's continued acceleration, each new step in automation is a bigger one, which increases the ambiguity of objectives and of paths to attaining them. It is no accident, therefore, that many of the cases referred to in the corporate culture literature are in the more technologically dynamic industries.

A key challenge in capitalizing on the new technological potential for flexibility is, therefore, to adapt organizational cultures to maximize spontaneous cooperation. When knowledge is the critical resource, economic theory[31] has shown with considerable analytic rigor that under any realistic set of assumptions there can be no optimal incentive structure: a market system can create incentives that encourage the production of knowledge, but these incentives will inhibit its distribution; a bureaucracy—central planning—can mandate the optimal pattern of distribution, but fails to provide the incentives for the production of new knowledge. A third form of organization—cooperation—premised on shared values is the only form which can hope to surmount these complementary weaknesses. The challenge, however, is to create and sustain it.

IMPLEMENTATION

The implementation of new automated capabilities creates unique problems, the solutions to which seem to make a cultural fabric of cooperation a necessity, not merely a humanistic ideal.

The new systems' flexibility is such that a commitment to learning needs to have been well established as a culture, with the supporting strategies, structures, procedures, and skills. Failing that, fear of system underutilization and vulnerability will deter the participants from pursuing the new technological development opportunities.

Jaikumar's survey of Flexible Manufacturing Systems[32] is striking in this regard, since most of the systems in use in the U.S. today show a remarkable lack of flexibility. The average number of parts being produced on FMSs in the U.S. is 8—compared to over 30 for comparable systems in Japan[33] and perhaps as many as 85 in Germany.[34] Many U.S. firms appear to be paying lip-service to the idea of flexibility, while in reality using the FMSs as automated assembly lines. History may yet vindicate them: when FMS technology gets cheaper, such an approach might indeed be viable, even optimal. In the meantime, however, it is a very expensive recipe for frustration.

The difficulty facing U.S. managers is to maintain the commitment to continual learning in the face of schedule constraints that are, in the U.S. context, rarely lifted for long enough to permit the development and debugging of the programs required for more than a minimal set of parts.

Ettlie has surveyed users of flexible, programmable automation and confirms the results of numerous other studies of the conditions of effective implementation.[35] The principal factors of success appear to be the following:

- *A close relationship with suppliers:* rather than an arm's-length transaction, the complexity and flexibility of the new technology call for an ongoing collaboration over an extended period of time. Many respondents refer to the relationship as a "marriage."
- *A good fit of the proposed new process technology and the product range:* too many firms like the idea of flexibility, but have not established the link between their range of products and the types and degrees of flexibility that would make business sense. They consequently often over-order then under-utilize.
- *A clear strategic vision on the part of the user:* to guide the user through this gold mine of opportunities that so easily turns into a minefield of problems, users need a long-term strategy for their process technology development path.
- *The training of operators:* the need to actively pursue development of the flexibility potential in-house calls for in-depth training of both a hands-on and a theoretical kind. Merely operational training is insufficient.

The concept that seems to be emerging—if only with considerable difficulty—is that of "planning for effective implementation." Traditionally, implementation has been a residual task; but with the new technologies, the learning process is both so lengthy (often three years or more for an FMS) and so critical (without it, little flexibility is realized) that implementation finally begins to assume importance amongst the competing priorities of management.

LABOR FORCE CONSIDERATIONS

Of all the implementation issues, labor requirements are among the least well-managed:

- Vendors have traditionally been rarely available for advice or help—although that is now beginning to change, since the new technologies can't even get off the ground without major vendor commitments to the implementation phase.
- Vendors are also a somewhat interested party, especially when the capital expenditure is going to be justified on labor cost savings: many users of numerically controlled machine-tools or of word processing equipment were overly impressed by advertisements promising to reduce

dramatically their dependence on skilled machinists and typists.

- Internally, the assessment of the skill impact of new technology is a low priority task; the choice of the new equipment is itself motivated by technical capabilities and cost savings: the skills required to make it function effectively are rarely examined. The idea that the work force capabilities are themselves a critical competitive resource may get a mention in the annual "Employee communications" sheet, but almost never plays the kind of role in strategic or even operating plans that it should command.
- Once the equipment arrives, skill and training issues often take second place to the more urgent task of debugging, getting the system up and running, and getting manufacturing back onto schedule. When Manufacturing's mission is defined in traditional, narrow terms, manufacturing managers have little room to move.
- When attention is finally paid to the question of the optimal skill mix, it is usually in firefighting mode: how to absorb the displaced headcount or how to deal (reactively) with union job classification grievances.

To the extent that the skill requirements are planned for, it is in a largely unconscious mode. The "fantasy" that governs much of this unconscious process is one I have called the "deskilling myth": received wisdom and wishful thinking encourage managers to believe that new technologies permit them to make do not only with proportionately *fewer* workers—but also with *less-skilled* workers.[36] Of course there are some cases where deskilling is possible, but the evidence, meager as it is, suggests strongly that the general effect of new technologies—over-riding the local effects of any particular management's philosophy—is an upgrading one.[37]

The problem is that firms too often "back into" this upgrading, losing many of the advantages that can be gained from a more proactive approach to the process of transforming their human resource profile.[38]

Wheelwright and Hayes offer an insightful characterization of the manner in which firms planning to make manufacturing a competitive weapon will view their work force.[39] When management believes its process capabilities must evolve in a dynamic manner, work-force management has to be focused on stimulating worker learning rather than on "command and control" for execution of the standard, stable set of procedures. The contribution of workers in these more dynamic environments is via their attention rather than mere effort, since the central task is one of problem solving. Whereas in the static model of manufacturing, direct supervisory control is sufficient, in the more dynamic environment, the process specifications are rarely stable enough to avoid relying on indirect modes of control via systems and values.

Wheelwright and Hayes do not link the need to move to a more dynamic model specifically to technological tendencies. But much of the research on the implementation of advanced, programmable automation points unambiguously in this direction, as the following brief survey of some recent research in Germany, Britain, and the U.S. will confirm.

FLEXIBLE AUTOMATION AND SKILLS: AN EMERGENT PARADIGM

Recently, several colleagues and myself have independently and informally been polling companies on the major "show-stoppers" in flexible automation projects. We were not surprised to discover that literally every company we talked to had their share of horror stories. We were, however, surprised to discover that in the overwhelming majority of cases, it was the human resource management issues that were the major stumbling blocks in implementing the new technologies. These experiences seem to be pushing managers to reconsider their human resource policies.

Recent research might help identify the optimal policies. Indeed, a surprising degree of convergence can be found in a series of studies conducted in numerous countries, all pointing to advanced automation's new and higher skill requirements. To the casual observer, unaware of the tenor of research on automation and skill over the last 10 or 15 years, such a consensus might appear entirely natural. But in reality, it constitutes a remarkable transformation of the dominant discourse within the research community.

The automation and work research of the 1950s and 1960s was dominated by authors like Blauner, Woodward, and Mallet, who—despite considerable nuance between them—all saw automation leading to a recomposition of jobs and an upgrading of skills relative to the limited job requirements of the assembly line.[40]

Partly because of the prominence of less-skilled workers in the resurgence of class conflict in the late 1960s, and partly because of the internal limits of the older research (the optimism of which seemed based almost exclusively on the narrow base of continuous process industries), the late 1960s and the 1970s saw a very different approach dominate. A series of studies originating in different countries expressed a striking convergence on the proposition that automation's potentially favorable effect on skill requirements would not be realized, since automation's mode of deployment was a reflection of its social context. Authors like Braverman in the U.S., Freyssenet in France, Beynon and Nichols in the U.S., Kern and Schumann in Germany, and Panzeri in Italy all argued various forms of a single thesis: capitalist societies would tend to deskill work in their constant search for lower production costs and greater control over the production process.[41]

That there was an element of polemical intent in these studies was fairly evident. Nevertheless, and in the absence of systematic statistics, case studies were used to great effect to show: a frequent gap between workers' capabilities and job requirements (underutilization); instances where efficiency did seem to call for deskilling; and other instances in which managerial ideologies led to deskilling at the expense of efficiency. The microdynamics of power relations within the firm became the focus of attention and the premise for extrapolations to overall skill requirement trends.

Even its partisans had some difficulties with this deskilling thesis—in particular the multitude of counter examples, the preponderance of statistical evidence pointing to a distinct if modest long-run rise in skill requirements, and the need to include the impact of worker resistance and specific market conditions on skill outcomes. Research therefore then veered away from the "big generalizations" and began to focus on the microdynamics of automation and skill in specific institutional and market settings. This generated very worthwhile research into the "social construction" of skill definitions, into the impact of market conditions on relative bargaining power, and so forth. The *Sociologie du Travail* group in Paris can, in this respect, be compared to the research of Edwards in the U.S. and Gallie in Britain, to name but a few examples.[42]

Over the last five years or so, a change of tone has become manifest. The dynamism of capitalist work reorganization efforts and the extensiveness of industrial restructuring seemed to call for new efforts to reach viable, if modest, generalizations. But this time, the cases cited are much more frequently those of skill upgrading.[43]

What is retained of the preceding generation's work is an appreciation of the fact that in a market economy, these tendencies will typically manifest themselves in a chaotic manner, often leaving pockets of deskilling and redundancies that may indeed call for policy remedies. This lack of personnel planning is obviously a source of concern, both for workers who suffer and for those worried about competitiveness. (Lund and Hansen, for example, point out the serious difficulties that can lie ahead for firms which so polarize their skill distributions that there are no longer any promotion prospects for lower-level personnel—creating a morale problem—nor any internal recruiting possibilities for higher-level positions—a problem in some industries where much of the requisite production knowledge is noncodified.)[44] But despite these local and short-term issues, it seems reasonable to suggest that in general and over the longer run capitalist competition forces firms to seek out more productive combinations of machine and human capacities, and in the process the spontaneous outcome is, more often than not, an upgrading of worker skill requirements. Better labor force planning would then complement, rather than hold in check, the principal spontaneous tendency of industry.

One of the key conceptual difficulties of the automation-skill issue—a difficulty encountered by managers, engineers, and theorists alike—resides in the fact that automation is changing our notion of skill. This change in the nature of skill has added a new layer of complexity to old debates, since prognoses of deskilling often refer to a type of skill quite different from that implicit in upgrading views.

"Skill" is not just a one-dimensional variable; jobs differ in the types of skills they require. But beneath the multitude of specific occupational skills, one can identify certain common dimensions: one approach would be to distinguish the amount of training, the frequency of training, and responsibility, expertise, and interaction requirements. Much of the evidence of the recent research points to important changes in the content of these categories under the impact of automation. One way of mapping the nature of these changes is suggested in Table 26.1.

When skills shift from behavioral, experientially-based, individual attributes that are acquired once-and-for-all to capabilities that are more attitudinal, cognitive, social, and in continual evolution, it is easy for the worker to feel as if something tangible has been lost. To take an example: when control over the cutting tool path shifts from the skilled machinist to the part programmer, the machinist might feel that numerical control has undermined his or her distinctive competence. By the same token, a plant manager might be tempted to think that the new NC technology will permit important savings in training and thus in hourly wage rates. But the research underlying the new model suggests that the payoff to that NC investment will depend critically on the workers' sense of responsibility, problem-solving abilities, teamwork strengths, and willingness to regularly expand their capabilities as the automation itself evolves.[45]

Flexible automation thus increases the upside opportunity for the plant that appropriately matches new technologies with new skills. Flexible automation thus also increases the opportunity cost associated with an all-too-common form of myopia in skills management and training effort.

FOCUSING ON THE RIGHT STABILITIES

In conclusion, it may be useful to highlight one overarching lesson which emerges from a comparison between the managerial and the economic approaches to flexibility. In elementary economics, it is assumed that flexibility—flexibility of prices, of quantities, of location, of assets, etc.—is *prima facie* "a good thing." Of course, in more sophisticated economic analysis, phenomena such as the

TABLE 26.1. Old and New Content of Work

Factor	Old Content	New Content
Training amount	Minimized Narrow and shallow	Investment Broad and tall
Training frequency	One Time One-time investment	Continual Frequent retraining
Responsibility	Behavioral Responsibility for effort Discipline	Attitudinal Responsibility for process integrity and results Disposition
Expertise	Experiential Manual or rote	Cognitive Identifying and solving problems
Interaction	Low Stand-alone or sequential	Systemic Interdependence Team-work Interfunctional cooperation

importance of standards and the inevitability of various externalities make certain rigidities or stabilities appear as potentially beneficial in a world of "second-best" options. But these rigidities tend to be viewed with suspicion, since their presence can stop the invisible hand of the market from producing an optimal outcome.

From the managerial point of view, things appear almost inverted. Stability is a fundamental value, whereas flexibility is difficult to manage and *ceteris paribus* more costly. For managers, flexibility is potentially advantageous—and indeed, only becomes meaningful as a concept—against a backdrop of stabilities. The managerial question is therefore not simply how to reduce rigidities, but how to find the right mix of stabilities and flexibilities. That is why the deliberate introduction of certain stabilities can be a powerful policy lever: the rigidity of the Just-in-Time inventory management regime helps to uncover the production inefficiencies hidden under inventory.

The difference in approach derives primarily from the fact that for the manager, the reality of externalities, standards, increasing returns, etc. is the stuff of daily competitive life; while for the economists, they are concepts that only become important as higher-level refinements of the basic theory. Indeed, the tension between the two perspectives can only grow as technology becomes more important to competitiveness and the proactive management of standards and knowledge-sharing become more central.[46]

Managers therefore need to focus on identifying the stabilities which offer the greatest leverage. Several candidates are:

- the stability of an explicit and credible long-term technology strategy can allow lower-level managers to be flexible in local innovation;
- employment stability can generate the good will needed to maintain a culture of cooperation;
- stable "marriages" between vendors and users of new equipment help maintain flexible collaboration in the development of new technologies;
- a stable development growth path for worker skills can ensure that the knowledge mix evolves in the way needed for effective deployment of the new technologies.

In this context, an issue which this article has not yet addressed becomes critical: Industrial Relations. The institutionalization of employment conditions and worker-management relations is often perceived as an impediment to the flexibility that many economists see as necessary for the invisible hand to do its magic in the labor market; in this domain at least, many managers share the economists' view, seeing institutionalized industrial relations as inimical to the maintenance of a personalized, individualized relationship with their workers.

Such approaches to the issue might no longer be adequate. Walton argues persuasively that business efficiency can be significantly enhanced by a shift away from the "control" model—with its associated adversarial labor relations and its corollary of a preference for union-free environment—toward a "commitment" model—in which, rather than attempting to decertify unions, managers attempt to establish mutuality and joint planning.[47] The commitment model uses a constructive Industrial Relations climate to derive competitive advantage from the strengths of a more stable, motivated, and loyal work force. As we have seen, such advantages might become more valuable as firms adopt more dynamic technological strategies and move towards flexible automation: the studies reviewed in this article consistently point to the growing importance of lower-level personnel's problem-solving capability—and their motivation to use it.

The recommendation seems so compelling that one has to ask why more firms don't adopt the commitment model. The problem is easily identified: in order for unions to play a constructive role, they cannot relinquish the functions that have pitted them against management in their traditional adversarial role. To remain significant partners, they must continue to express and represent workers' interests. Experience, moreover, teaches that these interests are not always congruent with those of the firm's managers and share-holders. In the absence of unions, some of these divergences might be less acute, or at least less apparent. But as Reisman and Compa argue, the sources of these divergences are real enough that "if the existing unions cannot help them accomplish [their] goals, workers will find other approaches and methods."[48]

To derive the benefits of the commitment model advocated by Walton would therefore seem to require the construction of a system of Industrial Relations in which Labor and Management can pursue collaborative efforts at the same time as they give organized expression to their inevitable conflicts. The culture of cooperation needed to sustain the development and effective deployment of flexible automation will need to be sufficiently robust to absorb the inevitable tensions between various stakeholders' interests.

The problem for managers is that it is difficult to construct a single island of new industrial relations in a sea whose principal currents are oriented toward the destruction of all "labor market rigidities." Not all the answers to U.S. industry's competitiveness in the use of flexible automation are in the manager's hands.

APPENDIX—WHAT IS FLEXIBILITY?

The new developments in technology have prompted efforts to clarify the significance of the concept "flexibility." At this stage of research, no one approach has gained widespread acceptance. The definitions of Gerwin, Mandelbaum, Buzacott, Zelenovic, Browne, and Jaikumar have a somewhat ad hoc, domain-specific flavor.[49] The economic approach focuses on another, equally specific set of issues.[50]

Attempting to synthesize these various notions, we find that the economic definition is the most generic; Zelenovic's distinction of design and adaptation flexibility is a little less so; the others are partially overlapping.

The conceptual difficulty appears to reside primarily in linking the two key dimensions of flexibility: process and product. In each dimension, we can identify successively broader system boundaries: the process dimension encompasses individual machines, then systems, and finally the overall plant; the product dimension encompasses product mix, then design changes, then new products, and finally new product generations. (See Figure 26.3.)

Mandelbaum's action and state flexibility might be thought of as another way of distinguishing process and product; but Gerwin, Buzacott, Jaikumar, and Browne mix both dimensions in their typologies. For example, Gerwin's mix, design change, and parts are on the product dimension, whereas his volume and routing dimensions are on the process dimension.

	Mandelbaum	Gerwin	Buzacott	Browne	Jaikumar
Product:	state		state		
product mix		mix		production	
design changes		design change			
new products within family		parts		product	product
new families					
Process:	action				
			machine job	machine process	
system		routing		routing operations	process
plant		volume		expansion volume	program

Fig. 26.3. Product and process dimensions of flexibility.

From the engineers' point of view, it is the process dimension that seems most exciting: one of the most promising aspects of the current phase of automation lies in our growing ability to design into machine systems sufficient flexibility and intelligence to make them far more robust relative to a broad spectrum of process contingencies. Whether it be relative to breakdowns or to finding a given machine unexpectedly unavailable, the ability to automatically reroute products offers the promise of higher utilization rates through expanded flexibility in the process dimension.

From the societal and managerial points of view, however, the bigger challenges and opportunities seem to derive from the development of flexibility relative to changes in the product dimension. The new technologies permit the flattening of the average cost curve relative to a number of key competitive dimensions—product change-over time, new product manufacturing ramp-up time, product development cycle time. This flattening has the potential of transforming the rules of the competitive game by undercutting the cost advantage of the standardized commodity produced in long runs.

REFERENCES

1. P. S. Adler, "When Knowledge is the Critical Resource, Knowledge Management is the Critical Task," forthcoming in *IEEE Transactions on Engineering Management.*
2. U.S. Office of Technology Assessment, *Computerized Manufacturing Automation: Employment, Education, and the Workplace* (Washington, DC: USGPO, 1984).
3. M. Piore and C. Sabel, *The Second Industrial Divide* (New York, NY: Basic Books, 1984).
4. J. P. Goldhar and M. Jelinek, "Plan for Economies of Scope," *Harvard Business Review* (November/December 1983), pp. 141–148.
5. G. Boothroyd, "Economics of Assembly Systems," *Journal of Manufacturing Systems,* 1/2 (1982).
6. G. K. Hutchinson and J. Holland, "The Economic Value of Flexible Automation," *Journal of Manufacturing Systems,* 1/2 (1982).
7. R. H. Hayes and W. J. Abernathy, "Managing Our Way to Economic Decline," *Harvard Business Review* (July/August 1980).
8. R. B. Reich, *The Next American Frontier* (New York, NY: Times Books, 1983).
9. E. S. Buffa, *Meeting the Competitive Challenge* (Homewood, IL: Dow Jones-Irwin, 1984).
10. J. Woodward, *Industrial Organization: Theory and Practice* (London: Oxford University Press, 1965).
11. R. H. Hayes and S. C. Wheelwright, "Link Manufacturing Process and Product Life Cycles," *Harvard Business Review* (January/February 1979).
12. Boothroyd, op. cit.
13. R. H. Hayes and S. C. Wheelwright, *Restoring Our Competitive Edge* (New York, NY: John Wiley & Sons, Inc., 1984).
14. K. Ferdows, "Technology-Push Strategies for Manufacturing." *Tijdschrift voor Economie en Management,* 28/2 (1983).
15. W. J. Abernathy, *The Productivity Dilemma* (Baltimore, MD: Johns Hopkins University Press, 1978).
16. D. H. Ginsburg and W. J. Abernathy, eds., *Government, Technology, and the Future of the Automobile* (New York, NY: McGraw-Hill, 1980).
17. W. J. Abernathy, K. B. Clark, and A. M. Kantrow, *Industrial Renaissance* (New York, NY: Basic Books, 1983).
18. N. Rosenberg and C. Frischtak, "Technological Innovation and Long Waves," *Cambridge Journal of Economics,* 8 (1984).
19. S. C. Wheelwright and R. H. Hayes, "Competing Through Manufacturing," *Harvard Business Review* (January/February 1985).
20. Ibid.
21. See Appendix; J. R. Meredith, ed., *Justifying New Manufacturing Technology* (Norcross, GA: Institute of Industrial Engineers, 1986).
22. Manufacturing Studies Board, National Research Council, *Computer Integration of Engineering Design and Production: A National Opportunity* (Washington, D.C.: National Academy Press, 1984), p. 16.
23. R. S. Kaplan, "Yesterday's Accounting Undermines Production," *Harvard Business Review* (July/August 1984), pp. 95–101.
24. B. Gold, "Strengthening Managerial Approaches to Improving Technological Capabilities," *Strategic Management Journal,* 4 (1983).
25. R. H. Hayes and D. Garvin, "Managing as if Tomorrow Mattered," *Harvard Business Review* (May/June 1982).
26. J. E. Hodder and H. E. Riggs, "Pitfalls in Evaluating Risky Projects," *Harvard Business Review* (January/February 1985).
27. McKinsey & Co., Inc., "Forging CAD/CAM into a Strategic Weapon," January 1984.
28. P. S. Adler and D. A. Helleloid, "Effective Imple-

mentation of Integrated CAD/CAM: A Model," *IEEE Transactions on Engineering Management* (May 1987); P. S. Adler, "Managerial Challenges of CAD/CAM Integration," Industrial Engineering and Engineering Management Department, Stanford University, 1988.
29. C. H. P. Pava, *Managing New Office Technology* (New York, NY: Free Press, 1983).
30. W. G. Ouchi, "Markets, Bureaucracies and Clans," *Administrative Science Quarterly* (March 1980).
31. K. Arrow and L. Hurwicz, eds., *Studies in Resource Allocation Processes* (Cambridge: Cambridge University Press, 1977).
32. R. Jaikumar, "Post-Industrial Manufacturing," *Harvard Business Review* (November/December 1986).
33. T. Ohmi and Y. Yoshida, "Flexible Manufacturing Systems in Japan—Present Status," in *Flexible Manufacturing Systems: Proceedings of the 1st International Conference, Brighton, U.K.* (Amsterdam: North-Holland, 1982).
34. G. Spur and K. Mertins, "Flexible Fertigungssytems." *Produktionsanlagen der Flexiblen Automatisierung ZwF*, 76/9 (1981). For more data on these international differences, see J. Bessant and B. Haywood, "The Introduction of Flexible Manufacturing Systems as an Example of Computer-Integrated Manufacturing," *Operations Management Review* (Spring 1986).
35. J. Ettlie, "The Implementation of Programmable Manufacturing Innovations," Industrial Technology Institute, March 1983.
36. P. S. Adler, "New Technologies, New Skills," *California Management Review* (Fall 1986).
37. K. Spenner, "Deciphering Prometheus," *American Sociological Review* (December 1983).
38. M. Blomberg and D. Gerwin ("Coping With Advanced Manufacturing Technology," *Journal of Occupational Behavior*, 5: 113–130) document a case of poor job design in an FMS. B. Jones and P. Scott ("Working the System: A Comparison of the Management of Work Roles in American and British Flexible Manufacturing Systems," paper presented to annual conference of the Operations Management Association of Great Britain, University of Warwick, January 2–3, 1986) document a case of backing into more appropriate job design in an FMS.
39. S. C. Wheelwright and R. H. Hayes, "Competing Through Manufacturing," *Harvard Business Review* (January/February 1985).
40. R. Blauner, *Alienation and Freedom* (Chicago, IL: University of Chicago Press, 1964); J. Woodward, *Management and Technology* (London: H. M. Stationary Office 1958); S. Mallet, *La Nouvelle Classe Ouvriere* (Paris: Seuil 1963).
41. H. Braverman, *Labor and Monopoly Capital* (New York, NY: Monthly Review Press, 1974); H. Freyssenet, *Le Processus de Dequalification-Surqualification de la Force de Travail* (Paris: Centre de Sociologie Urbane, 1974); H. Beynon and J. Nichols, *Living with Capitalism* (London: Routledge-Kegan Paul, 1977); H. Kern and M. Schumann, *Industriearbeit und Albeiterbewusstsein* (Frankfurt A. M.: Europaische Verlagsanstalt, 1972); R. Panzeri, "Sullinso Capitalistico della Macchine," *La represa del Marxismo-Leninismo in Italia* (1972).
42. R. Edwards, *Contested Terrain* (New York, NY: Basic Books, 1979); D. Gallie, *In Search of the New Working Class* (Cambridge: Cambridge University Press, 1978).
43. R. E. Walton and G. I. Susman, "People Policies for the New Machines," *Harvard Business Review* (March/April 1987); R. Schultz-Wild and C. Kohler, "Flexible Manufacturing Systems: Manpower Problems and Policies," *Journal of Manufacturing Systems*, 4/2; H. Kern and M. Schumann, "Limits of the Division of Labour," *Economic and Industrial Democracy*, 8 (1987); A. Sorge, G. Hartmann, M. Warner, and I. Nicholas, *Microelectronics and Manpower* (Berlin: Gower, 1983); P. Senker, *Towards the Automatic Factory?* (Berlin: IFS [Publications] Springer-Verlag, 1986); L. Hirschhorn, *Beyond Mechanization* (Cambridge, MA: MIT Press, 1984); R. T. Lund and J. A. Hansen, *Connected Machines, Disconnected Jobs* (Cambridge, MA: MIT Center for Policy Alternatives, 1983).
44. Lund and Hansen, op. cit.
45. P. S. Adler and B. Borys, "Automation and Work: The Machine Tool Case," Stanford University, 1986.
46. Some economists, recognizing this problem, are generating important results from more complex models (see, for example, P. A. David, "Some New Standards for the Economics of Standardization in the Information Age," *Stanford Center for Economic Policy Research, Working Paper No. 11*, 1986). The issue in the future may well become whether formal economic models can handle the complexity of pervasive externalities and standards without losing analytic tractability. See P. S. Adler, "When Knowledge is the Critical Resource, Knowledge Management is the Critical Task," Stanford University, 1985.
47. R. E. Walton, "From Control to Commitment in the Workplace," *Harvard Business Review* (March/April 1985).
48. B. Reisman and L. Compa, "The Case for Adversarial Unions," *Harvard Business Review* (May/June 1985), p. 36.
49. D. Gerwin, "The Do's and Don't's of Computerized Manufacturing," *Harvard Business Review*, 60/2: 67–116; M. Mandelbaum, "Flexibility in Decision

Making: An Exploration and Unification," Ph.D. dissertation, Dept. of Industrial Engineering, University of Toronto, Ontario, Canada, 1978; J. A. Buzacott, "The Fundamental Principles of Flexibility in Manufacturing Systems," *Proceedings of the First International Conference on Flexible Manufacturing Systems,* Brighton, U.K., 1982; D. M. Zelenovic, "Flexibility—A Condition for Effective Production Systems," *International Journal of Production Research,* 20/3 (1982); J. Browne, "Classification of Flexible Manufacturing Systems," *The FMS Magazine* (April 1984); R. Jaikumar, "Flexible Manufacturing Systems: A Managerial Perspective," Harvard Business School, 1984.

50. R. A. Jones and J. M. Ostroy, "Flexibility and Uncertainty," *Review of Economic Studies* (1984).

27

A Perspective on Quality Activities in American Firms

NORIAKI KANO

Since 1963, when I was an undergraduate being supervised by Professor Kaoru Ishikawa at the University of Tokyo, I have been a student of quality management and engineering. Over the years, I have done research and taught in the field, widely disseminating its findings. I visited the U.S. for the first time in 1977. In over thirty subsequent visits, I have engaged in plant visits, seminars, conferences, and presentations; counseled nearly twenty firms; and taught and lectured at several universities. This article draws upon these experiences in evaluating recent quality trends in the United States.

MY PERSONAL VIEW ON TQM

In order to evaluate TQM in the U.S., I need to clarify how I arrive at my judgments. The following three sections lay out my analytical framework.[1]

Generalizing an Individual Observation

It is reasonable to ask how an observer synthesizes a specific observation from Workplace A, Factory B, or Company C into a view that allows generalization from the specific area to the whole unit. I adopted the following approach for generalizing my observations. I would explain to the managers of the factory what I observed in a workplace and what generalizations I derived from these observations. Then I would ask them whether each of these views was applicable only to the workplace I visited or whether it applied to the factory as a whole. If the views coincided, then I would conclude that they were characteristic not only of specific workplaces but also of the entire factory.

For example, in one case I observed that in the final inspection area, all the inspections were properly conducted, with the detected defectives being properly reworked. Then I visited the production workplace where one of the defectives had been produced. While the workplace supervisor had been informed of the defectives, I observed that the supervisor provided little more than a warning to the worker who had assembled the defective component. Moreover, he failed to analyze why this defective had been produced. As a result, the worker had difficulty knowing what preventive action to take in the future. I generalized these observations into a view, noting that the final inspection had been properly conducted, with the detected defectives being properly reworked, but the feedback was not properly analyzed to prevent future defects. If the managers agreed with my view that this was a common problem, then it could be accepted as a generalized pattern for the factory.

I would then sum up the views of the factories that I visited in a particular area, such as a region or country, and would incorporate their common elements into a hypothesis, a tentative general view, for the area. I would then ask the local quality specialists whether the hypothesis was generalizable to the factories in the area or applied only to the specific factories visited. If they agreed with me about the common elements, the generalization would be considered valid for that area.

The views selected did not necessarily encompass every feature of the area. It can be said, however, that these features were at least partially represented in each portion. The degree to which this is true can be determined by local experts with a thorough knowledge of the factories in the area.

The House of TQM: Structure of TQM[2]

I use "The House of TQM," which shows the structure of TQM, as the touchstone for identifying the evolution of TQM's implementation in the United States (see Figure 27.1). The portion from the floor up to the roof is TQM, where the floor symbolizes "motivational approach" and the roof shows "customer satisfaction/quality assurance"—which is the purpose of TQM. The base is the "intrinsic technology," which refers to the driving technology specific to an industry. For example, electrical engineering is the technology intrinsic to the electric industry; chemical engineering is intrinsic to the chemical industry. Intrinsic technology provides the necessary foundation upon which TQM is built.

Given the existence of intrinsic technology, it is still necessary to carry out "sweating work" (hard work such as promoting standardization, educating and training, and collecting and analyzing data) to obtain good quality. The problem is this: how to create the conditions that will impel management and employees to take up such sweating work. This "motivational approach" is shown as the floor in Figure 27.1.

"Concepts," one of the pillars of the house, shows how to proceed from a particular perspective when a given intrinsic technology and motivation already exist. "Concepts" consists of both a theory of quality (such as "quality is the satisfaction of the customer" or "the next processes are our customers") and a theory of management (such as the "PDCA Cycle" [Plan, Do, Check, Act], "build quality into the processes," or "management by facts"). Figure 27.1 shows this as the column on the left.

When actual activities based on these concepts begin, some "techniques" (the second pillar) for collecting and analyzing data become necessary. The seven QC tools and "The QC Story" procedure are typical techniques for this purpose, and statistical methods can also be applied here.

At this point, some methods for effectively and efficiently promoting all of these activities within the organization become necessary. Management by Policy, Daily Management, and

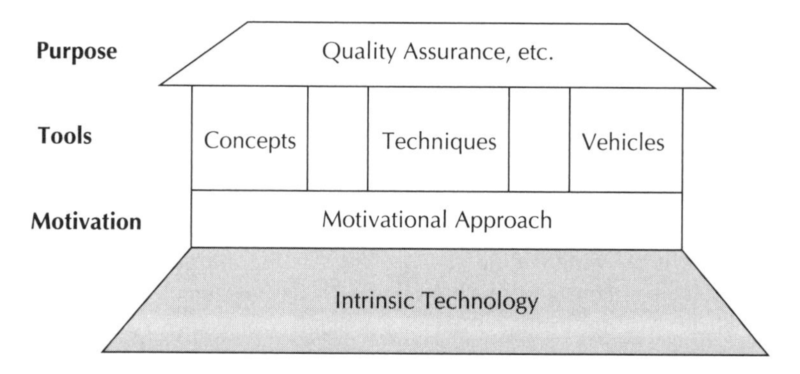

Fig. 27.1 House of TQM—structure of TQM.

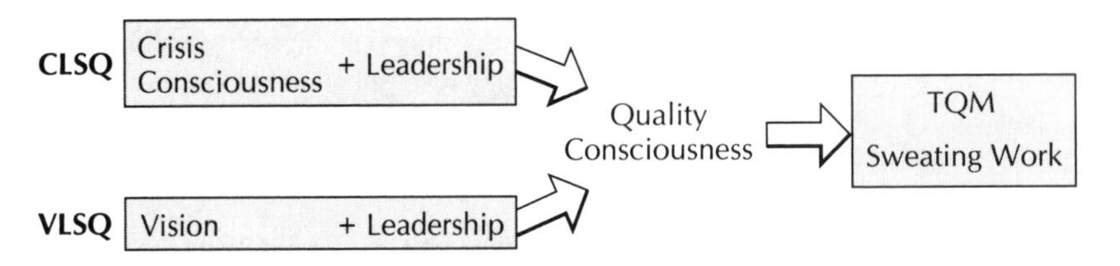

Fig. 27.2 Model for TQM introduction based on quality sweating theory.

QC Circles are methods that can be called "vehicles" (the third pillar) since they quicken and facilitate promotion.

Quality Sweating Theory

The process of introducing TQM can be explained by the "Quality Sweating Theory."[3] It states that TQM is an effective tool for improving quality, but its success depends on making many employees sweat. The quality sweating theory, encompasses two alternative approaches: CLSQ (Crisis consciousness and Leadership make people Sweat for Quality) and VLSQ (Vision and Leadership encourage people to Sweat for Quality). (See Figure 27.2.)

CHANGE OF QUALITY ACTIVITIES IN THE U.S. OVER THE LAST DECADE[4]

In 1980, NBC aired a T.V. program, "If Japan can, why can't we?" It might be said that this program symbolized the historical shift of the American quality movement. Around 1980, quality circle activities began booming in the United States. By the middle of the 1980s, however, the quality movement had come to emphasize the necessity of management commitment, stressing the importance of top management's leadership. Subsequently, many companies tried to promote company-wide quality activities by motivating executives and managers. This contrasts with past American practice where quality professionals played the primary role in quality activities. In recent years, this trend has been increasing. Consequently, some companies have successfully introduced and implemented Japanese TQC. A nation-wide quality improvement campaign has been spearheaded by the Malcolm Baldrige National Quality Award (MBNQA). "TQM" is called "TQC" in Japan although its activities can be better explained by TQM (Total Quality Management) than by TQC (Total Quality Control) because both "management" and "control" are foreign terms for the Japanese, who understand them as synonymous.[5] Hence, in this article, the term TQM is used in order to avoid confusion except when specifically discussing Japan, in which case the term TQC will be used.

Where They Were 10 Years Ago

On my desk, I have my 1982 lecture notes on "Overseas Quality Activities"[6] as well as an article of mine based on an interview with Dr. J. M. Juran at his New York office during July 1979.[7] They show the state of quality activities in the U.S. during that latter half of the 1970s and the beginning of the 1980s. My views on that period are as follows:

- The education of top management is less focused on quality.
- Some proponents of the quality circle movement simplistically believe that quality circles can solve all the quality problems of a company. This characteristic tends to make management neglect the necessity of committing itself to quality.

- Cross-functional communication is so weak that it is difficult to solve quality problems, which demand cooperation among departments.
- The cost of quality is too often over-emphasized.
- Strong resistance to the introduction of quality improvement activities is manifested by the "NIH" syndrome (resistance to that which was "Not Invented Here") even though these quality improvement activities seem to be effective in solving quality problems.
- The "AQL" (Acceptable Quality Level) practice of parts procurement as well as a short-term view for "achieving best quality at lowest cost" become obstacles to improving quality. Instead, managers need to make purchase decisions based on a long-term approach to continuous improvement.

Changes over the Last Decade: Two Episodes

In 1977, I visited the United States for the first time, staying for a month. In the production shop of a certain factory well-known for its good quality management, I found a percent-defective graph on the bulletin board. As I hadn't seen such a graph in any other factory which I had visited during the trip, I had a closer look at it. I found that the last data point plotted was in July 1976, eight months before my visit to the factory.

It was not until 1984, when I visited an American factory of a multinational corporation, that I felt things had begun to change in the United States. To my surprise, a considerable number of quality slogans and quality indicator graphs were displayed throughout the factory. One of the graphs carried data points for a control indicator at a certain process plotted hourly; the last data point was plotted at 11:00 of the day I was visiting. (I happened to see that graph around 11:30.)

Several other factories that I visited later also had many quality indicator graphs, including control charts. They were practicing "SPC" (Statistical Process Control), and data points were accurately plotted. Of course, I am not going to say that simply plotting data on graphs makes quality better, but this observation does illustrate a change in quality consciousness in the U.S. by the mid-1980s.

The Impact of the Malcolm Baldridge National Quality Award (MBNQA)

It could be said that the MBNQA precipitated the broad-based promotion of quality improvement within American companies. More importantly, it presented national quality leaders with opportunities to discuss the direction of the American quality activity systems through the formulation or revision of the guidelines for the MBNQA. They could do this during both the intensive training course for examiners and during the examiners' site visits as they audit the finalists. The successive revisions of the guidelines illustrate this development. Through this process, I believe that the institutional consensus among the experts is deepening and the American approach to quality is solidifying. Before the MBNQA was established, the consensus for nation-wide quality activities had neither been very strong nor well-organized, even though there were many excellent quality experts.

Change in the Quality System

In any country or organization, total quality activity can begin only if top management is conscious of the critical need for company-wide commitment to quality and its own responsibility for introducing such activity. As testimony to the leadership that top management is exercising today in some American companies, in lectures at quality-related conventions, top managers have begun to include specific examples of their own practices, instead of just sprinkling their speeches with such phrases as "quality first," "customer satisfaction," or "do it right the first time." Also, the top managers of leading U.S. companies were influential in establishing the MBNQA. Furthermore, although formerly only a few companies had Vice Presidents in charge of quality, in recent years their number has increased. In some companies, Senior Vice Presidents and Executive Vice Presidents in charge of quality have appeared. Finally, top managers of many American companies are visiting Japan to study TQM, a reflection of their positive attitude toward the study of quality. In summary, over the last decade in the U.S., what has changed most is the leadership of top management. Whereas before, quality activities were only implemented by the quality department, top management is now initiating corporate-wide quality activities.

As mentioned, a criterion that formerly dominated quality efforts in the U.S. was quality cost. Today, this criterion is increasingly less relevant and is being superseded by "customer satisfaction." In the audit for the MBNQA, this has become evident. Although "customer satisfaction" is only one of many audit categories, it is given the largest weight of all: 30%.

In U.S. companies that have introduced quality activities in accordance with Japanese methods, their implementation differs slightly from the original Japanese versions because of their different backgrounds. For example, there is an emphasis on team activities in the U.S. version, which can be seen as an effort to break through departmental barriers resulting from a strong chain of command. The term "cross-functional," which is frequently heard in the U.S., reflects the consciousness of barriers between departments.

GENERAL TRENDS OF QUALITY ACTIVITIES IN THE UNITED STATES[8]

Based on the House of TQM, the current status of quality activities in the U.S. can be analyzed using the categories of motivation, concepts, techniques, and promotion.

Motivation

As Dr. Juran said in 1979, "Top management believes that the situation is not so bad yet, and that is a problem of American firms."[9] After the economic recession of the early 1980s, however, America came to face growing trade and financial deficits. This seems to have given many people, especially top management in some leading companies, a crisis consciousness. As a result, top management has begun to play an important role in leading the quality movement. This explains why American firms have become so enthusiastic about quality in recent years.

Concepts

There are four criteria for implementing TQM:

- *Customer satisfaction; putting quality first.*
- *The PDCA cycle; Process-oriented production; Doing it right the first time.* Although the PDCA Cycle is well-known, many people have difficulty applying it to their jobs. Therefore, many facilitators try to implement this concept via management by policy and daily management. Often, when they do this, employees ask managers to improve their own way of working. Some resistant employees play with the acronym PDCA, turning it into "Please don't change anything!"
- *Emphasis on the use of data.* Along with the diffusion of the SPC method, the concept of management based on data is widely used. It is easy just to collect data and apply the method mechanically. In each case, however, what physical phenomena the data represent and what the analytical results actually mean need to be individually investigated and discussed, to determine whether the method used adequately reflects the significance of the data. In general, people are not taught to recognize the tangible phenomena based on the data. In most areas, managers and workers alike tend to do their jobs depending on their gut reactions, not on the data.

  Consider the following example. On my first trip to Florida Power and Light, we discussed the reduction of service interruptions. The FPL people made a presentation with a Pareto diagram of interruptions by causes showing that lightning was the largest cause. To my question of why the groundings or arresters had not prevented the interruptions, they answered that Florida had the most frequent and severe lightning attacks in the United States and that groundings or arresters would not have prevented service interruptions produced by strong lightning. I again asked, "Please show me the data which explain that intense lightning caused interruptions so severe that groundings or arresters could not prevent them." They could not show me any data, but promised to collect data by my next visit. Nearly half a year later when I visited again, they had collected the data and found that interruptions occurred even when they did not have such strong lightning. In addition, they found a certain percentage of poles with absent or insufficient groundings, which they had not recognized until they collected the data.

- *Employees' commitment; management is not a monopoly of managers.*

Techniques

The most popular techniques are the Pareto diagram, cause & effect diagram, and the histogram which are among the "Seven QC Tools." This is the same phenomenon which I observed in Japan. Some American companies are also starting to show an interest in applying the "QC Story," which is a standard procedure for problem-solving developed in Japan.

Due to the popularity of "SPC seminars," production lines are often flooded with control charts. In many cases, however, I observed that actions had not always been taken when the charts showed out-of-control situations. When I asked supervisors how they examined such facts and what actions they had taken, in most cases they could not explain their specific measures. Nevertheless, I have heard from the quality facilitators of the companies that have introduced control charts in their workplaces that their quality has been rapidly improved by the introduction of SPC. One explanation for this is that the very process of collecting data and plotting points on the charts causes foremen and workers to pay more attention to quality. Thus, even poorly used control charts can contribute to the process of improving quality by making workers more conscious of quality.

In addition to control charts, other statistical methods such as analysis of variance and multiple regression analysis have been taught to engineers and managers. I have not, however, found many successful applications of these techniques in U.S. firms, yet. In the factories I observed, a remarkable number of problems were solved by combining intrinsic technology with simple methods such as the Seven QC Tools.

In the Toyota Motor Company, which is well-known for its enthusiastic devotion to quality, when management discusses how much quality has been improved and how much they should stress further improvement, it is often said, "People make efforts to squeeze a dry towel drier." This towel metaphor means that in its early stages, a company with a dripping towel gets rid of a lot of water with just a little squeeze, without any tools. As the company evolves, a wringer might be needed. In an advanced stage, an electric vacuum drier is required. If asked how much the U.S. has improved its quality so far, I explain that at present the U.S. is using a wringer, such as the Seven QC Tools, to bring about effective results. I assume that more sophisticated methods, on the level of the electric vacuum drier—such as analysis of variance or multiple regression analysis—shall be required for U.S. firms to progress further.

The Taguchi Method and "QD" (Quality Deployment, sometimes called "QFD," Quality Function Deployment) also seem to have become somewhat popular. People are also beginning to show an interest in the "Seven New Management Tools for QC." Some people seem to find somewhat flamboyant ways to demonstrate certain tools, rather than using them to solve problems. This is a common phenomenon at the early stage of TQM's introduction into any country. As TQM develops further in the U.S., this phenomenon should disappear. It is nothing to worry about.

Promotional Vehicles

Promotional Organization

Who shall ultimately be responsible for promoting company-wide quality? The answer to that question is the key factor to successful implementation, especially in the United States. In companies which promote company-wide quality activities, I find the rank of the executive responsible for this job differs by company. In one company, the Executive Vice President is in charge; in another it is the Vice President or Director. Those companies which elevate someone from personnel to become the responsible Director seem to have a great deal of difficulty. The new Director faces such problems as how to persuade high-level line managers such as the Vice President of Manufacturing or the Division General Manager to take necessary action. I recall the late Dr. Kaoru Ishikawa mentioning that he used to think that the staff function remains strong in Western companies. But, in reality, the situation is the same as in Japanese companies, where the line function is stronger than the staff function, so that inter-departmental barriers are very thick.

When it comes to promoting quality in implementation units—such as a division, factory, and branch office—successful results depend on whether or not their heads, general staff, and managers are actively leading the promotion. On this point, I found no differences between American and Japanese companies.

Education and Training

In classrooms and workshops, American companies promote very extensive, and sometimes very intensive, training programs in quality management philosophies, management tools, and statistical methods. It seems to me, however, that they have difficulty in adapting these philosophies and techniques to their workplaces. This is partially due to the shortage of able and seasoned quality experts who can support and encourage both executives and managers to promote company-wide quality management by counseling them in actual situations as they occur.

In the U.S., the exchange of ideas between companies has increased along with the number of companies which promote company-wide quality activities. Given the brief history of company-wide quality promotion, dependence on information from other U.S. companies suffices neither qualitatively nor quantitatively. I assume, however, that the exchange of quality information will improve in the future.

Quality Circles

In the mid-1970s, Mr. Wayne Rieker at Lockheed Aircraft Co. first introduced quality circles to the United States. By the 1980s, the "IAQC" (International Association for Quality Circles), which was organized to promote the concept, was actively diffusing it. At the end of 1987, IAQC was reorganized and renamed "AQP" (Association for Quality & Participation). In addition to promoting quality circles, it holds numerous seminars on matters relating to quality and participatory management.

Quality circle activities seemed to peak around the middle of the 1980s. This peak points to the end of one phase of the quality movement. Companies had initially defined quality problems in relation to workers' morale and workmanship, so they believed those problems could be solved by quality circles. In the early 1980s, many American management study teams came to Japan. Most of them were interested in nothing more than quality circles. Gradually, however, many of them learned that promoting quality circles alone was not sufficient to achieve quality. So they began, sometimes hesitatingly, to emphasize company-wide implementation of quality activities.

These movements may also have been affected by the article "Quality Circles After the Fad," published in the *Harvard Business Review* by Prof. Edward Lawler III and associates.[10] The authors sharpened my own reflections on what happens to a company if it introduces quality circles without promoting company-wide quality activities.

Although the quality circle boom has been declining, it is clear that group activities of workers, such as quality circles, are essential to a comprehensive implementation of quality. I believe that U.S. companies will need to reactivate quality circles as part of the future development of TQM.

Team Activity

There are several different types of team activities in the United States. One type is similar to quality circle activity and involves a voluntary problem-solving group made up of workers in the same workplace. Upon completing one theme, the same team will work on the next one in an on-going activity. These teams are sometimes called functional teams since each group at the bottom of the organizational hierarchy shares one operational function.

In a second type of team activity, a Japanese and American joint venture company in the U.S. defines the team as the unit to which a set of jobs is assigned. In typical American companies, a job is assigned to each worker. In this case, a team of 10 workers is given a set of jobs to be shared, flexibly, among team members. In the unionized sector, generally, a detailed job classification is developed jointly with the labor union, and it is not unusual to see over 100 different kinds of jobs classified in just one factory. To adopt an alternate type of team approach, job classifications need to be reduced to allow workers to do multiple jobs. Since this change is linked to the wage system, it necessitates negotiating with the unions. Because these teams are formed primarily to build quality into a product during the assembling process as specified, rather than to improve it, they can be called "quality building-in process teams." Several workers in a plant where this team method has been adopted told me that their transformation from single-skill to multiple-skill workers has made them more interested in the job and more aware of quality.

A third approach involves initiating a team comprising staff or line managers, usually formed on orders from management. It is often called a task team or a project team. Once the task is completed, the team is disbanded. Participation in such teams, just like participation in quality circles, depends in principle on the free will of individual employees. Such teams are often formed across different departments; the nature of the task given sometimes demands the creation of two other types of teams: a problem-solving team (a team to address a particular problem arising in the process of promoting quality activity) and a task-achieving team (a team formed to implement a particular project, such as developing a new product).

Of the three types, the third one is the most standard team activity. It seems to me, however, that its content and nature varies from company to company, or even from department to department within the same company. In some cases, members of the team work on the task part-time while they do their regular job back in their own section. In other cases, members work full-time on the team activity. I also heard that the freedom to go beyond departmental barriers to do the activity required by the objective helped revitalize both staff and line managers who participated. This method is certainly useful in temporarily removing the barriers between departments. It could, however, produce new problems, such as a need for accumulated technical expertise and communication between different teams. Whatever the merits or demerits may be, this type of teamwork should not be overlooked when talking about quality activity in the U.S. today.

Management by Policy (Hoshin Kanri)

Of the various elements of Japanese TQC, what most interests American management is customer satisfaction activity and management by policy. "MBO" (Management by Objective) has long held a central place for American managers. But, in actuality, in many companies it has become a mere skeleton, without much substance. In contrast, management by policy looks very fresh. It has the following features:

- *The Four Stages of Management By Policy:* policy setting, policy deployment, policy implementation, and evaluation and feedback.

- *Linkage of Annual Company Policy with Longer-Term Plans:* Annual company policy is linked with long-term and mid-term plans.
- *Policy Elements:* Each policy consists of an objective and the strategies to achieve it.
- *Setting Up Annual Company Policy:* Annual company policy (frequently called presidential policy in Japan) is nominally a top-down approach. In reality, however, top management collects the opinions of people inside the company about chronic major problems and their wishes for their company's future and uses this information to formulate the annual plan.
- *Policy Deployment Conducted before the Fiscal Year Starts:* Each annual company policy is incorporated into each department manager's annual policy. That way, implementing departmental policies helps to realize the company policy. Additionally, in each department, the manager's policies are coordinated with those of its related functions. Whether the deployment mentioned above is vertical (between a superior and his or her subordinates) or horizontal (cross-functional), communication within the organization should be fluid.
- *Policy Implementation:* After company policy is deployed throughout each department, each function prepares the program and schedule needed to implement specific items to achieve those targets assigned to it. After the fiscal year begins, the items are implemented according to schedule. This process is conducted through the PDCA cycle, usually quarterly or monthly, but occasionally weekly or even daily.
- *Evaluation and Feedback:* At the end of the fiscal year, the implementation results are evaluated for each function, at each department and company level. This is done whether or not the implementation items have been realized and the targets achieved, and whether or not department managers' policies and company policies have been achieved. If they haven't been achieved at a particular level, the causes should be investigated. The results should be reported to policy makers. That would allow some compensatory action and prevent the recurrence of similar failures in the coming year.

This system was developed to implement the concept of PDCA cycle for company policies. Before introducing the system, a company's policies are often dream-like, with no systematic actions taken. In the 1960s, the "Integrated Management" system was created by Nippon Denso, Sumitomo Electric Industries, Nippon Kayaku, and Komatsu. It came to be named "Management by Policy" and was initiated in earnest by Bridgestone Co. around 1967.

In some American companies, Management by Policy has already been introduced and implemented. I observe that American top executives are involved with it in the following ways:

- *Two Categories of Strategy:* So far, U.S. executives have been involved in preparing various strategies. But when asked to implement them, they have not been necessarily successful. Gradually, these executives have begun to understand that excellent strategies are not always successful. Hence, in addition to a system for preparing strategies, they need to install a system for realizing them. In general, executives deal with two kinds of strategies. The first kind, which is effective immediately after decision making, involves personnel, budgeting, or merger and acquisition. The second kind is effective only with a company-wide effort, such as a quality strategy. Management by policy is the ideal vehicle for realizing the second kind of strategy. When it comes to the introduction and implementation of this second kind of strategy, however, U.S. companies have encountered a variety of difficulties inherent in management by policy, just as many Japanese companies did at a similar stage.
- *English Translation Problems:*[11] American companies have encountered the problem of how to translate "*Hoshin Kanri.*" The Japanese Society for Quality Control committee for the English translation of Japanese TQC terminol-

ogy, which I chaired, decided to translate the Japanese term "*Hoshin Kanri*" into "Management by Policy." But this translation alone is insufficient. What makes an exact translation difficult is that a considerable semantic gap exists between the English word "policy" and the Japanese word "*hoshin*." In English, "policy" implies something rather permanent. The phrase "annual policy" makes English speakers uncomfortable, for it implies that a policy might change from year to year. A detailed explanation is also in order because of the common assumption that "*hoshin*" or "policy" implies an objective and its methods. Because of this confusion, some American companies such as Hewlett Packard have begun using the Japanese word "*hoshin*." They feel it is better to use the original word untranslated and free from the misleading connotation of "policy." Some of the other names selected by American companies which have introduced management by policy are "policy deployment" and "hoshin process."

- *Chase Too Many Rabbits:* It is very common for a company introducing Management by Policy to set up too many policies to implement, and I often caution companies about this. This caution, however, is generally neglected or forgotten during the first year; and as a consequence, the results are inadequate. Then management pays attention to focusing its efforts from the second year on.
- *Inadequate Analysis of Data about the Current Status, A Preference for Pursuing Dreams:* When top executives decide company policy, they are apt to chase their dream rather than analyze data about the current status. Dreams are necessary, but top executives should understand that it takes many years to realize them. I recommend that part of the introductory year's annual policy include overcoming weaknesses—reducing failure cost and market claims, for example—rather than enhancing strengths, such as further increasing market share.
- *Insufficient Cross-Functional Coordination Can Undermine Company Policy:* One company, for example, set up a medium-term policy of developing a new product along with specific target for sales and profit. At the end of this period, however, the company achieved its sales target, but not its profit target. This was because of the unexpectedly high warranty costs incurred by the new product. The warranty problems resulted from poor design review and poor reliability which, in turn, arose from product failures under certain usage conditions, which had not been predicted. If service engineers had participated in the design review, the problems in all likelihood would have been minimized because these engineers would have recognized similar kinds of failures in the past. This product development policy was deployed under the auspices of the product development department, which had no contact with the service department. This is an example of poor policy deployment.
- *Development of a Project:* I have noticed that in American organizations, policy is deployed down to lower levels and finally, in most cases, to project teams. At the implementation stage for each project, analysis is conducted to solve problems, countermeasures are developed, and the results are replicated at different departments or sections. This process seems appropriate for American culture. In some companies, in fact, the task teams execute these projects. In the case of Japanese companies, however, the team projects are deployed within the ordinary organization; the team members come from departments and sections whose chiefs have key roles in the promotion of the policy objective.
- *Presidential Diagnosis:* Some companies have introduced presidential diagnosis—sometimes called "presidential audit" or "presidential review"—to ensure that top management's policy is deployed and implemented at each department, and to provide top management with an opportunity to learn what is happening on the first line of operations. In introducing presidential diagnosis, one company made the following argument:

> Top management's role is to make decisions on various kinds of important issues, including policy making. Implementation is mainly the job of managers who report the results to top management through the chain of command. Top management, therefore, macroscopically knows what is happening in the field. Under such a system, of what use is a presidential review? Would not top management's visit at each department or workplace create confusion in the chain of command?

Once such an argument was overcome, however, and the process implemented, the company discovered the merits of the diagnosis. In particular, top management became aware of very different kinds of microscopic information. This company still continues to use the presidential diagnosis. In another company, a division general manager's review was implemented on a trial basis for about two years before the process became company-wide and was eventually upgraded to presidential diagnosis.

- *Few Companies Are Conducting a Combined Evaluation for Objectives and Methods of Implementation:* In the evaluation stage, we need to check whether objectives are achieved as well as how they are realized. The evaluation results can be categorized into the following four cases:

	Objective	Method	Comment
Case A	OK	OK	Very Happy
Case B	OK	No Good	God Blesses the Company
Case C	No Good	OK	Hopeless
Case D	No Good	No Good	Still Hopeful

These cases can be compared to students who: Case A—fully attended the class and passed the exam; Case B—rarely attended the class but passed the exam; Case C—fully attended the class but failed the exam; Case D—rarely attended the class and then failed the exam. Student A and Student D are easily understood, and Student D may be successful in the next semester if he fully attends the class, so he is hopeful about next semester. Student B and Student C are not as easily understood. Student B is overjoyed that God blessed him, but he needs to be careful, for God will not always bless him. Student C may be hopeless next semester, even though he fully attends the class as he did last semester. Student C should analyze why he failed even though he fully attended the class. He needs to improve his approach to learning in order to pass the exam.

Similarly, in evaluating Management by Policy at the end of the fiscal year, both its objectives and the methods for realizing objectives should be evaluated. The reasons for each individual case should be analyzed. In many cases, the problems come from the mode of policy deployment.

The above combination of evaluating both the objective and its methods of implementation is an orthodox approach. Yet in the U.S., there are very few companies that conduct such an evaluation. In addition, some companies emphasize objectives; others, on the contrary, emphasize only methods of implementation (process improvements).

Daily Management

This approach is less marked among American companies than is Management by Policy, but it is so important that it could be defined as fundamental management. It is encouraging to find a few companies in the U.S. that have introduced this system in addition to Management by Policy, for it strengthens the capabilities of the first-line managers and supervisors. Florida Power and Light calls this system "Quality in Daily Work" (QIDW). In this activity, these are the major tools: a control item or a performance indicator for a job, process standards (including process flow charts and forms/files), and control charts/control graphs.

American companies seem to have the following strengths and weaknesses in this activity:

- In regard to daily management, the U.S. should be proud of its extensive use of computer systems, although there is still room for improvement from the standpoint of their effective utilization. I have been impressed by the considerable number of computer terminals that have been installed and the reduction of paper documents in offices. In contrast to this, in the offices of Japanese companies, many documents are still piled up on desks. Japanese companies are over 10 years behind the U.S. when it comes to promoting office automation.

- Process standards have been prepared by staff departments in headquarters in the United States. Therefore when something needs to be improved in the standards, the workplace manager cannot revise them by himself/herself. In Japan, many companies operate under a corporate standardization rule. It classifies standards into categories according to the responsible department or managerial level and authorizes each of the responsible departments or management levels to prepare and revise the category of standards they are responsible for. Workplace management can revise most operational standards on their own, and this helps management to rotate the PDCA cycle quickly.
- In the U.S., the training in SPC or control charts is not being linked with the training for process standardization. SPC is taught by statisticians who have limited experience in workplace management. They base their training on the notion of population, which is just a hypothetical statistical model. What prerequisites are essential to apply this model to process control in the workplace? The answer is process standardization, which bridges the statistical model and the actual job process in the workplace. Without specific standards and their operation, we can neither identify the population nor use the information provided by control charts such as "under control" or "out of control."
- It is encouraging to see that the judgement "out of control" is based on control charts. This shows that people have based their actions on "fact," rather than "intuition." In the process of finding the causes of "out of control," however, many people still depend on "intuition" over "data." In order to recover from this dependence, companies need executives and managers who earn the nickname "Mr. Show-Me-Data."
- One of the popular activities observed in American companies is the implementation of "5 WHY's"; namely, if a problem occurs, employees are encouraged to ask "Why" five times, on the assumption that the answer to this question will unearth a root cause for the problem. Although this is a very good practice, in many workplaces, I found employees who ask the questions without fully understanding the nature of the problem. Therefore, I recommend asking "5 WHAT's" before asking "5 WHY's," repeating "What is our problem?" five times.

QUALITY ACTIVITIES IN EACH STAGE OF THE NEW PRODUCT DEVELOPMENT, PRODUCTION, AND MARKETING PROCESSES

New Product Development

I was very curious about the process U.S. companies use to develop epoch-making new products. I found that they take a "genius approach," in which a company, when it discovers a very outstanding person, perhaps a genius, provides him/her with the necessary resources, conditions, and work environments. The company makes it easy for the person to do his/her best, without bringing into play age, wealth, education, or experience, as is popularly done in Japanese companies. In contrast to this, the method for new product development in Japanese companies which implement TQM could be termed the "systematic approach."

We are frequently surprised with the outstanding originality at the core of American new products—the result of giving a genius free rein—but we sometimes encounter problems with the subsidiary parts of the product which are handled by ordinary employees. In contrast to this, Japanese products may have fewer problems as well as less originality, although Japanese companies have made great efforts to downplay this weakness. To sum up, America is strong in its genius approach, but poor in its systematic approach. Japan is just the opposite. Companies in both countries need to overcome their weakness and further develop their strength. Although some companies are already examining this issue, it might take a considerable number of years to achieve this state. (I assume it could be done very well if a Japanese-American development center were established somewhere in the middle of the Pacific Ocean.)

Production

The discipline on the shop floor in American firms seems to have been improved considerably during the last decade. People work harder and housekeeping has been much improved. Quality circles, SPC, and team activities are becoming common tools for quality improvement. Some companies are eager to promote JIT as well.

Marketing Share

The management of American companies characteristically puts a great deal of emphasis on "amount of profit," rather than on "turnover growth," "profit ratio to sales turnover," or "market share." Based on this concept, U.S. manufacturing firms have cut off the less profitable business lines or transferred their manufacturing base overseas. It might be said that this is one of the negative elements that caused the "hollowing out" of U.S. manufacturing firms. Recently, however, some companies have begun to become more aware of market share and have made comparative studies of their competitors' product quality.

CONCLUSION

With regard to the promotion of quality, there appear to be American companies that have been implementing quality activities enthusiastically and effectively during the last decade and are beginning to take further steps for their steady promotion. Although the pace of their efforts is sometimes slow, they are having successful results.

At the beginning of 1993, as usual, various statistical reports about the past year were released by the Japanese press. Three of the articles caught my attention. One, dealing with semiconductor production, said that NEC, which had been in first place for the past seven years, was replaced by INTEL, whose sales had increased to five billion dollars (up 26 percent from the previous year). The second article, relating to total car sales, showed the Ford Taurus surpassing the Honda Accord in the U.S. as the best selling passenger car. The third article, using data from Dataquest, deals with personal computers and reports that, worldwide, American companies (IBM, Apple, and others) are regaining market share. These facts point to a trend: during the past ten years, American improvement efforts, including TQM, have gradually led to good results in some industries. We seem to be at the beginning of a new era in which international competition is based on quality. These facts are indeed impressive, and I want to keep my eye on how things develop after this.

Because America is such a big country, it will take much more time until it gets macroeconomic results, such as contributing to the elimination of the trade deficit. In the case of Japan, it took about 20 years after the introduction of QC before the trade balance turned positive.

As I have repeatedly said, similar problems arise in Japanese and American firms when they first implement TQM. I have often heard some American specialists insisting that it is difficult to implement TQM in the U.S. due to its culture. I disagree. A more important factor, I would suggest, is the degree of top management's motivation to be part of TQM, and the greatest enemy of TQM is top management's doubt. "Is there a need to do so much?" or "Do I have to do so much?" This shows that they still base their professional self-evaluation too much on traditional measures of their companies' standing. Quality will be improved if top management will assert its leadership to smoothly implement TQM corporate-wide—and from the long-term viewpoint.

When I consider how American firms will develop TQM activities in the future, one influence is mass media. It tends to discuss the short-term effects of quality. In addition, American consumers tend to be uncritical of poor quality products and services. Let me elaborate on both these points.

One characteristic in the development of Japan's TQC is that Japanese mass media paid little attention to the TQC movement, both its merits and demerits, until around 1980. As a result, Japanese top management could concentrate on TQC from a long-term perspective without worrying about mass media commentary. A drawback is that it took so much time to diffuse TQC activities throughout the country.

In the case of the U.S., at the beginning of the 1980s, American mass media began to pay attention to the quality movement, so information about it has been diffused rapidly. A considerable number of articles about it appeared in various media. There are some firms, however, that implement TQM simply because it is the fashion. So quality activities in such companies tend to be perfunctory or short-sighted. It might be said of these companies that they promote TQM for the sake of TQM. So it is not surprising that the media can find easy targets to criticize.

A friend of mine, a leading person in the American quality world, once showed me a critical article that appeared in one of the leading newspapers and asked for my comment. The article criticized the absolute amount of funds the firm expended for its TQM implementation. I answered that my friend should calculate the average cost per employee per year as well as the rate per gross sales. Typically, such articles give little consideration to the company's size and time over which the budget was spent; rather, they take the total amount as a reference, even when it is budgeted over five years. This often shocks people into thinking, "How expensive!"

According to my experience, it costs approximately 0.1–0.3% of gross sales, referring to total expenses, to run a department for education, training, and quality promotion. The cost can vary depending on how intensively a firm implements it or what related costs are included. So in the case of a mammoth enterprise whose gross sales amount to ten billion dollars, the total costs could be between ten and thirty million dollars a year. In a word, that's how much cost should be permitted when a firm implements quality activity to improve its product/service and to realize the use of scientific methods.

It seems to me that American consumers are quite sensitive to the prices of products. On the other hand, they are charitable about quality unless a product becomes a product liability issue. People seem to feel that it is no use complaining because the country is so large and there is too little competition. For example, this past summer when I crossed the continent on a red-eye flight, my plane took off from San Francisco Airport one hour behind schedule. The delay was due to the late arrival of a pilot. As a result, I missed my connection at Chicago and wasted three hours. I did not notice any passengers complaining to the ground staff of the airline at the Chicago terminal. Since the delay was caused not by weather but by the airline's lack of attention, one might assume that the ground service staff in Chicago would have been waiting for us at the arrival gate, eager to offer service for in-transit passengers. On that occasion, however, I could not find any ground service staff at the gate. The airline's indifference led me to wonder if passively accepting the world as it is undermines the quality improvement of American products and services.

In conclusion, there are many similarities between the development of quality activities in American TQM and Japanese TQC. Although some differences exist, they are not critical. Therefore it permits me to assert that the experiences and knowledge of both countries is to some extent interchangeable.

Acknowledgment: I am grateful to Prof. Robert E. Cole for helping to clarify the English language usage in this article as well as the content of various ideas and also to Ms. Noriko Kusaba for assisting me in preparing this article.

REFERENCES

1. N. Kano, "Business Trip to Europe after an Interval of Six Years," *Hinshitsu* [*Quality,* Journal of the Japanese Society for Quality Control, in Japanese], 12 (1982): 24–28.
2. N. Kano and K. Koura, "Development of Quality Control Seen Through Companies Awarded the Deming Prize," *Rep. Stat. Appl. Res.,* JUSE, 37 (1990–91): 79–105.
3. N. Kano, "Quality Sweating Theory: Crisis Consciousness, Vision and Leadership," *Hinshitsu* [*Quality*, Journal of the Japanese Society for Quality Control, in Japanese], 19 (1989): 32–42.
4. Ibid.
5. N. Kano, "English Translation of the Technical Terms of TQC Implementation," *Hinshitsu* [*Quality,* Journal of the Japanese Society for Quality Control, in Japanese], 22/3 (1992): 91–99 and 22/4 (1992): 87–93.
6. N. Kano, "A Comparative Study between the Western and Japanese QC," memorandum, the 207th JSA (Japanese Standards Association) COSCO Research Meeting [in Japanese], 1982.
7. N. Kano, "The Situation Is Not So Bad Yet, and That Is a Problem of American Firms," *Hinshitsu* [*Quality,* Journal of the Japanese Society for Quality Control, in Japanese], 10 (1980): 227–232 and 11 (1980): 87–93.
8. Kano (1989), op. cit.
9. Kano (1980), op. cit.
10. E. Lawler III and S.A. Mohrmann, "Quality Circles After the Fad," *Harvard Business Review,* 63 (January/February 1985): 65–71.
11. Kano (1992), op. cit.

CHAPTER SIX

MANAGING LINKAGES

Many individual components come together in a successful innovative firm; there is no single, powerful idea, technique, or way of organizing that dwarfs all others in importance. Innovation requires balancing strong opposing tendencies, for example, long-term versus short-term optimization, independence versus coordination, and the natural inclination of a highly competent specialist to believe his or her task is the most important contributor to success. Consequently, managing organizational linkages at several different conceptual levels is extremely important.

The first part of this section builds on the discussion of functional areas in section V, exploring how to link and integrate strong disciplines, which tend toward independence. Clark and Wheelwright direct our attention to cross-functional development teams, showing how to manage "heavyweight" structures that seem to synthesize functional skills especially well. Caldwell and Ancona explore another aspect of cross-functional teams, the way in which successful teams manage the boundary between the group and the outside world. Nonaka suggests that both redundancy and variety characterize successful Japanese teams that integrate various disciplines. Next, we move outside the team, focusing on linkages between subunits within an organization. Bartlett and Ghoshal describe one way to coordinate the development efforts of groups located in different companies. In their "transnational" design, each country group is a leader in some areas of a technology, and a follower in others. Taylor's interview with Percy Barnevik, the CEO of ABB, extends this notion, illustrating how a genuinely global company operates as a federation. Nonaka discusses linkages up and down organizational hierarchies, showing how Japanese middle managers generate innovation under the umbrella of a vision generated at the top. Quinn and his coauthors examine linkages in "network" organizations, showing how they manage intellectual capital by effecting five different types of coordination appropriate to different kinds of innovation problems. Finally, we explore linkages between different organizations. Kanter et al. describe how established firms manage the paradox of balancing efficiency and innovation by creating entrepreneurial ventures that are part of the parent while operating in a fairly independent fashion. Roberts and Berry pursue this theme further, developing a matrix that suggests when corporate venturing is the best solution for entering new businesses. Both articles describe important lessons of experience in managing the linkages between quasi-independent ventures and their parents. Doz and Hamel conclude this section with their article delineating the skills and techniques needed to manage successful strategic alliances.

28

Organizing and Leading "Heavyweight" Development Teams

KIM B. CLARK
STEVEN C. WHEELWRIGHT

Effective product and process development requires the integration of specialized capabilities. Integrating is difficult in most circumstances, but is particularly challenging in large, mature firms with strong functional groups, extensive specialization, large numbers of people, and multiple, ongoing operating pressures. In such firms, development projects are the exception rather than the primary focus of attention. Even for people working on development projects, years of experience and the established systems—covering everything from career paths to performance evaluation, and from reporting relationships to breadth of job definitions—create both physical and organizational distance from other people in the organization. The functions themselves are organized in a way that creates further complications: the marketing organization is based on product families and market segments; engineering around functional disciplines and technical focus; and manufacturing on a mix between functional and product market structures. The result is that in large, mature firms, organizing and leading an effective development effort is a major undertaking. This is especially true for organizations whose traditionally stable markets and competitive environments are threatened by new entrants, new technologies, and rapidly changing customer demands.

This article zeros in on one type of team structure—"heavyweight" project teams—that seems particularly promising in today's fast-paced world yet is strikingly absent in many mature companies. Our research shows that when managed effectively, heavyweight teams offer improved communication, stronger identification with and commitment to a project, and a focus on cross-functional problem solving. Our research also reveals, however, that these teams are not so easily managed and contain unique issues and challenges.

Heavyweight project teams are one of four types of team structures. We begin by describing each of them briefly. We then explore heavyweight teams in detail, compare them with the alternative forms, and point out specific challenges and their solutions in managing the heavyweight team organization. We conclude with an example of the changes necessary in individual behavior for heavyweight teams to be effective. Although heavyweight teams are a different way of organizing, they are more than a new structure; they represent

Reprinted with permission of The Free Press, a division of Simon & Schuster, from *Revolutionizing Product Development: Quantum Leaps to Speed, Efficiency and Quality* by Wheelwright, Steven C. and Kim B. Clark.

a fundamentally different way of working. To the extent that both the team members and the surrounding organization reorganize that phenomenon, the heavyweight team begins to realize its full potential.

TYPES OF DEVELOPMENT PROJECT TEAMS

Figure 28.1 illustrates the four dominant team structures we have observed in our studies of development projects: functional, lightweight, heavyweight, and autonomous (or tiger). These forms are described below, along with their associated project leadership roles, strengths, and weaknesses. Heavyweight teams are examined in detail in the subsequent section.

Functional Team Structure

In the traditional functional organization found in larger, more mature firms, people are grouped principally by discipline, each working under the direction of a specialized subfunction manager and a senior functional manager. The different subfunctions and functions coordinate ideas through detailed specifications all parties agree to at the outset, and through occasional meetings where issues that cut across groups are discussed. Over time, primary responsibility for the project passes sequentially—although often not smoothly—from one function to the next, a transfer frequently termed "throwing it over the wall."

The functional team structure has several advantages, and associated disadvantages. One strength is that those managers who control the project's resources also control task performance in their functional area; thus, responsibility and authority are usually aligned. However, tasks must be subdivided at the project's outset, i.e., the entire development process is decomposed into separable, somewhat independent activities. But on most development efforts, not all required tasks are known at the outset, nor can they all be easily and realistically subdivided. Coordination and integration can suffer as a result.

Another major strength of this approach is that, because most career paths are functional in nature until a general management level is reached, the work done on a project is judged, evaluated, and rewarded by the same subfunction and functional managers who make the decisions about career paths. The associated disadvantage is that individual contributions to a development project tend to be judged largely independently of overall project success. The traditional tenet cited is that individuals cannot be evaluated fairly on outcomes over which they have little or no control. But as a practical matter, that often means that no one directly involved in the details of the project is responsible for the results finally achieved.

Finally, the functional project organization brings specialized expertise to bear on the key technical issues. The same person or small group of people may be responsible for the design of a particular component or subsystem over a wide range of development efforts. Thus the functions and subfunctions capture the benefits of prior experience and become the keepers of the organization's depth of knowledge while ensuring that it is systematically applied over time and across projects. The disadvantage is that every development project differs in its objective and performance requirements, and it is unlikely that specialists developing a single component will do so very differently on one project than on another. The "best" component or subsystem is defined by technical parameters in the areas of their expertise rather than by overall system characteristics or specific customer requirements dictated by the unique market the development effort aims for.

Lightweight Team Structure

Like the functional structure, those assigned to the lightweight team reside physically in their functional areas, but each functional organization designates a liaison person to "represent" it on a project coordinating committee. These liaison representatives work with a "lightweight project manager," usually a design engineer or product marketing manager, who coordinates different

1. Functional Team Structure

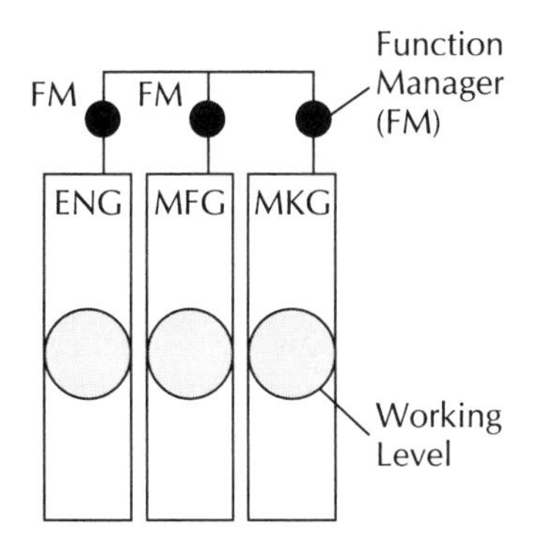

2. Lightweight Team Structure

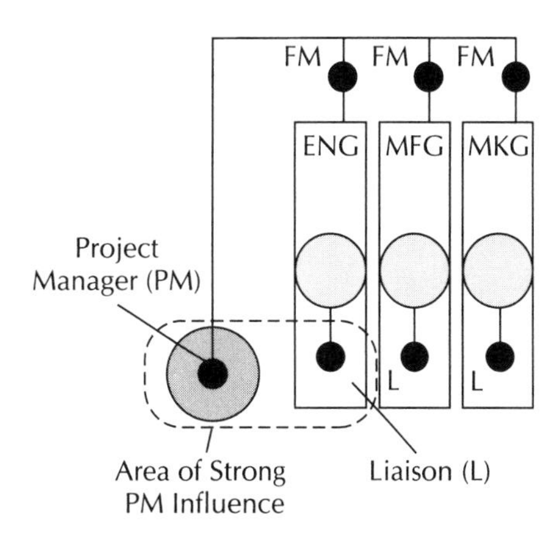

3. Heavyweight Team Structure

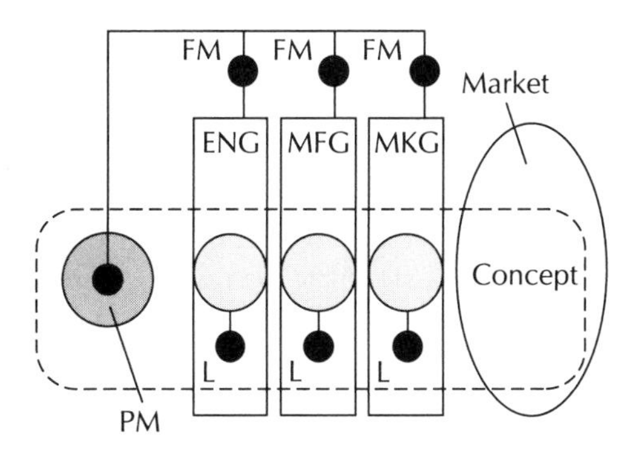

4. Autonomous Team Structure

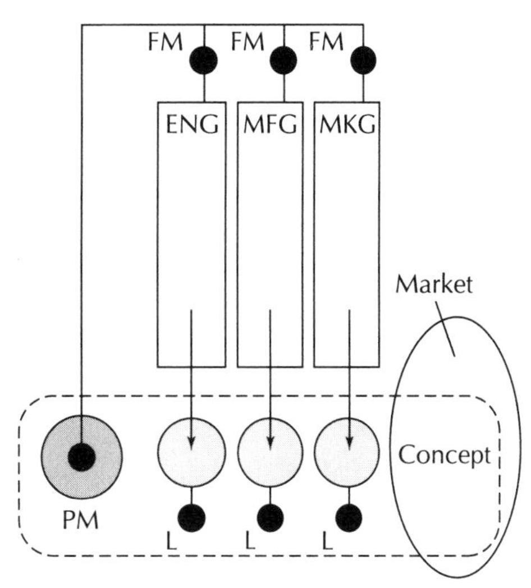

Fig. 28.1. Types of development teams.

functions' activities. This approach usually figures as an add-on to a traditional functional organization, with the functional liaison person having that role added to his or her other duties. The overall coordination assignment of lightweight project manager, however, tends not to be present in the traditional functional team structure.

The project manager is a "lightweight" in two important respects. First, he or she is generally a middle- or junior-level person who, despite considerable expertise, usually has little status or influence in the organization. Such people have spent a handful of years in a function, and this assignment is seen as a "broadening experience," a chance for them to move out of that function. Second, although they are responsible for informing and coordinating the activities of the functional organizations, the key resources (including engineers on the project) remain under the control of their respective functional managers. The lightweight project manager does not have power to reassign people or reallocate resources, and instead confirms schedules, updates time lines, and expedites across groups. Typically, such project leaders spend no more than 25% of their time on a single project.

The primary strengths and weaknesses of the lightweight project team are those of the functional project structure. But now at least one person over the course of the project looks across functions and seeks to ensure that individual tasks—especially those on the critical path—get done in a timely fashion, and that everyone is kept aware of potential cross-functional issues and what is going on elsewhere on this particular project.

Thus, improved communication and coordination are what an organization expects when moving from a functional to a lightweight team structure. Yet, because power still resides with the subfunction and functional managers, hopes for improved efficiency, speed, and project quality are seldom realized. Moreover, lightweight project leaders find themselves tolerated at best, and often ignored and even preempted. This can easily become a "no-win" situation for the individual thus assigned.

Heavyweight Team Structure

In contrast to the lightweight set-up, the heavyweight project manager has direct access to and responsibility for the work of all those involved in the project. Such leaders are "heavyweights" in two respects. First, they are senior managers within the organization; they may even outrank the functional managers. Hence, in addition to having expertise and experience, they also wield significant organizational clout. Second, heavyweight leaders have primary influence over the people working on the development effort and supervise their work directly through key functional people on the core teams. Often, the core group of people are dedicated and physically co-located with the heavyweight project leader. However, the longer-term career development of individual contributors continues to rest not with the project leader—although that heavyweight leaders makes significant input to individual performance evaluations—but with the functional manager, because members are not assigned to a project team on a permanent basis.

The heavyweight team structure has a number of advantages and strengths, along with associated weaknesses. Because this team structure is observed much less frequently in practice and yet seems to have tremendous potential for a wide range of organizations, it will be discussed in detail in the next section.

Autonomous Team Structure

With the autonomous team structure, often called the "tiger team," individuals from the different functional areas are formally assigned, dedicated, and co-located to the project team. The project leader, a "heavyweight" in the organization, is given full control over the resources contributed by the different functional groups. Furthermore, that project leader becomes the sole evaluator of the contribution made by individual team members.

In essence, the autonomous team is given a "clean sheet of paper"; it is not required to follow existing organizational practices and procedures, but allowed to create its own. This includes establishing incentives and rewards as well as norms for behavior. However, the team will be held fully accountable for the final results of the project: success or failure is its responsibility and no one else's.

The fundamental strength of the autonomous team structure is focus. Everything the individual team members and team leader do is concentrated on making the project successful. Thus, tiger teams can excel at rapid, efficient new product and new process development. They handle cross-functional integration in a particularly effective manner, possibly because they attract and select team participants much more freely than the other project structures.

Tiger teams, however, take little or nothing as "given"; they are likely to expand the bounds of their project definition and tackle redesign of the entire product, its components, and subassemblies, rather than looking for opportunities to utilize existing materials, designs, and organizational relationships. Their solution may be unique, making it more difficult to fold the resulting product and process—and, in many cases, the team members themselves—back into the traditional organization upon project completion. As a consequence, tiger teams often become the birthplace of new business units or they experience unusually high turnover following project completion.

Senior managers often become nervous at the prospects of a tiger team because they are asked to delegate much more responsibility and control to the team and its project leader than under any of the other organization structures. Unless clear guidelines have been established in advance, it is extremely difficult during the project for senior managers to make midcourse corrections or exercise substantial influence without destroying the team. More than one team has "gotten away" from senior management and created major problems.

THE HEAVYWEIGHT TEAM STRUCTURE

The best way to begin understanding the potential of heavyweight teams is to consider an example of their success, in this case, Motorola's experience in developing its Bandit line of pagers.

The Bandit Pager Heavyweight Team

This development team within the Motorola Communications Sector was given a project charter to develop an automated, on-shore, profitable production operation for its high-volume Bravo pager line. (This is the belt-worn pager that Motorola sold from the mid-1980s into the early 1990s.) The core team consisted of a heavyweight project leader and a handful of dedicated and co-located individuals, who represented industrial engineering, robotics, process engineering, procurement, and product design/CIM. The need for these functions was dictated by the Bandit platform automation project and its focus on manufacturing technology with a minimal change in product technology. In addition, human resource and accounting/finance representatives were part of the core team. The human resource person was particularly active early on as subteam positions were defined and jobs posted throughout Motorola's Communications Sector, and played an important subsequent role in training and development of operating support people. The accounting/finance person was invaluable in "costing out" different options and performing detailed analyses of options and choices identified during the course of the project.

An eighth member of the core team was a Hewlett Packard employee. Hewlett Packard was chosen as the vendor for the "software backplane," providing an HP 3000 computer and the integrated software communication network that linked individual automated workstations, downloaded controls and instructions during production operations, and captured quality and other operating performance data. Because HP support was vital to the project's success, it was felt essential they be represented on the core team.

The core team was housed in a corner of the Motorola Telecommunications engineering/manufacturing facility. The team chose to enclose in glass the area where the automated production line was to be set up so that others in the factory could track the progress, offer suggestions, and adopt the lessons learned from it in their own production and engineering environments. The team called their project Bandit to indicate a willingness to "take" ideas from literally anywhere.

The heavyweight project leader, Scott Shamlin, who was described by team members as "a crusader," "a renegade," and "a workaholic," became the champion for the Bandit effort. A hands-on manager who played a major role in stimulating and facilitating communication across functions, he helped to articulate a vision of the Bandit line, and to infuse it into the detailed work of the project team. His goal was to make sure the new manufacturing process worked for the pager line, but would provide real insight for many other production lines in Motorola's Communications Sector.

The Bandit core team started by creating a contract book that established the blueprint and work plan for the team's efforts and its performance expectations; all core team members and senior management signed on to the document. Initially, the team's executive sponsor—although not formally identified as such—was George Fisher, the Sector Executive. He made the original investment proposal to the Board of Directors and was an early champion and supporter, as well as direct supervisor in selecting the project leader and helping get the team underway. Subsequently, the vice president and general manager of the Paging Products division filled the role of executive sponsor.

Throughout the project, the heavyweight team took responsibility for the substance of its work, the means by which it was accomplished, and its results. The project was completed in 18 months as per the contract book, which represented about half the time of a normal project of such magnitude. Further, the automated production operation was up and running with process tolerances of five stigma (i.e., the degree of precision achieved by the manufacturing processes) at the end of 18 months. Ongoing production verified that the cost objectives (substantially reduced direct costs and improved profit margins) had indeed been met, and product reliability was even higher than the standards already achieved on the off-shore versions of the Bravo product. Finally, a variety of lessons were successfully transferred to other parts of the Sector's operations, and additional heavyweight teams have proven the viability and robustness of the approach in Motorola's business and further refined its effectiveness throughout the corporation.

The Challenge of Heavyweight Teams

Motorola's experience underscores heavyweight teams' potential power, but it also makes clear that creating an effective heavyweight team capability is more than merely selecting a leader and forming a team. By their very nature—being product (or process) focused, and needing strong, independent leadership, broad skills and cross-functional perspective, and clear missions—heavyweight teams may conflict with the functional organization and raise questions about senior management's influence and control. And even the advantages of the team approach bring with them potential disadvantages that may hurt development performance if not recognized, and averted.

Take, for example, the advantages of ownership and commitment, one of the most striking advantages of the heavyweight team. Identifying with the product and creating a sense of esprit de corps motivate core team members to extend themselves and do what needs to be done to help the team succeed. But such teams sometimes expand the definition of their role and the scope of the project, and they get carried away with themselves and their abilities. We have seen heavyweight teams turn into autonomous tiger teams and go off on a tangent because senior executives gave insufficient direction and the bounds of the team were only vaguely specified at the outset. And even if the team stays focused, the rest of the organization may see themselves as "second class." Although the core team may not make that distinction explicit, it happens because the team has responsibilities and authority beyond those commonly given to functional team members. Thus, such projects inadvertently can become the "haves" and other, smaller projects the "have-nots" with regard to key resources and management attention.

Support activities are particularly vulnerable to an excess of ownership and commitment. Often the heavyweight team will want the same control over secondary support activities as it has over the primary tasks performed by dedicated team members. When waiting for prototypes to be constructed, analytical tests to be performed, or quality assurance procedures to be conducted, the team's natural response is to "demand" top priority from the support organization or to be allowed to go outside and subcontract to independent groups. While these may sometimes be the appropriate choices, senior management should establish make-buy guidelines and clear priorities applicable to all projects—perhaps changing service levels provided by support groups (rather than maintaining the traditional emphasis on resource utilization)—or have support groups provide capacity and advisory technical services but let team members do more of the actual task work in those support areas. Whatever actions the organization takes, the challenge is to achieve a balance between the needs of the individual project and the needs of the broader organization.

Another advantage the heavyweight team brings is the integration and integrity it provides through a system solution to a set of customer needs. Getting all of the components and subsystems to complement one another and to address effectively the fundamental requirements of the core customer segment can result in a winning platform product and/or process. The team achieves an effective system design by using generalist skills applied by broadly trained team members, with fewer specialists and, on occasion, less depth in individual component solutions and technical problem solving.

The extent of these implications is aptly illustrated by the nature of the teams Clark and Fujimoto studied in the auto industry.[1] They found that for U.S. auto firms in the mid-1980s, typical platform projects—organized under a traditional functional or lightweight team structure—entailed full-time work for several months by approximately 1500 engineers. In contrast, a handful of Japanese platform projects—carried out by heavyweight teams—utilized only 250 engineers working full-time for several months. The implications of 250 versus 1500 full-time equivalents (FTEs) with regard to breadth of tasks, degree of specialization, and need for coordination are significant and help explain the differences in project results as measured by product integrity, development cycle time, and engineering resource utilization.

But that lack of depth may disclose a disadvantage. Some individual components or subassemblies may not attain the same level of technical excellence they would under a more traditional functional team structure. For instance, generalists may develop a windshield wiper system that is complementary with and integrated into the total car system and its core concept. But they also may embed in their design some potential weaknesses or flaws that might have been caught by a functional team of specialists who had designed a long series of windshield wipers. To counter this potential disadvantage, many organizations order more testing of completed units to discover such possible flaws and have components and subassemblies reviewed by expert specialists. In some cases, the quality assurance function has expanded its role to make sure sufficient technical specialists review designs at appropriate points so that such weaknesses can be minimized.

Managing the Challenges of Heavyweight Teams

Problems with depth in technical solutions and allocations of support resources suggest the tension that exists between heavyweight teams and the functional groups where much of the work gets done. The problem with the teams exceeding their bounds reflects in part how teams manage themselves, in part, how boundaries are set, and in part the ongoing relationship between the team and senior management. Dealing with these issues requires mechanisms and practices that reinforce the team's basic thrust—ownership, focus, system architecture, integrity—and yet improve its ability to take advantage of the strengths of the supporting functional organization—technical depth, consistency across projects, senior management direction. We have grouped the mechanisms and problems into six categories of management action: the project charter, the contract, staffing, leadership, team responsibility, and the executive sponsor.

The Project Charter

A heavyweight project team needs a clear mission. A way to capture that mission concisely is in an explicit, measurable project charter that sets broad performance objectives and usually is articulated even before the core team is selected. Thus, joining the core team includes accepting the charter established by senior management. A typical charter for a heavyweight project would be the following.

> The resulting product will be selected and ramped by Company X during Quarter 4 of calendar year 1991, at a minimum of a 20% gross margin.

This charter is representative of an industrial products firm whose product goes into a system sold by its customers. Company X is the leading customer for a certain family of products, and this project is dedicated to developing the next generation platform offering in that family. If the heavyweight program results in that platform product being chosen by the leading customer in the segment by a certain date and at a certain gross margin, it will have demonstrated that the next generation platform is not only viable, but likely to be very successful over the next three to five years. Industries and settings where such a charter might be found would include a microprocessor being developed for a new computer system, a diesel engine for the heavy equipment industry, or a certain type of slitting and folding piece of equipment for the newspaper printing press industry. Even in a medical diagnostics business with hundreds of customers, a goal of "capturing 30% of market purchases in the second 12 months during which the product is offered" sets a clear charter for the team.

The Contract Book

Whereas a charter lays out the mission in broad terms, the contract book defines, in detail, the basic plan to achieve the stated goal. A contract book is created as soon as the core team and heavyweight project leader have been designated and given the charter by senior management. Basically, the team develops its own detailed work plan for conducting the project, estimates the resources required, and outlines the results to be achieved and against which it is willing to be evaluated. (The table of contents of a typical heavyweight team contract book are shown in Table 28.1.) Such documents range from 25 to 100 pages, depending on the complexity of the project and level of detail desired by the team and senior management before proceeding. A common practice following negotiation and acceptance of this contract is for the individuals from the team and senior management to sign the contract book as an indication of their commitment to honor the plan and achieve those results.

TABLE 28.1. Heavyweight Team, Contract Book—Major Sections

Executive summary
Business plan and purposes
Development plan
Schedule
Materials
Resources
Produce design plan
Quality plan
Manufacturing plan
Project deliverables
Performance measurement and incentives

The core team may take anywhere from a long week to a few months to create and complete the contract book; Motorola, for example, after several years of experience, has decided that a maximum of seven days should be allowed for this activity. Having watched other heavyweight teams—particularly in organizations with no prior experience in using such a structure—take up to several months, we can appreciate why Motorola has nicknamed this the "blitz phase" and decided that the time allowed should be kept to a minimum.

Staffing

As suggested in Figure 28.1, a heavyweight team includes a group of core cross-functional team members who are dedicated (and usually physically co-located) for the duration of the development effort. Typically there is one core team member from each primary function of the organization; for instance, in several electronics firms we have observed core teams consisting of six functional participants—design engineering, marketing, quality assurance, manufacturing, finance, and human resources. (Occasionally, design will be represented by two core team members, one each for hardware and software engineering.) Individually, core team members represent their functions and provide leadership for their function's inputs to the project. Collectively, they constitute a management team that works under the direction of the heavyweight project manager and takes responsibility for managing the overall development effort.

While other participants—especially from design engineering early on and manufacturing later on—may frequently be dedicated to a heavyweight team for several months, they usually are not made part of the core team though they may well be colocated and, over time, develop the same level of ownership and commitment to the project as core team members. The primary difference is that the core team manages the total project and the coordination and integration of individual functional efforts, whereas other dedicated team members work primarily within a single function or subfunction.

Whether these temporarily dedicated team members are actually part of the core team is an issue firms handle in different ways, but those with considerable experience tend to distinguish between core and other dedicated (and often co-located) team members. The difference is one of management responsibility for the core group that is not shared equally by the others. Also, it is primarily the half a dozen members of the core group who will be dedicated throughout the project, with other contributors having a portion of their time reassigned before this heavyweight project is completed.

Whether physical colocation is essential is likewise questioned in such teams. We have seen it work both ways. Given the complexity of development projects, and especially the uncertainty and ambiguity often associated with those assigned to heavyweight teams, physical colocation is preferable to even the best of on-line communication approaches. Problems that arise in real time are much more likely to be addressed effectively with all of the functions represented and present than when they are separate and must either wait for a periodic meeting or use remote communication links to open up cross-functional discussions.

A final issue is whether an individual can be a core team member on more than one heavyweight team simultaneously. If the rule for a core team member is that 70% or more of their time must be spent on the heavyweight project, then the answer to this question is no. Frequently, however, a choice must be made between someone being on two core teams—for example, from the finance or human resource function—or putting a different individual on one of those teams who has neither the experience nor stature to be a full peer with the other core team members. Most experienced organizations we have seen opt to put the same person on two teams to ensure the peer relationship and level of contribution required, even though it means having one person on two teams and with two desks. They then work diligently to develop other people in the function so that multiple team assignments will not be necessary in the future.

Sometimes multiple assignments will also be justified on the basis that a function such as finance does not need a full-time person on a project. In most instances, however, a variety of potential value-adding tasks exist that are broader than finance's traditional contribution. A person largely dedicated to the core team will search for those opportunities and the project will be better because of it. The risk of allowing core team members to be assigned to multiple projects is that they are neither available when their inputs are most needed nor as committed to project success as their peers. They become secondary core team members, and the full potential of the heavyweight team structure fails to be realized.

Project Leadership

Heavyweight teams require a distinctive style of leadership. A number of differences between lightweight and heavyweight project managers are highlighted in Figure 28.2. Three of those are particularly distinctive. First, a heavyweight leader managers, leads, and evaluates other members of the core team, and is also the person to whom the core team reports throughout the project's duration. Another characteristic is that rather than being either neutral or a facilitator with regard to problem solving and conflict resolution, these leaders see themselves as championing the basic concept around which the platform product and/or process is being shaped. They make sure that those who work on subtasks of the project understand that concept. Thus they play a central role in ensuring

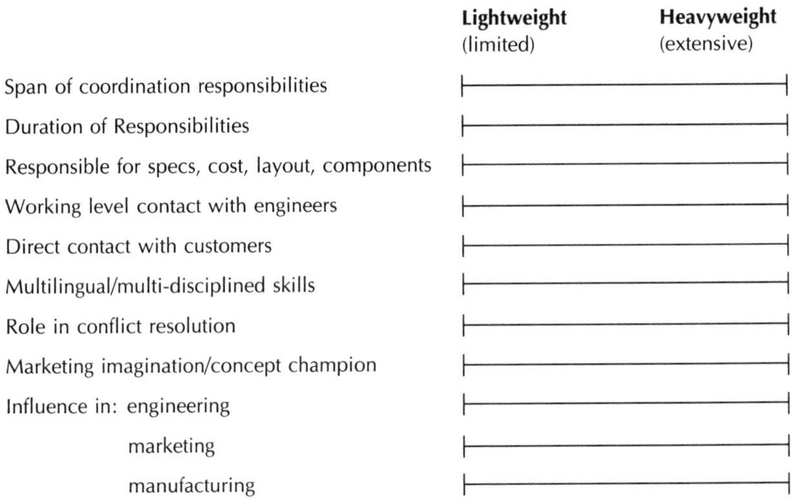

	Lightweight (limited)	Heavyweight (extensive)
Span of coordination responsibilities		
Duration of Responsibilities		
Responsible for specs, cost, layout, components		
Working level contact with engineers		
Direct contact with customers		
Multilingual/multi-disciplined skills		
Role in conflict resolution		
Marketing imagination/concept champion		
Influence in: engineering		
marketing		
manufacturing		

Fig. 28.2. Project manager profile.

the system integrity of the final product and/or process.

Finally, the heavyweight project manager carries out his or her role in a very different fashion than the lightweight project manager. Most lightweights spend the bulk of their time working at a desk, with paper. They revise schedules, get frequent updates, and encourage people to meet previously agreed upon deadlines. The heavyweight project manager spends little time at a desk, is out talking to project contributors, and makes sure that decisions are made and implemented whenever and wherever needed. Some of the ways in which the heavyweight project manager achieves project results are highlighted by the five roles illustrated in Table 28.2 for a heavyweight project manager on a platform development project in the auto industry.

The *first role* of the heavyweight project man-

TABLE 28.2. The Heavyweight Project Manager

Role	Description
Direct market interpreter	First hand information, dealer visits, auto shows, has own marketing budget, market study team, direct contact and discussions with customers
Multilingual translator	Fluency in language of customers, engineers, marketers, stylists; translator between customer experience/requirements and engineering specifications
"Direct" engineering manager	Direct contact, orchestra conductor, evangelist of conceptual integrity and coordinator of component development; direct eye-to-eye discussions with working level engineers; shows up in drafting room, looks over engineers' shoulders
Program manager "in motion"	Out of the office, not too many meetings, not too much paperwork, face-to-face communication, conflict resolution manager
Concept infuser	Concept guardian, confronts conflicts, not only reacts but implements own philosophy, ultimate decision maker, coordination of details and creation of harmony

ager is to provide for the team a direct interpretation of the market and customer needs. This involves gathering market data directly from customers, dealers, and industry shows, as well as through systematic study and contact with the firm's marketing organization. A *second role* is to become a multilingual translator, not just taking marketing information to the various functions involved in the project, but being fluent in the language of each of those functions and making sure the translation and communication going on among the functions—particularly between customer needs and product specifications—are done effectively.

A *third role* is the direct engineering manager, orchestrating, directing, and coordinating the various engineering subfunctions. Given the size of many development programs and the number of types of engineering disciplines involved, the project manager must be able to work directly with each engineering subfunction on a day-to-day basis and ensure that their work will indeed integrate and support that of others, so the chosen product concept can be effectively executed.

A *fourth role* is best described as staying in motion: out of the office conducting face-to-face sessions, and highlighting and resolving potential conflicts as soon as possible. Part of this role entails energizing and pacing the overall effort and its key subparts. A *final role* is that of concept champion. Here the heavyweight project manager becomes the guardian of the concept and not only reacts and responds to the interests of others, but also sees that the choices made are consistent and in harmony with the basic concept. This requires a careful blend of communication and teaching skills so that individual contributors and their groups understand the core concept, and sufficient conflict resolution skills to ensure that any tough issues are addressed in a timely fashion.

It should be apparent from this description that heavyweight project managers earn the respect and right to carry out these roles based on prior experience, carefully developed skills, and status earned over time, rather than simply being designated "leader" by senior management. A qualified heavyweight project manager is a prerequisite to an effective heavyweight team structure.

Team Member Responsibilities

Heavyweight team members have responsibilities beyond their usual functional assignment. As illustrated in Table 28.3, these are of two primary types. Functional hat responsibilities are those accepted by the individual core team member as a representative of his or her function. For example, the core team member from marketing is responsible for ensuring that appropriate marketing expertise is brought to the project, that a marketing perspective is provided on all key issues, that project sub-objectives dependent on the marketing function are met in a timely fashion, and that marketing issues that impact other functions are raised proactively within the team.

But each core team member also wears a team hat. In addition to representing a function, each member shares responsibility with the heavyweight project manager for the procedures followed by the team, and for the overall results that those procedures deliver. The core team is accountable for the success of the project, and it can blame no one but itself if it fails to manage the project, execute the

TABLE 28.3. Responsibilities of Heavyweight Core Team Members

Functional Hat Accountabilities
Ensuring functional expertise on the project
Representing the functional perspective on the project
Ensuring that subobjectives are met that depend on their function
Ensuring that functional issues impacting the team are raised pro-actively within the team
Team Hat Accountabilities
Sharing responsibility for team results
Reconstituting tasks and content
Establishing reporting and other organizational relationships
Participating in monitoring and improving team performance
Sharing responsibility for ensuring effective team processes
Examining issues from an executive point of view (Answering the question, "Is this the appropriate business response for the company?")
Understanding, recognizing, and responsibly challenging the boundaries of the project and team process

tasks, and deliver the performance agreed upon at the outset.

Finally, beyond being accountable for tasks in their own function, core team members are responsible for how those tasks are subdivided, organized, and accomplished. Unlike the traditional functional development structure, which takes as given the subdivision of tasks and the means by which those tasks will be conducted and completed, the core heavyweight team is given the power and responsibility to change the substance of those tasks to improve the performance of the project. Since this is a role that core team members do not play under a lightweight or functional team structure, it is often the most difficult for them to accept fully and learn to apply. It is essential, however, if the heavyweight team is to realize its full potential.

THE EXECUTIVE SPONSOR

With so much more accountability delegated to the project team, establishing effective relationships with senior management requires special mechanisms. Senior management needs to retain the ability to guide the project and its leader while empowering the team to lead and act, a responsibility usually taken by an executive sponsor—typically the vice president of engineering, marketing, or manufacturing for the business unit. This sponsor becomes the coach and mentor for the heavyweight project leader and core team, and seeks to maintain close, ongoing contact with the team's efforts. In addition, the executive sponsor serves as a liaison. If other members of senior management—including the functional heads—have concerns or inputs to voice, or need current information on project status, these are communicated through the executive sponsor. This reduces the number of mixed signals received by the team and clarifies for the organization the reporting and evaluation relationship between the team and senior management. It also encourages the executive sponsor to set appropriate limits and bounds on the team so that organizational surprises are avoided.

Often the executive sponsor and core team identify those areas where the team clearly has decision-making power and control, and they distinguish them from areas requiring review. An electronics firm that has used heavyweight teams for some time dedicates one meeting early on between the executive sponsor and the core team to generating a list of areas where the executive sponsor expects to provide oversight and be consulted; these areas are of great concern to the entire executive staff and team actions may well raise policy issues for the larger organization. In this firm, the executive staff wants to maintain some control over:

- resource commitment—head count, fixed costs, and major expenses outside the approved contract book plan;
- pricing for major customers and major accounts;
- potential slips in major milestone dates (the executive sponsor wants early warning and recovery plans);
- plans for transitioning from development project to operating status;
- thorough reviews at major milestones or every three months, whichever occurs sooner;
- review of incentive rewards that have company-wide implications for consistency and equity; and
- cross-project issues such as resource optimization, prioritization, and balance.

Identifying such areas at the outset can help the executive sponsor and the core team better carry out their assigned responsibilities. It also helps other executives feel more comfortable working through the executive sponsor, since they know these "boundary issues" have been articulated and are jointly understood.

THE NECESSITY OF FUNDAMENTAL CHANGE

Compared to a traditional functional organization, creating a team that is "heavy"—one with effective leadership, strong problem-solving skills and the ability to integrate across functions—requires basic changes in the way development works. But it also requires change in the fundamental behavior of engineers, designers, manufacturers, and marketers in their day-to-day work. An episode in a computer company with no previous experience with heavyweight teams illustrates the depth of change required to realize fully these teams' power.[2]

Two teams, A and B, were charged with development of a small computer system and had market introduction targets within the next twelve months. While each core team was co-located and held regular meetings, there was one overlapping core team member (from finance/accounting). Each team was charged with developing a new computer system for their individual target markets but by chance, both products were to use an identical, custom-designed microprocessor chip in addition to other unique and standard chips.

The challenge of changing behavior in creating an effective heavyweight team structure was highlighted when each team sent this identical, custom-designed chip—the "supercontroller"—to the vendor for pilot production. The vendor quoted a 20-week turnaround to both teams. At that time, the supercontroller chip was already on the critical path for Team B, with a planned turnaround of 11 weeks. Thus, every week saved on that chip would save one week in the overall project schedule, and Team B already suspected that it would be late in meeting its initial market introduction target date. When the 20-week vendor lead time issue first came up in a Team B meeting, Jim, the core team member from engineering, responded very much as he had on prior, functionally structured development efforts: because initial prototypes were engineering's responsibility, he reported that they were working on accelerating the delivery date, but that the vendor was a large company, with whom the computer manufacturer did substantial business, and known for its slowness. Suggestions from other core team members on how to accelerate the delivery were politely rebuffed, including one to have a senior executive contact their counterpart at the vendor. Jim knew the traditional approach to such issues and did not perceive a need, responsibility, or authority to alter it significantly.

For Team A, the original quote of 20-week turnaround still left a little slack, and thus initially the supercontroller chip was not on the critical path. Within a couple of weeks, however, it was, given other changes in the activities and schedule, and the issue was immediately raised at the team's weekly meeting. Fred, the core team member from manufacturing (who historically would not have been involved in an early engineering prototype), stated that he thought the turnaround time quoted was too long and that he would try to reduce it. At the next meeting, Fred brought some good news: through discussions with the vendor, he had been able to get a commitment that pulled in the delivery of the supercontroller chip by 11 weeks! Furthermore, Fred thought that the quote might be reduced even further by a phone call from one of the computer manufacturer's senior executives to a contact of his at the vendor.

Two days later, at a regular Team B meeting, the supercontroller chip again came up during the status review, and no change from the original schedule was identified. Since the finance person, Ann, served on both teams and had been present at Team A's meeting, she described Team A's success in reducing the cycle time. Jim responded that he was aware that Team A had made such efforts, but that the information was not correct, and the original 20-week delivery date still held. Furthermore, Jim indicated that Fred's efforts (from Team A) had caused some uncertainty and disruption internally, and in the future it was important that Team A not take such initiatives before coordinating with Team B. Jim stated that this was particularly true when an outside vendor was involved, and he closed the topic by saying that a meeting to clear up the situation would be held that afternoon with Fred from Team A and Team B's engineering and purchasing people.

The next afternoon, at his Team A meeting, Fred confirmed the accelerated delivery schedule for the supercontroller chip. Eleven weeks had indeed been clipped out of the schedule to the benefit of both Teams A and B. Subsequently, Jim confirmed the revised schedule would apply to his team as well, although he was displeased that Fred had abrogated "standard operating procedure" to achieve it. Curious about the differences in perspective, Ann decided to learn more about why Team A had identified an obstacle and removed it from its path, yet Team B had identified an identical obstacle and failed to move it at all.

As Fred pointed out, Jim was the engineering manager responsible for development of the supercontroller chip; he knew the chip's technical requirements, but had little experience dealing with chip vendors and their production processes. (He had long been a specialist.) Without that experience, he had a hard time pushing back against the vendor's "standard line." But Fred's manufacturing experience with several chip vendors enabled him to calibrate the vendor's dates against his best-case experience and understand what the vendor needed to do to meet a substantially earlier commitment.

Moreover, because Fred had bought into a clear team charter, whose path the delayed chip would block, and because he had relevant experience, it did not make sense to live with the vendor's initial commitment, and thus he sought to change it. In contrast, Jim—who had worked in the traditional functional organization for many years—saw vendor relations on a pilot build as part of his functional job, but did not believe that contravening standard practices to get the vendor to shorten the cycle time was his responsibility, within the range of his authority, or even in the best long-term interest of his function. He was more concerned with avoiding conflict and not roiling the water than with achieving the overarching goal of the team.

It is interesting to note that in Team B, engineering raised the issue, and, while unwilling to take aggressive steps to resolve it, also blocked others' attempts. In Team A, however, while the issue came up initially through engineering, Fred in manufacturing proactively went after it. In the case of Team B, getting a prototype chip returned from a vendor was still being treated as an "engineering responsibility," whereas in the case of Team A, it was treated as a "team responsibility." Since Fred was the person best qualified to attack that issue, he did so.

Both Team A and Team B had a charter, a contract, a co-located core team staffed with generalists, a project leader, articulated responsibilities, and an executive sponsor. Yet Jim's and Fred's understanding of what these things meant for them personally and for the team at the detailed, working level was quite different. While the teams had been through similar training and team startup processes, Jim apparently saw the new approach as a different organizational framework within which work would get done as before. In contrast, Fred seemed to see it as an opportunity to work in a different way—to take responsibility for reconfiguring tasks, drawing on new skills, and reallocating resources, where required, for getting the job done in the best way possible.

Although both teams were "heavyweight" in theory, Fred's team was much "heavier" in its operation and impact. Our research suggests that heaviness is not just a matter of structure and mechanism, but of attitudes and behavior. Firms that try to create heavyweight teams without making the deep changes needed to realize the power in the team's structure will find this team approach problematic. Those intent on using teams for platform projects and willing to make the basic changes we have discussed here, can enjoy substantial advantages of focus, integration, and effectiveness.

REFERENCES

1. See Kim B. Clark and Takahiro Fujimoto, *Product Development Performance* (Boston, MA: Harvard Business School Press, 1991).
2. Adapted from a description provided by Dr. Christopher Meyer, Strategic Alignment Group, Los Altos, CA.

29

Making Teamwork Work: Boundary Management in Product Development Teams

DEBORAH GLADSTEIN ANCONA
DAVID F. CALDWELL

> The product team consisted of some of the best engineers in the company and had the support of top management. Members of the group worked long hours and quickly became a cohesive and motivated team. The team leader established common goal statements with manufacturing and marketing around product functionality, cost, and quality. About six months after the team began its work, coordination began to break down. Disputes between the team and the marketing and manufacturing functions became more frequent and were regularly elevated to top management for resolution. After months of delay, division management reorganized the team and replaced a number of members, including the team leader. While the product was ultimately successful, the delay of time in getting the product to market was extremely costly. The original manager of the team summarized his experiences by saying "I try to deal with things in a rational way, but that's not necessarily the way other groups operate around here. Many things that were planned for the product six, eight, ten months ago weren't ever really committed to by manufacturing."

Despite substantial resources, motivated and talented people, and even the ability to work together effectively, some teams fail because they are unable to develop a new product that meets the expectations of others in the organization. Developing a product that will be supported throughout the organization requires more than the use of concurrent engineering techniques or a cross-functional team. Rather, what is required is that a team learn to manage across traditional lines of function and authority. This paper summarizes an investigation of the activities product development teams undertake in managing relations with other groups and links the pattern of those activities to the teams' overall success.

Rapid changes in technology and markets are the reality for an increasing number of industries. Because of this, speed to market is critically important in determining whether or not a new product is successful. Perhaps the most common approach for responding to increasing technical complexity and need for speed is the use of a development team to design the product rather than assigning development to a single individual or even a small group of individuals from one function. Such teams have the potential of speeding development by improving interunit coordination, allowing for the use of a parallel as opposed to a sequential design process, and reducing delays due to the failure to include the necessary information from throughout the organization (Kazanjian, 1988; House and Price, 1991).

However, if teams are to fulfill their promise of shortening the product development cycle, they must develop the ability to collect information and resources from a variety of sources—inside and outside the organization—and interact with others in the organization to negotiate deadlines and specifications, coordinate workflow, obtain support for the product idea, and smoothly transfer the product to those groups that will ultimately manufacture, sell, and service it.

A number of studies have shown the importance of communication across group boundaries in R&D laboratories and have identified some of the roles individuals play in communication. What much of this research has shown is that particular patterns of communication are related to the performance of technical teams. Allen (1971, 1984) showed that in successful R&D projects, some individuals serve as technological gatekeepers. These individuals provide an important link between other team members and the outside environment. Specifically, gatekeepers have broad connections to external, technical information sources and serve to refine that knowledge and direct it to other members of the team. These technological gatekeepers provide links between the team, and the broad technical environment outside the organization. Tushman (1977, 1979) expanded on these ideas and identified two other boundary roles, organization liaisons and laboratory liaisons, who serve to link the team to the organization and other units in the R&D function through the same two-step process of collecting and reviewing outside information and then communicating it within the team. His work also showed that the fit between the team's communication networks and the demands of the project will be related to the team's performance.

A somewhat broader framework is described by Roberts and Fusfeld (1981). They argue that successfully completing technical projects requires five work roles: idea generating (developing an idea for a new product or procedure); championing (gaining formal management support for the new idea); project leading (coordinating the activities and people necessary to develop the idea into a product); gatekeeping (collecting outside information and channeling it into the team); and sponsoring (providing resources and support for the project). Roberts and Fusfeld hypothesize that the importance of these roles changes throughout the product development process and that if these roles are not filled, the team is not likely to be successful. Although this classification covers a wide range of roles related to the product development process, two of them, championing and gatekeeping directly address the importance of the group's interactions with outsiders.

Collectively, these studies clearly show the importance of bringing technical information into the group and of linking the product development team to others in the organization. The studies also describe the various roles individuals occupy in this process. However, these studies have not investigated the full range of activities these teams take in dealing with others. Developing an understanding of the things team members do to establish and maintain connections with others can do two things. First, it can clarify the specific actions that contribute to filling specific roles. Second, it may provide some direction as to what the team can do to both acquire the information necessary for successful innovation and build the necessary bridges with others in the organization to ensure that the work of the team is supported. Thus, the purpose of this paper is to explore a fuller set of activities product develop teams take in dealing with other groups and to investigate how those activities contribute to the success or failure of the team. Our goal is to identify the specific things that product development team members do to manage their relations with outside groups and determine whether or not engaging in these activities enhances team performance.

RESEARCH SETTING AND METHODS

An important goal of our research was to describe the types of things group members do when they interact with others outside the group, identify patterns among these activities, and determine how these activities are related to the overall performance of the team. To identify this set of activities, we interviewed thirty-eight experienced product development team managers. The duration of the interviews averaged three hours. In the interviews, we asked each individual to describe the activities that the manager and team members carried out with people outside the group boundary. Questions dealt with the timing, frequency, target, and purpose of interactions across the entire life of the product development process. Through a review of these interviews we tried to develop an inclusive set of the things team members do in dealing with others. Based on that, we identified twenty-four specific activities that represented the complete set of boundary actions the managers described.

The thirty-eight interviews provided a rich set of data about how teams deal with other groups; however, because our goal was to provide a more direct empirical test of the impact of boundary activities on team operations, we collected systematic data from a set of product development teams, unrelated to those described by the interviewees. Thus, the conclusions we report are drawn from a study of forty-five product development teams in five high-technology companies in the computer, analytic instrumentation, and photographic industries. Each team was responsible for developing a prototype product for manufacture and sale. For example, one team was developing a product to automate the sampling process used in liquid chromatography, and another was developing a device that combined photographic and computer imaging processes. To ensure some compatibility across the teams, all the projects represented new product development (a major extension to an existing product line or the start of a new product line) as opposed to either basic research or the enhancement of features of an existing product, and all had a time frame of between one and one-half and three years.

We collected data about each of the teams from three sources: (1) an interview with the team leader regarding the history and nature of the project; (2) interviews with division or corporate management regarding the performance of the teams; and (3) a questionnaire survey of all members of the team including the leader.

Surveys were completed by 409, or 89 percent, of the members of the product development teams. The survey contained sections regarding the frequency with which team members communicated with others and assessments of the team's group process. In addition, the survey asked each team member to report how responsible he or she was for engaging in each of the twenty-four boundary activities we identified through the interviews described earlier. Team members were also asked to rate their own performance.

Executives in each firm rated the teams' performance at two points in time. The first was when the teams were about halfway through their lives. The same executives rated the teams again, after they had completed their work. At both points in time, executives assessed the teams in terms of effectiveness of their operations and their overall level of innovation.

The combination of surveys, interviews, and performance ratings allowed us to look systematically at how team members deal with others and determine whether or not the ways teams interacted with outside groups were related to assessments of performance.

PATTERNS OF BOUNDARY ACTIVITIES

The first issue we explored was the pattern of activities in which individuals engaged to manage the boundary of the group. Because the questionnaire asked individuals to rate their responsibility on twenty-four separate actions, we used a principal component analysis to identify the smaller set of dimensions or "themes" that underlie the specific actions. This analysis showed that the items relating to how team members dealt with others could be combined into three dimensions.

- *Ambassador*—One set of activities included those aimed at representing the team to others and protecting the team from outside interference. Examples of the actions making up this dimension included "Prevent outsiders from 'overloading' the team with too much information or too many requests" and "Persuade other individuals that the team's activities are important."

- *Task Coordination*—The second set of actions we identified were aimed at coordinating the team's efforts with others. Examples include discussing design problems with others, obtaining feedback about the team's progress, and coordinating and negotiating with other groups. Specific items included "Resolve design problems with other groups" and "Negotiate with others for delivery deadlines."
- *Scouting*—The third set of activities represented general scanning for ideas and information about the market, the competition, or the technology. In contrast to the other sets, these actions were less specifically focused and more directed to building a general awareness and knowledge base than addressing specific issues. Examples included "Find out what competing firms or groups are doing on similar projects" and "Collect technical information/ideas from individuals outside the team."

The nature of these activity sets represents the complexity of the boundary management tasks of technical teams. Scouting provides the team with the broad technical and market information that is necessary for product development. These activities are primarily lateral and involve investigating markets, technologies, and competition; in short, bringing a great deal of data into the team. Task Coordination represents the lateral connections across functions that are necessary to deliver a product that meets the expectations of other functional groups. Teams that excel at this activity bargain with other groups, trade services or essential resources, and get feedback from other groups. In contrast to the other activity sets, Ambassador activities are primarily, although not exclusively, vertical. That is, they are aimed at securing effective sponsorship for the team, ensuring that the team has necessary resources, aligning the new product to the strategic direction of the organization, and in general, creating a favorable impression of the team and its efforts.

BOUNDARY ACTIVITIES AND THE PRODUCT DEVELOPMENT TEAM

While it is useful to know the nature of the activities team members undertake to manage their relations with other groups, it is important to understand whether or not these activities are related to the performance of the team and how other factors may influence the way these activities are completed by teams. What follows are conclusions we have drawn from our research.

The Nature of External Activities Is Related to Teams' Performance

The ways teams managed their boundaries were strongly related to their performance. However, these relationships were not simple, rather they are dependent on both the frequency of specific activities and when performance is measured. The extent to which team members engaged Task Coordination activities was directly related to top management's evaluation of team performance; however, the relationship between the frequency of these activities and team success was much stronger when the evaluation was obtained *after* the team completed its work than if the evaluation was made while the team was still working. In other words, the frequency that team members engaged in things such as problem solving with other groups was related to the ultimate success of the team's efforts but not to interim evaluations of the team's progress.

A surprising relationship existed between Scout activities and performance. Those teams with high levels of Scout activities were rated substantially lower in performance at all points in time than those teams who engaged in fewer of these activities. At first glance, these results seem contrary to previous work on teams and to common sense. Why would teams who spend a great deal of effort understanding technical and market trends have poorer performance than those that don't? In looking specifically at teams we followed, some never got beyond exploring possibilities and moving to exploiting a technology and market niche. That is, some teams were never able to commit to a particular design. Rather, any new information would lead the team to reconsider the decisions it had already made. While bringing general information into the group is critical, it seems that it cannot become an enduring pattern. In fact, our findings are consistent with the notion that scanning the environment for broad, competitive, technical, and market information is critical *before* the team begins to work, but potentially disruptive once the team has begun to make product decisions.

A different pattern emerged for the relationship between Ambassador activity and performance. Teams with high levels of Ambassador activity received very positive evaluations from executives during the team's operations. However, when performance was assessed after the project was completed, the relationship between these activities and the success of the team was much lower. It seems that teams that devoted a great deal of effort to "managing upward" were initially viewed as successful by top management—but the impact of these activities declined over time.

Combining these results suggests that there is a particular timeliness to boundary activities. Linking boundary activities to management's evaluation of teams' performance suggests that Scouting should be done very early, perhaps even before the full team has begun operations; Ambassador activities are most necessary early in the group's process; and Task Coordination is important throughout the life of the team. This sequence of activities is similar to relationships between the phase of a project and the importance of roles hypothesized by Roberts and Fusfeld (1982).

The way groups manage their boundaries is clearly related to how top management in the firm evaluates the group's performance. However, a different pattern occurs when individual team members are asked to assess the performance of their own teams. As part of the questionnaire, team members were asked to rate the performance of their teams and to evaluate the extent to which the team members had developed a good internal process for working with each other (e.g., defining goals, prioritizing work, etc.) and good interpersonal relationships (e.g., getting along with each other, sticking together, etc.). A number of factors are worth note. First, top management's ratings of the teams were essentially uncorrelated with the team's assessment of its own performance. Second, neither the internal process teams developed, nor interpersonal relations were related to management's ratings of performance but both were strongly related to the *teams' own* assessments of their performances. Thus, company executives and team members seemed to have different models of what led to high performance. Team members associated high levels of performance with effectiveness in working with one another while management's assessments were driven by how effective the team was in working with outsiders.

Neither Frequent Communications nor Cross-Functional Teams Are a Substitute for Effective Boundary Management

Our results show that how teams interact with outsiders is related to their performance. The next question is whether or not other factors may serve as substitutes for these activities. Two other factors may influence the extent to which teams are able to develop positive relations with other functional groups. The first of these is the frequency with which team members interact with outsiders. The second represents the question of how the team is formed. That is, is placing individuals who represent different functions on the team a replacement for good boundary management?

The frequency of communication between the team and outsiders was much less predictive of performance than the type of activities. The total amount of communication had a small positive relationship with initial assessments of performance—but none at all with the final evaluations of the teams' efforts. What this shows is that it is not simply the amount team members interact with others that influences performance, rather it is the nature of those interactions. In other words, frequent communication with outsiders is necessary but not sufficient to build the cross-functional relations necessary for successful new product introduction. What is required is careful attention to the lateral negotiations represented by the Task Coordination activities and those early Ambassador activities that both shape others' perceptions of the team and buffer the team from interference.

Although at first glance the lack of a relationship between frequency of communication and team performance may seem counterintuitive, our findings are very consistent with previous studies of research teams (Allen, 1984; Katz and Tushman, 1979). As studies of boundary spanning roles have shown, effective teams do not rely on extensive external communication by all members, but instead have individuals (gatekeepers or liaisons) who collect, interpret, and triage information from sources outside the team or organization.

A popular approach to facilitate coordination is to form product development teams of individuals who represent the functional areas necessary to not only design, but also to build and market, and service the product. Although the conclusions we can draw must be somewhat tentative, the data from our teams show that simply using cross-functional teams is not the answer. In fact, among the teams we studied, there was very little difference in performance between those with individuals from many functions compared to those made up exclusively of engineers. What we did find was that teams made up of a diverse set of individuals had more frequent communication with outsiders, but that communication alone is not enough. Rather, external communication must be carefully managed to ensure that effective boundary tasks are accomplished. In addition, our results also showed that cross-functional teams had more difficulty in coordinating some aspects of their internal processes than did teams that were not so diverse.

What our research showed was that neither the frequency with which the team communicates with others nor the functional diversity of team, in and of themselves, were strongly related to team performance. Rather, if correctly managed, these factors may help provide some basis for helping the team effectively complete the Ambassador and Task Coordination activities that facilitate team performance.

The Importance of Boundary Management Is Independent of Most Characteristics of the Project

Although our sample of teams were all involved in the development of sophisticated products, there were substantial differences in the characteristics of the products. During the interviews with team leaders, we asked them to rate their projects in terms of: (1) the extent to which it used a "revolutionary" technology; (2) the experience of the company in developing similar products; (3) the degree of competition anticipated for the new product; (4) the extent to which resources were available to the product development team; and (5) the predictability of the market for the product. There were very few relationships between the characteristics of the product under development and the extent to which members of the product teams undertook the various boundary activities. Because all of our teams were developing products in high technology companies, there may be more similarities than differences among them. However, among teams of this type, specific characteristics of the product or environment had little effect on the extent to which individual team members felt responsible for Ambassador, Task Coordination, or Scout activities.

Related to this is the question of whether or not the nature of boundary activities change under the life of the project. Based on our data, the answer is partially "yes." Ambassador activities were highest among teams that were either beginning a project or were nearing project conclusion, that is, at the point where "ownership" of the new product was about to be transferred to others. These activities were lower among those teams in the middle of the process.

Finally, to what extent do individual characteristics relate to responsibility for boundary management activity? Not surprisingly, the answer is "somewhat." Team leaders and individuals who had experience in the marketing function were more likely to take on responsibilities for Ambassador activities than people without these characteristics. The lateral cooperation reflected in the Task Coordination activities is slightly more likely to be the responsibility of the team leader, employees who have been in the company a long time, and individuals with experience in the manufacturing function than of those who do not possess these characteristics. Individuals with experience in the marketing function report that they assume more responsibility for Scout activities than individuals without that experience. In thinking about how individual characteristics affect the extent boundary management activities, what must be kept in mind is that personal characteristics we measured—assigned leadership role, tenure in the company, experience on other teams, and experience in functional areas—had relatively little effect on what individuals reported they were responsible for. At most, these characteristics explained about 20 percent of the variance in individual levels of these activities.

Teams Have Distinct Strategies for Performing Boundary Activities

Our results show that there are distinct activities in which team members engage in dealing with others and that the frequency of these activities is related to the performance of the team. What was clear from our initial interviews was that not all teams exhibited all of these activities. Rather, just as teams develop unique styles of working internally, teams seemed to develop distinct strategies for approaching their environment; some seemed to specialize in particular sets of activities, others were generalists, and others did not seem to engage in much activity at all. To empirically identify those strategies among the teams we surveyed, we used a type of factor analysis to "cluster" teams into groups based on the frequency of their use of the different boundary activities. We found that teams used one of four different strategies: Ambassadorial, Researchers, Isolationists, and Comprehensives. The Ambassadorial teams engaged in very low levels of Task Coordination and Scouting but high levels of Ambassador activity. Researchers displayed high levels of Scouting, moderate levels of Task Coordination, and low levels of Ambassador activities. Isolationists were low on all the boundary activities. The Comprehensives combined high levels of Ambassador activities and Task Coordination with a low level of Scouting. In our study, approximately an equal number of teams used each of the four strategies.

The Ambassadorial teams seemed to specialize in working "vertically," that is, they concentrated on developing and maintaining relations with upper management. Relative to teams using other strategies, they communicated less frequently with outsiders. In terms of performance, these teams were only moderately effective. During the course of their work, top management rated them relatively high, particular in terms of meeting budgets and schedules. However, at the conclusion of the project, the outcomes of these teams were not highly rated.

The Researcher teams concentrated on acquiring broad technical and market information. Although they had relatively high levels of communication with outsiders, they devoted low effort toward cultivating relations with upper management or establishing links with other functional groups. The performance ratings of these teams were low, particularly in terms of meeting budgets and schedules. Much of the reason for this seemed to lay in the inability of these teams to commit to and follow through on a course of action. Perhaps because of their constant search for "new" information, the members of these teams reported difficulty in establishing goals and developing priorities.

The Isolationist teams attempted to be self-contained. Compared to other groups, they were low in their interactions with others. They generally reported that they worked well together, yet the evaluations they received from top management were very low. Of particular interest, the products these teams developed received very low ratings in terms of how innovative they were. This finding is one more piece of evidence that outside input, rather than simply the ability of a group of technologists to work together well, is necessary for innovation within an organization.

The highest performing teams combined high levels of Task Coordination and Ambassador activity. These Comprehensive teams were highly rated throughout the life of their projects. In addition, the products these groups developed were rated as much more innovative than those developed by teams using other strategies. Not surprisingly, these teams engaged in frequent communication with others. What is surprising is that these did not develop the same level of internal interaction and cohesiveness than did the less successful teams who used different strategies for dealing with others.

MANAGING THE TEAM'S BOUNDARY

The notion of managing beyond a team's borders is counterintuitive to many team managers. Most of our models of management suggest that team building begins with setting goals and priorities and having team members get to know and trust one another. Little in these models suggests that what a team also needs to do early in its life is link team priorities to corporate objectives or get feedback from other groups. Even fewer models suggest that teams may need to work to actively shape the expectations others hold for the team and product. Thus, the first step in successfully carrying out boundary management is educating team members about its importance.

Simply building an awareness of the importance of boundary management is not enough. The team needs to be organized to take on these responsibilities. The team might begin by developing a list of those individuals or groups within the organization who have the information, expertise, or resources the team might need. Parallel lists could identify the concerns other groups might have and the issues that need to be successfully resolved to gain commitment from the other groups that will take over the product. With this as a start, teams may then be able to decide how to allocate the work of managing external relations. As our research shows, the overall level of the responsibility for Task Coordination and Ambassador activity are related to team performance, with Task Coordination most directly related to final evaluations of the groups' work and Ambassador activity related to earlier evaluations. This suggests that teams that begin to build bridges with other groups, before specific coordination problems arise, may have the greatest opportunity for long-term success. Along these same lines, teams may use this exercise to determine when particular types of boundary activity need to be curtailed. For example, too much Scout activity, particularly once the team has committed itself to specifications, may be counterproductive.

Effective boundary management places special demands on the team leader. Our research shows that team leaders take on a large amount of this responsibility, yet our results also show that nature of these relations is such that the team leader cannot do it alone. Thus, the team leader may need to monitor the entire web of relations between the team and outsiders and ensure that these important tasks are completed. This may mean that the leader does many of these activities himself or herself, particularly those up and down the hierarchy, and develops a plan to ensure that lateral activities are completed by those who have the relevant technical knowledge. At a less obvious level, our results present another set of challenges for the leader. We found that the performance of the team was at least partially dependent on its relations with others. Yet, team members' models of what leads to effective performance do not include these variables. The leader must look beyond the ability of the group to work together and get along with one another to develop benchmarks for team performance that reflect how effectively the team is in working with outsiders.

At a broader level, what must an organization do to make it more likely that teams will develop effective boundary management strategies? When we presented the results of our research to executives in the companies we studied, they identified a number of things they might do to improve the ability of groups to deal with others. Most of the ideas they generated had implications for how team members are chosen and rewarded.

Selection and Development

In addition to making assignments to teams based on technical skills and the ability to work as part of a group, an explicit criterion for team assignment should be boundary management. That is, does the team have individuals on it who can effectively deal with outsiders? Are there individuals assigned to the team who can perform the activities necessary to serve as a gatekeeper or liaison? Not all team members need to be skilled in these external activities, yet the team's effectiveness depends on having sufficient resources in this area. This implies that management ensure that teams have individuals who are skilled at both Ambassador and Task Coordination activities and are capable of managing and focusing Scouting. This may require providing the training and feedback to develop individuals that can take on these activities.

Part of this training may involve describing a more complete model of group effectiveness to team members. Cross-functional teams and those that focus on building effective relations with other groups are inherently difficult to manage. In fact, encouraging teams to spend time working with nonteam members is counter to what frequently is seen as necessary to build a smoothly operating group. In fact, among the teams we studied, those that had the *least* communication with outsiders became the most cohesive and rated their own performance higher than other groups. For a team manager, this is something of a paradox. Encouraging the groups to build effective relations with outsiders contributes to high performance but at the same time may reduce the cohesiveness and sense of accomplishment of the group. If team managers have an expanded model of what constitutes an effective group, they may be less likely to concentrate solely on what is going on within the group.

Reward Systems

If teams are to develop an external focus, systems for supporting that will have to be developed. Particularly important may be the reward and career development systems. If individuals are evaluated solely by their performance in a function or if product team members are accountable to a functional manager rather than a team manager, it will be difficult for individuals to take on unfamiliar responsibilities—especially when it takes them outside their own expertise. If an organization wants to encourage cross-functional work, it must reward it. This does not mean that functional skills can be ignored in evaluations, rather it means that if the product development process requires cooperation and problem solving across functional lines, individuals possessing those skills must be rewarded.

BEYOND PRODUCT DEVELOPMENT TEAMS

The importance of how teams manage their boundaries with other groups is not confined to product development. Rather, boundary management will become a critical factor for success in many organizations. Organizations are changing; the flat, flexible, networked, global organization is often team based. Teams in these organizations cannot work the ways they have in the past. Success depends on developing linkages with others in the organization to get work done. The teams that can best pull together the expertise of the firm and move their ideas and products quickly through the organization are those that will succeed.

Note

A more complete description of this research is reported in: Bridging the boundary: external activity and performance in organizational teams. *Administrative Science Quarterly,* 37, 4 (1992); and Demography and design: predictors of new product Team Performance. *Organization Science,* 3, 3 (1992).

REFERENCES

Allen, T. J. (1971). Communication networks in R&D laboratories. *R&D Management,* 1, 14–21.

Allen, T. J. (1984). *Managing the flow of technology:*

Technology transfer and the dissemination of technical information within the R&D organization. Cambridge, MA: MIT Press.

House, C. H. and Price, R. L. (1991). The return map: Tracking product teams. *Harvard Business Review,* 69, 92–100.

Katz, R. and Tushman, M. (1979). Communication patterns, project performance, and task characteristics. *Organizational Behavior and Human Performance,* 23, 139–62.

Kazanjian, R. K. (1988). Relation of dominant problems to stages of growth in technology-based new ventures. *Academy of Management Journal,* 31, 257–79.

Roberts, E. and Fusfeld, A. (1981). Staffing the innovative technology-based organization. *Sloan Management Review,* 22(3), 19–34.

Roberts, E. and Fusfeld, A. (1982). Critical functions: Needed roles in the innovation process. In R. Katz (ed.), *Career issues in human resource management.* Englewood Cliffs, NJ: Prentice-Hall.

Tushman, M. (1977). Special boundary roles in the innovation process. *Administrative Science Quarterly,* 22, 587–605.

Tushman, M. (1979). Work characteristics and subunit communication structure: A contingency analysis. *Administrative Science Quarterly,* 24, 82–98.

30

Redundant, Overlapping Organization: A Japanese Approach to Managing the Innovation Process

IKUJIRO NONAKA

Innovation is a product of the interaction between necessity and chance, order and disorder, continuity and discontinuity. Innovation is the result not only of the planned allocation of resources to meet some predetermined clear objective, but also of some difficult to predict or duplicate redundancy, chance, uncertainty, or even chaos. It is not unusual to discover information and knowledge born of the development process that did not sequentially follow the innovators' original intent.

Most of the research to date has focused only on the analytical aspects of innovation through the hypothesizing of "problem solving" or "information processing" models.[1] In this process one first determines, based on market research, what kind of products to produce and at what price and quality levels to produce them—that is, one clearly defines the "problem" (largely ignoring how the problem is created). This problem is then efficiently disposed of by breaking it down into hierarchical units and analyzing and solving those units in a sequential, orderly fashion. In this way, once the problem is given, innovation moves forward in time through the sequential reduction of the information and decision burden. In problem solving theory, knowledge is accumulated in the problem solver's long-term memory and the problem solver searches through relevant items in this memory in order to efficiently process information. In this almost routinized computing process, the problem solver is considered to be following and calculating through clear, predetermined algorithms—in this case, the decision premises given.[2]

Yet, even if this theory can explain problem solving as the organizational phenomenon of "a reiteration of order" following such preset algorithms, it cannot explain the emergence of innovation as the organizational phenomenon of "the creation of new order through the acting upon of redundancy and chaos." This chaos refers not to a static state of disorder, but rather to the dynamic state of order without the periodicity and recursiveness of the emerging chaos theory.[3] "In real world practice, problems do not present themselves to the practitioner as givens. They must be constructed from the materials of problematic situations which are puzzling, troubling, and uncertain."[4] In Japanese organizations, the process of generating innovation takes place through the use of chaos to internally give birth to new problems, a process through which knowledge and information are created through the use of redundancy. Unlike the problem solving or information processing model, the innovation generation process of Japanese organizations is a "problem generating" or "information creating" model. From this viewpoint, innovation can be understood more dynamically, leading to the possibility that different and new patterns of managing the innovation process can be proposed.[5]

INFORMATION REDUNDANCY

The cases of innovation gathered for this study cover the entire product life cycle and include the personal computer (introduction), the printer (introduction), the copier (growth), the automobile (maturity), the camera (maturity), and consumer electronics (maturity).[6] Through these cases I shall explain the concept of "information redundancy"—a factor I postulate as fundamental to the innovation process.

Information redundancy refers to a condition where some types of excess information are shared in addition to the minimal amount of requisite information held by every individual, department (group), or organization in performing a specific function. While this excess information could be considered needless or superfluous from a standpoint of efficiently processing information in quantity, from a qualitative standpoint this excess information enriches the meaningful functions of the organization. When excess information is shared within the organization, it clarifies the meaning of the specific requisite information held by distinct individuals and groups. In addition, this excess information both increases the reliability and induces an expansion of the significance of such requisite information. Information redundancy stimulates the creative powers of information and is linked to the generation of information with new meanings.

Redundancy in the Project Team

The reasoning behind project teams is that a group can produce ideas in larger quantity and quality than those produced by the average individual. This is especially true in autonomous groups comprised of members with different functions and perceptive and behavioral patterns. For this reason, project teams with a cross-functional diversity are often used at every phase of innovation generation.

The development of the Honda "City" automobile is a case in point. During the development phase, project teams were formed mainly from R&D, production, and sales/marketing personnel who cooperatively analyzed the automobile market and competitive conditions, targeted market segments, established sales quantities and prices, and so on. This was accomplished rapidly and with much interaction, which promoted flexibility.

In another case, in Matsushita Electric's automatic "Home Bakery" project, not only R&D personnel but also members from the sales department and the manufacturing department were involved. At an early stage, the basic objectives for product design decisions, stabilization of quality, and cost reductions were established, but through grappling with these objectives their plans were fundamentally overturned and a new method of breadmaking was born.

At Epson, groups that cut across the hierarchy to encompass general managers, section managers, and group leaders are often employed in order to mobilize enough corporate power. In the case of the personal computer, there are a number of groups in charge of the same model's circuitry, casing, operating system, and language. Naturally, some conflict occurred in every group over specifications, costs, and deadlines. However, the Epson-style method of conflict resolution states that "the best way (to achieve an objective) is through opposition and conflict." Through rigorous self-assertion, a feeling of group identity in creating an object is born and the group is able to overcome difficult barriers. The EP101 dot-matrix printer was developed from the conflict within such a diverse, mixed team.

As Table 30.1 shows, the project teams in the innovation generation process consist of between 10 and 30 members and are composed of personnel with diverse backgrounds including: R&D, planning, production, quality control, sales/marketing, services, etc. In the innovation generation process of Japanese organizations, this kind of di-

TABLE 30.1. Functional Background of Major New Product Development Team Members

Company (product)	Functional Background							
	R&D	Production	Sales Marketing	Planning	Service	Quality Control	Other	Total
Fuji Xerox (FX3500)	5	4	1	4	1	1	1	17
Honda (City)	18	6	4	—	1	1	—	30
NEC (PC8000)	5	—	2	2	2	—	—	11
Epson (EP-101)	10	10	8	—	—	—	—	28
Canon (AE-1)	12	10	—	—	—	2	4	28
Canon (Mini Copier)	8	3	2	1	—	—	1	15
Mazda (New RX7)	13	6	7	1	1	1	—	29
Matsushita Electric (Automatic Home Bakery)	8	8	1	1	1	1	—	20

versity is a notable characteristic and is realized through the multi-functional nature of the project's members.

Redundancy in the Interdepartmental Development Process

In the innovation generation process of Japanese organizations, the concept of the division of labor, in the sense that areas of activities are clearly specialized, is not completely adopted. Rather, in what is also termed "a shared division of labor,"[7] every phase of innovation generation is loosely connected and overlaps, expanding and contracting with the unrestricted elasticity of diversity. Through each phase being autonomous yet loosely linked, interaction between each phase is induced and an abundant sharing of information is promoted in the innovation generation process, which then becomes a system totally sensitive to changes in its market.

More concretely, the special characteristic of Japanese enterprise is that, rather than clearly dividing each phase and operating remotely, every phase is made to overlap in a process that moves through the joint efforts of the participants. Thus, if the individuals in charge are able to carry out their tasks to the next phase, they can continue on and, in some cases, contribute throughout the entire development and production process.

For example, in the development of the Fuji Xerox FX3500, the NASA-style phased program was improved, with each phase from planning to production overlapping in a parallel development method termed the "sashimi approach." (See Figure 30.1.) As a result, the 38-month period previously required with a linear and sequential development process was compressed to 24 months. Behind this was the frequent personal contact in which the development-related staff was extensively involved. As shown by Figure 30.2, the distribution of time for daily duties amounted to 39.7% for design and drawings, 31.6% for meetings and discussions, 16.1% for search and production of materials, and 12.6% for other activities. Personal contact between the development project members amounted to about one-third of the total time. Had each phase been completely subdivided and the objective of each function made extremely specialized, each function would have been isolated. Activities exceeding the artificial boundaries on interaction created by such a division of labor would have been impossible. Frequent co-

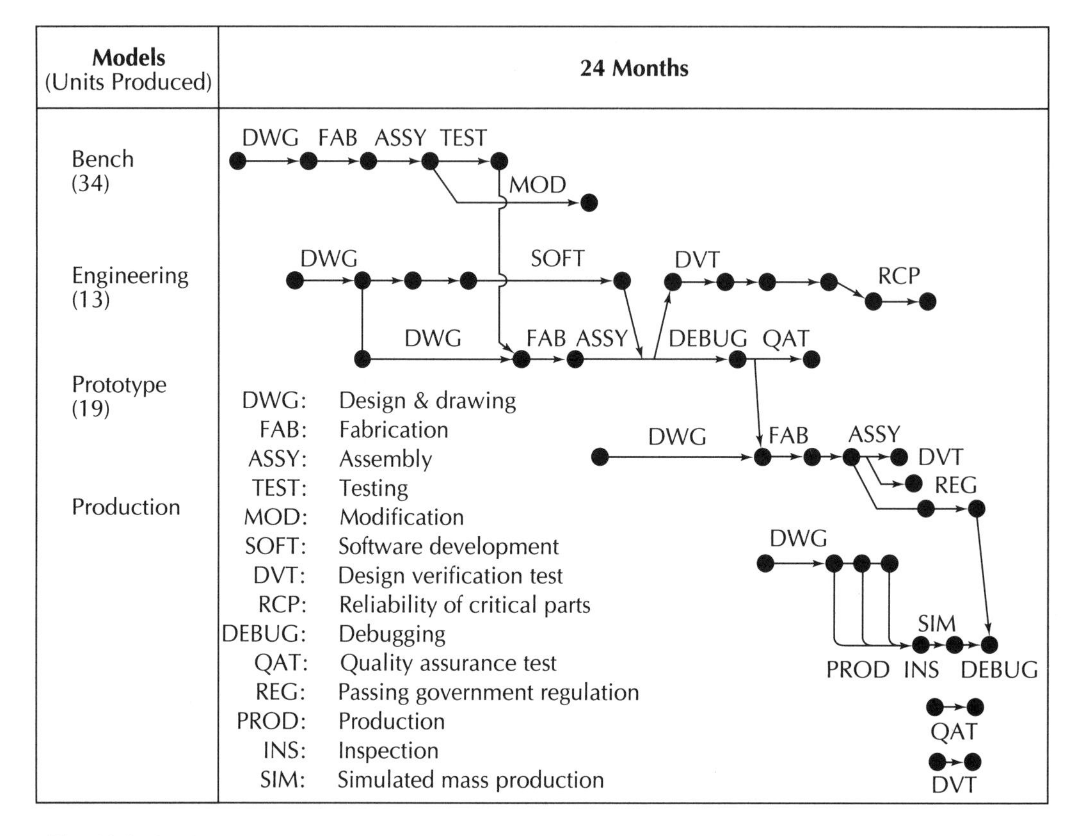

Fig. 30.1. Parallel development of the Fuji Xerox FX 3500. (*Source:* Ken-ichi Imai, Ikujiro Nonaka, and Hirotaka Takeuchi, Managing the New Product Development Process: How Japanese Companies Learn and Unlearn: In Kim B. Clark, Robert H. Hayes, and Christopher Lorenz, eds., *The Uneasy Alliance* (Boston, MA: Harvard Business School Press, 1985). p. 352. Reprinted by permission.)

operative interaction among project members, however, makes parallel development possible.

In the development of Canon's Autoboy Camera, the division of labor in each phase was extremely vague. One of the leaders on the Autoboy project noted:

> The idea that (the planners) were to continue looking after their creation until it became really good was always in the planners' minds. The fabricators too, in implementing the planners' idea, would confront them with "Why did you think that?" or "How come it has got to be done that way?" I think that though there is usually a division of labor, a mutual intrusion into each other's territory for the better emerges and it has a very good effect.

At Canon, the traditional concept of the division of labor was discarded. While such a situation creates much conflict, it also promotes a beneficial redundancy of information through the increased group interaction.

Redundancy in Inter-Organizational Relations with Suppliers

The overlapping of functional groups can also be observed in the organizational network of parent and affiliated or subcontracting companies. Affiliated companies have neither a restricted nor isolated division of labor in their relationship with parent companies; each affiliate, while autonomous, has loose yet closely interdependent ties with the parent.

In the development of Mazda's New RX7 (P747), a planner from Mazda's head office went to Hiroshima Aluminum, an affiliate supplying engine and brake parts, and conducted detailed meetings from the very start of parts development. A project leader at Hiroshima Aluminum observed:

> With this car, we had a designer from Mazda come in from the first stages. Usually corporate secrets are not allowed outside, but in this case we worked closely together from the start. Until now, we would receive the plans and work from those; this time, however, the design people and the manufacturing people formed a team and worked together. One question that comes up is why Mazda's people would go so far as to want to do something like this. There's a big gap between planning and manufacturing, and because this method covers that gap, I think it was very good. We can't understand things like the origins of the shapes we work with just from the drawings. Oh, the shapes come out clearly, but the reasons behind them are not communicated just from the prints.

Fuji Xerox has a similar relationship with its affiliates. For example, according to one Xerox planning manager:

> Early in the development process we try to have the vendor come to our factory and work with us. Vice versa, if we go to their plant, they accept us there. I think there's a relationship between this kind of mutual exchange and openness of information and flexibility in the development process. If one just puts out a purchase order when the plans are completed, those involved are divided clearly

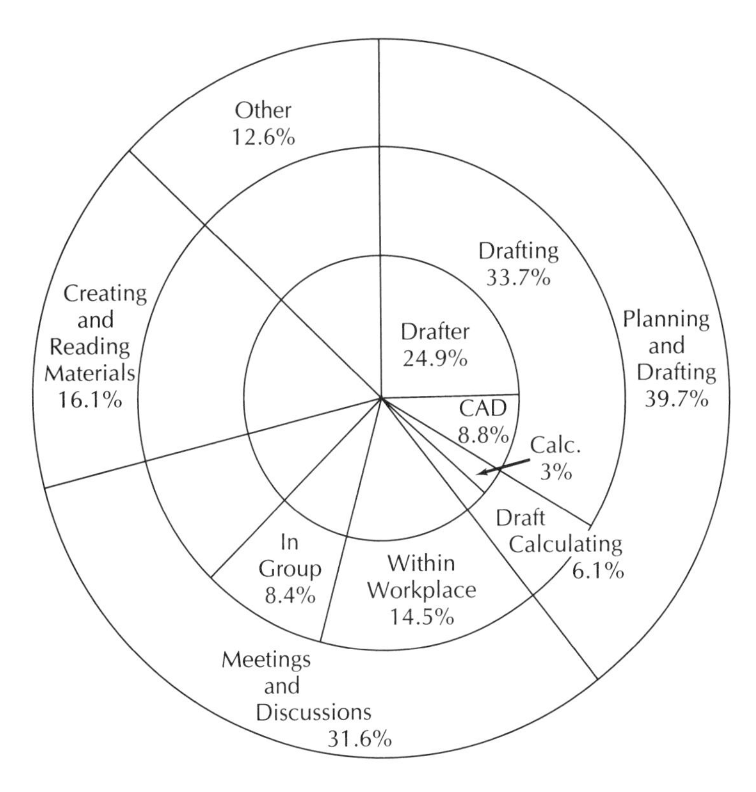

Fig. 30.2. FX3500 development leaders' (nineteen leaders) allocation of normal working hours (one month before completion of design for production).

> into the role of seller and buyer. However, by consulting with the vendor in advance, even if one places a purchase order for only one item, the vendor can make that item with the understanding of where it fits within the whole product. To go one step further, even at the rough sketch stage, if the subcontractor has some special capability, they will sense what I want and bring in several different kinds of things that take advantage of their strength, and the process becomes one where our planner simply has to approve their plans.

At Fuji Xerox, it is recommended that development project members call on vendors during the project. Vendors as well are encouraged to visit their subcontractors. In this way, Fuji Xerox is loosely connected to its subcontractors and its subcontractors' subcontractors, and through cooperative interaction information redundancy is promoted and the entire development can be carried out with flexibility and speed.

The Promotion of Mutual Investigation

When information is redundant—i.e., excess information is held—it becomes possible for one to enter the territory or specialty of another. In this process, a project member contributes some supplementary information that impinges on another member's specialty. By providing such extra information, the project member is able to point out problems which another member, lacking such information, might miss. There are many such cases in which previously undetected problems are brought up by group members from other domains. If information was non-overlapping, it would be impossible to enter other functional areas of disciplines to make adjustments of this sort. Behind the emergence of successive new problems in innovation generations is the principle that each individual has supplemental information which must be shared with other group members.

At the heart of the innovation generation process of Japanese organizations is this kind of cooperative utilization and learning through problem generation (learning by intrusion). Creating or defining a problem is prerequisite to problem solving and, consequently, information.

Furthermore, intense interactions where information is redundant facilitate the transfer of tacit knowledge among team members.[8] Since members share overlapping information, they can sense what others are trying to articulate. Especially in the concept development stage, it is critical to articulate images rooted in tacit knowledge. Meaningful information arises as a result of the conversion of tacit knowledge into articulable knowledge. Redundancy of information facilitates this process.

The Redundancy of Potential Command

In Japanese organizations, every member involved has the potential to create and suggest a solution to the problems at hand regardless of status within the hierarchy. The acquisition of great amounts of information becomes more important than the relationships among the various individuals, and the individual who brings forward the most critical solution has the authority to give orders. MacCulloch refers to this special characteristic as the "principle of redundancy of potential command."[9] The value of each part is potentially equal and arises from the nature of information in which the importance of each component contributes in its relation to the whole. In other words, in the redundancy of information, all of the parts can be potential leaders and the concept of rank no longer holds much meaning. The assumption that the top level is always the leader does not exist and all members—whether from top, middle, or lower levels—have the potential to play a leadership role depending on the context. While organizations have a hierarchical structure in their composition, they do not necessarily limit themselves to a similar chain of command or structural rank with regard to information generation.

Building Trusting Relationships

When information redundancy is present among individuals, groups, or organizations, the possibility of loyalty and trust forming among them heightens. When supplemental information is shared within a cooperative, tight-knit relationship, questioning normally perceived as destructive or hostile can be accepted with security and without a sense of crisis or stress. It becomes a foundation for making an imaginative place that nurtures dreams and continual challenge.

Trust is an important lubricant in the social system and has great significance in areas demanding imagination such as innovation generation. Without loyal relationships, it is hardly possible to have "synergetics"[10] where all can move up and down, left and right in unison to surpass the boundaries of their various areas of responsibility to create information. By nurturing such loyal and trusting relationships, information redundancy eliminates "deception," "cheating," or "blame." Williamson asserts that, compared to market transactions, the possibility of generating opportunism in intra-organization transactions is lessened;[11] the essence of this lies in information redundancy within the organization. In other words, through the sharing of information within a close, interactive group, information redundancy gives rise to loyal relationships which function to suppress the generation of opportunism.

PROBLEMS WITH INNOVATION GENERATION IN JAPANESE ENTERPRISES

Compromising in Group Think

Information redundancy can potentially nurture "group think."[12] While information redundancy can translate into flexible and rapid innovation generation, it can also give rise to compromise in the pursuit of innovation to its ultimate, potential limits. Even within the most heterogeneous of groups, information redundancy implies that there are no eccentric ideas and even that, from time to time, there is some hesitation in the submission of creative ideas.

This is particularly so in cases where the project team or task force has a tight deadline for innovation generation. Yamanouchi Akio, former head of Canon's Technology Development Center stated that:

> The task force or project team is an excellent way to conduct creative activities, but generally deadlines are clearly set which bring into the process various limitations that oppose project objectives. Sometimes, somewhere in the process there is compromise and it becomes impossible to pursue the outer limits of potential technology.

It is perhaps more difficult for this to occur in an innovation style like that of the United States where, in the pursuit of outer boundaries, designated areas are thoroughly dealt with in depth by individual specialists.

High Costs and Human Exhaustion

Another problem of information redundancy in the innovation generation process is the usually high cost of regularly generating problems and their solutions. One of the members of the "City" development team at Honda said the following about the costs of innovation:

> It often happens that no matter what, inevitably, work gets done because we ask for the impossible even though it is clearly impossible, and there are quarrels there and quarrels here. Because of this, the waste of energy and the waste of money is so great one is almost tempted to boast about it.

Also, innovation generation backed by information redundancy requires more social interaction than usual and when a problem comes up, all the members are subject to tense working relationships. This leads to human exhaustion. In a number of cases, the overtime of those in charge of development planning was, at its peak, over 100 hours per month and averaged 50 hours of overtime per month. One project team member commented:

> There's no difference here from the Japanese Army during World War II. What has impacted me deeply is how can Japanese enterprise keep winning through this. Sometimes we lose a battle and all of us that are mentally struggling worry that we're just doing the same old things.

This kind of overwork is not only limited to the extraordinary effort by the organization's members, but extends to the extraordinary effort necessary

TABLE 30.2. Education, Major, and Corporate Career of Core Members of the FX 3500 Development Team

Name	Education and Major	Career Path within Fuji Xerox
Yoshida Hiroshi	University, Education	Technical Service Staff → Personnel → Product Planning → Program Management
Fujita Ken'ichiro	University, Commerce	Marketing Staff → Product Planning → Program Management
Suzuki Masao	University, Mechanical Engineering	Design → Research → Design
Kitajima Mitsutoshi	University, Electrical Engineering	Technical Service Staff → Quality Guarantee → Production

among affiliated companies as well. The head of one affiliate made the following point:

> The way we do things is not by calmly stepping through various technical patterns, but by doing them frantically. At such a time, it doesn't matter if you're general manager or not, you start doing overtime as if it were normal. In that sense you find yourself in a situation where, say, you stop following normal protocol. You no longer pay attention to your appearance and you just do it.

THEORETICAL IMPLICATIONS

Information redundancy is the essential element that characterizes the innovation generation process in Japanese enterprise. Innovation generation in the Japanese enterprise should be viewed not as an information processing model for reaching solutions but rather as an information creating model.[13] Information redundancy, while holding the dangers of chaos and randomness, creates order in Japanese organizations because the corporate vision is clear and the individual team members share a sense of direction.[14] The establishment of deadlines functions as a means of focusing problems or information emission.

The competitive conditions characteristic of Japanese organizations have significance in prompting this order creation. One of the major reasons relates to the level of built-in "multi-functionality" of each group member in the Japanese enterprise. As Table 30.2 shows, members engaged in the development of the Fuji Xerox FX3500 combined within one group abilities relating to several different areas. Because of their diverse backgrounds, they not only bore information pertaining to their specific functions, but bore information pertaining to other functions as well. In addition, such prior job rotation insured that if the group member did not hold certain requisite information, he at least knew where to get it. In other words, the group participant need only "know where" information necessary to the group project is—not "know how" regarding the project itself. This provides the ability to process the significant information burden born from the intensive interaction among project members and focuses information creation at a fixed level. This can be called the concept of requisite variety in semantics as it prescribes the level of redundancy. Ashby's requisite variety concept in organizational theory says that the construction of information channels will match the volume of information generated in the environment.[15] Though mainly developed from the aspect of formal information processing, this concept can be expanded to cover the aspect of significant information.

Furthermore, overlapping within and across team members' functions encourages meaning creation. Once semantic information is created, it naturally seeks similar meaning in an elastic effort to self-organize. Thus, information generates information to both increase entropy and reduce entropy in a dynamic process. This "overlapping" is often referred to as "equipotentiality," which is "the overlap that permits the organism to exhibit a high degree of adaptability, i.e., to change its behavior in accordance with change in stimuli."[16] Overlapping functions in the same manner as parallel distributed processing in the field of computer and cognitive science, which exploits brain-style processing and is "a system that is flexible, yet rigid."[17] Learning takes place by changes in the system itself to adapt flexibly to a changing environment and the major determinant of the system's learning capability is the number of possible connections generated by a redundancy of functions.

In sum, the innovation generation process in Japanese enterprise demonstrates how the concept of information redundancy, along with the concept of requisite variety, is an indispensable factor in constructing a theory of information and knowledge-creating organization.

Note

The author would like to thank Professors Ken-ichi Imai, Francesco M. Nicosia, David C. Mowery, and Robert G. Harris, as well as the author's research assistants Shigemi Yoneyama (Hitotsubashi University) and Jeffrey Daggett (University of California at Berkeley) for their insightful comments and assistance.

REFERENCES

1. E.g., T.J. Allen, "Study of the Problem-Solving Process in Engineering Design," *IEEE Transactions on Engineering Management* (June 1966); M. L. Tushman, "Special Boundary Roles in the Innovation Process," *Administrative Science Quarterly,* 22 (December 1977).
2. H.A. Simon, "Rational Decision Making in Business Organization," *American Economic Review,* 69 (1979).
3. E.g., I. Prigogine and I. Stegners, *Order Out of Chaos* (Toronto: Bantam Books, 1984); H. Haken, "Can Synergetics Be of Use to Management Theory?," in H. Ulrich and G. J. B. Probst, eds., *Self-Organization and Management of Social Systems* (Berlin: Springer-Verlag, 1984); J. Gleick, *Chaos,* (London: Penguin Books, 1987).
4. D. A. Schön, *The Reflective Practitioner* (New York, NY: Basic Books, 1983), p. 40.
5. E.g., I. Nonaka, "Managing the Organizational Information Creation Process," in K. Imai, ed., *Innovation and Organization* (Toyo Keizai Shinposha, 1986 [in Japanese]); I. Nonaka, "An Organization Theory of Information and Knowledge Creation," *Organizational Science,* 22/4 (1989) [in Japanese]; I. Nonaka and M. Kenney, "Innovation as an Organizational Information Creation Process: A Comparison of Canon Inc. and Apple Computer," working paper, 1989; I. Nonaka and A. Yamanouchi, "Managing Innovation as a Self-Renewing Process," *Journal of Business Venturing,* 4/5 (1989).
6. Some of these studies have been published in K. Imai, I. Nonaka, and H. Takeuchi, "Managing the New Product Development Process: How Japanese Companies Learn and Unlearn," in K. Clark, R. Hayes, and C. Lorenz, eds., *The Uneasy Alliance,* (Boston, MA: Harvard Business School Press, 1985); H. Takeuchi and I. Nonaka, "The New New Product Development Game," *Harvard Business Review,* 64 (January/February 1986); Nonaka and Yamanouchi, op. cit.
7. Imai et al., op. cit.
8. M. Polanyi, *Personal Knowledge* (New York, NY: Harper & Row, 1962).
9. W.S. MacCulloch, "A Hierarchy of Values Determined by the Topology of Nervous Nets," *Embodiment of Mind* (Cambridge, MA: MIT Press, 1965).
10. Haken, op. cit.
11. O.E. Williamson, *Market and Hierarchies: Antitrust Implications* (New York, NY: The Free Press, 1975).
12. I. Janis, *Victims of Group Think* (Boston, MA: Houghton Mifflin, 1972).
13. Nonaka (1986), op. cit.
14. I. Nonaka, "Creating Organizational Order Out of Chaos," *California Management Review,* 30/3 (Spring 1988).
15. W. Ashby, *An Introduction to Cybernetics,* (London: Chapman and Hall, 1955).
16. M. Landau, "Redundancy, Rationality, and the Problem of Duplication and Overlap," *Public Administration Review,* 14/4 (1969).
17. D.A. Norman, "Reflections on Cognition and Parallel Distributed Processing," in J.L. McClelland and D.H. Rumelhard, eds., *Parallel Distributed Processing,* Volume 2 (Cambridge, MA: MIT Press, 1986).

31

Managing Innovation in the Transnational Corporation

CHRISTOPHER A. BARTLETT
SUMANTRA GHOSHAL

Philips has long held a leading position in the mature but profitable electric shaver business. Its Philishave is sold world-wide, but has particularly strong sales in Europe and North America. In recent years, this rather traditional and stable product category has been shaken up by several product innovations brought to market primarily by Japanese companies trying to fight their way into an established business with new clearly differentiated product ideas. The most important innovations have been in the area of miniaturization, rechargeability, and most radically, the concept of wet shaving with an electric razor.

For Philips, the new technologies and product market concepts were important trends that had to be fully understood and closely monitored, and in this task Philips' Japanese subsidiary had a critical role. However, the subsidiary had neither the technological capabilities nor the marketing expertise to develop the appropriate response to these events, and had to pass its data on to the product group and development laboratories in Europe to decide what actions Philips should take and to undertake the actual technological development. Finally, while the company had to consider introducing a new product in Japan, its competitive exposure was far greater in Europe and the United States. It was vital to introduce the new product in these locations and the co-operation of the local subsidiaries was necessary to protect Philips' global leadership position in this business.

This example, one of the thirty-eight cases that we documented in some detail in the course of a large research project involving extensive discussion with over 230 managers in nine large companies,[1] illustrates the complexity of managing the innovative process in today's multinational corporations. Not only must the company have sensing, responding and implementing capabilities around the globe, it must be able to co-ordinate those activities and link them in a flexible yet efficient manner, despite the fact that interdependent units may be separated by enormous physical and cultural distances.

INNOVATIONS AS THE KEY SOURCE OF GLOBAL COMPETITIVE ADVANTAGE

Scholarly research on international corporations has long identified innovations as the *raison d'être* of multinationals. A firm invests abroad to

 First appeared in Bartlett, Doz, and Hedlund, Eds. *Managing the Global Firm*, 1990. London, New York: Routledge, pp. 215–255.

make more profits out of innovations embodied in the products it has developed for the domestic market. The ticket for a firm to invest and manage its affairs in many different countries is its ability to innovate, i.e., to develop new products and processes, and to create an organization through which it can appropriate the benefits of its innovations more advantageously by operating as a multinational rather than selling or licensing its technology.[2]

In the current international environment, a company's ability to innovate is rapidly becoming the primary source of its ability to compete successfully. While once MNCs gained competitive advantage by exploiting global scale economies or arbitraging imperfections in the world's labour, materials or capital markets, such advantages have tended to erode over time. In most industries today, MNCs no longer compete primarily with numerous national companies, but with a handful of other giants who tend to be comparable in terms of size and geographic diversity. In this battle, having achieved global scale, international resource access and world-wide market position is no longer sufficient—many other MNCs will match such assets. The new winners are the companies that are sensitive to market or technological trends no matter where they occur, creatively responsive to the opportunities and threats they perceive world-wide, and able to exploit their new ideas and products globally in a rapid and efficient manner. Meanwhile, companies that are insensitive, unresponsive, or slow are falling victim to the rising costs of R&D, the narrowing technology gap between countries and companies and the shortening of product life cycles.

The Different Multinational Innovation Processes

World-wide competition has not only made innovations more important for MNCs, it has also made it necessary for them to find new ways of creating innovations. Traditionally, most MNCs have adopted one or both of two classic innovation processes. The first of these is what we describe as the 'centre-for-global' innovation process: sensing a new opportunity in the home country, using the centralized resources of the parent company to create a new product or process, and exploiting it world-wide. The other traditional process we have labelled 'local-for-local' innovation process: national subsidiaries of MNCs using their own resources and capabilities to create innovations that respond to the needs of their own environments.

While most MNCs have tried to develop elements of both processes, the tension that exists between them normally means that one will become the primary source of innovations in any particular company. Quite naturally, the centre-for-global process has been dominant in MNCs adopting what has been described elsewhere as the 'centralized hub' mode of operations while local-for-local innovations have been common in the 'decentralized federation' organizations.[3]

These traditional innovation processes also reflect the traditional mentalities of these two types of companies. The centre-for-global process tends to be associated with what may be described as the extreme global mentality which sees the diversity of international environments as an inconvenience whose effects must be minimized. Thus, these organizations modify their centre-for-global innovations reluctantly and only minimally to meet specific needs. On the other hand, the archetypal multinational mentality underlies the local-for-local innovation process wherein complete conformity to local needs is seen as an unavoidable price of admission to the market. To somewhat caricature these multinational mentalities, if the first reflects minimum compromise to meet local needs, the second implies unquestioning capitulation to the whims and fancies of local customers.[4]

In recent years, however, some of these traditional management attitudes have been changing. As a result, the innovative processes in some MNCs have been evolving. In the course of documenting innovation cases in our nine core companies, we have seen how successful MNCs are developing and managing some new ways of creating new products, technologies, and even administrative systems. These new approaches tend to fall into two broad categories which we have labelled 'locally-leveraged' and 'globally-linked' innovation processes. The first involves utilizing the resources of a national subsidiary to create innovations not only for the local market but also for exploitation on a world-wide basis. The second pools the resources and capabilities of many different components of the MNC—at both the headquarters and the subsidiary level—to create and implement an innovation jointly. In this process, each unit contributes its own unique resources to develop a truly collaborative response to a globally perceived opportunity.

Both of these new and much less frequently implemented innovative processes imply a very different attitude on the part of management—an attitude that constitutes what we call the transnational mentality. Rather than viewing the differences among international environments and the diversity of local demands as liabilities, managers with the transnational mentality see them as one of the company's greatest assets. By being exposed to a diversity of consumer needs, market trends, technological breakthroughs, and government demands, a company has a greater chance of stimulating new ideas internally.[5] By operating in different environments world-wide, the company has access to a broader range of resources and capabilities (particularly the scarcest resource of all—creative people) and is able to enhance its capacity for innovative response. Many companies in our sample tried to create new products, processes, and administrative systems through the two new innovation processes but it seemed that in the absence of this transnational mentality, such efforts rarely succeeded.

Some of the companies we studied, however, were in the midst of a transition to this transnational mentality. They were beginning to synthesize their learning from exposure to diverse environments. The new mentality led them to a new set of management processes that could integrate and focus their dispersed organizational capabilities, and thereby unlock the power of the new innovation processes. Why and how they have succeeded while many others have failed is the main topic of our discussion in this chapter.

MANAGING INNOVATIONS IN MNCS: THE MANAGEMENT CHALLENGE

While locally-leveraged and globally-linked innovations are becoming increasingly important, they are not substitutes for the more traditional innovation processes. In a competitive environment in which the ability to innovate is becoming the critical differentiating capability between the winners and losers, companies are recognizing the need to maximize the number of ways in which they can develop innovative products and processes. But as they try to develop all four of these innovative processes, managers have become aware of the advantages and disadvantages of each, and also of their mutually debilitating characteristics.

Centre-for-Global Innovations: Risk of Market Insensitivity

Centre-for-global innovations are necessary because certain key capabilities of the MNC must, of necessity, remain at the headquarters both because of the administrative need to protect certain core competencies of the company, and also to achieve economies of specialization and scale in the R&D activity. A good example is provided by Ericsson's development of the AXE digital switch.

Impetus for this came from early sensing of both shifting market needs and emerging technological changes. The loss of an expected order from the Australian Post Office combined with the excitement generated in a trade show by the new digital switch developed by CIT-Alcatel, then a small French competitor virtually unknown outside its home country, set in motion a formal review process within Ericsson's headquarters. The review resulted in a proposal for developing a radically new switching system based on new concepts and a new technology. The potential for such a product was high, but the costs and risks were also enormous. The new product was estimated to require over £50 million and about 2,000 man-years of development effort and take at least five years before it could be offered in the market. Even if the design turned out to be spectacular, diverting all available development resources during the intervening period could erode the company's competitive position beyond repair.

In sharp contrast to almost all the 'Principles of Innovation' proposed by Peter Drucker in his book *Innovation and Entrepreneurship*, corporate managers of Ericsson decided to place their bet on the proposal for the AXE switch, as the new product came to be called. They provided full authority and all resources so that Ellemtel, the R&D joint venture of Ericsson and the Swedish telecommunications administration, could develop the product as quickly as possible. For over four years, the technological resources of the company were devoted exclusively to this task. The development was carried out entirely in Sweden, and by 1976 the company had the first AXE switch in operation. By 1984, the system was installed in fifty-nine countries around the world.

The need to centralize the development process was driven by three main forces. First, management wanted to have control over a technology that was going to be at the very core of the company's long-term competitiveness. The cost of unco-ordinated or duplicated development in such a product is astronomical. Second, the effort required close integration between hardware and software development, and subsequently between the development and manufacturing functions. Such close co-ordination could best be provided at a central location. Finally, in a rapidly changing competitive environment, Ericsson knew it had to develop its new switch rapidly if it was to respond to tenders that had begun to specify digital capabilities. Centralizing development reduced the time and inefficiencies associated with more dispersed efforts.

The major risk of such a centralized development process is that the resulting innovations may be insensitive to market needs and may also be difficult to implement because of resistance from the subsidiaries in accepting a central solution. Ericsson was able to avoid many of these problems by seconding to the development a group of engineers from their Australian subsidiary who had recent and direct market experience in responding to the demands of one of the first tenders in the world to specify digital switching capabilities. Other companies, however, were not so fortunate and experienced some of the problems of centre-for-global innovation. For example, NEC designed NEAC 61 as a global digital switch, but with the primary objective of meeting the requirements of the US market. However, while the Japanese engineers at the corporate headquarters had excellent technical skills, they were not totally familiar with the highly sophisticated and complex software requirements of the independent telephone operating companies in the US. The result was that while everyone applauded the switch for the capabilities of its hardware, early sales suffered since the software did not meet some specific needs of end users that were significantly different from those of Japanese customers.

Local-for-Local Innovations: Risk of Needless Differentiation

Local-for-local innovations are essential for responsiveness to the unique attributes of each of the different national environments in which the MNC operates. Current fascination with globalization of markets tends to overlook the fact that while the forces of globalization have certainly strengthened in many industries, the need for responsiveness to national demands and local differences has not disappeared, and often has increased. For example, Unilever faces numerous pressures to develop globally standardized products. The cost and sophistication of R&D is increasing, economies of globally integrated operations are available, and competitive battles with major MNCs like P&G are forcing global responses. Yet, Unilever's ability to sense and respond in innovative ways to local needs and opportunities has been a major corporate asset. While advanced laundry detergents did not sell well in huge markets like India, where much of the laundry was done in streams, a local development that allowed synthetic detergents to be compressed into solid tablet form gave the company a product that could capture much of the bar soap market. Similarly, in Turkey, while the company's margarine products did not sell well, an innovative application of Unilever's expertise in edible fats allowed the local company to develop a product that competed with the traditional local clarified butter product, ghee.

But, on the negative side, such innovations may also reflect the efforts of national subsidiaries to differentiate themselves to retain their identity and autonomy, without any real need, and may impose differentiation costs without any significant benefits. Also, they may lead to considerable reinvention of the wheel as each subsidiary finds its own solution to common problems. In the course of interviewing managers in the different companies, we came across scores of instances of such limitations of local-to-local innovations. In Philips, for example, the British subsidiary spent a large amount of resources to create a new TV chassis that would be specially suitable for the local market. The final product was almost indistinguishable from the standard European chassis that headquarters managers were trying to implement, and resulted in the company having to operate five instead of four television set factories in Europe. While this may have been an objective of the managers in the national organization, it clearly compromised Philips' overall efficiency and competitiveness.

Locally Leveraged Innovations: NIH Risk

Locally-leveraged innovations permit management to take the most creative resources and innovate developments from its subsidiaries worldwide, and allow the whole company to benefit from them. In so doing, the company is often able to take the responses to market trends that emerge in one location and use them to lead similar trends in other locations. This kind of innovation requires management to develop and control a process of worldwide learning, but in so doing it allows the company substantially to leverage its world-wide innovative resources.

For example, Proctor and Gamble created the fabric softener product category with a brand called Downy in the US and Lenor in Europe. Unilever's entrant in this fast-growing segment was Comfort, but after years of effort it had done little to shake P&G's dominant first mover advantage. Then, its German subsidiary developed a new brand with a product position and marketing strategy that proved enormously successful in gaining market share rapidly. Management soon recognized that at the heart of the success of Kusshelveich (literally Teddy soft) was the bear that the Germans had developed as the product's symbol and identity. Consumer research showed that not only did the bear do an excellent job in communicating the desired image of softness, it also evoked strong recognition and trusting association in consumers that gave the advertising promise great credibility. The German brand (appropriately translated) and their product market strategy were successfully transferred to other markets throughout Europe and eventually to the US where the teddy bear spokesman rapidly helped Snuggle to build a 25 per cent share in P&G's home market where Unilever's Comfort had been struggling for a decade.

Yet local innovations are not always so easily transferred. The main impediments include attempts to transfer products or processes that are unsuited to the new environment, the lack of suitable co-ordinating and transfer mechanisms (a particular problem with much of the technology transfer), and the barriers presented by the NIH syndrome.[6]

Despite the outstandingly successful transfer of its bear fabric softener, Unilever management was unable to transfer a zero phosphate detergent product developed by its German subsidiary to other European subsidiaries. Insisting that its market needs were different, the French subsidiary proceeded with its own zero-P project. Product co-ordination managers at the central office believed that a NIH attitude was at least as important a factor, particularly in an environment where national companies were struggling to maintain their local R&D budgets against pressures from the centre for more co-ordination.

Globally Linked Innovation: Co-ordination Cost

The globally-linked innovation process is the one most suited to an environment in which the stimulus for an innovation is distant from the company's response capability, or where the resources and capabilities of several organizational units can contribute to developing the most innovative response to a sensed opportunity. By creating flexible linkages that allow the efforts of multiple units to be combined, a company can create synergies that can significantly leverage its innovation process. Like locally-leveraged innovations, the globally-linked process captures the MNC's potential scope economies and harnesses the benefits of world-wide learning.

One of the best examples we observed of this mode of innovation was the way in which Proctor and Gamble developed its global liquid detergent. When Unilever's US success with Wisk demonstrated the potential of the heavy duty liquid detergent category, P&G and Colgate rushed to the market with limited competitive products (Era and Dynamo respectively), but with limited success. All three companies tested their products in Europe, but due to different washing practices and the superior performance of European powder detergents which contained levels of enzymes, bleach and phosphates not permitted in the US, the new liquids, failed in all these test situations.

But P&G's European scientists remained convinced they could enhance the performance of the liquid to match the local powders. After seven years of work they developed a bleach substitute, a fatty acid with water softening capabilities equivalent to phosphate, and a means to give enzymes stability in liquid form. Their new product beat the leading powder in blind tests, and the product was launched as Vizir, establishing the heavy duty liquid segment in Europe.

Meanwhile, researchers in the US had been working on a new liquid to replace Era which had failed to establish a satisfactory share against Wisk. The challenge for liquids in the US was to deal with the high-clay soil content in dirty clothes, and this group was working on improving builders, the ingredients that prevent redisposition of dirt in the wash. Also during this period, the company's International Technology Co-ordination group was working with P&G's subsidiary in Japan and had developed a more robust surfacant (the ingredient that removes greasy stain) making the liquid more effective in the cold water washes that were common in Japan. Each unit had developed effective responses to its local needs, yet none of them was co-operating to share its breakthroughs.

When the company's head of R&D for Europe was promoted to the top corporate research job, one of his primary objectives was to develop more co-ordination and co-operation among the diverse local-for-local development efforts. Through several important organizational changes, he was able to develop the means for co-operation, and the world liquid project became a test case. Plans to launch Omni, the new liquid the US group had been working on, were shelved until the innovations from Europe and Japan could be incorporated. Similarly, the Japanese and Europeans picked up on the new developments from the other laboratories. The result was the launch of Liquid Tide in the US (a brand that was able to challenge market leader Wisk), the successful launch of Liquid Cheer in Japan, and Liquid Ariel in Europe. All of these products incorporated the best of the developments created in response to European, American, and Japanese needs.

But this process also has its own limitations. It requires a degree of internal co-ordination that may be extremely expensive and wasteful. The complex interlinkages among different organizational components that are necessary to facilitate this process can also overwhelm a company because of ambiguity and excessive diffusion of authority. One of the companies participating in our study estimated that a new system of periodic meetings that had been instituted for more effective integration of its European production plants resulted in company managers having to spend 2,581 person-days in one year just on travel and in being physically present to attend the meetings. Similarly, ITT faced enormous problems in attempting to develop its System 12 digital telecommunications switch through a collaborative effort of its different European subsidiaries. Trying to co-ordinate the efforts of the different units that were responsible for developing different components of the switch proved to be extremely time consuming and costly, leading to delays and budget overruns. This effort to create a globally-linked innovation may well have been responsible for the failures that led to the company's exit from this business which has traditionally been its primary activity.

The Management Challenge

The challenge for MNC managers, therefore, it not one of promoting one or the other of these different innovation processes, but to find organizational systems and processes that will simultaneously facilitate all the different processes. In other words, they must, at one and the same time, enhance the effectiveness of central innovations, improve the efficiency of local innovations, and create conditions that will make the newer forms of transnational innovations feasible. To do so, however, requires that the companies overcome two related but different problems. First, for each innovation process, they must avoid the different pathologies we have described. Second, to develop innovations simultaneously through all the different processes, they must find ways to overcome the contradictions among the organizational factors that facilitate these processes.

None of the companies in our sample had solved both problems fully. However, some of them had developed special competencies in managing one or the other of the different innovation processes and in overcoming the pathologies of those processes. Their experiences suggest some ways in which MNC managers can overcome the first problem and these are described in the next part of this chapter.

A few of these companies had also made some progress in overcoming the inherent contradictions among the different processes. A special system of internal differentiation in the roles and responsibilities of different organizational units appeared to lie at the core of the solution they were in the process of developing. Instead of finding ways so that each unit could contribute equally to all the different processes, those companies were systematically differentiating among the units based on their capabilities and needs, and were creating an internally differentiated organization so that each unit could have attributes that facilitated its participation in the particular innovation process to which it could make the greatest contribution. In the concluding part of the chapter, we draw some lessons from the emerging practices of these companies to suggest how multinational managers can build an overall organizational system for creating and exploiting the local, central, and two transnational innovation processes.

MAKING CENTRAL INNOVATIONS EFFECTIVE: LESSONS FROM MATSUSHITA

The key strength on which Japan's Matsushita Electric Company has built the global leadership position of its well-known Panasonic and National brands in the highly competitive consumer electronics industry is its ability to create central innovations and to exploit them quickly and efficiently throughout its world-wide operations. This is not to say that it does not employ some of the other modes of innovation, but, of all the companies we surveyed, Matsushita is the champion manager of central innovations. As we tried to identify the organizational mechanisms that distinguish Matsushita's way of managing central innovations from those of the others, three factors stood out as the most important explanations of its outstanding success in managing this innovation process: gaining the input of subsidiaries into the process, ensuring development efforts are linked to market needs, and managing responsibility transfers from development to manufacturing and to marketing.

Gaining Subsidiary Input: Multiple Linkages

The two most important problems facing a company innovating centrally are that those developing the new product or process may not understand market needs, or that those required to implement the new product introduction are not committed to it. (Philips learned both lessons very well when it tried to introduce its technologically superb V2000 video recorder in competition with Matsushita's VHS system and Sony's Beta format.) Matsushita managers are very conscious of this problem and spend a great deal of time building multiple linkages between headquarters and overseas subsidiaries designed not only to give headquarters managers a better understanding of country-level needs and opportunities, but also to give subsidiary managers greater access to and involvement in headquarters' product development processes.

Matsushita recognizes the importance of market sensing as a stimulus to innovation and does not want its centrally driven development process to reduce its environment sensitivity. Rather than trying to limit the number of linkages between headquarters and subsidiaries, or focus them through a single point, as many companies do for the sake of efficiency, Matsushita tries to preserve the different perspectives, priorities, and even prejudices, of its diverse groups world-wide, and ensure that they have linkages to those in the headquarters who can represent and defend their views.

The organizational systems and processes that connect different parts of the Matsushita organization in Japan with the video department of MESA, the US subsidiary of the company, are illustrative of these multifaceted interlinkages. The Vice President in charge of this department has his roots in Masushita Electric Trading Company (METC), the organization which has overall responsibility for Matsushita's overseas business. Although formally posted to the United States, he continues to be a member of the senior management committee of METC and spends about a third of his time in Japan. This allows him to be a full member of the top management team of METC that finalizes overall product strategy for the US market, including priorities for new product development. In his role as the V.P. of MESA, he ensures that the local operation implements the agreed video strategy effectively. The General Manager of this department is a company veteran who has worked for fourteen years in the video product division of the corporate headquarters of Matsushita Electric, the production and domestic marketing company in Japan. He maintains strong connections with the central product division and acts as its link to the local US market. Two levels below the department's general manager is the Assistant Product Manager, the junior-most expatriate in the organization. Having spent five years in the company's main VCR plant in Japan, he acts as the local representative of the factory and handles all day-to-day communication with factory personnel.

None of these linkages is accidental. They are deliberately created and maintained and they reflect the company's open acknowledgment that the parent company is not one homogeneous entity, but a collectivity of different constituencies and interests, each of which is legitimate and necessary. Collectively, these multiple linkages enhance the subsidiary's ability to influence key headquarters decisions relating to its market, and particularly decisions about product specifications and design. The multiple links not only allow local management to reflect its local market needs, they also give headquarters managers the ability to coordinate and control implementation of their strategies and plans, including those of implementing their innovations.

Linking Development to Needs: Market Mechanisms

But Matsushita's efforts to ensure innovations are linked to market needs does not stop at the input stage. The company has created an integrative process that ensures that the researchers and technologists are not sheltered from the pressures, constraints and demands felt by managers in the front line of the operations. One of the key elements in achieving this difficult organizational task is the company's willingness to employ internal 'market mechanisms' for directing and regulating the activities of central researchers and development engineers. Because the system is unique, we will describe some of its key characteristics.

Research projects undertaken by the central research laboratories (CRL) of Matsushita can be categorized into two broad groups. The first group consists of 'company total projects' which involve developing technologies that are important for Matsushita's long-term strategic position and that may be applicable across many different product divisions. Such projects are decided jointly by the research laboratories, the product divisions, and top management of the company, and are funded directly by the corporate Board. The second group of CRL research projects consists of relatively smaller projects which are relevant to the activities of particular product divisions. The budget for such research activities, which amounts to approximately half of the total research budget of the company, is allocated not to the research laboratories but to the product divisions. This creates an interesting situation in which technologically driven and market-led ideas can compete for attention. Each year, the product divisions suggest a set of research projects that they would like to sponsor. At the same time, the various research laboratories hold annual exhibitions and meetings and also write specific proposals to highlight research projects that they would like to undertake. The Engineering and Development groups of the product divisions mediate the subsequent contracting and negotiation process through which the expertise and interests of the laboratories and the needs of the product divisions are finally matched. Specific projects are sponsored by the divisions and are allocated to the laboratories or research groups of their choice, along with requisite funds and other resources.

The system creates intense competition for projects (and the budgets that go with them) among the research groups, and it is this mechanism that forces researchers to keep a close market orientation. At the same time the product divisions are conscious that it is their money being spent on product development and they become less inclined to make unreasonable or uneconomical demands of R&D.[7]

The market mechanism also works to determine annual product styling and features. Each year the company has its merchandising meetings, which are in effect, giant internal trade shows. Senior marketing managers from Matsushita's sales companies worldwide visit their supplying divisions and see on display the proposed product line for the new model year. Relying on their understanding of their individual markets, these managers pick and choose among proposed models, order specific modifications for their local markets, or simply refuse to take products they feel are unsuitable. Individual products or even entire lines might have to be redesigned as a result of input from the hundreds of managers at the merchandising meeting.

Managing Responsibility Transfer: Personnel Flow

In local-for-local innovations, the task of transferring responsibility from research to manufacturing and finally to marketing is facilitated by the smaller size and closer proximity of the units responsible for each stage of activity. This is not so where large central units take the lead role in the development of new products and processes, and Matsushita has built some creative means for managing these transitions. The systems rely heavily on the transfer of people. First, the careers of research engineers are structured so as to ensure that a majority of them spend about five to eight years in the central research laboratories engaged in pure research, then another five years in the product divisions in applied research and development, and finally in a direct operational function, usually production, wherein they take up line management positions for the rest of their working lives.[8] More importantly, each engineer usually makes the transition from one department to the next along with the transfer of the major project on which he has been working.

In other companies we surveyed it was not uncommon for research engineers to move to development, but not with their projects, thereby depriving the companies of one of the most important and immediate benefits of such moves. We also saw no other examples of engineers routinely taking the next step of actually moving to the production function. This last step, however, is perhaps the most critical in integrating research and production both in terms of building a network that connects managers across these two functions, and also for transferring a set of common values that facilitates implementation of central innovations.

Another mechanism that integrates production and research in Matsushita works in the opposite direction. Wherever possible the company tries to identify the manager who will head the production task for a new product under development and makes him a full-time member of the research team from the initial stage of the development process. This system not only injects direct production expertise into the development team, but also facilitates transfer of the innovation, once the design is completed. Matsushita also uses this mechanism as a way of transferring product expertise from headquarters to its world-wide sales subsidiaries. Although this is a common practice among many multinationals, in Matsushita it has additional significance because of the importance of internationalizing management as well as its products.

MAKING LOCAL INNOVATIONS EFFICIENT: LESSONS FROM PHILIPS

If Matsushita is the champion manager of central innovation, Philips, its arch rival in the consumer electronics business, is the master of local innovations. Again, this does not imply that they have not been successful at central innovations. Indeed, the company has a long list of products and processes that were developed in their Central Research Laboratories that extends from their earliest ventures in light bulbs to today's latest innovations in laser disc technology. In the present context, however, we would like to focus attention on why and how Philips has been able to foster a process of innovation at the national organization level to a degree unmatched by any other company of comparable size, diversity and maturity.

The first colour TV set of the company was produced and sold not in Europe, where the parent company is located, but in Canada, where the market had closely followed the US lead in introducing colour transmission. The K6 chassis introduced in Canada was designed in the company's central research laboratory in Holland, but the local subsidiary had played a major role in the development process and had an even greater input in designating the production system. The first stereo colour TV set of the company was developed by the Australian subsidiary; teletext TV sets were created by its British subsidiary; 'smart cards' by its French subsidiary, a programmed word-processing typewriter by North American Philips; the list of local-for-local innovations in Philips is endless.

Philips' ability to create such local innovations has been due in part to the administrative history that shaped the company's growth, and in part to a strong philosophy and explicit strategic choice to respond to the local market needs. The net result has been that, over a period of time, the company has accumulated substantial resources in its different national organizations which, in conjunction with the relatively high level of decentralization of authority, has made dispersed entrepreneurship one of its key organizational assets. Out of the many different factors that have facilitated local-for-local innovations in Philips, there are three that appear to have been the most significant—the company's use of a cadre of entrepreneurial expatriates, an organization that forced tight functional integration within a subsidiary, and the historical dispersion of resources and authority.[9]

A Cadre of Entrepreneurial Expatriates

Expatriate positions, particularly in the larger subsidiaries, have been very attractive for Philips managers for several reasons. With only 6 to 8 per cent of its total sales coming from Holland, many of the different national subsidiaries of the company have contributed much larger shares of the company's total revenues than the parent company. As a result, foreign operations have enjoyed relatively high organizational status compared to most companies of similar size with headquarters in the United States, Japan, or even the larger countries in Europe. Further, because of the importance of its foreign operations, the formal management development system of Philips has always required considerable international experience as a prerequisite for top corporate positions. Finally, Eindhoven, the small town in a rural setting that serves as the corporate headquarters of the company, is far from the sophisticated and cosmopolitan world centres that host many of its foreign subsidiaries. After living in London, New York, Sydney, or Paris, many managers find it hard to return to Eindhoven. Collectively, all these factors have led to the best and the brightest of Philips managers spending most of their careers in different national operations. This cadre of entrepreneurial expatriate managers has been an important facilitator of local-for-local innovations in the company.

Further, unlike companies such as Matsushita or NEC where an expatriate manager spends a tour of duty of three to five years in a particular national subsidiary and then returns to the headquarters, expatriate managers in Philips spend a large part of their careers abroad continuously, working for two to three years each in a number of different subsidiaries. This difference in the career systems results in very different attitudes on the part of these managers. In Philips, the expatriate managers follow each other into assignments and develop close relations among themselves. They tend to identify strongly with the national organization's point of view, and this shared identity makes them part of a distinct subculture within the company. In companies such as Matsushita, on the other hand, there is very little interaction among the expatriate managers in the different subsidiaries, and they tend to see themselves as part of the parent company temporarily on assignment in a foreign company. One result of these differences is that expatriate managers in Matsushita are far more likely to take a custodial approach which resists any local changes to standard products and policies, while expatriate managers in Philips, despite being just as socialized into the overall corporate culture of the company, are much more willing to be advocates of local views and to defend against the imposition of corporate ideas on national organizations.[10] This willingness to 'rock the boat' and openness to experimentation and change is the characteristic that fuels local innovations.

Furthermore, by creating this kind of environment in the national organization, Philips has had little difficulty in attracting very capable local management. In contrast to the experience in many Japanese companies where local managers have felt excluded from a decision-making process that encompasses headquarters management and the local expatriates only, local managers in Philips feel their ideas are listened to and defended in headquarters. This too, creates a supportive environment for local innovations.

Integration of Technical and Marketing Functions within Each Subsidiary

Historically, the top management in all national subsidiaries of Philips consisted not of an individual CEO but a committee made up of the heads of the technical, commercial, and finance functions. This system of three-headed management had a long history in Philips, stemming from the functional independence of the two Philips brothers, one an engineer and the other a salesman. Although this management philosophy has recently been modified to a system which emphasizes individual authority and accountability, the long tradition of shared responsibilities and joint decision-making has left a legacy of many different mechanisms for functional integration at multiple levels. These integrative mechanisms within each subsidiary enhance the efficiency of local-for-local innovations in Philips[11] just the same way that various means of cross-functional integration within the corporate headquarters facilitates centre-for-global innovations in Matsushita.

In most subsidiaries, these integration mechanisms exist at three organizational levels. First, for each project, there is an article team that consists of relatively junior managers belonging to the commercial and technical functions. It is the responsibility of this team to evolve product policies and to prepare annual sales plans and budgets. At times, sub-article teams may be formed to supervise day-to-day working, and to carry out special projects, such as preparing capital investment plans should major new investments be felt necessary for effectively manufacturing and marketing a new product.

A second tier of cross-functional co-ordination takes place at the product group level, through the group management team, which again consists of both technical and commercial representatives. This team meets once a month to review results, suggest corrective actions, and resolve any inter-functional differences. Keeping control and conflict resolution at this low level facilitates sensitive and rapid responses to initiatives and ideas generated at the local level.

The highest level co-ordination forum within the subsidiary is the senior management committee (SMC) consisting of the top commercial, technical, and financial managers in the subsidiary. Acting essentially as a local board, the SMC provides an overall unity of effort among the different functional groups within the local unit, and ensures that the national unit retains primary responsibility for its own strategies and priorities. Again, the effect is to provide local management with a forum in which actions can be decided and issues resolved without escalation for approval or arbitration.

Dispersed Resources and Decentralized Authority

Finally, perhaps the most important facilitator of local innovations in Philips is the dispersal of its organizational assets and resources, and the very high level of decentralization of authority which respectively enable and empower subsidiary managers to experiment and to seek novel solutions to local problems.

The decentralized organization structure and management philosophy have deep roots. From its inception in 1891, Philips has recognized the need to expand its operation beyond its small domestic market. In those early days, however, transport and communications barriers forced management to decentralize operations and delegate responsibilities within its far-flung empire. The forces of decentralization were reinforced by the protectionist pressures of the 1930s that made it practically impossible to ship products or components across different countries within Europe. During World War II, even R&D capabilities were dispersed to avoid the possibility of their falling into enemy hands and many corporate managers left Holland reducing the parent company's control of the world-wide operations. For all these historical reasons, Philips' national organizations developed a degree of autonomy and self-sufficiency that was rare among companies of its size and complexity.

It is said that for an innovation to arise, two factors are necessary. First, the innovation must be desirable for the local managers. Second, it must be feasible for them to create it.[12] Dispersed managerial and technological resources and effective integration among them have made local innovations feasible for Philips subsidiaries. Local autonomy and decentralized control over those resources, coupled with the leadership of a highly entrepreneurial group of expatriate managers have made such indications desirable for them.

MAKING TRANSNATIONAL INNOVATIONS FEASIBLE: LESSONS FROM L. M. ERICSSON

Innovations are created by applying required resources to exploit an opportunity, or to overcome a threat. In multinational corporations, however, the location of the opportunity (or threat) is often different from the location of the appropriate response resources. The transnational innovation processes (locally-leveraged or globally-linked) use linkages among different units of the organization to leverage existing resources and capabilities, irrespective of their locations, to exploit any new opportunity that arises in any part of the dispersed multinational company.

Among the companies we studied, there were several that were in the process of developing such organizational capabilities. A few appeared to have become quite effective in managing the required linkages and processes, and we were able to identify three organizational characteristics that seemed most helpful in facilitating the new integrated innovation processes. The first was an interdependence of resources and responsibilities among organizational units, the second was a set of strong cross-unit integrating devices, and the last was a pervasive management attitude of strong corporate identification and well-developed worldwide perspectives.

Interdependence of Resources and Responsibilities

Perhaps the most important requirement for facilitating global innovations is a need for the organizational configuration to be based on a principle of reciprocal dependence among units. Such an interdependence of resources and responsibilities breaks down the hierarchy between local and global interests by making the sharing of resources, ideas and opportunities, a self-enforcing norm. To illustrate how such a basic characteristic or organizational configuration can influence a company's process of innovation, let us contrast the way in which ITT, NEC and L. M. Ericsson developed the electronic digital switch that would be the core product for each company's telecommunications business in the 1980s and beyond.

From its beginnings, in 1920, as a Puerto Rican telephone company, ITT built its world-wide operation on an objective described in the 1924 annual report as being 'to develop truly national systems operated by the nationals of each company'. For half a century ITT's national 'systems houses', as they were called within the company, committed themselves to meeting local interests and market needs. All but the smallest systems houses were established as fully integrated, self-sufficient units with responsibility for developing, manufacturing, marketing, installing and servicing their own products. All major innovations had their origins in the powerful and independent national companies, and even a product as important to the company's world-wide success as the highly regarded Pentaconta electromechanical switch was developed by ITT's French subsidiary.

These powerful, independent, and entrepreneurial national companies became the source of many important innovations, but management had never been able to co-ordinate or integrate the diverse efforts very effectively. After enormous frustrations in trying to co-ordinate development efforts and technical standards on the company's production in the electromechanical then early electronic (SPC) exchange switches, management recognized that establishing collaborative processes and co-ordinated efforts was all but impossible in an organization in which the key units were so strongly independent and autonomous. Yet the increasing cost of developing a new switch, and the shortening life cycle of successive generations of technology were forcing the company to take a more integrated approach.

The first sign that exchange switches built on a digital technology might replace recently introduced analog signal processing products occurred in the United States in the late 1970s when Northern Telecom's switch started creating what became referred to as 'digital fever' in North America. Despite the fact that ITT's British and French companies held the original patents for the process of sampling analog signals to convert them into digital form, European markets were being converted to ITT's existing analog switch and the local subsidiaries showed no interest in further developing and applying their pioneering digital research. At headquarters, the company's general technical director saw this as an opportunity to seize the initiative and create a company standard for this product that could then be developed in a co-ordinated global fashion. Consequently he assigned a team at the company's small Connecticut research centre to work on the project. But within a year, the Europeans were interested (and concerned) enough to send their own team of engineers to work with the headquarters group on the new System 12 switch. Their market knowledge and development experience allowed them to bend the researchers' global specifications better to reflect European needs. Soon after they convinced the company to take the responsibility for System 12 out of the hands of 'theoretical researchers in Connecticut' and transfer it to 'practical engineers close to the market'.

But the company wanted to ensure that the original work was not dissipated, diverted, and adapted by local systems houses, and created an International Telecommunications Center (ITC) in Brussels to lead, co-ordinate and control the development process of the System 12 switch. But the newly formed group and the staff managers responsible for it soon found they were no match for the powerful systems houses and the well-entrenched line managers who ran them. Although ITC management was able to allocate some development tasks and keep control over some standards, the large systems houses generally refused to rely on others for the development of critical parts of the system or accept standards that did not fit with their view of local needs. As a result, duplication of effort and divergence of specifications began to emerge, and the cost of developing the switch ballooned to over US $1 bn. The biggest problems appeared when the company decided to take System 12 into the US market. In true ITT tradition, the US business wanted to assert its independence and launched a major new R&D effort, despite appeals from the chief technological officer that they risked developing System 13. After years of effort and hundreds of millions of dollars in costs, ITT acknowledged it was withdrawing from the US market. The largest and most successful international telecommunications company in the world was blocked from its home country by the inability to transfer and apply its leading edge technology in a timely fashion. It was a failure that eventually led to ITT's withdrawal from direct involvement in telecommunications.

If effective global innovation was blocked by the extreme independence of the organizational units in ITT, it was impeded in NEC by the strong dependence of national subsidiaries on the parent company. Like ITT, NEC managers first detected 'digital fever' in the US market. The Japanese manager in charge of the US company recognized the importance of this trend early but did not have the resources, the capability or the authority to take much action. His role was one of selling corporate products and developing a beachhead in the US market. In Japan, technical managers were wary about a supposed trend to digitalization that they saw nowhere else world-wide (they called it a passing fad). They were sceptical about the claims of digital's technological superiority, and they were hesitant about beginning developmental work on a new switch that would compete with NEC's existing electromechanical and electronic products.

When the US managers finally were able to elicit sufficient support, the new NEAC 61 digital switch was developed almost entirely by headquarters personnel. Even in deciding which features to design into the new product, the central engineering group tended to discount the requests of the North American sales company and rely on data gathered in their own staff's field trips to US customers. Although the NEAC 61 was regarded as having good hardware, its software was thought to be unadapted to US needs. Sales did not meet expectations.

Both ITT and NEC have recognized the limitations of their independent and dependent organization systems and have begun to adapt them. But the process of building organizational interdependence is a slow and difficult one that must be constantly monitored and adjusted. In our sample of companies, L. M. Ericsson was by far the most consistent and experienced practitioner of creating and managing a delicate balance of inter-unit interdependency. The way in which it did so suggests the value of a constant readjustment of responsibilities and relationships as a way of adapting to changing strategic needs while maintaining a dynamic system of mutual dependence.

Like ITT, Ericsson had built during the 1920s and 1930s a substantial world-wide network of operations sensitive and responsive to local national environments; but like NEC, it had a strong home market base and a parent company technological, manufacturing, and market capability that was available to support those companies. Keeping the balance between and among these units has been a consistent company objective. In the late 1930s, when management became concerned that the growing independence of its offshore companies was causing divergence in technology, duplication of effort, and inefficiency in the sourcing patterns, they pulled sales and distribution control to headquarters and began consolidating responsibilities under product divisions. In the early 1950s, when these divisions showed signs of isolation and short-term focus, corporate staff functions were given more power, particularly in R&D. This led to the company's development of a crossbar switch that was an industry leader. As the product design and manufacturing technology for this product became well understood and fully documented, Ericsson management was able to respond to the demands of increasingly sophisticated and aggressive host governments, and transfer more manufacturing capacity and technological know-how abroad. Because assembly of crossbar switches was so labour intensive, it could often be done more efficiently abroad, and offshore sourcing of components and subassemblies increased.

Following half a century of constant ebb and flow in the centralization and decentralization of various responsibilities, Ericsson had no hesitation and little difficulty in adjusting tasks and responsibilities in response to the coming of electronic switching in the 1970s. Development efforts and manufacturing responsibilities were pulled back to Sweden, but where national capabilities, expertise or experience could be useful in the corporate effort, the appropriate local personnel were seconded to headquarters. In this way, the work the Australian subsidiary had done on a digital group selector was incorporated into the company's AXE digital switch. The AXE was designed knowing that local modifications would be necessary. As a modular system with very clear specifications, it allowed national companies to make necessary adaptations without compromising the integrity of the system. Similarly, in manufacturing, Ericsson's global computer-aided design and manufacturing system has allowed the delegation of component design and production to any national unit without the parent losing control.

With such central control, Ericsson has been willing to delegate substantial design, development, and manufacturing responsibilities to its subsidiaries and in recent years, the interdependence of units has increased. Primary responsibility for global development of peripheral products has been delegated (for example, Italy is the centre for transmission system development, Finland has mobile telephones, and Australia develops the rural switch). Further, headquarters has given some of these units responsibility for handling certain export markets (e.g., Italy's role in Africa). Increasingly, the company is moving even advanced software development offshore to subsidiary companies with access to more software engineers than it has in Stockholm.

By changing responsibilities, shifting assets and modifying relationships in response to evolving environmental demands and strategic priorities, Ericsson has maintained a dynamic interdependence between its operating units that has allowed it to develop entrepreneurial and innovative subsidiary companies that work within a corporate framework defined by a knowledgeable and creative headquarters group. This kind of interdependence, while hard to achieve rapidly, is the basis for global innovations.

Inter-unit Integrating Devices

We have shown how central innovations require strong integrating mechanisms at headquarters, and how local innovations are facilitated by co-ordinative capabilities within national units. The two newer global innovation processes need a different kind of integration process—one that operates across units. Such organizational positions and devices are more difficult to develop and manage than the inter-unit mechanisms, and an examination of the approach taken by L. M. Ericsson may provide some insight into how it can be achieved.

Unlike ITT, where relationship among national companies was often competitive and where headquarter—subsidiary interactions were often of an adversarial nature, L. M. Ericsson has been able to develop an organizational climate that is more co-operative and collaborative. This is essential if units are to work together to develop and implement innovations, but is not easily achieved. There are three important pillars to Ericsson's success in inter-unit integration—a clearly defined and tightly controlled set of operating systems, a people-linking process employing such devices as temporary assignments and joint teams, and inter-unit decision forums, particularly subsidiary boards, where views can be exchanged and differences resolved.

Ericsson management feels strongly that its most effective integrating device is strong central control over key elements of its strategic operation. Unlike ITT, Ericsson has not had strong or sophisticated administrative systems (it only introduced strategic plans in 1983), but its operating systems have been carefully developed. As indicated earlier, AXEs product specifications are tightly controlled and the CAD/CAM systems allow close central co-ordination of manufacturing. Rather than causing a centralization of decision-making, management argues that these strong operating systems allow them to delegate much more freely, knowing that local decisions will not be inconsistent with or detrimental to the overall interests.

But, in addition to strong systems, inter-unit co-operation requires good interpersonal relations, and Ericsson has developed these with a long-standing policy of transferring large numbers of people back and forth between headquarters and subsidiaries.[13] It differs from the more common transfer patterns in both direction and intensity, as a comparison with NEC's transfer process will demonstrate. Where NEC may transfer a new technology through a few key managers, Ericsson will send a team of 50 or 100 engineers for a year or two; while NEC's flow is primarily from headquarters to subsidiary, Ericsson's is a balanced two-way flow with people coming to the parent not only to learn but also to bring their expertise; and while NEC's transfers are predominantly Japanese, Ericsson's multidirectional process involves all nationalities.

Australian technicians seconded to Stockholm in the mid-1970s to bring their experience with digital switching into the development of AXE developed enduring relationships that helped in the subsequent joint development of a rural switch in Australia a decade later. Confidences built when an Italian team of 40 spent 18 months in Sweden to learn about electronic switching in the early 1970s provided the basis for greater decentralization of AXE software development and a delegated responsibility for developing the transmission systems.

But any organization in which there are shared tasks and joint responsibilities will require additional decision-making and conflict-resolving forums. In Ericsson, often divergent objectives and interests of the parent company and the local subsidiary are exchanged in the national company's board meetings. Unlike many companies whose local boards are pro forma bodies whose activities are designed solely to satisfy national legal requirements, Ericsson uses its local boards as legitimate forums for communicating objectives, resolving differences and making decisions. At least one and often several senior corporate managers are members of each board and subsidiary board meetings become an important means for co-ordinating activities and channelling local ideas and innovations across national lines.

National Competence, World-Wide Perspective

If there is one clear lesson from ITT's experience with the development of its System 12, it is that a company cannot innovate globally if its managers identify primarily with local parochial interests and objectives. But NEC's experience shows that when management has no ability to defend national perspectives and respond to local opportunities, global innovation is equally difficult. One of the important organizational characteristics Ericsson has been able to develop over the years has been a management attitude that is simultaneously locally sensitive, and globally conscious.

At the Stockholm headquarters, managers will emphasize the importance of developing strong country operations, not only to capture sales that require responsiveness to national needs, but also to tap into the resources that are available through world-wide operation. Coming from a small home country where it already hires over a third of the graduating electrical and electronics engineers, Ericsson is very conscious of the need to develop skills and capture ideas wherever they operate in the world. But, at the same time, local managers see themselves as part of the world-wide Ericsson group rather than as independent, autonomous units. Constant transfers and working on joint teams over the years has helped broaden many manager's perspectives from local to global, but giving local units systemwide mandates for products has confirmed their identity with the company's world-wide operations.

MANAGING INNOVATION IN THE TRANSNATIONAL

As we highlighted earlier in the chapter, the challenge of managing innovation in world-wide organizations is two-fold: first, management must enhance the efficiency and effectiveness of each of the different innovation processes and, second, it must create conditions that allow innovations to come about through all the different processes simultaneously.

In the preceding part of the chapter, we have described some of the ways in which managers can achieve the first of these two tasks. However, to benefit from these specific suggestions we have made, managers must also ensure that efforts to strengthen one of the innovation processes do not drive out the others. And this task of achieving simultaneity of the different innovation processes often proves to be the great challenge because of the dilemma that organizational attributes which facilitate one of the innovation processes often tend to impede the others.

The Organizational Dilemma

Consider, for example, the organizational attributes that are required to facilitate local-for-local and centre-for-global innovations. As we have illustrated, the former process requires that national subsidiaries have certain slack resources, and the requisite autonomy for deployment of those resources for creating local innovations. But such independent and resource-rich subsidiaries also tend to become victims of what Rosabeth Kanter, the Harvard sociologist and author of best selling *Changemasters* calls the 'entrepreneurial trap'—a mentality in which 'the need to be the source, the originator, leads people to push their own ideas single-mindedly'. This mentality impedes the subsidiary's ability and willingness to adopt centre-for-global innovations. Philips has long suffered from this problem, just as companies like Matsushita have suffered from the reverse problem of sheer incapability or lack the motivation wherein perceived or actual scarcity of slack resources and local authority in the national organizations have led to efficient adoption of centre-for-global innovations, but have constrained the company's ability to facilitate local-for-local innovations.

Overcoming the Dilemma: The Transnational Organization

At the core of this dilemma lies an assumption that managers of most multinational companies make about their international organization: they assume that organizational structure and management processes must by symmetric and homogeneous. This assumption is common to companies with what we have described as the global and the multinational mentalities and appears to be extremely widespread in practice.

Although there are wide differences in importance of operations in major markets like Germany, Japan, or the United States compared with subsidiaries in Argentina, Malaysia, or Nigeria, for example, most multinationals treat their foreign subsidiaries in a remarkably uniform manner. One executive we talked to termed this approach 'the UN model of multinational management'. While the functions carried out by the different subsidiaries may be different, they are administratively created as similar and equal. Thus, it is common to see managers express subsidiary roles and responsibilities in the same general terms, apply their planning and control systems uniformly world-wide, involve country managers to a like degree in product development, and evaluate them against standard criteria. This norm of symmetry and uniformity is inherent in the family metaphor that multinationals, irrespective of their origin, use persistently: Subsidiaries are children of the parent, and therefore there should be no discrimination—read differentiation—among them.

This symmetrical and homogeneous organizational approach encourages management to envision two roles in the multinational company: a local role for each subsidiary, and a global role for the headquarters. As a result, the relationship between the headquarters and the subsidiaries is also viewed in unidimensional terms. It is assumed that the relationship must be based on either dependence or independence of the subsidiary and, therefore, on either local autonomy or central control.

These assumptions and their administrative consequences constrain the flexibility of most multinational companies and imprison them into an either/or choice between central and local innovations. The very simplicity and clarity in these traditional organizational systems prevent the companies from developing the relatively more complex global innovation processes we have described and from achieving the even more difficult task of facilitating all the different local, central, and global innovation processes simultaneously.

These limitations of the symmetrical and homogeneous mode of operations have become increasingly clear to multinational corporations, and in many of the companies we surveyed, we found managers experimenting with alternative ways to manage their world-wide operations. And as we reviewed these various approaches, we saw a new pattern emerging that suggested a significantly different model of international organization based on some important new assumptions and beliefs. This organizational model we call the transnational and we have described the key characteristics of such organizations in a separate paper.[14] However, one attribute of the transnational—its ability and willingness to explicitly differentiate the roles and responsibilities of its different national subsidiaries—is of particular importance to our present discussions since this attribute appears to allow such companies to break out of central/local dilemma and to create an organizational infrastructure for managing central, local, and global innovations simultaneously and effectively.

Differentiated Subsidiary Roles for Different Innovation Processes

In most industries, a few key markets lead the industry's evolution. They are often the largest, most sophisticated and most competitive markets in which the nature of impending global changes is first mirrored. Results of competitive battles in such markets usually have a great deal of influence on the future world-wide competitive positions of firms. In the telecommunications switching business, for example, the United States is perhaps the principal lead market in the world. In the consumer electronics industry, in contrast, Japan, the United States, and a few of the major European markets share the lead position.

These are the markets that provide the stimuli for most global products and processes of a multinational company. Local innovations in such markets become useful elsewhere as the environmental characteristics that simulated such innovations diffuse to other locations. Similarly, the technological, competitive, and market-sensing processes that are required as inputs for centre-for-global and globally-linked innovations must also be provided by local operations in these lead markets. Relatively speaking, the sensing task for global innovations is much less intense in other 'follower' markets in which the company operates.

While the sensing opportunities lie outside the company and are determined by the strategic importance of different national environments, capabilities required to respond to the stimuli through development of new products or processes lie inside the organization and are determined by the company's administrative history. Further, while environmental opportunities are footloose, shifting from location to location, organizational resources are not easily transferable within the same company due to various administrative, regulatory and other reasons. The result is a situation of environment-resource mismatches: The company accumulates excessive resources in some environments that are relatively non-critical, and very limited or even no resources in some critical lead markets that offer the greatest opportunities and challenges.

Such environment-resource mismatches are pervasive in multinationals. Ericsson has significant technological capabilities in Australia and Italy—relatively insignificant markets in the global telecommunications business—but almost no presence in the United States which not only represents almost 40 per cent of the world's telecommunications equipment demand, but is the source of much of the new technology. Procter and Gamble is strong in the United States and Europe, but not in Japan where important consumer product innovations have occurred recently and where a major global competitor is emerging. Matsushita has appropriate technological and managerial resources in Japan and the United States, but not in Europe—a huge market, and home of arch rival, Philips.

These differences in external environments and internal capabilities imply some significant differences in the contributions that different subsidiaries can potentially make to the different innovational processes we have described. Thus, instead of either choosing to facilitate innovativeness or adoption in all subsidiaries, or seeking to find the non-existent grand compromise between the two, managers of transnational companies allocate different roles to their different operating units, based on the contributions the units can make to the different innovation processes. While none of the companies we studied had developed an explicit set of criteria for allocation of such differentiated roles, Figure 31.1 shows a simple framework for such differentiation that reflects some of the norms that appeared implicitly to guide their various approaches.[15]

Some subsidiaries which are located in challenging and stimulating environments and which possess high levels of technological and managerial capabilities, are allocated the role we label as strategic leaders and these subsidiaries serve as the transnational's innovative spark plugs.[16] They create local-for-local innovations, many of which are subsequently diffused to the rest of the organization as locally-leveraged innovations as the new technologies, tastes or business practices that first emerge in their environment diffuse around the world. Similarly, like the German organization of P&G that led the development of Vizir, they play key roles in creating globally-linked innovations. They also contribute to centre-for-global innovations, providing headquarters managers with the early warning signals of an emerging market need or opportunity and by serving as a learning ground for central managers to test their ideas and fine-tune their responses.

If, however, the subsidiary in a critical mar-

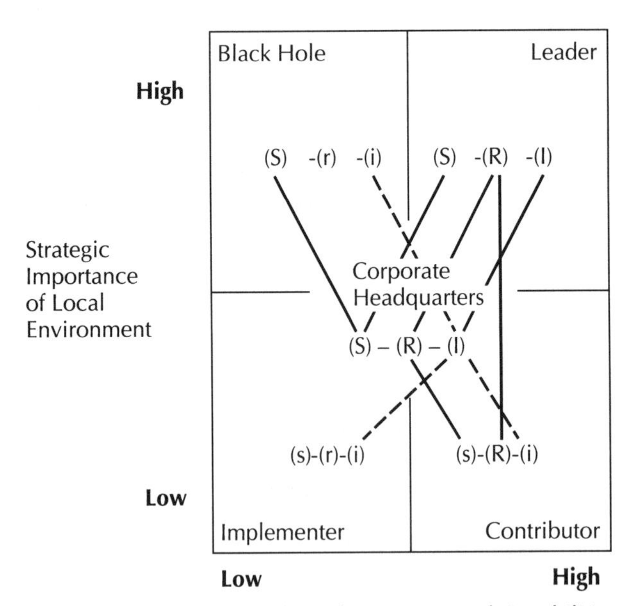

Fig. 31.1. Managing innovations in the transnational organization.

ket does not have adequate internal capabilities to exercise the lead role, it can be designated the sensor role of serving as a scanning and monitoring unit to direct the global innovation processes. Such, for example, has been the role played by the Japanese subsidiaries of many North American companies, and of the US subsidiaries of many Japanese companies.

National subsidiaries with relatively high levels of resources in nonstrategic markets can, in the traditional organization forms of the multinational, become a major source of disruption for the global strategies of the companies. The normal tendency in such units is to utilize the excess resources to demand and justify greater local differentiation and autonomy.[17] In the transnational organization, these subsidiaries are allocated the contributor role through which these excess resources are channelled to global rather than local tasks. Specific tasks with regard to centre-for-global or globally-linked innovations are farmed out to them and their primary contributions to the company's innovation processes becomes one of response rather than sensing.

Finally, organizations with relatively low levels of resources in relatively non-strategic markets are allocated the implementer role, in which their principal task is seen as one of adapting and implementing central and global innovations in the local context, creating in this process such local innovations as the adaptation task may require.

It is important to note, however, that these roles apply to specific businesses, functions, and product lines only, and not to all the activities of the subsidiary. Thus, a particular subsidiary may play a lead role for a specific product and an implementer or contributor role for others. For example, the UK subsidiary of Philips plays the lead role for the company's teletext TV sets, but an implementer role in the compact disc player business.[18]

Recognition of these differences in the potentials of its different organizational units to contribute to the different innovation processes lead to a set of clear norms in the transnational for deciding where and at what scale they should build the sensing, response, and implementation capabilities that are required to support the different innovation processes. In Figure 9.1 we provide a schematic representation of how such capabilities may be configured and linked within the transnational organization so as to enhance the overall innovation capacity of the company.

In the implementer subsidiaries, each of these capabilities may be built only to the extent that they are required to execute local tasks, including the task of creating and implementing local-for-local innovations. Such local level capabilities are represented by small letters in the diagram with (s), (r), and (i) standing for sensing, response, and implementation, respectively.

In the leader subsidiaries, in contrast, each of these capabilities is required in global scale to support its global roles. Recall, however, that these roles apply to specific businesses, functions, or product lines only, and not for all the activities of the subsidiary, and the global scale sensing, response, and implementation capabilities (represented by capital letters, (S), (R), and (I) respectively) are built in these units only with regard to the specific activities for which they carry the leadership role. Thus, if the French national organization in Philips is given this role for the 'smart card' business, the unit is made responsible for scanning the world-wide technological, competitive, and market environments of this business and not just the environment in France. Similarly, adequate development resources are also provided within the subsidiary to support its 'world product mandate', and appropriate production and marketing capabilities are also established locally so that the unit can serve as a world-scale source for the product as well as the driver for the company's global strategy for this business.

The sensor is not a viable strategic role, and the company's long-term goal should be to build up adequate resources and capabilities in such subsidiaries so that they can assume a leadership role in the relevant business. This process, however, is difficult and requires a long period of time. In the meanwhile such subsidiaries can and must serve as key sources of intelligence; collecting, interpreting, and circulating critical technological and competitive information on developments in its environment. Given the strategic importance and complexity of its local environment, this task requires a high level of sensing capability. Thus, while its response and implementation capabilities may be built up over time, advanced scanning capabilities must be built in such locations on high priority to support their global sensor roles.

The innovation process, however, requires that these dispersed and differentiated capabilities be interlinked so that the sensing, response, and implementation processes can function both sequentially and interactively. Thus, the sensing capabilities of the leader and the sensor subsidiaries must be linked by establishing mechanisms to ensure that intelligence acquired by the latter is passed on to the former. Similarly, the response capabilities of the contributor must also be integrated with those of the leader so as to capture the excess slack in the contributor organization to support the global level tasks that are carried out by the leader.[19] Even the implementation capabilities of the different subsidiaries may be interlinked, or they may be connected via the headquarters since all subsidiaries other than the leader carry only local-level responsibilities for this task.

In our description of the system we have assumed a situation where a national subsidiary plays the leadership role. However, this is not always necessary and the headquarters can and must play the leadership role for certain businesses, and in those cases the linkage among the dispersed sensing, response, and implementation capabilities must be established with the headquarters playing the nodal role. Even otherwise, in most companies, the centre retains the responsibilities for overall strategic direction and co-ordination, and has advanced sensing, response, and implementation capabilities either by design or as a result of historical evolution. Therefore, even for those activities for which the leadership role is allocated to a national subsidiary, the central capabilities must be linked with those of the leader to sustain the overall co-ordination role of the headquarters.

Let us illustrate how such a system might function. In L. M. Ericsson, the Australian subsidiary carries the leadership role for small rural switches built and marketed by the company. For this business, both the headquarters and the Italian subsidiary of the company act as sensors and as contributors. To support this way of operations, the Australian subsidiary has built a highly developed scanning capability and closely monitors worldwide developments in this particular business. The company also has a centralized scanning unit in Stockholm, and this unit passes on to Australia any information on the rural switch that may come to its attention. Similarly, some of the development engineers in Italy are allocated specific tasks for designing rural switches and they, in effect, work for the Australian subsidiary, despite being located physically in Italy. The development process is coordinated through the company's CAD/CAM system which serves as the linking device for the response task. The headquarters co-ordinates the design activities being carried out in Italy and Australia, but the Australian subsidiary drives the entire process and considers itself responsible for the overall task of designing and developing new products to support this small but profitable business of the company.

CONCLUSION: ORGANIZATIONAL CAPABILITY IS KEY

In the course of our study, we found that managers in many companies see the task of managing innovations as one of committing the greatest possible resources for creating new products and processes. This view implied that creating innovations was a specialized function performed in isolation from the mainstream of company activities and that the role of top management was one of resource allocation. Innovation could be improved if R&D budgets were increased, or, if the allocation of funds was made more efficient.

The emerging transnational companies, on the other hand, take a very different view of this task. More than increasing resources or developing new capabilities, we found that managers in these companies consider the challenge to be one of enhancing and leveraging the company's existing assets. Rather than focusing the innovative responsibility on a specialized group, they see the task to be one of involving all units of the organization in the process; and instead of developing one highly efficient process of innovation, they recognize the need to manage and integrate multiple processes.

In short, the challenge is much more one of developing organizational capabilities than one related to technological skills or resource allocation. Indeed, the two new transnational innovative processes are based on the ability to leverage existing innovative resources and capabilities by capturing synergies in their combined application or gaining scale and scope economies through broader exploitation of innovations. And to provide just one dramatic illustration of how organizational capabilities may triumph over resource commitments, recall that spending well over US $1 bn on the development of its digital switch did not assure ITT success over L. M. Ericsson who spent only a third of that amount.

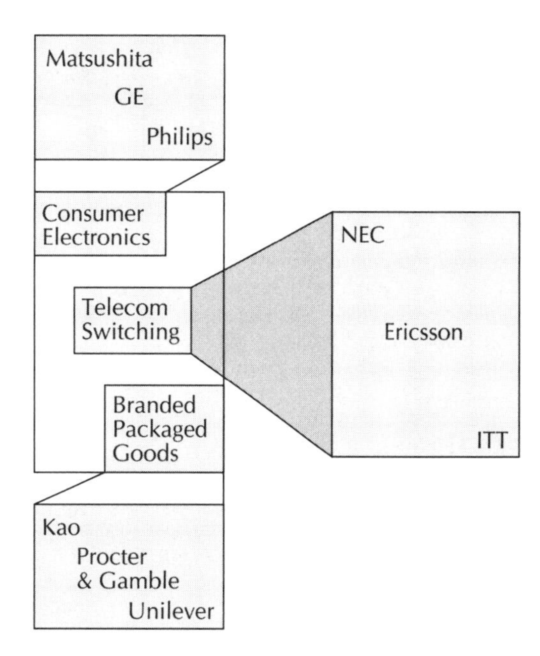

Fig. 31.2. Sample: choice of industries and companies.

Notes

1. This research project consisted of three phases. The first aimed at identifying and describing the key challenges that are being faced by managers of multinational corporations and to document 'leading practice' in coping with these challenges. This was also the hypothesis-generating phase and the sample was selected to represent the greatest variety of strategic and organizational situations. We chose three industries: consumer electronics, branded packaged products, and telecommunications switching. Each of these businesses was highly international but represented a very different set of key strategic demands. The first offered the greatest benefits of globalization, the forces of national responsiveness were particularly strong in the second, and the third represented the situation where both global and local forces were strong. Within each industry, we selected a group of firms that represented the greatest variety of administrative heritages including differences in nationality, internationalization history, and corporate culture. Philips, Matsushita, and GE in consumer electronics; Kao, Procter and Gamble, and Unilever in branded packaged products, and ITT, NEC, and L. M. Ericsson in telecommunications switching were the obvious choices.

Figure 31.2 provides a schematic representation of our sample in terms of the strategic characteristics of the industries, and the competitive postures of the firms. In this representation, we adopt the Global Integration—National Responsiveness framework proposed by Prahalad (1975) and subsequently developed by Doz (1979), Prahalad and Doz (1981), and Doz, Bartlett, and Prahalad (1981). For each box, the vertical axis represents the strength of globalizing forces in the industry or the extent of global integration in the company's strategic posture, and the horizontal axis represents the need for national responsiveness in the business or the extent of local differentiation in the company's overall competitive strategy (for a more detailed description of the strategic demands of these industries and administrative heritages of the companies, see Bartlett and Ghoshal 1987:a).

In each of these companies, we tried to identify as many specific cases of innovations as possible, and to document the participants' views on the organizational factors that facilitated or impeded the innovation process. To this end, we interviewed about 230 managers in these companies, both at their corporate headquarters and also in their national subsidiaries in the US, UK, Germany, France, Italy, Taiwan, Singapore, Japan, Australia, and Brazil. None of the interviews lasted less than an hour, and some took as long as three to five hours. We also collected and analysed internal documents relating to the histories of these innovations. This effort led to identification of thirty-eight innovation cases which form the key data base for this article (these innovations have been listed and briefly described in Ghoshal 1987b).

In the next phase of the project, we conducted detailed questionnaire surveys in three of the nine companies. The principle objective of the survey was to carry out a preliminary test of some of the hypotheses that were generated from the first phase of clinical research, also to define the hypotheses more precisely, and to develop suitable instruments for testing them more rigorously. Approximately 160 managers from NEC, Philips, and Matsushita participated in the survey, the results of which broadly confirmed most of the ideas that were generated from the first phase of the study (see Ghoshal and Bartlett 1987a and b).

Finally, in the third phase of the study, the hypotheses were tested through a large sample mailed questionnaire survey that yielded usable data from sixty-six of the largest US and European multinational corporations.

The overall findings of the project are being reported in our book, *Managing Across Borders: The Transnational Solution* (Harvard Business School Press, 1989).

2. This argument is implicit in the Product Cycle Theory proposed by Vernon (1966). It has been stated more explicitly in the internationalization and appropriability theories of foreign direct investment: see Buckley and Casson (1976), Rugman (1982), and Calvet (1981).

3. See Bartlett (1986) for descriptions of centralized hub and decentralized federation modes of multinational operations. In the centralized hub mode, the key assets and capabilities of the company and key value-adding activities, such as product development and manufacturing, are retained at the centre, or are tightly controlled from the centre. The national subsidiaries of centralized hub organizations act primarily as delivery pipelines to supply centrally manufactured global products to the local markets, and their roles are limited to local sales and service tasks. In contrast, in companies operating in the decentralized federation mode, key assets and capabilities of the multinational are dispersed among the different subsidiaries, each of which is allowed to develop as a self-contained and autonomous operation that is able to respond to local demands and opportunities. In a loose sense, the centralized hub mode supports what Porter (1986) calls a 'pure global' strategy, while the decentralized federation corresponds to what he describes as the 'multidomestic' strategy.

4. See Levitt (1983) for an interesting, if somewhat provocative, discussion of these two mentalities.

5. Vernon (1980) has highlighted the potential advantages that could accrue to a multinational through such a global scanning capability. The same point has also been made by Kogut (1983) and a number of other scholars in the field of international management.

6. The Not-Invented-Here (NIH) syndrome refers to the resistance of managers to accept ideas or solutions that have been generated elsewhere, and not by them. For a discussion and elaboration of this syndrome, see Katz and Allen (1982).

7. Westney and Sakakibara (1985) have observed a similar system of internal quasi-markets governing the interface between R&D and operating units in a number of other Japanese companies.

8. See Westney and Sakakibara (1985) for a more detailed discussion of these transitions in the careers of research personnel in Japanese companies, and for a comparison of American and Japanese practices in this regard.

9. The organizational form we have described as the decentralized federation shares many common features with the mother-daughter organization that Franko (1976) described as representative of many large European multinationals. Hedlund (1984) showed that national subsidiaries of such companies typically develop strong and entrepreneurial local management teams and also accumulate relatively high levels of local resources—two attributes that we propose are key to the ability of such companies to foster local-for-local innovations.

10. See Van Mannen and Schein (1979) for a rich and theory-grounded discussion on how such differences in socialization processes and career systems can influence manages' attitudes towards change and innovation.

11. Much of the earlier research on organizational innovation has been focused on innovations that are conceived, created, and implemented within individual subunits of large and multi-unit organizations. Burns and Stalker (1961), for example, state this explicitly: 'The twenty concerns which were subject of these studies were not all separately constituted business companies . . . (some of them) were small parts of the parent organization. . . . This is why we have used "concern" as a generic term.' Other researchers have similarly observed a district sales office of General Electric, or a department in the headquarters of 3M, or a divisional data processing office of Polaroid, but not the overall configuration of any of these companies. In essence, therefore, these studies have focused on what we call local-for-local innovations. And most of them have identified internal integration, local slack and decentralized authority as key factors that facilitate such innovations in organizations that have been variously described as 'organic', 'integrative', or simply 'excellent' (see, for example, Burns and Stalker 1962; Peters and Waterman 1982; Kanter 1983). To this extent, our findings regarding the factors that make local innovations efficient fully conform to the conclusions reached by earlier researchers. A major point of departure in our study, however, is that we view such local innovations as one of many different processes through which innovations may come about in complex organizations and suggest that the organizational factors that facilitate the different innovation processes are not only different, but may also be mutually contradictory.

12. See Mohr (1969).

13. The use of personnel transfers as an integration mechanism in multinational companies has been highlighted by many scholars, most notably by Edstrom and Galbraith (1977).

14. See Bartlett (1986) for a broad description of the transnational organization. Subsequently, some key attributes of such organizations have been described and illustrated in Bartlett and Ghoshal (1987:b). Based on their own research in a wide range of companies, a number of other researchers have also identified this emerging trend of a very different organizational model being adopted by some major multinational companies around the world—a model that has been variously described as the multi-focal organization (Prahalad and Doz 1987), the hierarchy (Hedlund 1986) and the horizontal organization (White and Poynter in this volume). While there are some considerable differences among these different organizational models, there are also some significant similarities. In essence, all these observations point towards the emergence of a more complex organizational form in multinational companies, with significantly higher levels of internal differentiation and integration that are supported through the simultaneous use of a wide variety of structures and management processes.

15. Refer to Bartlett and Ghoshal (1986) for a more detailed description of these different roles of national subsidiaries and for some suggestion on how these roles can be allocated and managed.

16. Rugman and Poynter (1982) have observed a similar phenomenon in the trend toward assigning 'global product mandates' to mature national subsidiaries.

17. The potential of such a disruptive role being played by resource-rich subsidiaries in noncritical environments has been highlighted by White and Poynter (1984) based on their study of the Canadian subsidiaries of a number of large multinational corporations.

18. This is an extremely important point that we have elaborated in Bartlett and Ghoshal (1986) and therefore do not discuss in this paper. If the roles are not differentiated by product lines or businesses, a few subsidiaries will come to play the lead role in general while others will become implementers for all tasks. This can have some adverse motivational consequences, not unlike those described by Haspeslagh (1982) for companies that embraced the portfolio concept uncritically and allocated star roles to some businesses and cash cow, question mark, or dog roles to others.

19. The response capability can be of many different kinds: research and development, manufacturing, marketing and general management competencies are but some of the capabilities that may be required to respond to a specific situation. In the specific context of R&D capabilities, the need for such linkages among dispersed facilities has been emphasized by Håkanson (see his paper, chapter 10 of this volume) and also by Ronstadt (1977).

REFERENCES

Bartlett, C. A. (1986) 'Building and managing the transnational: the new organizational challenge', in M. E. Porter (ed.), *Competition in Global Industries*, Boston: Harvard Business School Press.

Bartlett, C. A. and Ghoshal, S. (1986) 'Tap your subsidiaries for global reach', *Harvard Business Review*, vol. 64, No. 6, November–December.

(1987a) 'Managing across borders: new strategic requirements', *Sloan Management Review*, Summer.

(1987b) 'Managing across borders: new organizational responses', *Sloan Management Review*, Fall.

Buckley, P. J. and Casson, M. (1976) *The Future of Multinational Enterprise*, London: Macmillan.

Burns, T. and Stalker, G. M. (1961) *The Management of Innovation*, London: Tavistock.

Calvert, A. L. (1981) 'A synthesis of foreign direct investment theories and theories of the multinational firm', *Journal of International Business Studies*, Spring–Summer.

Doz, Y. L. (1979) *Government Control and Multinational Strategic Management*, New York: Praeger.

Doz, Y. L., Bartlett, C. A., and Prahalad, C. K. (1981) 'Global competitive pressures and host country demands: managing tensions in MNCs', *California Management Review*, Spring.

Drucker, P. F. (1985) *Innovation and Entrepreneurship*, New York: Harper & Row.

Edstrom, E. and Galbraith, J. R. (1977) 'Transfer of managers as a coordination and control strategy in multinational organizations', *Administrative Science Quarterly*, June.

Franko, L. G. (1976) *The European Multinationals*, London: Harper & Row.

Ghoshal, S. and Bartlett, C. A. (1987a) 'Innovation processes in multinational corporations', unpublished manuscript, Fontainebleau, France: INSEAD.

(1987b) 'Creation, adoption, and diffusion of innovations by subsidiaries of multinational corporations', unpublished manuscript, Fontainebleau, France: INSEAD.

Haspeslagh, P. (1982) 'Portfolio planning: uses and limits', *Harvard Business Review,* (January–February): 58–73.

Hedlund, G. (1984) 'Organization in-between: the evolution of the mother-daughter structure of managing foreign subsidiaries in Swedish MNCs', *Journal of International Business Studies,* Fall, vol. 15, No. 2.

(1986) 'The hypermodern MNC—a heterarchy?', *Human Resource Management,* Spring, vol. 25, No. 1.

Kanter, R. M. (1983) *The Change Master,* New York: Simon & Schuster.

Katz, R. and Allen, T. J. (1982) 'Investigating the not invented here (NIH) Syndrome: a look at the performance, tenure, and communication patterns of 50 R&D project groups', *R&D Management,* 12.

Kogut, B. (1983) 'Foreign direct investment as a sequential process', in C. P. Kindelberger and D. B. Audretsch (eds), *The Multinational Corporation in the 1980s,* Cambridge, MA: MIT Press.

Levitt, T. (1983) 'The globalization of markets', *Harvard Business Review,* May–June: 92–102.

Mohr, L. B. (1969) 'Determinants of innovation in organizations', *American Political Science Review,* 63: 111–28.

Peters, T. J. and Waterman, R. H. (1982) *In Search of Excellence,* New York: Harper & Row.

Porter, M. E. (1986) 'Competition in global industries: a conceptual framework', in M. E. Porter (ed.), *Competition in global industries,* Boston: Harvard Business School Press.

Prahalad, C. K. (1975) 'The strategic process in a multinational corporation', unpublished doctoral dissertation, Graduate School of Business Administration, Harvard University, Boston.

Prahalad, C. K. and Doz, Y. L. (1981) 'An approach to strategic control in MNC's, *Sloan Management Review,* 22, No. 4, Summer.

(1987) *The Multinational Mission: Balancing Local Demands and Global Vision,* New York: Free Press.

Ronstadt, R. C. (1977) *Research and Development Abroad by US Multinationals,* New York: Praeger.

Rugman, A. M. and Poynter, T. A. (1982) 'World product mandate: how will multinationals respond?' *Business Quarterly,* October.

Van Mannen, J. and Schein, E. H. (1979) 'Toward a theory of organizational socialization', in B. Shaw (ed.), *Research in Organizational Behavior,* JAI Press.

Westney, D. E. and Sakakibara, K. (1985) 'The role of Japan-based R&D in global technology strategy', *Technology in Society,* No. 7.

White, R. E. and Poynter, T. A. (1984) 'Strategies for foreign-owned subsidiaries in Canada', *Business Quarterly,* Summer.

Vernon, R. (1966) 'International investment and international trade in the product cycle', *Quarterly Journal of Economics,* May.

Vernon, R. (1980) 'Gone are the cash cows of yesteryear', *Harvard Business Review,* November–December.

32

The Logic of Global Business:
An Interview with ABB's Percy Barnevik

WILLIAM TAYLOR

NOTE: Percy Barnevik, president and CEO of ABB Asea Brown Boveri, is a corporate pioneer. He is moving move aggressively than any CEO in Europe, perhaps in the world, to build the new model of competitive enterprise—an organization that combines global scale and world-class technology with deep roots in local markets. He is working to give substance to the endlessly invoked corporate mantra, "Think global, act local."

Headquartered in Zurich, ABB is a young company forged through the merger of two venerable European companies. Asea, created in 1890, has been a flagship of Swedish industry for a century. Brown Boveri, which took shape in 1891, holds a comparable industrial status in Switzerland. In August 1987, Barnevik altered the course of both companies when he announced that Asea, where he was managing director, would merge with Brown Boveri to create a potent new force in the European market for electrical systems and equipment.

The creation of ABB became a metaphor for the changing economic map of Europe. Barnevik initiated a wrenching process of consolidation and rationalization—layoffs, plant closings, product exchanges between countries—that observers agreed will one day come to European industries from steel to telecommunications to automobiles. And soon more than a metaphor, Barnevik's bold moves triggered a wholesale restructuring of the Continent's electrical power industry.

The creation of ABB also turned out to be the first step in a trans-Atlantic journey of acquisition, restructuring, and growth. ABB has acquired or taken minority positions in 60 companies representing investments worth $3.6 billion—including two major acquisitions in North America. In 1989, ABB acquired Westinghouse's transmission and distribution operation in a transaction involving 25 factories and businesses with revenues of $1 billion. That same year, it spent $1.6 billion to acquire Combustion Engineering, the manufacturer of power-generation and process-automation equipment.

Today ABB generates annual revenues of more than $25 billion and employs 240,000 people around the world. It is well balanced on both sides of the Atlantic. Europe accounts for more than 60% of its total revenues, and its business is split roughly equally between the European Community countries and the non-EC Scandinavian trading bloc. Germany, ABB's largest national market, accounts for 15% of total revenues. The company also generates annual revenues of $7 billion in North America, with 40,000 employees. Although ABB remains underrepresented in Asia, which accounts for only 15% of total revenues, it is an important target for expansion and investment. And ABB's business activities are not limited to the industrialized world. The company has 10,000 employees in India, 10,000 in South America, and is one of the most active Western investors in Eastern Europe.

In this interview, Percy Barnevik, 49, offers a detailed guide to the theory and practice of building a "multi-domestic" enterprise. He explains ABB's matrix system, a structure designed to leverage core technologies and global economies of scale without eroding local market presence and responsiveness. (See the insert "The Organizing Logic of ABB.") He describes a new breed of "global managers" and explains how their skills differ from those of traditional managers. He reckons candidly with the political implications of companies such as ABB.

The interview was conducted at ABB's Zurich headquarters by HBR associate editor William Taylor.

HBR: Companies everywhere are trying to become global, and everyone agrees that ABB is more global than most companies. What does that mean?

Percy Barnevik: ABB is a company with no geographic center, no national ax to grind. We are a federation of national companies with a global coordination center. Are we a Swiss company? Our headquarters is in Zurich, but only 100 professionals work at headquarters and we will not increase that number. Are we a Swedish company? I'm the CEO, and I was born and educated in Sweden. But our headquarters is not in Sweden, and only two of the eight members of our board of directors are Swedes. Perhaps we are an American company. We report our financial results in U.S. dollars, and English is ABB's official language. We conduct all high-level meetings in English.

My point is that ABB is none of those things—and all of those things. We are not homeless. We are a company with many homes.

Are all businesses becoming global?

No, and this is a big source of misunderstanding. We are in the process of building this federation of national companies, a multidomestic organization, as I prefer to call it. That does not mean all of our businesses are global. We do a very good business in electrical installation and service in many countries. That business is superlocal. The geographic scope of our installation business in, say, Stuttgart does not extend beyond a ten-mile radius of downtown Stuttgart.

We also have businesses that are superglobal. There are not more than 15 combined-cycle power plants or more than 3 or 4 high-voltage DC stations sold in any one year around the world. Our competitors fight for nearly every contract—they battle us on technology, price, financing—and national borders are virtually meaningless. Every project requires our best people and best technology from around the world.

The vast majority of our businesses—and of most businesses—fall somewhere between the superlocal and the superglobal. These are the businesses in which building a multidomestic organization offers powerful advantages. You want to be able to optimize a business globally—to specialize in the production of components, to drive economies of scale as far as you can, to rotate managers and technologists around the world to share expertise and solve problems. But you also want to have deep local roots everywhere you operate—building products in the countries where you sell them, recruiting the best local talent from the universities, working with the local government to increase exports. If you build such an organization, you create a business advantage that's damn difficult to copy.

What is a business that demonstrates that advantage?

Transportation is a good one. This is a vibrant business for us, and we consider ourselves number one in the world. We generate $2 billion a year in revenues when you include all of our activities: locomotives, subway cars, suburban trains, trolleys, and

The Organizing Logic of ABB

ABB Asea Brown Boveri is a global organization of staggering business diversity. Yet its organizing principles are stark in their simplicity. Along one dimension, the company is a distributed global network. Executives around the world make decisions on product strategy and performance without regard for national borders. Along a second dimension, it is a collection of traditionally organized national companies, each serving its home market as effectively as possible. ABB's global matrix holds the two dimensions together.

At the top of the company sit CEO Percy Barnevik and 12 colleagues on the executive committee. The group, which meets every three weeks, is responsible for ABB's global strategy and performance. The executive committee consists of Swedes, Swiss, Germans, and Americans. Several members of the executive committee are based outside Zurich, and their meetings are held around the world.

Reporting to the executive committee are leaders of the 50 or so business areas (BAs), located worldwide, into which the company's products and services are divided. The BAs are grouped into 8 business segments, for which different members of the executive committee are responsible. For example, the "industry" segment, which sells components, systems, and software to automate industrial processes, has 5 BAs, including metallurgy, drives, and process engineering. The BA leaders report to Gerhard Schulmeyer, a German member of the executive committee who works out of Stamford, Connecticut.

Each BA has a leader responsible for optimizing the business on a global basis. The BA leader devises and champions a global strategy, holds factories around the world to cost and quality standards, allocates export markets to each factory, and shares expertise by rotating people across borders, creating mixed-nationality teams to solve problems, and building a culture of trust and communication. The BA leader for power transformers, who is responsible for 25 factories in 16 countries, is a Swede who works out of Mannheim, Germany. The BA leader for instrumentation is British. The BA leader for electric metering is an American based in North Carolina.

Alongside the BA structure sits a country structure. ABB's operations in the developed world are organized as national enterprises with presidents, balance sheets, income statements, and career ladders. In Germany, for example, Asea Brown Boveri Aktiengesellschaft, ABB's national company, employs 36,000 people and generates annual revenues of more than $4 billion. The managing director of ABB Germany, Eberhard von Koerber, plays a role comparable with that of a traditional German CEO. He reports to a supervisory board whose members include German bank representatives and trade union officials. His company produces financial statements comparable with those from any other German company and participates fully in the German apprenticeship program.

The BA structure meets the national structure at the level of ABB's member companies. Percy Barnevik advocates strict decentralization. Wherever possible, ABB creates separate companies to do the work of the 50 business areas in different countries. For example, ABB does not merely sell industrial robots in Norway. Norway has an ABB robotics company charged with manufacturing robots, selling to and servicing domestic customers, and exporting to markets allocated by the BA leader.

There are 1,100 such local companies around the world. Their presidents report to two bosses—the BA leader, who is usually located outside the country, and the president of the national company of which the local company is a subsidiary. At this intersection, ABB's "multidomestic" structure becomes a reality.

the electrical and signaling systems that support them. We are strong because we are the only multidomestic player in the world.

First, we know what core technologies we have to master, and we draw on research from labs across Europe and the world. Being a technology leader in locomotives means being a leader in power electronics, mechanical design, even communications software. Ten years ago, Asea beat General Electric on a big Amtrak order for locomotives on the Metroliner between New York and Washington. That win caused quite a stir; it was the first time in one hundred years that an American railroad bought locomotives from outside the United States. We won because we could run that track from Washington to New York, crooked and bad as it was, at 125 miles an hour. Asea had been pushing high-speed design concepts for more than a decade, and Brown Boveri pioneered the AC technology. That's why our X2 tilting trains are running in Sweden and why ABB will play a big role in the high-speed rail network scheduled to run throughout Europe.

Second, we structure our operations to push cross-border economies of scale. This is an especially big advantage in Europe, where the locomotive industry is hopelessly fragmented. There are two companies headquartered in the United States building locomotives for the U.S. market. There are three companies in Japan. There are 24 companies in Western Europe, and the industry runs at less than 75% of capacity. There are European companies still making only 10 or 20 locomotives a year! How can they compete with us, when we have factories doing ten times their volume and specializing in components for locomotives across the Continent? For example, one of our new plants makes power electronics for many of the locomotives we sell in Europe. That specialization creates huge cost and quality advantages. We work to rationalize and specialize as much as we can across borders.

Third, we recognize the limits to specialization. We can't ignore borders altogether. We recently won a $420-million order from the Swiss Federal Railways—we call it the "order of the century"—to build locomotives that will move freight through the Alps. If we expect to win those orders, we had *better* be a Swiss company. We had better understand the depth of the Swiss concern for the environment, which explains the willingness to invest so heavily to get freight moving on trains through the mountains to Italy or Germany and off polluting trucks. We had better understand the Alpine terrain and what it takes to build engines powerful enough to haul heavy loads. We had better understand the effects of drastic temperature changes on sensitive electronics and build locomotives robust enough to keep working when they go from the frigid, dry outdoors to extreme heat and humidity inside the tunnels.

There are other advantages to a multidomestic presence. India needs locomotives—thousands of locomotives—and the government expects its suppliers to manufacture most of them inside India. But the Indians also need soft credit to pay for what is imported. Who has more soft credit than the Germans and the Italians? So we have to be a German and an Italian company, we have to be able to build locomotive components there as well as in Switzerland, Sweden, and Austria, since our presence may persuade Bonn and Rome to assist with financing.

We test the borderlines all the time. How far can we push cross-border specialization and scale economies? How effectively can we translate our multidomestic presence into competitive advantages in third markets?

Is there such a thing as a global manager?

Yes, but we don't have many. One of ABB's biggest priorities is to create more of them; it is a crucial bottleneck for us. On the other hand, a global company does not need thousands of global managers. We need maybe 500 or so out of 15,000 managers to make ABB work well—not more. I have no interest in making managers more "global" than they have to be. We can't have people abdicating their nationalities, saying "I am no longer German, I am international." The world doesn't work like that. If you are selling products and services in Germany, you better be German!

That said, we do need a core group of global managers at the top: on our executive committee, on the teams running our business areas (BAs), in other key positions. How are they different? Global managers have exceptionally open minds. They respect how different countries do things, and they have the imagination to appreciate why they do them that way. But they are also incisive, they push the limits of the culture. Global managers don't passively accept it when someone says, "You can't do that in Italy or Spain because of the unions," or "You can't do that in Japan because of the Ministry of Finance." They sort through the debris of cultural excuses and find opportunities to innovate.

Global managers are also generous and patient. They can handle the frustrations of language barriers. As I mentioned earlier, English is the official language of ABB. Every manager with a global role *must* be fluent in English, and anyone with regional general management responsibilities must be competent in English. When I write letters to ABB colleagues in Sweden, I write them in English. It may seem silly for one Swede to write to another in English, but who knows who will need to see that letter a year from now?

We are adamant about the language requirement—and it creates problems. Only 30% of our managers speak English as their first language, so there is great potential for misunderstanding, for misjudging people, for mistaking facility with English for intelligence or knowledge. I'm as guilty as anyone. I was rushing through an airport last year and had to return a phone call from one of our managers in Germany. His English wasn't good, and he was speaking slowly and tentatively. I was in a hurry, and finally I insisted, "Can't you speak any faster?" There was complete silence. It was a dumb thing for me to say. Things like that happen every day in this company. Global managers minimize those problems and work to eliminate them.

Where do these new managers come from?

Global managers are made, not born. This is not a natural process. We are herd animals. We like people who are like us. But there are many things you can do. Obviously, you rotate people around the world. There is no substitute for line experience in three or four countries to create a global perspective. You also encourage people to work in mixed-nationality teams. You *force* them to create personal alliances across borders, which means that sometimes you interfere in hiring decisions.

This is why we put so much emphasis on teams in the business areas. If you have 50 business areas and five managers on each BA team, that's 250 people from different parts of the world—people who meet regularly in different places, bring their national perspectives to bear on tough problems, and begin to understand how things are done elsewhere. I experience this every three weeks in our executive committee. When we sit together as Germans, Swiss, Americans, and Swedes, with many of us living, working, and traveling in different places, the insights can be remarkable. But you have to force people into these situations. Mixing nationalities doesn't just happen.

You also have to acknowledge cultural differences without becoming paralyzed by them. We've done some surveys, as have lots of other companies, and we find interesting differences in perception. For example, a Swede may think a Swiss is not completely frank and open, that he doesn't know exactly where he stands. That is a cultural phenomenon. Swiss culture shuns disagreement. A Swiss might say, "Let's come back to that point later, let me review it with my colleagues." A Swede would prefer to confront the issue directly. How do we undo hundreds of years of upbringing and education? We don't, and we shouldn't try to. But we do need to broaden understanding.

Is your goal to develop an "ABB way" of managing that cuts across cultural differences?

Yes and no. Naturally, as CEO, I set the tone for the company's management style. With my Anglo-Saxon education and Swedish upbringing, I have a certain way of doing things. Someone recently asked if my ultimate goal is to create 5,000 little Percy Barneviks, one for each of our profit centers. I laughed for a moment when I thought of the horror of sitting on top of such an organization, then I realized it wasn't a silly question. And the answer is no. We can't have managers who are "un-French" managing in France because 95% of them are dealing every day with French customers, French colleagues, French suppliers. That's why global managers also need humility. A global man-

Power Transformers—The Dynamics of Global Coordination

ABB is the world's leading manufacturer of power transformers, expensive products used in the transmission of electricity over long distances. The business generates annual revenues of $1 billion, nearly four times the revenues of its nearest competitor. More to the point, ABB's business is consistently and increasingly profitable—a real achievement in an industry that has experienced 15 years of moderate growth and intense price competition.

Power transformers are a case study in Percy Barnevik's approach to global management. Sune Karlsson, a vice president of ABB with a long record in the power transformer field, runs the business area (BA) from Mannheim, Germany. Production takes place in 25 factories in 16 countries. Each of these operations is organized as an independent company with its own president, budget, and balance sheet. Karlsson's job is to optimize the group's strategy and performance independent of national borders—to set the global rules of the game for ABB—while allowing local companies freedom to drive execution.

"We are not a global business," Karlsson says. "We are a collection of local businesses with intense global coordination. This makes us unique. We want our local companies to think small, to worry about their home market and a handful of export markets, and to learn to make money on smaller volumes."

Indeed, ABB has used its global production web to bring a new model of competition to the power transformer industry. Most of ABB's 25 factories are remarkably small by industry standards, with annual sales ranging from as little as $10 million to not more than $150 million, and 70% of their output serves their local markets. ABB transformer factories concentrate on slashing throughput times, maximizing design and production flexibility, and focusing tightly on the needs of domestic customers. In short, the company deploys the classic tools of flexible, time-based management in an industry that has traditionally competed on cost and volume.

As with many of its business areas, ABB built its worldwide presence in power transformers through a series of acquisitions. Thus one of Karlsson's jobs is to spread the new model of competition to the local companies ABB acquires.

"Most of the companies we acquired had volume problems, cost problems, quality problems," he says. "We have to convince local managers that they can run smaller operations more efficiently, meet customer needs more flexibly—and make money. Once you've done this 10 or 15 times, in several countries, you become confident of the merits of the model."

Karlsson's approach to change is in keeping with the ABB philosophy: show local managers what's been achieved elsewhere, let them drive the change process, make available ABB expertise from around the world, and demand quick results. A turnaround for power transformers takes about 18 months.

In Germany, for example, one of the company's transformer plants had generated red ink for years. It is now a growing, profitable operation, albeit smaller and more focused than before. The work force has been slashed from 520 to 180, throughput time has been cut by one-third, work-in-process inventories have decreased by 80%. Annual revenues have fallen $70 million per year to a mere $50 million—but profits are up substantially. Today the German manager who championed this company's changes is in Muncie, Indiana, helping managers of a former Westinghouse plant acquired by ABB to reform their operation.

ager respects a formal German manager—Herr Doktor and all that—because that manager may be an outstanding performer in the German context.

Let's talk about the structures of global business. How do you organize a multidomestic enterprise?

ABB is an organization with three internal contradictions. We want to be global and local, big and small, radically decentralized with centralized reporting and control. If we resolve those contradictions, we create real organizational advantage.

That's where the matrix comes in. The matrix

ABB's global scale also gives it clout with suppliers. The company buys up to $500 million of materials each year—an enormous presence that gives it leverage on price, quality, and delivery schedules. Karlsson has made strategic purchasing a priority. ABB expects zero-defect suppliers, just-in-time deliveries, and price increases lower than 75% of inflation—major advantages that it is in a position to win with intelligent coordination.

Sune Karlsson believes these and other "hard" advantages may be less significant, however, than the "soft" advantages of global coordination. "Our most important strength is that we have 25 factories around the world, each with its own president, design manager, marketing manager, and production manager," he says. "These people are working on the same problems and opportunities day after day, year after year, and learning a tremendous amount. We want to create a process of continuous expertise transfer. If we do, that's a source of advantage none of our rivals can match."

Creating these soft advantages requires internal competition and coordination. Every month, the Mannheim headquarters distributes detailed information on how each of the 25 factories is performing on critical parameters, such as failure rates, throughput times, inventories as a percentage of revenues, and receivables as a percentage of revenues. These reports generate competition for outstanding performance within the ABB network—more intense pressure, Karlsson believes, than external competition in the marketplace.

The key, of course, is that this internal competition be constructive, not destructive. Since the creation of ABB, one of Sune Karlsson's most important jobs has been to build a culture of trust and exchange among ABB's power transformer operations around the world and to create forums that facilitate the process of exchange. At least three such forums exist today:

- The BA's management board resembles the executive committee of an independent company. Karlsson chairs the group, and its members include the presidents of the largest power transformer companies—people from the United States, Canada, Sweden, Norway, Germany, and Brazil. The board meets four to six times a year and shapes the BA's global strategy, monitors performance, and resolves big problems.
- Karlsson's BA staff in Mannheim is not "staff" in the traditional sense—young professionals rotating through headquarters on their way to a line job. Rather, it is made up of five veteran managers each with worldwide responsibility for activities in critical areas such as purchasing and R&D. They travel constantly, meet with the presidents and top managers of the local companies, and drive the coordination agenda forward.
- Functional coordination teams meet once or twice a year to exchange information on the details of implementation in production, quality, marketing, and other areas. The teams include managers with functional responsibilities in all the local companies, so they come from around the world. These formal gatherings are important, Karlsson argues, but the real value comes in creating informal exchange throughout the year. The system works when the quality manager in Sweden feels compelled to telephone or fax the quality manager in Brazil with a problem or an idea.

"Sharing expertise does not happen automatically," Karlsson emphasizes. "It takes trust, it takes familiarity. People need to spend time together, to get to know and understand each other. People must also see a payoff for themselves. I never expect our operations to coordinate unless all sides get real benefits. We have to demonstrate that sharing pays—that contributing one idea gets you 24 in return."

—William Taylor

is the framework through which we organize our activities. It allows us to optimize our businesses globally *and* maximize performance in every country in which we operate. Some people resist it. They say the matrix is too rigid, too simplistic. But what choice do you have? To say you don't like a matrix is like saying you don't like factories or you don't like breathing. It's a fact of life. If you deny the formal matrix, you wind up with an informal one—and that's much harder to reckon with. As we learn to master the matrix, we get a truly multidomestic organization.

Can you walk us through how the matrix works?

Look at it first from the point of view of one business area, say, power transformers. The BA manager for power transformers happens to sit in Mannheim, Germany. His charter, however, is worldwide. He runs a business with 25 factories in 16 countries and global revenues of more than $1 billion. He has a small team around him of mixed nationalities—we don't expect superheroes to run our 50 BAs. Together with his colleagues, the BA manager establishes and monitors the trajectory of the business.

The BA leader is a business strategist and global optimizer. He decides which factories are going to make what products, what export markets each factory will serve, how the factories should pool their expertise and research funds for the benefit of the business worldwide. He also tracks talent—the 60 or 70 real standouts around the world. Say we need a plant manager for a new company in Thailand. The BA head should know of three or four people—maybe there's's one at our plant in Muncie, Indiana, maybe there's one in Finland—who could help in Thailand. (See the insert "Power Transformers—The Dynamics of Global Coordination.")

It is possible to leave the organization right there, to optimize every business area without regard for ABB's broad collection of activities in specific countries. But think about what we lose. We have a power transformer company in Norway that employs 400 people. It builds transformers for the Norwegian market and exports to markets allocated by the BA. But ABB Norway has more than 10,000 other employees in the country. There are tremendous benefits if power transformers coordinates its Norwegian operation with our operations in power generation, switchgear, and process automation: recruiting top people from the universities, building an efficient distribution and service network across product lines, circulating good people among the local companies, maintaining productive relations with top government officials.

So we have a Norwegian company, ABB, Norway, with a Norwegian CEO and a headquarters in Oslo, to make these connections. The CEO has the same responsibilities as the CEO of a local Norwegian company for labor negotiations, bank relationships, and high-level contacts with customers. This is no label or gimmick. We *must* be a Norwegian company to work effectively in many businesses. Norway's oil operations in the North Sea are a matter of great national importance and intense national pride. The government wouldn't—and shouldn't—trust some faraway foreign company as a key supplier to those operations.

The opportunities for synergy are clear. So is the potential for tension between the business area structure and the country structure. Can't the matrix pull itself apart?

BA managers, country managers, and presidents of the local companies have very different jobs. They must understand their roles and appreciate that they are *complementing* each other, not competing.

The BA managers are crucial people. They need a strong hand in crafting strategy, evaluating performance around the world, and working with teams made up of different nationalities. We've had to replace some of them—people who lacked vision or cultural sensitivity or the ability to lead without being dictators. You see, BA managers don't own the people working in any business area around the world. They can't order the president of a local company to fire someone or to use a particular strategy in union negotiations. On the other hand, BA managers can't let their role degrade into a statistical coordinator or scorekeeper. There's a natural tendency for this to happen. BA managers don't have a constituency of thousands of direct reports in the same way that country managers do. So it's a difficult balancing act.

Country managers play a different role. They are regional line managers, the equivalent of the CEO of a local company. But country managers must also respect ABB's global objectives. The president of, say, ABB Portugal can't tell the BA manager for low-voltage switchgear or drives to stay out of his hair. He has to cooperate with the BA managers to evaluate and improve what's happening in Portugal in those businesses. He should be able to tell a BA manager. "You may think the plant in Portugal is up to standards, but you're being too loose. Turnover and absenteeism is twice the Portugese average. There are problems with the union, and it's the managers' fault."

Now, the presidents of our local companies—ABB Transformers in Denmark, say, or ABB Drives in Greece—need a different set of skills. They must be excellent profit center managers. But they must also be able to answer to two bosses effectively. After all, they have two sets of responsibilities. They have a global boss, the BA manager, who creates the rules of the game by which they run their businesses. They also have their country boss, to whom they report in the local setting. I don't want to make too much of this. In all of Germany, where we have 36,000 people, only 50 or so managers have two bosses. But these managers have to handle that ambiguity. They must have the self-confidence not to become paralyzed if they receive conflicting signals and the integrity not to play one boss off against the other.

Isn't all this much easier said than done?

It does require a huge mental change, especially for country managers. Remember, we've built ABB through acquisitions and restructurings. Thirty of the companies we've bought had been around for more than 100 years. Many of them were industry leaders in their countries, national monuments. Now, they've got BA managers playing a big role in the direction of their operations. We have to convince country managers that they benefit by being part of this federation, that they gain more than they lose when they give up some autonomy.

What's an example?

Finland has been one of our most spectacular success stores, precisely because the Finns understood how much they could gain. In 1986, Asea acquired Strömberg, the Finnish power and electrical products company. At the time, Strömberg made an unbelievable assortment of products, probably half of what ABB makes today. It built generators, transformers, drives, circuit breakers—all of them for the Finnish market, many of them for export. It was a classic example of a big company in a small country that survived because of a protected market. Not surprisingly, much of what it made was not up to world-class standards, and the company was not very profitable. How can you expect a country with half the population of New Jersey to be profitable in everything from hydropower to circuit breakers?

Strömberg is no longer a stand-alone company. It is part of ABB's global matrix. The company still exists—there is a president of ABB Strömberg—but its charter is different. It is no longer the center of the world for every product it sells. It still manufactures and services many products for the Finnish market. It also sells certain products to allocated markets outside Finland. And it is ABB's worldwide center of excellence for one important group of products, electric drives, in which it had a long history of technology leadership and effective manufacturing.

Strömberg is a hell of a lot stronger because of this. Its total exports from Finland have increased more than 50% in three years. ABB Strömberg has become one of the most profitable companies in the whole ABB group, with a return on capital employed of around 30%. It is a recognized world leader in drives. Strömberg produces more than 35% of all the drives ABB sells, and drives are a billion-dollar business. In four years, Strömberg's exports to Germany and France have increased ten times. Why? Because the company has access to a distribution network it never could have been built itself.

This sounds enormously complicated, almost unmanageable. How does the organization avoid getting lost in the complexity?

The only way to structure a complex, global organization is to make it as simple and local as possible. ABB is complicated from where I sit. But on the ground, where the real work gets done, all of our operations must function as closely as possible to stand-alone operations. Our managers need well-defined sets of responsibilities, clear accountability, and maximum degrees of freedom to execute.

I don't expect most of our people to have "global mind-sets," to do things that hurt their business but are "good for ABB." That's not natural.

Take Strömberg and drives in France. I don't want the drive company president in Finland to think about what's good for France. I want him to think about Finland, about how to sell the hell out of the export markets he has been allocated. Likewise, I don't expect our profit center manager in France to think about Finland. I expect him to do what makes sense for his French customers. If our French salespeople find higher quality drives or more cost-effective drives outside ABB, they are free to sell them in France so long as ABB gets a right of first refusal. Finland has increased its shipments to France because it makes economic sense for both sides. That's the only way to operate.

But how can an organization with 240,000 people all over the world be simple and local?

ABB is a huge enterprise. But the work of most of our people is organized in small units with P&L responsibility and meaningful autonomy. Our operations are divided into nearly 1,200 companies with an average of 200 employees. These companies are divided into 4,500 profit centers with an average of 50 employees.

We are fervent believers in decentralization. When we structure local operations, we always push to create separate legal entities. Separate companies allow you to create *real* balance sheets with *real* responsibility for cash flow and dividends. With real balance sheets, managers inherit results from year to year through changes in equity. Separate companies also create more effective tools to recruit and motivate managers. People can aspire to meaningful career ladders in companies small enough to understand and be committed to.

What does that mean for the role of headquarters?

We operate as lean as humanly possible. It's no accident that there are only 100 people at ABB headquarters in Zurich. The closer we get to top management, the tougher we have to be with head count. I believe you can go into any traditionally centralized corporation and cut its headquarters staff by 90% in one year. You spin off 30% of the staff into free-standing service centers that perform real work—treasury functions, legal services—and charge for it. You decentralize 30% of the staff—human resources, for example—by pushing them into the line organization. Then 30% disappears through head count reductions.

These are not hypothetical calculations. We bought Combustion Engineering in late 1989. I told the Americans that they had to go from 600 people to 100 in their Stamford, Connecticut headquarters. They didn't believe it was possible. So I told them to go to Finland and take a look. When we bought Strömberg, there were 880 people in headquarters. Today there are 25. I told them to go to Mannheim and take a look at the German operation. In 1988, right after the creation of ABB, there were 1,600 people in headquarters. Today there are 100.

Doesn't such radical decentralization threaten the very advantages that ABB's size creates?

Those are the contradictions again—being simultaneously big and small, decentralized and centralized. To do that, you need a structure at the top that facilitates quick decision making and carefully monitors developments around the world. That's the role of our executive committee. The 13 members of the executive committee are collectively responsible for ABB. But each of us also has responsibility for a business segment, a region, some administrative functions, or more than one of these. Eberhard von Koerber, who is a member of the executive committee located in Mannheim, is responsible for Germany, Austria, Italy, and Eastern Europe. He is also responsible for a worldwide business area, installation materials, and some corporate staff functions. Gerhard Schulmeyer sits in the United States and is responsible for North America. He is also responsible for our global "industry" segment.

Naturally, these 13 executives are busy, stretched people. But think about what happens when we meet every three weeks, which we do for a full day. Sitting in one room are the senior managers collectively responsible for ABB's global strategy and performance. These same managers individually monitor business segments, countries, and staff functions. So when we make a decision—snap, it's covered. The members of the executive committee communicate to their direct reports, the BA managers and the country managers, and the implementation process is under way.

We also have the glue of transparent, centralized reporting through a management information system called Abacus. Every month, Abacus collects performance data on our 4,500 profit centers and compares performance with budgets and forecasts. The data are collected in local currencies but translated into U.S. dollars to allow for analysis across borders. The system also allows you to work with the data. You can aggregate and disaggregate results by business segments, countries, and companies within countries.

What kind of information does the executive committee use to support the fast decision making you need?

We look for early signs that businesses are becoming more or less healthy. On the tenth of every month, for example, I get a binder with information on about 500 different operations—the 50 business areas, all the major countries, and the key companies in key countries. I look at several parameters—new orders, invoicing, margins, cash flows—around the world and in various business segments. Then I stop to study trends that catch my eye.

Let's say the industry segment is behind budget. I look to see which of the five BAs in the segment are behind. I see that process automation is way off. So I look by country and learn that the problem is in the United States and that it's poor margins, not weak revenues. So the answer is obvious—a price war has broken out. That doesn't mean I start giving orders. But I want to have informed dialogues with the appropriate executives.

Let's go back to basics. How do you begin building this kind of global organization?

ABB has grown largely through mergers and strategic investments. For most companies in Europe, this is the right way to cross borders. There is such massive overcapacity in so many European industries and so few companies with the critical mass to hold their own against Japanese and U.S. competitors. My former company, Asea, did fine in the 1980s. Revenues in 1987 were 4 times greater than in 1980, profits were 10 times greater, and our market value was 20 times greater. But the handwriting was on the wall. The European electrical industry was crowded with 20 national competitors. There was up to 50% overcapacity, high costs, and little cross-border trade. Half the companies were losing money. The creation of ABB started a painful—but long overdue—process of restructuring.

That same restructuring process will come to other industries: automobiles, telecommunications, steel. But it will come slowly. There have been plenty of articles in the last few years about all the cross-border mergers in Europe. In fact, the more interesting issue is why there have been so *few*. There should be *hundreds* of them, involving *tens of billions* of dollars, in industry after industry. But we're not seeing it. What we're seeing instead are strategic alliances and minority investments. Companies buy 15% of each other's shares. Or two rivals agree to cooperate in third markets but not merge their home-market organizations. I worry that many European alliances are poor substitutes for doing what we try to do—complete mergers and cross-border rationalization.

What are the obstacles to such cross-border restructuring?

One obstacle is political. When we decided on the merger between Asea and Brown Boveri, we had no choice but to do it secretly and to do it quickly, with our eyes open about discovering skeletons in the closet. There were no lawyers, no auditors, no environmental investigations, and no due diligence. Sure, we tried to value assets as best we could. But then we had to make the move, with an extremely thin legal document, because we were absolutely convinced of the strategic merits. In fact, the documents from the premerger negotiations are locked away in a Swiss bank and won't be released for 20 years.

Why the secrecy? Think of Sweden. Its industrial jewel, Asea—a 100 year-old company that had built much of the country's infrastructure—was moving its headquarters out of Sweden. The unions were angry: "Decisions will be made in Zurich, we have no influence in Zurich, there is no codetermination in Switzerland."

I remember when we called the press conference in Stockholm on August 10. The news came as a complete surprise. Some journalists didn't even bother to attend; they figured it was an announcement about a new plant in Norway or something. Then came the shock, the fait accompli. That started a communications war of a few weeks where we had to win over shareholders, the public, governments, and unions. But strict confidentiality was our only choice.

Are there obstacles besides politics?

Absolutely. The more powerful the strategic logic behind a merger—the greater the cross-border synergies—the more powerful the human and organizational obstacles. It's hard to tell a competent country manager in Athens or Amsterdam, "You've done a good job for 15 years, but unfortunately this other manager has done a better job and our only choice is to appoint your colleague to run the operation." If you have two plants in the same country running well but you need only one after the merger, it's tough to explain that to employees in the plant to be closed. Restructuring operations creates lots of pain and heartache, so many companies choose not to begin the process, to avoid the pain.

Germany is a case in point. Brown Boveri had operated in Germany for almost 90 years. Its German operation was so big—it had more than 35,000 employees—that there were rivalries with the Swiss parent. BBC Germany was a technology-driven, low-profit organization—a real underperformer. The formation of ABB created the opportunity to tackle problems that had festered for decades.

So what did you do?

We sent in Eberhard von Koerber to lead the effort. He made no secret of our plans. We had to reduce the work force by 10%, or 4,000 employees. We had to break up the headquarters, which had grown so big because of all the tensions with Switzerland. We had to rationalize the production overlaps, especially between Switzerland and Germany. We needed lots of new managers, eager people who wanted to be leaders and grow in the business.

The reaction was intense. Von Koerber faced strikes, demonstrations, barricades—real confrontation with the unions. He would turn on the television set and see protesters chanting, "Von Koerber out! Von Koerber out!" After a while, once the unions understood the game plan, the loud protests disappeared and our relationship became very constructive. The silent resistance from managers was more formidable. In fact, much of the union resistance was fed by management. Once the unions got on board, they became allies in our effort to reform management and rationalize operations.

Three years later, the results are in. ABB Germany is a well-structured, dynamic, market-oriented company. Profits are increasing steeply, in line with ABB targets. In 1987, BBC Germany generated revenues of $4 billion. ABB Germany will generate twice that by the end of next year. Three years ago, the management structure in Mannheim was centralized and functional, with few clear responsibilities or accountability. Today there are 30 German companies, each with its own president, manufacturing director, and so on. We can see who the outstanding performers are and apply their talents elsewhere. If we need someone to sort out a problem with circuit breakers in Spain, we know who from Germany can help.

What lessons can other companies learn from the German experience?

To make real change in cross-border mergers, you have to be factual, quick, and neutral. And you have

Change Comes to Poland—The Case of ABB Zamech

Last May, Zamech, Poland's leading manufacturer of steam turbines, transmission gears, marine equipment, and metal castings began a new life as ABB Zamech—a joint venture of ABB (76% ownership), the Polish government (19% ownership), and the company's employees (5% ownership). ABB Zamech employs 4,300 people in the town of Elblag, outside Gdánsk. In September, two more Polish joint ventures became official—ABB Dolmel and Dolmel Drives. These companies manufacture a wide range of generating equipment and electric drives and employ some 2,400 workers.

The joint ventures are noteworthy for their size alone. ABB has become the largest Western investor in Poland. But they are perhaps more significant for their managerial implications, in particular, how ABB is revitalizing these deeply troubled operations. The company intends to demonstrate that the philosophy of business and managerial reform it has applied in places like Mannheim, Germany and Muncie, Indiana can also work in the troubled economies of Eastern Europe. That philosophy has at least four core principles:

1. Immediately reorganize operations into profit centers with well-defined budgets, strict performance targets, and clear lines of authority and accountability.
2. Identify a core group of change agents from local management, give small teams responsibility for championing high-priority programs, and closely monitor results.
3. Transfer ABB expertise from around the world to support the change process, without interfering with it or running it directly.
4. Keep standards high and demand quick results.

Barbura Kux, president of ABB Power Ventures, negotiated the Polish joint ventures and plays a lead role in the turnaround process. "Our goal is to make these companies as productive and profitable as ABB's operations worldwide," she says. "We don't make a 'discount' for Eastern Europe, and we don't expect the change process to take forever. We provide more technical and managerial support than we might to a company in the United States, but we are just as demanding in terms of results."

ABB Zamech has come the furthest to date. The change program began immediately after the creation of the joint venture. For decades, the company had been organized along functional lines, a structure that blurred managerial authority, confused product-line profitability, and slowed decision making. Within four weeks, ABB Zamech was reorganized into discrete profit centers. There are now three business areas (BAs)—the casting foundry, turbines and gears, and marine equipment—as well as a finance and administration department and an in-house service department. Each area has a leadership team that generates the business plans, budgets, and performance targets by which their operations are judged. These teams made final decisions on which employees would stay, which would go, what equipment they would need—tough-minded business choices made for the first time so as to maximize productivity (employee and capital) and business area profitability.

The reorganization was a crucial first step. The second big step was installing ABB's standard finance and control system. For decades, Zamech had been run as a giant overhead machine. Roughly 80% of the company's total costs were allocated by central staff accountants rather than traced directly to specific products and services. Managers had no clear idea what their products cost to make and thus no idea which ones made money. Tight financial controls and maximum capital productivity are critical in an economy with interest rates of 40%.

Formal reorganization and new control systems, no matter how radical, won't have much of an effect without big changes in who is in charge, however. ABB made two important decisions. First, there would be no "rescue team" from Western Europe. All managerial positions, from the CEO down, would be held by Polish managers from the former Zamech. Second, managers would be selected without regard to rank or seniority; indeed, there would be a premium on young, creative talent. ABB was looking for "hungry wolves"—smart, ambitious change agents who would receive intense training and be the core engine of Zamech's revival.

Most of the new leaders came from the ranks of middle management. The company's top executive, general manager Pawel Olechnowicz, ran the steel castings department prior to the joint venture's creation—a position that put him several layers below the top of the 15-layer management hierarchy. Employees had already elected him general manager shortly before the creation of ABB Zamech, so he looked like a good choice. The marine BA leader had been a production manager in the old Zamech, another low-level position, and the turbines and gears BA manager had been a technical director.

"We put in place a management team that lacked the standard business tools," Kux explains. "They didn't know what cash flows was, they didn't understand much about marketing. But their ambition was incredible. You could feel their hunger to excel. When we began the talent search, we told our Zamech contacts that we wanted to see the 30 people they would take along tomorrow if they were going to open their own business."

Next came the process of developing a detailed agenda for reform. The leadership team settled on 11 priority issues, from reorganizing and restraining the sales force to slashing total cycle times and redesigning the factory layout. Each project was led by a champion—some from top management ranks, some from the other "hungry wolves." A steering committee made up of the general manager, the deputy general manager, the business area managers, and Kux meets monthly to review these critical projects.

To support the change initiatives, ABB created a team of high-level experts from around the world—authorities in functional areas like finance and control and quality, as well as technology specialists and managers with heavy restructuring experience. Team members do not live in Poland. Kux says it is unrealistic to expect top people to spend a year or two in the conditions they would find in Elblag. But they visit frequently and stay updated on progress and problems.

The logistics of expertise transfer are more complicated than they sound. For example, most of the Polish managers spoke little or no English—a serious barrier to effective dialogue. So ABB began intensive language training. "If Polish managers want to draw from the worldwide ABB resource pool, they *must* speak English," Kux emphasizes. "Most communication doesn't happen face-to-face where you can have an interpreter. Last May, I couldn't simply pick up the phone and talk to the general manager. Today we speak in English on the phone almost every day."

Of course, speaking on the telephone in English assumes a working telephone system—a dangerous assumption in the case of Poland. Thus another prerequisite for effective expertise transfer was creating the infrastructure to make it possible. ABB has linked Zamech and Dolmel by satellite to its Zurich headquarters for reliable telephone and fax communications. (It is now easier to communicate between Zamech and Zurich and Dolmel and Zurich than it is between Zamech and Dolmel.) In January, ABB Zamech began electronically transferring three monthly performance reports to Zurich—another big step to make communications more intensive and effective.

Once it created the communications infrastructure, however, ABB had to reckon with a second language barrier—the language of business. To introduce ABB Zamech's "hungry wolves" to basic business concepts and to enable them to transfer these concepts into the ranks, ABB created a "mini MBA program" in Warsaw. The program began in September, covers five key modulates (business strategy, marketing, finance, manufacturing, human resources) and is taught by faculty members of INSEAD, the French business school. Sessions run from Thursday evening through Saturday noon, use translated copies of Western business school cases, and closely resemble what goes on in MBA classes everywhere else.

The change program at ABB Zamech has been under way for less than a year, and much remains to be done. But it is already generating results. The company is issuing monthly financial reports that conform to ABB standards—a major achievement in light of the simple systems in place before the joint venture. Cycle times for the production of steam turbines have been cut in half and now meet the ABB worldwide average. A task force is implementing a plan to reduce factory space by 20%—an important step in streamlining the operation. ABB will draw on the Zamech experience as it begins the reform process at Dolmel and Dolmel Drives.

"You *can* change these companies," Kux says. "You *can* make them more competitive and profitable. I can't believe the quality of the reports and presentations these people do today, how at ease they are discussing their strategy and targets. I have worked with many corporate restructurings, but never have I seen so much change so quickly. The energy is incredible. These people really want to learn; they are very ambitious. Basically, ABB Zamech is their business now."

—William Taylor

to move boldly. You must avoid the "investigation trap"—you can't postpone tough decisions by studying them to death. You can't permit a "honeymoon" of small changes over a year or two. A long series of small changes just prolongs the pain. Finally, you have to accept a fair share of mistakes. I tell my people that if we make 100 decisions and 70 turn out to be right, that's good enough. I'd rather be roughly right and fast than exactly right and slow. We apply these principles everywhere we go, including in Eastern Europe, where we now have several change programs under way. (See the insert "Change Comes to Poland—The Case of ABB Zamech.")

Why emphasize speed at the expense of precision? Because the costs of delay are vastly greater than the costs of an occasional mistake. I won't deny that it was absolutely crazy around here for the first few months after the merger. We *had* to get the matrix in place—we couldn't debate it—and we *had* to figure out which plants would close and which would stay open. We took ten of our best people, the superstars, and gave them six weeks to design the restructuring. We called it the Manhattan Project. I personally interviewed 400 people, virtually day and night, to help select and motivate the people to run our local companies.

Once you've put the global pieces together and have the matrix concept working, what other problems do you have to wrestle with?

Communications. I have no illusions about how hard it is to communicate clearly and quickly to tens of thousands of people around the world. ABB has about 15,000 middle managers prowling around markets all over the world. If we in the executive committee could connect with all of them or even half of them and get them moving in roughly the same direction, we would be unstoppable.

But it's enormously difficult. Last year, for example, we made a big push to squeeze our accounts receivable and free up working capital. We called it the Cash Race. There are 2,000 people around the world with some role in accounts receivable, so we had to mobilize them to make the program work. Three or four months after the program started—and we made it very visible when it started—I visited an accounts receivable office where 20 people were working. These people hadn't even *heard* of the program, and it should have been their top priority. When you come face-to-face with this lack of communication, this massive inertia, you can get horrified, depressed, almost desperate. Or you can concede that this is the way things are, this is how the world works, and commit to doing something about it.

So what do you do?

You don't inform, you *overinform*. That means breaking taboos. There is a strong tendency among European managers to be selective about sharing information.

We faced a huge communications challenge right after the merger. In January 1988, just days after the birth of ABB, we had a management meeting in Cannes with the top 300 people in the company. At that meeting, we presented our policy bible, a 21-page book that communicates the essential principles by which we run the company. It's no glossy brochure. It's got tough, direct language on the role of BA managers, the role of country managers, the approach to change we just discussed, our commitment to decentralization and strict accountability. I told this group of 300 that they had to reach 30,000 ABB people around the world within 60 days—and that didn't mean just sending out the document. It meant translating it into the local languages, sitting with people for a full day and hashing it out.

Cannes and its aftermath was a small step. Real communication takes time, and top managers must be willing to make the investment. We are the "overhead company." I personally have 2,000 overhead slides and interact with 5,000 people a year in big and small groups. This afternoon, I'll fly up to Lake Constance in Germany, where we have collected 35 managers from around the world. They've been there for three days, and I'll spend three hours with them to end their session. Half the executive committee has already been up there. These are active, working sessions. We talk about how we work in the matrix, how we develop people, about our programs around the world to cut cycle times and raise quality.

I'll give a talk at Lake Constance, but then we'll focus on problems. The manager running high-voltage switchgear in some country may be unhappy about the BA's research priorities. Someone may think we're paying too much attention to Poland. There are lots of tough questions, and my job is to answer on the spot. We'll have 14 such sessions during the course of the year—one every three weeks. That means 400 top managers from all over the world living in close quarters, really communicating about the business and their problems, and meeting with the CEO in an open, honest dialogue.

Let's discuss the politics of global business. For senior executives, the world becomes small every day. For most production workers, though, the world is not much different from the way it was 20 years ago, except now their families and communities may depend for jobs on companies with headquarters thousands of miles away. Why shouldn't these workers worry about the loss of local and national control?

It's inevitable that a global business will have global decision centers and that for many workers these decision centers will not be located in their community or even their country. The question is, does the company making decisions have a national ax to grind? In our case the answer is no. We have global coordination, but we have no national bias. The 100 professionals who happen to sit in Zurich could just as easily sit in Chicago or Frankfurt. We're not here very much anyway. So what does it mean to have a headquarters in Zurich? It's where my mail arrives before the important letters are faxed to wherever I happen to be. It's where Abacus collects our performance data. Beyond that, I'm not sure if it means much at all.

Of course, saying we have no national ax to grind does not mean there are any guarantees. Workers will often ask if I can guarantee their jobs in Norway or Finland or Portugal. I don't sit like a godfather, allocating jobs. ABB has a global game plan, and the game plan creates opportunities for employment, research, exports. What I guarantee is that every member of the federation has a fair shot at the opportunities.

Let's say you're a production worker at ABB Combustion Engineering in Windsor, Connecticut. Two years ago, you worked for a company that you knew was an "American" company. Today you are part of a "federation" of ABB companies around the world. Should you be happy about that?

You should be happy as hell about it. A production worker in Windsor is probably in the boiler field. He or she doesn't much care about what ABB is doing with process automation in Columbus, Ohio, let alone what we're doing with turbines outside Gdańsk, Poland. And that's fair. Here's what I would tell that worker: we acquired Combustion Engineering because we believe ABB is a world leader in power plant technology, and we want to extend our lead. We believe that the United States has a great future in power plants both domestically and on an export basis. Combustion represents 80 years of excellence in this technology. Unfortunately, the company sank quite a bit during the 1980s, like many of its U.S. rivals, because of the steep downturn in the industry. It had become a severely weakened organization.

Today, however, the business is coming back, and we have a game plan for the United States. We plan to beef up the Windsor research center to three or four times its current size. We want to tie Windsor's work in new materials, emissions reduction, and pollution control technology with new technologies from our European labs. That will let us respond more effectively to the environmental concerns here. Then we want to combine Combustion's strengths in boilers with ABB's strengths in turbines and generators and Westinghouse's strengths in transmission and distribution to become a broad and unique supplier to the U.S. utility industry. We also have an ambition for Combustion to be much more active in world markets, not with sales agents but through the ABB multidomestic network.

What counts to this production worker is that we deliver, that we are increasing our market share in the United States, raising exports, doing more R&D. That's what makes an American worker's life more secure, not whether the company has its headquarters in the United States.

Don't companies like ABB represent the beginning of a power shift, a transfer of power away from national government to supranational companies?

Are we above governments? No. We answer to governments. We obey the laws in every country in which we operate, and we don't make the laws. However, we do change relations *between* countries. We function as a lubricant for worldwide economic integration.

Think back 15 years ago, when Asea was a Swedish electrical company with 95% of its engineers in Sweden. We could complain about high taxes, about how the high cost of living made it difficult to recruit Germans or Americans to come to Sweden. But what could Asea do about it? Not much. Today I can tell the Swedish authorities that they must create a more competitive environment for R&D or our research there will decline.

That adjustment process would happen regardless of the creation of ABB. Global companies speed up the adjustment. We don't create the process, but we push it. We make visible the invisible hand of global competition.

33

Toward Middle-Up-Down Management: Accelerating Information Creation

IKUJIRO NONAKA

The concepts of "top-down" and "bottom-up" management pervade management research and the popular business literature. Both center on information flow and information processing. Top-down management emphasizes the process of implementing and refining decisions made by top management as they are transmitted to the lower levels of the organization. Bottom-up management emphasizes the influence of information coming up from lower levels on management decision making. The management styles of individual firms are usually seen as located somewhere on the continuum between these two types.

However, organizations must not only process information; they must also create it. If we look closely at R&D activities, we find a pattern in some firms that does not fit on the continuum between top-down and bottom-up. It is a process that resolves the contradiction between the visionary but abstract concepts of top management and the experience-grounded concepts originating on the shopfloor by assigning a more central role to middle managers. This process, which is particularly well suited to the age of fierce market competition and rapid technological change, I call *middle-up-down management*. The development of the Honda "City" exemplifies this management style.

DEVELOPMENT OF HONDA CITY

In 1978, the management of Honda Motors was worried that the company was losing its vitality. Its basic models, such as the Civic and the Accord, had reached "middle age" at a time of major generational shift in the Japanese market. For the first time, Japanese born after World War II outnumbered those born before and during the war. Honda's top management decided to let the younger design engineers develop something for their own generation. The project began with the catch-phrase, *Boken o yatte miyo*—literally, "Let's make an adventure."

The youngest people on Honda's design staff were selected as members of the City development team. Their average age was only twenty-seven. Former president Kiyoshi Kawashima and other top managers promised there would be no interference with the team's operation. One of the team members recalls, "It was surprising and wonderful that the company dared to entrust younger staff like us with the design and development of a new model." Even so, the company did not just abandon them, but rather sought to impress them with a high degree of responsibility.

Mr. Nobuhiko Kawamoto, vice president in

Reprinted from "Toward Middle-Up-Down Management: Accelerating Information Creation" by Nonaka, I., *Sloan Management Review*, (Spring 1988), pp. 9–18, by permission of publisher.

charge of R&D, describes how Honda's top managers approach research projects:

> We usually control the tasks of researchers quite tightly, and then loosen the control from time to time. Ideas with great potential often emerge from this process. What we have to do is to watch these creative spurts carefully, notice them when they are good, and then develop them. Of course, too much freedom doesn't work. But sometimes we take the chance of giving researchers basic goals and responsibilities, and then letting them go by themselves. In other words, we put them upstairs, remove the ladders, and say, "You have to jump from there. We don't care if you can't." I think human beings display their greatest creativity under pressure.

Some say Honda not only puts researchers upstairs without a ladder, but also "sets the first floor on fire."

Mr. Kawamoto appointed Mr. Hiroo Watanabe, a chief researcher who was thirty-five years old at the time, to be the leader of the team. Mr. Kawamoto explains, "It's hard to get enough cooperation if you only have young staff members. Therefore I chose a skilled senior as their leader."

The autonomous team faced some challenging goals. The overall assignment from headquarters was to "create something different from the existing concept." This involved two major targets: creating a popular, fuel-efficient car (a top management goal) and designing a low-priced but not cheap model (a self-imposed team objective). The team eventually saw two ways to achieve these goals.

They first attempted to develop a "mini Civic" by shortening the car 100mm in front and back. Although the original instructions from top management—on a single sheet of paper—had been general and ambiguous, management resolutely rejected the compromise. Team leader Watanabe recalls, "No matter how we refined development plans of this kind, Mr. Kawamoto rejected them again and again, saying we had to start all over from the very beginning. We didn't know what to do, but we couldn't tell them that a fresh start was absolutely impossible. So we were finally persuaded to try it." Top management thus forced the team forward but maintained the loose power balance between the two groups. Mr. Watanabe, who had reached a conceptual dead end, went to Europe, where he was inspired by Austin Mini to lead the team to create a "luxurious mini" model.

The second means of reaching the team's goal involved challenging an idea that was dominant in the automobile industry: that a car should be "horizontally long and vertically short." At the time, most automobile manufacturers were looking for fuel economy with lightweight materials and aerodynamic designs. Consequently vertically short designs were prevalent. After a month of heated discussion, the team began to move in another direction. Their new policy involved what Mr. Watanabe called "automobile evolution theory":

> The theory is that the ultimate form of the automobile requires us to maximize human space by minimizing machine parts. We decided that the automobile was evolving toward this ideal form, and that Honda should lead this evolutionary movement. The first step was to design a tall model that would challenge the "common sense of Detroit," which permitted car designers to pursue beautiful forms at the expense of space for human use. . . . It takes a technological perspective to see the significance of a taller model. . . . It is a cube that can be more lightweight, lower priced, and more solid.

Based on this theory, the development team chose a "horizontally short and vertically tall" car as its target. The "luxurious mini" became the core of their design policy. The challenge was to design a machine-minimum (minimum space devoted to the machine) and man-maximum (maximum space for human use) automobile. These seemingly contradictory goals required the creation of a new viewpoint.

One of the team members recalls, "I feel, however illogical it sounds, that the success of this project owes a lot to the very wide gap between the ideal and the actual. We could not achieve the ideal goal by an incremental improvement of the actual. Revolutionary reformation was necessary, and, in order to achieve this, new technologies and concepts were generated one after another."

The group, then, was given autonomy and was simultaneously forced to challenge long-held assumptions. These alone were not sufficient for the realization of creative concepts, however. The importance of also having a group of people with heterogeneous backgrounds cannot be overstated. Any group can create a wider variety of concepts than the average individual. This potential is best realized when group members are extremely heterogeneous with respect to jobs, orientations, and behavior. In the phase of concept formation, the City development team was a hybrid, except in that all of the members were genuine car maniacs. In fact, many worried that the team would be unable to reach consensus.

The group needed to introduce, challenge, process, and integrate a wide range of information and ideas. One of the methods used was *tamadashi kai*, or meetings to create and share information. *Tamadashi kai* are not formal, and every meeting was held at a different time and place. Each was entirely devoted to discussion and aimed at clearing some hurdle the project had run into. The participants in these meetings were not just the team members. When necessary, the staffs of related departments were invited as well. All participants were required to be equally involved and absorbed in the lively discussion regardless of titles or qualifications. However, criticism was taboo. As Mr. Watanabe says, "To criticize is ten times easier than to propose a constructive opinion. The discussion would have been useless if participants had remained silent for fear of being criticized." Several times, Mr. Watanabe held day-and-night discussion sessions, most often in rooms at small taverns or inns near the research lab instead of at luxurious hotels. In ceaseless discussion from morning to midnight, participants had to use all their wits to challenge existing paradigms.

During the development process, the project team comprised people from the development, production, and sales departments. The team made use of a quick information creation system based on a hybridization of the three departmental positions. The system performed the following functions: procuring personnel, facilities, and budget figures for the production plant; analyzing the automobile market and the competition; and setting a market target and determining sale price and production quantity. This system, which collects, creates, and implements information, is called the "SED system." It is also used to develop about forty-five new motorcycle models each year. The SED system was established when Mr. Kawashima was president of Honda. Its initial aim was to manage development activities more systematically by integrating the knowledge and wisdom of ordinary people instead of relying on a "hero" like founder Soichiro Honda.

The operation of the system is quite flexible. The three areas, Sales, Engineering, and Development (SED), are nominally distinct, but there is a built-in learning process that encourages invasion into others' areas. The actual work flow requires researchers to collaborate with their colleagues. Mr. Watanabe comments, "I am always telling the team members that our work is not a relay race—that's my work, and yours starts here. Every one of us should run all the way from the start to the finish. Like in a rugby game, all of us should run together, passing the ball left and right, reaching the goal as one united body."

This process leads to a high level of information sharing. In a project team, members share a huge amount of managerial information received through conversations with top management; they also share market information concerning the competition. Moreover, as the division of labor is rather unclear and flexible, members can meddle wherever they like. There is thus no information lag between top management and the team leader. Also, since whatever the various members are doing is out in the open, each knows almost instinctively what his or her coworkers have in mind. This information sharing among project members reflects a distinctive aspect of Honda's broad corporate culture. The company is already famous as an originator of the "large room" system, and all its meeting rooms are glass-plated so that what is going on can be seen from the outside. As the degree of information sharing increases, individuals identify themselves with the team as a whole, and begin "self-controlling." "Basically speaking," says Vice President Kawamoto, "organizational management is no longer necessary if each individual properly performs what is expected. Once a goal is given and roles are specified to a certain extent, our staff works quite well."

Every participating member grows up by casting off his older skin in the process of successful project development. Prominent success helps other members to follow, saying, "Hey! Look at what those guys have done." After the successful completion of a project, participants are assigned to other projects so that the knowledge they have acquired can be transferred throughout the organization. An engineer comments, "I think it's pretty difficult to articulate really meaningful know-how in text, figures, or other measurable forms. The knowledge is alive because . . . it changes continuously. . . . The best way to transfer it is through human interaction."

On the other hand, Honda dislikes easy imitations of recent successes, and sometimes even goes so far as to destroy its own accumulated knowledge. "The most severe criticism for us is to say, 'It looks like something else,' " states Mr. Hiroshi Honma, a chief engineer. The concept of "Tall Boy" (the nickname for the City model) was created by destroying the concepts that dominated Honda and the automobile industry in general at the time. A special engine, suspension, and radial tire were developed exclusively for the new car. All other parts were also designed for the City, to avoid giving customers an image of a "mini Civic." About ninety patents were applied for during the project, which clearly indicates the "unlearning" of accumulated knowledge and the acquisition of new knowledge.

Middle management plays a key role in this process of abandoning the old and generating the new. The Honda City case clearly shows the critical importance of Mr. Watanabe, the middle manager selected as project leader. His role had several key aspects: providing direct information links to top management; transforming top management's general vision into directions for the team's activities and for pursuing the creation of meaning; managing "chaos" and keeping it within tolerable limits; and providing the context for integration across specialties. While Honda pays tribute to the energy and drive of the young researchers who generated the new product idea, top management clearly recognizes the strategic role of its middle management project leaders as well.

PRODUCT DEVELOPMENT AND INFORMATION CREATION

If we wish to understand how the City development project was managed, the familiar information-processing paradigm is not especially helpful. For innovation is not so much a process of gradually reducing uncertainty (processing information) in moving toward a prescribed goal. Rather, it is a process through which uncertainty is intentionally *increased* when circumstances demand the generation of chaos from which new meaning can be created. This process is full of discovery, surprise, and redundancy. Senior management at Honda, in forcing the project members to challenge their most deeply held assumptions, forced them not to process information but to create it. To do this successfully, project members had to confront ambiguity, contradiction, and failure.

The information-processing paradigm emphasizes the structure of the organization. The information creation paradigm, in contrast, stresses the process of creating meaningful information through personal interaction.[1] The *quality* of information becomes more important than the quantity. Inductive, synthetic, and holistic methodologies become more useful than the deductive, analytic, and reductionistic ones used in information processing.

If we view the organization as a three-tiered structure—composed of the individual, the group, and the organization as a whole—then we can pinpoint the specific characteristics that are important

TABLE 33.1. Levels of Organizational Information Creation

Level	Emergent Property	Factors Related to Information Creation
Organization	Structure	Competitive resource allocation
Group	Interaction	Direct dialogue
Individual	Autonomy	Action and deliberation

to information creation at each tier of the organization (see Table 33.1).

The Individual Level

The emergent, or critical, property of information creation at the individual level is autonomy. This level is characterized by action and deliberation: Only here is it possible to deliberate and act autonomously. Autonomy begins to be realized when individuals are given the freedom to combine thought and action at their own discretion, and are thereby able to guarantee the unity of knowledge and action.

The methodology for information creation can be looked at in a number of ways—deductive or inductive, analytic or synthetic, reductionistic or holistic, etc. If we approach information creation with the idea that a new point of view will emerge as facts and knowledge are related, then our approach is essentially inductive. An inductive style is more likely to produce innovation, because the knowledge that forms the basis for information creation is often inarticulate. (Michael Polanyi has called this tacit knowledge—intuitive knowledge that cannot be completely expressed either in writing or in speech.[2]) The essential elements of tacit knowledge—the creation, learning, or recognition of the new and the unknown—are particularly important in the process of information creation that leads to innovation. Overt knowledge is important, but intuitive knowledge is essential, because it reflects an internalized understanding gained from previous experience. In any event, the interaction between perception and action is the methodological basis for individual information creation.

This methodology is characterized by a never-ending rephrasing of the basic questions one is dealing with. However, this type of self-reflection far more often leads to failure than to success. Therefore, if one does not have the freedom to fail as well as the freedom to succeed, then one is not truly autonomous.

THE GROUP LEVEL

The emergent property at the group level is interaction—more concretely, open and frank dialogue. Human interaction is best realized within the organization at the group level.

The creation of information is the creation of a new perspective. The dynamic, complementary process that results in a shift to a new point of view requires interaction—a dialogue or debate—among people. The process is convoluted, involving a cycle of affirmation, denial, and resolution before new information is created. Since the significance of information is elastic during this process, individuals have the opportunity to interpret and reinterpret for themselves; this freedom allows group members to organize information individually. Unity and coherence are born from this group action. Coherence itself, however, can serve both to promote and to hinder the creation of information. Coherence often produces a pressure for conformity, and differing opinions are confined or limited with the birth of what Janis calls "group think."[3] However, this tendency must be balanced against the fact that trust is the precondition for creative dialogue, as well as for the open exchange and cooperative possession of information.

THE ORGANIZATIONAL LEVEL

The emergent property of the organization as a whole is structure. An organization's structure regulates the depth of the relationship between groups (sections) involved in information creation. From a macro perspective, structure produces the means for the distribution of resources among the various groups in the organization, and thereby contributes to a greater competitive capability. The structure of an organization is designed to be able to mediate between the desires of the group and of the individual in relation to information creation. It thus addresses the problem of allocating resources properly among competing interests.

The key resources in information creation are people, things, money, and information. One of the objects of competitive resource allocation is the proper integration of these limited business resources. An organizational structure based on the logic of competitive resource allocation regulates the ability of a group or individual to create information. The structure thereby provides basic direction, in terms of breadth and depth, to individual acts of information creation over a specified period of time.

This discussion of methodology at the individual, group, and organizational levels suggests some differences between the levels with respect to information creation. It is important to recognize, however, that these methodologies can either promote progress or generate contradictions. For example, even when individual autonomy exists, there may be no constructive dialogue at the group level. Or if there is constructive dialogue at the group level, but no competitive allocation of resources at the organizational level, then productive information creation cannot be realized in a competitive market. And even if inductive and deductive methodologies were unified so that information creation could be managed in a single process, the breadth and depth of the organization's information creation might thereby be lost. I therefore suggest that the special characteristics of these distinctive methodologies be allowed to coexist.

METHODOLOGIES OF ORGANIZATIONAL INFORMATION CREATION

Top-down management is essentially deductive; bottom-up management is essentially inductive. Let us briefly consider how these two managerial styles affect the "emergent properties" of resource allocation, interaction, and autonomy. Later we will propose middle-up-down management (seen in the development of the Honda City) as a methodology for information creation that can incorporate the strengths of both inductive and deductive management.

Deductive Management

Resource Allocation

The management methods used in deductive corporations are premised on the belief that information creation occurs mainly at the top. The role of top management is to clarify decision premises and to design organizational structures that can reduce individual information and decision burdens. Top management also allocates resources using sophisticated analytical techniques. Since decision making is concentrated at headquarters, a common set of clear-cut and measurable criteria that transcends the specific requirements of the various divisions is needed. ROI is typically used as such a criterion, with cash flow within and across individual strategic business units becoming the major concern with respect to resource allocation.

The underlying principle supporting such a management approach is the information-processing paradigm. But the hierarchy designed by top management in a deductive manner is not suited to allow organizational members at lower levels to create information in a flexible manner.

Interaction

Top-down, strong leadership is the basic policy adopted by deductive management. Information is processed; it moves from the upper levels to the lower levels, and variety reduction is the keystone. The elimination of "noise," "fluctuation," and "chaos" is the paramount concern. Information creation at the lower levels proceeds with great difficulty.

Information activity between divisions has a sequential relay pattern; work completed by one division is passed on to another division.

There is a tendency for the transformation of information into knowledge to occur with great intensity within the narrow areas of labor divisions. However, the amount of semantic information and knowledge absorbed and accumulated by the lower levels of the organization is small because of the lack of personal interaction.

AUTONOMY

Top managers and corporate staff possess the greatest autonomy. They are likely to adopt a hands-off, deductive methodology rather than a hands-on one. Consequently their information creation activities sometimes move far from the individual, shopfloor viewpoint. However, there is a potential for creating visionary concepts at the organizational level that could not be reached based on individual experience.

Inductive Management

RESOURCE ALLOCATION

Inductive management maintains that the organizational creation of information begins with the vision of the individual—the entrepreneurial individual—and that people who have an interest in a project will become the core of any long-term effort.

Technology is seen as the interaction between people and systems of information or knowledge. Thus the concept of synergy is basic to inductive management. Resources are allocated in a way that encourages interaction, allowing new concepts and theories to develop in the most natural way possible. The ideal inductive organization is "self organizing." Autonomous information creation takes place by expanding from the individual level to the group level and then to the organizational level. At 3M, for example, a project can become a department and then a division if it is sufficiently successful.

INTERACTION

A supportive leadership that moves in step with the individual, the group, and the organization is necessary for information creation in an inductive-management organization. The support of an influential leader is necessary for individuals or self-organizing groups that have vision, since they will need help overcoming opposition from within the organization.

The need for a supporting sponsor to assist the intracompany entrepreneur is particularly emphasized at 3M. Before a daring and promising idea can stand on its own, it must be defined and supported by a sponsor willing to risk his or her reputation in order to advance or support changes in intracompany values. The leadership style of the sponsor can be summed up in the unspoken maxim, "The captain bites his tongue until it bleeds." On the basis of past experience, the leader relies on his or her own criteria (consciously and unconsciously) to guide the creation of new information.

AUTONOMY

Autonomy is given to those working as entrepreneurs at every organizational level. In many cases such individuals create meaningful information in the midst of interactive, tense relations, by testing and deepening their intuitive understanding through practice. Their information creation may be based on hunches or intuition, or on the ability to recognize the essence holistically in a moment.

Since the individual internalizes a great deal of tacit understanding, a career-path personnel policy that stresses promotions and transfers is used to support the organizational transfer of understanding. On the other hand, since the unlearning of acquired personal experience is difficult, inductive management may be unsuitable in instances where there are frequent large-scale reorganizations or replacements due to acquisitions or divestitures.

SYNTHESIZING INDUCTIVE AND DEDUCTIVE MANAGEMENT

Today, the intensity of market competition and the speed required for efficient information creation suggest a need to synthesize these two managerial styles. This synthesis involves the conceptualization of symbiotic management,[4] or what I call compressive management. The development of the Honda City is a good example of this methodology, which can also be called middle-up-down management. The core of this managerial style is not the top managers or the entrepreneurial individuals, but rather the middle managers.

Middle management occupies a key position; it is equipped with the ability to combine strategic macro (context-free) information and hands-on micro (context-specific) information. In other words, middle management is in a position to forge the organizational link between deductive and inductive management.

Middle management is able to most effectively eliminate the noise, fluctuation, and chaos within an organization's information creation structure by serving as the starting point for action to be taken by upper and lower levels. Therefore, middle managers are also able to serve as the agent for change in the organization's self-renewal process.

Resource Allocation

Top management is responsible for determining the overall direction of the company and for establishing the time limits on realizing that vision. Time is the key resource. Each individual performing day-to-day tasks has his or her own vision. It is the middle manager who works, within a certain time limit, as a "translator" in charge of unifying individual visions and creating a larger vision, which will in turn be reflected in future individual visions. The group functions as the field for the realization of this process. In order to achieve this vision, middle managers work with upper- and lower-level personnel. However, it is the top that selects the middle, and selecting the right people becomes the most important foundation of an effective corporate strategy. In addition to deciding who will formulate and implement a strategy, the top serves as a catalyst that creates fluctuation or chaos.

Consequently, in compressive management, the entrepreneurial middle receives broad direction from the top and begins the process of information creation within the group, working to involve relevant individuals and carrying out information creation intensively within a compressed period of time. Through interaction with top management, middle management secures the resources required to achieve its vision. In this process, both deductive strategic planning and inductive emanation of information from the needs of the market are integrated to establish a definite direction for resource deployment and to create a practical concept which follows that direction.

The unit for resource allocation should be designed by the top so that the middle can create meaningful concepts. The structure of this unit can take a variety of forms, but usually consists of a multidisciplinary team led by middle management.

Interaction

Before the entrepreneurial middle can realize its vision, it must first confront and survive the criticism of other members of the group through intensive communication. As a result of this criticism, a more concrete concept will be formed. In order to realize a vision, an idea must successfully challenge the stability of the organization, involving people from both top and bottom, left and right.

This process often involves the following steps. The first stage is establishing creative chaos.[5] Top management offers a challenging goal and creates tension. As the organization moves in the direction of innovation, creative chaos is amplified to focus on specific contradictions in order to solve the problem. These contradictions produce a demand for a new perspective, speeding up information creation activity. This approach is exemplified by the Honda R&D manager's statement, "Creativity is born by pushing people against the wall and pressuring them almost to the extreme."

The second stage involves the formation of a self-organizing team that tries to create a new order (meaning) out of the chaos. This self-organizing group has the following characteristics: it is autonomous; it is multidisciplinary, so as to encourage cross-fertilization among its members; and it creates challenging goals that force it to transcend the existing contradictions. This team forms the core for an intense level of activity and works independently of other divisions within the corporation.

The third stage is the synchronization of concept creation. This stage is the embodiment of the spiral in which information creation moves from middle management to the top and bottom. These movements resemble the punting and passing that occur in a rugby match as the opposing teams attempt to win ground.[6] The realization of a concept is made possible by the intraorganizational divisions pulling together in a "shared division of labor" and by promoting "active cooperative phenomena."[7]

The fourth stage involves the transfer of learning and unlearning. Innovation that aims at a distant and vaguely defined goal goes through apparently redundant phases of shared division of labor. The natural consequence of this process is to activate the information creation activities at all levels of the organization. The successful innovation generates a new order, and gives birth to organizational learning and unlearning.

AUTONOMY

A group is given both autonomy (freedom) and a time limitation (constraint). Middle management becomes the logical center for the fusion of the deductive and inductive styles of management. Although it may be possible to balance the use of stored syntactic information and of tacit understanding, the need for a rapid response to changing conditions will not allow middle management to concentrate exclusively on the creation of information. The requirement to simultaneously expand the knowledge base and process information may eventually place an excessive burden upon the middle management group. If these people are not allowed to recharge their batteries from time to time, the long-term capacity for organizational information creation will weaken.

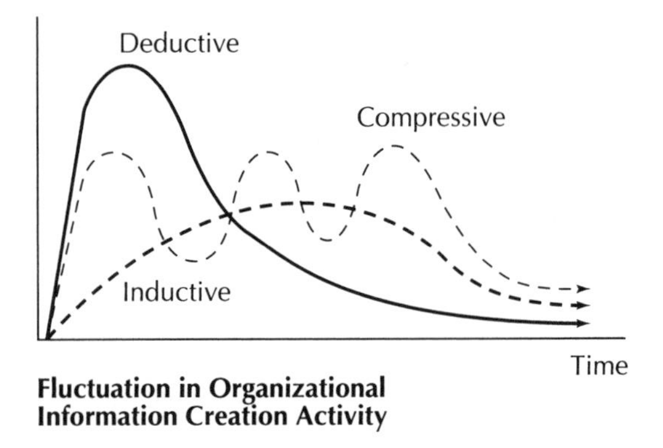

Fig. 33.1. A comparison of organizational information creation patterns.

PROPER MANAGEMENT OF ORGANIZATIONAL INFORMATION CREATION

I have spoken of three methodologies for information creation—deductive, inductive, and compressive. Their approximate patterns are sketched in Figure 33.1 and Table 33.2.

One cannot make an unqualified choice of methodology until one has considered the special

TABLE 33.2. Comparison of Methodology of Organizational Information Creation

	Deductive Management	Inductive Management	Compressive Management
Resource Allocation			
Key resource	Money	People	Time
Time management	Periodical planning	Self-management	Deadline
Unit of resource allocation	SBU	Individual	Self-organizing team
Interaction			
Top management	Leader	Sponsor	Catalyst
Context of interaction	Within headquarters	Among voluntary individuals	Among designated individuals within the group
Direction	Top-down	Bottom-up	Middle-up-and-down
Autonomy			
Methodology	Deductive, hands-off	Inductive, hands-on	Hands-on and -off
Knowledge	Articulate	Tacit	Articulate/tacit
Problem	Analysis paralysis	Inductive ambiguity	Exhaustion

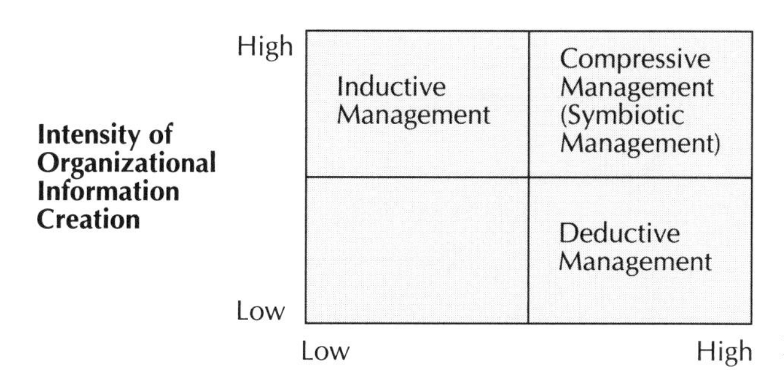

Fig. 33.2. The relationship between organizational information creation and environmental characteristics.

environmental characteristics present. The relationship between the environment and the appropriate management methodology is perhaps best illustrated in Figure 33.2.

As environmental uncertainty increases, the organization can adapt itself more effectively with a high level of information creation occurring at all levels of the organization, rather than with a low level of information creation. In this sense, as the need for information creation increases, companies will probably make a shift from deductive management to inductive or compressive management, which have higher information creativity.

In the meantime, as market reactions speed up as a result of intense competition, companies will likely shift from inductive or deductive management to compressive management to cope with that problem. However, compressive management must come to grips with the problem of placing a great deal of pressure on middle management to process an expanding base of information within a limited time period. Therefore, whether or not information creation that is both high in quality and well coordinated can occur will depend largely on how entrepreneurial middle management really is.

CONCLUSION

The essential logic of compressive management is that top management creates a vision or dream, and middle management creates and implements concrete concepts to solve and transcend the contradictions arising from gaps between what exists at the moment and what management hopes to create. In other words, top management creates an overall theory, while middle management creates a middle-range theory and tests it empirically within the framework of the entire organization.

Mr. Tadashi Kume, president of Honda, expresses the role of middle management as follows: "I continually create dreams, but people run in different directions unless they are able to directly interact with reality. Top management doesn't know what bottom management is doing. The opposite is also true. For example, John at Honda Ohio is not able to see the company's overall direction. We at corporate headquarters see the world differently, think differently, and face a different environment. It is middle management that is charged with integrating the two viewpoints emanating from top and bottom management. There can be no progress without such integration."

Honda is, in fact, a company continually creating contradictions. Their company principle is, "Maintaining an international viewpoint, we are dedicated to supplying products of the highest efficiency at a reasonable price for worldwide customer satisfaction." Simultaneously, the company stresses the need to "be in touch with the reality of local conditions." Consequently, the company constantly generates conflict between dreams and reality. In order to resolve these contradictions, middle management must create and implement business and product concepts capable of being tested empirically. This is a never-ending process, illustrated in Figure 33.3.

Mr. Katsutoshi Wada, head of the Human

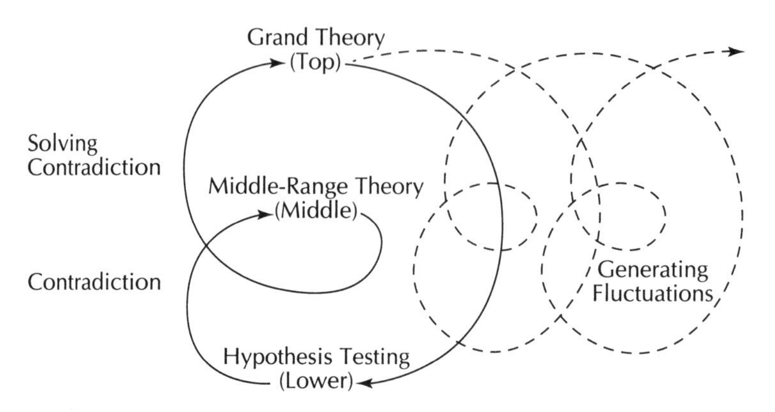

Fig. 33.3. Middle-up-down information creating process.

Resources Development Center at Honda, characterizes Honda management as follows:

> Mr. Soichiro Honda, the founder of our company, did not articulate this, but we now realize that an organization has a kind of culture, with dreams on the one hand and a realistic methodology on the other. Taken together, the dream is always larger than the reality, qualitatively as well as quantitatively. The solution to the gap between the two comprises the essential mission of Honda management. We are perpetually engaged in an effort to mediate and solve the contradictions generated by a naive romanticism and a hard realism. Without coupling such romanticism and realism, it would be difficult to manage our company well. Neither a dreaming child nor an adult who has lost all his dreams can produce good management. Good management can be achieved only by matching the dream of a child with the realism of an adult.

It is middle management's role to create and realize verifiable business concepts for the creative solution of contradictions and gaps between the ideal and the actual. The development of the Honda City illustrates this process concretely. The project was initiated with a top management dream to "create on original car of high energy and resource efficiency." This dream was then brought to the middle, which created and realized a more concrete concept—and added its own vision—using a self-organizing group as the core of the project.

The process of organizational information creation can also become a process of organizational theory development—in other words, the language and assumptions that underlie the corporation's existence can be transformed if information creation is sufficiently innovative. One of the significant products of the Honda City project, for example, was the use of such metaphors as "automobile evolution" and "cube." Concepts are created through the spiral loop of recognition and practice. The metaphor can leverage movement toward a companywide change of perspective.

Middle-up-down management is a type of organizational information creation that involves the total organization. It may best embody the essence of an organization spontaneously surviving in the business environment's ceaseless generation of changes.

Note

The author wishes to acknowledge the assistance of Professor D. Eleanor Westney of M.I.T. and of Professor Hirotaka Takeuchi and Tsuyoshi Numagami of Hitotsubashi University.

References

1. I. Nonaka, *Corporate Evolution* (in Japanese) (Tokyo: Nikkei Shimbun-sha, 1985).
2. M. Polanyi, *Tacit Dimension* (New York: Doubleday, 1966).

3. I. Janis, *Victims of Group Think* (Boston: Houghton Mifflin, 1972).
4. T. Kagono, I. Nonaka, K. Sakakibara, and A. Okumura, *Strategic vs. Evolutionary Management: A U.S.-Japan Comparison of Strategy and Organization* (Amsterdam: North Holland, 1985).
5. I. Nonaka, "Creating Organizational Order out of Chaos: Self-Renewal in Japanese Firms," *California Management Review*, forthcoming.
6. H. Takeuchi and I. Nonaka, "The New New Product Development Game," *Harvard Business Review*, January–February 1986.
7. K. Imai, I. Nonaka, and H. Takeuchi, "Managing the New Product Development Process: How Japanese Companies Learn and Unlearn," in K. B. Clark, R. H. Hayes, and C. Lorenz eds., *The Uneasy Alliance* (Boston: Harvard Business School Press, 1985).

34

Managing Intellect

JAMES BRIAN QUINN
PHILIP ANDERSON
SYDNEY FINKELSTEIN

With rare exceptions, the economic and producing power of a modern corporation or nation lies more in its intellectual and systems capabilities than in its hard assets—raw materials, land, plant, and equipment. Intellectual and information processes create most of the value added for firms in the large service industries—like software, medical care, communications, education, entertainment, accounting, law, publishing, consulting, advertising, retailing, wholesaling, and transportation—which provide 79 percent of all jobs and 76 percent of all U.S. GNP. In manufacturing as well, intellectual activities—like R&D, process design, product design, logistics, marketing, marketing research, systems management, or technological innovation—generate the preponderance of value added. By the year 2000 McKinsey & Co. estimates that 85 percent of all jobs in America and 80 percent of those in Europe will be knowledge based.

The capacity to manage intellect and to convert it into useful outputs has become the critical executive skill of the era. Yet few managements have systematic answers to even these basic questions:

- What is intellect? Where does it reside? How do we capture it? Leverage it?
- What special skills are needed to manage professional vs. creative intellect? How can a firm measure the value of its intellect? How can managers leverage their firm's intellectual resources to the maximum?

WHAT IS INTELLECT?

Webster's Dictionary defines intellect as "knowing or understanding; the capacity for knowledge, for rational or highly developed use of intelligence." The intellect of an organization—in order of increasing importance—includes: (1) *cognitive knowledge* (or know what); (2) *advanced skills* (know how); (3) *system understanding* and *trained intuition* (know why); and (4) *self-motivated creativity* (care why). Intellect clearly resides inside the firm's human brains. The first three levels can also exist in the organization's systems, databases, or operating technologies. If properly nurtured, intellect in each form is both highly leverageable and protectable. *Cognitive knowledge* is essential, but usually far from sufficient for success. Many may know the rules for performance—on a football field, piano, or accounting ledger—but lack the higher skills necessary to make money at it in competition.

Similarly, some possess advanced skills but lack system understanding; they can perform selected tasks well, but often do not fully understand how their actions affect other elements of the organization or how to improve the total entity's effectiveness. Similarly, some people may possess both the knowledge to perform a task and the *advanced skills* to compete, but lack the will, motivation, or adaptability for success. Highly *motivated* and *creative* groups often outperform others with greater physical or fiscal endowments.

The value of a firm's intellect increases markedly as one moves up the intellectual scale from cognitive knowledge toward motivated creativity. Yet, in a strange and costly anomaly, most enterprises reverse this priority in their training and systems development expenditures, focusing virtually all their attention on basic (rather than advanced) skills development and little or none on systems, motivational, or creative skills (see Figure 34.1). The result is a predictable mediocrity and loss of profits.

CHARACTERISTICS OF INTELLECT

The best managed companies avoid this by exploiting certain critical characteristics of intellect at both the strategic and operational levels.

EXPONENTIALITY

Properly stimulated, knowledge and intellect grow exponentially. All learning and experience curves have this characteristic. As knowledge is captured or internalized, the available knowledge

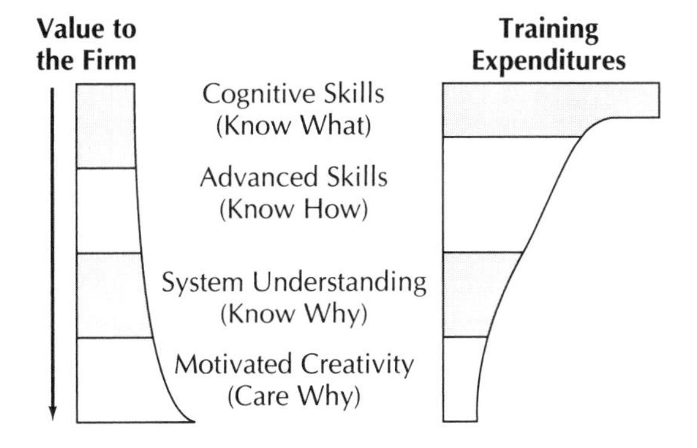

Fig. 34.1. Skills value vs. training expenditures.

base itself becomes higher. Hence a constant percentage accretion to the base becomes exponential total growth, as documented by many studies of organizational learning curves. The effect accelerates as higher levels of knowledge allow the organization to attack more complex problems and to interrelate with other knowledge sources it earlier could not access. For example, Microsoft has moved from single operations systems or applications programming to an environment in which it must integrate all applications into a uniform graphics-based environment and hook many different problem modules together across a variety of computing platforms. To meet these challenges, both Microsoft's knowledge base and rate of intellectual growth have had to expand dramatically in the last five years.

The strategic consequences of exploiting exponentiality are profound. Once a firm obtains a knowledge-based competitive edge, it becomes ever easier to maintain its lead and harder for competitors to catch up. The most serious threat is that through complacency intellectual leaders may lose their knowledge advantage. This typically occurs when they fail to adapt to changing external conditions, particularly new concepts that obsolesce their earlier skills—as molecular design techniques are obsoleting blind screening in pharmaceuticals today. This is why the highest level of intellect, self-motivated creativity, is so vital. Firms that nurture "care why" are able to thrive on today's rapid changes and simultaneously refresh their cognitive, advanced skills, and systems-knowledge bases.

Because a firm is a leader it can attract better talent than competitors. The best people want to work with the best. These people can then perceive and solve more complex and interesting customer problems, make more profits as a result, and attract even more talented people to work on the next round of complexity. Driving and capturing individual's exponential learning has been the key to strategic success for most intellectual enterprises—from Bell Labs or Intel, to Microsoft, McKinsey, or the Mayo Clinic. For example:

- Microsoft, realizing that software design is a highly individualistic effort, interviews hundreds of candidates to find the few most suited to write its advanced operating systems. It then places its new members directly onto small (3- to 7-person) teams under experienced mentors to design complex new software systems at the frontier of user needs. Microsoft's culture drives everyone with the unstated expectation of 60- to 80-hour weeks on intensely competitive projects. The best commercial programmers seek out and stay with Microsoft largely because they believe "Microsoft will determine where the industry moves in the future" and they can share the excitement and rewards of being at that frontier. Each Microsoft success, in turn, builds the experience base and recruitment attractiveness for the next wave of challenges.

Sharing

Another important characteristic is that knowledge is one of the few assets that grows most—also usually exponentially—when shared. Communication theory states that a network's potential benefits grow exponentially as the nodes it can successfully interconnect expand numerically. As one shares knowledge with other units, not only do those units gain information—linear growth—they share it with others and feed back questions, amplifications, and modifications, which add further value for the original sender, exponential total growth. Proper leveraging through external knowledge bases—especially those of specialized firms, customers, and suppliers—can create even steeper exponentials. There are, however, some inherent risks and saturation potentials in this process. Determining what knowledge should be concentrated and protected, what may be decentralized and shared, and how to share and sift knowledge from data are critical elements in intellectual strategies. For example:

• The core intellectual competency of many financial firms—like Fidelity, State Street Boston, or Aetna—lies in the human experts and the systems software that collect and analyze the data surrounding their investment specialties. Access to the internals of these centralized software systems is tightly limited to a few specialists working at headquarters. Here they share and leverage their own specialized analytical skills through close interactions with other financial specialists, "rocket scientist" modelers, and the unique access the firm has to massive transactions data. These companies further leverage their systems' outputs as broadly as possible through their extensive brokerage outlets, which, in turn, yield more information. Nevertheless, the structure of sharing must be carefully controlled. For security and competitive reasons, sales brokers cannot have access to their corporate system's analytics, and corporate analysts must be kept out of brokers' individual customer files. Yet for maximum impact the system itself must capture and manipulate data aggregates from both sources. The detail in which this sharing can occur—without compromise—is a major source of competitive edge as is its integration into the software and support systems of the firm. The latter provide the key to converting intellect into a proprietary asset.

Both genuine intellectual synergies and proprietary protection derive from so embedding and sharing the firm's crucial knowledge capacities in its other systems that isolated elements of the firm's intellect will lose significant value if they are detached from the whole of the organization's competencies. For example, when the trust department of a major bank developed a unique set of analytics to support the mortgage-backed securities market, a rival lured away the key programmers who had developed the system. However, the competitive loss turned out to be minor. In a different corporate culture with different management, databases, support systems, and customer relationships, the pillager was unable to implement the analytics effectively.

Expandability

Unlike physical assets, intellect: (a) increases in value with use; (b) tends to have much underutilized capacity; (c) can be self-organizing; and (d) is greatly expandable under pressure. How can a company exploit these characteristics?

Arthur Andersen & Company (AA&Co) offers some interesting insights.

• Andersen attempts to electronically interlink more than 51,000 people in 243 offices in 54 countries. Its AANET, a T-1 level, private, all-fiber-optics network connects most major offices, through data, voice, and video interlinks. AANET allows AA&Co specialists—by posting problems on electronic bulletin boards and following up with visual and data contacts—to instantly self-organize around a customer's problem anywhere in the world. It thus taps into otherwise dormant capabilities, and vastly expands the energies and solution sets available to customers. The capacity to share in AA&Co's worldwide variety of problems and unique solutions is enhanced through centrally collected and carefully indexed subject, customer reference, and resource files accessible directly via AANET or from CD-ROMs, distributed to all offices. These, in turn, expand the intellectual capabilities AA&Co field personnel have available to add value for future customers. The key intellectual asset is not AA&Co's database per se. What matters most is the ability to motivate use of the knowledge, which depends more on AA&Co's unique training, culture, mentoring, and worldwide customer base. Many of the files could be duplicated, but the dynamics permitting their effective extension and use could not.

Effective managers of intellectual processes consciously harness all these characteristics. The processes they use resemble successful coaching more than anything else. The critical activities are: (1) recruiting the right people; (2) stimulating them to internalize the information knowledge, skills, and attitudes needed for success; (3) creating systematic technological and organizational structures to capture, focus, and leverage intellect to the greatest possible extent; and (4) demanding and rewarding top performance from all players. Easier said than done. But much can be learned from how successful (and failing) practitioners have handled these issues. Our conclusions draw on an extensive literature search, hundreds of personal interviews, and numerous published case studies of the leading professional and innovative companies of the United States, Europe, and Japan.[1]

PROFESSIONAL INTELLECT

There are important differences between managing professional versus creative intellect. Although much attention has been given to managing creativity, little has been written about managing professionals. Yet professionals are the most important source of intellect for most organizations. For every truly creative organization, there are probably twenty to 100 professional groups creating high value deep within integrated firms or directly for customers. What characterizes such professionals?

Perfection, Not Creativity

While no precise delineation applies in all cases, most (90 to 98 percent) of a typical professional's activity is directed at perfection, not creativity.[2] The true professional commands a complete body of knowledge—a discipline—and updates that knowledge constantly. In most cases, the customer wants the knowledge delivered reliably with the most advanced skill available. Although there is an occasional call for creativity, the preponderance of work in actuarial units, dentistry, hospitals, accounting units, opera companies, universities, law firms, aircraft operations, equipment maintenance, etc. requires the repeated use of highly developed skills on relatively similar, though complex, problems. People rarely want their surgeons, accountants, airline pilots, maintenance personnel, or nuclear plant operators to be very creative, except in emergencies. While managers clearly must prepare their professionals for these special circumstances, the bulk of their attention needs to be on delivering consistent, high-quality intellectual output. What are the critical factors?

HYPER-SELECTION

The leverage of intellect is so great that a few top-flight professionals can create a successful organization or make a lesser one billow. Marvin Bower created McKinsey & Co.; Robert Noyce and Gordon Moore spawned Intel: William Gates and Robert Allen built Microsoft; Herb Boyer and Robert Swanson made Genentech; Einstein put

Princeton's Institute of Advanced Studies on the map; and so on. Finding and developing extraordinary talent is thus the first critical managerial prerequisite. McKinsey long focused on only the top 1 percent of graduates from the top five business schools and screened heavily from these. Microsoft interviews hundreds of highly recommended people for each software designer hired, after a withering selection process that tests not just their cognitive knowledge but their capacity to think about new problems under high pressure. Similarly—recognizing talent and commitment as their most critical elements for success—experienced venture capitalists spend as much time on relentlessly pursuing and selecting top people as on the quantitative aspects of projects.

Intense Training, Mentoring, and Peer Pressure

These factors literally force professionals to the top of their knowledge ziggurat. The best students go to the most demanding schools. The top graduate schools—whether in law, business, engineering, science, or medicine—further reselect and drive these students with greater challenges and with 100-hour work weeks. Then upon graduation the best of the survivors go back to even more intense "boot camps" in medical internships, law associate programs, or other outrageously demanding training situations as pilots, consultants, or technical specialists. The result is to drive the best professionals up a learning curve that is steeper than anyone else's. People who go through these experiences quickly move beyond those in less demanding programs, becoming noticeably more capable—and valuable—within six months to a year than those other organizations challenge in a lesser way. The keys are forcing professional trainees' growth with constantly heightened (preferably customer-induced) complexity, thoroughly planned mentoring, high rewards for performance, and strong stimuli to understand, systematize, and advance their professional disciplines. The great intellectual organizations all seem to develop deeply ingrained cultures around these points. Most others do not.

Constantly Increasing Challenges

Intellect grows most when challenged. Hence, heavy internal competition and constant performance appraisal are common. Leaders tend to be demanding, visionary, and intolerant of half-efforts. At Bell Labs, 90 percent of its carefully selected basic researchers moved on (voluntarily or through "stimulated changes") within seven years. Microsoft tries to force out the lowest performing 5 percent of its highly screened talent each year. Leaders often set almost impossible "stretch goals" as HP's Bill Hewlett (improve performance by 50 percent), Intel's Gordon Moore (double the componentry per chip each year), or Motorola's Galvin (six sigma quality) did. Top professionals like to be evaluated, to compete, to know they have indisputably excelled against their peers. The best organizations constantly push their professionals beyond the comfort of their catalogued book knowledge, simulation models, and controlled laboratories. They relentlessly drive associates to deal with the more complex intellectual realms of live customers, real operating systems, and highly differentiated external environments and cultural differences. They insist on and actively support mentoring by those nearest the top of their fields. And they reward associates for their competencies. Mediocre organizations do not.

Managing an Elite

Each profession tends to regard itself as an elite. Members look to their profession and to their peers to determine codes for behavior and acceptable performance standards. They often disdain the values and evaluations of those "outside their discipline." This is the source of many professional organizations' problems. Professionals tend to surround themselves with people having similar backgrounds and values. Unless consciously fractured, these discipline-based cocoons quickly become inward-looking bureaucracies, resistant to change, and detached from customers. Because professionals' knowledge is their power base, many are reluctant to share it with others unless there are powerful inducements. Even then the different values of other groups—although supposedly seeking the same goal—become points of conflict. Hence, in manufacturing researchers disdain product designers, who disdain engineers, who disdain line personnel, who disdain staff, and so on. In medical care, medical researchers disdain MDs (because "they don't understand causation"). MDs disdain both researchers (who "don't understand practical variations among real patients") and nurses (who "don't understand the discipline"). And nurses disdain both doctors and researchers (as "lacking true compassion" for the patient's discomfort and fears). All disdain administrators (as "nonproductive bureaucrats"). Professional values—and the elitism they inculcate if not consciously ameliorated—are the cause of so much of the contentiousness in professional organizations.

Because they have unique knowledge and have been trained as an elite, professionals tend to regard their judgment in all realms as sacrosanct. Professionals hesitate to subordinate themselves to others to support organizational goals not completely congruent with their special viewpoint. This is why most professional firms operate as partnerships and not hierarchies, and why it is so hard for them to adopt a distinctive strategy. It also provides the imperative for evaluating professionals by three separate methods: (1) by peers for professionalism; (2) by customers for relevance; and (3) by enterprise norms for net value.[3]

Few Scale Economies?

Yet many enterprises seem to overlook or violate all these critical characteristics in developing, leveraging, and measuring professionals' capabilities. One reason; conventional wisdom has long held that there are few scale economies—which allow leverages—in professional activities. A pilot can only handle one aircraft; a great chef can cook only so many different dishes at once; a top researcher can conduct only so many unique experiments, a doctor can only diagnose one patient's illness at a time, and so on. In such situations, adding professionals at a minimum multiplies costs at the same rates as outputs. In fact, most often, growth brought *diseconomies* of scale as the bureaucracies coordinating, monitoring, or supporting the professionals actually expanded faster than the professional base. Universities, hospitals, personnel, accounting groups, and consultancies seemed to suffer alike. For years, the only ways firms found to create leverage were (1) to push their people through more intense training or work schedules than competitors; or (2) to increase the number of "associates" supporting each professional. The latter even became the accepted meaning of the term "leverage" in the legal, accounting, and consulting fields.

But new technologies and management approaches now enable firms to develop, capture, and leverage intellectual resources to much higher levels. The keys are: (1) to design organizations and technology systems around *intellectual flows* rather than command and control concepts; (2) to develop performance measurements and incentive systems that reward managers for developing intellectual assets and customer value—and not just for producing current profits and using physical assets more efficiently. Companies as diverse as Sony, AT&T, Merck, Rockwell, State Street Bank, and Microsoft have found effective ways to do this.

CORE INTELLECTUAL COMPETENCIES

The crux of leveraging intellect is to focus one's own resources on those things—important to customers—where the company can create uniquely high value for its customers. Conceptually, this means disaggregating both corporate staff activities and the value chain into manageable intellectual clusters. On close examination, these turn out to be "intellectually-based service activities" (see Figure 34.2). Such activities can either be performed internally or outsourced depending on one's own relative costs and competencies. For maximum effectiveness, a company should concentrate its own resources and executive time on those few activities where it performs at "best-in-world" levels. These are typically clusters of intellectually based skills, knowledge bases, or systems—not products and physical assets—that enable the firm to pro-

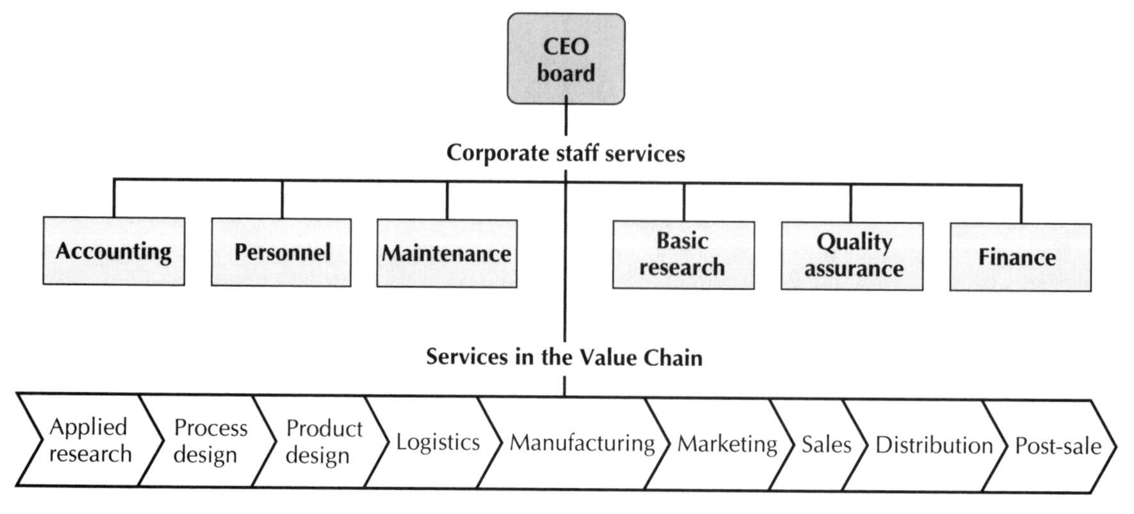

Fig. 34.2. Your core service competency.

duce higher value outputs per unit cost than its competitors. By developing these unique core competencies in depth as a secure "strategic block" between its suppliers and its customers, the company can more aggressively outsource or enter alliances with the world's most effective external suppliers to leverage its resources in other areas.

External Leveraging

Through this process managers can substantially decrease investments and risks, focus their own time and resources, improve the intellectual resources at their disposal, flatten their organizations, enhance their strategic flexibility, innovativeness, and response times—and serve customers better. Many entrepreneurial ventures like Apple Computer, Silicon Graphics, Nike, or Novellus have started in this highly concentrated, but heavily outsourced fashion, leveraging their fiscal capital by factors of three or more—and their intellectual capital by tens to hundreds—as compared to integrated companies. If a company is not "best in world" at an activity (including all transaction costs) it gives up competitive edge by performing that activity in-house. Today, most venture capitalists realize they can only obtain bank returns on capital invested in fixed assets (unless these embody the company's core competency). They seek to invest in and leverage the unique intellectual resources of the company and not to undertake the investments and risks of activities others could perform better. Larger companies—from Continental Bank to Wal-Mart, MCI, Honda, and Boeing—operate in this fashion too.

- When interviewed, MCI had only 1,000 full-time technical people internally, but had 20,000 professionals working for it full time on vendor premises. MCI not only did not have to invest in facilities, overhead, and benefits costs for these people on a permanent basis, "many outside professionals were higher quality than those MCI could hire internally." These specialized suppliers could afford to invest in more elaborate facilities and support systems to attract top talent to work with their firms on the frontiers of their particular specialties. In addition to many software, construction, and system maintenance activities outsourcing, MCI also actively sought out and exploited the innovations of thousands of small firms with new hardware or services to attach to its core software and electronic hardware or "pipeline" capabilities, thus leveraging its own innovative capabilities by hundreds of times.

Once strategists define each activity in their value chain or staff services as "intellectually based services," opportunities abound for a company to restructure its strategy, internal activities, and external coalitions for greater fiscal and intellectual leverage. Generally, there are many external providers who, by specializing, can provide greater depth in that service and produce it with greater quality or lower cost than any internal department could. No company can hope to be better than all outside specialists in all the elements of its value chain. And new technologies and management systems have dramatically shifted the balance between what it pays to outsource and what the company can effectively produce internally.

In today's hypercompetitive climate, "core competency with outsourcing" strategies allow companies to leverage their intellectual assets to the greatest possible extent. They can simultaneously be the lowest cost, broadest line, most flexible, and most highly differentiated producer in their markets—yet have the needed depth in selected skills to be strong alliance partners and to move rapidly in emerging new markets.[4] No other strategy supports efficiency (through focus), innovative flexibility (through multiple sourcing), and stability (through market diversity) to the same extent. Many enterprises in financial services (State Street Boston or Continental Bank), retailing (L. L. Bean or Toys "R" Us), or communications, entertainment, and lodging services, (Paramount, Turner, or Marriott) operate in this fashion. So do industrial giants like the oil majors, Boeing, Sony, or 3M. "Integrated" oil companies often outsource many intellectually based service elements in their value chains to great advantage. (See Figure 34.3). Similarly,

- Boeing outsources many parts of its commercial airliners to those who have greater skills in specialized areas. Internally it produces mainly those portions of the craft that contain the critical flight control and power plant interfaces. It concentrates its own intellectual capabilities on understanding aircraft technologies and its customers' needs and focuses its operations on the design, logistics, and flexible assembly processes necessary to coordinate and control the quality and performance of the aircraft.
- 3M's extensive growth has rested on its R&D skills in four related "historic technologies": abrasives, adhesives, coating-bondings, and nonwoven technologies. In each it has developed

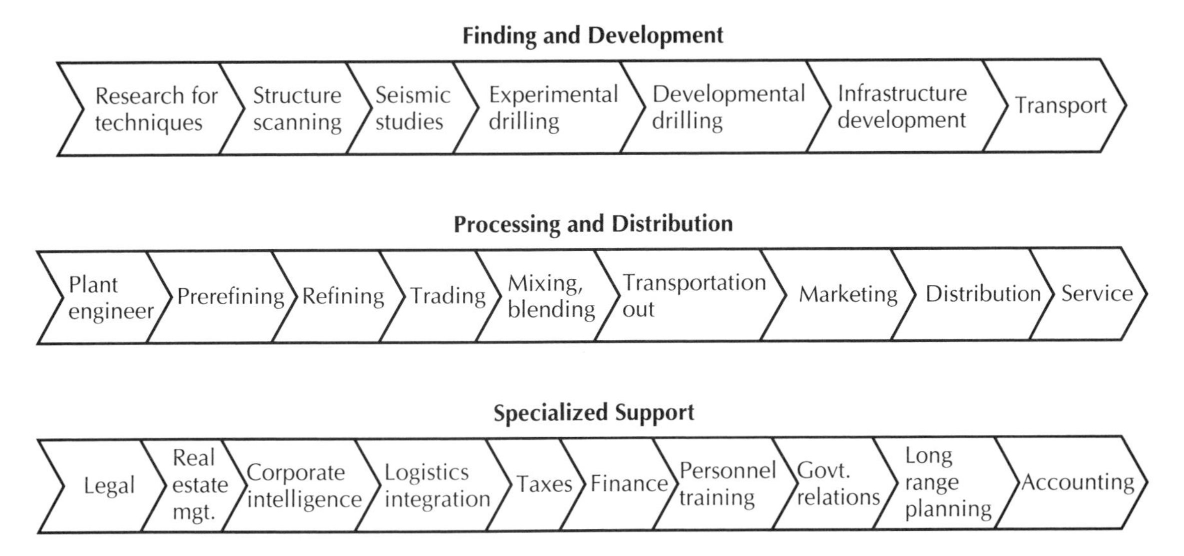

Fig. 34.3. "Integrated" oil companies.

knowledge bases and skill depths exceeding those of its major competitors. When combined with two other core intellectual competencies—its remarkable "entrepreneurial-innovation system" and its strong, broad-based marketing distribution system—these "historic technologies" have allowed 3M to create and support over 50,000 products for a variety of markets, to maintain a flexible innovative culture, and to sustain continuous growth for six to seven decades at compounded rates of over 10 percent.

ORGANIZING AROUND INTELLECT

Exploiting these new intellectually based strategies often calls for new organization concepts. In the past—to enhance efficiencies—most companies formed their organizations around dominating process investments, product clusters, geographic needs, or specialized management functions. These clearly optimized the capacity of power holders to direct and control the organization. However, the highly varying demands of customers, the unwillingness of professionals to work in such hierarchical situations, the rigidity and lack of responsiveness of these organizations, and the extended capabilities new technologies present for managing highly disaggregated organizations enable—and require—entirely new structures. The term "network organizations" has been widely—and confusingly—used to embrace a variety of these new forms varying from extremely flat, to "horizontal matrix," to alliance, to cross-disciplinary team, to holding-companies structures that merely finance a number of unrelated divisions self-coordinating on an ad-hoc basis.[5] This overly broad categorization reveals little about how the various forms differ, when to use them, or how to manage them for maximum effect.

The main function of organization in today's hypercompetitive environment is to develop and deploy—i.e., attract, harness, leverage, and disseminate—intellect effectively. Each of these organization forms does this in its own way (see Table 34.2), and should be used only for those particular purposes it handles best. Not only is no one form a panacea, many different structures can be used to advantage within the same company. Because they are useful for certain purposes, hierarchies will doubtless continue in many situations. But we expect much greater use of four other basic organizational forms that leverage professional intellect uniquely well. These are the infinitely flat, inverted, starburst, and spider's web forms. Table 34.1 summarizes the primary differences in these organization forms and their utility in leveraging intellect. The key variables in choosing among these new forms are:

- *Locus of intellect:* Where deep knowledge of the firm's particular core competencies primarily lies.
- *Locus of customization:* Where intellect is converted to novel solutions.
- *Direction of intellectual flow:* The primary direction(s) in which value-added knowledge flows.

TABLE 34.1. Outline of Four Forms of Organizing

	Infinitely Flat	Inverted	Starburst	Spider's Web
Definition of node	Individual	Individual	Business units	Individual
Locus of intellect	Center	Nodes	Center and nodes	Nodes
Locus of customization	Nodes	Nodes	Center and nodes	Project
Direction of flow	Center to nodes	Nodes to center	Center to nodes	Node to node
Method of leverage	Multiplicative	Distributive	Additive	Exponential
Example	Brokerage firms, aircraft operations	Hospitals, construction engineering	Major movie studios, mutual fund groups	Internet, SABRE

- *Method of leverage:* How the organization leverages intellect.

All the forms tend to push responsibility outward to the point at which the company contacts the customer. All tend to flatten the organization and remove layers of hierarchy. All seek faster, more responsive action to deal with the customization and personalization an affluent and complex marketplace demands. All require breaking away from traditional thinking about: lines of command, one person–one boss structures, the center as a directing force, and management of physical assets as the key to success. But each differs substantially in its purposes and management. And each requires very different nurturing, balancing, technologies, and support systems to achieve its performance goals.[6]

"Infinitely Flat" Organizations

In infinitely flat organizations—so called because there is no inherent limit to their span—the primary locus of intellect is at the center, e.g., the operations knowledge of a fast-food franchising organization or the huge body of data and analysis capability at the center of a brokerage firm. The nodes of customer contact are the locus of customization. Intellect flows primarily one way, from the center to the nodes. The leverage is arithmetic—the amount of leverage equals the value of the knowledge times the number of nodes using it. Single centers in such organizations presently can coordinate anywhere from 20 to 18,000 individual nodes (see Figure 34.4). Common examples include highly dispersed fast-food, brokerage, shipping, mail order, or airline flight operations.

Several other characteristics are also important. The nodes themselves rarely communicate with each other, operating quite independently. The center rarely needs to give direct orders to the line organization. Instead, it is primarily an information source, a communications coordinator, and a reference desk for unusual inquiries. Lower organizational levels generally connect into the center to obtain information to improve their performances,

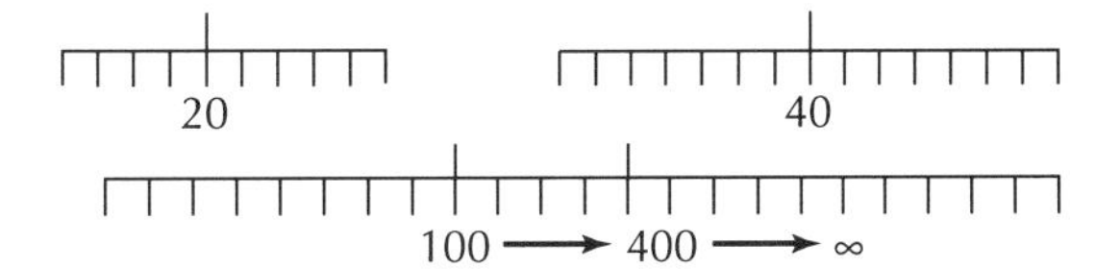

Fig. 34.4. Infinitely flat organizations.

rather than for instructions or specific guidance. Most operating rules are programmed into the system and changed automatically by software. Many operations may even be monitored electronically. These organizations are essentially single-layer hierarchical structures where electronics has replaced human command-control procedures. They are neither "horizontal" nor "networks" in a true sense.

- For example, Merrill Lynch's 480 domestic brokerage offices each connect directly into the parent's central information office to satisfy the bulk of their information and analytic needs. Although regional marketing structures exist, business is conducted as if each of Merrill Lynch's 17,000 + branch-office contact people reported directly to headquarters, with their only personal oversight being at the local level. Technology permits the overall company to capture data and develop information with the full power and scale economies available only to a major enterprise. Yet local brokers manage their own small units and accounts as independently as if they alone provided the total service on a local basis.

Infinitely flat organizations operate best when the activity at the node can be broken down and measured to the level of its minimum repeatable transaction elements (as for example, the cooking and operating details in fast-food chains of the basic components of financial transactions in brokerage operations). Response times can be nearly instantaneous, and the organizations can support any desired degree of centralized or decentralized autonomy. Control can be exercised at the most detailed level—yet, if desired, systems can eliminate most of the routine in jobs, free up employees for more personalized or skilled work, and allow tasks to be very decentralized, individually challenging, and rewarding. Under proper circumstances, the electronics systems of such organizations capture the aggregate experience curve of the entire enterprise, allowing less trained people quickly to achieve performance levels ordinarily associated with much more experienced personnel. They also allow market knowledge to be captured, mixed, and matched at the greatest possible level of detail to enable identification and exploitation of new opportunities. Well-designed systems simultaneously offer highest responsiveness, maximum levels of efficiency, and greatest consistency in quality. Such has been their effect in firms like Fidelity Securities, Federal Express, Benneton's, Merrill Lynch, or Domino's Pizza.

Infinitely flat organizations present certain inherent management problems. Lower level personnel wonder how to advance in a career path when there is no "up." Those at the center require totally different skills from those at the nodes. Traditional job evaluation systems break down, and new compensation systems based on professional capability, individual performance, and customer satisfaction become imperative. Reward systems need to include a great variety of titles, intangible performance measurements and rewards, and constant training and updating by advisory teams from the central office. There is a tendency for systems to rigidify with time, if companies continue use of the same measurement and control systems. Consequently, external scanning systems, constant customer sampling, and personal observation systems are essential to supplement the structured "hard information" linkages of these very flat organizations. But the essence of their management is capturing, analyzing, and disseminating the most detailed possible level of customer-relevant information from the center to the contact nodes.

The Inverted Organization

In the inverted form, the major locus of *both* corporate intellect and customization is at the nodes contacting customers, not at the center. Hospitals or medical clinics, therapeutic care-giving units, or consulting-engineering firms provide typical examples. The nodes tend to be highly professional and self-sufficient. Accordingly, there is no need for direct linkage between the nodes. When critical knowledge about operations diffuses, it usually does so informally from node to node—or formally from node to center—the opposite of the infinitely flat organization. The leverage of this form is "distributive," i.e., the support organization efficiently distributes logistics, analysis, or administrative support to the nodes. But it does not give orders to them.

In inverted organizations, the former line hierarchy becomes a "support" structure, only intervening in extreme emergencies—as might the CEO of a hospital or the chief pilot of an airline (see Figure 34.5). The function of "line" managers becomes bottleneck breaking, culture development, communication of values, developing special studies, consulting on request, expediting resource movements, and providing service economies of scale. Hierarchy may exist within some groups because members of this "support" structure must ensure consistency in the firm's application of their specialized knowledge (for example, about government regulations or accounting rules). Generally, however, what was "line" (order giving) management now performs essentially "staff" (analytical or support) activities.

• A well-known example of an inverted organization was SAS after Jan Carlzon became CEO. He utilized the concept of inverting to empower SAS's contact people in ground and flight operations to bypass its heavily entrenched bureaucracies. Another is Nova Care, the largest provider of rehabilitation care in the United States and one of the fastest growing health care compa-

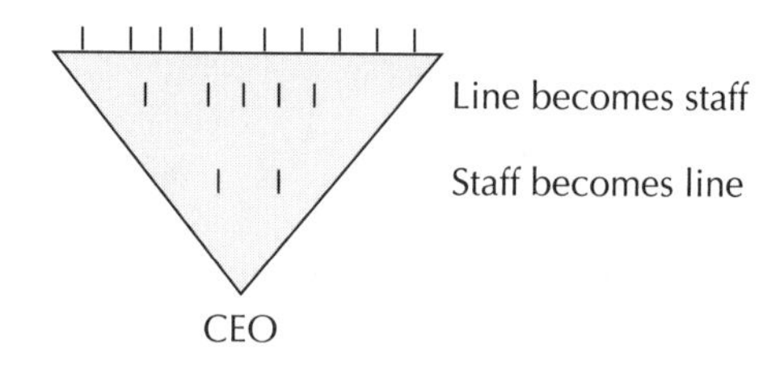

Fig. 34.5. The inverted organization.

nies of the last decade. With its central resource—well-trained physical, occupational, and speech therapists in short supply—NovaCare provides the business infrastructure for over 4,000 therapists, arranging and merging contracts with nursing homes and chains, handling accounting and credit activities, providing training updates, and stabilizing and enhancing therapists' earnings. However, the key to performance is the therapists' knowledge and their capacity to deliver this individually to patients.

The inverted organization works well when: (1) servicing the customer at the point of contact is the most important activity in the enterprise; *and* (2) the person at the point of contact has more information about the individual customer's problem and its potential solutions than anyone else. Experience suggests that because they present unique problems, inverted organizations should be used sparingly, and not as gimmicks to improve perceived empowerment. While seeming to diminish line authority, line members' roles often increase in importance as they perform more influential activities—such as strategic analysis, resource building, or public policy participation—once freed from their traditional information-passing routines.

The inverted organization poses certain unique challenges. The apparent loss of formal authority can be very traumatic for former "line managers." Given acknowledged formal power, contact people may tend to act ever more like specialists with strictly "professional" outlooks, and to resist any set of organization rules or business norms. Given their predilections, contact people often don't stay current with details about the firm's own complex internal systems. And their empowerment without adequate information and controls (embedded in the firm's technology systems) can be extremely dangerous. A classic example is the rapid decline of People Express, which enjoyed highly empowered and motivated point people, but lacked the systems or computer infrastructures to enable them to self-coordinate as the organization grew.

A frequent cause of failure in this and the infinitely flat nodes is inadequate segmentation and dissemination of information into detailed enough elements to monitor and support individuals' actions at the nodes. Many such organizations also fail because they attempt to transition from their more traditional hierarchies to the new forms without thoroughly overhauling their measurement and support systems. In our recent study of over 100 such situations in major service organizations, less than twenty percent had changed their performance measurement systems significantly; and only about 5 percent had yet changed their reward systems. The complications were predictable. People continued to perform against traditional norms and with bureaucratic slowness.

The Starburst

Another highly leverageable form, the starburst, serves well when there is highly specialized and valuable intellect at *both* the nodes and the center. Starbursts are common in creative organizations that constantly peel off more permanent, but separate, units like shooting stars from their core competencies (see Figure 34.6). These spin-offs remain partially or wholly owned by the parent, usually can raise external resources independently, and are controlled primarily by market mechanisms. Some common examples of this organizing form and its different internal relationships include: movie studios, mutual fund groups or venture cap-

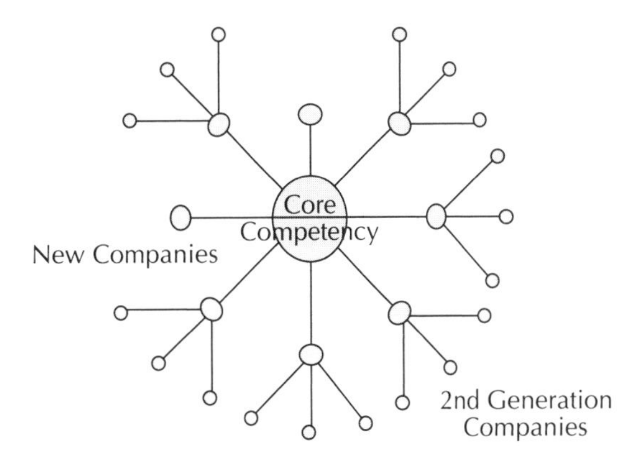

Fig. 34.6. The starburst organization.

italists (among service firms), and Thermoelectron, TCG, or Nypro (among industrials).

In operation, the center retains deep knowledge of some common knowledge base (e.g., specialized plastic molding technology for Nypro, managing no-load funds for Vanguard, or risk-taking and resource-assembly skills for movie studies). Unlike holding companies, starbursts contain some central core of intellectual competency. They are not merely banks or portfolio managers. The nodes—essentially separate permanent business units, not individuals or temporary clusters—have continuing relationships with given marketplaces and are the locus of important, specialized, market, or production knowledge. The nodes may in time spin out further enterprises from their core. The flow of intellect is typically from the center toward the outer nodes. The organization rarely transfers knowledge from one node laterally to another. But nodes do feed back to the core-specialized information that others nodes may find useful (without direct competition with the source's marketplace). The nodes both multiply the number of outlets using the firm's core competency and leverage it: (1) through their access to specialized market expertise; and (2) through the independent relationships and external financing they can generate as separated entities.

Starburst organizations work well when the core embodies an expensive or complex set of competencies and houses a few knowledgeable risk takers who realize they cannot micromanage the diverse entities at the nodes. They work well in very ambiguous environments where it is difficult to estimate outcomes without undertaking a specific market test. Usually they occur in environments where entrepreneurship—not merely flexible response—is critical. To be effective the nodes require the economies of scale and new opportunity spin-offs that only a large, specialized, knowledgeable base can provide. The center usually maintains and renews its capacities to develop these opportunities by charging the market units a fee or taking a share of their equity. In addition to maintaining the core competency, the corporate center generally manages the culture, sets broad priorities, selects key people, and raises resources more efficiently than could the nodes. Unlike conglomerates, starbursts maintain some cohesive, constantly renewed, and critical intellectual competencies at their center.

The classic problem of this organizational form is that managements often lose faith in their freestanding "shooting stars." After some time they try to consolidate functions in the name of efficiency or economies of scale—as some movie studios, HP, TI, and 3M did to their regret—and only recover by reversing such policies. Starbursts also encounter problems if their divisions move into heavy investment industries, or such capital-intensive mass production activities that one unit's needs can overwhelm the capacity of the core—as HP's computer division overwhelmed its Test Equipment groups. In most starburst environments, the nodes are so different that even sophisticated computer systems cannot provide or coordinate all the information needed to run these firms from the center. Rather than try, managers must either live with quasimarket control or spinoff the subsidiary entirely. Starbursts tend to work extremely well for growth by innovating smaller scale, discrete (product or service) lines positioned in diverse marketplaces.

The "Spider's Web"

The spider's web form is a true network. (The term "spider's web" avoids confusion with other "network-like" forms, particularly those that are more akin to holding company or matrix organizations.) In the spider's web there is often *no* intervening hierarchy or order-giving center among the nodes. In fact it may be hard to define where the center is. The locus of intellect is highly dispersed, residing largely at the contact nodes (as in the inverted organization). However, the point of customization is a project or problem that requires the nodes to interact intimately or to seek others who happen to have the knowledge or special capabilities that a particular problem requires.

- The purest example of a spider's web is the INTERNET, which is managed by no one. Other

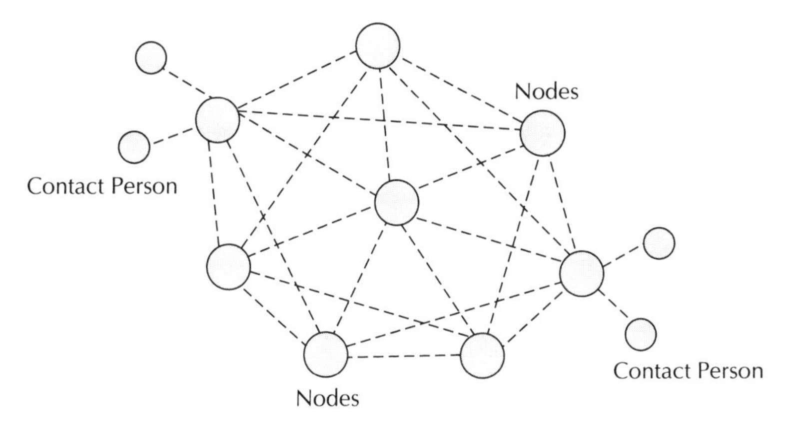

Fig. 34.7. Spider's web organizations.

common examples include most open markets, securities exchanges, library consortia, diagnostic teams, research, or political action groups.

The organization's intellect is essentially latent, underutilized until a project forces it to materialize through connections people make with each other. Information linkages are quite complex; intellect flows from many nodes to many others. Nodes typically collaborate only temporarily in delivering a specific service or solution in project form to a particular customer. The intellectual leverage is exponential—growing at some power $2^{(n-1)}$ greater than the number of nodes—as defined by interactive learning or network theories. Even with a modest number of collaborating nodes (eight to ten), the number of connections and leverage of knowledge capabilities multiplies by hundreds of times (see Figure 34.7).

Individual nodes may operate quite independently, when it is not essential to tap the knowledge of other sources to solve a problem efficiently. On a given project there may or may not be a single authority center. Often decisions will merely occur through informal processes if the parties agree. Occasionally, however, the various nodes may need to operate in such a highly coordinated fashion that they delegate temporary authority to a project leader—as when widely dispersed researchers present a contract proposal or an investment banking consortium services a multinational client.

The spider's web form has existed for centuries (among universities and scientists, or within trading groups), but these were overlooked in the mass production era that sought the greater stability, predictability, and control that military or clerical hierarchies seemed to offer. In today's highly competitive environments they offer unique advantages because they can simultaneously support high specialization, multiple geographic locations, and a disciplined focus on a single problem or customer set. They are particularly useful in situations where problem sites are very dispersed and radically different specialties need to be tapped. Often no one person in a given organization knows exactly what the problem's dimensions are, where issues may be located, or who may have potential solutions. This form of organizing releases the imaginations of many different searchers in diverse locations, multiplies the numbers of possible opportunity encounters, and encourages the formation of entirely new solutions from a variety of disciplines.

While they are usually effective for problem finding and analysis, spider's webs present important challenges when used for decision making. Dawdling is common, as nodes work on refining their specialist solutions instead of solving the complete problem together. Assigning credit for intellectual contributions is difficult, and cross-competition among nodes can inhibit the sharing on which such networks depend. Extreme overload can emerge as networks become jammed with trivia. Significant changes at both the network and local levels are usually essential to make spider's webs effective in professional situations. First among these is creating a culture for communication and willing sharing. How groups communicate and what they are willing to communicate are as important as the knowledge they have. And overcoming professionals' natural reluctance to share their most precious asset, knowledge, usually presents some common, and difficult, challenges. One enterprise exemplifies many patterns:

• Arthur Andersen & Co. (AA&Co) found that its AANET required major incentive shifts. Profit sharing had to be shifted away from local profit centers toward more partnership-wide sharing. To encourage use of the system, a person's participation on the Net had to be considered in all promotion and compensation reviews. Initially AA&Co. spent large sums to encourage people to follow up on network exchanges and to meet personally on important problems. To stimulate wider use of the system, senior partners purposely posed questions into employees' e-mail files each morning "to be answered by 10 o'clock" and followed up on queries in their own units. Until AA&Co. made these and other supporting structural changes to achieve desired cultural modifications, AANET—despite its technological elegance—was less than successful.

Because spider's webs are so dependent on individual goals and behavior, there is no "best way" to manage them except to stimulate a sense of interdependency and identity with the problem at hand. Since each network is unique, no one management approach will work for all forms. Shared interest of participants, shared value systems, and mutual personal gains for members are, of course, the essential starting points for any network relationship. However, research suggests effective network managers generally: (1) force team overlaps to increase learning and shared information; (2) purposely keep hierarchical relations ill-defined; (3) constantly update and reinforce project goals; (4) avoid overelaborate rules for allocating profits to individual nodes; (5) develop continuous mechanisms for updating information about the external environment (e.g., tax code changes, customer needs, or scientific results); (6) involve both clients and peers in performance evaluations; and (7) provide node members with personal and team rewards for participation. These kinds of active management interactions are usually needed to avoid the most common failures and frustrations. The other key leverage factor, technology, requires special attention.

INTEGRATING DIVERSE FORMS

Each organization form performs best for a specific set of intellectual tasks. Hence most large enterprises will require a mixture of these basic building blocks, combined with more traditional hierarchical structures. The critical nonorganizational element in integrating and leveraging these diverse forms within the same firm is a software framework that can mix, match, and manipulate "the smallest replicable units" of tasks, customized data, or transactions across all units. What does this involve?

The Smallest Replicable Level

In earlier years, the smallest manageable unit for organizations and data seemed to be at individual store, office, or franchise level. Later, as volume increased, it often became possible for a parent company to manage and measure critical performance variables at individual departmental, sales counter, activity, or stock-keeping-unit (SKU) levels. Then the successful formula approaches of McDonald's, Pizza Hut, H & R Block, and the insurance companies pushed the repeatability unit to even smaller "micromanagement levels" such as cooking processes, detailed work schedules, maintenance cycles, accounting transactions, or contract phrases. Proper management around these permitted franchises, agents, and headquarters to guarantee quality and desired service levels continuously and to make corrections within minutes when something went wrong.

TABLE 34.2. How Different Organizing Forms Develop Intellect

Type of Intellect	Infinitely Flat	Inverted	Starburst	Spider's Web
Cognitive (Know-what)	Deep knowledge and information at center	Primary intellect at nodes, support services from center	Depth at center (technical) and (markets) at the nodes	Dispersed, brought together for projects
Advanced Skill (Know-how)	Programmed into systems	Professionalized skills informally transferred node to node	Transferred from center to node, then node to node via the core	Latent until a project assembles a skill collection
Systems Knowledge (Know-why)	Systems experts at the center. Customer knowledge at the nodes	Systems and customer expertise at the nodes	Split: between central technical competency at the core, systematic market knowledge at nodes	Discovered in interaction or created via search enabled by the network
Motivated Creativity (Care-why)	Frees employees from routine for more skilled work	Great professional autonomy	Entrepreneurial incentives	Personal interest, leveraged through active interdependence stimulation

Now in some industries—like banking, publishing, communications, structural design, entertainment, or medical research—it has become possible to disaggregate the critical units of service production into digitized sequences, electronic packets, data blocks, or "bytes" of information that can be endlessly combined or manipulated for new effects and to satisfy individual customer and operating needs. These microunits permit the highest possible degrees of segmentation, strategic fine-tuning, value-added, and customer satisfaction—at lowest cost—in the marketplace. The larger the organization, the more refined are these replicability units, and the greater their leverage for creating value-added.

In keeping with the exponentiality of knowledge, systems that capture such data from the outset can build up a detailed information base that later provides an insurmountable competitive edge. By constantly updating, analyzing, and sharing data patterns from all its detailed sources, successful IT systems automatically capture and disseminate the highest level of experience available within the enterprise. They also provide the basis for laterally integrating the wide range of different organization forms now available. For example:

- American Airlines' (AA) well-known SABRE and associated operating systems interlink to provide consistent data for AA's very different reservation service (spider's web) flight operations (infinitely flat), financial controls (conservative bureaucratic), ground operations and maintenance (decentralized bureaucratic), personnel training (specialized functional), and other operating modes. The organization structure, operating style, and culture within each unit can remain unique and appropriate to its tasks. Yet the software system ensures coordination among them, minimizes costs, and ensures that desired service levels and consistency are delivered to customers.
- Similarly NovaCare (described above) keeps track of all its 4,000 therapists' activities in 15 minute "units" of detail. These provide the basis for scheduling, compensation, billing, and follow-up on all therapies. They enable NovaCare to ensure that all its customers (patients, nursing homes, hospitals, hospital directors, doctors, nursing directors, payors, and regulating bodies) are properly served, charged, and compensated. The system serves a variety of centralized functional (accounting), geographical hierarchical (hospital), inverted (therapy), and spider's web (professional

knowledge exchange) structures. Though collecting information in immense detail, the system has unburdened care-givers (who hated detailed paperwork), and freed up regional coordinators (who earlier had spent enormous time in collecting, analyzing, and relaying activity reports) for more personalized and higher value-added activities. The same detailed information about individual operations and transactions also helped to coordinate the purchasing, logistics, financial, and regulatory compliance groups which support these operations.

Far from depersonalizing operations, such measures and software provide a framework that allows professionals to behave independently, responsibly, and consistently even though they cannot understand major portions of the system outside their specialties. With the proper information at their fingertips, contact people can provide greater quality, personalization, and value-added for customers. With details handled by the electronics system, they can concentrate on the more conceptual, personalized, and human tasks that provide most satisfaction for employees and customers. Such systems are the glue that welds together highly dispersed service delivery centers and leverages the critical intellect of professional organizations.

MEASURING OUTPUTS

The same systems also serve a central role in the performance tracking and measurement systems that make any complex organization function properly. Direct sensing already measures many important aspects of intellectual and service output in real time for major communications, airline, retailing, wholesaling, banking, health care, fast-foods, electric power, etc. operations. They also handle many aspects of quality—e.g., signal quality, signal strength, power variations, fluid flows, service cycle times, error rates, delays, downtimes, credit worthiness, variability costs, inventory levels, environmental and operating conditions, "vital systems" performance in health care, and so on. Software systems, for capturing microdetails about operations, customers, and the environment are rapidly becoming the most valuable competitive tools of intellectual management.

However, effective systems for managing outputs typically must go well beyond just software. They must encompass at least four dimensions: (1) peer review of professional performance; (2) customer appraisal of outputs received; (3) business evaluation of efficiency and effectiveness; and (4) appraisal of the changes in intellectual assets created. While such measurements would appear critical, surprisingly few firms perform all four. Merck's and AT&T's approaches to measuring intellectual outputs have been widely publicized elsewhere. Without attempting a detailed treatment here, a few other examples will suggest how certain leading enterprises attack individual aspects of this problem.

- After each project, a major investment banking concern asks each team member, its customer group, and the team head to rank all important participants on the project in terms of their demonstrated professional knowledge, specific project contributions, and team support. Customers rate their overall satisfaction with the project and with the firm as a service supplier. Annual surveys, which rank the company against all competing firms' performance on twenty-eight critical dimensions, supplement this. Costs and profits are collected for each project, with the latter allocated among participating groups on a simple formula basis. Annually, for each division, the company calculates the net differential between the market value of each division (if sold) and its fixed asset base. This net "intellectual value" of the unit is tracked over time as a macromeasure of how well management is growing its intellectual assets.
- McKesson Drugs, at the strategic level, emphasizes five strategic "themes" for competitiveness: customer-supplier satisfaction, people development, market positioning, relative net delivered cost, and innovation. For each factor it uses internal and external metrics to track its position relative to competitors. It has forty-two "customer satisfactors" that it measures on a routine basis—including a seven-page questionnaire that goes to

more than 1,000 customers each year, with quarterly updates on a smaller set of factors it considers most important. Through its ECONOMOST system, which links its warehouses directly to PCs on its druggist customers' desktops, it can track all cost and measurable (delivery) quality features, like fulfillment accuracy, delivery times, or stock-outs at both customer and warehouse levels. The same system tracks detailed costs and profits by SKU, and uses market feedback data to advise customers on how they can improve their own productivity. The profits from such services—derived from ECONOMOST—and measures of McKesson's probable costs and profits without its various ECONOMOST software packages give the company a measure of this system's enormous value as intellectual property.

SUMMARY

Not only must a company's strategic, recruiting, organizing, and technological systems be designed in ways that focus and leverage its intellectual capabilities, it must also develop measurement and reward infrastructures to make sure these intentions are honored across all units. Too few firms pursue their intellectual management through all these levels. Based on careful research, we have tried to provide some practical guidelines suggesting how successful enterprises have designed their strategies, organizations, training, and measurement systems to maximize the value of their most critical asset, intellect.

Notes

1. Specific examples are expanded at length in Quinn, *Intelligent Enterprise,* New York, Free Press, 1992; National Research Council, *Technology in the Service Society,* 1993; and Quinn and Mintzberg, *The Strategy Process,* Englewood Cliffs, NJ, Prentice-Hall, 1995.

2. D. Schon, *The Reflective Practitioner,* New York, Basic Books, 1983.

3. J. B. Quinn and F. Hilmer, "Strategic Outsourcing," *Sloan Management Review,* July 1994, amplify those concepts.

4. R. Miles and C. Snow, "Organizations: New Concepts for New Forms," *California Management Review,* spring 1986.

5. J. B. Quinn, P. C. Paquette, "Technology in Services: Creating Organizational Revolutions," *Sloan Management Review,* winter 1990, first described these different organization forms.

6. H. Mintzberg, *Mintzberg on Management: Inside Our Strange World of Organizations,* New York, Free Press, 1989.

35

Engines of Progress: Designing and Running Entrepreneurial Vehicles in Established Companies; Raytheon's New Product Center, 1969–1989

ROSABETH MOSS KANTER
JEFFREY NORTH
LISA RICHARDSON
CYNTHIA INGOLS
JOSEPH ZOLNER

The problem of generating and sustaining entrepreneurship in established companies contains a major paradox: how to *routinize and manage* a process often assumed to be *spontaneous and opportunistic*. Scholars of entrepreneurship and innovation stress the emergent, chaotic, and unpredictable character of new ventures and innovation projects (e.g., Quinn 1985; Kao 1989)—termed "newstream" activities (Kanter 1989), or activities producing a new revenue flow. Such newstream chaos and uncertainty pose a sharp contrast to the more orderly management process for on-going operations in the organizational "mainstream," because of necessary differences in outlook between entepreneurs and administrators/bureaucrats as well as differences in the kinds of tasks they face (Kanter 1985; Stevenson and Gumpert 1986; Kanter 1989, pp. 201–214).

Thus, research has tended to demonstrate that entrepreneurship is difficult for established companies to tolerate, let alone manage, for more than a short period of time. Of course, established companies can organize in ways that promote more innovation in their mainstream businesses, and even occasionally spawn new businesses (Kanter 1983, Peters and Waterman 1982); and an occasional major, strategic new venture with large investment can succeed (Burgelman and Sayles 1986). But many documented efforts to systematically produce new ventures on a large scale over a long period of time, such as new venture departments, have been disappointing to their corporate sponsors and, hence, short-lived (Fast 1979, Sykes 1986).

Is it possible to *routinize the unpredictable*—that is, to design a mechanism that will reliably and continuously produce newstreams?

The Harvard Business School Research Program on Entrepreneurship in Established

Reprinted by permission of the publisher from "Engines of Progress: Designing and Running Entrepreneurial Vehicles in Established Companies" by Kanter, R., J. North, L. Richardson, C. Ingols, and J. Zolner, from *Journal of Business Venturing*, Vol. 6, pp. 145–163.

Companies undertook to address this question by examining a variety of entrepreneurial vehicles (programs driving the creation of new revenue streams through new ideas) in large North American corporations. Between 1986 and 1989, the research team studied eight corporate venturing programs in depth, in a range of industries, to compare and contrast the ways in which companies could organize to find, nurture, and use newstream activities as well as the viability of each approach. The eight sites were selected to maximize differences in strategic intent and operations (see Kanter 1989a on methodology; Kanter 1989b on theory; Kanter 1990). Further differences emerged from the dynamic nature of the vehicles themselves, which had complicated histories before the research began, changed considerably in form and intent during the study period, and were continuing to evolve as the formal research was completed. The dynamism was a result, not only of the normal flux and change in corporate life, but also of the additional problem of defining the appropriate relationship between mainstream and newstream.

A newstream program is an attempt to "bureaucratize entrepreneurship," that is, subject entrepreneurship to routines and controls, in order to capture its benefits without losing its essence. Such a program is an attempt to harness creativity and transform it into new ventures—without diluting creativity by the addition of commercialization requirements or sacrificing relevance on the altar of creativity. The effective juggling of both creativity and relevance, producing a regular flow of new revenue sources at a modest level—and doing this for 20 years—is the central theme of the Raytheon New Products Center case.

Raytheon's New Products Center (NPC) demonstrates how newstream development efforts can be effectively linked to, and contribute to the revitalization of, mainstream businesses. It also shows how the creativity and entrepreneurial spirit that can conceive of new opportunities and develop them can coexist with disciplines and controls. As Angle (1989) has observed, creativity is not necessarily innovation. The issue for organizations is not necessarily finding creative people, because the potential for creativity may be widespread in the population (e.g., Campbell 1960, Langer 1989) and may be a matter of education or situation as well as individual traits. Subject-matter expertise, social or political skill, and contacts outside the field of primary focus—as a stimulus to new thinking—have been correlated with the creativity of research scientists, for example (Amabile 1988). At the same time, however, some personal attributes often associated with creativity have led analysts to assume that creative people are "hard to manage"; for example, self-motivation or risk-orientation. And creativity has been assumed to be a spontaneous, unpredictable process not subject to the usual managerial controls.

Therefore, the major organizational issue is the *integration* of the creative process into the organizational mainstream to produce innovation. Clearly, managers influence creative tasks by defining them and setting expectations, framing the problem, and offering parameters such as deadlines, uses, etc. Michael Eisner, president of Walt Disney, linked discipline and creativity, remarking that "the only people I let roll around the floor having a tantrum are my three-year-olds" (Kao 1989, pp. 23–24). As to whether or not an organization can have creativity on demand, George Freedman, founder of the Raytheon New Products Center, told the field team that the simplest way to "manage" creativity is to provide a deadline.

The Raytheon NPC case illustrates how creativity can be routinely produced to serve corporate purposes. It also illustrates the political and organizational side of corporate entrepreneurship, because effective management of the organizational context was responsible for this newstream program's success. The Raytheon NPC counts as one of two long-lived entrepreneurial vehicles among the eight corporate newstream programs studied. As such, it has been the subject of much interest from visitors eager to start similar programs; founder Freedman wrote a book (Freedman 1989; see also Freedman 1986, Kanter 1990). But the very reasons for its longevity—modest expectations, involvement in prototype development rather than commercialization—make it seem less "entrepreneurial" to some observers.

RAYTHEON'S NPC, 1969–1989

Raytheon's NPC, formally established in 1969 as part of the Microwave and Power Tube Division, has been an independent facility since 1976. Operating on an eventual budget of approximately $3 million per year, and in 1986 moving to a newly dedicated facility in a Burlington, MA technology park, the New Products Center assisted Raytheon companies in domestic, commercial, and industrial markets, but primarily in the major appliance area, to develop new products, product extensions, and new lines of business. The NPC worked on as many as 60 projects at one time, and a critical part of the director's job was assigning priorities to projects and ensuring that their resources were being effectively deployed. The NPC did not charge its "clients" (the Raytheon product divisions) for its services; this fostered greater acceptance and use of the facility. Some issues and problems were routed to the NPC from the head office; NPC administrators developed a network of contacts throughout the company which allowed them quickly to direct the problems to the appropriate problem solver.

The NPC was purposely kept small. The stated NPC philosophy was that many of the best innovations come from constant interaction and exchange of ideas among a small group; 50 people was thus held to be an upper limit, to avoid segmenting the staff into separate departments, which would jeopardize the creative process that results from interaction among different disciplines (Kanter 1983). Between 1969 and 1989, over $50 million had been invested in the NPC. During that time, in excess of 50 products developed at the NPC were accepted by the clients; estimates of revenues generated by these projects were in the range of several hundred million dollars per year. In addition, at least one stand-alone business was developed by NPC staff, the industrial micro-waving unit.

While the NPC did research, it was a business generator rather than an R&D lab. Raytheon divisions each retained their own captive R&D organizations. The NPC served mainly the commercial divisions, but not in the same way as the traditional R&D did. The NPC stressed development—innovation in the interests of new products and business lines. Thus, the staff demonstrated competence in market assessment and manufacturing process as well as the technologies needed for new products.

THE RAYTHEON COMPANY

Founded in Cambridge, MA in 1922, by 1989 Raytheon was a diversified, high-technology company ranked among *Fortune*'s 100 largest industrial corporations. Before a major diversification program that began in 1964, Raytheon was primarily a producer of transmitting tubes, radar systems, and missile guidance systems for the United States government. In 1964, when sales amounted to $454 million, 83% of this total was accounted for by government contracts. By 1985, when sales had grown to $6.4 billion, government business accounted for only half of this amount.

Raytheon's primary businesses involved electronics. Whereas the relative importance of military contracts had declined by 1989, Raytheon remained the Department of Defense's ninth largest prime contractor. Raytheon's other business areas included Aircraft Products, Major Appliances, and Energy Services (see Exhibit 1). In 1989, Raytheon employed more than 77,600 people in 80 facilities in 26 states, and another 24 facilities in Canada, Europe, and Japan. With 31,000 employees in Massachusetts alone, Raytheon ranked as the largest industrial employer in the state.

Raytheon was a research-intensive organization with a history of innovation. Since 1977, annual expenses for research and development fluctuated between 3 and 4% of sales. Major Raytheon innovations included:

- A rectifier tube that changed the shape and size of home radios

Exhibit 1. Raytheon's Business Segments

ELECTRONICS

Largest business area: missile systems, surveillance radars, air-traffic control systems, military communications equipment, electronic and microwave components; semiconductor devices and other components; commercial products, including medical and marine equipment.

AIRCRAFT PRODUCTS

Beech Aircraft Corporation, a leading supplier of aircraft for general aviation: single- and twin-engine piston and turboprops, and business jets for corporations, commuter airlines, government and military organizations, private pilots; and missile targets for the government.

MAJOR APPLIANCES

Amana Refrigerators and *Caloric Corporation* (including *Modern Maid, Glenwood,* and *Sunray* brands) full-line appliance companies. *Speed Queen Company:* laundry equipment for home and commercial use.

ENERGY SERVICES

United Engineers and Constructors: design, construction, and maintenance of electric power and industrial plants. *Seismograph Service Corporation:* geophysical exploration service for oil and gas. *The Badger Company:* design and construction of petroleum, petrochemicals, and chemical processing plants.

OTHER LINES

D.C. Heath and Company: educational publishing, including *Caedmon* spoken-word recordings and *Arabesque* classical recordings. *Cedarapids:* equipment for the road-building and construction industry. *Raytheon Service Company:* technical service and logistic support for government and commercial customers.

Source: *Raytheon 1986 Annual Report.*

- A mass-production breakthrough on magnetron power tubes, which are the heart of radar systems
- The first guided missile to intercept and knock down an aircraft in flight
- The first interception and destruction of a ballistic missile by a guided missile
- The world's first electronic depth sounder
- The first application of microwave energy to the heating of food and the innovation and commercial introduction of microwave ovens.

The first microwave oven for home use, trademarked the Radarange, was introduced by Raytheon's Amana division in 1967. Amana, along with Speed Queen and Caloric, formed Raytheon's Major Appliance group. The group was the fifth largest major appliance company in the United States, with 1986 sales nearing $900 million. The most recent thrust of the group was to manufacture products, particularly for microwave ovens, for the mid-priced marked segment, for both first-time and replacement buyers.

Throughout the 1980s, all of the appliance group units looked for new markets and new lines of business. Amana, for example, added dishwashers as a step in offering a complete appliance line; the group was then able to bid on construction contracts that called for complete appliance packages. Speed Queen introduced the "no-coin" alternative to coin-operated washers and dryers. A dormitory at Bentley College in Waltham, MA was the site of a pilot project to test an apparatus that allows the customer to use a laundry card (similar to an ATM card) rather than coins. Although its original strength was in gas ranges, Caloric introduced the first electric range featuring Ultra-Ray broiler technology, something that has been available on gas ranges for some time.

The NPC was a vital link in the new business development process. It contributed to developing the laundry card and the technology for electric ranges, and it was instrumental in Caloric's entry into the field of microwave cooking when it developed industry-leading combination ovens, which added microwave-cooking capability to conventional stoves. In addition, the NPC received funding from the Laser Products division, a non-consumer unit, to support development work on lasers. Raytheon's industrial microwave operation was a new business unit formed and staffed by NPC entrepreneurs, who left the NPC to run the business after they started it.

EVOLUTION OF RAYTHEON'S NPC

In the mid-1960s, George Freedman, a materials engineer with Raytheon, and a group of his colleagues, asked the company to allow them to form a separate new products group.

One impetus was shared aptitude for and love of creating new products; another might have been a movement among some peace-oriented engineers to resist working on war-related products. The engineers were often frustrated when too often they were reduced to "bootlegging"—working on new product projects on the sly, using funds appropriated from officially sanctioned projects. Because that money was, by definition, so scarce and piecemeal, the bootleg projects were carefully and efficiently managed (Freedman 1988). Whereas Freedman believed that bootlegging was an inappropriate practice for the development of new product ideas at Raytheon, the Raytheon culture made it hard for innovations to emerge outside of the rigid R&D structures already in place. Earlier in his 44-year career at Raytheon, Freedman had left to start his own venture. "That experience rendered me poor rather than rich (the exact opposite of my original intention)," he later wrote, "but I don't regret it because it enriched my level of knowledge in the field of new ventures. . ." (Freedman 1988, p. 14). Thus, seeking to found a new products center was an outlet for Freedman's entrepreneurial desires, but in a protected corporate environment.

It took three years of persuasion and lobbying before the NPC was established. Four factors led to the ultimate decision by Raytheon to support an NPC:

1. Over a period of several years, Freedman had become friendly with people throughout Raytheon who had problems in the materials area. He soon had a portfolio of ideas and problems that could begin to justify the ongoing support of an innovation unit by Raytheon.
2. The microwave oven, which was introduced in 1967, was a huge success that proved the value of innovation and creativity. Such a visible success made Freedman's search for a corporate "champion" much simpler.
3. Freedman found a champion in Palmer Derby, a former MIT classmate who had risen to one of Raytheon's 25 vice-presidencies and was known as a frustrated inventor.
4. Freedman threatened to leave Raytheon.

Even these conditions were not sufficient to translate support into an actual center. Feeling that he had reached an impasse, Freedman contacted a friend in the public relations department, who showed him some products that had been successfully developed, and suggested that perhaps this would make an interesting news story. His colleague agreed. When the news broke in the newspapers, it appeared that the NPC had been established before it had actually been approved. Immediately the materials group began to receive calls from inside and outside of the company. After the story had been on television, Raytheon's stock rose by four points, an event that countered senior management annoyance about the inappropriate news release. Derby, then Assistant General Manager of Raytheon's Microwave and Power

Tube Division, approved the NPC. Derby and Freedman had to act quickly, before senior management support waned, so they rented two "portable office" trailers to house the NPC.

The NPC began as an applied research group tied to a single business. It started under the aegis of the Microwave and Power Tube Division because of the hope that many innovations could be created from the microwave technology that Raytheon had pioneered, and, probably more important, because of Derby's sponsorship. The NPC's stay in the Power and Tube Division was short-lived, however, because of technological limitations. The NPC attempted to provide new sockets for microwave tubes; it did not produce a cooking tube to compete with Japan. Instead, the NPC started looking at different kinds of equipment, including heat-transfer units, sand molds, commercial pass-through cookers, and technology for tempering frozen beef—none of them tubes. Eventually, in 1976, the NPC became part of the corporate office. Still, the early show-of-faith sponsorship was essential to getting started, just as the research literature has shown (e.g., Souder 1981), because it permitted development of a track record to attract other support.

The NPC occupied its trailers for nearly a year, then moving to a small, shabby building in Waltham, MA, filled with gutted washing machines, sophisticated computer systems, and a microwave oven that doubled as a tool cabinet to save space. Freedman strongly believed that a development group needed the autonomy of its own space (Freedman 1988; see Kanter 1989, on newstream pressures for autonomy). There was no real plan, but since the company expected the NPC to be self-supporting, Freedman's "portfolio of problems" soon came into play. For the first seven years, the NPC led a hand-to-mouth existence by providing engineering support that would *occasionally* lead to the development of new products. The ratio of engineering support to new product development was about 5 to 1 until about 1976. The ratio reversed when the NPC moved to its corporate base and became a resource to Amana, Speed Queen, and Caloric (which a decade later accounted for nearly $1 billion of the corporation's $7 billion in sales).

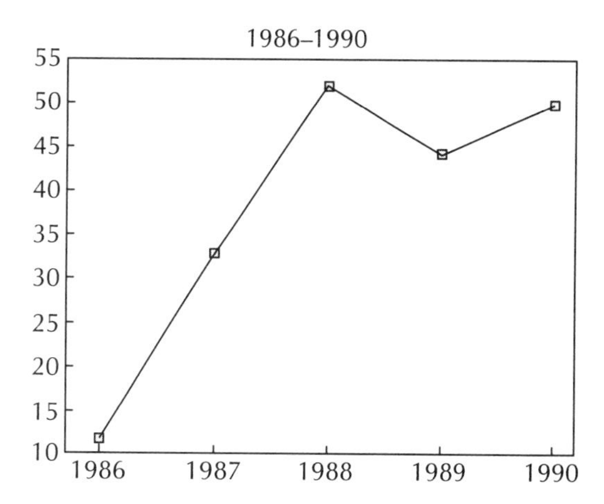

Fig. 35.1. Number of patents, NPC engineers, 1986–1990.

Raytheon's insistence that the NPC be self-supporting was complicated by the unwillingness of managers to risk their budgets on an untested group of innovators. Engineers preferred to solve their own problems, so their acceptance of the NPC's services was additionally hindered by resistance on the part of mainstream business to outside influences, to anything "not invented here." Until Raytheon's top management was convinced by its track record to furnish the NPC with its own budget, seven years after the NPC was founded, Freedman found himself looking for outside funding, such as grants from trade associations, to support the NPC.

Staffing the NPC at the outset was not difficult. Freedman had instinctively hired people like himself—innovators and mavericks—so he already had five or six people in place when the NPC was created. Subsequently, staffing consisted of "picking up some orphans": finding the "parents" of ideas that had been shelved for one reason or another. This is similar to one of the approaches adopted at General Electric Medical Systems Group, where innovations were based on adoption of a task suggested or tried by an earlier manager who had been blocked, had lost interest, or had been transferred (Kanter 1983).

By 1989, the NPC operated as a discretionary expense center on a $3.1 million budget. The director of the NPC reported to the head of strategic

Exhibit 2. A Tour of the New Products Center, 1988

In the heart of an industrial park in Burlington, MA, stood a single-story 28,000-square-foot building dedicated to innovation and creativity. The opening of the new facility in 1986 to house Raytheon's NPC marked both the culmination of NPC founder George Freedman's 44-year career with Raytheon, and his 18 years nurturing and championing the NPC. Since its creation in 1969, the NPC moved from a pair of trailers to modest office and lab space in Waltham, MA in 1970, and finally to the new Burlington facility in 1986, a building specifically designed to house the NPC.

The NPC was part laboratory and part corporate showcase. Engineers and secretaries alike were accustomed to looking up from their work and seeing a delegation from a Raytheon division or customer group touring the facility. In the lobby, mauve couches line two walls of the reception area; in front of them sat a large crimson table and coordinating floral arrangement. The walls, carpets, and reception desk were all coordinating shades of gray. On the left, in an office where meetings with clients were held, the walls were covered with framed advertisements of successful products that originated within the NPC. Leading back from the lobby were a string of offices, all containing stylish black furniture: the offices were both functional and fashionable. Computer workstations sat on the desks; purchased and leased artwork graced the walls; and all offices had glass interior walls.

A broad doorway from the administration area led into the main conference room, which doubled as a functional work area and a shrine to the NPC's successes. Two alcoves abutted the main conference area. The first of these contained plaques commemorating each of the patents earned by NPC engineers. The other contained special industry awards that the NPC won over the years. The room was equipped with a projection room and a long table seating 14. The room connected with the NPC's test kitchen, and acted as a corridor between the administration area and the engineering labs.

Beyond the administration area lay the engineers' offices, which were a series of large cubicles in a sunny and spacious part of the building. Each cubicle had a different size and shape. The

planning and met annually with the president of Raytheon to present the NPC's budget proposal for the upcoming year. Most of its funding came from corporate sources, but it also received support from some divisions (e.g., Laser Products) for specific development work, and the center raised funds through revenues for licensing.

The NPC developed for 10 years without significant change. In 1984, Freedman began talking about his retirement (which took place in February 1987). In the summer of 1986, the subject of a change of management was raised for the first time. The NPC also moved from Waltham to a new facility in Burlington in this period (see Exhibit 2), and a group executive-vice-president was appointed to coordinate the efforts of Raytheon's appliance companies and develop strategies to increase their competitiveness.

In late 1986, Robert Bowen became the director of the NPC. He had been with Raytheon for over 30 years and had been part of the NPC from the beginning, acting as manager of technology transfer under Freedman. Immediately following Freedman's retirement, Bowen and Wes Teich took on joint responsibility for the NPC with another old hand, who soon retired, leaving Bowen in charge. Like Freedman, Robert Bowen was a would-be entrepreneur who stayed at Raytheon only because of the NPC. Bowen emphasized communication and mainstream integration. He saw his job as making sure that the vice president of Strategic Planning had no surprises.

Under Bowen, interaction with Raytheon businesses increased; for example, he sent staff into the divisions on "sales" calls so that they could gain familiarity with the mainstream lines of business.

engineers were given small budgets with which to decorate their own spaces, making each office a custom-designed workspace. Rooms were not numbered at the NPC's Burlington facility; the 35 employees move about easily and the individual touches adequately identify workspaces. The flow of people throughout the building suggested that doors were merely a means of ensuring quiet during meetings.

From the engineers' area, the tour led to the technicians' bench area. The bench area is both spacious and well-organized, even though the NPC worked on as many as 60 projects at a time. Tour leaders make a particular point of introducing visitors to the technicians, who explain their work on a particular project. The technicians have smaller cubicles, less luxurious than the engineers', but comfortable places nonetheless for them to retreat for small meetings or individual work.

Within the ring created by the administration area, engineers' offices, and bench area were a variety of larger rooms where larger projects were underway. These range from a project to develop improved insulation for refrigerators, to a more intensive x-ray, to a piece of emulsifying equipment being prepared for a local teaching hospital. Another room contained a complete machine shop with metal, plastic, and woodworking capability, where, according to George Freedman, "anything can be made." Adjacent to it was a room that accommodated the drafters. There was also a company library, which had the latest issues of *Good Housekeeping, Adhesives Age,* and at least 60 other trade related periodicals. There was also a terminal designed for access to a technical data bank.

The tour ended in a large well-appointed test kitchen that led into the main conference room. A wide variety of experimental appliances were installed as part of a fully operational kitchen. Designers' sketches on the wall illustrated the aesthetic commitment the NPC offered to its clients. Designers developed the sketches, which showed how new appliances can be integrated into a typical kitchen. The NPC also employed a home economist who interacted with clients and engineers to develop new inventions, and tested many of the smaller products, such as microwave cookware and packaging, in the lab that looked like a kitchen.

Organization

During the study period, the NPC had three loosely defined departments and about 35 people. The first was the new concepts group, considered the "real" inventors and experts in their fields. The second group was composed of product engineers, who were also very innovative, and who turned the concepts into physical models that the clients could see and touch. Like the inventors, they tended to be generalists, but they were more adept at interacting with the people at the client's company. The third group was the specialized microelectronics and computer group. Because almost all appliances used controls that were increasingly computer-driven, the NPC maintained its own computer specialists. Indeed, new products sometimes came from redesigning old appliances with modern controls, like substituting plastic cards with magnetic strips for coin mechanisms in washers and dryers.

Everyone at NPC working in an engineering capacity met formally twice a month (informal interaction occurred continuously) to review the progress made on various projects and to prioritize work depending on client needs and the probability for success. At any one time, the NPC would be working on over 60 new product programs at various stages. By the end of 1989, the NPC had successfully shepherded over 100 new products through the early development phase; they went into production at a rate of approximately two per year, becoming parts of the product lines of eight of Raytheon's businesses.

The ideas for development programs came from three sources:

- About 30% from a request from a client. The client would have done a lot of preliminary work and would have a strong concept of what was being asked for. These requests enjoyed virtually total success (in terms of being developed further by the client).
- About 25% from more general requests for technical help. These requests, in other industries, might be handled by a technology center.
- About 45% of product ideas were internally generated. As long as the allocation of funds to "pure" innovation was not unreasonable, it was permitted. About one-third of the product ideas generated in this manner were accepted by a client.

Progressing from idea creation to a prototype was an *iterative process,* with constant involvement from the client marketing and manufacturing people. Marketing people determined whether there would be a market for the new product with expected opportunities, whereas manufacturing concerns revolved around the ability to make the product profitably with given constraints. It was often said by NPC staff that making a product was the difficult part—innovation came easily. The NPC's task was selling the idea to the client, convincing the client to take the risk, and then getting both manufacturing and marketing managers on board—"to get them to adopt another child."

Equally difficult was maintaining a balance between the "creators" who loved to tinker endlessly with a concept, and the "doers," who were motivated by taking a concept from the conceptual stage to the point where it could be turned over to the client. The popular phrase, "to shoot the engineer," described the occasional need to pry a project away from its creator in order to expedite development rather than perfecting the minute details. The NPC's responsibility usually ended once a prototype had been delivered to the client. It was then the client's responsibility to manufacture and market the product. This practice of transfer and hands-off after the development stage was designed to meet the different management and investment requirements at later points in the entrepreneurial process, similar to the way in which IBM reintegrated developments from its Independent Business units into mainstream divisions (Alterowitz 1989). Because of the close working relationship with the mainstream throughout the process, hands-off was much smoother than in situations in which the corporate entrepreneurs were isolated (Kanter 1988).

A key to NPC's success in converting ideas to products was getting the clients to believe that they invented the new product. Ideas generated solely within the NPC were less likely to succeed than those with a champion for a new product concept in the mainstream client unit. If a product was based on a perception of need identified solely within the NPC, then NPC managers had to devote a great deal of effort to selling the idea. Nevertheless, the NPC still had to expend a portion of its effort in selling ideas to the clients.

There was wide agreement among NPC staff and the strategic planning office to which it reported about time allocation at the NPC during the study period:

- 80% on only what the NPC understood the clients wanted them to do.
- 15% responding to the "fire drills": because of NPC specialties and technical knowledge they could help clients solve ongoing problems, particularly in production. This was not part of the NPC charter but it made political and organizational sense for the NPC to do it.
- 5% for ideas generated at the NPC. NPC managers had to approve of the idea—the clients had not identified the need but there was the outside chance that something would come along that would interest the client.

This last category of time allocation was justified by NPC's immersion in both the marketplace and the technologies, as described below. Sometimes, the appliance companies would not have identified a need because they were unaware that one existed or that a practical solution to a problem could be developed.

Exhibit 3. How a New Products Center Should Operate

1. THE GOAL

There can be only one goal: to aid the company's growth and profit. It does this by developing new products for selected profit centers within the company. These new products are worthless if they do not generate increased revenues and profits—no matter how brilliant the concepts involved. Thus, success is not based on published papers, patents, or just work that was "fun." The criterion of success can only be the number of new products that enter the marketplace successfully. If they do not enter the marketplace and make money, they do not succeed.

2. THE INGREDIENTS

The first required ingredient for a new products center is the right people; the second is the right environment. There should be creative engineers whose talents for invention are reinforced by an atmosphere designed for that purpose (as a new products center is).

These people should include generalists who are capable of and stimulated by working in a variety of fields simultaneously. Raytheon's NPC generalists were trained as mechanical and electrical engineers, physicists, materials engineers, and the like, but they cross each other's boundaries with ease.

Then there are the specialists, people who confront unique problems as only specialists can. At the NPC, there are computer experts who design the electronic controls for the products that the generalists develop. Another specialist group is the materials engineers. Everything we sell is made of materials, except software: if the product fails in the field it is usually because materials have fatigued, fractured, or corroded. Too many companies neglect materials innovation as a way to achieve radically new products. At Raytheon, we have tried to rectify that oversight—and it has stood us in good stead.

Finally, there is the product engineering group. Their task and specialty is product transfer to the client after successful models have been demonstrated. They understand tooling, manufacturing, and costs much better than the new concepts group does. Furthermore, client manufacturing managers prefer to work with clients.

3. RELATIONSHIPS WITH CLIENTS

This is the make-or-break issue. Success can come only with the establishment of effective channels of communication, leading to an accumulation by the NPC staff of understanding of all elements of client businesses. It also requires mutual trust. This can take years to achieve, and once done, it may have to be done all over again when personnel change. The point is, the new products center must become part of each client's team—as though the NPC staff were employees of the client, with the same loyalties.

Since every relationship is based on a different culture, a new products center must give much of its energy to developing sympathy with and sensitivity to them. Only then will there be real agreement on which market area to challenge, which product to develop, the scale of the company's investment, the role of each person in each group, the credit to be assigned to individuals, or the fanfare and publicity that will attend to that credit.

Source: *George Freedman,* Management Review, *December 1986.*

Sources of new product ideas were diverse. Some came from the military side of Raytheon, which sought commercial applications for new technologies. Other ideas came from listening to television ads, reading trade magazines, going to trade shows, or simply creative genius. Brainstorming was another idea source. Outside experts in relevant fields were on call. In addition to consulting with mainstream businesses to determine their specific needs, NPC engineers and technicians kept in close contact with their counterparts in the client companies in order to keep up with their technological advances. NPC staff were expected to understand the entire business development spectrum from market sensitivity to the business needs of the receiving units. While "autonomous" in one sense, they were far from isolated. Staff traveled widely, visitors to the NPC were common, and at one point the NPC communicated with the receiving businesses by frequent videotapes.

Managing and Motivating Creative People

Freedman was considered the quintessential representative of the NPC, serving as director through 1986. He often said that the best inventors are totally without principle because they steal ideas wherever they see them; his willingness to break corporate rules to get results was a demonstration of this.

Freedman was known to be good at picking the "right" people (important in entrepreneurial projects)—(Alterowitz 1989), selecting for enthusiasm for doing new things as well as a sense of humor. But, as important as technical competence was, good communication skills were even more essential for the "salesmanship" and inter-Raytheon diplomacy that the NPC process required. The NPC's 35 engineers, technicians, and administrative staff members were thus a small, self-selected group, especially when compared with the 77,000 employed by Raytheon as a whole.

The NPC enjoyed a very low turnover rate: of the six people who started the Center in 1969, for example, four remained in 1989. Of those employees who departed over the years, the most notable was the big group that developed the industrial microwave business unit out of NPC roots. No one had ever left for a rival company. A few people had left to start their own businesses, a few retired, but for the most part people stayed at the NPC.

What was behind this success at retaining people? The answer to this question may lie in the make-up of the employees. They were variously described as people who are nonconformists, generalists, very versatile, who did not like to work normal hours or follow set procedures, and were often "difficult to manage."

The NPC's management philosophy, however, succeeded in producing creativity on demand. There was an emphasis on giving staff every possible ego boost. The directors fought for financial recognitions in the form of bonuses and stock options, but these were token awards (e.g., a $500 patent bonus) compared with the lucrative financial deals other companies were pressed to offer their internal entrepreneurs. NPC people were advertised outside the NPC as bright "hotshot" types. Publicity was used often to boost people's reputations—a key motivational tool (Kanter 1989c). Articles about NPC products and patent awards were posted throughout the NPC building.

The NPC represented for many an escape from the more bureaucratic structures in Raytheon's research and development facilities. At the NPC, the engineers worked on up to four or five projects at once, usually rotating among them at will. Specific definition of tasks and deadlines made it possible to produce innovation reliably, though creative breakthroughs were still serendipitous. When scheduling became tight, it was occasionally necessary to drag an engineer away from a project because s/he was obsessed with perfecting the device assigned to another one. A mild renegade spirit existed at the NPC, a spirit that sometimes did not fit precisely with the Raytheon system.

TYPES OF NEWSTREAM SUCCESSES

There were three principal ways that the NPC contributed to new ventures and new revenue streams for Raytheon. Each involved a different kind of relationship between the NPC and the rest of the corporation.

New Products to Serve Mainstream Client Needs

Most of the NPC's activity fell into this category. The following example is typical. A vice president of Amana came to the NPC with a specific request. Amana was the largest producer of microwave ovens, but competition was increasing. The vice president wanted the NPC to create a free item to go with its ovens, replacing a small trivet then used as a give-away. NPC engineers worked on and developed a coffee brewer, but its impact on sales was not impressive. They switched their efforts to designing a microwave popcorn maker. Designing a popper that would heat every kernel evenly was a formidable task. It took the NPC 10 months and 12 separate patents to perfect it. Raytheon had licensed six other companies to produce it. By the time the promotion was over, Amana had given away more than a million popcorn makers and microwave sales had increased 10%.

NPC-Generated New Ventures

A second type of newstream success was much less common but had larger dollar impact: the development of a new business out of ideas generated by the NPC. Industrial microwaving was the best example.

The development of the microwave oven was closely linked to the workings of the NPC. Microwave technology for military applications of radar was pioneered at Raytheon in the 1940s and '50s. Later, much of this research was funded by NASA. Raytheon and NASA scientists worked toward the development of microwave-powered helicopters and trains, linear accelerators, and platforms in space that would convert solar energy to microwaves and beam the energy back to earth.

Because of the NPC, Raytheon became the dominant force in the market for large industrial microwaves. The business started after an NPC member was visiting a Hormel meat processing plant and noticed a room full of 100-pound boxes of frozen meat. The meat had been shipped frozen from Australia and had to thaw before it could be processed into hamburger patties. Not only did the process take five days, but the floors were covered with blood. The 100-pound block of meat became 97 pounds, valuable protein was lost through blood loss, and expensive pollution controls had to be installed to deal with the blood. Furthermore, a potentially unhealthy condition was created as the meat hung and thawed.

The NPC engineer immediately saw that all the costs and problems associated with thawing 100-pound blocks of meat could be solved with microwave technology. A group of engineers went on to develop large microwave ovens for the meat-packing industry that could thaw those same blocks of meat in 30 minutes instead of five days with zero blood-drip loss. Eventually, large ovens were in virtually every meat processing plant in the country. Other uses for industrial microwave include pre-cooking bacon for fast-food restaurants, and curing rubber products like huge tires for earth-moving machines. From the small group of engineers who were originally part of the NPC, the Industrial Products Group, which makes these microwave products, became a separate operation accounting for $10 million in annual sales by 1988 and employing 50 people.

Technology Leverage and Cross-Fertilization

The NPC also contributed to Raytheon in the related areas of technical leverage and licensing revenues. Some of the NPC's technical products had significant leverage effects on other company businesses regardless of modest dollar volume. For example, a new magnet material developed by the NPC was first announced in 1970 as "the world's most powerful magnet." Although it never exceeded $1 million in annual sales, it gave Raytheon an advantage over its competitors when it was first incorporated into traveling-wave tubes. Eventually, put into an electronic component in missiles, the magnets had a major impact on the company's largest segment, its missile systems business (Freedman 1986).

Other products that were developed at the NPC but rejected by the client, or which otherwise did not fit into the company's businesses, were licensed for use outside the company. Originally, the NPC was told by superiors not to get into licensing. Yet the NPC did so anyway, "working undercover for three years," as a staff member put it, until eventually it was discovered and officially approved because there was so much money coming in. By 1988, annual licensing revenues amounted to about $1 million (the equivalent of the pre-tax profits of a typical Raytheon $10 million business). The NPC generally assigned the rights to the licensing revenues to the mainstream organization whose projects stimulated the development of the product.

In addition to its role in developing newstreams for mainstream businesses, the NPC also facilitated internal technology transfer. This was possible because of its independent role, straddling or bridging several businesses. For example, NPC received funding from the Laser division to develop several products that allowed the group to achieve major competitive advances. The NPC could subsequently apply what was learned about laser technology to work being done at Amana, Caloric, or Speed Queen, where familiarity with laser technology was absent. It was widely believed that the innovation process could be stimulated by seemingly random connections between projects that might become important breakthroughs (Kanter 1983, 1988).

The process also worked in reverse. For example, when the NPC was working to reduce the level of vibration in a refrigerator, they called in assistance from engineers in another division who worked on vibration problems in a missile system. If outside expertise was needed, there were 50 Raytheon plants within a half-hour drive of Boston. Thus, a group of 35 people could draw on thousands of experts and make connections across businesses.

The NPC's View of Factors in Its Success

Whether the product was one that has been specifically requested by a client or developed from within the NPC, the NPC staff believed that certain factors made it more likely that the product would be successfully developed (Freedman 1986, 1988).

- *A "receiving" mainstream champion is needed.* Someone in the client organization who enthusiastically supports the development and devotes a significant amount of time selling it within his or her own organization is invaluable. If a Raytheon division isn't interested in the new product or doesn't want to manufacture it, the idea will probably be dropped.
- *The client should become involved in the development process at an early stage.* If the division's manufacturing people can have input toward the design, and their marketing people can give advice on the aesthetics before the product get too far along, they are much more likely to support its development.
- *The product should become tangible as early as possible.* The people at the client organization want to be able to see and touch the product in the prototype stage. If the computer control that runs the appliance has not yet been incorporated into the product itself, then the NPC people know that it should be hidden under the table during the product demonstration so the client does not get the incorrect impression that the new product comes with a bulky control box sitting on top of it. "You have to do a lot of that with clients," one project manager observed. "Give them a prototype which looks pretty much like it would in the real world, otherwise they get hung up on the cosmetic details and don't see what the new product concept is really about."
- *The product must be priced right.* When tinkering with new technology and product ideas, it is easy to get carried away. Because the NPC supports the consumer appliance businesses at Raytheon, the technology and the new ideas have to be affordable.
- *The clients should be made to believe that they invented the product.*
- *The client's design engineers and R&D people who have the biggest "not invented here" problems should be kept out of the picture.*

RESULTS

The more than 50 successful new products and businesses developed over a 15-year period resulted in over $300 million in incremental sales and licensing revenues during the 1980s—all for an initial investment of under $20 million. Despite the high percentage return, however, in absolute terms the NPC was only a very small part of Raytheon. In a company the size of Raytheon, a facility with a $3.1 million budget and fewer than 40 employees could be easy to miss. Nevertheless, the NPC tried to maintain a high profile. The NPC's existence was justified by the flow of products through its doors. The threat lay in the danger of being perceived to be in competition with the traditional R&D structures, even though NPC administrators have successfully positioned the Center as a supplement to, not a replacement for, "traditional" R&D channels. On the other hand, since the NPC was funded by the corporation and not the individual divisions, and since it was so small, it was less likely that its budget would be cut during business downturns. Thus, the NPC deliberately stayed small.

The large size of Raytheon made it easy to insulate the NPC from the rest of the company when appropriate. In the area of human resource management, the environment of the NPC made it an attractive workplace for individuals who might otherwise chafe under the more bureaucratic structures in the rest of the company, particularly the military contracting side. Whereas the NPC was subject to the reward structure of the entire company, NPC leaders fought successfully to secure compensation and bonuses for the staff.

The funding structure was particularly critical to the NPC success. The NPC was part of an entrepreneurial process with minimal risk. First, the investment in the NPC itself was exceedingly modest. Then, once products were designed, subsequent financial risk was mitigated by leaving the development decisions in the hands of the client companies. The NPC was neither budgeted for nor had to concern itself with the funds necessary to take a newly developed product from the laboratory to the marketplace. Whereas NPC personnel were consulted during the process that followed the departure of the product from its auspices, additional responsibility was assumed by the mainstream businesses. This clear delineation of development and production permitted the NPC to concern itself with purely developmental issues. Since the client companies were involved with the process throughout and made the final go/no-go decisions, the success rate of development ideas was boosted.

As in any innovation process, there were failures in the NPC portfolio. Some ideas may be killed before they have reached the production stage. The most common causes of idea death were:

- *A product that would be too expensive to produce,* such as a clothes dryer that saved energy by dehumidifying clothes in an air-conditioning cycle rather than by heating them.
- *A technical imperfection,* such as a heating unit built to fit inside a fireplace, which fell apart after a year because of a corrosive gas it gave off internally.
- *The departure of a key player,* particularly the champion, from a client organization.

Other failures stemmed from unexpected events and environmental shifts occurring in the middle of the development process (see Kanter 1988, for a discussion of the vulnerability of "middle" period). Although not a failure in technical terms, the NPC's cook-by-weight microwave oven, an effort to create "one-touch cooking," was perceived as a failure because it performed below expectations. Items to be cooked were placed on a built-in scale, and a computer control would determine the proper cooking time automatically. The idea had come from the interaction between the NPC and the Industrial Microwave Division; to heat large blocks of frozen meat required cooking times based on weight. In a meeting to discuss industrial microwaves, someone asked whether or not the same concept could be applied to home use if a scale could be built into home microwave ovens. The idea was a good one, but the timing was bad. Designing the scale and computer controls took longer than expected. One of the reasons for the delay was an accident that occurred when the oven and its then-separate computer control was being field tested at the home of an engineer. The engineer's cat apparently liked the heat that the computer control generated, so it spent hours lying on top of it. One day the cat got violently ill—into the computer—and destroyed several months' worth of work and components costing $64,000. Even if the cat had not delayed the project, the market timing simply would not have worked out. Amana was introducing this expensive, top-of-the-line oven just as the Japanese and Koreans were hitting the market with low-priced models. The opportunity was lost, and instead of becoming a big seller, the cook-by-weight oven was only a marginal success, with sales of only about 1500 units a year.

Overall, development successes far outnumbered failures. The NPC had an enviable track record as an entrepreneurial vehicle.

THE NEW PRODUCTS CENTER AS AN ENTREPRENEURIAL VEHICLE: ISSUES IN PERSPECTIVE

The NPC was successful at managing many of the paradoxes of entrepreneurship in established companies. The NPC had many of the elements often found to be associated with the creativity and idea-generation phases of innovation (Amabile 1988; Kanter 1988) along with excellent links to the mainstream businesses to ensure further investment in the NPC's creative outputs. The NPC was small enough to allow technical specialists to exchange ideas freely and to stimulate each other's creativity. Its staff was loosely managed and encouraged to work on several projects at a time, allowing discoveries and insights from one project to be applied to other projects. As a corporate entity that was centrally funded, the NPC had the independence to work in the most promising areas and to work through the appliance divisions, getting to know their businesses, their markets, and their resources. Not being part of any one division, the NPC also brought a fresh point of view to any project while having immunity from the daily problems of that division. At the same time, it maintained close working relationships with the established businesses, thus supporting their needs.

Because the NPC's responsibility ended when it handed over a working prototype to a division, its staff had the relative luxury of being the blue-sky conceptualizers without the responsibility for actually manufacturing and marketing the product—and without threatening the incumbents in the mainstream businesses.

It is a truism of entrepreneurship research that entrepreneurship requires "patient money." So does an entrepreneurship *program*. By beginning modestly and peripherally, with low expectations and shoestring funding but a protective sponsor, the NPC could take the time to experiment with its own organization and operations. It would wait to become visible until it found the appropriate balance between independence and integration and developed a track record. The track record of success would then create legitimacy and widespread support, reducing its dependence on a single sponsor. Whereas much of the popularized literature on innovation has stressed the single sponsor to back up a single product champion, innovation efforts are more likely to be sustained when they have multiple sponsors at many levels of management (Schroeder et al. 1989, pp. 129–130).

Whereas the program was too peripheral to be a threat or to face unmeetable expectations for performance, the long period of development of the NPC concept allowed it to gradually gain acceptance by the mainstream. In contrast, some of the new venture or innovation groups in other companies began with great fanfare and were cancelled within a *shorter* time frame than it took for the NPC to find its ultimate niche. The Eastman Kodak program, for example, promised much greater payoff to the company than the Raytheon NPC, as it was undertaken on a much larger scale and attempted to build stand-alone new businesses; but it also sank faster, with greater losses. Other programs hope to be role models for "transferring the corporate culture" but then forget to serve the businesses, generating political battles, jealousies, and tensions that result in program elimination.

The Raytheon NPC began as a "skunkworks" (Peters and Waterman 1982) in an almost literal sense, as it was founded by a maverick with bootlegged funds and staffed with people who could not fit into the mainstream company. But from the beginning the NPC signaled its intention to be useful to the mainstream, not opposed to it. By concentrating on new opportunities for existing businesses and the use of latent Raytheon technology and expertise, the NPC avoided the diversification trap—a tendency to equate business revitalization (and thus acquisition and entrepreneurship) with moving *away* from the company's areas of competence (Biggadike 1979).

The NPC was small in size but not self-contained. It was clearly *not* a "reservation" that had to exist on isolation in order to remain creative (Galbraith 1982); its creativity derived not only from staffing and internal management of NPC activities, but also from its very linkages with the Raytheon mainstream. It drew technological resources from all over Raytheon, including R&D groups in client organizations and hundreds of engineers on large defense projects. It stayed closely linked with the mainstream operations that would receive and exploit its new ventures, referring to them as "clients." It applied technology from one area to another, helping to build synergies by serving as a bridge between the businesses it served.

Overall, it was too small to be a threat, and too integrated to become peripheral or irrelevant.

The Raytheon NPC, in short, had modest goals: new products for existing businesses. It did not seek to generate enormous financial returns, to transform the corporate culture, or to "save" mature businesses by finding large new markets. But the new products it developed have become successful lines equivalent to new ventures, spawning at least one stand-alone new business unit, and the licensing revenues it earns are substantial.

Thus, the paradox of entrepreneurship in established companies *can* be resolved. Innovation can be routinized. Mainstream and newstream can be linked. Indeed, one can conclude from the Raytheon NPC case that the success of a new venture program depends on just such linking. If successful entrepreneurship is a matter of "coordinating independence" (Kao 1989), then successful corporate entrepreneurship derives from stressing the coordination even more than the independence.

NOTE

1. This report covers one of eight programs examined by Rosabeth Moss Kanter's research team in 1986–1989. Joseph Zolner, Cynthia Ingols, Lisa Richardson, and Jeffrey North contributed to the case report, and Gina Quinn helped edit it. This report is based on interviews in 1986, 1987, 1988, and 1989 and a review of company documents and published materials. Thanks are due to the Harvard Business School Division of Research for research support.

REFERENCES

Alterowitz, R. 1988. *New Corporate Ventures: How to Make Them Work.* New York: John Wiley and Sons.

Amabile, T.M. 1988. A model of creativity and innovation in organizations. In B.M. Staw and L.L. Cummings, eds., *Research in Organizational Behavior,* Vol. 10. Greenwich, CT: JAI Press.

Angle, H.L. 1989. Psychology and organizational innovation. In A.H. Van de Ven, H.L. Angle, and M.S. Poole, eds., *Research on the Management of Innovation: The Minnesota Studies.* New York: Ballinger, pp. 135–170.

Biggadike, R. May–June 1979. The risky business of diversification. *Harvard Business Review* 57:103–111.

Burgelman, R., and Sayles, L. 1986. *Inside Corporate Innovation.* New York: Free Press.

Campbell, D.T. 1960. Blind variation and selective retention in creative thought processes. *Psychological Review* 67:380–400.

Freedman, G. December 1986. R&D in a diverse company: Raytheon's new product center. *Management Review* 75:40–45.

Freedman, G. 1988. *The Pursuit of Innovation.* New York: AMACOM.

Galbraith, J. Summer 1982. Designing the innovating organization. *Organizational Dynamics* 10:5–25.

Kanter, R.M. 1983. *The Change Masters.* New York: Simon and Schuster.

Kanter, R.M. 1985. Supporting innovation and venture development in established companies. *Journal of Business Venturing* 1(1):47–60.

Kanter, R.M. 1988. When a thousand flowers bloom: Structural, collective, and social conditions for innovation in organization. *Research in Organizational Behavior* 10:169–211.

Kanter, R.M. 1989a. Three tiers for innovative research. *Communication Research* 15:509–523.

Kanter, R.M. 1989b. *When Giants Learn to Dance.* New York: Simon and Schuster.

Kanter, R.M. November/December 1989c. The new managerial work. *The Harvard Business Review* 67:85–92.

Kanter, R.M. 1990. *Rosabeth Moss Kanter on Synergies, Alliances, and New Venture.* Harvard Business School Video Series. Boston: Harvard Business School and Nathan-Tyler Productions.

Kao, J.J. 1989. *Entrepreneurship, Creativity, and Organization.* Englewood Cliffs, NJ: Prentice-Hall.

Langer, E. 1989. *Mindfulness.* Reading, MA: Addison-Wesley.

Peters, T., and Waterman, R. 1982. *In Search of Excellence.* New York: Harper and Row.

Quinn, J.B. May–June 1985. Managing Innovation: Controlled Chaos. *Harvard Business Review* 63:73–84.

Schroeder, R.G., Van de Ven, A., Scudder, G.D., and Polley, D. 1989. The development of innovation ideas. In A.H. Van de Ven, et al., same as above, pp. 107–134.

Souder, W.E. May 1981. Encouraging entrepreneurship in the large corporations. *Research Management* 24:18–22.

Stevenson, H., and Gumpert, D. 1986. The Heart of Entrepreneurship. *Harvard Business Review* 64:84–94.

36

Entering New Businesses: Selecting Strategies for Success

EDWARD B. ROBERTS
CHARLES A. BERRY

Entry into new product-markets, which represents diversification for the existing firm, may provide an important source of future growth and profitability. Typically, such new businesses are initiated with low market share in high growth markets and require large cash inflows to finance growth. In addition, many new product-market entries fail, draining additional cash resources and incurring high opportunity costs to the firm. Two basic strategic questions are thus posed: (1) Which product-markets should a corporation enter? and (2) How should the company enter these product-markets to avoid failure and maximize gain?

Although these questions are fundamentally different, they should not be answered independently of one another. Entering a new business may be achieved by a variety of mechanisms, such as internal development, acquisition, joint ventures, and minority investments of venture capital. As Roberts indicates, each of these mechanisms makes different demands upon the corporation.[1] Some, such as internal development, require a high level of commitment and involvement. Others, such as venture capital investment, require much lower levels of involvement. What are the relative benefits and costs of each of these entry mechanisms? When should each be used?

This article attempts to analyze and answer these questions, first by proposing a framework for considering entry issues, second by a review of relevant literature, third by application of this literature to the creation of a matrix that suggests optimum entry strategies, and finally by a test of the matrix through a case analysis of business development decisions by a successful diversified corporation.

ENTRY STRATEGY: A NEW SELECTION FRAMEWORK

New business development may address new markets, new products, or both. In addition, these new areas may be ones that are familiar or unfamiliar to a company. Let us first define "newness" and "familiarity":

Newness of a Technology or Service: the degree to which that technology or service has not formerly been embodied within the products of the company.

Reprinted from "Entering New Businesses: Selecting Strategies for Success" by Roberts, E. and C. Berry, *Sloan Management Review* (Spring 1985), pp. 3–17, by permission of the publisher.

Newness of a Market: the degree to which the products of the company have not formerly been targeted at that particular market.

Familiarity with a Technology: the degree to which knowledge of the technology exists within the company, but is not necessarily embodied in the products.

Familiarity with a Market: the degree to which the characteristics and business patterns of a market are understood within the company, but not necessarily as a result of participation in the market.

If the businesses in which a company presently competes are its *base* businesses, then market factors associated with the new business area may be characterized as *base, new familiar,* or *new unfamiliar.* Here, "market factors" refers not only to particular characteristics of the market and the participating competitors, but also includes the appropriate pattern of doing business that may lead to competitive advantage. Two alternative patterns are performance/premium price and lowest cost producer. Similarly, the technologies or service embodied in the product for the new business area may be characterized on the same basis. Figure 36.1 illustrates some tests that may be used to distinguish between "base" and "new" areas. Table 36.1 lists questions that may be used to distinguish between familiar and unfamiliar technologies. (Equivalent tests may be applied to services.) Questions to distinguish between familiar and unfamiliar markets are given in Table 36.2.

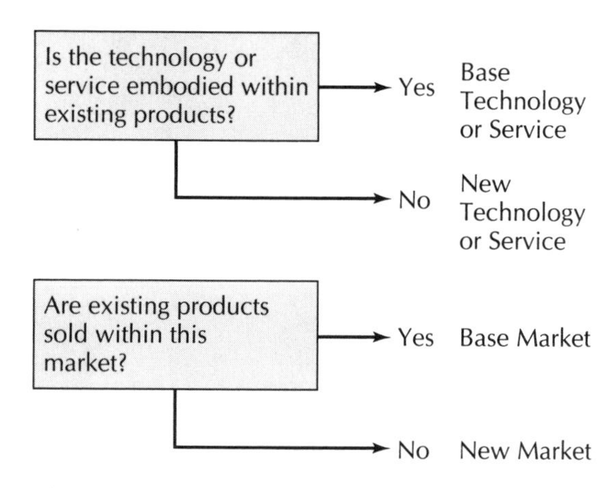

Fig. 36.1. Tests of newness.

The application of these tests to any new business development opportunity enables it to be located conceptually on a 3 × 3 technology/market

TABLE 36.1. Tests of Technological Familiarity

	Decreasing Familiarity
1. Is the technological capability used within the corporation without being embodied in products, e.g., required for component manufacture (incorporated in processes rather than products)?	↓
2. Do the main features of the new technology relate to or overlap with existing corporate technological skills or knowledge, e.g., coating of optical lenses and aluminizing semiconductor substrates?	
3. Do technological skills or knowledge exist within the corporation without being embodied in products or processes, e.g., at a central R&D facility?	
4. Has the technology been systematically monitored from within the corporation in anticipation of future utilization, e.g., by a technology assessment group?	
5. Is relevant and reliable advice available from external consultants?	

TABLE 36.2. Tests of Market Familiarity

	Decreasing Familiarity
1. Do the main features of the new market relate to or overlap existing product markets, e.g., base and new products are both consumer products?	↓
2. Does the company presently participate in the market as a buyer (relevant to backward integration strategies)?	
3. Has the market been monitored systematically from within the corporation with a view to future entry?	
4. Does knowledge of the market exist within the corporation without direct participation in the market, e.g., as a result of previous experience of credible staff?	
5. Is relevant and reliable advice available from external consultants?	

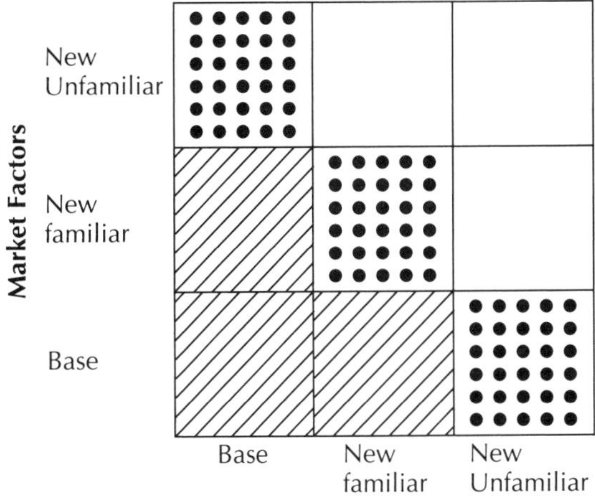

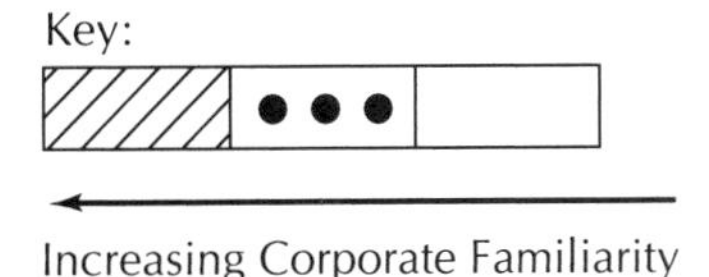

Fig. 36.2. The familiarity matrix.

familiarity matrix as illustrated in Figure 36.2. The nine sectors of this matrix may be grouped into three regions, with the three sectors that comprise any one region possessing broadly similar levels of familiarity.

LITERATURE REVIEW: ALTERNATIVE STRATEGIES

Extensive writings have focused on new business development and the various mechanisms by which it may be achieved. Much of this literature concentrates on diversification, the most demanding approach to new business development, in which both the product and market dimensions of the business area may be new to a company. Our review of the literature supports and provides details for the framework shown in Figure 36.2, finding that familiarity of a company with the technology and market being addressed is the critical variable that explains much of the success or failure in new business development approaches.

Rumelt's pioneering 1974 study of diversification analyzed company performance against a measure of the relatedness of the various businesses forming the company.[2] Rumelt identified nine types of diversified companies, clustered into three categories: dominant business companies, related business companies, and unrelated business companies. From extensive analysis Rumelt concluded that related business companies outperformed the averages on five accounting-based performance measures over the period 1949 to 1969.

Rumelt recently updated his analysis to include Fortune 500 companies' performances through 1974 and drew similar conclusions: the related constrained group of companies was the most profitable, building on single strengths or resources associated with their original businesses.[3] Rumelt, as well as Christensen and Montgomery, also found, however, that the performance in part reflected effects of concentrations in certain categories of industrial market clusters.[4] While some (e.g., Bettis and Hall)[5] have questioned Rumelt's earlier conclusions, still others (e.g., Holzmann, Copeland, and Hayya)[6] have supported the findings of lower returns by unrelated business firms and highest profitability for the related constrained group of firms.

Peters supports Rumelt's general conclusions on the superior performance of related business companies.[7] In his study of thirty-seven "well-managed" organizations, he found that they had all been able to define their strengths and build upon them: they had not moved into potentially attractive new business areas that required skills that they did not possess. In their recent book Peters and Waterman termed this "sticking to the knitting."[8]

Even in small high technology firms similar effects can be noted. Recent research by Meyer and Roberts on ten such firms revealed that the most successful firms in terms of growth had concentrated on one key technological area and introduced product enhancements related to that area.[9] In contrast, the poorest performers had tackled "unrelated" new technologies in attempts to enter new product-market areas.

TABLE 36.3. Entry Mechanisms: Advantages and Disadvantages

New Business Development Mechanisms	Major Advantages	Major Disadvantages
Internal developments	Use existing resources	Time lag to break even tends to be long (on average eight years) Unfamiliarity with new markets may lead to errors
Acquisitions	Rapid market entry	New business area may be unfamiliar to parent
Licensing	Rapid access to proven technology Reduced financial exposure	Not a substitute for internal technical competence Not proprietary technology Dependent upon licensor
Internal ventures	Use existing resources May enable a company to hold a talented entrepreneur	Mixed record of success Corporation's internal climate often unsuitable
Joint ventures or alliances	Technological/marketing unions can exploit small/large company synergies Distribute risk	Potential for conflict between partners
Venture capital and nurturing	Can provide window or new technology or market	Unlikely alone to be a major stimulus of corporate growth
Educational acquisitions	Provide window and initial staff	Higher initial financial commitment than venture capital Risk of departure of entrepreneurs

The research discussed above indicates that in order to ensure highest performance, new business development should be constrained within areas related to a company's base business—a very limiting constraint. However, no account was taken of how new businesses were in fact entered and the effect that the entry mechanism had on subsequent corporate performance. As summarized in Table 36.3, the literature identifies a wide range of approaches that are available for entering new business areas, and highlights various advantages and disadvantages.

NEW BUSINESS DEVELOPMENT MECHANISMS

INTERNAL DEVELOPMENTS

Companies have traditionally approached new business development via two routes: internal development or acquisition. Internal development exploits internal resources as a basis for establishing a business new to the company. Biggadike studied Fortune 500 companies that had used this approach in corporate diversification.[10] He found that, typically, eight years were needed to generate a positive return on investment, and performance did not match that of a mature business until a period of ten to twelve years had elapsed. However, Weis asserts that this need not be the case.[11] He compared the performance of internal corporate developments with comparable businesses newly started by individuals and found that the new independent businesses reached profitability in half the time of corporate effort—approximately four years versus eight years. Although Weiss attributes this to the more ambitious targets established by independent operations, indeed the opposite may be true. Large companies' overhead allocation charges or their attempts at large-scale entry or objectives that preclude early profitability may be more correct explanations for the delayed profitability of these ventures.

Miller indicates that forcing established attitudes and procedures upon a new business may severely handicap it, and suggests that success may not come until the technology has been adapted, new facilities established, or familiarity with the new markets developed.[12] Miller stresses this last factor as very important. Gilmore and Coddington also believe that lack of familiarity with new markets often leads to major errors.[13]

Acquisitions

In contrast to internal development, acquisition can take weeks rather than years to execute. This approach may be attractive not only because of its speed, but because it offers a much lower initial cost of entry into a new business or industry. Salter and Weinhold point out that this is particularly true if the key parameters for success in the new business field are intangibles, such as patents, product image, or R&D skills, which may be difficult to duplicate via internal developments within reasonable costs and time scales.[14]

Miller believes that a diversifying company cannot step in immediately after acquisition to manage a business it knows nothing about.[15] It must set up a communication system that will permit it to understand the new business gradually. Before this understanding develops, incompatibility may exist between the managerial judgment appropriate for the parent and that required for the new subsidiary.

Licensing

Acquiring technology through licensing represents an alternative to acquiring a complete company. J.P. Killing has pointed out that licensing avoids the risks of product development by exploiting the experience of firms who have already developed and marketed the product.[16]

Internal Ventures

Roberts indicates that many corporations are now adopting new venture strategies in order to meet ambitious plans for diversification and growth.[17] Internal ventures share some similarities with internal development, which has already been discussed. In this venture strategy, a firm attempts to enter different markets or develop substantially different products from those of its existing base business by setting up a separate entity within the existing corporate body. Overall the strategy has had a mixed record, but some companies such as 3M have exploited it in the past with considerable success. This was due in large part to their ability to harness and nurture entrepreneurial behavior within the corporation. More recently, IBM's Independent Business Units (especially its PC venture) and Du Pont's new electronic materials division demonstrate the effectiveness of internal ventures for market expansion and/or diversification. Burgelman has suggested that corporations need to "develop greater flexibility between new venture projects and the corporation," using external as well as internal ventures.[18]

The difficulty in successfully diversifying via internal ventures is not a new one. Citing Chandler,[19] Morecroft comments on Du Pont's failure in moving from explosive powders to varnishes and paints in 1917:

> [C]ompeting firms, though much smaller and therefore lacking large economies of scale and production, were nonetheless profitable. . . . Their sole advantage lay in the fact that they specialized in the manufacture, distribution, and sale of varnishes and paints. This focus provided them with clearer responsibilities and clearer standards for administering sales and distribution.[20]

Joint Ventures or Alliances

Despite the great potential for conflict, many companies successfully diversify and grow via joint ventures. As Killing points out, when projects get larger, technology more expensive, and the cost of failure too large to be borne alone, joint venturing becomes increasingly important.[21] Shifts in national policy in the United States are now encouraging the formation of several large research-based joint ventures involving many companies. But the traditional forms of joint ventures, involving creation of third corporations, seem to have limited life and/or growth potential.

Hlavacek et al. and Roberts[22] believe one class of joint venture to be of particular interest—"new style" joint ventures in which large and small companies join forces to create a new entry in the market place. In these efforts of "mutual pursuit," usually without the formality of a joint venture company, the small company provides the technology, the large company provides marketing capability, and the venture is synergistic for both parties. Recent articles have indicated how these large company/small company "alliances," frequently forged through the creative use of corporate venture capital, are growing in strategic importance.[23]

Venture Capital and Nurturing

The venture strategy that permits some degree of entry, but the lowest level of required corporate commitment, is that associated with external venture capital investment. Major corporations have exploited this approach in order to become involved with the growth and development of small companies as investors, participants, or even eventual acquirers. Roberts points out that this approach was popular as early as the mid-to-late 1960s with many large corporations, such as Du Pont, Exxon, Ford, General Electric, Singer, and Union Carbide.[24] Their motivation was the so-called "window on technology," the opportunity to secure closeness to and possibly later entry into new technologies by making minority investments in young and growing high-technology enterprises. However, few companies in the 1960s were able to make this approach alone an important stimulus of corporate growth or profitability. Despite this, ever increasing numbers of companies today are experimenting with venture capital, and many are showing important financial and informational benefits.

Studies carried out by Greenthal and Larson[25] show that venture capital investments can indeed provide satisfactory and perhaps highly attractive returns, if they are properly managed, although Hardymon et al.[26] essentially disagree. Rind distinguishes between direct venture investments and investment into pooled funds of venture capital partnerships.[27] He points out that although direct venture investments can be carried out from within a corporation by appropriate planning and organization, difficulties are often encountered because of a lack of appropriately skilled people, contradictory rationales between the investee company and parent, legal problems, and an inadequate time horizon. Investment in a partnership may remove some of these problems, but if the investor's motives are something other than simply maximizing financial return, it may be important to select a partnership concentrating investments in areas of interest. Increasingly, corporations are trying to use pooled funds to provide the "windows" on new technologies and new markets that are more readily afforded by direct investment, but special linkages with the investment fund managers are needed to implement a "window" strategy. Fast cites 3M and Corning as companies that have invested as limited partners in venture capital partnerships.[28] This involvement in business development financing can keep the company in touch with new technologies and emerging industries as well as provide the guidance and understanding of the venture development process necessary for more effective internal corporate venturing.

In situations where the investing company provides managerial assistance to the recipient of the venture capital, the strategy is classed "venture nurturing" rather than pure venture capital. This seems to be a more sensible entry toward diversification objectives as opposed to a simple provision of funds, but it also needs to be tied to other company diversification efforts.

Educational Acquisitions

Although not discussed in the management literature, targeted small acquisitions can fulfill a role similar to that of a venture capital minority investment and, in some circumstances, may offer significant advantages. In an acquisition of this type, the acquiring firm immediately obtains people familiar with the new business area, whereas in a minority investment, the parent relies upon its existing staff to build familiarity by interacting with the investee. Acquisitions for educational purposes may therefore represent a faster route to familiarity than the venture capital "window" approach. Staff acquired in this manner may even be used by the parent as a basis for redirecting a corporation's primary product-market thrust. Harris Corporation (formerly Harris-Intertype) entered the computer and communication systems industry using precisely this mechanism; it acquired internal skills and knowledge through its acquisition of Radiation Dynamics, Inc. Procter & Gamble recently demonstrated similar behavior in citing its acquisition of the Tender Leaf Tea brand as "an initial learning opportunity in a growing category of the beverage business."[29]

One potential drawback in this educational acquisition approach is that it usually requires a higher level of financial commitment than minority investment and therefore increases risk. In addition, it is necessary to ensure that key people do not leave soon after the acquisition as a result of the removal of entrepreneurial incentives. A carefully designed acquisition deal may be necessary to ensure that incentives remain. When Xerox acquired Versatec, for example, the founder and key employees were given the opportunity to double their "sellout" price by meeting performance targets over the next five years.

Though not without controversy, major previous research work on large U.S. corporations indicated that the highest performers had diversified to some extent but had constrained the development of new business within areas related to the company's base business. The range of mechanisms employed for entering new businesses, previously displayed in Table 36.3, is divided in Figure 36.3 into three regions, each requiring a different level of corporate involvement and commitment. No one mechanism is ideal for all new business development. It may be possible, therefore, that selective use of entry mechanisms can yield substantial benefits over concentration on one particular approach. If this presumption is valid, then careful strategy selection can reduce the risk associated with new business development in unrelated areas.

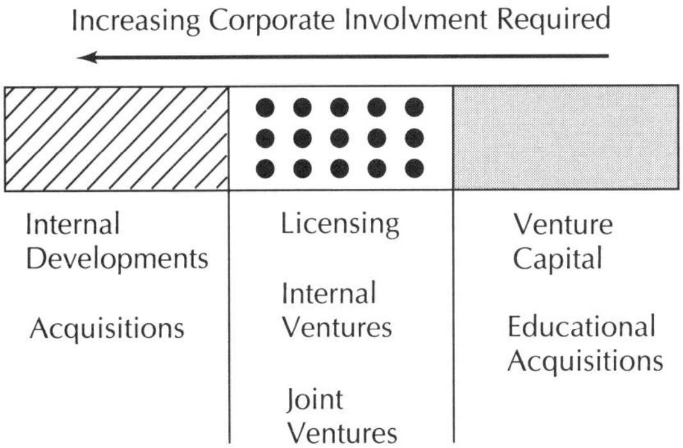

Fig. 36.3. Spectrum of entry strategies.

DETERMINING OPTIMUM ENTRY STRATEGIES

How can the entry strategies of Figure 36.3 be combined with the conceptual framework of Figure 36.2? Which entry strategies are appropriate in the various regions of the familiarity matrix? The literature provides some useful guides.

In his discussion of the management problems of diversification, Miller proposes that acquisitive diversifiers are frequently required to participate in the strategic and operating decisions of the new subsidiary before they are properly oriented towards the new business.[30] In this situation the parent is "unfamiliar" with the new business area. It is logical to conclude that if the new business is unfamiliar *after* acquisition, it must also have been unfamiliar *before* acquisition. How then could the parent have carried out comprehensive screening of the new company before executing the acquisition? In a situation in which familiarity was low or absent, preacquisition screening most probably overlooked many factors, turning the acquisition into something of a gamble from a business portfolio standpoint. Similar arguments can be applied to internal development in unfamiliar areas and Gilmore and Coddington specifically stress the dangers associated with entry into unfamiliar markets.[31]

This leads to the rather logical conclusion that entry strategies requiring high corporate involvement should be reserved for new businesses with familiar market and technological characteristics. Similarly, entry mechanisms requiring low corporate input seem best for unfamiliar sectors. A recent discussion meeting with a number of chief executive officers suggested that, at most, 50 percent of major U.S. corporations practice even this simple advice.

The three sections of the Entry Strategy Spectrum in Figure 36.3 can now be aligned with the three regions of the familiarity matrix in Figure 36.2. Let us analyze this alignment for each region of the matrix, with particular regard for the main factors identified in the literature.

Region 1: Base/Familiar Sectors

Within the base/familiar sector combinations illustrated in Figure 36.4, a corporation is fully equipped to undertake all aspects of new business development. Consequently, the full range of entry strategies may be considered, including internal development, joint venturing, licensing, acquisition, or minority investment of venture capital. However, although all of these are valid from a corporate familiarity standpoint, other factors suggest what may be the optimum entry approach.

The potential of conflict between partners may reduce the appeal of a joint venture, and minority investments offer little benefit since the investee would do nothing that could not be done internally.

The most attractive entry mechanisms in these sectors probably include internal development, licensing, and acquisition. Internal development may be appropriate in each of these sectors, since the required expertise already exists within the corporation. Licensing may be a useful alternative in the base market/new familiar technology sector since it offers fast access to proven products. Acquisition may be attractive in each sector but, as indicated by Shanklin, may be infeasible for some companies in the base/base sector as a result of antitrust legislation.[32] For example, although IBM was permitted to acquire ROLM Corporation, the Justice Department did require that IBM divest ROLM's MIL-SPECS Division because of concern for concentration in the area of military computers.

It may therefore be concluded that in these base/familiar sectors, the optimum entry strategy range may be limited to internal development, li-

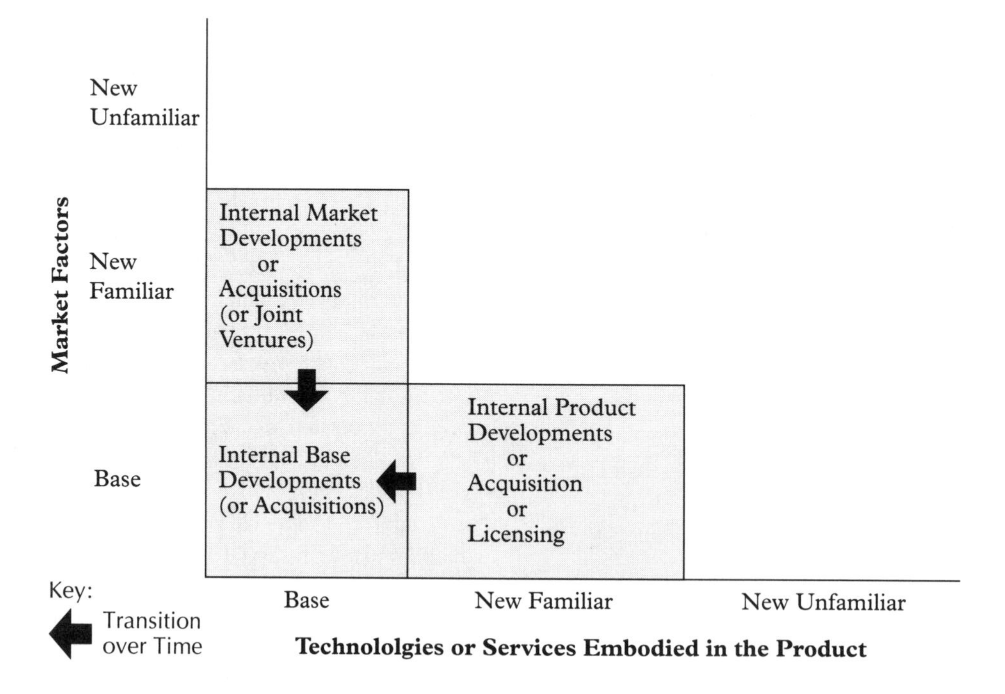

Fig. 36.4. Preferred entry mechanisms in base/familiar sectors.

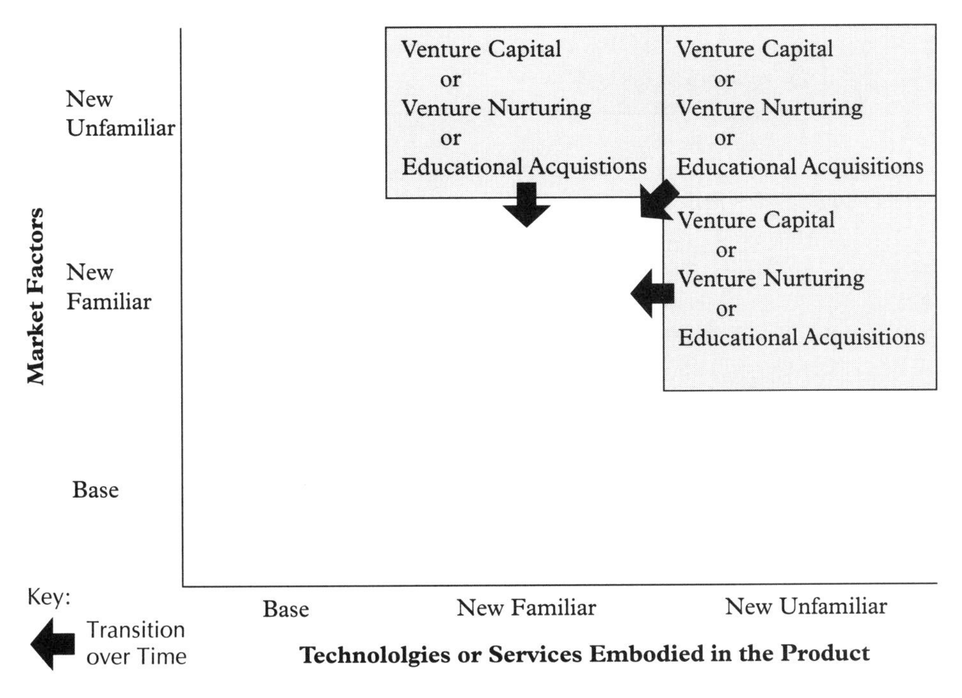

Fig. 36.5. Preferred entry mechanisms in familiar/unfamiliar sectors.

censing, and acquisition as illustrated in Figure 36.4. In all cases a new business developed in each of these sectors is immediately required to fulfill a conventional sales/profit role within the corporate business portfolio.

Finally, since new businesses within the base market/new familiar technology and new familiar market/base technology sectors immediately enter the portfolio of ongoing business activities, they transfer rapidly into the base/base sector. These expected transitions are illustrated by the arrows in Figure 36.4.

Region 2: Familiar/Unfamiliar Sectors

Figure 36.5 illustrates the sectors of lowest familiarity from a corporate standpoint. It has already been proposed that a company is potentially competent to carry out totally appropriate analyses only on those new business opportunities that lie within its own sphere of familiarity. Large-scale entry decisions outside this sphere are liable to miss important characteristics of the technology or market, reducing the probability of success. This situation frequently generates unhappy and costly surprises. Furthermore, if the unfamiliar parent attempts to exert strong influence on the new business, the probability of success will be reduced still further.

These factors suggest that a two-stage approach may be best when a company desires to enter unfamiliar new business areas. The first stage should be devoted to building corporate familiarity with the new area. Once this has been achieved, the parent is then in a position to decide whether to allocate more substantial resources to the opportunity and, if appropriate, select a mechanism for developing the business.

As indicated earlier, venture capital provides one possible vehicle for building corporate familiarity with an unfamiliar area. With active nurturing of a venture capital minority investment, the corporation can monitor, firsthand, new technologies and markets. If the investment is to prove worthwhile, it is essential for the investee to be totally familiar with the technology/market being monitored by the investor. The technology and market must therefore be the investee's base business. Over time active involvement with the new investment can help the investor move into a more familiar market/technology region, as illustrated in Figure 36.5, from which the parent can exercise appropriate judgment on the commitment of more substantial resources.

Similarly, educational acquisitions of small young firms may provide a more transparent window on a new technology or market, and even on the initial key employees, who can assist the transition toward higher familiarity. It is important, however, that the performance of acquisitions of this type be measured according to criteria different from those used to assess the "portfolio" acquisitions discussed earlier. These educational acquisitions should be measured initially on their ability to provide increased corporate familiarity with a new technology or market, and not on their ability to perform immediately a conventional business unit role of sales and profits contributions.

Region 3: Marginal Sectors

The marginal sectors of the matrix are the two base/new unfamiliar combinations plus the new familiar market/new familiar technology area, as illustrated in Figure 36.6. In each of the base/new unfamiliar sectors, the company has a strong familiarity with either markets or technologies, but is totally unfamiliar with the other dimension of the new business. In these situations joint venturing may be very attractive to the company and prospective partners can see that the company may have something to offer. However, in the new familiar technology/market region the company's base business strengths do not communicate obvious familiarity with that new technology or market. Hence, prospective partners may not perceive that a joint venture relationship would yield any benefit for them.

In the base market/new unfamiliar technology sector the "new style" joint venture or alliance seems appropriate. The large firm provides the mar-

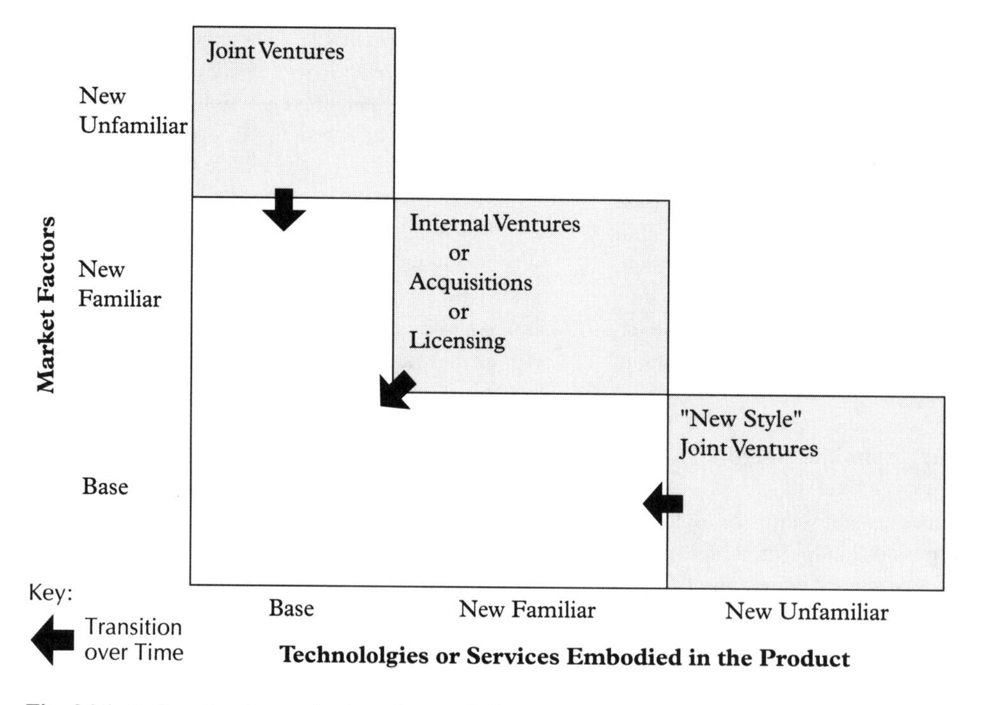

Fig. 36.6. Preferred entry mechanisms in marginal sectors.

Market Factors	Base	New Familiar	New Unfamiliar
New Unfamiliar	Joint Ventures	Venture Capital or Venture Nurturing or Educational Acquistions	Venture Capital or Venture Nurturing or Educational Acquisitions
New Familiar	Internal Market Developments or Acquisitions (or Joint Ventures)	Internal Ventures or Acquisitions or Licensing	Venture Capital or Venture Nurturing or Educational Acquisitions
Base	Internal Base Developments (or Acquisitions)	Internal Product Developments or Acquisition or Licensing	New Style Joint Ventures

Technololgies or Services Embodied in the Product

Fig. 36.7. Optimum entry strategies.

keting channels and the small company provides the technological capability, forming a union that can result in a very powerful team.[33] The complement of this situation may be equally attractive in the new unfamiliar market/base technology sector, although small companies less frequently have strong marketing/distribution capabilities to offer a larger ally.

The various forms of joint ventures such as these not only provide a means of fast entry into a new business sector, but offer increased corporate familiarity over time, as illustrated in Figure 36.6. Consequently, although a joint venture may be the optimum entry mechanism into the new business area, future development of that business may be best achieved by internal development or acquisition as discussed in the earlier base/familiar sectors section of this article.

In the new familiar market/new familiar technology sector, the company may be in an ideal spot to undertake an internal venture. Alternatively, licensing may provide a useful means of obtaining rapid access to a proven product embodying the new technology. Minority investments can also succeed in this sector but, since familiarity already exists, a higher level of corporate involvement and control may be justifiable.

Acquisitions may be potentially attractive in all marginal sectors. However, in the base/new unfamiliar areas this is dangerous since the company's lack of familiarity with the technology or market prevents it from carrying out comprehensive screening of candidates. In contrast, the region of new familiar market/new familiar technologies does provide adequate familiarity to ensure that screening of candidates covers most significant factors. In this instance an acquisitive strategy is reasonable.

Sector Integration: Optimum Entry Strategies

The above discussion has proposed optimum entry strategies for attractive new business opportunities based on their position in the familiarity matrix. Figure 36.7 integrates these proposals to form a tool for selecting entry strategies based on corporate familiarity.

TESTING THE PROPOSALS

In testing the proposed entry strategies, Berry studied fourteen new business development episodes that had been undertaken within one highly successful diversified technological corporation.[34] These episodes were all initiated within the period 1971 to 1977, thus representing relatively recent activity while still ensuring that sufficient time had elapsed for performance to be measurable.

The sample comprised six internal developments (three successful, three unsuccessful); six acquisitions (three successful, three incompatible); and two successful minority investments of venture capital. These were analyzed in order to identify factors that differentiated successful from unsuccessful episodes, measured in terms of meeting very high corporate standards of growth, profitability, and return on investment. Failures had not achieved these standards and had been discontinued or divested. The scatter of these episodes on the familiarity matrix is illustrated in Figure 36.8. Internal developments are represented by symbols A to F, acquisitions by symbols G to L, with symbols M and N showing the location of the minority investments.

The distribution of success and failure on the matrix gives support to the entry strategy proposals that have been made in this article. All high corporate involvement mechanisms (internal development and regular "portfolio" acquisitions) in

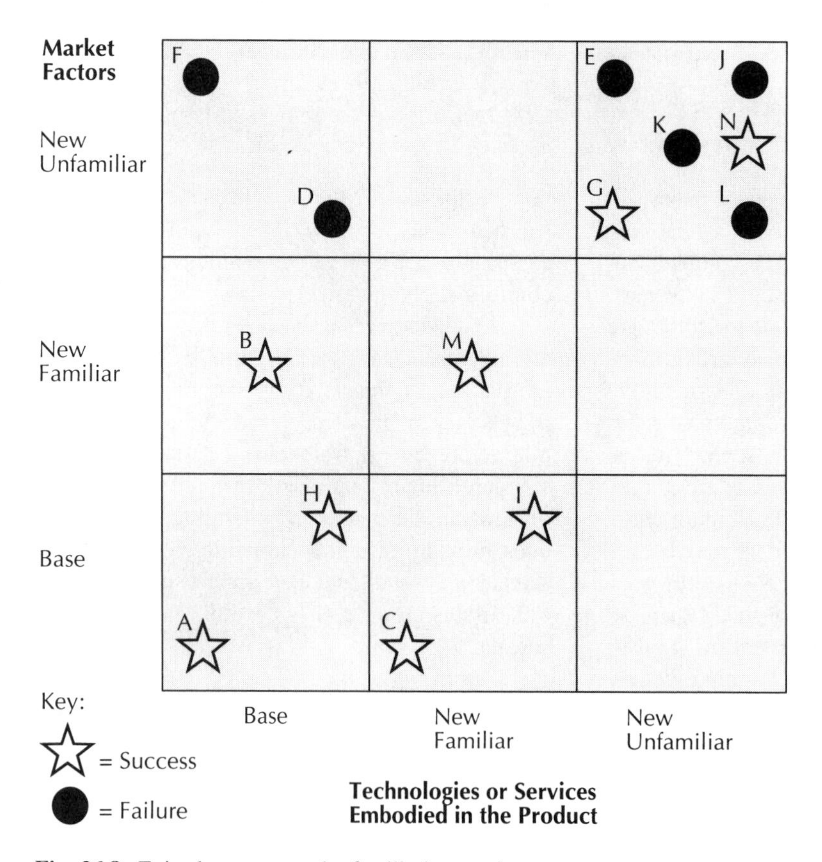

Fig. 36.8. Episode scatter on the familiarity matrix.

familiar sectors were successful. However, in unfamiliar areas, only one of this category of entry mechanism, acquisition G, succeeded. This acquisition was a thirty-year-old private company with about 1,000 employees, producing components for the electronics and computer industries. Company G was believed to offer opportunities for high growth although it was unrelated to any area of the parent's existing business. The deal was completed after an unusually long period of two years of candidate evaluation carried out from within the parent company. The only constraint imposed upon Company G following acquisition was the parent's planning and control system, and in fact the acquired company was highly receptive to the introduction of this system. This indicated that Company G was not tightly integrated with the parent and that any constraints imposed did not severely disrupt the established operating procedures of the company.

All factors surrounding the acquisition of Company G—its size, growth market, low level of constraints, and low disruption by the parent—suggest that Company G might have continued to be successful even if it had not been acquired. Representatives of the parent agreed that this might be the case, although they pointed out that the levels of performance obtained following acquisition might not have occurred if Company G had remained independent. Hence, if an acquired company is big enough to stand alone and is *not* tightly integrated with the parent, its degree of operational success is probably independently determined.

It is important to point out that despite the success that occurred in instance G, an acquisition of this type in unfamiliar areas must carry some degree of risk. The parent is liable to overlook many subtle details while screening candidates. Furthermore, when an established company is acquired and continues to operate with a high degree of independence, identification of synergy becomes difficult. Synergy must exist in any acquisitive development if economic value is to be created by the move.[35] Consequently, an acquisition of this type not only carries risk, but may also be of questionable benefit to shareholders, especially if a high price was required because of an earlier good performance record.

The other success in an unfamiliar area, episode N, is a minority investment of venture capital. By the very nature of minority investments, corporate involvement is limited to a low level. Although some influence may be exerted via participation on the board of directors of the investee, again, the investee is not tightly bound to the parent. Consequently, the success of the investee tends to be determined to a large extent by its own actions.

Detailed examination of episodes G and N suggests a good reason for the subject companies' success despite their location in unfamiliar sectors—the companies did not require significant input from the unfamiliar parent in the decision-making process. This suggests that new business development success rates in unfamiliar areas may be increased by limiting corporate input with the decision-making process to low levels until corporate familiarity with the new area has developed. These experiences support the entry proposals already outlined in this article.

Some companies have already adopted entry strategies that seem to fit the proposals of this article, and Monsanto represents one of the best examples. Monsanto is now committed to significant corporate venturing in the emerging field of biotechnology. Its first involvement in this field was achieved with the aid of its venture capital partnership Innoven, which invested in several small biotechnology firms, including Genentech. During this phase Monsanto interacted closely with the investees, inviting them in-house to give seminars to senior management on their biotechnology research and opportunities. Once some internal familiarity with the emerging field had developed, Monsanto decided to commit substantial resources to internal research-based ventures. Monsanto used venture capital to move from an unfamiliar region to an area of more familiar technology and market, and is currently continuing those venture capital activities to seek new opportunities in Europe. Joint ventures with Harvard Medical School and Washington University of St. Louis are further enhancing its familiarity with biotechnology, while producing technologies that Monsanto hopes to market. Contract research leading to licenses from small companies, primarily those in which it holds minority investments, is another strategy Monsanto is employing. Although the outcome is far from determined, Monsanto seems to be effectively entering biotechnology by moving from top right to bottom left across the familiarity matrix of Figure 36.7.

CONCLUSION

A spectrum of entry strategies was presented in this article, ranging from those that require high corporate involvement, such as internal development or acquisition, to those that require only low involvement, such as venture capital. These were incorporated into a new conceptual framework designed to assist in selecting entry strategies into potentially attractive new business areas. The framework concentrates on the concept of a corporation's "familiarity" with the technology and market aspects of a new business area, and a matrix was used to relate familiarity to optimum entry strategy.

In this concept, no one strategy is ideal for all new business development situations. Within familiar sectors virtually any strategy may be adopted, and internal development or acquisition is probably most appropriate. However, in unfamiliar areas these two high involvement approaches are very risky and greater familiarity should be built *before* they are attempted. Minority investments and small targeted educational acquisitions form ideal vehicles for building familiarity and are therefore the preferred entry strategies in unfamiliar sectors.

Early in this article, research results from the literature were outlined which indicated that in order to ensure highest performance, new business development should be constrained within areas related to a company's base business. However, this research did not account for alternative entry mechanisms. This article proposes that a multifaceted approach, encompassing internal developments, acquisitions, joint ventures, and venture capital minority investments, can make available a much broader range of business development opportunities, at lower risk than would otherwise be possible.

REFERENCES

1. See E.B. Roberts, "New Ventures for Corporate Growth," *Harvard Business Review,* July–August 1980, pp. 134–142.
2. See R.P. Rumelt, *Strategy, Structure and Economic Performance* (Harvard Business School, Division of Research, 1974).
3. See R.P. Rumelt, "Diversification Strategy and Profitability," *Strategic Management Journal* 3 (1982): 359–369.
4. See H.R. Christensen and C.A. Montgomery, "Corporate Economic Performance: Diversification Strategy versus Market Structure," *Strategic Management Journal* 2 (1981): 327–344.
5. See R.A. Bettis and W.K. Hall, "Risks and Industry Effects in Large Diversified Firms," *Academy of Management Proceedings '81,* pp. 17–20.
6. See O.J. Holzmann, R.M. Copeland, and J. Hayya, "Income Measures of Conglomerate Performance," *Quarterly Review of Economics and Business* 15 (1975): 67–77.
7. See T. Peters, "Putting Excellence into Management," *Business Week,* 21 July 1980, pp. 196–205.
8. See T.J. Peters and R.H. Waterman, *In Search of Excellence* (New York: Harper and Row, 1982).
9. See M.H. Meyer and E.B. Roberts, "New Product Strategy in Small High Technology Firms: A Pilot Study" (MIT Sloan School of Management Working Paper #1428-1-84, May 1984).
10. See H.R. Biggadike, "The Risky Business of Diversification," *Harvard Business Review,* May–June 1979, pp. 103–111.
11. See L.A. Weiss, "Start-Up Businesses: A Comparison of Performances," *Sloan Management Review,* Fall 1981, pp. 37–53.
12. See S.S. Miller, *The Management Problems of Diversification* (New York: John Wiley & Sons, 1963).
13. See J.S. Gilmore and D.C. Coddington, "Diversification Guides for Defense Firms," *Harvard Business Review,* May–June 1966, pp. 133–159.
14. See M.S. Salter and W.A. Weinhold, "Diversification via Acquisition: Creating Value," *Harvard Business Review,* July–August 1978, pp. 166–176.
15. See Miller (1963).
16. See J.P. Killing, "Diversification through Licensing," *R&D Management,* June 1978, pp. 159–163.
17. See Roberts (1980).

18. See R.A. Burgelman, "Managing the Internal Corporate Venturing Process," *Sloan Management Review,* Winter 1984, pp. 33–48.
19. See A.D. Chandler, *Strategy and Structure* (Cambridge, MA: MIT Press, 1962).
20. See J.D.W. Morecroft, "The Feedback Viewpoint in Business Strategy for the 1980s" (MIT Sloan School of Management, Systems Dynamics Memorandum D-3560, April 1984).
21. See J.P. Killing, "How to Make A Global Joint Venture Work," *Harvard Business Review,* May–June 1982, pp. 120–127.
22. See: J.D. Hlavacek, B.H. Dovey, and J.J. Biondo, "Tie Small Business Technology to Marketing Power," *Harvard Business Review,* January–February 1977; Roberts (1980).
23. See: *Business Week,* "Acquiring the Expertise but Not the Company," 25 June 1984, pp. 142B–142F; *Inc.,* "The Age of Alliances," February 1984, pp. 68–69.
24. See Roberts (1980).
25. R.P. Greenthal and J.A. Larson, "Venturing into Venture Capital," *Business Horizons,* September–October 1982, pp. 18–23.
26. See G.F. Hardymon, M.J. Denvino, and M.S. Salter, "When Corporate Venture Capital Doesn't Work," *Harvard Business Review,* May–June 1983, pp. 114–120.
27. See K.W. Rind, "The Role of Venture Capital in Corporate Development," *Strategic Management Journal* 2 (1981): 169–180.
28. See N.D. Fast, "Pitfalls of Corporate Venturing," *Research Management,* March 1981, pp. 21–24.
29. See Procter & Gamble Company, *1983 Annual Report* (Cincinnati, OH: 1984), p. 5.
30. See Miller (1963).
31. See Gilmore and Coddington (1966).
32. See W.L. Shanklin, "Strategic Business Planning: Yesterday, Today and Tomorrow," *Business Horizons,* October 1979, pp. 7–14.
33. See: *Business Week* (25 June 1984); Hlavacek, Dovey, and Biondo (1977); *Inc.* (February 1984).
34. C.A. Berry, "New Business Development in a Diversified Technological Corporation" (MIT Sloan School of Management/Engineering School Master of Science Thesis, 1983).
35. See Salter and Weinhold (July–August 1978).

37

The Use of Alliances in Implementing Technology Strategies

YVES DOZ
GARY HAMEL

INTRODUCTION

Strategic alliances are in vogue. Many new strategic alliances are being created every year, most of them with a substantial technological content, or even a specific R&D focus (Doz, 1992). While the phenomenon is not new (for example, alliances existed at the beginning of the 20th Century in such industries as non-ferrous metals and chemicals) the pace of alliance formation has accelerated since the late 1980s.

This paper first analyzes the reasons for the recently observed surge in the number of strategic alliances. The intensification and sectoral spread of global competition is one central driving force, calling for accelerated/quicker permutations of ever more complex skill sets. Few firms have all the in-house capabilities required to compete globally: market access, management and operational disciplines, critical technologies, and creative product and market ideas. Most need to complement internal capabilities. The first section of this paper analyzes what additional capabilities companies seek as they are confronted by growing, often technology-based, global competition, and why alliances may give them access to or internalization of these capabilities.

1. "NO FIRM IS AN ISLAND"

In most industries, companies are now racing towards what they see as the future high ground of industry leadership. While a few global competitive battles featuring daily in the media, such as those for high definition television, notebook computers, AIDS treatment, civilian aircraft, or motor cars, capture a disproportionate share of the attention of researchers and policy makers alike, many more such global races are going on, for instance in the areas of new materials, voice recognition computing, electric vehicles, communication networks, membrane separation, plant protection and growth regulation.

Some of the contestants are global companies that seek to expand their leadership in their traditional industry, while others are smaller companies hoping to exploit significant innovations. Yet other companies, seeking an escape from declining industries, are rushing into new fields in which they are comparatively small and inexperienced, and where their existing understanding, skills and other intangible resources may be only tangentially useful. Further, new fields often result from "hybrid" technologies bringing several disciplines together (e.g., bio-electronics), or from creative ways of combining product innovation and process capabilities, or from the integration of diverse technologies into complex systems. In both cases, the range of required technologies broadens. To win the race to extend industry horizons and extract the benefits, most companies increasingly find they are in need of an array of "complementary assets" (Teece, 1987).

Complementary Capabilities

Even large, well-established firms, which have the strongest heritage to defend, see technology as the main determinant of competitive success in their core business. As a result, they may increasingly need access to new technology developed by other firms. Only with new technologies can they leverage the market access capabilities they built (e.g., distribution infrastructure and global brands), capitalize on the management disciplines they cultivate (such as quality management, value-engineering, or project management skills) and continue to exploit and strengthen their reputation and brand image. Market access skills and management disciplines are of little value unless they can be applied to a constant stream of new products and technologies. Existing companies need innovative products to exploit existing distribution networks and to capitalize on the management disciplines they have cultivated; innovation thus becomes the key to protecting the value and future longevity of past efforts.

Historically, skills have been unevenly distributed between firms and between regions of the world (Hamel, 1990). European companies may create science-based inventions, but often suffer from a poor understanding of markets and thus fail to bring those inventions to new markets (Doz, 1994; Hamel and Prahalad, 1991). American companies may succeed at fundamental market innovation, but fail to buttress initial successes with fast-paced improvements in cost and quality. Japanese companies design production and product development processes for speed and efficiency (De Meyer, 1991). By pursuing manufacturing flexibility and agility, they are able to engage in "trial-and-error" product development and introduction (Nevens, Summe and Uttal, 1990; Hamel and Prahalad, 1991). Japanese companies have also often developed a capability to blend various technologies into collective *know-how* that results in deeply rooted and widely held competencies, for example, in joint product and process development or in commercialization skills (Prahalad and Hamel, 1990; Fruin, forthcoming; Nonaka and Takeuchi, 1995). Yet, Japanese companies sometimes have to rely on others for market access or fundamental innovations in sciences.

While significant productivity gaps still result from historically different approaches and priorities toward management disciplines (Womack et al., 1990), approaches to these disciplines now tend to converge within an industry, and the productivity differences between the survivors are narrowing. It is therefore unlikely that differences in management disciplines—no matter how wide a gap they may create at some point in time between competitors—will result in sustainable long-term competitiveness differences between survivors.[1] While constant improvements in management disciplines can exhaust some weaker "follower" companies, and lead to their demise, survivors are increasingly evenly matched in costs, quality, and cycle time. Further, these improvements may only generate diminishing returns: beyond "6 sigma" quality, for example, what is the competitive advantage to be gained from further improvement?

Similarly, as European, American, and Japanese companies rush to build marketing and manufacturing networks across the triad's markets, market access asymmetries may disappear and competitors with more or less matching worldwide market access capabilities emerge. They will "cover the world" in approximately equal terms.

In summary, increased global competition has forced companies, which historically developed different capabilities and skills, into a process of rapid convergence. In this process, they are increasingly trying both to close skill gaps vis-à-vis their competitors and to develop new competencies and skills that their competitors will find difficult to imitate and match.

The Race to the Future

This convergence process takes place along several dimensions: global market access infrastructure and skills; management disciplines that improve the integrity and the value of the products; and competencies and technologies that allow product innovation and market creation. Yet, the speed at which a company can achieve convergence varies, with more rapid progress in some dimensions. Increasing familiarity with world markets has made competition for market access less dependent on skills and more on resources. Management disciplines become increasingly well-codified, and thus more easily transferable, although their successful implementation may still benefit from a mentor. Competencies are much more difficult to codify, and thus to transact. Nevertheless, competitive victory will belong to those who build new competencies the fastest.

The pace at which a company can move toward convergence thus depends on the type of new capabilities it requires. Market access may require greater or fewer inputs of efforts and local knowledge, depending on *which* market is to be penetrated. The structure of distribution, the type of products sold, and possible protection policies all play a role. Developing channels alone to penetrate a complex market in which protectionism may arise is a difficult, slow, and risky process. Alliances with distributors can accelerate market penetration. The early reliance of Japanese companies on U.S. mass merchandisers in the consumer electronics industry gave the Japanese manufacturers the scale needed for automating production—and achieving very high quality—without having to invest in market development (Hamel and Prahalad, 1985). Similarly, Japanese companies now use alliances with local companies to learn about markets for more sophisticated products such as computer systems and telecom networks. Indeed, historically, market access alliances have generally been the most widespread form of strategic alliances and joint ventures (e.g., Franko 1973; Harrigan 1986). Better understanding and longer experience of global markets may reduce the need for market access alliances, and global competition increases the pressure for manufacturers to directly control their distribution. It is thus not surprising to observe a decline of market access alliances, relative to other alliances (Doz, 1992).

Why Technology Alliances?

Through patents, licensing rights, and technology transfer agreements, the market for technology may be relatively efficient. The market for competencies may be much less efficient. Developing a complex, partly tacit competence is a problem of example as much as of cash: in the absence of a mentor with whom to co-practice the competence as in an apprenticeship, it may be impossible to develop the competence autonomously. At best, independent development may be a long "time-incompressible" process (Dierickx and Cool, 1994). Alliances may allow one company to intercept the skills of another and give it the opportunity to close skill gaps much faster than internal development would allow. For example, General Motors used its joint venture with Toyota to learn "lean" manufacturing processes and new labor relation practices. Similarly, Thomson, the French consumer electronics group, relied on its alliance with JVC to learn to mass produce the micromechanic subsystems key to successful video cassette recorder production (Hamel and Doz, 1990).

Many of the most critical current technologies are increasingly embedded, i.e., firm- and context-specific: It is not the abstract theoretical understanding of technologies but their effective day-to-day practice that matters. Take VLSI memories: understanding their technology is not conceptually difficult, but designing actual memories, and running a high-yield memory factory is harder! The hurdle *is* practice, not theory. Embedded technologies, based on systemic skills and tacit collective process know-how, are less easily imitable than the increasingly well-codified management disciplines such as total quality management and concurrent engineering, and harder to match than market access skills (Prahalad and Hamel, 1990; Dierickx and Cool, 1994; Itami and Roehl, 1987).

Related to the embedded nature of technology we observe another difficulty: the growing interdependence between previously stand-alone technologies and the need to incorporate a wider range of technologies into complex systems and integrated solutions (Cantwell, 1989). As the range of technologies to be integrated widens, the required combination of technologies is increasingly difficult for single firms to master. Managers also believe that the integration capabilities to bring these technologies together will establish a new hierarchy within industries. System integrators, who are close to top management issues and end-users, will define solutions; prime contractors will lead in their development, and subcontractors will follow in their implementation. The current convergence of information technology services, software development and strategy consulting, or the concerns of aerospace companies for system and mission integration skills bear witness to this trend, at least in some industries.

The embedded and systemic characteristics of new technologies, which makes them more akin to competencies, have significant implications for alliances. The more embedded a technology, the closer the cooperation needed to share and transfer it. As a result, we are likely to witness an increasing number of alliances driven by the motives to learn new skills and more attempts at more intimate "closer" alliances.

In sum, alliances may provide vehicles for faster, less costly, and more permanent skill borrowing and internalization (the ability to use and improve the new skill independently from the partner) than other approaches to developing new competencies, particularly when these skills are embedded and largely tacit (Hamel, 1990). In that sense alliances may be most useful at the "cutting edge" of the learning agenda, to access and internalize technologies and know-hows that are embedded, largely tacit, uncodified, and thus difficult to access via contractual approaches that do not involve a close collaboration between the partners. This is the case of many new technologies, which are themselves a growing source of competitive advantage when compared to market access and management disciplines.

Co-specialization or Internalization

Embeddedness beyond a point can become a deterrent to internalization. The very tacitness and embeddedness of a technology make it difficult to learn, even in the co-practice apprenticeship that characterizes many technology alliances. The required learning effort may exceed the resources and the capabilities of the firm. The risks involved may also be seen as too high, because of uncertain imitability. The outcome of the learning process may not be easily anticipated.

Access to the benefits provided by the use of the partner's skills, rather than internalization, may thus be chosen. It may not be necessary to learn the partner's skills when access is achieved "by proxy." In fact internalization objectives may interfere with effective complementation and interweaving of skills. Visible internalization attempts are likely to raise the partner's suspicion, which may erode goodwill and willingness to contribute the skills required for successful collaboration.

The relative balance in priorities between technology and skill complementation, on the one hand, and technology and skill internalization on the other, is likely to depend on various factors. First, it depends partly on the scope of the learning opportunity. Whether the skills to be exercised in the alliance extend to other activities is the key issue here. For instance, when Thomson Consumer Electronics started its alliance for VCR manufacturing with JVC (back in 1983) it was clear to Thomson's management that the comanufacturing of VCRs provided an opportunity to learn a set of micromechanical precision manufacturing skills that would be useful not just for the joint production of VCRs, or even for other VCRs made by

Thomson, but for a whole range of new products, where precision manufacturing was also important, such as CD video disks and players. This provided a strong motivation for Thomson to learn JVC's skills. Conversely, Corning, in an optical fiber alliance with Siemens, did not see much value in trying to internalize Siemen's telecommunication networking skills, as they are not pertinent to Corning's other businesses.

Second, the will to learn skills may depend on the perception by the partners of each other's positions. It was clear to Thomson Consumer Electronics, for example, that while it cooperated with JVC in Europe, it also competed against JVC at the level of distribution, brand image, and market share in other regions, and to some extent in Europe too. Indeed, the partner who took most of the JV's production—i.e., competed most effectively in the end market—also gained more influence on the JV as its economics became more dependent on the partner with higher sales. Conversely, both General Electric's and SNECMA's managers attributed a good part of the success of their very successful jet engine alliance—CFM—to the fact that they did not compete against each other in any segment of the civilian jet engine markets.[2] Mutual interest was also reinforced by competing against common rivals, Pratt & Whitney and Rolls Royce.

Third, the nature of the skills themselves may facilitate or hinder cooperation. The more distinct and different, the more difficult the learning, and the more cospecialized—in the sense of jointly creating value that would not be available to the partners without these combined skills—the more the continued goodwill of the partners is essential, hence internalization is both more difficult and less attractive. Similarly, the more skills are embedded in other businesses, which are not themselves part of the alliance, the more difficult will be attempts to internalize. If Siemens' competence in transmission systems is more a result of its overall range of telecom equipment and software skills than the result of the cooperation with Corning, Corning will be quite satisfied to access this cluster of competencies, knowing full well that trying to encroach into them or substitute for them would be senseless. Siemens may reach a similar conclusion if it observes that Corning's excellence in optical fibers is rooted in a cluster of competencies that are nurtured by many other Corning businesses. While what is "exchanged" through the alliance is limited, what is created in the alliance is important. Conversely, one can consider NUMMI, the famous joint venture between General Motors and Toyota. What was "created" in the NUMMI alliance is relatively minor: a modest number of small cars assembled in a joint plant. In GM, and even Toyota in the United States, the production of that plant is dwarfed by each partner's own production. Yet, what is exchanged through the joint venture is critical: a first-hand experience of co-practicing Toyota's manufacturing methods for GM, a direct experience of managing U.S. unionized labor for Toyota.[3] Alliances that are fora for exchange rather than crucibles of new value creation are likely to be intrinsically more difficult for the partners to manage successfully.

Single Relationships or Networks

Some companies' alliance network is but the sum of individual discrete alliances, with little coordination and cumulativeness between alliances. Although the company may end up with many different alliances, it may have entered each on its own merits, as if other alliances did not exist, and also manage each of them as a discrete operation. While this is a viable approach where external constraints provide particularly strong incentives for cooperation—as in the European aerospace industry—the discrete treatment of each partnership may also constitute a missed opportunity for leverage between alliances. Corning, for example, is very clear about the nodal nature of its position, at the center of a network of alliances, each capitalizing on a cluster of core technologies in the inorganic chemistry area, leveraged via focused alliances into a number of application domains. Technologies and competencies likely to be applicable to several alliances—across both geographies and products—are consciously cultivated. Other companies use a virtual network of alliances over time as a ladder to learn from one alliance to the next, and use the learning from each alliance to bargain for a stronger position in subsequent alliances.

While each alliance still needs to be managed effectively for its own sake, on its own terms, the fact that each belongs to a network opens new avenues for more ambitious technology strategies that build on the whole network, and guide its further development in a strategically coordinated manner. To engage in a network of alliances without managing them in a coordinated fashion results in missed strategic opportunities.

In sum, we can categorize the management issues faced in technology alliances along two dimensions:

1. Whether the main logic of the alliance is to create new value within the alliance proper, mostly through technology and skill complementation and cospecialization, or to use the alliance as a conduit to exchange skills, and thus to transfer value between the partners, mostly through skill learning and internalization.
2. Whether the alliance should be seen, and managed, as a discrete relationship, a single alliance as if other alliances did not exist, or whether the management challenge is more accurately captured in terms of complementarity between alliances and evolution of a dynamic network of alliances over time.

These categories are summarized in Figure 37.1, below. While many alliances obviously straddle several categories, we will first consider single alliances that fit clearly in one category and network with a clear purpose. We will then analyze more complex and more ambiguous alliances, which straddle multiple categories.

	Individual Alliance	Network of Alliance
Capability Complementation	GE-SNECMA	Corning Glass
Capability transfer	Thomson JVC	Fujitsu

Fig. 37.1. Technology alliance management categories.

The alliance between General Electric and SNECMA in the jet engine business mentioned earlier provides an interesting example of successful capability complementation, whereas the Thomson JVC alliance provides an example of a company—Thomson—trying to use the alliance to appropriate and internalize some of its partner's capabilities. Corning Glass provides an example of a network of alliances conceived and managed as a series of opportunities to leverage a common cluster of competencies across many alliances. The history of Fujitsu's relationships with its partners, complemented by some other examples, will show the value of a nodal position in a skill acquisition network.

2. MANAGING VARIOUS TYPES OF TECHNOLOGY ALLIANCES

Complementary Capabilities in an Individual Alliance

The General Electric-SNECMA alliance, which serves as an example of successful capability complementation in an individual alliance, had many reasons to succeed. Some were extrinsic to the alliance management itself, others the result of the alliance design and management process.

In the early 1970s, SNECMA had envisioned the need for a "10 ton" thrust engine for the civilian market, but lacked the experience and credibility to sell to airlines and maintain engines on a worldwide basis. It turned to GE, which had independently identified the same market segment, had a military engine that could be adapted, but lacked the resources needed.

The collaboration started in 1973, following the removal of a Department of Defense bar on the export of engine cores by GE to France. Fortunately it was possible to mate GE's core with the SNECMA-developed systems (mainly the main bypass fan) without divulging GE-specific information other than the size and functional specifications of the core—allowing GE to get U.S. government approval for the collaboration. The task interdependence between the partners was kept very low with "black box" cores shipped to France, but very little joint work between the two companies. The engine was aimed at a market segment, the so-called "10 ton" engine (referring to the engine unit thrust), which everyone expected to be important, but failed to materialize. Although prototypes of the new engine, coded CFM56, ran well as early as 1976, no sales were made. The program was on hold, to the deep embarrassment of the partners. It was nearly cancelled by the French government and GE's corporate management in 1979, before the first sales were finally achieved, to reengine older civilian transport planes and military tankers in the United States. In retrospect, managers of both partners saw a hidden benefit to this delay: It had given them the time to proceed with the project, and learn to work together in an unhurried fashion, and it had tested their commitment to the project.

In sum, despite the delay in the market development, the alliance offered great value creation potential to both companies. It allowed them to enter the civilian markets and to become challengers to Pratt & Whitney and Rolls Royce—from which they ultimately took market leadership in the 1980s—in a way they could not have achieved on their own, given limited skills and credibility, particularly after sales service (SNECMA) and limited resources (GE).

The partners were also highly strategically compatible, in fact, sharing a similar objective.[4]

Further, in the case of GE-SNECMA, the partners were also organizationally compatible. Both companies shared similar technical values, and their top management knew and respected each other. The fact that the jet engine world is relatively small, and that both the head of GE's jet engine business, Mr. Neuman, and the CEO of SNECMA, Mr. Ravaud, were not conventional executives steeped in corporate hierarchies, but also deeply committed to the jet engine business, facilitated the cooperation.

Each partner undertook significant efforts to adjust to the other. A joint venture was created, as a small (thirty or so person) group with its headquarters in Cincinnati, where GE's aircraft engine business is centered, but headed by a Frenchman, and more or less equally staffed by secondees from both partners, as a way to facilitate the relationship with the more geographically distant French partner. SNECMA sent its managers to participate in GE's internal management development programs, while GE executives stationed in France made efforts to blend with the culture and learn the language of SNECMA. SNECMA also reorganized some aspects of its operations so their structure would match that of GE, establishing direct counterparts in each organization.

Following the initial development period, and in particular once the engine entered service and the lifting, in the early 1980s, of certain U.S. Department of Defense restrictions on technology sharing, the two partners worked closely on improving the basic engine and developing new versions (for new aircraft programs this time) and started to learn from each other—particularly on manufacturing process, materials technologies, and quality management. The fact that the cooperation had been established on the basis of workshare, limiting the haggling over the respective values of their contributions to one momentous but early negotiation process, made further collaboration and exchange of technical data easier.

Although both companies had strong, proud cultures, they were also receptive, and their managers were willing to learn from each other. Further, the high stakes of the alliance, and the joint "crossing of the desert" in the late 1970s, cemented stronger bonds between them. The workshare split, following the initial negotiation, also facilitated co-operation by "designing out" of the relationship many of the more contentious issues that might have marred co-operation. The two organizations were thus able to learn to work together, and to learn from each other unhampered by organizational discrepancies or by recurring conflicts in the collaboration process.

This situation can be contrasted with that of a less successful alliance: the AT&T–Olivetti attempt to work together in the mid-1980s. The two organizations were hardly compatible. Olivetti was a nimble organization, used to competing in a very tough IBM-compatible computer market, and to making quick, sometimes intuitive, decisions in that fast-changing competitive environment. AT&T was emerging from a long period as a telephone monopoly, in which it had set its own pace and created its own markets. Its decision-making processes were slow and explicit, relying on strategic planning procedures, and a wide basis of consensus, or at least coalition, among product line, geographic, and functional executives. In contrast, Olivetti was led by a very small team grouped around Carlo de Benedetti. Further, the AT&T–Olivetti partnership started not with a joint venture and a broad once-and-for-all agreement, but as a series of periodically renewed supply contracts between the partners. The renegotiation of these contracts threw the partners back into a conflictual mode every six months or so, while the deep differences in their organizations, and decision-making styles and speeds made a joint approach to problems—or even a mutual understanding of issues—extremely difficult. Predictably, the alliance never developed into a true collaboration, and fell in disarray when irreversible commitments were called for (to a new joint product line in minicomputers) or when economic circumstances became more difficult (with the fall in the value of the dollar for personal computers and typewriters).

As the relationship between SNECMA and GE strengthened, the initial cospecialization decreased. As the engine went into service, and the development priorities shifted to improving its reliability and durability, the clear-cut division of tasks between the partners that shaped the early phases of the alliance could not be maintained. Improvements and reactions to problems faced by engines in use needed closer and less predictable patterns of collaboration and information exchange. Improvements interactively affected all subsystems, irrespective of which partners had designed and supplied them. This led to closer collaboration between them, and decreased specialization.

By 1991, the skill bases had converged very significantly, leading even to a role reversal in the development of some new products. SNECMA, for example, is developing the high pressure "core" of a new generation version of the CFM engine, while GE develops the low pressure fan. Although one can thus argue that skill internalization took place quite extensively over time, leading to a possible weakening of the alliance, the partners were kept together by other forces. As mentioned earlier, they have carefully avoided competition in the civilian markets—by being partners in all their programs, albeit not necessarily through the CFM joint venture—and limited competition in the military markets.[5] Further, the extremely high costs of new product development, as well as the limited market size for new engines, now strongly encourage further cooperation, almost irrespective of technology and skill complementarity. In other words, the level of mutual benefits (versus the dissolution of the alliance) as well as that of trust and commitment between the partners is now such that the alliance is likely to continue and further deepen in the future.[6]

In planning for or assessing individual skill-based alliances, it is important, as illustrated by the analysis of the GE-SNECMA collaboration—in contrast to the AT&T–Olivetti attempts—to focus on a few critical issues:

1. *How will the alliance create value?* In particular, what is the amount of cospecialization between the technologies and skills of the partners, offering the potential of value creation not otherwise available to the partners? It is important, also, to consider the risk of each partner encroaching on the technology and competence domain of the other. The more limited that risk, the more genuinely cooperative, rather than potentially competitive, the alliance. Lastly, it is useful to compare the alliance, as a means to combine cospecialized skills to alternatives such as acquisitions, mere supplier–customer agreements, licensing and technology transfer agreements, and internal development.

2. *What is the strategic compatibility between the partners?* Even when the alliance creates value, it is possible for the partners to disagree on priorities and strategies, and thus fail to reach a lasting

agreement. External factors, totally independent from the alliance, may also destabilize it.[7] It is thus essential to assess the potential compatibility between partners' priorities, and the resilience of the alliance to external factors.

3. *How compatible are the partners' organizations and cultures?* As we discussed, SNECMA and GE were compatible, and SNECMA took practical measures to make itself even more compatible, whereas AT&T and Olivetti were essentially incompatible, but recognized this only slowly. Exposing the partner organizations to each other led to convergence in the GE-SNECMA cases—as the starting positions were already close enough to allow the management of both companies to quickly reorganize areas of required improvement. AT&T's and Olivetti's starting positions were so remote that even identifying areas and approaches for improvement proved difficult.

4. *How to make the process of collaboration convergent?* When the two organizations and cultures are deeply different, it is important to pay explicit attention to the design of the collaboration process. This might mean matching the sequence of mutual commitments demanded from each partner with the level of understanding and trust achieved together.[8] Far-reaching collaborative agreements may not be easily obtained too early in an alliance, until partners are close enough. Hence, rather than focus on the up-front resolution of all differences, it may be more practical to design an accelerated convergence process to foster mutual understanding between the partners.

5. *How effective is the design of the alliance?* Here again, the contrast between CFM and AT&T–Olivetti is instructive. The structure of the CFM alliance effectively "designed out" many of the potentially contentious areas between the partners, such as transfer pricing. With these out of the way, collaboration could proceed unimpeded. Conversely, the design of the AT&T–Olivetti alliance as a series of contracts led to divisive discussions being re-entered periodically.

6. *How balanced are contributions and benefits over time?* In principle, alliances where benefits accrue early to one partner and late to the other are perfectly feasible, and often very useful. In practice, however, they are rife with tension—unless the level of mutual trust is very high, and the later benefits are not substantially more uncertain than the earlier ones.

7. *How strong are the expectations of future benefits?* Successful cospecialization alliances have a tendency to grow and continue over time, creating a constantly extended "shadow of the future," which itself contributes to current trust and cooperation.[9] The more the time horizon of the collaboration can be extended, the more likely the collaborative behavior in the shorter term.

When partners attend carefully to the above issues, alliances based on technology and skill complementation and cospecialization are the most robust alliances, since the benefits they provide their partners remain dependent on the continuation of the alliance.

One can observe, for example, that Corning has relied extensively on such alliances in its many businesses, and that most of these alliances have proved very resilient over time. Corning uses a set of alliance selection and management principles that consistently addresses the issues raised above. In addition, however, Corning faces a set of distinct issues that stem from its reliance on not just a few individual alliances, but on a whole network.

Managing a Network of Complementary Capability Alliances

Corning Glass is unique, among major corporations, in deriving the majority of its turnover from joint ventures and alliances, as a matter of its own choice.[10]

From the outset, Corning had been a technology-driven specialty glass company. It established an R&D lab as early as 1908. Early on, Corning's management had entered a series of joint ventures to combine its technologies with partner companies offering complementary technologies, skills, and capabilities for market access and understanding. In 1937 an alliance with PPG gave it access to the building industry, in 1938 an alliance with Owens

Illinois provided an entry into glass fibers, and in 1943 Dow Corning was created to develop applications for the silicon products invented by Corning.

Over time, building on these early experiences, Corning developed a strategic approach to technology alliances. It sought partners providing a high degree of complementarity and cospecialization—mainly by bringing application, end product, and user market knowledge and skills—while minimizing the threat of encroachment posed by these partners. For example, in a more recent alliance with Siemens to develop the optical communication fiber business, Corning provided the fiber development and manufacturing skills. Yet it relied on Siemens expertise in communication networks to develop the market, and to help guide Corning's technology development efforts. Corning was not about to develop expertise in communication systems, while Siemens was not interested in, nor able to, contribute much to the development and manufacturing of the fiber itself. The two firms were complementary and cospecialized, and their overlap was limited to fiber connectors and splicers, that both made.

Although one could argue that mere supply contracts, rather than joint ventures, might suffice, Corning preferred equity joint ventures, and treated these as strategic partnerships rather than mere investments. Corning's top management felt that the early uncertainties about the new markets it sought to develop, and the initial information asymmetries between partners—Corning knowing component material and process technology, the partner knowing system technologies and market and customer characteristics—were best overcome via equity alliances pooling its interests and that of its partner in a single entity. Corning was also increasingly concerned not just with application development and market access, but also with combining its technologies with that of its partners, and with developing markets rapidly ahead of competition.[11] Hence the market access speed provided by partners already engaging the appropriate market linkages was essential to Corning.

In each of its many alliances, Corning applied these principles of complementarity, cospecialization, and rapid application development, while minimizing overlaps and risks of encroachment between the partners and Corning. In addition, Corning had developed an expertise in managing partnerships, and a commitment to their success among its top management, which allowed Corning to be much more successful than less experienced companies in the management of alliances.

Yet, Corning's identity as a network of alliances raised several difficult issues. First, alliance partners had a tendency to pull Corning in too many directions. Serving a variety of strategic alliances ran the risk of technology dispersion, spreading limited R&D resources and capabilities too thinly. Corning's implicit strategic concept of being a cluster of core competencies leveraged via a series of alliances into discrete business domains made this dispersion pressure all the more worrisome. Corning was an outstanding partner—able to contribute a lot to individual alliances—only insofar as it achieved synergies across alliances. If each alliance had an independent technology base, economies of scale in core technologies and economies of scope across alliances would be lost, making Corning a much less attractive partner. In nearly all its alliances, Corning's value to the partner—and hence Corning's main source of influence and bargaining power in the alliance—was rooted in its technology and competence leadership. Synergies were essential to maintain that leadership and its affordability. In a situation in which Corning was often allied with partners of larger size or greater resources (notably Siemens, Asahi Glass) affordability remained essential, but difficult.

This had led Corning to divest from some of its more mature partnerships, and to partner some of its core activities. TV tubes provide an example of this approach. As their technology stabilized, and as margins declined, this became a less attractive business for Corning. In Europe, Corning divested from its joint venture with Thomson, Videocolor, while in the United States it sold part of the equity of the TV Glass business to Asahi. Establishing a domestic alliance in one of its historically core businesses was a difficult decision for Corning, but one it justified by the need to transfer resources to more promising higher technology, faster growing opportunities, such as porous glass for flat screens used in professional goods and portable computers.

Selectivity in new alliances was also difficult to achieve. For example, Corning had entered one alliance with Ciba-Geigy in 1985 in the medical diagnostics business. It subsequently discovered that Ciba was willing to invest significantly, across a wide front, in the medical diagnostic business, and to use the joint venture as the vehicle to carry out that effort, while Corning's interests were much more narrowly focused. Corning decided not to commit the scale of resources and the diversity of efforts required to match Ciba's ambition, and the partners split after a few years, Corning emphasizing its own development in the diagnostic laboratory business.

The ability to maintain strategic control in a network of alliances with strong partners where the core technologies and core competencies' synergies across alliances are weak is highly questionable. For a company such as Corning, or for other companies heavily involved in a network of alliances, this may be the central strategic dilemma.

We can thus summarize the critical issues that a network of technology-leveraging alliances brings, in addition to the issues present in single alliances we discussed in the previous sections, in the following way.

1. *How broad is the scope of application of existing core competencies and technologies across the alliance network?* Obviously, the broader the application scope, relative to the resource commitment required to maintain the technology or the competence, the better the "hub" (or nodal) position in the network. Related applications provide economies of scale and scope by giving the opportunity to practice and exercise competencies and technologies common to several alliances. Conversely, as we argued earlier, an alliance network with no synergies across the competencies and technologies required in each alliance would be of limited value.[12]

2. *Can core technology and core competence leadership be maintained by the nodal partner?* As we argued, Corning remains valuable to its partners only insofar as it maintains leadership in the crucial core competence areas.

3. *Are the various partners compatible?* By and large, Corning's partners have been in different "downstream" industries; their offerings are not substitutes of each other. As a result, there are few interdependencies, if any, between Corning's various partners. Where Corning entered multiple partnerships in different parts of the world in the same business—say TV tubes—problems emerge with industry globalization, in particular when the partner is a downstream end-product supplier. The partnership with Asahi, or that with Thomson, may well have cost Corning other potential customers, such as Sony, or encouraged some downstream competitors to remain or become vertically integrated upstream—such as Philips producing its own TV tube glass envelopes. Having partners that are separate from each other in product and market positions generally facilitates the role of the nodal partner.

4. *Are the partners equally able and willing to invest in business growth?* The nodal partner can maintain its leadership position only insofar as its technical and financial resources are not stretched too thin by multiple partners who can outspend (or outsmart) the nodal partner in its various application and market areas.

5. *Is the managerial capability of the nodal partner up to the task?* Singly, each of the alliances in a network raise the partner compatibility and process issues discussed in the previous sub-section. The diversity of partners is both a source of difficulty, and, potentially, of learning, as the diversity calls for more generalized alliance management skills. Originally, Corning had treated each alliance separately, and the gentlemanly but loose "country club" atmosphere of Corning in the 1960s had allowed the management of each alliance to flexibly adjust to the requirements of the task and of the partner organization. In the 1980s, the strategy of the company became tighter and Corning started to manage the various alliances as

part of an overall strategic design. Yet, apparently, by the late 1980s, Corning had developed enough expertise in selecting and managing strategic alliances to be able to rein in its various alliances without serious conflicts with the partners (except when conflicts originated in genuine differences of strategic priority, as with Ciba-Geigy).

Transferring Capabilities via an Alliance

Corning was careful to maximize stable complementation and cospecialization between its capabilities and that of its partners, and to minimize the actual capability overlap, and potential encroachment on its part and that of its partners. Alliances can also provide a vehicle to reduce cospecialization and complementarity as a way to erode the influence of a partner in the alliance, and possibly to regain competitive independence when the skills of the partner have been successfully absorbed. Taking the JVC–Thomson example, we have already shown why the motives to learn the partner's skills may be important. In this section we consider the "how" of the learning process, building further from the same example.[13]

In 1983, at the start of the alliance, Thomson's objective was to learn micromechanical mass manufacturing skills from JVC. The mechanical system of the VCR, called a "mechadeck," is made of the extremely precise assembly of a series of rotating heads and drums, including the main drum around which the tape from the cartridge rotates in a helicoidal pattern, where tolerances are extremely tight. The precision assembly of the mechadeck, and the quality of the design and the manufacturing of its components largely determine the quality of the VCR to the customer, in particular its image sharpness and its reliability.

The initial design of the alliance, originally including Thorn as a third partner, planned that JVC would provide components, process equipment, and would license product designs to a joint venture located in Europe (with plants mainly in Germany and Britain), staffed by workers provided by Thomson and Thorn and comanaged by JVC and the European partners. Products would be made with both the JVC brand and the brands of the European partners, and sold separately in their respective channels. Thomson and Thorn would also become allies of JVC in promoting the VHS standard and in dealing with regulatory authorities in the EEC. In particular, the joint venture (called J2T) was a way for JVC to overcome the import restrictions imposed by the French, and by other European countries on VCRs from Japan.

By deciding products and processes, by providing components, and by setting productivity standards, JVC exercised almost full control over the cost levels of J2T. By deciding which models were made in Europe (as against supplied directly from Japan), and by setting its own pricing policies, JVC also influenced end product prices, and thus, the margins of its European partners in Europe.

Initially, following the nationalization of the company by the French Socialist government in 1981, Thomson had been barred from participating in the joint venture with JVC. In 1983, it bargained its entry in the alliance (allowing it broader market access in Europe) against a commitment from JVC to increase the European content of the product, which also involved setting up a component and subsystem plant in France, outside of J2T, to be owned and run by Thomson with JVC's technical assistance. This provided more freedom to Thomson to develop its own skills in micromechanics.

To learn these skills from JVC, Thomson engaged in several complementary approaches.

First, Thomson engineers analyzed both the product and the process, and broke down the capability to design and manufacture VCRs into several complementary steps:

- Capability to assemble JVC-supplied subassemblies using JVC-supplied equipment and process controls
- Capability for autonomous improvements in assembly efficiency through improvements to JVC-specific assembly line layout and procedures

- Capability for specifying and bringing on stream European process equipment to supplant that supplied or specified by JVC
- Capability for developing advanced product features independently from JVC
- Capability for separate manufacture and assembly of precision components
- Capability for design and manufacture of a VCR independent of JVC
- Capability for simultaneous and closely coupled advances in both product design and product manufacturing processes, independent of J2T
- Design and manufacturing capability for next generation product (e.g., hand-held mini video-camera).

This breakdown was important insofar as it facilitated the "journey" to develop a micromechanical capability. Each stage was carefully defined, with a practical milestone. In contrast, we observed alliances where the will to learn was strong, but where the partner failed to decompose the learning task into a series of steps. When the initial distance between the partners was substantial, this led to a permanent learning gap. The will to learn was not operationalized into practical, achievable, and measurable steps, and the learning agenda appeared so daunting that it in effect discouraged any serious attempt to learn.

Identifying the learning objectives precisely, and disaggregating the resulting learning agenda into specific achievable steps, however, is not enough: This only points the way toward a feasible journey. To actually carry out the learning requires several other approaches.

First, Thomson structured the alliance to allow these steps to be taken. As local content had to be increased to meet the EEC-wide agreement on Japan, Thomson and JVC planned a sequence of local content increases which clearly paralleled the learning agenda developed by Thomson, and allowed successive steps to be taken. The alliance was also physically structured to allow learning. In the Berlin assembly plant, for example, Thomson co-located J2T's assembly lines with some lines fully under Thomson's control. These latter lines were first used to experiment processes and to develop experience in manufacturing, and then to manufacture Thomson's own VCR types, rather than the ones licensed from JVC. Workers, who were supplied by Thomson, were rotated between the J2T lines and Thomson's own, allowing for an efficient and immediate transfer of learning. At the beginning, JVC underestimated Thomson's ability to learn, and was not too concerned with Thomson's learning efforts. Later on, market access and good relationships with the European governments became important enough to JVC to trade these implicitly for support to Thomson's learning. JVC was not too concerned since its rate of skill renewal was comparable to, or better than, Thomson's rate of skill catch-up. For example, Thomson was not able to independently develop and produce the miniaturized VCR or the smaller and cheaper video cameras using the new VHS-C or Video8 formats.[14]

Second, Thomson worked hard to improve its receptivity to JVC's skills. Beyond the co-location of activities it used other approaches. For example, it hired former watchmakers from the Jura mountains, who had been put out of business by the Japanese watch industry, lured by the dual prospect of exercising their skills and getting even. Although VCRs are made to tighter tolerance than clocks, and many watches, the technologies and mindsets required to make both types of products are relatively close. Thomson also engaged in Japanese-style small group activities to articulate, exchange, and diffuse the learning occurring in the plants. This took place both at the level of individual steps, and, more importantly, at that of understanding the interactions between these steps. As Thomson's capabilities increased, the interdependency between product features, process characteristics, and manufacturability and quality, for example, became clearer and clearer. Also key to improving receptivity were the constant reminders from top management that the most important goal was to learn, not the economic and financial performance of the alliance. The priority given to the learning agenda was clearly communicated and repeated over time to all involved.

Third, the learning agenda was transformed into a learning calendar. For the first three years (1983–86) Thomson concentrated on mastering assembly technologies with sufficient precision. In parallel, Thomson had started developing its own models and testing their manufacturability and reliability. It also started to substitute manufacturing equipment sourced in the EEC for the equipment provided by JVC, and using that European equipment, quickly surpassed the productivity targets set by JVC. By 1987, Thomson was able to make its first full VCR, using both a Thomson product design, and processes developed by Thomson, and not relying on Japanese process equipment. Thomson also started purchasing components directly in the Far East to substitute for those supplied by JVC. This often allowed significant reductions in component costs. By 1988, Thomson was fully able to design and engineer both the product and the process for VCRs, and to purchase the required components, independently from JVC.

Fourth, opportunities to leverage the newly acquired skills were increased through acquisitions: acquiring General Electric's consumer electronics operations gave Thomson an opportunity to enter the American and Asian markets, while Thorn EMI's consumer electronics business provided a stronger position in the UK and Northern Europe. The latter also gave Thomson manufacturing plants in Singapore. In 1989 Thomson entered into an alliance with Toshiba, with a much more balanced starting position, to comanufacture VCRs in Singapore for mass markets worldwide. The learning resulting from the J2T experience proved quite useful in negotiating, setting up, and operating this new alliance.

In sum, although the J2T alliance continued in the early 1990s, and was used as a sourcing operation by both JVC and Thomson to serve the European markets, it had fulfilled its more strategic function of providing Thomson with enough of an apprenticeship of JVC's manufacturing processes to put Thomson at the more advanced end of consumer electronics. The cost to Thomson, though, had been high. The cumulative learning effort was costly, more in expenditure than in investment, and Thomson was only able to play an independent role in the VCR business when very keen Korean competition had driven most of the profits out of that business. Was it a phyrric victory? Perhaps not. Thomson could not afford not to be in the VCR business, nor to be in it in a permanent dependence on Japanese competitors. Yet the micromechanical technologies central to VCR quality have applications beyond VCRs in products and businesses that are of interest to Thomson.

We can summarize the key issues raised in individual alliances whose main purpose is skill learning and capability transfer around a few questions.

1. *How clear is the learning agenda?* A broad learning goal is of little help unless it is translated into a specific learning agenda, with well-defined capabilities, rooted in the mastery of practical tasks, and preferably arranged in a logical sequence of capability development, or a "learning ladder." Internalizing new skills requires the operationalization of the learning goal.

2. *How open are we to learn?* Both the J2T story and the above plea for clarity and practicality in the learning agenda may understate the difficulty of the issue, because they ignore the "unlearning" barriers. In some skill learning alliances, focusing too early on a clear learning agenda may be counter-productive. The initial mindset of the partner may make it oblivious to the more important learning, and lead to a focus on learning which does not require a deep shift in mindset. For example, General Motors probably lost several years, and many billions of dollars in its alliance with Toyota, NUMMI, by focusing initially on the plant layout, material flows, equipment and the like. These were consistent with the mindset of GM's management, but did not, by themselves, explain NUMMI's remarkable performance. Less visible to GM, but more critical, was the whole system of labor relations in the plant, and the spirit a different system fostered. It took several years for GM to discover that it had been focusing on only a small subset of Toyota's competencies. Full realization only came after GM had spent too many billions

of dollars to reconfigure its own plants, patterned on Toyota's, to no avail (Adler and Cole, 1993). Conversely, in the example we have discussed extensively, Thomson was fortunate not to have to unlearn much about micromechanics. The technology area was essentially new to Thomson, at least insofar as high volume, high precision manufacturing was concerned.

3. *What is the nature of the capabilities we need to develop?* Understanding the nature of required capabilities is essential to designing the interface between the partners, or even in deciding in favor of a partnership over other firms of collaboration. The more the capabilities belong to the realm of competencies, (tacit know-how in action, embedded organizationally, systemic in interaction and cultivated through learning by doing), the more co-practice is essential. Indeed, without the NUMMI experience, GM might have permanently missed the essence of Toyota's management processes, despite the rather abundant literature available on them. Where the capabilities are more explicit and less like competencies, co-practice and apprenticeship become less important, and the capabilities may be transferred by other means (typically technology licensing, perhaps supported by some specified technical assistance). Even in the case of an alliance, a less intimate form of cooperation may be sufficient.

4. *How should the learning process be operationalized, once the capabilities to be developed have been ascertained?* We dwelled at some length on this question in our description of the J2T alliance. The key points to recall are: breaking down the learning agenda into a sequence of practical steps (with each new step building on the previous ones); maximizing managerial receptivity; and establishing a learning schedule.

5. *How should we establish the boundaries between the partners to allow observation and co-practice?* The partner who wants to learn, and has the capability to do so, is likely to try to establish porous boundaries and to leave the interface loosely defined. In the early phases of its collaboration with ICL, as a way to learn about users, for example, Fujitsu tried to convince ICL's management to let Fujitsu engineers accompany account managers on visits to their clients in the UK. At the same time, ICL was defending a much more arm's length relationship, in which it would provide technical specs for Fujitsu to develop products against these specs. Conversely, in the J2T case, dedicating only part of the Berlin plant to J2T, and rotating the workforce between sections of the plant provided a crude but effective mechanism for both co-practice and independent replication and experimentation. Similarly GM has used NUMMI as a learning and training ground for its production managers, who were transferred back to GM's own operations, after having been seconded to NUMMI for a while.

6. *How can we learn through practice?* Practice alone is no guarantee of learning, unless the right mindset and skills are present. To some extent in learning situations "believing is seeing" (Crossan and Inkpen, 1994). Grasping the more subtle competencies can only come to the trained eye and the trained hand. Thomson saw its use of watchmakers, engineers, and technicians as a cornerstone in its learning process. Too much distance between the learning starting point and the new required skills, and the implicit mindset they call for, makes learning impossible. Such a gap diminishes the chance that the partnership will allow the lagging partner not just to copy the skills of the leading partner, but also to independently develop these skills further.

7. *How compatible are skill learning and internalization with the continuation of the alliance?* To a large extent JVC was "locked in" by the time its management realized the extent of Thomson's learning. Thomson's very incompetence at the beginning of the alliance served Thomson well: it was easy to learn from an unsuspecting JVC who had discounted Thomson's skills. As the extent of Thomson's learning became visible, JVC was too dependent on the JVC-given European protectionism and the political salience of the venture to pull out or even to trigger a visible crisis with Thomson. JVC did tighten up Thomson's access to its plants in Japan (labs had always been off-limits), but relatively late in the learning process. Similarly, al-

though obviously displeased, Toyota was hard put to prevent GM from appointing to the Saturn project the managers who had been posted at NUMMI.

8. *How can we benefit from the learning investment?* Clearly, there are several ways to benefit from the learning. One, already discussed at some length, involves widening the scope of application of the newly learned technologies and competencies, thus increasing the leverage from the investment. Another is to use the results from the learning to renegotiate a more favorable alliance with the partner, for instance by rebalancing financial flows in favor of the partner who has learned. A look at the history of royalty payments between Xerox and Fuji-Xerox is instructive in that regard: one can observe a progressive rebalancing of the relationship over thirty years, increasingly favorable to Fuji-Xerox, as it assumes a growing product development and sourcing platform role for the whole Xerox group (Gomez Casseres, 1991). A third benefit is to use the learning from one alliance to improve one's position in the next, as Thomson did in its alliance with Toshiba, subsequent to J2T. This positioned each alliance in the broader context of an intertemporal learning network, where the learning ladder extends not just between steps in an alliance, but also in a more ambitious and longer-term process, from alliance to alliance.

Integrating a Network of Competence-Acquisition Alliances

Alliance networks can play many different roles. In our discussion of Corning we stressed one of these roles: the competence leverage multiplier role. Networks can also allow one to build a stronger competitive coalition. This can help establish a standard: JVC's desire to secure the leadership position of the VHS standard did play a role in its initial willingness to cooperate with Thomson and then other independent consumer electronics suppliers, Thorn and Telefunken in 1977–78. Alternatively, the network can help achieve a sufficient global presence to challenge the leader in an industry: this was one of General Electric's motives in initiating the relationship with SNECMA, and a key interest of Fujitsu in its collaboration with Amdahl and ICL early on.

Contrary to other nodal partners we observed, Fujitsu was able to coordinate its policies and to integrate its learning across alliances in a very effective way. A central corporate alliance staff group, headed by a senior executive, was able to provide expertise and guidance to the various sub-units of Fujitsu involved in various alliances and to encourage the sharing of learning between alliances. Fujitsu's nodal position in a wide range of alliances was actively exploited by that central group.

Beyond the interdependencies between co-existing alliances, one can also consider an intertemporal network of alliances over time, each contributing to the development of competencies on the part of the company. For example, Siemens had fallen behind in semiconductor technology when it concluded, in the early 1980s, that it could not be fully dependent on competitors, or even on independent merchant suppliers such as INTEL, for such critical components. Playing a catch-up game was relatively difficult though, in an industry where the technology moves rapidly and where process skills are of the essence.

Siemens first entered an alliance with Toshiba, which was also trying to catch up with industry leaders, for similar reasons to Siemens. Toshiba brought process manufacturing skills that Siemens lacked, and a good experience of managing technology alliances. Siemens brought other skills, such as product development. Strengthened by this initial alliance, Siemens then allied with Philips, first in the "Mega" project to develop 1 megabit memories. While Siemens focused on DRAMs, Philips focused on SRAMs, thus minimizing the overlap in product, and the potential competition, while allowing a common investment in production skills and capabilities. The project was launched in 1984 by Siemens, and Philips joined in, under the auspices of ESPRIT in 1985. By 1986, Siemens bought technology from Toshiba, which had moved fast to develop its own 1 megabit memories. The relative closeness of Philips and Siemens in their strategic interest, in their technologies, cultures, and in geography facilitated a rapid start of the collaboration process. The extensive use of parallel development was one other way to speed up the process. By 1988, Siemens had a brand new facility in Regensburg, successfully producing 1 megabit DRAM using a mix of Toshiba's technology, of Siemens' own, and of the joint results from the Mega investment. In 1989, the "Mega" project was followed up by the Joint European Submicron Silicon Initiative (JESSI). JESSI was aimed to complement the "Mega" project as an investment into the overall European capabilities in semiconductors, from material and manufacturing equipment providers to semiconductor users. However, the largest participants were still to be Siemens and Philips. In 1990, Siemens entered an alliance with IBM, to develop further the submission technology and build the next generation of memories. The alliance was subsequently joined by Toshiba, which in the meantime had captured the leadership position for 1 megabit DRAMs. For Siemens, one can summarize these ten years of effort as a way to reenter the race for advanced microelectronics.

Other companies had followed similar approaches. For example, following its demise at the end of World War II, the German Aerospace industry engaged in a series of alliances, starting in the early 1960s. Playing an increasingly equal role to that of its partners, it skillfully used the learning resulting from each alliance to gain strength in the subsequent ones. Specific technologies appropriated in one alliance were used as bargaining chips to gain a better role in the next.

A network of alliances can also be used, in certain junctures, as a way of maintaining options in the face of uncertain technologies and markets. The response of major pharmaceutical firms to the AIDS crisis, for example, fits that pattern. Merck provides an interesting illustration. It maintained a whole series of alliances to explore the various approaches to AIDS, treatment, vaccines, better tests, and to pursue various options in each approach.

In sum, networks of technology alliances provide for a series of potential advantages, which can be summarized around key issues.

1. *Can a network of alliances be used to make industry structures, and terms of competition evolve toward one's own strength?* The most obvious example is provided by the VCR story, where by quickly establishing a series of alliances to promote the VHS standard, the JVC–Matsushita group not only gained the advantage over Sony and scuppered Philips' ambition to enter the VCR business on its own terms, but also "froze" the technology for a number of years, and shifted the competitive arena toward rapid production scale up, mass manufacturing and distribution, and cost reduction, all areas in which Matsushita excels. In general, where standards play a role, and where the competencies and skills of competitors differ widely, the need for alliance networks to set the parameters of competition, and to influence them in the interest of the focal company is significant.

2. *Can a nodal network position be used to cumulate and leverage competencies across alliances?* While Corning was essentially a competence developer, building its own competencies to be applied in its various alliances, Fujitsu, as we discussed, was partly a competence developer, partly a competence exchanger, especially insofar as it could use the same competencies across alliances. By combining the learning from several technologically related alliances with different partners, the nodal partner can normally build competencies faster, and maintain or increase its leadership in the coalition.[15]

3. *Can learning be cumulated over time between alliances and used as a source of growing bargaining position from alliance to alliance over time?* The experience of both Siemens and DASA (the consolidation of the German aerospace industry) suggests that this can be done, given two conditions: first, the technological trajectory of the industry must be clear enough; and second the strategic architecture of the company must be tight enough to inspire confidence in long-term competence development efforts. Both semiconductors and aerospace are indeed relatively stable, if technologically very demanding industries. In more volatile environments, learning cumulation may be less easily achievable.

4. *Can a network allow you to keep open a multiplicity of options at an affordable cost?* In highly uncertain technological areas, like AIDS prevention and treatment, an approach built on a network of relationships, and constructing a type of "bet" focusing on individual technologies and research areas makes sense. It acts both as an information network, and as a portfolio of options.

In sum, networks of alliances can provide many advantages, but only provided the network is actively managed from the center. In other words, it must not be run merely like a collection of alliances. Active management, as we discussed briefly in the case of Fujitsu, is not easy though.

Conclusion

The main conclusions of this section are presented in summary form in Figure 37.2.

3. TECHNOLOGY ALLIANCES AS LEARNING PROCESSES

The preceding discussion of various types of alliances, singly or in the context of an alliance net-

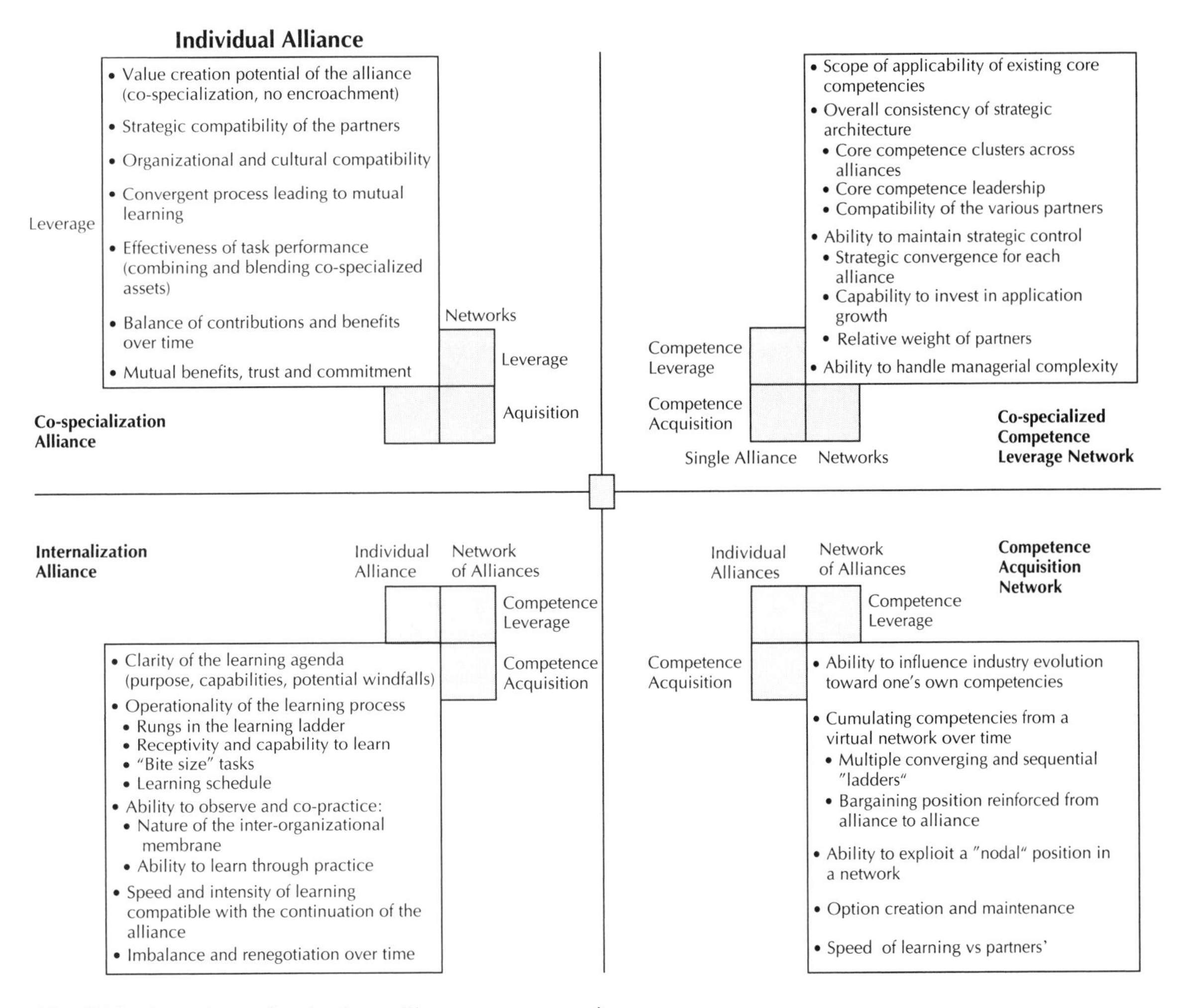

Fig. 37.2. A typology of technology alliance management issues.

work may convey a sense of the management of alliances being a deterministic and analytical process, where the appropriate questions can be raised, and answers provided at the outset of the relationship. This is rarely the case.

Alliances are often pursued as ways to explore new applications, and new technologies, or the match between the two. By their very nature they confront uncertainty and information asymmetry, indeed they are sometimes designed to overcome information asymmetry between the partners. Further, although a full explicit agreement is often seen as desirable, seeking such an agreement early on can foreclose the possibility of collaboration, given uncertainty, information asymmetry and lack of familiarity that exists between the partners. The need to learn in alliances—and not just in terms of skills—should therefore not be underestimated. And nor should the potential for dysfunctional learning be underestimated. As an illustration of the need for learning, and the difficulty of learning, this section starts with the description of the alliance between Alza and Ciba-Geigy, from which we draw various implications on the nature of technology alliances as learning processes.

Ciba-Geigy and Alza

Alza was started in 1968 by Alex Zaffaroni, a medical doctor and Syntex executive. He had become convinced that the administration of drugs to patients could be considerably improved by better controlled release devices, such as transdermal patches, osmotic pills, implantable slow release polymers, or implantable minipumps.

Over the first ten years of its existence, Alza gathered a first-rate group of scientists and technicians of drug delivery technologies and identified highly promising work. Yet its products failed in the market (they were more costly and harder to use than conventional products, and addressed markets of limited size) and by 1977 Alza was finding itself in an increasingly precarious financial position. Ciba-Geigy's pharmaceutical division, which was short of really new products, was keen to differentiate itself from competitors and wished to extend the life of some of its products, or to use known chemical substances, but ones whose use was difficult in conventional forms and dosages. In 1976, Ciba-Geigy had carried out a preliminary study of new drug delivery methods, and identified Alza as a source of potential technology. When Ciba-Geigy found out that Alza was looking for funds and for marketing expertise, it signaled interest.

The relationship between Ciba-Geigy and Alza started in 1977, when Ciba-Geigy reached a research agreement with Alza, and also took a significant equity position in the company. Alza was to develop for Ciba-Geigy applications of advanced drug delivery systems on which Alza had been working since its founding in 1968, in particular transdermal (TTS) and oral (OROS) slow release systems. These were to be combined with active ingredients that were difficult to use in other forms (e.g., nitroglycerine for angina pectoris) or proprietary substances of Ciba-Geigy. The patented life of these proprietary substances was to be extended by the use of the advanced delivery system. The R&D priorities were set at the beginning in the course of a "joint research conference" to be followed about twice yearly by a session of a joint research board to review progress and reassess priorities. Priorities included both the development of new technologies (e.g., implantable polymers and minipumps) and the adaptation to Ciba-Geigy's needs of technologies already largely developed by Alza. Coordination was to be ensured by a series of complementary approaches: scientific liaison desks were set up in all three locations (including Ciba-Geigy-U.S.), project sponsors were appointed (but only on a part-time basis on top of existing responsibilities in Ciba-Geigy), a business liaison desk was established in Ciba-Geigy, and internal coordination within Ciba-Geigy was to be achieved via an "advisory team" in Basle, regrouping all key persons involved in the interface with Alza and in the work on the advanced delivery systems, and via a specialized R&D group in Ciba-Geigy-U.S. Ciba-Geigy and Alza expected to obtain, within two to three years, a first wave of applications both for TTS and OROS technologies and to research further products and systems. Sixty percent of the joint R&D budget was devoted to short-term projects, the balance to new technologies.

The alliance was thus initially focused on capability complementation. Alza was providing delivery system technologies, while Ciba-Geigy would provide substances, and manufacturing, registration and marketing expertise for the new products.

By 1979 some difficulties started to develop. While projects made good progress, it became clear that the OROS technologies needed more application-specific development, and that progress on pharmacokinetics was required to guide this development. In the absence of joint Ciba-Geigy-Alza R&D teams, it was difficult to perform this interactive application-specific development between the system features and the substance characteristics. The problem was compounded by the need to involve the manufacturing function, which had been traditionally quite separate from research at Ciba-Geigy. These difficulties made OROS a less immediate alternative to slow release technologies already developed within Ciba-Geigy. TTS projects made good progress largely because they found maverick sponsors within Ciba-Geigy who developed strong personal commitments to the projects and to their counterparts at Alza. Most Ciba-Geigy sponsors, though, had many other commitments, no specific budget, and no ability to modify or bypass the usual slow sequential development process whereby new drugs move from one test to another. Further, contacts were poor between Ciba-Geigy-Basle and Ciba-Geigy-U.S. Their relationship suffered from: the rivalry between the subsidiary; a struggle for independence, and the consideration of the relationship with Alza in that context; and the intent of divisional headquarters on achieving more international coordination within R&D. A tradition of consensual decision making did not facilitate the resolution of such tensions within Ciba-Geigy. As a result, Alza sometimes found itself involved in parallel relationships on the same topic with both Ciba-Geigy-Basle and Ciba-Geigy-U.S., in which the two Ciba-Geigy subunits did not communicate.

Faced with these delays and frustrations, Alza's management reacted in several ways. First, the problems reinforced the view that Ciba-Geigy was perhaps not the ideal partner over the long run, and that the cooperation was better seen as transitory, with Alza regaining its independence after a few years. Second, in an attempt to get faster action on the part of Ciba-Geigy, Alza's managers and scientists started to "walk the corridors" in Ciba-Geigy, trying to get quick decisions. This behavior violated the accepted norms of interaction in Ciba-Geigy, and triggered adverse reactions on the part of Ciba-Geigy's executives and personnel, in particular as Alza's executives had a tendency to leverage their access to Ciba-Geigy's top management. This access was easier for them than for Ciba-Geigy's own employees, given the hierarchical nature of Ciba-Geigy's processes. TTS was favored by the relatively high informal status, and willingness to bypass rules of its sponsors, as compared to a more "usual" process for OROS. Third, Alza started to look for third party research projects, as a way to generate revenues quickly and to make full use of its R&D capability. Alza's management was also concerned that unless it maintained a steady research effort upstream, the value of the partnership with Alza to Ciba-Geigy, and Alza's bargaining power in it, would decline over time.

Within Ciba-Geigy, those who had doubted the value of new drug delivery systems, and of Alza's innovations, saw slow progress as confirmation of their negative prior judgment. Ciba-Geigy's executives also started to realize that in the absence of third party contracts, Alza's losses would mount, leading to a cumulative negative cash flow of $57 million by 1986. Exercising its power of shareholder, Ciba-Geigy imposed on Alza drastic savings, which affected mostly the marketing and manufacturing functions, but left R&D relatively unscathed. The Alza plant was sold, and proceeds used to pay back loans. Alza's management felt these cuts were inescapable, if only to restore their partner's confidence.

In 1979–80, Ciba-Geigy reorganized its development activities by therapeutic area teams to speed up product development, as other pharmaceutical companies were doing too. At the same time, although Alza was authorized by Ciba-Geigy to look for third party contracts, a slow and complicated vetting process by Ciba-Geigy deterred potential third party partners from entering an agreement with Alza. Increasingly frustrated with Alza's "intrusions" in their decision making, Ciba-Geigy executives resorted more and more to "premeetings" where they would reach decisions on a common Ciba-Geigy position to be presented to, and if need be, imposed on Alza. Alza's managers obviously felt railroaded. Openness in communication between the partners started to break down. Ciba-Geigy was wary that information on development and manufacturing could leak to competitors, while Alza started to suspect that Ciba-Geigy wanted to develop its own in-house advanced drug delivery system capability, including research. As a result, communication exchanges were reduced to their contractual minimum, and the development of some applications further suffered. The Ciba-Geigy researchers, frustrated by the lack of open communication, came to believe in 1980–81 that they had obtained from Alza about all there was to obtain, and that neither further process know-how nor major breakthroughs were likely to result from the partnership.

At the end of 1981, Ciba-Geigy decided to divest from Alza and phase out the research agreement over the following three years. The cooperation agreement was revised to give Alza more generous royalties on the products it had developed, and less exclusivity to Ciba-Geigy, facilitating Alza's search for multiple third party agreements. Although Ciba-Geigy remained a shareholder of Alza for a while, and keeps cooperating with Alza on specified products, the alliance was dissolved. It was not seen by either party as a failure, though. Several TTS-based products introduced in the early '80s have become major successes for Ciba-Geigy, while Alza, once out of the alliance with Ciba-Geigy, was able to quickly reach a number of third party agreements and to succeed as a contract researcher for a variety of health care companies.

The Nature of Learning in the Alza-Ciba-Geigy Alliance

Contrary to the simpler alliances discussed in the earlier section, it is difficult to pigeon-hole the Ciba-Geigy-Alza alliance. Part of the difficulty stems from its evolution. During the first two years (1977–79), the alliance was clearly seen as a capability complementation one. However, over time, it became clear that, for OROS at least, capability complementation was actually rather difficult. The task was just more difficult than expected, and required more application-specific work than was provided for in the initial agreement. By 1979, these difficulties were recognized, but this sometimes led to more friction. For example, the more active involvement of Alza's managers in Basle led to adverse reactions on the part of Ciba-Geigy's personnel. Alza failed to achieve effective action, by Ciba-Geigy's rules, and failed to change these rules, or obtain a way to override them.

During the first two years of the alliance, thus, learning took place on several fronts. The partners learned about the common task, and found it more difficult than anticipated. The partners also learned how difficult it was to work together successfully. A set of differences of a somewhat similar nature to those observed in the relationship between AT&T and Olivetti stood in the way of successful collaboration between Alza and Ciba-Geigy. They also learned about their own goals, and their partner's, and the fact that these goals were not always easily reconcilable. For example, the head of Ciba-Geigy's pharmaceutical division had entered the alliance with both concern to not stifle entrepreneurship and innovation at Alza, and a desire to use these as a role model for Ciba-Geigy's own R&D. The first concern called for an arm's length relationship, in which Alza would be protected from Ciba-Geigy, the latter called for more exposure and joint work. Similarly, Alza's management was keenly aware that its long-term value to Ciba-Geigy, and potentially to third parties, was critically dependent on new research. But the agreements with Ciba-Geigy made Alza's future dependent on royalty payments stemming from the early introduction of already developed products. As a result of these dual priorities, the focus of Alza's technical skills, and the balance between advanced research and product development support became difficult issues.

The partners learned not only about their own sets of goals, and their potential internal contradictions, but about each other's. Ciba-Geigy discovered that Alza had not given up its ambition to become a fully fledged autonomous pharmaceutical company, while Alza discovered that Ciba-Geigy had not necessarily forsaken its own R&D on advanced drug delivery systems. This led to a heightened sense of possible encroachments.

Not all learning was constructive, in that useful learning occasionally triggered dysfunctional behavior. As discussed in the summary description of the Alza-Ciba-Geigy alliance, Alza's understanding of the complexity of its partner's decision making processes led to an active "lobbying" and "corridor management" attitude, which aggravated a good part of Ciba-Geigy's staff who were unused to such intrusive behavior. This, in turn, led some, within Ciba-Geigy to decommit personally from the relationship, not the outcome sought by Alza.

Learning also led to modify the interface, or the internal organization of the partners. Although there were various reasons for Ciba-Geigy's reorganization by therapeutic area development teams in 1979, the desire to facilitate the relationship with Alza, and to speed up the development process, did play a role.

Beyond 1979, the awareness of competition grew. Alza was increasingly worried about its technology lead being eroded by Ciba-Geigy's perceived slowness and by the rapid catch-up of competitors, especially that of Key Pharmaceuticals.

Learning in Technology Alliances

This example illustrates the fact that technology alliances are evolutionary processes where the initial conditions are frequently questioned by intermediate learning outcomes of the very process of collaboration. In managing technology alliances it is important to be attentive to at least five domains of learning:

Environment learning. Results from a better appreciation of the environment of the partnership. This better appreciation may just result from action getting feedback from the environment, leading the partners to revise or sharpen their initial assumptions, or from changes in the environment in the course of the partnership, making some of the initial assumptions obsolete. The internal environment of the partnership, within each partner, may also change as a function of other opportunities developing or closing, or of internal organizational and political aspects, such as the partnership gaining or losing powerful sponsors for reasons that have little to do with its evolution itself. As partners may have more or less similar perceptions of the external environment, and more or less compatible evolutions of their internal environment, environmental learning may lead to convergence or divergence between them, and to increasingly similar or different assumptions about the future environment.[16]

Task learning. Starting the cooperation work may also lead to learning about the task, in particular when the task was broadly defined at the outset, and/or when it is complex or uncertain. Depending on the novelty of the task of the partnership, more or less task learning may be required. When partners start with information asymmetry, complementary but not very differentiated skills, and use the alliance to develop new opportunities, the task is unlikely to be defined accurately or precisely at the outset, calling for more precise definition through extensive task learning as the partnering process unfolds.

Process learning. Similarly, the very process of collaboration itself can be an object of learning. Partners discover *each other.* Recognition of their differences in structure, processes, action routines and the like may lead partners to learn to overcome, or even to exploit these differences in the collaboration process. Some of the noise and ambiguity which are inherent to collaboration between dissimilar organizations may be eliminated over time as the partners learn to relate to each other more clearly and to cooperate more effectively.

Skill learning. Some alliances may also offer "windows" into the partners' respective skills and opportunities to the partners to learn new skills from each other (Hamel, 1990, 1991). Skill learning may allow the partners to cooperate more closely as their respective skills converge, but may also decrease the value of future cooperation between them (in particular if skill learning ranked high on the partners' hidden agenda) and make current cooperation rife with tension if skill learning is asymmetric between partners.

Goal learning. The collaboration process itself may also reveal the partners' goals to one another. The behavior of each partner in the alliance may provide clues about their hidden agenda and true motives beyond, or beside, the explicitly shared goals of the collaboration. Further, and more interestingly, the collaboration process itself may lead the partners to clarify, revise, or refocus their own goals. New unseen benefits to the cooperation may become visible, or new outcomes may now be expected, given the partner's interpretation of the evolution of the partnership.

The observation common to all five dimensions of action learning outlined above is that the determinants of partnership outcomes cannot be reduced to partnership starting conditions. Initial conditions obviously matter in setting and construing the learning processes, but the outcome of these learning processes is not fully determined by their starting position.

Learning results in the revision of the initial conditions of the alliance, such as task definitions, or interface structures and processes, and may even lead partners to reform their own organizations and decision making processes, as discussed in the GE-SNECMA case.[17] For obvious reasons, commitments are also affected by the tacit or explicit revaluations of expected outcomes that result from learning in the course of the alliance.

It is therefore useful to consider alliances not as "one-off" agreements to be implemented, but as iterations of interactive cycles of cooperation over time. This entails ongoing learning in several domains, punctuated by periodic, but not necessarily fully explicit nor shared, reassessments, leading on the part of each partner to changes in commitment to the alliance, task definition, interface mechanisms and approaches, and, possibly, in each party's organization and management processes.

CONCLUSION

This chapter has argued that alliances have a key role to play in the implementation of technology strategies, and more generally of corporate strategies. This is especially the case when the skill bases of partners are different, and where the technologies involved are more tacit and systemic, rather like competencies and less easily manageable and learnable than explicit and discrete technologies.

Second, we have argued that it is important to distinguish alliances whose purpose is to combine cospecialized technologies between the partners, in which each retains a discrete set of technologies, from alliances whose goal is, explicitly or implicitly, to learn and internalize the partner's skills via the alliance. They raise very different issues, and the strategic and managerial priorities are quite different between the two types of alliances.

Thirdly, we have argued that the management of technology alliances is exercised at least at two levels: individual alliances and a network of alliances. It is important to manage actively both levels, and the issues are quite different. It is also important to recognize that alliance networks have to be managed differently—like individual alliances—depending on their main purpose.

Finally, we have argued that alliances, by their very nature cannot be planned and decided once and for all at the outset. Instead they should be considered as multiple cycles of learning processes, resulting in periodic reassessment and readjustment to the relationship.

It is perhaps worth ending with a word of caution. While alliances often provide an attractive option, partly because they are a vehicle to match commitments to learning, as uncertainty and information asymmetries are removed, their cost should not be underestimated. Their costs in terms of management time and energy, organizational stress, and possible constraints on strategic freedom far exceed the visible costs on the P&L statements or balance sheets. Inexperienced managers should be wary of entering alliances without a full appreciation of these less discernible costs that manifest themselves only over time as the relationship develops. These conclusions are not meant to dissuade companies from entering into alliances, simply to encourage them to manage those alliances more actively.

Notes

1. The current superiority of "lean" over "mass" manufacturing disciplines in the car industry may force the elimination of some of the weaker car makers in Europe and the United States, and their replacement with Japanese transplants. Over the longer run, however, several U.S. and European car manufacturers will adjust, and regain competitiveness against the Japanese, at least in this particular dimension.

2. Although CFM did not cover the high end of the thrust range, and thus the larger jet engines, nor necessarily new innovative projects, such as propfan engines, GE had always offered SNECMA the opportunity to codevelop any new products, and to participate significantly in all its existing programs.

3. In the early 1990s, Toyota succeeded in excluding the United Auto Workers from NUMMI. In addition, Toyota gets public relation benefits from NUMMI's success, which points to mismanagement, rather than an intrinsically undertrained and restive workforce as the cause of the decline of the U.S. car industry's competitiveness.

4. Although this particular case illustrates strategic similarity and alignment between partners sharing a common ambition, such commonality should not be taken as a precondition for success, which is why we use the word compatibility rather than commonality. Partners pursuing different but compatible objectives in an alliance may actually find agreement sometimes easier as the benefits they expect from the alliance are not in competition with each other. Reaching a "package deal" solution, in which each partner finds satisfaction vis-à-vis different objectives, may be easier.

5. With the exception of the re-engining of old Mirage planes, their products are not interchangeable, but military aircraft producers do compete with, for example, the Dassault Mirage 2000 planes using SNECMA engines, and their U.S. competitors, the F16 in particular, using GE engines.

6. It is possible, though, that a change in status of one or the other of the partners would lead to a restructuring of the alliance. For example, SNECMA's privatization, or a divestiture by GE of its aero-engine group could lead to imbalance in feasible financial commitment between the partners, possibly leading to a restructuring of the current relationship.

7. For example, the fluctuations in the value of the dollar, its deep fall of the late 1980s in particular, did create tensions in the transatlantic alliances we researched. Jet engines being sold in dollars, SNECMA had to make tremendous productivity efforts, and to engineer smart hedging deals, to avoid a conflict with GE on pricing and market development policies.

8. It is important to note here that GE had already relied on SNECMA as a risk sharing subcontractor for the CF6 engine, to be installed on the Airbus A300, before it engaged into the CFM alliance. The experience of the CF6 project had allowed the two companies to get to know each other in the context of a collaboration where mutual commitments were lesser than in the "50–50" CFM alliance. In contrast again, AT&T and Olivetti had to rush into trying to make momentous commitments, such as a new joint computer product line, from a situation in which they knew and trusted each other relatively little.

9. For obvious reasons, the more partners expect to continue cooperation over the long term, the more cooperative and truthful they are likely to be in the short term. Negotiation theorists call this phenomenon the shadow of the future. One of the more salient consequences of this is that partnerships that have a set date of termination—or renewal—are likely to grow increasingly contentious as they get closer to that date. This calls for renegotiation and renewal dates well in advance of the expiration date.

10. There are companies such as Aérospatiale, Deutsche Aerospace, or British Aerospace where alliances and partnerships account for a higher share of their total sales. However, many of these firms' alliances have been mandated by external forces, such as military and civilian administrations, in the various European states. The management of most of these companies deplores rather than welcomes at least some of the alliances in which they are forced to participate.

11. Historically, Corning had missed several opportunities by failing to transform its innovations into products fast enough, from electric bulbs where it lost out to General Electric, to semiconductors where its subsidiary, Signetics, had been an also-ran, sold to Philips in 1975, and to ophthalmic glass where Corning lost out to flexible contact lenses.

12. This is a close analogy to the argument in favor of related diversification over unrelated diversification and is not exclusive to alliances, although the questions of why other companies see Corning as an attractive partner and of how Corning maintains its bargaining influences over these partners, add an alliance-specific twist to it.

13. For a very detailed treatment of the example, see Hamel, 1990 (Appendix) and Hamel and Doz, 1990.

14. Some cynics at JVC argued that Thomson was exhausting itself in a catch-up race it could not win, and that new product development capabilities and speed at JVC provided a good "safety margin" against Thomson's efforts.

15. In some cases, by serving more protected markets, and having its partners compete on more broadly contested ones, the nodal partner may maintain better margins than its partners and be able to use the resources it generates to build into competencies and transform itself into a true competence developer rather than a competence exchanger. This seems to have been Fujitsu's evolution vis-á-vis both Amdahl and ICL.

16. Although different assumptions may make the initial negotiation between the partners easier, the quality of the collaboration process may be weakened if these stay very different.

17. Similarly, following its unfortunate experience with Olivetti, AT&T transformed itself and the way it managed alliances, which proved quite useful in further cooperative efforts, such as the one with Italtel in the late 1980s.

REFERENCES

Adler, P.S., & R.E. Cole. 1993. "Designed for Learning: A Tale of Two Auto Plants." *Sloan Management Review 34,* 3, 85–94.

Cantwell, J. 1989. *Technological Innovation and Multinational Corporations.* Oxford: Basil Blackwell.

Crossan, M. and A. Inkpen. 1994. "Promise and Reality of Learning through Alliances." *International Executive 36,* 3, 263–73.

De Meyer, A. 1991. "Removing the Barriers in Manufacturing: The 1990 European Manufacturing Futures Survey." *European Management Journal 9,* 1, 22–29.

Dierickx, I. and K. Cool. 1994. "Competitive Strategy, Asset Accumulation and Firm Performance," in H. Daems and H. Thomas (eds.), *Strategic Groups, Strategic Moves and Performance.* Oxford: Pergamon Press.

Doz, Y. 1994. "New Product Development Effectiveness: A Triadic Comparison in the Information Technol-ogy Industry," in T. Nishiguchi (ed.), *Managing Product Development.* New York: Oxford Univers-ity Press.

Doz, Y. 1992. "The Role of Partnerships and Alliances in the European Industrial Restructuring," in K. Cool, D. Neven, and I. Walter (eds.), *European Industrial Restructuring in the 1990s.* London: Macmillan.

Franko, L. 1973. *Joint Venture Survival in Multinational Corporations.* New York: Praeger.

Fruin, M. (forthcoming). *Knowledge Works.* New York: Oxford University Press; and "Innovation and Organisation Design: Organisational and Property Rights and Permeability in Product Development Strategies in Japan," in Y. Doz (ed.), *Managing Technology and Innovation for Corporate Renewal.* New York: Oxford University Press.

Gomez Casseres, B. 1991. *Xerox and Fuji-Xerox.* Harvard Business School case.

Hamel, G. 1990. "Competitive Collaboration: Learning, Power and Dependence in International Strategic Alliances." Ph.D. dissertation, University of Michigan.

Hamel, G. and Y. Doz. 1990. "The Video Cassette Recorder Industry (C)." INSEAD–London Business School Case Series.

Hamel, G. and C.K. Prahalad. 1991. "Corporate Imagination and Expeditionary Marketing." *Harvard Business Review,* 69, 81–92.

Hamel, G. and C.K. Prahalad. 1985. "Do You Really Have a Global Strategy?" *Harvard Business Review,* 63, 139–148.

Harrigan, K. 1986. *Managing for Joint Venture Success.* Lexington, MA: Lexington Books.

Itami, H. and T. Roehl. 1987. *Mobilizing Invisible Assets.* Boston, MA: Harvard Business School Press.

Nevens, T.M., G.L. Summe, and B. Uttal. 1990. "Commercializing Technology: What the Best Companies Do." *McKinsey Quarterly 4,* 3–22.

Nonaka, I. and Hirotaka Takeuchi. 1995. "The Knowledge-Creating Company. New York: Oxford University Press.

Prahalad, C.K. and G. Hamel. 1990. "Core Competence and the Concept of the Corporation." *Harvard Business Review,* 68, 79–91.

Teece, D. 1987. "Profiting from Technological Innovation: Implications for Integration," in D.J. Teece (ed.), *The Competitive Challenge: Strategies for Industrial Innovation and Renewal.* Cambridge, MA: Ballinger.

Womack, J.P., D.T. Jones, and D. Roos. *The Machine That Changed the World.* New York: Macmillan.

CHAPTER SEVEN

EXECUTIVE LEADERSHIP AND MANAGING INNOVATION AND CHANGE

We have explored in detail how successful firms surmount the problem of managing inconsistent requirements across cycles of technological change.

What glue binds an "ambidextrous" organization that is capable of leading and adapting to conflicting demands in turbulent environments? Executive leaders hold such enterprises together and direct them to capture value from innovation. They balance continuity and change, guiding companies through long periods of gradual adaptation, punctuated by dramatic transformations that take place in relatively brief intervals.

Tushman et al. begin this section by describing this pattern of convergence and upheaval. Their model provides concrete advice for managers who must decide how to manage both types of change, and must know when each type is appropriate. Nadler and Tushman discuss how top executives actually implement the difficult, yet necessary process of fundamentally transforming an organization's architecture. Kolesar provides a detailed case study of how one executive effected a seismic change in a traditional industrial giant. Step by step he describes how the CEO instilled a total quality culture in a firm whose institutional resistance to such a metamorphosis was rooted in its proud heritage and the lessons learned from decades of dominating its industry. Howard concludes this section by providing a concrete example that illustrates how a CEO leads a traumatic reorganization that fundamentally alters his company's identity. Changing his firm's definition of its position in the world not only alters its strategy, but requires a shift in its deep structure.

38

Convergence and Upheaval: Managing the Unsteady Pace of Organizational Evolution

MICHAEL L. TUSHMAN
WILLIAM H. NEWMAN
ELAINE ROMANELLI

A snug fit of external opportunity, company strategy, and internal structure is a hallmark of successful companies. The real test of executive leadership, however, is in maintaining this alignment in the face of changing competitive conditions.

Consider the Polaroid or Caterpillar corporations. Both firms virtually dominated their respective industries for decades, only to be caught off guard by major environmental changes. The same strategic and organizational factors which were so effective for decades became the seeds of complacency and organization decline.

Recent studies of companies over long periods show that the most successful firms maintain a workable equilibrium for several years (or decades), but are also able to initiate and carry out sharp, widespread changes (referred to here as reorientations) when their environments shift. Such upheaval may bring renewed vigor to the enterprise. Less successful firms, on the other hand, get stuck in a particular pattern. The leaders of these firms either do not see the need for reorientation or they are unable to carry through the necessary frame-breaking changes. While not all reorientations succeed, those organizations which do not initiate reorientations as environments shift underperform.

This article focuses on reasons why for long periods most companies make only incremental changes, and why they then need to make painful, discontinuous, system-wide shifts. We are particularly concerned with the role of executive leadership in managing this pattern of convergence punctuated by upheaval.

Here are four examples of the convergence/upheaval pattern:

> Founded in 1915 by a set of engineers from MIT, the General Radio Company was established to produce highly innovative and high-quality (but expensive) electronic test equipment. Over the years, General Radio developed a consistent organization to accomplish its mission. It hired only the brightest young engineers, built a loose functional organization dominated by the engineering department, and developed a "General Radio culture" (for example, no conflict, management by consensus, slow growth). General Radio's strategy and associated structures, systems, and people were very successful. By World War II, General Radio was the largest test-equipment firm in the United States.

After World War II, however, increasing technology and cost-based competition began to erode General Radio's market share. While management made numerous incremental changes, General Radio remained fundamentally the same organization. In the late 1960s, when CEO Don Sinclair initiated strategic changes, he left the firm's structure and systems intact. This effort at doing new things with established systems and procedures was less than successful. By 1972, the firm incurred its first loss.

In the face of this sustained performance decline, Bill Thurston (a long-time General Radio executive) was made President. Thurston initiated system-wide changes. General Radio adopted a more marketing-oriented strategy. Its product line was cut from 20 different lines to 3; much more emphasis was given to product-line management, sales, and marketing. Resources were diverted from engineering to revitalize sales, marketing, and production. During 1973, the firm moved to a matrix structure, increased its emphasis on controls and systems, and went outside for a set of executives to help Thurston run this revised General Radio. To perhaps more formally symbolize these changes and the sharp move away from the "old" General Radio, the firm's name was changed to GenRad. By 1984, GenRad's sales exploded to over $200 million (vs. $44 million in 1972).

After 60 years of convergent change around a constant strategy, Thurston and his colleagues (many new to the firm) made discontinuous system-wide changes in strategy, structure, people, and processes. While traumatic, these changes were implemented over a two-year period and led to a dramatic turnaround in GenRad's performance.

Prime Computer was founded in 1971 by a group of individuals who left Honeywell. Prime's initial strategy was to produce a high-quality/high-price minicomputer based on semiconductor memory. These founders built an engineering-dominated, loosely structured firm which sold to OEMs and through distributors. This configuration of strategy, structure, people, and processes was very successful. By 1974, Prime turned its first profit; by 1975, its sales were more than $11 million.

In the midst of this success, Prime's board of directors brought in Ken Fisher to reorient the organization. Fisher and a whole new group of executives hired from Honeywell initiated a set of discontinuous changes throughout Prime during 1975–1976. Prime now sold a full range of minicomputers and computer systems to OEMs and end-users. To accomplish this shift in strategy, Prime adopted a more complex functional structure, with a marked increase in resources to sales and marketing. The shift in resources away from engineering was so great that Bill Poduska, Prime's head of engineering, left to form Apollo Computer. Between 1975–1981, Fisher and his colleagues consolidated and incrementally adapted structure, systems, and processes to better accomplish the new strategy. During this convergent period, Prime grew dramatically to over $260 million by 1981.

In 1981, again in the midst of this continuing sequence of increased volume and profits, Prime's board again initiated an upheaval. Fisher and his direct reports left Prime (some of whom founded Encore Computer), while Joe Henson and a set of executives from IBM initiated wholesale changes throughout the organization. The firm diversified into robotics, CAD/CAM, and office systems; adopted a divisional structure; developed a more market-driven orientation; and increased controls and systems. It remains to be seen how this "new" Prime will fare. Prime must be seen, then, not as a 14-year-old firm, but as three very different organizations, each of which was managed by a different set of executives. Unlike General Radio, Prime initiated these discontinuities during periods of great success.

The Operating Group at Citibank prior to 1970 had been a service-oriented function for the end-user areas of the bank. The Operating Group hired high school graduates who remained in the "back-office" for their entire careers. Structure, controls, and systems were loose, while the informal organization valued service, responsiveness to client needs, and slow, steady work habits. While these patterns were successful enough, increased demand and heightened customer expectations led to ever decreasing performance during the late 1960s.

In the face of severe performance decline, John Reed was promoted to head the Operating Group. Reed recruited several executives with production backgrounds, and with this new top team he initiated system-wide changes. Reed's vision was to transform the Operating Group from a *service*-oriented back office to a *factory* producing high-quality products. Consistent with this new mission, Reed and his colleagues initiated sweeping changes in strategy, structure, work flows, controls, and culture. These changes were initiated concurrently throughout the back office, with very little participation, over the course of a few months. While all the empirical performance measures improved substantially, these changes also generated substantial stress and anxiety within Reed's group.

For 20 years, Alpha Corporation was among the leaders in the industrial fastener industry. Its reliability, low cost, and good technical service were important strengths. However, as Alpha's segment of the industry matured, its profits declined. Belt-tightening helped but was not enough. Finally, a new CEO presided over a sweeping restructuring: cutting the product line, closing a plant, trimming overhead; then focusing on computer parts which call for very close tolerances, CAD/CAM tooling, and cooperation with customers on design efforts. After four rough years, Alpha appears to have found a new niche where convergence will again be warranted.

These four short examples illustrate periods of incremental change, or convergence, punctuated by discontinuous changes throughout the organization. Discontinuous or "frame-breaking" change involves simultaneous and sharp shifts in strategy, power, structure, and controls. Each example illustrates the role of executive leadership in initiating and implementing discontinuous change. Where General Radio, Citibank's Operating Group, and Alpha initiated system-wide changes only after sustained performance decline, Prime proactively initiated system-wide changes to take advantage of competitive/technological conditions. These patterns in organization evolution are not unique. Upheaval, sooner or later, follows convergence if a company is to survive; only a farsighted minority of firms initiate upheaval prior to incurring performance declines.

The task of managing incremental change, or convergence, differs sharply from managing frame-breaking change. Incremental change is compatible with the existing structure of a company and is reinforced over a period of years. In contrast, frame-breaking change is abrupt, painful to participants, and often resisted by the old guard. Forging these new strategy-structure-people-process consistencies and laying the basis for the next period of incremental change calls for distinctive skills.

Because the future health, and even survival, of a company or business unit is at stake, we need to take a closer look at the nature and consequences of convergent change and of differences imposed by frame-breaking change. We need to explore when and why these painful and risky revolutions interrupt previously successful patterns, and whether these discontinuities can be avoided and/or initiated prior to crisis. Finally, we need to examine what managers can and should do to guide their organizations through periods of convergence and upheaval over time.

THE RESEARCH BASE

The research which sparks this article is based on the abundant company histories and case studies. The more complete case studies have tracked individual firms' evolution and various crises in great detail (e.g., Chandler's seminal study of strategy and structure at Du Pont, General Motors, Standard Oil, and Sears[1]). More recent studies have dealt systematically with whole sets of companies and trace their experience over long periods of time.

A series of studies by researchers at McGill University covered over 40 well-known firms in diverse industries for at least 20 years per firm (e.g., Miller and Friesen[2]). Another research program conducted by researchers at Columbia, Duke, and Cornell Universities is tracking the history of large samples of companies in the minicomputer, cement, airlines, and glass industries. This research program builds on earlier work (e.g., Greiner[3]) and finds that most successful firms evolve through long periods of convergence punctuated by frame-breaking change.

The following discussion is based on the history of companies in many different industries, different countries, both large and small organizations, and organizations in various stages of their product class's life-cycle. We are dealing with a widespread phenomenon—not just a few dramatic sequences. Our research strongly suggests that the convergence/upheaval pattern occurs within departments (e.g., Citibank's Operating Group), at the business-unit level (e.g., Prime or General Radio), and at the corporate level of analysis (e.g., the Singer, Chrysler, or Harris Corporations). The problem of managing both convergent periods and upheaval is not just for the CEO, but necessarily involves general managers as well as functional managers.

PATTERNS IN ORGANIZATIONAL EVOLUTION: CONVERGENCE AND UPHEAVAL

Building on Strength: Periods of Convergence

Successful companies wisely stick to what works well. At General Radio between 1915 and 1950, the loose functional structure, committee management system, internal promotion practices, control with engineering, and the high-quality, premium-price, engineering mentality all worked together to provide a highly congruent system. These internally consistent patterns in strategy, structure, people, and processes served General Radio for over 35 years.

Similarly, the Alpha Corporation's customer driven, low-cost strategy was accomplished by strength in engineering and production and ever more detailed structures and systems which evaluated cost, quality, and new product development. These strengths were epitomized in Alpha's chief engineer and president. The chief engineer had a remarkable talent for helping customers find new uses for industrial fasteners. He relished solving such problems, while at the same time designing fasteners that could be easily manufactured, The president excelled at production—producing dependable, low-cost fasteners. The pair were role models which set a pattern which served Alpha well for 15 years.

As the company grew, the chief engineer hired kindred customer-oriented application engineers. With the help of innovative users, they developed new products, leaving more routine problem-solving and incremental change to the sales and production departments. The president relied on a hands-on manufacturing manager and delegated financial matters to a competent treasurer-controller. Note how well the organization reinforced Alpha's strategy and how the key people fit the organization. There was an excellent fit between strategy and structure. The informal structure also fit well—communications were open, the simple mission of the company was widely endorsed, and routines were well understood.

As the General Radio and Alpha examples suggest, convergence starts out with an effective dovetailing of strategy, structure, people, and processes. For other strategies or in other industries, the particular formal and informal systems might be very different, but still a winning combination. The formal system includes decisions about grouping and linking resources as well as planning and control systems, rewards and evaluation procedures, and human resource management systems. The informal system includes core values, beliefs, norms, communication patterns, and actual decision-making and conflict resolution patterns. It is the whole fabric of structure, systems, people, and processes which must be suited to company strategy.[4]

As the fit between strategy, structure, people, and processes is never perfect, convergence is an ongoing process characterized by incremental change. Over time, in all companies studied, two types of converging changes were common: fine-tuning and incremental adaptations.

Converging Change: Fine-Tuning

Even with good strategy-structure-process fits, well-run companies seek even better ways of exploiting (and defending) their missions. Such effort typically deals with one or more of the following:

- *Refining* policies, methods, and procedures.
- Creating *specialized units and linking mechanisms* to permit increased volume and increased attention to unit quality and cost.
- *Developing personnel* especially suited to the present strategy—through improved selection and training, and tailoring reward systems to match strategic thrusts.
- Fostering individual and group *commitments* to the company mission and to the excellence of one's own department.
- Promoting *confidence* in the accepted norms, beliefs, and myths.
- *Clarifying* established roles, power, status, dependencies, and allocation mechanism.

The fine-tuning fills out and elaborates the consistencies between strategy, structure, people, and processes. These incremental changes lead to an ever more interconnected (and therefore more stable) social system. Convergent periods fit the happy, stick-with-a-winner situations romanticized by Peters and Waterman.[5]

Converging Change: Incremental Adjustments to Environmental Shifts

In addition to fine-tuning changes, minor shifts in the environment will call for some organizational response. Even the most conservative of organizations expect, even welcome, small changes which do not make too many waves.

A popular expression is that almost any organization can tolerate a "ten-percent change." At any one time, only a few changes are being made; but these changes are still compatible with the prevailing structures, systems, and processes. Examples of such adjustments are an expansion in sales territory, a shift in emphasis among products in the product line, or improved processing technology in production.

The usual process of making changes of this sort is well known: wide acceptance of the need for change, openness to possible alternatives, objective examination of the pros and cons of each plausible alternative, participation of those directly affected in the preceding analysis, a market test or pilot operation where feasible, time to learn the new activities, established role models, known rewards for positive success, evaluation, and refinement.

The role of executive leadership during convergent periods is to reemphasize mission and core values and to delegate incremental decisions to middle-level managers. Note that the uncertainty created for people affected by such changes is well within tolerable limits. Opportunity is provided to anticipate and learn what is new, while most features of the structure remain unchanged.

The overall system adapts, but it is not transformed.

Converging Change: Some Consequences

For those companies whose strategies fit environmental conditions, convergence brings about better and better effectiveness. Incremental change is relatively easy to implement and ever more optimizes the consistencies between strategy, structure, people, and processes. At AT&T, for example, the period between 1913 and 1980 was one of ever more incremental change to further bolster the "Ma Bell" culture, systems, and structure all in service of developing the telephone network.

Convergent periods are, however, a double-edged sword. As organizations grow and become more successful, they develop internal forces for stability. Organization structures and systems become so interlinked that they only allow compatible changes. Further, over time, employees develop habits, patterned behaviors begin to take on values (e.g., "service is good"), and employees develop a sense of competence in knowing how to get work done within the system. These self-reinforcing patterns of behavior, norms, and values contribute to increased organizational momentum and complacency and, over time, to a sense of organizational history. This organizational history—epitomized by common stories, heroes, and standards—specifies "how we work here" and "what we hold important here."

This organizational momentum is profoundly functional as long as the organization's strategy is appropriate. The Ma Bell and General Radio culture, structure, and systems—and associated internal momentum—were critical to each organization's success. However, if (and when) strategy must change, this momentum cuts the other way. Organizational history is a source of tradition, precedent, and pride which are, in turn, anchors to the past. A proud history often restricts vigilant problem solving and may be a source of resistance to change.

When faced with environmental threat, organizations with strong momentum

- may not register the threat due to organization complacency and/or stunted external vigilance (e.g., the automobile or steel industries), or
- if the threat is recognized, the response is frequently heightened conformity to the status quo and/or increased commitment to "what we do best."

For example, the response of dominant firms to technological threat is frequently increased commitment to the obsolete technology (e.g., telegraph/telephone; vacuum tube/transistor; core/semiconductor memory). A paradoxical result of long periods of success may be heightened organizational complacency, decreased organizational flexibility, and a stunted ability to learn.

Converging change is a double-edged sword. Those very social and technical consistencies which are key sources of success may also be the seeds of failure if environments change. The longer the convergent period, the greater these internal forces for stability. This momentum seems to be particularly accentuated in those most successful firms in a product class (for example, Polaroid, Caterpillar, or U. S. Steel), in historically regulated organizations (for example, AT&T, GTE, or financial service firms), or in organizations that have been traditionally shielded from competition (for example, universities, not-for-profit organizations, government agencies and/or services).

On Frame-Breaking Change

FORCES LEADING TO FRAME-BREAKING CHANGE

What, then, leads to frame-breaking change? Why defy tradition? Simply stated, frame-breaking change occurs in response to or, better yet, in anticipation of major environmental changes—changes which require more than incremental adjustments. The need for discontinuous change springs from one or a combination of the following:

- *Industry Discontinuities*—Sharp changes in legal, political, or technological conditions shift the basis of competition within industries. *Deregulation* has dramatically transformed the financial services and airlines industries. *Substitute product technologies* (such as jet engines, electronic typing, microprocessors) or *substitute process technologies* (such as the planar process in semiconductors or float-glass in glass manufacture) may transform the bases of competition within industries. Similarly, the emergence of industry standards, or *dominant designs* (such as the DC-3, IBM 360, or PDP-8) signal a shift in competition away from product innovation and towards increased process innovation. Finally, *major economic changes* (e.g., oil crises) and *legal shifts* (e.g., patent protection in biotechnology or trade/regulator barriers in pharmaceuticals or cigarettes) also directly affect bases of competition.
- *Product-Life-Cycle Shifts*—Over the course of a product class life-cycle, different strategies are appropriate. In the emergence phase of a product class, competition is based on product innovation and performance, where in the maturity stage, competition centers on cost, volume, and efficiency. Shifts in patterns of demand alter key factors for success. For example, the demand and nature of competition for minicomputers, cellular telephones, wide-body aircraft, and bowling alley equipment was transformed as these products gained acceptance and their prod-

uct classes evolved. Powerful international competition may compound these forces.

- *Internal Company Dynamics*—Entwined with these external forces are breaking points within the firm. Sheer size may require a basically new management design. For example, few inventor-entrepreneurs can tolerate the formality that is linked with large volume; even Digital Equipment Company apparently has outgrown the informality so cherished by Kenneth Olsen. Key people die. Family investors may become more concerned with their inheritance taxes than with company development. Revised corporate portfolio strategy may sharply alter the role and resources assigned to business units or functional areas. Such pressures especially when coupled with external changes, may trigger frame-breaking change.

Scope of Frame-Breaking Change

Frame-breaking change is driven by shifts in business strategy. As strategy shifts so too must structure, people, and organizational processes. Quite unlike convergent change, frame-breaking reforms involve discontinuous changes throughout the organization. These bursts of change do not reinforce the existing system and are implemented rapidly. For example, the system-wide changes at Prime and General Radio were implemented over 18–24-month periods, where as changes in Citibank's Operating Group were implemented in less than five months. Frame-breaking changes are revolutionary changes *of* the system as opposed to incremental changes *in* the system.

The following features are usually involved in frame-breaking change:

- *Reformed Mission and Core Values*—A strategy shift involves a new definition of company mission. Entering or withdrawing from an industry may be involved; a least the way the company expects to be outstanding is altered. The revamped AT&T is a conspicuous example. Success on its new course calls for a strategy based on competition, aggressiveness, and responsiveness, as well as a revised set of core values about how the firm competes and what it holds as important. Similarly, the initial shift at Prime reflected a strategic shift away from technology and towards sales and marketing. Core values also were aggressively reshaped by Ken Fisher to complement Prime's new strategy.
- *Altered Power and Status*—Frame-breaking change always alters the distribution of power. Some groups lose in the shift while others gain. For example, at Prime and General Radio, the engineering functions lost power, resources, and prestige as the marketing and sales functions gained. These dramatically altered power distributions reflect shifts in bases of competition and resource allocation. A new strategy must be backed up with a shift in the balance of power and status.
- *Reorganization*—A new strategy requires a modification in structure, systems, and procedures. As strategic requirements shift, so too must the choice of organization form. A new direction calls for added activity in some areas and less in others. Changes in structure and systems are means to ensure that this reallocation of effort takes place. New structures and revised roles deliberately break business-as-usual behavior.
- *Revised Interaction Patterns*—The way people in the organization work together has to adapt during frame-breaking change. As strategy is different, new procedures, work flows, communication networks, and decision-making patterns must be established. With these changes in work flows and procedures must also come revised norms, informal decision-making/conflict-resolution procedures, and informal roles.
- *New Executives*—Frame-breaking change also involves new executives, usually brought in from outside the organization (or business unit) and placed in key managerial positions. Commitment to the new mission, energy to overcome prevailing inertia, and freedom from prior obligations are all needed to refocus the organization. A few exceptional members of the old

guard may attempt to make this shift, but habits and expectations of their associations are difficult to break. New executives are most likely to provide both the necessary drive and an enhanced set of skills more appropriate for the new strategy. While the overall number of executive changes is usually relatively small, these new executives have substantial symbolic and substantive effects on the organization. For example, frame-breaking changes at Prime, General Radio, Citibank, and Alpha Corporation were all spearheaded by a relatively small set of new executives from outside the company or group.

Why All At Once?

Frame-breaking change is revolutionary in that the shifts reshape the entire nature of the organization. Those more effective examples of frame-breaking change were implemented rapidly (e.g., Citibank, Prime, Alpha). It appears that a piecemeal approach to frame-breaking changes gets bogged down in politics, individual resistance to change, and organizational inertia (e.g., Sinclair's attempts to reshape General Radio). Frame-breaking change requires discontinuous shifts in strategy, structure, people, and processes concurrently—or at least in a short period of time. Reasons for rapid, simultaneous implementation include:

- *Synergy* within the new structure can be a powerful aid. New executives with a fresh mission, working in a redesigned organization with revised norms and values, backed up with power and status, provide strong reinforcement. The pieces of the revitalized organization pull together, as opposed to piecemeal change where one part of the new organization is out of synch with the old organization.
- *Pockets of resistance* have a chance to grow and develop when frame-breaking change is implemented slowly. The new mission, shifts in organization, and other frame-breaking changes upset the comfortable routines and precedent. Resistance to such fundamental change is natural. If frame-breaking change is implemented slowly, then individuals have a greater opportunity to undermine the changes and organizational inertia works to further stifle fundamental change.
- Typically, there is a *pent-up need for change*. During convergent periods, basic adjustments are postponed. Boat-rocking is discouraged. Once constraints are relaxed, a variety of desirable improvements press for attention. The exhilaration and momentum of a fresh effort (and new team) make difficult moves more acceptable. Change is in fashion.
- Frame-breaking change is an inherently *risky and uncertain venture*. The longer the implementation period, the greater the period of uncertainty and instability. The most effective frame-breaking changes initiate the new strategy, structure, processes, and systems rapidly and begin the next period of stability and convergent change. The sooner fundamental uncertainty is removed, the better the chances of organizational survival and growth. While the pacing of change is important, the overall time to implement frame-breaking change will be contingent on the size and age of the organization.

Patterns in Organization Evolution

This historical approach to organization evolution focuses on convergent periods punctuated by reorientation—discontinuous, organization-wide upheavals. The most effective firms take advantage of relatively long convergent periods. These periods of incremental change build on and take advantage of organization inertia. Frame-breaking change is quite dysfunctional if the organization is successful and the environment is stable. If, however, the organization is performing poorly and/or if the environment changes substantially, frame-breaking change is the only way to realign the organization with its competitive environment. Not all reorientations will be successful (e.g., People Express' expansion and up-scale moves in 1985–86). However, inaction in the face of performance crisis and/or environmental shifts is a certain recipe for failure.

Because reorientations are so disruptive and fraught with uncertainty, the more rapidly they are implemented, the more quickly the organization can reap the benefits of the following convergent period. High-performing firms initiate reorientations when environmental conditions shift and implement these reorientations rapidly (e.g., Prime and Citibank). Low-performing organizations either do not reorient or reorient all the time as they root around to find an effective alignment with environmental conditions.

This metamorphic approach to organization evolution underscores the role of history and precedent as future convergent periods are all constrained and shaped by prior convergent periods. Further, this approach to organization evolution highlights the role of executive leadership in managing convergent periods *and* in initiating and implementing frame-breaking change.

EXECUTIVE LEADERSHIP AND ORGANIZATION EVOLUTION

Executive leadership plays a key role in reinforcing system-wide momentum during convergent periods and in initiating and implementing bursts of change that characterize strategic reorientations: The nature of the leadership task differs sharply during these contrasting periods of organization evolution.

During convergent periods, the executive team focuses on *maintaining* congruence and fit within the organization. Because strategy, structure, processes, and systems are fundamentally sound, the myriad of incremental substantive decisions can be delegated to middle-level management, where direct expertise and information resides. The key role for executive leadership during convergent periods is to reemphasize strategy, mission, and core values and to keep a vigilant eye on external opportunities and/or threats.

Frame-breaking change, however, requires direct executive involvement in all aspects of the change. Given the enormity of the change and inherent internal forces for stability, executive leadership must be involved in the specification of strategy, structure, people, and organizational processes *and* in the development of implementation plans. During frame-breaking change, executive leadership is directly involved in *reorienting* their organizations. Direct personal involvement of senior management seems to be critical to implement these system-wide changes (e.g., Reed at Citibank or Iacocca at Chrysler). Tentative change does not seem to be effective (e.g., Don Sinclair at General Radio).

Frame-breaking change triggers resistance to change from multiple sources change must overcome several generic hurdles, including:

- Individual opposition, rooted in either anxiety or personal commitment to the status quo, is likely to generate substantial individual resistance to change.
- Political coalitions opposing the upheaval may be quickly formed within the organization. During converging periods a political equilibrium is reached. Frame-breaking upsets this equilibrium; powerful individuals and/or groups who see their status threatened will join in resistance.
- Control is difficult during the transition. The systems, roles, and responsibilities of the former organization are in suspension; the new rules of the game—and the rewards—have not yet been clarified.
- External constituents—suppliers, customers, regulatory agencies, local communities, and the like—often prefer continuation of existing relationships rather than uncertain moves in the future.

Whereas convergent change can be delegated, frame-breaking change requires strong, direct leadership from the top as to where the organization is going and how it is to get there. Executive leadership must be directly involved in: motivating constructive behavior, shaping political dynamics, managing control during the transition period, and managing external constituencies. The executive team must direct the content of frame-breaking change *and* provide the energy, vision, and resources to support, and be role models for, the new order. Brilliant ideas for new strategies, structures, and processes will not be effective unless they are coupled with thorough implementation plans actively managed by the executive team.[6]

When to Launch an Upheaval

The most effective executives in our studies foresaw the need for major change. They recognized the external threats and opportunities, and took bold steps to deal with them. For example, a set of minicomputer companies (Prime, Rolm, Datapoint, Data General, among others) risked short-run success to take advantage of new opportunities created by technological and market changes. Indeed, by acting before being forced to do so, they had more time to plan their transitions.[7]

Such visionary executive teams are the exceptions. Most frame-breaking change is postponed until a financial crisis forces drastic action. The momentum, and frequently the success, of convergent periods breeds reluctance to change. This commitment to the status quo, and insensitivity to environmental shocks, is evident in both the Columbia and the McGill studies. It is not until financial crisis shouts its warning that most companies begin their transformation.

The difference in timing between pioneers and reluctant reactors is largely determined by executive leadership. The pioneering moves, in advance of crisis, are usually initiated by executives within the company. They are the exceptional persons who combine the vision, courage, and power to transform an organization. In contrast, the impetus for a tardy break usually comes from outside stakeholders; they eventually put strong pressure on existing executives—or bring in new executives—to make fundamental shifts.

Who Manages the Transformation

Directing a frame-breaking upheaval successfully calls for unusual talent and energy. The new mission must be defined, technology selected, resources acquired, policies revised, values changed, organization restructured, people reassured, inspiration provided, and an array of informal relationships shaped. Executives already on the spot will probably know most about the specific situation, but they may lack the talent, energy, and commitment to carry through an internal revolution.

As seen in the Citibank, Prime, and Alpha examples, most frame-breaking upheavals are managed by executives brought in from outside the company. The Columbia research program finds that externally recruited executives are more than three times more likely to initiate frame-breaking change than existing executive teams. Frame-breaking change was coupled with CEO succession in more than 80 percent of the cases. Further, when frame-breaking change was combined with executive succession, company performance was significantly higher than when former executives stayed in place. In only 6 of 40 cases we studied did a current CEO initiate and implement multiple frame-breaking changes. In each of these six cases, the existing CEO made major changes in his/her direct reports, and this revitalized top team initiated and implemented frame-breaking changes (e.g., Thurston's actions at General Radio).[8]

Executive succession seems to be a powerful tool in managing frame-breaking change. There are several reasons why a fresh set of executives are typically used in company transformations. The new executive team brings different skills and a fresh perspective. Often they arrive with a strong belief in the new mission. Moreover, they are unfettered by prior commitments linked to the status quo; instead, this new top team symbolizes the need for change. Excitement of a new challenge adds to the energy devoted to it.

We should note that many of the executives who could not, or would not, implement frame-breaking change went on to be quite successful in other organizations—for example, Ken Fisher at Encore Computer and Bill Poduska at Apollo Computer. The stimulation of a fresh start and of jobs matched to personal competence applies to individuals as well as to organizations.

Although typical patterns for the when and who of frame-breaking change are clear—wait for a financial crisis and then bring in an outsider, along with a revised executive team, to revamp the company—this is clearly less than satisfactory for a particular organization. Clearly, some companies benefit from transforming themselves before a crisis forces them to do so, and a few exceptional executives have the vision and drive to reorient a business which they nurtured during its preceding period of convergence. The vital tasks are to manage incremental change during convergent periods; to have the vision to initiate and implement frame-breaking change prior to the competition; and to mobilize an executive which can initiate and implement both kinds of change.

CONCLUSION

Our analysis of the way companies evolve over long periods of time indicates that the most effective firms have relatively long periods of convergence giving support to a basic strategy, but such periods are punctuated by upheavals—concurrent and discontinuous changes which reshape the entire organization.

Managers should anticipate that when environments change sharply:

- Frame-breaking change cannot be avoided. These discontinuous organizational changes will either be made proactively or initiated under crisis/turnaround condition.
- Discontinuous changes need to be made in strategy, structure, people, and processes concurrently. Tentative change runs the risk of being smothered by individual, group, and organizational inertia.
- Frame-breaking change requires direct executive involvement in all aspects of the change, usually bolstered with new executives from outside the organization.
- There are no patterns in the sequence of frame-breaking changes, and not all strategies will be effective. Strategy and, in turn, structure, systems, and processes must meet industry-specific competitive issues.

Finally, our historical analysis of organizations highlights the following issues for executive leadership:

- Need to manage for balance, consistency, or fit during convergent period.
- Need to be vigilant for environmental shifts in order to anticipate the need for frame-breaking change.
- Need to effectively manage incremental as well as frame-breaking change.
- Need to build (or rebuild) a top team to help initiate and implement frame-breaking change.
- Need to develop core values which can be used as an anchor as organizations evolve through frame-breaking changes (e.g., IBM, Hewlett-Packard).
- Need to develop and use organizational history as a way to infuse pride in an organization's past and for its future.
- Need to bolster technical, social, and conceptual skills with visionary skills. Visionary skills add energy, direction, and excitement so critical during frame-breaking change.

Effectiveness over changing competitive conditions requires that executives manage fundamentally different kinds of organizations and different kinds of change. The data are consistent across diverse industries and countries, an executive team's ability to proactively initiate and implement frame-breaking change *and* to manage convergent change seem to be important factors which discriminate between organizational renewal and greatness versus complacency and eventual decline.

NOTE

The authors thank Donald Hambrick and Kathy Harrigan for insightful comments and the Center for Strategy Research and the Center for Research on Innovation and Entrepreneurship at the Graduate School of Business, Columbia University, for financial support.

REFERENCES

1. A. Chandler, *Strategy and Structure* (Cambridge, MA: MIT Press, 1962).
2. D. Miller and P. Friesen, *Organizations: A Quantum View* (Englewood Cliffs, NJ: Prentice-Hall, 1984).
3. L. Greiner, "Evolution and Revolution as Organizations Grow," *Harvard Business Review* (July/August 1972), pp. 37–46.
4. D. Nadler and M. Tushman, *Strategic Organization Design* (Homewood, IL: Scott Foresman, 1986).
5. T. Peters and R. Waterman, *In Search of Excellence* (New York, NY: Harper and Row, 1982).
6. Nadler and Tushman, op. cit.
7. For a discussion of preemptive strategies, see I. MacMillan, "Delays in Competitors' Responses to New Banking Products," *Journal of Business Strategy,* 4 (1984): 58–65.
8. M. Tushman and B. Virany, "Changing Characteristics of Executive Teams in an Emerging Industry," *Journal of Business Venturing* (1986).

39

Implementing New Designs: Managing Organizational Change

DAVID A. NADLER
MICHAEL L. TUSHMAN

CASE

Orient Oil Corporation is a major force in the energy industry. About ten years ago, Orient senior management decided to diversify out of petroleum-based industries into other ventures, some related to the basic business and some completely unrelated. For a while, this seemed to work out. After the first three years of the acquisition strategy, Orient faced problems in managing the different companies within its structure, so it called in a consulting firm, which did an organization study and recommended the establishment of twenty-three individual strategic business units, each with its own president and a complete set of staff functions. The business units were in turn grouped into four major sectors, each headed by a senior corporate executive.

About four years into the strategy, things started to go sour. Many of the management systems, approaches, and methods that had worked so well in the oil business seemed to lead to one disaster after another. For a while, the huge profits of the oil business could be used to cover up the continuing stream of catastrophes in the other units, but finally it became obvious to both insiders and outsiders that Orient was incapable of effectively managing businesses outside its basic industry. A strategic decision was made to start divesting or liquidating the acquisitions and to move back to the base business.

A very senior group was convened to work intensively for several weeks to develop a plan to reorganize the company. They produced a top secret document that proposed reorganizing the company into two major operating units—one focusing primarily on energy exploration and production and the other focusing on refining, distribution, and sales. A third group would hold the non-energy-related business but would be chartered to do itself out of business in two years through divestiture or liquidation.

The policy committee of the corporation met for a full-day session to hear the report of the design task force and, after many hours of debate, decided to go ahead with the group's recommendations. Having worked hard for several hours, the group took a short break at about 4:00 in the afternoon and decided to reconvene at 4:30 to "tie up the loose ends."

As the meeting started again, the discussion moved toward the issue of announcements. Many members of the group were pushing for an immediate announcement. Rumors were flying around the company that something was up, and the group members were concerned about the consequences of possible information leaks. In the midst of this discussion, one of the group turned to the task force chairman and said, "Once we make the announcement, what do you want people to do then?" The task force chairman looked around the room for help and saw none. He responded, "Well, there's a lot we haven't figured out yet. We'll just tell people not to worry about it." The questioner came back, "Aren't we just creating problems, then, by announcing this thing? We're just going to disturb people, and nobody will be doing any work." The room was silent.

INTRODUCTION

As with Orient Oil, many managers see their organization design work as completed when "the announcement" is made. Because so much energy may have been expended on reaching an agreement on a design, little thought may have been given to what will happen next. As a result, after the announcement is made, managers suddenly begin to think about how to manage the implementation of the change in design.

In fact, implementing a new design is difficult, as is the implementation of any major change within an organization. Design changes are particularly problematic because it seems so easy to create a design on paper that managers often overlook how truly difficult it is to install a new design and make it work effectively. Truly effective implementation is difficult and often takes a good deal of time.

Many design failures—in which everyone agrees that the reorganization was a disaster—are not failures because of a technically inadequate design but rather are failures of implementation. In practice, an adequate or even mediocre design, if implemented well, can be effective, while the most elegant and sophisticated of designs poorly implemented will fail.

This chapter, then, will be devoted to the question of implementation of organization designs. The underlying issue in design implementation appears to be one of managing organizational change. We will therefore start by providing a way of thinking about changes in organization. Next, we will point out some of the very predictable problems that one encounters when attempting to bring about change. Finally, we will discuss some implications for managing change and outline some specific techniques and action areas for enhancing the implementation of organization design changes.

CRITERIA FOR ORGANIZATIONAL CHANGE

During the past decade, there has been increasing interest in the subject of managing organizational change.[1] One approach to thinking about change that many have found useful was originally proposed by Richard Beckhard and Reuben Harris. They saw the implementation of a change, such as a new organization design, as the moving of an organization toward a desired future state. They saw changes in terms of transitions (see Figure 39.1). At any time, an organization exists in a **current state** (A). The current state describes how the organization functions prior to a change. In terms of our congruence model, we can think of the current state as a particular configuration of the strategy, task, individual, and formal and informal organizations. A change involves movement toward a desired **future state** (B), which describes how the organization should function after the change. In a design, the full set of design documents (strategic design, impact analysis, operational design, and so on) provides a written description of the intended future state.

The period between the current state (A) and the future state (B) can be thought of as the **transition state** (C). In the most general terms, then, the effective management of change involves de-

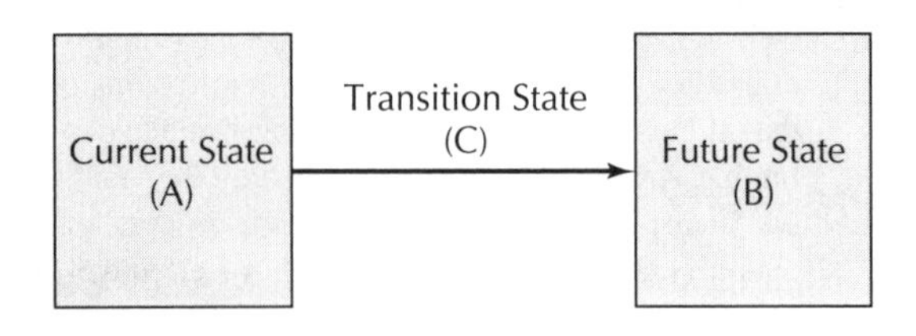

Fig. 39.1 Organization change as transition.

veloping an understanding of the current state, developing an image of the desired future state, and moving the organization through a transition period. In design, we deal with the first two of these steps. Implementation concerns the moving of the organization through the transition period. Typically, as much care needs to be taken in designing the transition as in designing the future state—both are critical.

Several criteria can be used to judge the effective management of transitions. Building on the transition framework just presented, an organizational change, such as the implementation of a new design, can be managed effectively when:

1. The organization is moved from the current state to the future state—in which the design is actually installed or implemented.
2. The functioning of the organization design in the future state meets expectations, or works as planned. In the case of design, this meets that the design in practice met the criteria that it was intended to satisfy.
3. The transition is accomplished without undue cost to the organization. This means that the design is implemented without significant disruptions to the business or damage to relationships with customers, suppliers, or regulators. While there is always some cost associated with implementation, the cost should be managed, predictable, and controlled consistent with the estimates done in the impact analysis. "Undue" cost is cost that is unplanned, unpredicted, or uncontrolled.
4. The transition is accomplished without undue cost to individual organization members. Here again, the key operative word is "undue" as defined by the original impact analysis. Much of the cost to individuals occurs more through the manner in which changes are made than through the change itself.

Of course, not every implementation of a new design can be expected to meet all of these criteria consistently, but such standards provide a target for planning implementation. The question is how to maximize the chances that the design will be implemented effectively.

PROBLEMS OF IMPLEMENTING ORGANIZATIONAL CHANGES

What are the issues that must be addressed if managers are to implement effectively? On the broadest level, there are two basic issues—what the change should be and how the change should be implemented. Throughout this book we have been dealing with the first issue as it relates to organization design. We have stressed the importance of diagnosis of problems and causes, followed by systematic work to develop and then choose from alternative solutions that will be responsive to those problems. The second question—how the changes are implemented—is the one on which we will focus now. Observations of changes seem to indicate that there are three types of problems encountered in some form whenever a significant organizational change is attempted.

The Problem of Power

Any organization is a political system made up of various individuals, groups, and coalitions competing for power. Political behavior is thus a natural and expected feature of organizations. Such behavior occurs during the current and future states. In the transition state, however, these dynamics become even more intense as an old design, with its political implications, is dismantled and a new design takes its place. Any significant change (and design changes clearly are significant in terms of power) poses the possibility of upsetting or modifying the balance of power among various formal and informal interest groups. The uncertainty created by change creates ambiguity, which in turn tends to increase the probability of political activity as people try to create some structure and certainty by attempting to control their environment.

Individuals and groups may take political action based on their perceptions of how the change will affect their relative power position in the organization. They will try to influence where they will sit in the organization (both formal and informal) that emerges from the transition and will be concerned about how the conflict of the transition period will affect the balance of power in the future state. Finally, individuals and groups may engage in political action because of their ideological position with regard to the change—the new design, strategy, or approach may be inconsistent with their shared values or their image of the organization.

The Problem of Anxiety

Change in organizations involves the movement from something that is known toward something that is unknown. Individuals naturally have concerns, such as whether they will be needed in the new organization, whether their skills will be valued, and how they will cope with the new situation. These concerns can be summarized in the question that is frequently voiced during a major organizational change—"what's going to happen to me?" To the extent that this question cannot fully be answered (such as in the Orient Oil case at the beginning of this chapter), individuals may experience stress and feel anxious.

As stress and anxiety increase, they may result in a variety of behavior or performance problems. For example, stress may result in difficulty in hearing or integrating information. It may lead people to resist changes that they might otherwise support or in the extreme, engage in irrational and even self-destructive acts. Resistance is a common occurrence, although in many large organizations people may not actively resist the change by openly refusing to implement the new organization design. What does occur is that people passively or subtly resist the change or act in ways that objectively do not appear to be constructive for either the individual or the organization.

The Problem of Organizational Control

A significant change in organization design tends to disrupt the normal course of events within the organization. Thus, it frequently undermines existing systems of management control, particularly those that are embedded in the formal organizational arrangements. An impending change may suddenly make control systems irrelevant or cause them to be perceived as "lame ducks." As a result, it is often easy to lose control during a change. As goals, structures, and people shift, it becomes difficult to monitor performance and make correct assumptions, as one would during a more stable period.

A related problem is that most of the formal organizational arrangements are designed either to manage the current state (the existing design) or to manage the future state (the proposed new design), but those same designs may not be adequate for the management of the transition state. In most situations, they are not appropriate for managing implementation, since they are steady state management systems designed to run organizations already in place. They are not transitional management devices.

IMPLICATIONS FOR CHANGE MANAGEMENT

Each of these three problems lead to some relatively straight-forward conclusions about actions needed to manage change (see Table 39.1). To the extent that a change presents the possibility of significant power problems, the management of the organization's political system must shape the political dynamics associated with the change, preferably prior to implementation. Second, to the extent that change creates anxiety and the associated patterns of dysfunctional behavior, it is critical to motivate individuals, through communications and rewards, to react constructively to the change. Finally, if a change presents significant control problems, this implication is the need to pay attention to the management of the transition state to ensure effective organizational control during the transition period. The question is how to do this. There appear to be some patterns in the effectively managed changes. While not universal principles,

TABLE 39.1. Change Problems and Implications

Problem		Implication
Power	→	Need to shape the political dynamics associated with change
Anxiety	→	Need to motivate constructive behavior in response to the change
Control	→	Need to systematically manage the transition state

they represent some relatively consistent differences between the actions that managers take in effective cases of change management and the actions taken in ineffective cases.

For each of the three implications for change management, there are four actions that appear to characterize effectively managed changes. On the following pages, each action area will be explained and a list of illustrative techniques associated with each area will be discussed. The summaries of these techniques are listed in Tables 39.2, 39.3, and 39.4, respectively.

Action Areas for Shaping Political Dynamics

The first set of practices concerns the organization as a political system. Any significant change usually involves some modification of the political system, thus raising issues of power. The implication is a need to shape and manage the political dynamics prior to and throughout the transition. This concept relates to four specific action areas.

The first action area involves getting the support of key power groups within the organization in order to build a critical mass in favor of the change.

TABLE 39.2. Shaping Political Dynamics

Action	Purpose	Technique
Get support of key power groups	Build internal critical mass of support for change	Identify power relationships Key players Stakeholders Influence relationships Use strategies for building support Participation Bargaining/deals Isolation Removal
Demonstrate leadership support of the change	Shape the power distribution and influence the patterns of behavior	Leaders model behavior to promote identification with them Articulate vision of future state Use reward system Provide support/resources Remove roadblocks Maintain momentum Send signals through informal organization
Use symbols	Create identification with the change and appearance of a critical mass of support	Communicate with Names/graphics Language systems Symbolic acts Small signals
Build in stability	Reduce excess anxiety, defensive reactions, and conflicts	Allow time to prepare for change Send consistent messages Maintain points of stability Communicate what will not change

TABLE 39.3. Motivating Constructive Behavior

Action	Purpose	Technique
Surface/create dissatisfaction with the current state	Unfreeze from the present state and provide motivation to move away from the present situation	Present information on Environmental impact Economic impact Goal discrepancies How change affects people Have organization members collect/ present information
Obtain the appropriate levels of participation in planning/ implementing change	Obtain the benefits of participation (motivation, better decisions, communication); control the costs of participation (time, control, conflict, ambiguity)	Create opportunities for participation Diagnosis Design Implementation planning Implementation evaluation Use a variety of participation methods Direct/indirect Information vs. input vs. decision making Broad vs. narrow scope Expertise vs. representation
Reward desired behavior in transition to future state	Shape behavior to support the future state	Give formal rewards Measures Pay Promotion Give informal rewards Recognition/praise Feedback Assignments
Provide time and opportunity to disengage from current state	Help people deal with their attachment and loss associated with change	Allow enough time Create opportunity to vent emotions Have farewell ceremonies

The organization is a political system with competing groups, cliques, coalitions, and interests, each with varying views on any particular change. Some favor the change. Some oppose it. Some may be disinterested. But the change cannot succeed unless there is a critical mass of support; several steps can be used to build that support. The first step is identifying the power relationships as a basis for planning a political strategy. This step may involve identifying the key players in the organization, or the individual and/or group stakeholders—the individuals who have a positive, negative, or neutral stake in the change. Frequently, drawing a diagram or creating a stakeholder or influence map may be useful in conceptualizing these relationships. This map should include not only the various stakeholders but their relationships to each other—who influences whom and what the stakes are for each individual.

Having identified the political topography of the change, the next step is to think about approaches for building support. There are several possible methods. The first is participation, which has long been recognized as a tool for reducing resistance to change and for gaining support. As individuals or groups become involved in a change, they tend to see it as *their* change, rather than one imposed on them.

Participation, while desirable, might not be feasible or wise in all situations. In some cases, participation merely increases the power of opposing groups to forestall the change. Thus, another approach may be bargaining with groups, or cutting

TABLE 39.4. Managing the Transition

Action	Purpose	Technique
Develop and communicate a clear image of the future state	Provide direction for management of transition; reduce ambiguity	Develop as complete a design as possible General impact statements Communicate Repeatedly Multiple channels Tell and sell Describe how things will operate Communicate clear, stable image/vision of the future
Use multiple and consistent leverage points	Recognize the systemic nature of changes and reduce potential for creating new problems during transition	Use all four organizational components Anticipate poor fits Sequence changes appropriately
Use transition devices	Create organizational arrangements specifically to manage the transition state	Appoint a transition manager Provide transition resources Design specific transition devices (dual systems, backup) Develop a transition plan
Obtain feedback about the transition state; evaluate success	Determine the progress of the transition; reduce dependence on traditional feedback processes	Use formal methods Interviews Focus groups Surveys/samples Use informal channels Use participation

deals. In this case, those favoring the change get the support of others by providing some incentive to comply.

A third step is isolation. There may be those who resist participation or bargaining and who persist in attempting to undermine the change. The goal in this situation is to minimize the impact of such individuals on the organization by assigning them to a position outside the mainstream.

In the extreme, the final step is removal. In some cases, individuals who cannot be isolated or brought into constructive roles may have to be removed from the scene through a transfer to another organization or by outplacement. Obviously, participation and bargaining are more desirable and leave a more positive aftermath; however, it would be naive to assume that these first two methods will be successful in all cases.

An important consideration in creating the political momentum and sense of critical mass is the activity of leaders. Thus, a second action area is leader behavior in support of the change. Leaders can greatly shape the power distribution and influence patterns in an organization. They can mold perceptions and create a sense of political momentum by sending out signals, providing support, and dispensing rewards.

Leaders can take a number of specific actions. First, they can serve as models; through their behavior, they provide a vision of the future state and a source of identification for various groups within the organization. Second, leaders can serve as important persons in articulating the vision of the future state. Third, leaders can play a crucial role by rewarding key individuals and specific types of behavior. Fourth, leaders can provide support through political influence and needed resources. Similarly, leaders can remove roadblocks and, through their public statements, maintain momentum. Finally, leaders can send important signals through the informal organization. During times of uncertainty and change, individuals throughout the organization tend to look to leaders for signals concerning appropriate behavior and the direction of movement in the organization. Frequently, potent signals are sent through such minor acts as patterns of attendance at meetings or the phrases and words used in public statements. By careful attention to these subtle actions, leaders can greatly influence the perceptions of others.

The third action area concerns the use of symbols associated with a change. Such symbols as language, pictures, and acts create a focus for identification and the appearance of a critical mass within the organization's political system. Symbols are used by public and social movements and are similarly relevant to dealing with the political system within an organization. A variety of devices can be used, such as names and related graphics that clearly identify events, activities, or organizational units. Language is another symbol; it can communicate a unique way of doing business. The use of symbols is a mundane behavior that can, however, have a powerful impact on the clarity of the informal organization. The more focused the informal organization, the less the political turbulence. For example, a particular promotion, a firing, the moving of an office, or an open door, all can serve to create and send important signals. These small but visible signals by the leaders (as mentioned above) can be important in providing a symbolic sense of political movement.

The final action area is that of building stability. Too much uncertainty can create excess anxiety and defensive reaction, thus heightening political conflict to a counterproductive level. The organization must provide certain "anchors" to create a sense of stability within the context of the transition. This can help limit the reverberations of the change and dampen counterproductive political activity. A number of steps, such as preparing people for the change by providing information in advance, can buffer them to a degree against the uncertainty that will occur. Secondly, some stability can be preserved—even in the face of change—if managers are careful to maintain the consistency of messages they convey to organization members throughout the period of change. Nothing creates more instability than inconsistent or conflicting messages. Thirdly, it may be important to maintain certain very visible aspects of the business, such as preserving certain units, organizational names, management processes, or staffing patterns or keeping people in the same physical location. Finally, it may help to communicate specifically what will not change—to mediate the fears that everything is changing or that the change will be much greater than what actually is planned.

In summary, the four action areas focus on identifying the political system and then developing a political strategy. Specific action includes using leadership and related symbols to maintain momentum and critical mass in support of the change and building stability to prevent the counterproductive effects of extreme anxiety.

Motivating Constructive Behavior

When a broad, significant change occurs in an organization, the first questions many people ask are "What's in it for me?" and "What's going to happen to me?" This is an indication of the anxiety that occurs when people are faced with the uncertainty associated with organizational change. Anxiety may result in a number of reactions, ranging from withdrawal to panic to active resistance. The task of management is to somehow relieve that anxiety and motivate constructive behavior through a variety of actions. Some actions are aimed at providing much needed information communicating the nature, extent, and impact of the change. Others are focused on providing clear rewards for required behavior, recognizing and dealing with some of the natural anxiety. There are four specific action areas.

The first action area is to surface or create dissatisfaction with the current state. Individuals may be psychologically attached to the current state, which is comfortable and known, compared to the uncertainty associated with change. A critical step, then, is to demonstrate how unrealistic it is to assume that the current state has been completely good, is still good, and will always remain good. The goal is to "unfreeze" people from their inertia and create willingness to explore the possibility of change. Part of their anxiety is based on fantasies that the future state may create problems, as well as on fantasies about how wonderful the current state is.

Techniques for dealing with this problem involve providing specific information, such as educating people about what is occurring in the environment that is creating the need for change. In addition, it is useful to help people understand the economic and business consequences of not changing. It may be helpful to identify and emphasize discrepancies—the discrepancy between the present situation and the situation as it should be. In critical cases, it may be necessary to paint a disaster scenario, in which people can see what would happen if the current state continued unchanged. It may be helpful to present a graphic image of how the failure to change would affect people. One manager for example, talked very graphically about what would happen if the division did not become successful within eighteen months: "They'll pull buses up to the door, close the plant, and cart away the workers and the machinery." The manager presented a highly graphic image of the consequences of not making the change. An alternative to management's presenting this kind of information may be to involve organization members in collecting and presenting their own perceptions. Participating in the collection and discovery process may make the information more salient, since it comes from peers in the work force.

There is a need to overcommunicate during change management efforts. Extreme anxiety impairs normal functioning; thus, people may be unable to hear and integrate messages effectively the first time. Therefore, it may be necessary to communicate key messages two, three, four, and even five times to individuals through various media.

The second action area for motivation is to obtain participation in planning and implementing change. Employee participation in the change process yields proven benefits. It tends to capture people's excitement. It may result in better decisions because of employee input, and it may create more direct communications through personal involvement. On the other hand, participation also has some cost. It takes time, involves giving up some control, and may create conflict and increase ambiguity. The question, then, is to choose where, how, and when to build in participation. People may participate in the early diagnosis of the problems, in the design or development of solutions, in implementation planning, or in the actual execution of the implementation. There are many options. Various individuals or groups may participate at different times, depending on their skills and expertise, the information they have, and their acceptance and ownership of the change. Participation can be direct and widespread or indirect through representatives. Representatives may be chosen by position, level, or expertise. Using some form of participation usually outweighs the costs of no involvement at all.

The third action area is to visibly reward the desired behavior in both the future and transition states. People tend to do what they perceive they will be rewarded for doing. To the extent that people see their behavior as leading to rewards or outcomes they value, they will tend to be motivated to perform as expected. It is important to realize that during implementation, the old reward system frequently loses potency and new rewards are not set up as an early step. This results in a situation in which an individual is asked to act in one way but has been rewarded for acting in another way. Sometimes people are punished by the existing measurement system for doing things that are required to make the change successful. Management needs to pay special attention to the indicators of performance, to the dispensation of pay or other tangible rewards, and to promotion during the transition. In addition, there are informal rewards, such as recognition, praise, feedback, or the assignment of different roles, and it is important to carefully manage these to ensure that they support constructive behavior during the transition. It is equally important to reestablish clearly an appropriate reward system for the future state.

The fourth action area directly affects individual anxiety. It is the need to provide time and opportunity to disengage from the current state. People associate a sense of loss with change. It is predictable that they will go through a process of "letting go of," or mourning, the old structure. Management, knowing that this is essential, can greatly assist in this process. A number of specific techniques are possible. One is to provide the appropriate time for letting go, while giving people enough information and preparation to work through their detachment from the current state. Another technique may be to provide the opportunity to vent emotions through an event similar to a wake. This can be done in small group discussions, in which people are encouraged to talk about their feelings concerning the organizational change. While this may initially be seen as promoting resistance, it can have the opposite effect. People will undoubtedly talk about these issues, either formally or informally. If management can recognize such concerns and encourage people to express their feelings, it may help them let go of them and move into constructive action. It may also be useful to create ceremony, ritual, or symbols, such as farewell or closing-day ceremonies, to help give people some psychological closure on the old organization.

Thus, there are four action areas in motivating constructive behavior. One concerns helping people detach themselves from the current state. The second concerns obtaining appropriate levels of participation in planning or implementing the change. The third concerns rewarding desired behavior during the transition, and the final action area has to do with helping people let go of their psychological attachment to the present situation.

Managing the Transition

The third implication concerns the actual and explicit management of the transition state, which is that time period between the current state and the implemented future state. It is frequently characterized by great uncertainty and control problems, because the current state is disassembled prior to full operation of the future state. Managers need to coordinate the transition with the same degree of care, the same resources, and the same skills as they manage any other major project. There are four specific action areas in which managers can work (see Table 39.4).

The first action area is to develop and communicate a clear image of the future state. The ambiguity of change without a focus produces major problems. It is difficult to manage toward something when people do not know what that something is. In the absence of a clear direction, the organization gets "transition paralysis," and activity grinds to a halt. This is caused by uncertainty over what is appropriate, helpful, or constructive behavior. Several specific practices are relevant in this situation. First, there is a need to develop as complete a design as possible for the future state. This may not always be feasible, but to the extent possible, it is important to articulate at least a vision ahead of time. Secondly, it may be useful to construct a statement that identifies the impact of the change on different parts of the organization. Thirdly, it is important to maintain a stable vision and to avoid unnecessary changes, extreme modification, or conflicting views of that vision during the transition.

Finally, there is a need to communicate. As previously indicated, it is important to communicate repeatedly and to use multiple media, be it video, small group discussions, large group meetings, or written memos. It is critical to think of this communication as both a telling and a selling activity. People need to be informed, but they also need to be sold on why the change is important. This may necessitate repeated explanations of the rationale for the change, the nature of the future state, and the advantages of the future. Finally, the future state must be made real, visible, and concrete. Communications should include information on future decision-making and operating procedures. The way in which this is communicated can help shape the vision of the future. For example, one company showed television commercials, both inside and outside its organization, demonstrating the specific types of customer service that it was attempting to provide. The commercials gave people clear, graphic, and memorable images of the future state.

The next area is to use multiple and consistent leverage points for changing behavior. This issue relates to the organizational model underlying this approach to change management. An organization is a system made of tasks, individuals, formal organizational arrangements, and informal organizational arrangements. During a transition, when certain aspects of the organization are being changed, there is a potential for problems arising from a poor fit. An organization works best when all elements fit smoothly. Managers need to use all of these levers for change. Specifically, managers need to think about modifications that need to be made in the work, individuals, formal structure, and informal arrangements. Secondly, there is a need to monitor and/or predict some of the poor fits that may occur when changing any of the organizational components. It is necessary to plan the changes to minimize poor fit among different elements of the organization.

The next action area involves using transition devices. The transition state is different from the current and future states; therefore, there may be a need to create organizational arrangements that are specifically designed to manage the transition state. These devices include: (1) a transition manager; (2) specific transition resources, including budget, time, and staff; (3) specific transition structures, such as dual management systems and backup support; and (4) a transition plan. All of these can be helpful in bringing needed management attention to the transition.

The final area is to obtain feedback and an evaluation of the transition state. The transition is a time when managers need to know what is going on in the organization. There is usually a breakdown in the normal feedback devices that managers use to collect information about how the organization is running. This is particularly serious during a period of change when there may be high anxiety and people hesitate to deliver bad news. Therefore, it is critical to build in various channels for feedback. Formal methods may include individual interviews, various types of focus-group data collection, surveys used globally or with select samples, or feedback gathered during a normal business meeting. Informal channels include senior managers's meetings with individuals or with groups, informal contacts, or field trips. Finally, feedback may be promoted through direct participation by representatives of key groups in planning, monitoring, or implementing the change.

In summary, the initial emphasis in transition management is on identifying a clear image of the future state. Secondly, there is a need to pay attention to the changing configuration of the organizational system and to develop—where needed—unique organizational arrangements to manage the transition period. Finally, there is a need to monitor progress through the development of feedback systems. All of these are important elements in managing a transition.

SUMMARY

OPENING CASE REVISITED

The reorganization at Orient Oil presents the potential for a full range of problems associated with change. This change requires concentrated attention from skilled managers. In Orient, the senior team was exposed to a speaker who raised some of the key issues in managing change and described some new and different methods of change management. As a consequence, the senior group at Orient decided to delay the announcement of the change and appointed a highly respected senior member to head a transition team. This team was dedicated full time to the transition management task. The first assignment was to develop an initial transition plan, which included specific sections dealing with communications; how to get participation in the change; how to sequence the individual changes; how to assign jobs and coordinate moves; how to make the changes in information and control systems; and, finally, what the senior team had to do to lead the change effectively. The senior team reviewed this plan several weeks later and approved it. The change was implemented, led by the senior team, and guided by the transition manager and the transition team. Within six months, all of the major elements of the transition plan had been accomplished. As the Orient Oil chairman reflected, "When I think about all of the things we changed and all of the things we had to accomplish, I never thought that we could have done this so well. I'm amazed."

Why have we placed such an emphasis on change management? The process of implementation is a critical determinant of the success of a new organization design. To develop a design and then not give significant thought, time, and effort to the planning and management of its implementation is to do only part of the job of design.

As we look at organizations over time, we find that changes in design are a normal part of life. They are not one-time events but an ongoing element in the development of an organization. Adaptive organizations are able to respond quickly and effectively to new conditions—they are able to reconfigure to support new strategies as needed. Therefore, organizations that remain effective over time have to develop the capacity to execute change and, in some cases, competently manage accelerated change.

This leads us to our final topic: organizations over time. What are the types of design changes that we might anticipate at different points in time? Moreover, how might managers think ahead and begin to plan necessary changes in organization design in a way that anticipates rather than responds to problems and opportunities? Can creative organization design done in anticipation be an effective competitive tool for the manager?

Note

1. R. Beckhard and R. Harris, *Organizational Transitions*, (Reading, MA: Addison-Wesley, 1977); M. Beer, *Organization Change and Development*, (Santa Monica, CA: Goodyear, 1980); N. Margulies and J. Wallace, *Organizational Change*, (Glenview, IL: Scott, Foresman, 1973); D. Nadler, *Feedback and Organization Development*, (Reading, MA; Addison-Wesley, 1977); N. Tichy, *Managing Strategic Change*, (New York: Wiley, 1983); M. Tushman, *Organization Change: An Exploratory Study and Case History*, (Ithaca, NY: NYSSILR, Cornell University, 1974).

40

Vision, Values, Milestones: Paul O'Neill Starts Total Quality at Alcoa

PETER J. KOLESAR

It is really quite impossible to be affirmative about anything which one refuses to question; one is doomed to remain inarticulate about anything which one hasn't by an act of the imagination made one's own.

—JAMES BALDWIN, *NOTES OF A NATIVE SON*

In November of 1987, Paul H. O'Neill, the new Chairman of the Board and Chief Executive Officer of the Aluminum Company of America (Alcoa), appointed a task force of a dozen senior managers to explore the issue of quality management at Alcoa and report to him and the company's senior management Operating Committee with concrete recommendations for change. Over a period of about six months, this Quality Task Force and the Operating Committee labored first to identify the challenges and opportunities, and then to design and begin implementation of Alcoa's "Excellence through Quality" initiative. This is a somewhat personal account of the start-up of this new total quality management initiative at Alcoa.[1]

BACKGROUND

The forces that prompted and shaped "Excellence through Quality" are deeply rooted in Alcoa history. Alcoa is the largest aluminum company in the world with fiscal 1987 total revenues of $7.8 billion dollars, shipments of 2.3 million metric tons of primary aluminum and fabricated aluminum products, and 55,000 employees. Alcoa was founded in Pittsburgh in 1888 by Charles Martin Hall, the inventor of the modern electrolytic aluminum smelting process, and was seeded with venture capital from the Mellon family. From its founding until 1948, when the U.S. Government forced Alcoa to sell off some of its plants to emerging competitors, the company held a virtually complete national aluminum monopoly. Earlier, the company had been moving toward a global dominance in aluminum, but in 1928, under the stress of becoming truly multinational from its domestic base and culture, Alcoa divested its international holdings. These became Alcan, the second largest aluminum company in the world. Over its century plus of life, the company has had an enviable record of success and adaptability. Starting from scratch it created the processes, the products, and the markets for aluminum. Alcoa meant aluminum. Alcoa meant quality.

Yet by the early 1980s, many inside and outside the company believed that the aluminum industry was now mature and the outlook poor. The reduction of trade barriers under GATT had opened North America to competition on aluminum more successfully than had the earlier anti-trust maneuvers of the U.S. Government. Third world governments that held the principle sources of raw material became themselves manufacturers of primary aluminum, often at sharp subsidies. Aluminum, a precious metal when Alcoa was founded, had now become a commodity and the international supply, demand, and pricing system was out of Alcoa's or anyone's control. Another troubling trend was seen: exotic substitute materials appeared in the market.

In the 1982 Alcoa Annual Report, then Chairman W. H. Krome George observed these dilemmas and suggested that effective responses must imply very big changes in "the Aluminum Company." Shortly thereafter, in 1983, his successor, Charles W. Parry, articulated a new diversification strategy that would take Alcoa into the production and sales of the emerging materials that many feared would replace aluminum—highly engineered laminates, polymers, ceramics, composites, and the like. Alcoa would become "The Engineered Materials Company" and, according to this corporate vision, by the turn of the century more than half the company's revenues would be from non-aluminum products. To carry off this strategy Parry dramatically increased and re-focussed the company's research and development expenditures and embarked on a program of acquisitions.[2]

Only four short years later, apparently unimpressed with their Chairman's ability to clarify and execute his diversification strategy, the Alcoa Board of Directors in a quiet internal revolution requested Mr. Parry's early retirement. Thus, Paul H. O'Neill, then President of International Paper, became the new Alcoa Chairman and CEO in April 1987. In his introductory communications to employees, to the financial community, and to stockholders, this first ever outside CEO spoke about safety, about quality, and about the people of Alcoa as a valued resource. He also signalled a fundamental shift of strategy: Further business diversification would be put on hold while he concentrated on improving the performance results of the base business. It was back to aluminum basics.

The options available to O'Neill and Alcoa in pushing an agenda of improving the base aluminum business would be shaped by two facts about Alcoa and the industry. First, truly *fundamental* process change had not happened in the industry and appeared to be infeasible. The smelting of aluminum was still done by the century old Hall-Heroult electrolytic process (notwithstanding very substantial efforts by Alcoa and Alcan to make major smelting innovations in the 1970s and early 1980s). To be sure there had been a century of continual refinements, and by many measures Alcoa was still a world leader in aluminum process technology. Second, *revolutionary* new aluminum products did not appear to be coming along quickly either. Alcoa was no Hewlett-Packard or 3M, companies for which very significant percentages of current revenues come from products that did not exist 5 years ago. As of 1987 the last significant new product introduction was still the all aluminum beverage can that had been introduced twenty-two years earlier—by Alcoa. Again, this product had been refined over the two decades since its introduction and by 1987 accounted for about a third of Alcoa's aluminum revenues. Where was the next such innovation coming from? Perhaps it would be the aluminum intensive automobile that Alcoa was working towards?

Thus, it made sense that a back to the *base business strategy* would also mean a back to the *basics of the business* strategy. While serving as International Paper's President, Paul O'Neill had been an observer and participant in the significant total quality management initiative undertaken by IP between 1985 and his departure for Alcoa in April 1987. Despite millions of dollars invested, a well-thought-out plan, attractive icons on the walls, thousands of hours of training, and the energies of many dedicated individuals, the results at IP failed to match the private expectations or public promises of the program's initiators. Neither Paul O'Neill, nor any number of other observant and frustrated IP employees were able to get IP's quality program on track. Nevertheless, O'Neill was still convinced that quality could be made a central part of his strategy at Alcoa. "But," he said, "it has to produce real value." How would the Alcoa quality effort be different? What follows is the story of what Alcoa did and why.

THE TASK FORCE STARTS WORK

In November 1987, Paul O'Neill and C. Fred Fetterolf, Alcoa's President, jointly commissioned the Quality Task Force to undertake a study of quality management and to recommend a course of action to them and to the Operating Committee (the seven senior Alcoa officers who, since O'Neill's installation as CEO, increasingly acted as his cabinet in running the company).[3] The 12-member Quality Task Force had six members of vice presidential rank and six members just one level below who were seen as "up-and-comers." As Alcoa was rather niggardly with the VP title—there were only 27 in the 55,000 employee company—this was a high horse-power group. The Quality Task Force Chair was the Vice President of Engineering, Thomas L. Carter.

At Paul O'Neill's suggestion, Tom Carter called on me in December to organize a seminar on the statistical design of experiments for the Task Force. It became clear that focussing on such complex statistical material for this group of high level executives at such an early stage in their explorations of quality management was premature. Together we formulated a new set of objectives for an initial quality *education* experience for the Task Force. We began early on to speak of education rather than *training* which connotes specific task oriented learning. The Task Force, at this stage the total quality initiative, did not yet need specific skills. They needed to acquire information and *understanding* on which to make strategic choices.

Tom Carter voiced his perception of the needs: "What I'd like is that within 3 weeks the Task Force should have a more sound factual and experiential basis for understanding TQC, understanding what it could be at Alcoa, and understanding the potential impact on the company. I'd like them to have the beginnings of a plan for how to get there." Then, in response to a question about the likely barriers, Carter ticked off, "They don't know what they don't know. They think it doesn't apply to us, and that at Alcoa we're different. They'll fear a loss of power and autonomy. Lastly, in reality the Alcoa culture is not very fact-driven. Decisions are made around here on the basis of decibel level."

We began by holding an intensive three-day introductory seminar away from the job site, Alcoa's Pittsburgh Corporate Headquarters. The material was presented in a manner that encouraged critical consideration and evaluation, and assessment of relevance to the realities at Alcoa. From the outset and repeatedly throughout the sessions, we stressed two essential points. First, we stated: "Around the world companies like yours are at this very time convened in similar seminars, digesting similar ideas, and are evolving similar quality strategies and systems. There are no secrets in total quality. In rather short order, everyone will know the essentials of what needs to be done. Success will go to those who are superior at implementation—to those who *execute* well." Second, "Diagnosis must precede and shape the specific actions and design to fit this company in this industry at this time. What are the specific organizational and performance problems at Alcoa for which total quality is an alleged solution? Any effective TQ design for Alcoa must be appropriate for the most important of these challenges, and must recognize this company's history and culture!"

Day 1 was an overview of the best thinking in quality management circa 1988. To set the stage historically and philosophically, it started with a summary of the opening day of W. Edwards Deming's now legendary 1950 Tokyo lectures to senior Japanese industrial executives and scientists.[4] Other material included various definitions of quality, the economic impact of quality and the concept of quality related costs, the strategies of prevention and continuous improvement, Juran's ideas on quality management implementation, and the philosophy and implementation of Kaizen in Japan.

After dinner on the first day, we performed Deming's famous red bead experiment. In this simulation, Dr. Deming illustrates the devastating impact of variation on managerial decision making through sampling beads from an urn. Red and white beads are mixed in the urn and the players sample beads by dipping a paddle into the urn. The white beads are acceptable production and the red beads are the defectives. Deming urges the employees (participants) to try not to produce red beads. Of course they can't—the red beads are already in the bowl. They are inherent in the system and try as one might red beads can't be avoided in the sampling. As the simulation proceeds, he harangues the workers, displays the data on an overhead projector and makes some telling points about "management by the numbers," and about the sources and solutions of quality problems. We played out the bead experiment with the Alcoa managers as the "willing workers" and with me as the "foreman," imitating Deming's unique style as closely as I could.[5] I fired employees, rewarded the employee of the month, and threatened to close the plant. A run chart of the red bead "defect" data was plotted on an overhead projector as we proceeded.

However, we added a special twist to the red beads. Prior to the seminar, I had requested from Mr. Carter some Alcoa time series data on product quality, process reliability, sales, safety performance, and the like. I had plotted these Alcoa data on a scale identical to that used in plotting the red bead data. My Alcoa plots were done in two versions, one labelled and one not. After the bead drawing, I displayed the unlabelled versions one at a time on one overhead projector while the red bead data plot was projected simultaneously on another. The question was put to the participants: "Here are some data on an important Alcoa process, over there are the red bead data. What is the difference?" In most cases, it was indeed impossible to see what the differences were, if any. When the identity of the Alcoa plots was revealed, the room echoed to some soft gasps. These data were all important Alcoa performance measures and the similarity to the red bead data was apparent. Most telling was the plot of safety data, which pushed an Alcoa hot button. This plot showed Alcoa's safety performance was quite variable and not satisfactory. Yet according to the control limits, it was in "statistical control" and was not improving despite the company's long-standing safety program. It must be remarked that Alcoa was then, and is now, the best performer on safety in its industry. But Paul O'Neill was not content and he was targeting safety levels equal to those of Dupont, the U.S. national benchmark on industrial safety. But, to reach that level would require a reduction in the Alcoa serious injury rate by nearly a factor of five. Here was an exhortation without much substance to the safety improvement program in place. The participants immediately saw the analogy between the Alcoa Corporate proclamations about improved safety performance and the exhortations of foreman Deming about "no red beads."

All in all, the red bead experiment proved to be an "aha" experience that transcended ordinary classroom learning. Over the course of the next nine months, this experiment was repeated in quality awareness training sessions around the company with data specific to each business unit. It entered into Alcoa corporate folklore, and managers would spontaneously remark, "there go the red beads again" or "no red beads"—a testimony to the appropriateness of this part of the Deming philosophy. Almost immediately the Quality Task Force took onto itself the assignment of looking into safety at Alcoa from a statistical process control and Deming Plan-Do-Check-Act (PDCA) problem-solving cycle point of view. That hands-on work would prove an important part of the learning and maturation of the group on quality management.

On the morning of day 2 of the seminar we continued with an overview of statistical control and process capability ideas, and with an introduction to Japanese style implementing PDCA problem-solving process. After lunch, the focus shifted dramatically as Columbia Professor Mike Tushman put the *content* of total quality aside for a moment and began to work on the *process of changing* to total quality. He introduced a framework for problem diagnosis and problem solving at a macro organizational level and made a direct analogy to our earlier discussions of the Deming style of analysis on the factory floor.[6] It was PDCA all over again, and we reminded the Task Force that though Deming's Cycle was first employed on the factory floor, by 1967 the Toyota Board of Directors had adopted it for analysis of all corporate problems at all levels. Later, in a case study of Toyota Machine Tool's march towards the Deming Prize they would see how the master planning process employed was an elaborate PDCA Cycle. The afternoon's work dealt with a number of case studies, and in each one, as soon as the main problem symptoms were displayed, these Alcoans would leap into action and propose solutions with little orderly analysis. When this tendency was contrasted to the orderly (and apparently slower) Deming PDCA cycle, the Task Force members reminded themselves, and us, of the Alcoa corporate slogan "We can't wait for tomorrow!" Our own joking rejoinder was to describe their actual implementation of the Alcoa slogan with an alternative: "Ready! Fire! Aim!" Along with "no red beads" this phrase would also enter the new Alcoa lexicon.

On the first evening, discontent and uneasiness among many Task Force members became apparent. It surfaced that perhaps a majority of the Task Force was uncertain about their mission, about why *they* had been chosen, and about what was actually expected of them. The day's course materials seemed quite interesting, most averred, but "to what purpose? why are we here?" O'Neill and Fetterolf joined the Task Force on the second evening for dinner. They were alerted to the emergent issue and encouraged to help get the Task Force (back) on the rails. The dozen Task Force members sat around the outside of a large U-shaped table and the Chairman and President took chairs in the opening of the U. Mr. O'Neill started rather softly, "Safety is my number one priority with quality right behind it . . . all-consuming quality. Quality in the broadest possible sense. I'm asking this group to help us think about it and shape what we as a company do about quality. Tell us candidly what we need to know so that together we can make quality a reality—and quickly. I want our relationship to be interactive." After about 10 minutes of remarks along these lines, President Fred Fetterolf spoke, "I fear what will happen if we don't do something dramatic on costs. Experts say that 25% cost reduction is available through quality. That seems credible to me . . . I became convicted of it on the Japanese trip.[7] It is an absolute necessity. We are getting a better picture of what is good and bad manufacturing. I'm concerned about the unevenness of commitment on quality across the company, about the lack of common language and understanding. Tell us what the impediments are. Tell us what the operating committee needs to know and needs to do. I believe we have no choice if we are to be the kind of company we want to be. We are willing to spend the time and money needed to do this and win."

For perhaps an hour Task Force members directed a series of questions at the Chairman and President: "What do you want to be different at Alcoa 5 years from now?" "Why isn't the Operating Committee doing this themselves, instead of us?" "Paul, what were the quality barriers at International Paper?" "Fred, what about timing?" and on and on with more questions and answers. And then the 64 million dollar question, "Will we still be focussing on our aluminum activities?" (Only one member of the Task Force was from a non-aluminum part of the business.) Paul O'Neill replied, "I've not seen any data that shows me that this industry is doomed. . . . You can make a great success by being the best in an industry in which others make excuses for their poor performances. The business that you are not in and looks easy to you is a business you don't know much about."

The performance of Alcoa's two most senior officers was ostensibly persuasive. They were clear about why they were interested in quality, they were clear that they saw a long process ahead and they were clear that they were willing to invest in it. They were also clear in their charge to the group that it investigate total quality on behalf of the

Operating Committee. However, after O'Neill and Fetterolf left and the Task Force debriefed what it had heard and how it now felt, some Task Force members were still uneasy about their mission. This discomfort had several sources including the members' unfamiliarity with O'Neill himself, an ongoing uneasiness about the basic strategic balance between the aluminum and non-aluminum parts of the business, and most importantly for those on the Task Force who reported directly to the four Alcoa Group Vice Presidents—the "barons" on the Operating Committee—an uncertainty about their bosses' support of the quality initiative. It was widely intimated among the group that the most powerful baron thought that total quality was bull, and that two others were at best neutral. Moreover, the Task Force commission had not been made all that explicit to anyone. The Task Force was working for and would report to the Chairman and President, yet most members still worked for the barons in whose hands their careers rested completely. It was also becoming clear that carrying out this Task Force mission would be a bigger job than they anticipated but nothing was being taken *off* their plates.

The seminar work continued all through day 3 with an analysis of Xerox's efforts to install its "Leadership Through Quality" version of TQM being our main vehicle. At this time, January 1988, the details of the development and implementation of the Xerox Corporation's version of total quality were only beginning to emerge in public. Tushman had done some work in the Xerox effort, and his intellectual partner David Nadler had been a central player in the Xerox "Leadership Through Quality" effort.[8] Tushman relayed to the Task Force the thinking and tactics employed by the senior management at Xerox in managing what was planned as a four-year massive change effort. We showed videotapes of two of the key players at Xerox: David Kearns, the CEO who instigated the total quality effort, and Fred Henderson, who was the first Xerox Corporate Vice President for Quality. We examined the general framework of the Xerox total quality implementation plan and the levers that senior management at Xerox proposed to use in managing the company-wide change effort—including how Xerox had used or failed to use them. These ranged from Xerox's cascading training program for all 200,000 employees worldwide, to micro management of senior managers' own personal behaviors (role modelling), to changes in the organization structure itself, and on to changes in the corporate reward system.

The seminar closed down with the Task Force assessing what had been of value and what had not, as well as their taking the first steps in what would eventually be a very detailed assessment of quality management at Alcoa. This assessment, once started, would take on a power of its own and propel the Task Force into an even broader critical inquiry into the general management of the company. To create a reference point, the participants made a checklist of the beliefs, behaviors, systems, and processes of a total quality enterprise. Many components on this list were at considerable variance with existing practices in most U.S. companies and specifically in Alcoa. The Task Force prioritized the list according to what appeared to be most important and appropriate for Alcoa. In addition, they made an appraisal of how big it felt the gaps were. Opinions varied, of course, but most in the group felt that the components of total quality were appropriate and that many of the gaps were substantial. This work, done in groups, helped consensus to emerge. Quality management, the Task Force concluded, was now looking much broader than statistical control, it was bigger than they had thought and could not be neatly separated from the rest of management.

BENCHMARKING

A week later the travelling began. Over a two-week period, the entire group visited companies that had acquired reputations as leaders in quality management, including Allen-Bradley, Corning Glass, Xerox, and Florida Power and Light. An internal visit was made to PEP Industries, an Alcoa subsidiary whose main business is manufacturing wiring harnesses for the automotive industry. PEP had been propelled and helped into Total Quality by Ford Motor Company and had been an early recipient of Ford's Q1 Award. A little later, there would also be influential visits to Preston Trucking and the Tennessee Eastman Division of Kodak.

All of this work was done before the Baldrige Award existed and before the term "benchmarking" had been popularized and so well codified. These highly structured benchmarking visits, organized and planned by Task Force Chairman Carter, were used to identify both overall quality management frameworks and specific quality management systems and tools that could be adopted and adapted at Alcoa. They were also used to test the reality of the theory. Carter's staff prepared a briefing book in advance of each visit containing basic background on the visited company and as many specifics on its quality management approaches as was publicly available. Using this material, the Task Force identified in advance the key issues to learn more about or to test at each visit. These objectives were communicated in advance to the host firm and specific Task Force members were assigned to acquire information on them during the site visit. Each Task Force member produced individual de-briefing notes on his specific assignment as well as on overall impressions. These data were assembled on the corporate plane on the way back to Pittsburgh and a typed summary report was frequently available the next day.

Among the general questions and issues explored on the benchmarking visits were the scope and definition of quality and quality management. Specifics that were asked included: How and why did the TQ program get started? What was the role of the corporate leaders? What were the specific TQ goals? What were the specifics on quality training, on the TQC organization, and on the external and internal resources required? What were the success stories, progress to date, and proof of success? A very important question on all visits was: "What would you now do differently and what would you advise Alcoa to do differently, and why?"

Now the Task Force work proceeded in high gear. In addition to their real jobs, members worked about half-time on the quality mission. Momentum and engagement were building. A report to the Chairman, President, and the Operating Committee was set for the end of February. In preparation, the Task Force also visited three other Alcoa facilities, and individual or pairs of Task Force members visited Ford Motor Company, LTV Steel, IBM, and AT&T at Oklahoma City. Input and data were solicited and obtained from 11 other Alcoa locations, and all Alcoa 1988 Business Unit Operating Plans were reviewed.

Two of the visits were particularly influential. The first was to Xerox, which we had already studied during the initial seminar. Xerox, like Alcoa was a manufacturer, had invented an industry and had held a monopoly. Xerox had been propelled into quality management in the early 1980s when, under attack from the Japanese, it had lost half its market share and saw complete disaster looming. While this had not (yet) happened to Alcoa, the Alcoa managers identified with and were moved both by the struggles of Xerox and by the apparent success of their TQM counterattack. Xerox—and this was two years prior to their winning the Baldrige Award—proclaimed itself the first American company to win back market share from the Japanese, and without government assistance. Alcoa was a supplier to Xerox of the aluminum cylinders—the photoreceptors which copy machines use to form an image. A year earlier, there had been quality and cost problems and a threat of loss of the account. But Alcoa had worked hard to improve and in the process Xerox had taught the Alcoa plant in question some statistical control tools.

Xerox had been motivated to use quality as its change and survival theme by the success some years earlier of its Japanese affiliate. The Task Force read and analyzed Fuji-Xerox President Kobayashi's recounting of his push for the Deming Prize.[9] Xerox had gone about its quality transformation in a particularly studied way. The role of Xerox Chairman David Kearns in the change process, the design of the "cascading" Xerox total quality awareness education, the parallel quality management organization that Xerox created and added to the existing corporate structure, the Xerox six-step version of the PDCA problem-solving process, the Xerox formalization and refinement of its competitive benchmarking process, and the overall Xerox total quality management change plan delineated in the famous "green book" that the Task Force examined but could not copy—all these elements had an influential impact on the ultimate Alcoa design. Even the Xerox name for its effort, "Leadership Through Quality," would be mirrored by the name Alcoa ultimately chose for its own program. None of this suggests that the Xerox story or program was bought uncritically. On some aspects, Xerox talked a better game than the evidence showed that they played. Xerox had also been an Alcoa supplier and inquiries around the Alcoa system on Xerox performance did not always match the "Leadership Through Quality" rhetoric. The theme was: learn from Xerox, applaud them, and do still better.

The visit to Florida Power and Light—eventually, Alcoans would make three visits there—was influential in different ways. The Task Force quickly saw that the quasi-military management style at FP&L, and its wholesale and rather slavish adoption of Japanese total quality control techniques, would not find ready acceptance at Alcoa. Yet a number of FP&L elements were adopted. The FP&L quality improvement story-board method of running improvement teams was much admired and would be licensed from FP&L. The Alcoa senior managers were quite impressed with FP&L's adoption of the Japanese style of Policy Deployment to give focus and accountability to the massive quality effort underway. There was a downside too. The thrust of much of the quality management effort at FP&L—then a year prior to their winning the first Deming Prize awarded outside of Japan—appeared to focus too hard on "winning the prize" and there were suspicions among some on the Task Force that FP&L was doing some things primarily to look good to the jury.

All in all, the benchmarking visits had an impact on the twelve Alcoa senior managers that no amount of reading, lectures by professors, or harangues by gurus could possibly equal. Whenever possible on these visits Task Force members paired off with their peers. (It became one of our standard requests prior to a visit.) They verified whether total quality had actually had an impact on their peers' lives and management styles. Later on, we would expose the Operating Committee to similar experiences for that very reason—we wanted person-to-person transfer of experience. In a typical commentary, a senior Alcoa manager remarked on a flight back to Pittsburgh from a visit, "You know, Professor, we didn't see anything different from what you had told us about weeks ago, or than what we'd read in Deming and Juran. But when Joe Blow, my peer at ABC Corporation, showed me what he's been doing, the light really went on for me. Before, I understood it with my mind. Now, I believe it in my gut."

THE SAFETY PROBLEM: LEARNING BY DOING

One other experience had a deep impact on the Task Force: their hands-on examination of the issue of safety at Alcoa. Recall that in the introductory seminar, I used Alcoa safety data displayed as a control chart. The data used were monthly serious accident rates over a 24-month period at a major Alcoa mill, and it had been impossible to tell the difference between it and the red bead plot. I had walked slowly over to the tray of beads, leaned over it, cupped my hands to my mouth, and stage-whispered into the tray, "No red beads! No lost time accidents!" The group decided on the spot that it wanted to examine the issue of safety at Alcoa from a total quality perspective. More safety data were collected and processed TQ style: Pareto charts, run and control charts, and the like. The Task Force spent a day defining specific safety problems and examined data linking cause and effect. The two-decade history of the safety program at Alcoa was recounted.

The Task Force seized on several points: how little data analysis was typically done on such problems; how improvements in safety had tapered off at Alcoa when management attention had waned;

how easy it is to talk about, but how hard it is to actually get to the root causes of a social and technical phenomenon as complex as safety; the futility of managing via exhortation. They experienced first-hand their own deeply ingrained tendency toward a "ready-fire-aim" style of management. They also experienced the frustration of trying to improve a process where the special causes had largely been eliminated and where the remaining faults "were in the system" (as Deming would put it) or were "designed that way" (as Juran would put it).

This analysis of the safety issue at Alcoa was particularly striking to the Task Force because O'Neill—since his first day on the job as CEO—had stated publicly that safety in the workplace was his number one priority. At a January meeting with financial analysts in New York, while the Task Force was in session, he had said, "Alcoa's most important human value is that its employees work safely. Some things are negotiable. Excellent safety performance is not. When you pay attention to all the details associated with achieving excellent safety performance and housekeeping, good economics will result." Similar remarks had been made continually inside the company since O'Neill's appointment in April of 1987. But it was now nearly the end of January 1988 and no systemic attack on safety had yet been mounted. This extended and painful self-analysis on the safety issue would color the thinking of the majority of the Task Force on quality. Several members recounted that some years back the safety program at Alcoa had produced real improvements and then had levelled off when an important leader retired. They mused: "What would keep a total quality program from duplicating this experience?" "What happens to quality if O'Neill leaves or if his attention gets refocussed elsewhere?"

REPORT TO THE OPERATING COMMITTEE

By February 18th, 44 days after the initial seminar, the Task Force was ready to fashion its conclusions and assemble a set of recommendations for O'Neill, Fetterolf, and the Operating Committee.

A particularly insightful aspect of the Task Force's company assessment was its review of the Alcoa 1988 Business Unit Operating Plans from a freshly acquired total quality perspective. This work defined serious gaps between where Alcoa was currently and where a total quality driven Alcoa would have to be. The Task Force found that quality was not an explicit component of the business unit mission statements, strategy, or philosophy. Overall, the understanding of quality was limited and, when it was included in planning, the focus was exclusively on a limited set of physical quality characteristics like surface finish or dimension. Tracking and monitoring of customer returns, product specs, and scrap were just beginning and measurements of rework, good parts per hour, and flow times were not routinely monitored. The business unit plans did not leverage improvements made at one plant across the system. Little evidence existed that suppliers were being brought into the Alcoa quality process. At most business units, specific quality goals were absent from the plans. Following on the century-old traditions of a capital-intensive industry, the Task Force observed that where there was a push for quality it was typically cast in terms of capital expenditures to replace older equipment. The business plans also revealed that generally quality improvement appeared to be a bottom-up process. Only in a few noteworthy cases was there a top-down process driven by senior management. Moreover, Alcoa appeared to be all over the map on philosophy and approach, with different units influenced by one of a number of outside gurus whose messages were at times in conflict.

Among the misconceptions about quality management that the Task Force found in the business plans were the following:

- Random production upsets that hurt Alcoa in the past will not happen in the future.
- The identification and prioritization of cost improvement projects can be done independently of quality considerations.

- Everyone *knows* that Alcoa's products are generally higher quality than the competition, and we can guarantee our quality without bringing our processes under control.
- SPC (statistical process control) can be "turned on" for specific customers who require it, and it is primarily a matter of running off some charts in order to help in marketing.
- Multiple sourcing by our customers is a way of life that our quality cannot change and our customers changing needs do not include higher quality from us.

In November 1987, just after the Task Force had been commissioned, Carter had taken a survey of the members' opinions on the level of quality in Alcoa. Graded on a scale of 1 to 10 (with 10 being excellent and 1 being terrible), the mean Alcoa overall score had been 4.6. Now, at the end of February, as the Task Force completed its initial quality education and benchmarking visits, they repeated the self-evaluation and the mean score moved down to 4.0. The individual evaluations also moved closer to one another. A framework for self and external evaluations was standardized and applied to several individual Alcoa business units, as well as to all the companies that had been benchmarked. In those pre-Baldrige days, the Alcoa Quality Task Force had to create their own framework for this evaluation.

Of the outside companies visited, Xerox and Florida Power and Light scored highest overall. Of the Alcoa business units studied, PEP Industries scored highest, nearly as high as the two leading outsiders. This fact was to prove important to many Alcoans as evidence that quality management at its best was demonstrably possible inside the company.

While the Task Force formulated its findings and report to O'Neill, Fetterolf, and the Operating Committee, the Task Force's thinking was dominated by possible reactions of the Operating Committee, their direct bosses in most cases. The Chairman and President already shared a strong commitment to a new quality initiative, but how to convince the Operating Committee weighed heavily on their minds. Ultimately, the strategy adopted was not to try too hard to convince them and to state essentially: "We have done what you asked and examined quality inside and outside Alcoa. We have also studied quality management theories and practices as described by leading thinkers and as practiced in Japan. It, '*quality management*,' is bigger and more important than we imagined when we started. It is vital to the future of Alcoa. Moreover, we have changed as a result of what we have seen and *others* need to change too. Although we can describe the main things we learned in the last two months, we do not expect our description to be convincing or sufficient. For you to understand what happened to us and what Alcoa should do, our most important recommendation is that you experience a process similar to what we just went through."

The body of the report was a series of points that summarized the activities of the Task Force and then listed the main findings in four categories:

- *Culture*: Quality is a rallying point around which an entire company can be energized. It is appealing to most people, building on their innate desire to excel and control their destiny. It is more inspiring than pure financial goals. Quality can be a win-win proposition vis-à-vis the union. The quality management view that 80% of the problems are with the "system" is a key contributor to this.[10] On the cautionary side, quality is not a quick fix and those companies that see it that way fail. It will not become "real" until it permeates the values, norms, and culture of the organization.
- *Impact*: Quality is a differentiator that can be harvested in price and market share if it is perceived by the *customer*. In the short term, there are costs to beginning a quality effort—e.g., training, facilitators. Getting beyond the symptoms of quality problems to root causes and solutions is critical. Quality emphasis and metrics are quickly moving beyond detection to full product life cycle indicators and into staff and support functions.
- *Enablers*: Implementation of quality requires substantial changes and additions to the organization including training, creation of top-

down commitment, and the creation of broad-based involvement. Distinct attitude changes are required, including: a focus on never-ending improvement against quantitative metrics; attention to and respect for detail at all organizational levels; a fact-based orientation in decision making; and granting appropriate authority to worker teams. Successful quality efforts seem to involve substantial investments in people versus things. A change is required in supplier relations with the burden on us to understand our own needs, a thrust toward fewer suppliers and shared gains, early supplier involvement in product and process design, and a need for a formal supplier certification program. There is probably an 80/20 rule on the impact of quality management tools.

- *Strategies*: Quality can be used as a weapon to gain share; to produce a unique product; to set a pace competitors cannot sustain; and to reduce cost. Quality can be used to reduce barriers to progress and improvement in the organization—horizontal, vertical, supplier/customer, union/management, and staff/line. Quality can be used as a vehicle for focussed top-down deployment of an overall strategy/policy and as a vehicle for overall change of a broader scope. Quality can be used to close the gap between our performance and our goals or values on things that matter to the organization, e.g., *safety*.

The Task Force saw that their most important mission was to get a gut commitment to quality by the Operating Committee, and second to continue their work as that commitment was developed. To these ends they recommended:

- The issue of quality is so central to the business and its overall management and the magnitude of the changes needed around it are so great that a corporate quality office and officer are needed.
- You, the Operating Committee, should get educated on total quality. We will design for you an intensive study and travel program based on the best of our experiences.
- In the interim, we will continue to serve as a design team and will develop a quality implementation plan and structure for the company. Some elements of the Alcoa quality initiative must remain your responsibility and we will call these to your attention. *One of these responsibilities which remains uniquely yours, is the development of a corporate strategic objective statement, values, and guiding principles. An effective quality initiative cannot be mounted without this. . . .*

THE OPERATING COMMITTEE GETS EDUCATED

The Operating Committee agreed to all the Task Force recommendations and prepared to begin their education while the Task Force itself continued its exploration and design work. Later on, O'Neill, Fetterolf, and the Operating Committee would admit that they undertook this education grudgingly as they did not understand "why they needed additional education on a topic to which the group was so committed." In mid-April, a three-day Operating Committee quality awareness seminar was held at a retreat in the mountains outside Pittsburgh. There was an atmosphere of high stakes and high expectations on the part of the Task Force and all involved in the design and delivery. These were the men who *ran* the company and the conventional wisdom of total quality ideology was that the success or failure of quality management at Alcoa would depend on their engagement.

For Paul O'Neill this was another opportunity to put his own stamp on the emerging quality initiative, but to do so without inhibiting his team's ability to discover it for themselves and thereby make it their own, as well as his. This group was his cabinet in formation, but they were not yet used to functioning as such for that was not the way Charlie Parry had run the company. Only nine months earlier, these same men had run the company under a different leader, with a very different style, and with a very different strategy. Only O'Neill was different. To date, not a single new face had been brought into the senior management ranks nor had anyone left or been removed. We knew of no precedent for O'Neill's strategy of changing the fundamentals of corporate mission and management style without at the same time changing at least some of the people at the top. Thus, this seminar and the field trips that followed were also supporting a parallel agenda of changing the outlook and style of the senior management team at the company—and of molding them into more of a team.

To start off the awareness seminar, the participants were asked to lay out concerns, issues, and questions about quality management they felt should be addressed. Several items on their list were the concerns of senior managers everywhere who face implementation of a quality initiative. One in particular was a most direct expression of a concern of many American managers. It was one of the Group Vice Presidents who asked, "How is quality different from metallurgy? I run a business in which metallurgy is at the core of all our products and processes. I know very little metallurgy and I have experts for that. It works well. But I'm told that quality is different and that I have to be directly involved. Is this true and if so, why? And exactly how do I have to be involved?" Another Operating Committee member asked, "I've been around this company long enough to have seen many programs come and go, including some quality programs. Most made sense on their face and had some initial impact, but for the most part the promises were never really fulfilled and eventually—in fact, not too long after their flamboyant introduction—these programs just fade away and we are back to the same old Alcoa we were before. What is going to make this 'total quality thing' any different?" A third member asked, "I've been to Japan and I've seen it in action there and I know it works, but that is a very different culture. Can we make it work here without becoming Japanese, without losing who we are and without losing our own distinct American and Alcoan advantages? I don't think we can or should try to make Alcoa into Toyota."

These issues and others were squarely faced and explored, but clearly not settled outright. The prime seminar objectives, however, were achieved: An ecumenical framework of contemporary quality management was developed and critiqued. The magnitude of the change from traditional management practices was articulated, and the process and problems of such a large scale organizational change discussed. At all points in the deliberations, the relevance to specific Alcoa situations was at the fore. We again did the red bead experiment and the Xerox case study. By the end of the three days, it was the judgment of our attending Task Force coach, and most importantly of our chief customer, Paul O'Neill, that the seminar—the first step in the Operating Committee's education process—had "worked." Now the whole Operating Committee—including O'Neill and Fetterolf and accompanied on each trip by several members of the Quality Task Force—went benchmarking to Tennessee Eastman, to Florida Power and Light, and to Alcoa's own PEP Industries. Paul O'Neill visited one-on-one with David Kearns, the CEO of Xerox.

A consensus was forming and enthusiasm was growing, and there was a sentiment that Alcoa "is probably doing this 'just in time' for we are at best half a step ahead of the competition, if that." In early May, the Operating Committee and the Quality Task Force participated in an event that would prove to have a very high impact and further solidify what the responsibilities of the senior management in leading quality management really were. It was an "Alcoa Quality Day" and it enabled the Operating Committee to generate a good part of their own answer to that month-old question, "How is quality different from metallurgy?" Tom Carter identified and brought to Alcoa's Pittsburgh headquarters eight successfully implemented quality improvement projects from around the company. The goal was to show some of the best of what was already happening inside Alcoa and to illustrate with homegrown Alcoa examples many of the quality improvement themes that had been discussed in the seminar, encountered in the readings of Deming, Juran, and Imai, or seen on external benchmarking visits. We included projects that spanned a broad range of applications: internal supplier-customer relations; cooperation with external suppliers; and satisfying an external customer. They had projects that utilized a complete set of problem-solving tools and methods ranging from the simplest of Pareto charts through more complex statistical process control and sophisticated designed experiments.

There was an air of excitement in the Alcoa Corporate Conference Room at Pittsburgh Headquarters that morning. A score of presenters had been flown in for the occasion from all over the Alcoa system. The first speaker, with a boxer-like build and ruddy complexion wearing a suit that was neither Brooks Brothers nor Armani, stood with very apparent nervousness in the corner of the room. The primary audience was more than a score of Alcoa's most senior managers including the CEO and President—enough to make anyone a bit nervous. The master of ceremonies welcomed everyone and then simply said, "We'll get started with the presentation by the Warrick Mill." Our man stepped to the podium and introduced himself: "I'm Ralph Box. I'm a baking room operator from the Warrick Mill and this is not my usual Monday morning." The room exploded with laughter, but quickly hushed attentively as Ralph detailed his role in a project that had made significant quality and financial improvements. He spoke with obvious knowledge of gauge capability, and interpreted complex statistical control charts. His business sense and his pride of these accomplishments were evident and impressive. This was high-power employee involvement already in place somewhere in Alcoa. And so it went: from improved process control in anode baking, to better relations with a major carbon supplier, to the elimination of over-control in smelter tapping, to use of complex statistically designed experiments, to reducing the cracking of ingots as they are cast.

After five hours of project presentations and follow up questions and answers, all the presenters and observers save the Operating Committee left the conference room. The Operating Committee debriefed. Several questions went up on the flip chart: "What are the most important things we have heard and learned?" "What does this imply for what quality management should be at Alcoa?" "What does this imply for *our* roles in quality management at Alcoa?" They homed in on several conclusions. First was that individually and collectively they, the Operating Committee, were remarkably ignorant that there was this level of quality improvement activity and competence—even in their own businesses. Second, it was obvious that Alcoa as a corporation had done little to create, sustain, and encourage this activity—to date quality at Alcoa was a bottom-up, grassroots, almost underground movement. Third, there was very erratic performance overall—excellent as these examples were, they were isolated pockets of excellence. Moreover, there had been no meaningful dissemination of the results of these improvement projects to other locations or businesses, not to speak of dissemination of the quality improvement methodology itself. Fourth, although several of the projects dealt with complex technologies, there had been no involvement of the Corporate R&D resources at any stage. A major conclusion that would shape the actions of the Operating Committee emerged from this several-hour discussion and working session: "*It is our responsibility to create an Alcoa in which this excellence is the rule rather than the exception—an Alcoa in which excellence is shared, systematized, and rewarded.*"

Bringing It All Together

With the first wave of the Operating Committee's education and benchmarking completed, and the Task Force coming to closure on its recommendations for specific quality management systems, tools, training, and the like, an extraordinary day-long joint meeting of both groups was held on May 27, 1988. The day was structured to bring the thinking of the Quality Task Force and Operating Committee together, to see if they were congruent, and to make important implementation decisions. Each group had a specific set of issues to report on and was to spend the morning by itself putting the final shape to its findings and recommendations. The Operating Committee was to present its long-term goals and its image of what it wanted quality management to be at Alcoa, while the Task Force was to provide the specific "enablers" it had been developing. The day would end with the Operating Committee again closeted to review and act on the day's proceedings.

The Operating Committee reported first. They were now deeply involved in quality and were particularly excited this morning, having just spent the previous two days on a variety of quality related tasks, including the last of their benchmarking visits. They were eager to share what they had done and learned, to hear from the Task Force, and to get on with the job. They now understood why the Task Force had been so insistent about the need for them to get educated and to travel. They too had changed. During their long morning session, the Operating Committee had framed their conclusions as follows: "What," they asked, "would we want a visitor from space, who descended into Alcoa five years from today, to see with respect to quality management?" Several flip charts were filled in freewheeling brainstorming style and these were then critiqued and organized. Their image of quality management was:

- A commitment to be the best. Aggressive benchmarking with trending. A consistent definition and processes across the company.
- A strong focus on the customer. An Alcoa that knows who its customers are, what they need, and how satisfied they are. Systems in place to elicit data and feedback on this.
- Quality processes, systems, and culture are institutionalized. These should be unifying and motivating and relate to both individual and team improvement efforts.
- A quality culture that produces 59,000 "Larry Birds" [referring to qualities that the Operating Committee saw epitomized by the great Boston Celtic basketball star] excellence, teamwork, selflessness, competitiveness, and hard work.
- There should be an Alcoa scientific problem-solving process deployed and used across the company—on the pattern of the PDCA, Xerox, or FP&L systems.
- Wide use of teams as appropriate. Teams should be fact-driven, flexible in structure and mission. Teams should be guided and managed. [There was a strong aversion to the concepts and name "Quality Circles."]
- A formalized Alcoa set of guiding principles is required, including Alcoa vision and values statements.

Having been "charged" by the Task Force with working on a set of guiding principles for Alcoa, the Operating Committee at this point simply acknowledged that they too shared in the assessment of how important this need was for a new focus and stated that they had begun the work. They closed their presentation with a list they generated of 14 prioritized characteristics of excellent companies. The Committee had also given their own appraisal of where they felt Alcoa stood. Thus, with this list they had come a long way toward jointly defining their problem. (Again, had the Baldrige framework existed at the time, some of this work might have been simpler.) All that remained would be execution.

The Task Force was pleased with the congruence in the Operating Committee's report and their own thinking on these issues. They could have written this report themselves. Now it was the Task Force's turn, and Tom Carter presented their recommendations. His opening remarks were passionate and startling to the Operating Committee. "We have worked diligently at this quality charge you put before us. We thank you for the opportunity. We have been changed and convinced as a consequence of what we have done. We are taking up this quality mission and you cannot take it away from us. We will proceed regardless of what you do!" The Operating Committee members glanced sheepishly and perplexedly at one another—who was denying anyone anything?

The content that followed that challenge was straightforward, given where everyone now stood:

- More education and training for both groups and the design of a quality awareness training for the "top 100 in the company." (This would soon become the top 300 or so.) Topics like benchmarking, team processes, reward systems, and the like remained to be explored in depth.
- The work of the Operating Committee on values and guiding principles was urged on, and it was recommended that it be augmented by an Alcoa Quality Policy.
- A senior Corporate Quality Officer was called for and it was recommended that the person be

TABLE 40.1. The Operating Committee's Characteristics of Excellence

Priority	Characteristic	Importance	Alcoa Score
1.	Being the Best	100	48
2.	Safety	95	64
3.	Quality Process	84	43
4.	Importance of the Individual	74	51
5.	Meaning to Employee	71	53
6.	Ethics and Integrity	65	81
7.	Excellence in Manufacturing	54	47
8.	Execution, Attention to Detail	54	37
9.	Good Corporate Citizenship	49	78
10.	Growth and Profits	46	32
11.	Technology	33	60
12.	Innovation	32	31
13.	Shareholder Value	26	30
14.	Knowledge and Education	20	42

a member of the Operating Committee. A list of desirable characteristics for this person was offered.

- Specification of top management roles and accountabilities were to be defined by the Operating Committee with primary attention given to the roles of the CEO, the COO, the Corporate Quality Officer, and the Group Vice Presidents. The Task Force offered recommendations on these roles.
- A set of quality management enabling mechanisms and tools were identified for development and deployment, these included a variety of quality related training, a problem-solving process, and an Alcoa Quality Book or master implementation plan (patterned on the "Green Book" quality process definition and implementation plan of Xerox).
- The Task Force declared their jobs done and resigned as a body, but they recommended that a standing-senior level Quality Steering Committee be created to serve as an advisory body on the implementation. They then all agreed to serve on this body, if appointed.

After some joint discussion, the Operating Committee went into session by itself to consider the recommendations. All were accepted in principle, although some needed to be worked on over the coming weeks and months. By the end of the afternoon they had selected Alcoa's new Vice President for Quality. He would be Thomas L. Carter, the Quality Task Force Chairman.

VISION, VALUES, AND MILESTONES

During their external benchmarking visits, the Task Force had been very impressed with how tightly the quality programs at the host firms were linked to a larger sense of corporate purpose. Whether it be Xerox, Toyota, Florida Power and Light, or Tennessee Eastman, the quality strategy, goals, and ethics were always contained within a larger all-encompassing mission and value statement. Such linking with a larger purpose also came up in their readings. Perhaps at first, reading Deming's opening comments to Japanese business leaders in Tokyo in 1950 may have seemed corny—"International trade is an essential component of peace and prosperity. International trade depends on quality . . . quality leads to productivity, to competitive position . . . to jobs, jobs, and more jobs for the Japanese people." His first point, "strive for constancy of purpose," may have seemed a platitude until CEO Donald Peterson of Ford stated that focussing on corporate purpose was the single most important thing that Deming had taught the world's second largest automaker. The Task Force quickly became convinced that the level of extraordinary quality achievement that they envisioned for Alcoa would simply not be attainable unless the quality effort was linked to a higher sense of purpose for their company as well.

There was, however, no Alcoa corporate mission statement that any Task Force member was aware of. On investigation they would later discover one tucked away in the archives and indeed, it seemed that every ten years or so a new Alcoa values statement would be articulated by someone at or near the top of the company and then filed away. Alcoans had long thought nobly of their company, of its ethical behavior, and of its contributions to society and to the nation. To many the company had been a kind of family, literally and figuratively—remarkably, one of the officers on the Task Force would shortly occupy the same vice presidential position as had his father. In only a few companies would employees at all levels refer to themselves with a term like "Alcoans" and really mean it. Though some said these feelings were weakening, and cynics said they were completely gone, one could feel how strongly the Task Force felt these sentiments. What's more, a psychological void had been created by the sudden transition from CEO Parry and his diversification strategy to CEO O'Neill and his back to aluminum strategy. The Task Force had not been pleased when, during their initial deliberations in January, they had seen a newspaper account of a speech O'Neill had given to New York security analysts expounding on Alcoa values—values that had not been discussed or affirmed within the company. The concern was less with the substance of the remarks as with a feeling that "nobody told us, nobody asked us" and a fear that a possibly fleeting opportunity would be missed to reach out to bring all Alcoans together again and get the company refocussed.

The Task Force recommendation met with a favorable response from the Operating Committee, and once the issue was put before them they too felt a strong need for clarification and focus. At meetings with the rank and file and mid-level managers, Operating Committee members were confronted with confusion about what parts of the old strategy were still valid, about how significant the changes really were, and about which of the old Alcoa values still held. Many Alcoans were saying, "We're confused, we need to know where we are going." The Operating Committee saw inconsistent behavior with respect to the company's implied values. And there was real concern, even fear, among some business unit managers about a new set of demanding financial targets that O'Neill had set for the company.

An enormous research, formulation, feedback, and revision effort was unleashed that would go on for months. Though the Operating Committee undertook the responsibility of formulating the mission statement, the Quality Task Force stayed involved doing background work, benchmarking the vision and mission statements of other leading firms, and serving both formally and informally as a sounding board for the long series of drafts that the Operating Committee would produce. By my tally, the Operating Committee spent at least 11 full days in plenary session on discussion, drafting, redrafting, and the like. And that does not count the very considerable time spent singly or in small groups working on support tasks.

As this work was going on, the Operating Committee was getting closer to its own articulation of what quality management should be at Alcoa. Indeed, the two tasks became inseparable. The work began in May and the final articulation of the "first draft" was not arrived at until September. With communication to other senior level managers and reaction to their feedback, the final product was not ready for roll-out until January 1989. During those months, there was hardly a time when the Alcoa mission statement was not being actively worked on by the Operating Committee. "Forged" is the operative expression, for every single word and phrase was hammered out with great care. When they were finished, this group of the nine most senior Alcoa managers would have a document that each had a tremendous investment in, one that they expected would be the compass to guide Alcoa into the future. It would be called the "Alcoa Vision, Values, and Milestones." The *Vision* was to be the target, a setting of the corporate sights on where Alcoa wanted to be in the future; the *Values* were to be the standards the company and its employees would live by; and the *Milestones* would be the measures and checkpoints along the way that would both mark out progress and specify what remained to be done. A year later, Paul O'Neill would refer to the Vision and Values document as Alcoa's enduring constitution that would take it into the next century. The final articulation, arrived at after many weeks of internal debate and several rounds of feedback from business unit and other upper level managers, was:

Alcoa's Vision

Alcoa is a growing worldwide company dedicated to excellence through quality—creating value for customers, employees, and shareholders through innovation, technology, and operational expertise.

Alcoa will be the best aluminum company in the world, and a leader in other businesses in which we choose to compete.

Alcoa's Values

Integrity: Alcoa's foundation is the integrity of its people. We will be honest and responsible in dealing with customers, suppliers, co-workers, shareholders, and the communities where we have an impact.

Safety and Health: We will work safely in an environment that promotes the health and well-being of the individual.

Quality and Excellence: We will provide products and services that meet or exceed the needs of our customers. We will relentlessly pursue continuous improvement and innovation in everything we do to create significant competitive advantage compared to world standards.

People: People are the key to Alcoa's success. Every Alcoan will have equal opportunity in an environment that fosters communication and involvement while providing reward and recognition for teams and individual achievement.

Profitability: We are dedicated to earning a return on assets that will enable growth and enhance shareholder value.

Accountability: We are accountable—individually and in teams—for our actions and results.

This was more a statement of intent and ambition than of fact. In fact, the company had not been growing for some time. While it had important but very selective international operations and relationships, it could not truly be called international or worldwide. The best aluminum company? Some in the company and outside clearly thought so, but as more cross-industry benchmark data on product and process performance and on management practices became available, the self-evaluations were becoming harsher. "We'll know we are really the best," said Fred Fetterolf, "when the only question left on the table is who is second, and when our performance level is world class, not just the best in our own industry." It was hoped by the Operating Committee that the phrase "leadership in other industries in which we choose to compete" would clarify the role and standards in Alcoa's emerging other businesses like ceramics in which it would simply not be credible for neophyte Alcoa to claim to be the best. "Be a player and leader now, then raise the stakes," said an Operating Committee member.

Who could disagree? Was there, is there substance here? These statements might appear at first reading like "motherhood and apple pie" platitudes. But not so to the Operating Committee who labored mightily over them and who proposed to run the company by them, to live by them, to be personally accountable on them. Each word and expression included was evaluated from the standard, "Can we fully commit to that?" As an example, the Integrity value at one time included the word *fair* as in "we will be fair and honest." There was a long debate on the point that while Alcoa might want to be fair, and indeed strive to be fair, it could not guarantee fairness. So, out went "fairness." At the end of this process, the Operating Committee had words to live by. The quality thrust that had initiated this work was further amplified by a quality policy statement which became an integral part of the package.

Alcoa's Quality Policy

We are committed to quality in everything we do. It is Alcoa's policy to:

- Provide products and services which consistently meet or exceed the needs of our external and internal customers through the efficient use of resources.
- Involve all Alcoans in never-ending improvement in the quality of products, processes, and services.
- Provide every Alcoan with the training and tools necessary to contribute to the quality effort.

Our success will be measured by the satisfaction of our customers.

Both the Task Force and the Operating Committee had been strongly influenced by the frequent appearance of the theme of management by fact, by the repeated emphasis placed on the importance of specific *measurable* goals at each of the companies benchmarked, and by the policy deployment technique that was used at some of the benchmarked firms to assist in answering the 64 million dollar question: "Alright, now that we have articulated these noble aspirations, how do we propose to make them happen?" They moved part way toward an Alcoa response with the development of the "Alcoa Milestones."

The work started from the question "How will we know when we are there?" An exercise they used to develop responses was to imagine that a visitor from outer space would descend on Alcoa in five years. "What would you expect him to see as he walked the halls of our offices, the aisles of our mills?" Such a visitor wouldn't be fooled, he would be naive but intelligent and perceptive. All agreed that, like Deming, this visitor wouldn't be impressed by slogans, ambitions, or other quality management "artifacts." He'd have to see specific behaviors, systems, and performance to demonstrate that Alcoa had already made progress towards the vision, values, and quality policy, and was really on the road to world-class excellence. In the spirit of the total quality ideology that states that if you can't measure it you can't manage it, he'd need measures of progress and performance—or in the jargon of total quality, "metrics." So began the development of "The Alcoa Five-Year Milestones."

While the Task Force played a key role in proposing them, the Operating Committee, after feedback from the middle management ranks, made the final definitions. (Actually, the Vision, Values, and Milestone statements were all rolled out simultaneously.) The Task Force laid important groundwork by defining a set of desirable characteristics for milestones which included that they should be: measurable, actionable, enduring, significant, encompassing a sense of "best" and of "stretch," limited in number, simple to understand and use, and owned. They should also be communicable, motivational, and *enabled*. This last word was meant to capture the idea that attainment of the milestones must be supported by the investments and actions of the company and its most senior management. (Echoes of Deming again. His Point 11 states: "Stop giving arbitrary targets, goals and quotas to the work force without the means to achieve them.") There was a strong aversion to throwing up a wish list to the rank and file and hoping for miracles to happen.

This work was tough going, the list of criteria was awfully long and the drafters were getting tired. After many rounds of discussion within both the Operating Committee and the successor to the Task Force, communications to selected members of the next level of Alcoa management, and revisions based on the feedback received, they produced the final Milestone list (see below). By now the Operating Committee had put in at least 22 days of work in plenary sessions—and still more on their own or in small subgroups—getting educated on quality, doing benchmarking, and designing the Alcoa quality constitution. It would be understandable that fatigue was setting in. The final list was arrived at perhaps as much by acquiescence as by active consensus.

Five-Year Milestones for the Alcoa Vision

Customer Commitment

- Establish interactive relationships with internal and external customers and suppliers based on understanding of real needs and performance to targets with evidence of continuous improvement.

Employee Involvement

- Everyone clearly understands their customer needs.

- Clear definition of accountability for each team and individual.
- A team approach to problem solving and continuous improvement.
- Provide the training and education to enable all Alcoans to excel in their jobs.

Excellence

- Over the next five years, achieve a 50% reduction in serious injury and lost workday incidence rates as we move toward our goal of an injury-free workplace.
- For each process, activity, and technology, benchmark our position against best in the world. Close the gap where advantage can be gained.
- Critical Processes in control and capable.

Financial and Growth

- Average 15% return on shareholders' equity, not less than 10% in any year (and corresponding return on assets and return on investment.)
- Achieve growth objectives through improved asset utilization and approved expansion plans and pursue additional corporate growth as required to realize a 20% real growth in revenue above 1988 levels while achieving corporate financial return objectives.

Five-Year Milestones for the Alcoa Values

Integrity

- Be completely aware and committed to our conduct and behavior. (High now, same in five years.)

Safety and Health

- A 50% reduction in lost work days and Serious Injury Frequency.
- Stress [reduction]
- Housekeeping better
- More involvement of hourly [employees]

Quality

- Know and meet more customers' needs
- More processes in control
- More benchmarking
- High level of awareness
- More partnerships with (less) suppliers
- Internal customer needs met
- Facts used and available

People

- More minority and females in management positions
- Higher selection standards
- More involved in High School preparation of students for Industrial Jobs
- Greater number of cross-functional teams and involvement
- More apprentice training and development

Excellence

- Objectively understand World Standards and gaps closed/closing
- Win the Malcolm Baldrige Award
- Be the Benchmark

Looking at the Milestones in the light of the Task Force's own criteria, what is an observer to make of them? On their face and as a package they are directionally appealing, yet individually some of them clearly violate the defining criteria. Importantly, only a few were measurable. A suggested solution was to put "metrics" on each—a complicating "fleas on the backs of fleas" suggestion. If these milestones were to be the central thrust of the company, there were some gaps that might be perilous: There were *no* milestones on innovation, *no* milestones on technology. What would be the impact on the morale and behavior of Alcoa managers. Some of the metrics were definitely "stretch objectives," so much so that some managers argued that they were demotivating. There was much inconclusive debate on this vital point, and two Milestones that got particular attention in this regard were O'Neill's 15% ROE target and the critical processes in control and capable milestone. Would the marketplace—the price of aluminum ingot on the London Metal Exchange—permit the former, and would the time span of 5 years and the resources in place enable the latter? Matters of opinion surely, and how this would play out remained to be seen. To one senior Alcoan, the milestones looked like a score of crushing new priority "ones," while to another they were his Magna Carta for the next five years.

So in January 1989, one year after the Task Force began its work, this first part of the job was done. The new Alcoa constitution was written with quality at its core and Paul O'Neill, Fred Fetterolf, and the Operating Committee felt that Alcoa was poised to *start* its quality journey. Many saw the Alcoa quality effort—at this time still unnamed as the company sought to avoid the "quality management is another project" stigma by refusing to have it labeled—as the key to its future. In a short while, it would be named and its spirit portrayed with a variant on the Alcoa stylized "A" logo. "Managing for Quality" sat at the apex of the A, with "Continuous Improvement" on the left leg of the A, and with "Employee Involvement" at the right leg. In the white space under the A's crossbar sat the "Customer."

Under this banner, the good ship Alcoa would sail off to the quality wars. A few on the renamed and restaffed Corporate Quality Steering Committee imagined that one destination along the way might be the Malcolm Baldrige National Quality Award.

They and the new Corporate Quality Group were already hard at work creating the enabling mechanisms and resources for the quality journey—designing an Alcoa Problem-solving Process, rolling out the first wave of quality awareness training to the "top 100," codifying a benchmarking process, and the like. As the process continued it was looking less and less simple, but more and more important to most Alcoans involved. There no longer appeared to be any serious question of whether Alcoa would do total quality or what the broad parameters would be.

EPILOGUE

If the replacement of CEO Parry by CEO O'Neill was the dropping of one shoe, the other shoe dropped at Alcoa on August 9, 1991 when on short notice Paul O'Neill summoned some 50 of Alcoa's senior executives to an extraordinary meeting at Pittsburgh Headquarters. No agenda had been announced, and tension and rumor were rife since all present knew that President Fred Fetterolf had resigned two weeks before "over differences in policy." No explanation of what those differences were had been offered.

Opening the proceedings, O'Neill announced that the meeting was to discuss "change, but that two things would not change—the Alcoa Vision and Values." The changes he then announced were sweeping. In addition to the departure of Fetterolf, three other members of the Operating Committee had taken early retirement along with the Vice President of Engineering. There would be no replacement for the departed President, and the Operating Committee would cease to exist, as would the Group Vice Presidencies. All 25 business-unit managers would be given broader authority and bigger responsibilities and would henceforth report directly to the CEO. Paul O'Neill envisioned the new organization as an inverted triangle or pyramid with the business units and their customer relationships—the value-creating activities—forming the broad base of the triangle at the top of his diagram, the pooled corporate resources that service the business units one level below, and the CEO below that, the apex of the triangle at the bottom of the diagram. Two layers of the organization had been cut away.

Equally striking, the complex and fuzzy milestones were replaced with three intense and focussed imperatives:

- Cause your operating assets to perform at world-leadership rates. If someone else is doing it better you are not meeting the standard.
- Live by the Alcoa Values and Policies.
- Adopt quantum-leap improvement objectives that will at a minimum close 80% of the gap between current performance and the world benchmark on those few measures critical to your business.

The last of the above objectives would be the one that O'Neill would lean on particularly heavily as he impelled Alcoa management to accelerate the pace of change. Ironically, it is less stringent in this articulation than in its original version in the "Milestones," but the charge now appeared more forceful when put in isolation from the rest and when the CEO made it his *primary* focus. It was as if Alcoa had said it before, but hadn't really meant it. How strongly he now felt about the benchmark performance issue was reinforced by the other central change in O'Neill's thinking. "I believe we have made a major mistake in our advocacy of the idea of continuous improvement," he said. "Continuous improvement is exactly the right idea if you are the world leader in every thing you do," he explained. "It is a terrible idea if you are lagging behind the world leadership benchmark. It is probably a disastrous idea if you are far behind the world standard. In too many cases, we fall in the second and third categories. In these cases, we need a rapid quantum-leap improvement . . . [else] we will never be the world leader."

What inspired these changes? Over the more than two and a half years since "Excellence Through Quality" and the "Vision, Values, and Milestones" had been rolled out, and while he was convinced that the Alcoa constitution and direction were sound, Paul O'Neill had become increasingly frustrated with the pace of improvement at the company. There had been real improvement and the financial markets and the business press had recognized it, but in private O'Neill's impatience showed through. He was not about to wait for wrenching change to be imposed on him and Alcoa by external forces, as had happened at Xerox in the early 1980s and as was happening now, in the early 1990s, at General Motors. O'Neill was aware of painful shortfalls in Alcoa performance relative to the best of its competitors—even on some processes that Alcoa had *invented*. He also knew of substantial process performance gaps between Alcoa mills carrying out identical operations. The Alcoa quality effort was not addressing these gaps fast enough, he thought, and it never would "if we persist in our use of the traditional command and control system of management where many thousands of people believe their only responsibility is to do what they are told to do."

At the August 9th meeting, Paul O'Neill drove home his points with sharp observations and some disquieting facts. He said, "I am no longer willing to accommodate myself to the pace and direction of the organization when my own observations and instincts tell me we should be doing something different."[12] The examples he cited began with the issues on which he had been focussing since his first morning as Alcoa CEO: safety, the environment, and customer satisfaction:

> "While we have reduced our serious injury rate from 5.48 in 1987 to 3.80 [currently]; Du Pont is at 1.08 and even more alarming is that 19 Alcoans have been killed over the last four and a half years [of my tenure as CEO.]"
>
> "While we have always been committed to environmental protection we have just paid a record $7.5 million fine to the State of New York."
>
> "We operate some of the lowest cost alumina refineries in the world, but the levels of fines and soda [a pair of chronic quality problems in that industry] are not providing customer satisfaction."

He also cited a number of other crucial operational and financial issues, one of which particularly merits attention here as it simultaneously illustrates an issue that had for some time been a key irritation to O'Neill, relates to one of the central of the "Alcoa Quality Milestones," and brings us full circle back to principles emphasized in Deming's 1950 Tokyo lectures. He said, "We have constructed more Hall cells[13] than any other company in the world, yet our pot life is below the industry average; and this process, which we invented and have operated for over 100 years, is not in [statistical] control and [is not] capable [of meeting specifications consistently]."

To many observers these announcements were jarring discontinuous changes—changes which some interpreted as O'Neill's repudiation of total quality. A red flag to orthodox TQMers was his rejection of that arch TQ theme "continuous improvement." But not so to the man himself. In a recent conversation, he said "Back there in '88 and '89, we were learning to do quality, now we are really doing it. We have gone through a shifting of the gears, and perhaps the last shift up was, well, quite a bit bigger than the others."

It seems to me that the package of changes Paul O'Neill implemented in August of 1991 is consistent with the philosophy and goals he had articulated from the outset, and with his own personal very hands-on style. While the changes also can be seen as fitting within the quality framework that Alcoa had developed, they do significantly bend that framework. And, there are issues here worthy of some reflection. One can question whether it was necessary—indeed, whether it was dysfunctional—

to dismiss and bad mouth "continuous improvement" in order to emphasize, as needed to be done, the requirement for breakthrough improvements. Some observers and students of change and innovation processes propose that continuous improvement and breakthrough change are ideas in conflict, that an organization that emphasizes the former will inhibit the latter. My colleague in much of the Alcoa work, Michael Tushman, is of this persuasion and we have argued the matter often. When Tom Carter raised the issue even before the Task Force work began, I took a "Japanese position"—agreeing with the Kaizen philosophy that continuous improvement should enhance a firm's ability to carry off breakthroughs by adding greatly to its storehouse of product, process, and customer knowledge, and by adding to its ability to implement and maintain the breakthrough changes.[14] If this is true, then a false choice has been posed, and one needs both. Moreover, the "Excellence Through Quality" initiative had invited all 60,000 Alcoans to participate in continuous improvement, had trained them to do so, and had at great effort organized countless improvement teams that were already working at the time of the August 9th announcements. What were these Alcoans to make of their Chairman's statement that this continuous improvement concept to which they had just been wooed was a "disastrous idea." Most of these people would have no opportunity to participate in a quantum-leap improvement on one of the "few measures critical to your business." In the total quality roll-out, they had been told that they were part of a new team approach. Was that part of the "Excellence Through Quality" pact still intact?

Second, in these August 9th changes at Alcoa, one can see an intensification of the tension between competing long- and short-term goals. Even before this meeting, there were a significant number of Alcoa managers who felt and acted as if the quality talk was well and good, "but when you get down to it, you'd better make your numbers." "Do I hit my 15% ROI target this year, or invest in my breakthrough improvements for two [and more] years out?" Answer, "YES." Managing the tension that exists here will test the maturity and patience of Paul O'Neill and Alcoa and the integrity of their quality process.

In closing, and still speaking to the last point, we observe that no American company we have studied closely has satisfactorily resolved a key TQM implementation problem—one which Alcoa diagnosed during its first benchmarking visit to Xerox. The dilemma was expressed nicely for the visiting Alcoans when their Xerox host said: "This store is open for business while under repair!" Or, in other words, "How do we do all this improvement 'stuff' and still make and sell product?" Xerox, its Baldrige Award notwithstanding, recognized the problem but never resolved it. It took Xerox senior managers years to recognize that major elements of its well-designed "Leadership Through Quality" program were not being effectively implemented, that it was not fulfilling on the grand design. Under pressure for quarterly results, many business unit managers at Xerox had gone back to business as usual and quality management processes had been in effect shelved. Xerox was finding that, strenuous as they are, quality program design and quality training are the easy parts of total quality. The implementation is what is really difficult. Four years after the roll-out, when Xerox did a rigorous self-assessment as part of its preparation for the Baldrige Award Application, it created an extensive list of required enhancements to its quality efforts. These have only partially been addressed, and yet Xerox is now moving "Beyond Total Quality." Perhaps this is the American way.

Up to the time of the August 9th changes and despite its awareness of some of Xerox's problems, its refinement of many of the Xerox processes, and its more than doubling of the amount of training given its senior managers over that of Xerox, Alcoa hadn't resolved these problems either. As noted above, one could also see Alcoa managers struggling, "Do we do the business or the quality stuff?" And they too were seeing that the quality stuff really was a lot harder than they'd like, and the payoffs were not immediate. The Alcoa Task Force had been very critical of Xerox for mounting such an elaborate quality effort and then waiting years to evaluate its effectiveness. Yet to date, Alcoa has not done its own broad self-evaluation. The deep culture, systems, and technology changes that the highest levels of quality management require do not appear to yield to the "quick study" and rapid implementation that the American corporate culture seems to demand. It really remains to be seen whether the August 9th changes, the "Excellence Through Quality" Architecture, and most importantly Alcoa's continuing follow-up will propel it to the world eminence and outstanding performance that are its CEO's goals.

Alcoa has never adopted any particular company as a TQM role model, but once when questioned about which other companies he specially admired, Tom Carter, the Alcoa Vice President for Quality said, "Well, I don't know if there are any 10s out there, but Toyota Machine Tool sure is a 9+." He went on to add, "And Toyota got to where they are by 40 years of relentless work. At Alcoa we don't have 40 years."

REFERENCES

1. For about an 18-month period, the author served as the chief outside consultant to both the Quality Task Force that designed Alcoa's quality initiative as well as to the senior management "Operating Committee" that directed the implementation. During much of the time he was joined in this work by his Columbia Business School colleague, Professor Michael Tushman. Peter Kolesar's research and teaching background is in engineering, operations management, and applied statistics. He entered the world of quality management through that avenue. Michael Tushman's work by contrast is concerned with organizational design and change, and the management of innovation.
2. For a detailed history of Alcoa up to this time, see George David Smith, *From Monopoly to Competition: The Transformation of Alcoa, 1888–1986* (New York, NY: Cambridge University Press, 1988). The articles of Michael Schroeder and Thomas Stewart carry the story forward to the appointment of O'Neill as Chairman and CEO and his redirection of the company. Michael Schroeder, "The Quiet Coup at Alcoa," *Business Week*, June 27, 1988, pp. 58–65; Thomas A. Stewart, "A New Way to Wake Up a Sleeping Giant," *Fortune*, October 22, 1990, pp. 90–100.
3. The Operating Committee included the Chairman and CEO, the President, the Chief Financial Officer, the General Counsel, the Vice President of Human Resources, and four Business Group Vice Presidents. When we write about the meetings and activities of this group we include all the aforementioned as participants.
4. The content of Deming's 1950 Tokyo lectures may be found in W. Edwards Deming, *Elementary Principles of the Statistical Control of Quality—A Series of Lectures* (Tokyo: Nippon Kaguku Gijutsu Rommei, 1951). This is an edited version of lecture notes transcribed by Japanese participants. Proceeds from sales of this book were donated by Dr. Deming to the Japanese Union of Scientists and Engineers, the group which had sponsored his visit. The funds were used to establish the Deming Prize.
5. For an accurate description of the Deming red bead experiment and his interpretation see Mary Walton, *The Deming Management Method* (New York, NY: Dodd Mead & Co., 1986), Chapter 4; and Andrea Gabor, *The Man Who Invented Quality* (New York, NY: Times Books, 1990).
6. The particular framework used for the diagnosis was an organizational model of Nadler-Tushman. David Nadler and Michael Tushman, "A Model for Organizational Diagnosis," *Organizational Dynamics* (Autumn 1980). During the seminar we made the point that alternative diagnostic frameworks existed and that it was more important that one be used than which one. Our main purpose was to methodically expose all the elements of the enterprise that affected quality and that conversely would be affected by a move toward total quality management.
7. Some months earlier, Mr. Fetterolf had been on a benchmarking visit to Japan to review advanced manufacturing techniques. While he and the other participants learned and were inspired by the experience, nothing concrete had resulted from the Advanced Manufacturing Task Force. This failure was indeed well known to, and on the minds of, several Quality Task Force members.
8. The history of the Xerox effort is now reasonably well documented in the public literature. See, for example, the story as recounted by two of the key participants, David Kearns and David Nadler, *Prophets in the Dark: How Xerox Reinvented Itself and Beat Back the Japanese* (New York, NY: Harper Business, 1992). Jacobson and Hillkirk describe in detail the challenge from Japan and some of the personalities involved. Gary Jacobson and John Hillkirk, *Xerox, American Samurai* (New York, NY: Collier Books, 1986). On the Xerox quality effort, see Chapter 7 in Gabor, op. cit., which is, by the way, much more critical of the Xerox approach.

9. Yotaro Kobayashi, "Quality Control in Japan: The Case of Fuji-Xerox," *Japanese Economic Studies* (Spring 1983), pp. 75–104.
10. Most of the quality gurus testify to the large proportion of problems that are "due to the system." See, for example, the books of Deming and Juran. There has been no scientific documentation of this contention. The author is tempted to make his estimate far higher in some industries and far lower in others. In part the issue is whether processes are "in control." W. Edwards Deming, *Out of the Crisis*, MIT-CAES, Cambridge, 1986. Joseph M. Juran, *Juran on Planning for Quality* (New York, NY: The Free Press, 1988).
11. A sense of the public response to the Alcoa quality initiative and to Paul O'Neill's leadership can be gotten from Stewart, op. cit.; Thomas F. O'Boyle and Peter Pae, "The Long View: O'Neill Recasts Alcoa With His Eyes Fixed On a Decade Ahead," *The Wall Street Journal*, April 9, 1990, pp. A1 and A4. Over this period, Alcoa was repeatedly listed as first among the Metals Industry in *Fortune* magazines's annual "America's Most Admired Corporations" beauty contest. (About which a senior Alcoa manager observed, "That's real nice, but the industry sucks.") Over the two-year period 1989 to 1991, Alcoa's common stock earnings were 28.5% as compared to −28.3% for Alcan and 12.2% for Reynolds.
12. Here, Paul O'Neill is predicting a dire future unless significant anticipatory action is taken. One is reminded of a 1939 address by Winston Churchill to the House of Commons castigating the British government for its tardy and ineffectual response to the threat of Nazi Germany. It was a much more crucial and historical issue that Churchill faced then, but his words ring true in this contemporary industrial context as well. He said, "When the situation was manageable it was neglected, and now that it is thoroughly out of hand, we apply too late the remedies which might then have effected a cure. There is nothing new in the story. It is as old as the Sibylline books. It falls into that long dismal catalogue of the fruitlessness of experience and the confirmed unteachability of mankind. Want of foresight, unwillingness to act when action would be simple and effective, lack of clear thinking, confusion of counsel until the emergency comes, until self-preservation strikes its jarring gong—these are the features which constitute the endless repetition of history."
13. A Hall cell is the electrolytic device in which the smelting of aluminum is done. It is essentially a large carbon lined bathtub. An operating efficiency issue is how long such a pot can be run before it must be shut down and relined with carbon—this time is called "pot life." Alcoa was founded by the cell's inventor, Charles Martin Hall.
14. For a development of this line of reasoning, see Masaaki Imai, *Kaizen: The Key to Japan's Competitive Success* (New York, NY: Random House, 1986).

41

The CEO as Organizational Architect:

An Interview with Xerox's Paul Allaire

ROBERT HOWARD

NOTE: As chairman and CEO of the Xerox Corporation, Paul Allaire leads a company that is a microcosm of the changes transforming American business.

With the introduction of the first plain-paper copier in 1959, Xerox invented a new industry and launched itself on a decade of spectacular growth. But easy growth led Xerox to neglect the fundamentals of its core business, leaving the company vulnerable to low-cost Japanese competition. By the early 1980s, Xerox's market share in copiers had been cut in half, and return on assets had shrunk to 8%.

In the mid-1980s, Xerox embarked on a long-term effort to regain its dominant position in world copier markets and to create a new platform for future growth. The first step: use the techniques of total quality to reengineer the fundamentals of the business—product development, manufacturing, and customer service. Thanks to the company's Leadership through Quality program, Xerox became the first major U.S. company to win back market share from the Japanese. And return on assets has increased to 14%.

Since becoming CEO in 1990, Allaire has moved quickly to take Xerox's corporate transformation to a new level. He has redirected the company's strategy to position Xerox as "the document company," at the intersection of the two worlds of paper-based and electronic information. And Allaire has guided the company through a fundamental redesign of what he calls the "organizational architecture" of Xerox's document-processing business, which accounts for roughly 77.5% of the company's $17.8 billion in 1991 revenues.

Many companies are reorganizing to cope with new competitive realities, but few CEOs have approached the process of organizational redesign as systematically and methodically as Allaire has. Over the past two years, he has created a new corporate structure that balances independent business divisions with integrated research and technology and customer operations organizations. He has redefined managerial roles and responsibilities, changed the way managers are selected and compensated, and renewed the company's senior management ranks. And he has articulated the new values and behaviors that Xerox managers will need to thrive in a more competitive and fast-changing business environment.

Reprinted by permission of *Harvard Business Review*. "The CEO as Organizational Architect: An Interview with Xerox's Paul Allaire" by Howard, R., (September/October, 1992).

Allaire comes to this challenge with a deep knowledge of the Xerox organization. Since joining the company as a financial analyst in 1966, he has served in a variety of senior management positions at both Rank Xerox, the company's European joint venture, and at Xerox headquarters in Stamford, Connecticut. In addition to his position as CEO, he was elected chairman of the board in May 1991.

The interview was conducted at Xerox's Stamford headquarters by HBR senior editor Robert Howard.

HBR: In the past decade, companies across the economy have been decentralizing, downsizing, and flattening. What makes Xerox's reorganization different from what other companies are doing—or, for that matter, different from what Xerox has done in the past?

Paul Allaire: When most companies reorganize, usually they focus on the formal structure of the organization—the boxes on the organizational chart.

Typically, top management just moves people around or tries to shake up the company by breaking up entrenched power bases. Rarely do senior executives contemplate changing the basic processes and behaviors by which a company operates.

Until recently, Xerox was no different. In the 1980s, we went through a number of reorganizations. But none of them got at the fundamental question of how we run the company.

The change we are making now is more profound than anything we've done before. We have embarked on a process to change completely the way we manage the company. Changing the structure of the organization is only a part of that. We are also changing the processes by which we manage, the reward systems and other mechanisms that shape those processes, and the kind of people we place in key managerial positions. Finally, we are trying to change our informal culture—the way we do things, the behaviors that drive the business.

In fact, the term "reorganization" doesn't really capture what we are trying to do at Xerox. We are redesigning the "organizational architecture" of the entire company.

What is happening in your business that makes this more fundamental change so necessary?

Our business is rapidly becoming more competitive—and more complicated. Xerox grew from a high-tech startup to one of the country's biggest corporations in large part because we were the only game in town. Then in the 1980s, we faced our first serious competition from the Japanese. This posed a real threat to the company, but thanks to our quality process, we have responded to that threat extremely well.

However, the competitive challenges don't stop there. Xerox is now in the midst of a technological transformation that is revolutionizing our business. It is changing the skills our employees need, the competitors we face—and, indeed, the very nature of the business we are in.

How so?

Our traditional business is light-lens copiers and duplicators, fairly sophisticated electro-optical mechanical devices. Increasingly, they incorporate many computer systems—to control for copy quality, for instance—but the guts of the machine are electro-optical and mechanical. These are stand-alone devices.

With the evolution of digital technology, however, and its rapid reduction in cost, the light-lens element in a copier can now be replaced with a scanning device that digitizes the information on the page. Once you've captured that image electronically, you can do all kinds of things with it in addition to just making a copy.

Take the example of our new digital color copier. Once you scan a document into the copier, you can change the colors on it. Perhaps it's an old or damaged document. Well, you can enhance it and clean it up. You can also edit the document and even add photographs. You can merge electronic information—coming from a personal computer or a mainframe or over the network—with paper information. And once it's complete, you can send your new electronic document to be printed somewhere else. You can store it electronically as well, so you no longer have to deal with cumbersome paper files. Our traditional "stand-alone" copying devices are fast becoming components of complex, digital document systems.

What does this technological transformation mean for the business?

Enormous opportunity—and equally enormous challenges. In this new technological environment, it's not enough anymore just to make and sell discrete products. Rather, we need to offer our customers distinctive capabilities that will have a major impact on the way they do business.

As the worlds of paper-based information and digital information merge, we have redefined our strategic direction to focus on "the document." We believe that documents—whether in electronic or paper form—are key to making organizations more efficient. By focusing on managing documents and using them more effectively, Xerox can help our customers improve productivity.

What are the implications of that strategy for Xerox as an organization?

We have to change. For one thing, our people need new skills. Product designers now have to be far more knowledgeable about computer systems and electronics. They have to know how to function in the new "systems" environment.

But probably the greatest impact is on our relationship with customers. As we move into this systems world, we aren't just making and selling boxes anymore. Increasingly, we are working with customers to design and redesign their basic business processes. In the future, Xerox won't just sell copiers. It will sell innovative approaches for performing work and for enhancing productivity.

But that means that our salespeople need to understand the customer's business, what the customer's real needs are, and how the customer is going to use our products. It's a partnership in which we take a more consultative approach.

Can you give an example?

Recently I was in Austin, Texas to make a call on a customer who until then had not wanted to have anything to do with us. The chief information officer at the company did not want any paper in his organization—and, therefore, he did not want any copiers because copiers generate paper.

When we finally got a meeting, our salesperson didn't say a word about copiers. He talked about product documentation. "Have you ever had any delays in getting your product out the door because your documentation wasn't ready?" Well, of course, the answer was yes. So our salesperson said, "We may have some capability to help you with that. We want to come in and look at the way you are doing documentation and see if we can figure out a better way to do it. If we can get your products out the door more rapidly, with more up-to-date documentation—and, by the way, more cheaply—would you be interested?"

Needless to say, he was. So our salesperson did a study to analyze how the company could organize its work more effectively. Not only that, he got the CIO a tour at another company down the road where we had already done a similar thing. And we made the sale.

You mentioned that the new business environment also means new competitors.

Our competitors are multiplying. Some of our Japanese rivals, like Canon, will certainly come along with us in this new digital world. But we will also be competing with companies that are now in the printing business or in the computer systems business. Increasingly, we are moving into the more complex business environment that a lot of computer companies already find themselves in—where we will compete against companies in some situations and cooperate with them in others.

What's more, we can't just say, "Out with the old and in with the new." Even as we develop new skills, define new ways of interacting with the customer, and confront new kinds of competition, we must still pursue our traditional business. We need to maintain and build on all the skills we have developed in the light-lens business. But at the same time, we must manage this technological transformation. We have to do two things at once.

So Xerox's new architecture is a direct response to these new business challenges.

Absolutely. Operating effectively in this more complex and volatile business environment requires the capacity to cope with change—and at a very rapid pace. The technology is changing quickly. The demands of the marketplace are also changing. What's more, they're both moving targets. They are going to continue to change. So we have to change the company itself.

On one level, that's simple: If we intend to sell customers our expertise as designers and implementers of new, more effective, and more productive business processes, then we had better make sure that our own organization is a showcase for better ways of working.

But even more important, we have to create a new organizational architecture flexible enough to adapt to change. We want an organization that can evolve, that can modify itself as technology, skills, competitors, and the entire business change.

What did Xerox's old architecture look like and why wasn't it adequate to these new challenges?

Until recently, Xerox had the structure, practices, and values of a classic big company. First, we were an extremely "functional" organization. If you were in manufacturing, you strived to make manufacturing as good as possible—and only secondarily to make the businesses that manufacturing affected work well. The same was true for sales, R&D, or any other function.

Second, we were staff-driven. Staff was primary, line businesses were secondary. I used to be the chief staff officer at Xerox, so I know what I'm talking about. At the time, in the mid-1980s, I viewed myself—quite inaccurately, I should add—as the number two person in the company. It certainly wasn't in terms of responsibility. There were people with billions of dollars worth of revenue and profit responsibility, whereas I didn't have any. And yet, I was the CEO's key adviser, and I ran the management process. I could wield a lot of weight without a lot of real responsibility.

That's how the power structure worked—and for good reason. The fact is, in a functional organization, you need many staff functions in order to make things work. To ensure that manufacturing, for example, was hooked to sales, both had to be hooked in to corporate.

What's wrong with a staff-driven functional organization in the new business environment that you describe?

It breeds dependence and passivity. In a functional organization, there is a natural tendency for conflicts to get kicked upstairs. People get too accustomed to sitting on their hands and waiting for a decision to come down from above. Well, sometimes the decision does come down. But sometimes it doesn't. And even when it does, often it comes too late, because market conditions have already changed or a more nimble competitor has gotten there first. Or maybe the decision is simply wrong—because the person making it is too far from the customer.

We used to go through a big exercise here every Fall that went by the grandiose name of "resource optimization." Basically, the top-management team, aided by the marketing staff, took our total R&D spending—some $800 million to $900 million—and asked, "How are we going to allocate that budget?" First, we figured out how much money we could spend; then, we set priorities among our R&D projects against these resources to decide what got funded and what didn't.

It was as if the people who actually had to spend the money were living in a welfare state. They told us what they needed—"I need $50 million," "I need $40 million," "I need $30 million." Then they waited. At some point, somebody would come back to them and say, "What will happen if you only get $30 million instead of $40 million?" And then they would revise their plans according to the new figure. Finally, somebody would come back yet again and say, "OK, you get $35 million." That's how the process worked. It was completely crazy.

The old model is all too familiar. But what is the new model?

We are trying to break the bonds that tie up energy

and commitment in a big company. Our goal is to make this $17 billion company more entrepreneurial, more innovative, and more responsive to the marketplace. In fact, we intend to create a company that combines the best of both worlds—the speed, flexibility, accountability, and creativity that come from being part of a small, highly focused organization, and the economies of scale, the access to resources, and the strategic vision that a large corporation can provide.

How are you going to accomplish that?

By redefining the three key components that make up any organizational architecture. The first, we call "hardware." These are the things managers usually think of as organizational structure, the formal processes by which we get things done: the business planning system, control mechanisms, measurement systems, reporting relationships, reward systems, and the like.

The second is "people"—the new skills managers need, including the kind of personality and character they must have to operate effectively in the new environment. Clearly, if you are running a functional organization, you go with functional experts and you manage with a lot of staff who hold the functions together. But in a decentralized, more entrepreneurial organization like we have now, you need a completely different kind of manager.

Finally, the third component of our new architecture is what we call "software." This is the most difficult to describe but probably the most important: the informal networks and practices linking people together, the value system, the culture. Any company that leaves these out of its organizational change effort is making a big mistake.

Why is that?

Think about how any organization works. The informal organization is often far more important than the shape of the organizational chart or who is in what box. Two companies that look exactly alike when you compare their organizational charts can turn out, in fact, to operate very differently.

A successful organization is one where the formal organization and the informal one work together, rather than always working against each other. When there isn't a good fit between them, too much human energy and creativity gets wasted simply in making the organization work. But when you align the formal and the informal, most of your energy can be externally focused on achieving the objectives of the business.

Let's start with "hardware." What does Xerox's new formal organization look like?

When we began thinking through our new design two years ago, we needed to solve a basic paradox. For all the reasons I've mentioned, we wanted to get rid of our traditional functional organization and create more entrepreneurial business units.

And yet, for a company like Xerox, just setting up a bunch of small autonomous units isn't enough. Most of our businesses rest on a common technology base. What's more, we are dealing with a market where in many cases, the same customer is buying a variety of different products from us. So it's important not to lose the advantages of a situation where a salesperson can say, "I am the single person who will represent Xerox to you. And if you have a problem anywhere in the world, we will take care of you."

Take, for example, our total satisfaction guarantee. Anybody who buys a product from Xerox, no matter who sells it or where the customer buys it, has a satisfaction guarantee good around the world. That is very, very powerful. So how do you have speed, flexibility, autonomy and still be able to give the customer that guarantee?

How have you resolved that paradox?

We have created an organization that by its very design forces managers to confront—and manage—the necessary tensions between autonomy and integration. The centerpiece of our new architecture is a set of nine relatively independent business divisions. These are stand-alone businesses organized around specific products and markets and with profit and loss responsibility. Each division consists of a number of business teams, smaller entities tied to the marketplace by a specific customer need. In fact, the business team leader is the new entry-level general manager job in the company.

But these divisions don't exist in a vacuum. They are linked to other parts of the company whose responsibility is to leverage corporatewide resources and relationships in support of the divisions.

In a sense, we have turned the traditional vertically organized company on its side. At one end is technology, and we have retained an integrated corporate research and technology organization. At the other end is the customer. We have organized our sales and service people into three geographic customer-operations divisions, so that we can keep a common face to the customer. Between these two poles are the new business divisions. Their purpose is to create some "suction" on technology and pull it into the marketplace.

Finally, we have created a new unit called "strategic services" that will provide support to the business divisions in areas such as specialized manufacturing and purchasing, where economies of scale still make a big difference. And we've established a six-person corporate office, served by a much leaner corporate staff, to make sure all these units fit together in a strategically coherent way.

A big advantage of this structure is that it is remarkably easy to adapt. When we see new markets emerge or new technologies that don't fit into our current structure, we can simply add another business team or even a whole new division. Similarly, we can split a business division—or even eliminate one altogether—without changing the basic architecture of the company.

But as you point out, structure alone isn't enough to make a company work. How do you make sure that people manage the tension between autonomy and integration?

First, we are making it very clear to the presidents of these divisions that their primary responsibility is to make a success of these businesses. We in the corporate office will provide them with some constraints, some boundaries. If they feel they absolutely must cross those boundaries, then we expect them to tell us. But otherwise, they have the freedom to run the business, to meet the objectives that they have communicated to us and to implement the strategy they have presented. Now of course, if we don't think their business plan is credible, then we'll object and talk with them about it. But the initiative is up to them.

Take the example of R&D spending I mentioned earlier. Before, setting priorities was a top-down process. Now, we're turning that process completely upside down. We're saying to the business division presidents, "You tell us what you are going to deliver and the resources you need to do it." They are setting the priorities now. And they should be because they are the people closest to the market.

See, they should be coming to corporate management the way an entrepreneur goes to a venture capitalist or banker. And we at the top should be looking at individual strategies and how those strategies come together for the company as a whole. We shouldn't be starting out with a fixed R&D budget. How the hell is anybody smart enough to say what the right amount should be?

But, of course, the business division presidents aren't really independent entrepreneurs. They're part of a big corporation.

That's true. Even as these division presidents develop their own businesses, they have to think with their corporate hat on as well. Again, technology provides a good example. One of the big challenges we face as we move into a technological environment founded on integrated systems is to create a common design architecture for our products. When all our products were stand-alone devices, a common architecture didn't much matter. The only goal was to optimize individual product design. As a result, we have copiers where the paper goes from left to right, and we have had printers where the paper goes from right to left.

But when it comes to designing an integrated system that includes all the copier components and all the printer components, obviously you can't have it both ways. So who decides what the unified architecture will be? I have no interest in making that decision alone. I am not anywhere near smart enough. And while we have created a technology architecture unit in our corporate research organization, I don't want our technologists deciding on their own either.

So, we have put a decision-making process in place that will directly involve the line managers of the business divisions. Of course, they will have to make compromises at times. There may well be an optimum technology for one division and a very different one for another. Since we can't afford to develop both, they will have to sit down and say, "What is the right technology for Xerox as a whole?"

So too at the customer end. We want the customer to integrate the company. We don't want to integrate the company and then go to the customer. The job of the customer operations divisions will be to build a relationship with the customer and then figure out with the business divisions how to integrate our offerings for the customer. To do this, the customer operations people will have to understand not only the customer but also the strategies and directions of the divisions.

It sounds like the tension between autonomy and integration is built in to the new managerial roles you have created. What kind of people do you need to perform those roles effectively?

We want people who can hold two things in their heads at the same time, who can think in terms of their individual organizations but also in terms of the company as a whole. Our architecture won't work if people take a narrow view of their jobs and don't work together.

To do that requires a total business focus—something that the narrow functional organization does not provide. This was a big issue when we were selecting our business division people. The fact is, we simply do not have as many general managers as we need who have that broad experience. There is no traditional career path inside Xerox to become a business team general manager or a business division president.

So how did you find the right people?

People may find this hard to believe, but we actually designed the entire organization without any particular individuals in mind. That goes for the key managerial positions in the business divisions. It even goes for my senior management team in the corporate office. We didn't discuss who would occupy those positions until we had decided what kind of structure we needed and wanted.

Once we had designed the new organization, however, we tried to be explicit about the type of people we need to operate within it. We started with a clean slate and defined the ideal characteristics of a manager in the new Xerox. We developed a whole new set of criteria for evaluating people—23 characteristics in all, 7 of which we decided were absolutely critical for running a business division or team. Many are things that, frankly, we never thought of as so important before.

Like what?

Take strategic thinking. In the old organization, that wasn't very critical for most people other than those at the very top of the company. But if you're a business division president in the new architecture, it's crucial.

Another example is strategic implementation. We have a lot of people who are strategists or planners but who never had any responsibility for implementing their plans. But knowing how to think strategically is no longer good enough. You must also know how to act strategically.

And then, clearly, the "software" items are critical: teamwork, the ability to delegate and to empower subordinates. Other criteria have more to do with personal character. For instance, one we call "personal consistency." That's a fancy way of saying integrity, personal courage, "grace under pressure." We don't want people in these senior management positions who are all over the map, or who are constantly holding their fingers up in the air to see which way the wind is blowing.

What did you do with these characteristics once you identified them?

I sat down with a few top managers, and we rated some 45 people we thought were candidates for roughly the top 25 jobs in the company, including the top position in all the business and customer operations divisions. First, each of us rated everybody individually. Then we compared and discussed our individual ratings of each candidate and agreed on a common rating. And we did it without regard to particular jobs. It was only afterward that we considered individuals for specific positions.

We selected some very nontraditional people based on this approach. People came to the surface who wouldn't have in the old system (see the insert, "Three Managers, Three Perspectives on the New Xerox"). For instance, 3 of our 9 new business division presidents have been with Xerox for less than a year. And about one-third of the top 40 people in the company have been at Xerox for less than three years.

What's more, we continued this rigorous selection process down one more level into the organization. Recently, the 12 business division and customer operations presidents used it to choose their business team general managers and support staff. They rated more than 100 people for about 50 jobs.

Assuming you've found the right people, how are you managing for the new behaviors you need?

One of the things I learned long ago is that if you talk about change and then leave the reward and recognition system exactly the same, nothing changes. And for good reason: people quite rationally say, "I hear what he is saying, but it's not what I get paid to do or what I get promoted for. So what's in it for me?" Therefore, if you are trying to change the way you run a company, one of the most visible things you have to change is the way you compensate, the way you reward and recognize people.

The compensation of senior executives has come in for a lot of criticism recently. Did that influence how you designed your new reward system?

There is a lot of emotion around the issue of executive compensation right now. But frankly, when we began to design our new compensation system, we were focused almost exclusively on our internal goals: to encourage the new kinds of behavior essential to making the new organizational architecture work and to achieving our business goals. As we developed our plan, we discovered that it also provides exactly what shareholders want. It aligns individual compensation with the strategic objectives of the company. It rewards good performance and penalizes poor performance.

How does the new plan work?

Like most companies, we had a long-term incentive plan based on stock options and a bonus. But the way it was structured, you were guaranteed a payout unless the company completely fell apart. On the other hand, if the company did extremely well, the payout didn't really go up a lot. So the plan didn't function effectively as an incentive for people to work together and aggressively drive the company forward.

We decided to add two new features to our incentive plan for the people in roughly the top 50 jobs in the company. First, while there continue to be stock options, we also have a program for awarding people what we call "incentive shares," outright grants of Xerox stock. However, these shares are earned only if the company meets certain hurdles fixed in advance and based on return on assets. Everybody has to work together to make those numbers, and the numbers don't get reset from year to year. If we miss it the first year, that means that next year's target is all that much tougher. We have to do better in 1992 than we did in 1991 and improve again in 1993 and in 1994.

What's more, we believe that to be fully committed to the long-term success of the company, you have to have some skin in it yourself. Therefore, we have made it a condition of participation in this plan that senior managers must actually purchase one year's salary worth of Xerox stock.

Compared with the old plan, there is potential for bigger payout, but there is also a lot more at risk. There is more risk because you have to buy stock. Second, if we do not make our return-on-assets targets, we receive no incentive shares at all. So now we have a situation where people could actually lose money if the company doesn't perform. But if we perform well, we can all do much better.

But that long-term incentive plan only affects the top people in the company. How are you changing reward systems for people further down in the organization?

We have also made important changes in our bonus plan, which covers the top 2,000 people in the company. In the old world, a bonus was "additive." Say, 30% was based on how the corporation did during the year. Maybe 50% was based on how your division did. And 20% was based on how you as an individual did.

But the effect of such a system is to encourage people to say, "Well, I don't have a big impact on the company, so I'll just optimize my individual objectives or my division's objectives. And whatever happens in the corporation will just be icing on the cake."

In our new reward system, we have what I call a "multiplicative" approach. The categories are the same—corporate performance, division performance, personal performance. But instead of adding them together, we multiply them against each other. So if a manager does well personally—say 120 on a scale of 100—but the company as a whole does poorly—say 70 on a scale of 100—his bonus will suffer. He will get only 84% of his intended bonus.

On the other hand, if a manager does well and through her efforts and those of many others the company also does well—say both score 120 on a scale of 100—then her bonus increases substantially to 144% of the intended bonus.

This changes the whole dynamic. The message to the individual manager—not only in terms of his or her self-interest but in terms of pressure from peers—is "you had better be contributing to the company as a whole."

Finally, we have set aside an extra pool of money—up to an additional 10% of the entire bonus pool—for the managers who show exemplary performance in these soft values I've been emphasizing. I have the option of rewarding these managers handsomely.

Why are you placing so much emphasis on the intangibles of "software"?

Because the hardest stuff is the soft stuff—values, personal style, ways of interacting. We are trying to change the total culture of the company. When you talk about that in general terms, everybody is all for it. But when you talk about it in terms of individuals, it is much tougher. And yet, if individuals don't change, nothing changes.

What holds people back?

Imagine someone who is 45 years old, who has been a manager for 20 odd years and a successful one, who has gotten to, say, the vice president level. All of a sudden, we are saying to that person, "We know you've been successful, but we're changing the rules. You have to do things differently." That triggers enormous insecurity. Because it is perfectly natural for the person to be thinking, "I am in this nice corner office because of the way I operate. And yet I hear all this stuff about change."

Maybe the individual even understands the need for change—intellectually. But the real problem is, "How am I going to act when I get in trouble? Am I really going to trust all these new approaches. Am I really going to be comfortable with delegating responsibility and authority? Am I really going to take full accountability and not cleverly spread it around so no one can throw stones at me?"

We are asking people to take initiative, make decisions, and stand by them. Well, that's great when things go well. But what about when they don't?

What do you say to somebody like that?

You just have to be as explicit as you can with everyone about the new criteria for success and then give people honest feedback about where they stand. For instance, when we rated managers against our new job selection criteria, we also told them, "Here are the three or four characteristics where you are weakest. It is important that you work on these weaknesses because part of the way we will evaluate your success is your ability to work in this new manner."

How have people reacted?

It's been something of a shock. In our old performance appraisal system, we had five levels—and good performers regularly scored fours or fives. But with the new system, we rate people on a seven-point scale. Because we were using this for selection purposes, we tried to be completely honest. So, many people are getting threes and twos on a specific characteristic. They look at it and say, "Wait a minute. I'm one of the people being selected?"

Now ideally, we would like to have a portfolio of players in these jobs who are scoring fives, sixes, and sevens. But realistically, we simply don't have many of those people yet. In the new system, the highest possible score is 161. Well, our top people are in the 50s, and the bottom people are in the 20s. Believe me, that has gotten people's attention.

But we're not doing it just to be perverse. The change process we have begun is extremely difficult. It's hard for any organization to get out of its history. We have to be very diligent so that old habits, practices, and behaviors don't sneak back in.

How common is that?

My favorite example comes from the management team that designed our new architecture. Here is a group of 20 people who, more than anybody in the company, have internalized the new approach to running things (see the insert, "Architecting a New Organization"). And yet, when it came to deciding how to organize the business divisions, this group recommended that we start with the business division staff—the head of personnel, head of finance, etc.—and that the staff help each division president choose his or her key operating people. In other words, they immediately fell back into the traditional top-down, functional approach.

I said, "Hold on. Don't we have it backwards? I thought we were trying to run this business by the line? Why would you select the staff people to help the presidents select the line? Why don't we do it just the reverse? Who are the key line managers required to run these businesses? We can help them figure that out. They don't need to hire individual staff people to do that. We can help them put their team leaders in place, then the presidents and their team leaders can decide together what kind of staff support they need."

At first, they were convinced I was wrong. But as we talked it through, you could see a shock of recognition go through the group. Everyone started to say, "Well, I think we just violated one of the principles with this recommendation." They began to realize that it would send exactly the wrong message to the organization. Finally, somebody said to the person who had made the original proposal, "I think you'd better back off on this. Not only is he the boss, but I think he is right on this one."

I keep reminding people that what we are doing isn't easy. In fact, if they are finding it easy, then they should be concerned because that's a sure sign they are not really succeeding at it!

Everybody knows about the benchmarking tradition at Xerox. Is there any company that you benchmark the entire organization against? Are there companies operating today as you want Xerox to be operating, say, five years from now?

Unfortunately, no. I say "unfortunately" because it would be a lot easier if there were. We are quite shameless in our willingness to steal good ideas from elsewhere. But there aren't any companies that operate as a company the way we intend to.

The important thing to remember is that this new organizational architecture is just the beginning. Our ultimate goal is to organize the entire company into self-managed work teams—or what we call "productive work communities."

I envision a time when this company will consist of many, many small groups of people who have the technical expertise and the business knowledge and the information tools they need to design their own work process and to improve and adapt that process continuously as business conditions change. These work groups will be tied directly to the customer. They will be working in a much less supervised environment. And they will have the resources they need to redesign their work environment and modify their behaviors as needed to achieve their objectives in the marketplace (see the in insert, "From Strategy to Product: The Story of Paper's works").

Put another way, Xerox's new architecture won't really be complete until what you are now doing at the top of the company is also taking place at the lowest levels—until everybody becomes an "organizational architect."

Right. Now, a lot of companies are experimenting with creating self-managed organizations. You can see a factory that operates that way or maybe a warehouse or service facility somewhere. Indeed, you can find pockets of organizational innovation at a lot of big companies. I think we've visited just about all of them.

But it doesn't provide the answer that we are looking for: How do you do it on the scale of an entire large corporation? There is no big company that is managed that way. There is no example—or even a very good theory—of how to run a multi-billion dollar company that way.

We intend to keep looking and keep experimenting. Because I firmly believe that is the future. And as an organization, we are committed to living in that future.

INDEX

AANET, 520
Abacus system, 487
ABB Asea Brown Boveri, 477–93
ABB Dolmel, 489–491
ABB Power Ventures, 489
ABB Zamech, 489–91
Abernathy/Utterback model, 46, 290
Acceptable Quality Level, 405
Accutron, 40n.10
Acquisitions, 545
Action, speed of, 210–12
Action requirements, 214
Activation theory, 197n.6
Adaptation patterns, 54–59, 61, 62–65, 162
Adaptionism, 68
Adaptive society, 130, 139–40
Adhocracies, 240
Administrative automation, 386
Advertising, 224–25; *See also* Marketing
Aerospace industry, 176, 251, 392, 560, 561–62, 572
Age factors, employee, 183–96, 331–32
Aircraft industries, 289, 527
Aiwa, 101, 104
Akio, Yamanouchi, 449
Alagasco, 207
Alberding, Dick, 242
Alcoa, 607–29
  background of, 607–9
  benchmarking at, 612–14
  operating committee of, 615–21
  Quality Task Force inception, 609–12
  safety problems, 614–15
  values and milestones of, 621–26
Allaire, Paul A., 159–60, 170–71, 631–41
Allen, Robert, 510
Allen, Tom, 194
Allen-Bradley, 612
Alliances. *See* Joint ventures; Technology alliances
Alpha Corporation, 585, 586, 590, 592
Alza, 574–77, 576–77
Amana, 148, 151, 527, 529, 534–35, 536
Amanuma, Aki, 115
Ambassador activities, 435–36, 437, 438, 439, 440
Ambidextrous organizations, 6, 12–17, 19–21
Amdahl, 334, 571, 580n.15
American Airlines, 521
American Locomotive, 146, 147, 149
Ampex Corporation, 78, 79, 87
Anchor fork, 39
Anthropology of work, 347
Anxiety and stress, 598, 602, 603–4
Apollo Computer, 584, 593
Apollo space program, 176
APOS system, 112
Apple Computer, 47, 179, 206
  compatability issues, 227
  contractural agreements, 295–96
  innovations, 288
  intellectual management, 512
  technology strategy/cycle, 283, 369
Appliance industry, 527
Applied Materials, 210
Appraisal and evaluation. *See* Performance appraisal and evaluation
Appropriability regimes, 288–89, 291–96, 302–3
Architectural innovation, 11, 12, 14–15, 17, 18, 19; *See also* Organizational architecture; Strategic architecture; Technological architecture
Arthur Andersen & Company, 509, 520
Asahi, 565, 566
Asea Brown Boveri. *See* ABB Asea Brown Boveri
Aspartame. *See* NutraSweet
Assemblers, 29
Assembly lines. *See* Flexible automation; *specific product*
ASUAG holding company, 27, 28, 29, 33, 34, 37
AT&T, 178–79, 357, 360, 371, 373n.18, 522
  organizational evolution, 589
  quality programs, 613
  technology alliances, 562–63, 564, 576–77, 579n.8, 580n.17
Australia, 461, 466, 467, 469, 472
Austria, 379, 380
Automation. *See* Computer industry; Flexible automation; Robotic manufacturing
Automotive industry, 122
  core competencies, 172, 179
  dominant designs, 289–90
  early history of, 293
  evolution of electric car concept, 51
  flexible automation, 388–89
  global study of, 353

Automotive industry (*Continued*)
Japanese approach to, 57–58, 353, 354
Japanese-U.S. competition, 414
platform projects, 425
quality programs, 201–2
technology alliances, 558, 579nn.1, 2
worker longevity, 189
Autonomous project teams, 422–23
Autonomy, 498, 500, 502, 637
Avis, 211
AXE digital switch, 466–67

Baldridge Award. *See* Malcolm Baldridge National Quality Award
Baldwin Locomotive, 149
BA managers, 484–85
Bandit pager, 423–24
Bandwagon effect, 78, 91
Bankers Trust, 209
Banking automation, 279
Bank One Corporation, 278
Barnevik, Percy, 477–93
Base/familiar sectors, 548–49
BASF products, 7, 10
Battelle, 34
Batteries, rechargeable, 103–4, 107
Bayer products, 7, 10
BE&K construction company, 211
Beckhard, Richard, 596
Behavior changes, 430–32, 598
Behavior motivation, 602–4
Behavior research, 60
Bell companies, 178–79, 510, 588
Benchmarking, 612–14, 618, 640
Benedetti, Carlo de, 563
Beresford, Charles, 132
Bessemer process, 138–39
Beta technology. *See* Videocassette recorder industry
BICC, 361
BIC pens, 151
Biological evolution. *See* Evolutionary theory and process
Blumenthal, Jay, 210
BOC gases company, 205
Body scanner, 361
Boeing, 293, 531–14
Boleky, Edward, 237
Booz Allen technology study (1985), 332
Bossidy, Larry, 215
Bottom-up management, 494, 499
Boundary management, 433–41
Bowen, Robert, 363, 530
Bower, Marvin, 510
Bowmar Instrument Corporation, 288
Box, Ralph, 619
Boyer, Herb, 510
BP Chemicals, 206
Brackett, Gary, 212
Branded packaged products, 473n.1; *See also specific brands and product types*
Breakthrough Illusion, The (Florida/Kenney), 353
Breast scanner, 359, 360, 361, 371
Brown Boveri, 34, 477; *See also* ABB Asea Brown Broveri
Bulova, 28, 31, 36
Burns, Tom, 235, 236, 237, 239
Buttons software, 349

C&C concept. *See* Computing and communications (C&C) concept
CAD/CAM integration, 386, 391, 392, 467, 472
Caile, Jim, 361
Calculators, electronic, 146, 148–49, 150, 151, 288
Caloric, 527–28, 529, 536
Cambria Company, 139
Canada, 461
Cannon, Walter, 132
Canon, Inc., 148, 149, 181, 295, 446, 633
Capabilities. *See* Core capabilities and competencies
Cardiac pacemakers, 294
Career management. *See under* Employees
Carlson, Chester, 344
Carlzon, Jan, 516–17
Carter, Thomas L., 609, 616, 618, 620, 621, 628, 629
Cartier, 40n.7, 41n.14
Casio (electronics firm), 30
Cassette recorders. *See* Sony Walkman
Caterpillar, 583
CD. *See* Compact disc
Celler, Emmanuel, 230
Cellular telephones, 356, 358, 359–62, 369–70, 371, 373–74n.20
Cement industry, 45, 47, 49, 50
Central coordination, 325
Central innovations, 458–61
Centre-for-global innovation, 453, 470
CFM, 560, 564, 579n.2, 8
Channel-driven changes, 112
Channel strategy issues, 294–301
Chaos theory, 443
Chaparral Steel, 211, 212, 258, 261, 262–63, 265, 266
Chemicals company, 259, 260, 261–62, 265
China, 32
Churchill, Winston, 630n.12
Ciba-Geigy, 7, 10, 15, 16, 18, 19, 566, 574–77
Cipher Data Products, Inc., 295
Citibank, 584–85, 589, 590, 592
Citizen (electronics firm), 30
City National Bank of Columbus, Ohio, 279
Civil rights movement, 230
Coca-Cola syrup, 289
Codified knowledge, 289
Cognitive knowledge, 506
Cognitive processes, 192–93
Colgate-Palmolive, 178, 457
Collective learning, 179
Combustion Engineering, 477, 486, 492–93
Commercialization, 126, 368
Commitment, corporate, 147–49
Communications industry, 513
Communications issues, 319–29
in boundary management, 434, 437–38
electronic, 326–29
for global businesses, 491–92
for organizational change, 604
organizational mechanisms for, 325–26, 329
rules and procedures for, 323–24
Compact disc (CD), 119, 471
Competence-acquisition alliances, 571–72
Competencies, 5, 48–49, 50, 51, 390, 467–68; *See also* Core capabilities and competencies
Competition. *See* Product competition
Competitive advantage, 3, 4, 7, 11, 20, 452–58
Competitive positioning, 280
Competitive space, 175, 181–82
Competitive vision, 17
Complementary assets, 290–91

Complementary capabilities, 557–58, 564–67
Complementary products, 78, 80–81, 89–91, 143
Comprehensive teams, 440
Computer industry, 16, 17, 215, 356
  compatability issues, 227
  competitiveness, 304
  congruence hypothesis and, 168
  core products, 180
  dominant designs, 289
  dynamic tension, 240
  joint projects, 337
  market share, 242
  new product development, 141–42, 259, 288, 291
  operating systems, 227, 288
  organizational evolution, 584
  organizational structure, 233, 250–51
  personal computers for, 300–301
  printers for, 124–25, 295–96
  profitability, 294
  semiconductor chips for, 121
  technology cycles of, 47, 49, 369
  technology strategy, 283
  See also Flexible automation; Hardware tools; Software development
Computer Integrated Manufacturing (CIM), 386
Computerized axial tomography. *See* CT scanners
Computing and communications (C&C) concept, 176–78
Concept testing, 358
Conceptual model, 160–61
Congruence model, 159–71, 200, 203, 206
Connelly, John, 278
Continental Bank, 512
Continuous-aim firing system, 130–40
Continuous improvement, 53, 57, 58
Contract book, 426
Contractural modes, 294–96, 297–300
Convergence process, 558
Convergence/upheaval pattern, 583–85, 586–91
Cooper, Marty, 359
Core capabilities and competencies, 4, 172–82, 255–68, 268n.6
  for enhancing development, 259–62
  genes analogy, 276
  history of, 255–59
  for inhibiting development, 262–65
  intellectual, 508, 511–14
  in new industies, 142
  product/process development and, 265–67
  for technology alliances, 557–58, 567–72
  technology and, 178, 275
Core products, 175, 180
Core rigidities, 262–67
Core technologies, 178, 281
Corning, Inc., 19, 165, 546, 612
  discontinuous innovation, 357, 359, 361, 362, 370, 371, 373n.18, 374n.22
  technology alliances, 560, 564–67, 571, 572, 579–80n.11, 580n.12
Corporate commitment, 147–49
Corporate Technology Unit, 335
Corporations. *See* Multinational corporations; Organizational structure; *specific companies and industries*
Corzine, Jon, 204
Cost issues
  for automated technologies, 386
  during innovation process, 449–50, 455, 537
  for personal stereos, 99, 101, 102, 109–10
  in product competition, 181
  for quality, 405, 415, 610
  for R&D, 308, 453, 455, 634, 636
  for videocassette recorders, 82, 87
  for watches, 28, 29, 31, 33, 35, 148
Crandall, L. S., 72
Cray, Seymour, 17
Cray Research, 207
Creative destruction, 45, 50, 154n.1
Creative tensions, 186
Creativity, 205–8, 212, 213, 525, 534
Crop fungicides, 7, 10, 15, 16, 17, 18, 19
Crown Cork and Seal, 278, 280
CT scanners, 141–42, 144, 155n.8, 282, 287, 300–301
  discontinuous innovation and, 356, 357–58, 359, 360, 361, 369, 370, 371
  See also Body scanner; Breast scanner
Cultural change, 70–71, 160, 166, 200–215, 392, 616
Current state, 596
Curtis Mathes, 89
CUSP programming language, 349
Customer-driven innovations, 181, 242, 351–52, 355
  lifestyle as focus, 112–16
Cycles, 61, 122–23; *See also specific industry and type*

Darwin, Charles, 69, 71
DASA (aerospace company), 572
Data collection. *See* Information gathering
Data General, 337
Data processing. *See* Computer industry; *specific companies*
David, Paul A., 70
Dayton-Hudson, 212
Decay component term, 194–95
Decentralized organization, 463
Decision flow chart, 297
Decision-making process, 222–25
Deductive management, 499, 500–502
De Havilland Comet, 288, 293
Delta Airlines, 207–8
Deming, W. Edwards, 609, 610, 618, 627
Deming Prize, 629n.4
Dependence patterns, 228
Derby, Palmer, 528–29
Deregulation, 588
Derivitive projects, 106, 107–8
Design. *See* Organizational change; Product design
DeSimone, L. D., 213
Deskilling, 395
Development cycles. *See* Product development; Project development
Diesel-electric engines, 149, 151
Diesel engines, 144, 146, 147
Digital Equipment Company, 589
Digital technology, 35, 464–65
Direct Numerical Control, 386
Discontinuous technology and innovation, 7, 9, 10, 14, 17–19
  core competencies and, 256
  effectiveness of, 59–61
  marketing and, 353–72
  model of adaptation, 57
  organizational evolution and, 588, 593

Discontinuous technology (*Continued*)
technology change and, 43, 45, 46, 47–51
Discrete manufacturing, 122
Distribution systems, 36–37, 38, 87–89
Diversification, 543, 545
Dolmel Drives (company), 489–491
Dominant design, 9–10, 11, 17, 18
for innovation, 289–90
organizational evolution and, 588
in technology cycle, 48, 49–50, 375n.35
Dominant standards, 77, 78–79
DOS. *See* MS-DOS software
Dotted-line organizations, 245–49
Douglas company, 293
Dow Corning, 565
DRAMs, 122, 571, 572
Drucker, Peter, 455
Drug industry. *See* Medical and drug industry
Duality management, 19, 20
Duke, David, 362
DuPont, 206, 545, 610
Durot, Marcel, 363
Dvorak, August, 70, 73
Dvorak Simplified Keyboard (DSK), 70
Dynamic networks, 304
Dynamics
internal company, 589
political, 599–602
Dynamic tension, 239–40, 245

Eastern Europe, 478, 491
Eastman Kodak, 176, 178, 538, 613
EBAUCHE S.A. trust, 27
Economic changes, 588
Economic dominance, 121–22
Edison, Thomas, 72
EDP. *See* Electronic Data Processing
Education. *See* Learning; Training programs
Educational acquisitions, 546–47
Efficiency, 14, 60, 234
E. F. Hutton, 226
Eindhoven, Netherlands, 461–62
Einstein, Albert, 510
Eisner, Michael, 525
Electrical Musical Industries. *See* EMI Ltd.
Electric auto, 51
Electric industry, 144
Electric light, 47
Electronic calculators. *See* Calculators, electronic
Electronic communication, 326–29
Electronic Data Processing (EDP), 279
Electronics company, 259, 261, 262, 263, 265
Electronics industry, 527; *See also specific companies and product types*
Electronics Software Applications (project), 260
Electronics Workstation, 260
Electronic watch, 32–33, 36, 37, 39, 143, 144, 148, 150, 152
Elf Acquitane, 205
Ellemtel, 455
Embedded technologies, 558–59
EMI Ltd., 287, 300–301, 373, 569
Employee participation groups (EPGs), 211
Employees
adapted, 189, 196n.1
age factors, 183–96, 331–32
anthropology of work and, 347
career management of, 183–96, 197n.5, 218, 340
companies owned by, 211, 261
evaluation of, 165, 193, 207, 339–40, 637–38
flexible automation and, 393–96
of global companies, 460–61
job descriptions, 237
job longevity, 183–96
lack of planning for, 395–96
organizational architecture and, 635
for project development, 426–27
recruitment of, 338
underutilization of, 395
See also Performance issues; Training programs; *specific industries*
Empowerment. *See* Power issues
Enablers, 616–17
Encore Computer, 593
Energy industries, 527
England. *See* United Kingdom
Entertainment industry, 513
Entrepreneurship, 16, 135, 461–62, 512, 518
Harvard Business School's Raytheon study, 524–39
Entry strategies, 547–52
Entry timing, 146–47
Environment, organizational, 163, 256
Environmental changes, 583, 587–88
Environmental concerns, 457
Environmental learning, 577
EPGs. *See* Employee participation groups
Epson printers, 125, 444
Equal (NutraSweet), 363
Equifinality, 162
Equilibrium, 162
Escapement mechanism, 8, 39
ESPRIT, 571
Essay on Man (Pope), 68–69
Estridge, Philip, 301
Europe
automotive industry, 58
electronics industry, 75
global businesses, 457, 461, 464, 470
global competition, 332
personal stereo industry, 102, 103–4, 111, 115
technological adaptations, 54–56
technology alliances, 565, 569
videocassette recorder industry, 75, 78, 84–85, 90
Xerox research lab, 349
See also East Europe; *specific countries*
Evaluation
employee, 165, 193, 207, 339–40, 637–38
project, 391
See also Feedback
Evans, Robley D., 135
Eversharp pens, 146
Evolutionary theory and process, 69, 70–71, 273–85
Executives, 19, 430, 589–90, 591–93, 638; *See also specific companies and personal names*
Exercise price, 309
Expandability, 508–9
Expectations, 264
Exponentiality, 507
Exports. *See* Global markets
Express project, 351–52
External need. *See* Situational control

Failure. *See* Mistakes and failure
Fairchild Semiconductor, 223
Familiarity and familiarity matrix, 541–43, 548, 549–51

FASELEC laboratory, 35
Fast product cycle industries, 99
FAX machines, 181
FDA. *See* Food and Drug Administration
Federal Express, 203, 206, 207, 210–11
Feedback, 81, 162, 410
Ferment, era of, 47, 48, 49, 375n.35
Fetterolf, C. Fred, 609, 611, 612, 615, 616, 617, 618, 623, 626, 629–30n.7
Field service, 283
Financial firms, 508, 513
Fine-tuning, 586
Finevest Foods, 207
Finland, 485, 486
First movers, 76–77, 83, 142, 214
Fisher, George, 424
Fisher, Ken, 584, 589, 593
Fiske, Bradley, 135
Five WHY's, 413
Flath, Gene, 237
Flexibility, 114, 117, 234
Flexible automation, 385–99
Flexible Manufacturing Systems (FMS), 58, 65, 386, 393
Florida Power and Light, 406–7, 412, 612, 614, 618
Fomon, Robert, 226
Food and Drug Administration (FDA), 360, 363
Ford Motor Company, 172, 260, 414
  new products, 258, 259, 290
  quality programs, 165, 201–2, 613
Formal organizations, 166, 200, 203, 253
Frame-breaking change, 588–89, 591, 592, 593
France, 40n.8, 456, 464, 471, 567
Franchises, 520
Freedman, George, 525, 528–29, 531
Fujisawa, Takeo, 224
Fujitsu, 334, 570, 571, 572, 573, 580n.15
Fuji-Xerox, 337, 445, 447–48, 450, 571, 613
Functional project teams, 420
Funding issues, 308
Furukawa, 361
Future state, 596

Galbraith, Jay, 162
Galileo, 133
Gardner, John, 219
Gaslight technology, 47
Gates, William, 510
GATT, 608
Gear ratio, 131, 132, 137
Genentech, 510, 553
General Electric, 89, 101, 150–51, 152, 155n.23
  CT scanner, 300, 357–58, 359, 360, 361, 369, 370, 373
  failures, 354, 480
  growth rate and size, 173, 374n.22
  technology alliances, 560, 561–62, 563, 564, 569, 570, 571, 578, 579–80n.22, 579nn.2, 5, 6, 7
  technology strategy, 281–82
General Foods, 360, 374n.22
General Motors, 4, 144, 146, 147, 149, 151
  as innovator, 354
  management, 172
  quality control, 201, 202
  technology alliances, 558, 560, 569, 570
General Radio Company, 583–84, 586, 588, 589, 590
General Watch Company, 28
Generational design changes, 106–10, 111–12
Generational innovation, 11
Generative mechanisms, 276–77
Genetic change, 71
GenRad, 584
George, W. H. Krome, 608
Germany, 25, 41n.15, 103, 379, 380
  engineer training, 323
  flexible automation, 393
  global businesses, 433, 456, 470, 477, 479, 482
  technology alliances, 572
Gilbreth, Frank B., 70
Glass industry, 46–47, 48, 49
Global businesses
  logic of, 477–93
  managing innovation in, 452–73
  skills distribution in, 557–58
  See also Multinational corporations
Global manager, 480–83
Global markets
  manufacturing competition, 303–5
  for personal stereos, 102, 115–16
  R&D communications, 319–29
  for watches, 28–37
Global Technology Unit, 335
Goal attainment, 165
Goal learning, 578
Goldman, Sachs (company), 209, 210
Goodale, Robert, 207
Governance process, 179–80
Government funding, 215
Granada, 90, 91
Great Britain. *See* United Kingdom
Grosseteste, Robert, 133
Group One, 193–95
Group think, 449
Grove, Andrew, 210, 214–15, 237
Growth, corporate, 144, 173–74
Growth theorists, 186–87
Grundig, 84–85
GTE (company), 173
Gunnery techniques, 129–40

Haggerty, Pat, 251
Hall, Charles Martin, 607
Hall cell, 630n.13
Hamilton Watch Company, 40n.6
Hardware tools, 13, 168, 291, 392, 635
Harris, Reuben, 596
Harris Corporation, 546–47
Harvard Business School, 524–25
Headphone design, 103
Hearing-aid production, 3–4, 5, 14–15, 16
Heavyweight project teams, 422, 423–30
Henderson, Fred, 612
Henson, Joe, 584
Hershey Foods, 206
Heskett, James, 213
Hetzel (Swiss inventor), 33
Hewitt Associates, 209
Hewlett, Bill, 510
Hewlett, David, 251
Hewlett-Packard, 5, 211, 288, 608
  core capabilities, 259, 261, 263, 266
  organizational structure, 234, 239, 241–42, 248, 250–53
  professional intellect at, 510
  quality programs, 246–47
  size of, 251
  support for Motorola, 423
  technical base, 261–62
  test marketing, 265
Hierarchical authority, 225–26, 236–37

Hilti AG, 378–84
Hiroshima Aluminum, 447
Hitachi, 58, 59, 81, 83, 84, 86, 90, 93n.30
 growth rate, 173–74
 R&D facilities, 333, 334
Homophyly, 192–93
Honda, Sochiro, 224, 504
Honda Corporation, 172, 179, 181
 automotive developments, 444, 449
 City development project, 494–97
 entry into U.S. market, 224–25
 innovations, 354
 U.S. competition with, 414
Hong Kong watch industry, 28, 30, 31–32, 40n.8, 13, 41n.14
Honma, Hiroshi, 497
Hoshin Kanri (Management by Policy), 409–12
Hospital employees, 186
Houghton, Amo, 370, 371
Hounsfield, Godfrey N., 287, 300
HP. *See* Hewlett-Packard
Human exhaustion, 449–50
Human interface, 168
Human resources. *See* Employees
Hunt, R. W., 139
Hyper-selection, 509–10

IBM Corporation, 4, 16, 17, 18, 72
 Apple Computer compared with, 227
 channel strategy, 295, 300–301
 growth rate, 173
 as innovator, 288, 354
 mistakes, 207
 new businesses, 545, 548
 personal computer development, 300–301
 reputation, 301
 technology alliances, 572
 typewriter production, 72, 146
IBM ProPrinter, 124–25
Ibuka, Masaru, 82, 100, 116
ICL (company), 334, 570, 571, 580n.15
Identifying process, 139
Implementation
 of flexible automation, 393
 by global businesses, 471
 innovation as outcome of creativity and, 212–13
 of organizational change, 597–98
 power issues and, 225–28
 strategic innovation based on, 20
Incentive systems, 261
Incremental changes and innovations, 10, 11, 12, 14, 17, 19, 20, 48
 organizational evolution and, 585, 587–88
 in personal stereo industry, 108, 112–16, 117
India, 455, 478, 480
Indiginous Technology Unit, 335
Individual control, 189–93
Inductive management, 500–502
Industrial designers. *See* Design
Industrial relations, 397–98
Industrial restructuring, 25
Industries, new. *See* New business development
Industry foresight, 13
Infinitely flat organizations, 515
Informal organizations, 166–67, 200, 253
Information creation, 494–504
Information gathering, 63, 65
Information processes, 191–92
Information redundancy, 444–49
Information sharing, 210
Innovation, 3, 7, 10, 11, 48
 benefits for users, 379
 contractural agreements, 295
 coproduction of, 350–51
 customer-driven, 112–16, 181, 242, 351–52, 355
 efficiency and, 388
 in global business, 452–73
 gunnery case study, 129–40
 Japanese management of, 443–51
 local, 346–50
 for managers, 176, 186–87, 188, 190
 new industry resulting from, 141–54
 organizational structure and, 233–53
 in personal stereo industry, 116
 process of, 132, 212
 promotion of, 204–12
 through social control, 200–215
 strategy and culture linked with, 214–15
 value from, 287–306
 See also Discontinuous technology and innovation; New products; *specific fields*
Innovation and Entrepreneurship (Drucker), 455
Innovation streams, 4, 5, 6, 11–12, 12–16, 18–21, 233
Input, organizational, 162–63, 169
Institute of Advanced Studies (Princeton), 510
Integration modes, 296–300, 637
Integrative mechanisms, 277–79
Intel, 203, 210, 214, 215, 234, 237, 571
 organizational changes, 240
 professional intellect at, 510
 quasi-structure, 248, 249
 sales, 414
 size of, 251, 252
 technology development, 238
Intellect, 506–23
 characteristics of, 507–9
 core competencies and, 511–14
 definition of, 506–7
 organization forms and, 520–22
 organizing around, 514–17
 output measurement and, 522–23
 professional, 509–11
 starburst organization and, 520
 See also headings beginning with Knowledge
Intellectual leadership, 172–73
Interaction, 499–500, 501–2
Interchangeability, 31
Interdepartmental development process, 445–46
Interdependence patterns, 228
Interface standards, 80
Inter-firm networks, 25
Internal business development, 544–45
Internal diversity, 6
Internal dynamics, 589
Internal interdependence, 161–62
Internalization, 559–60
Internal ventures, 545
International markets. *See* Global markets
International Paper, 608–9
International Telecommunications Center, 464–65
INTERNET, 519
Interpretation control, 189–93
Inter-unit integration, 466–67
Intrinsic technology, 403
Inventors. *See* Innovation
Inverted organization, 516–17

Investments, 304, 310–12, 314
Ishikawa, Kaoru, 408
Isolation, 601
Isolationist teams, 439–40
Italy, 467, 472
Iteration, 361–63
ITT (company), 173, 357, 457, 464–67

Jacobs, John, 221
Jacobsen, Alan, 213
Japan
automotive industry, 57–58
computer industry, 58–59, 65
core competencies, 173, 255, 557
electronics industry, 75
flexible automation, 393
global businesses, 331–41, 452, 455, 457, 467, 469, 470, 471
importance of manufacturing, 126
innovation management, 443–51
personal stereo industry, 99–119
platform projects, 425
quality programs, 201, 404, 618
steel industry, 64–65
technical strength of, 124, 126, 128
technology acquisition, 282
technology alliances, 557, 558
U.S. competition with, 245, 304, 353, 354, 389, 391
U.S. joint ventures, 409
videocassette recorder industry, 18, 75–92, 93n.30
watch industry, 28, 30–31, 32, 34, 37, 40n.8
See also specific companies
Japan Victor Company. *See* JVC
Jephthah story, 68
Job descriptions, 237
Job longevity, 183–96
Jobs. *See* Employees
Johnson, Andrew, 221
Johnson, Lyndon, 230–31
Johnson, Philip, 69
Johnson & Johnson, 207, 211
Joint European Submicron Silicon Initiative (JESSI), 572
Joint ventures, 489–91, 545–46, 551; *See also* Technology alliances
Jones, Jim, 221
J. P. Morgan, 209
Juran, J. M., 404, 406, 618
Jura region (Switzerland), 24, 33
JVC (Japan Victor Company), 18, 75–92, 93n.30, 180
technology alliances, 558, 559–60, 567–68, 569, 571, 572

Kaizen philosophy, 58
Kanban system, 65
Kanter, Rosabeth, 220, 468
Karlsson, Sune, 482, 483
Kawamoto, Nobuhiko, 494–95, 497
Kawashima, Kihachiro, 224
Kawashima, Kiyoshi, 494, 496
Kearns, David, 159, 612, 613, 618
Kelleher, Herb, 207
Kennedy, John F., 230
K. Hattori/Seiko watches, 152; *See also* Seiko watches
Kissinger, Henry, 222
Knowledge. *See* Intellect
Knowledge building, 256, 289, 309, 313–14
Knowledge intensity, 385
Knowledge sharing, 508
Kodak. *See* Eastman Kodak
Koerber, Eberhard von, 479, 486, 488
Kokkonen, Kim, 237, 238
Kolesar, Peter, 629n.1
Korea, 304
Kotter, John, 213
Kume, Tadashi, 503
Kux, Barbura, 489, 490

Labor force. *See* Employees
Ladder paradigm, 121–22, 126, 127
Lambert, Tom, 357
Language issues, 478, 481, 490, 492, 602
Laserwriter, 295–96
Lauderdale, Mike, 211
Laundry products, 455, 456, 457
LCD display watches, 144
Leadership, 172–74, 219, 280; *See also* Management
Lead user method, 376–84
Learning, 179, 225, 275
in technology alliances, 569, 571, 572, 573–78
See also Intellect; Probe and learn process; Training programs
Leavitt, Harold, 162
LED display watches, 32, 145
Legal shifts, 588
Lehigh University industry study, 47
Leonardo da Vinci, 132
Leveraging, 176–78, 512–14, 535–36
Lexus, 181
Licensing, 535–36, 545, 551, 558
Lightweight project teams, 420–22
Lincoln, Abraham, 221–22
Litton/Monroe, 148, 149, 152
L. M. Ericsson, 5, 463–68, 472
Local area networks, 265, 357
Local-for-local innovations, 453, 455–56, 460–63, 470
Locomotives, 479–80
Lodging services, 513
Longley (touch-typing method developer), 73, 74
LTV Steel, 613
Lucy, Charles, 359, 360–61

Machine that Changed the World (Womack, Jones, and Roos), 353
Magnavox, 89
Magnetic Video Corporation, 90
Mahan, Alfred Thayer, 136–37, 138
Malaysia, 111
Malcolm Baldridge National Quality Award, 159, 353, 404, 405–6, 626, 628
Mammography. *See* Breast scanner
Management
conceptual models for, 160–61
core capabilities and, 256–57, 260, 263
of creative people, 534
of discontinuous innovation, 354–56, 372
of flexible automation, 385–99
of global businesses, 458, 468–72, 480–83
of information creation, 502–3
of Japanese innovation, 443–51
middle-up-down, 494–504
of organizational change, 598–605
organizational strategy and, 163, 171, 253
of Polish joint venture, 489–490
power issues for, 218, 228–31
of product design, 118
R&D and, 314–16, 319–29
scorecard, 172–74
in self-designing organization, 245

Management (*Continued*)
 served market orientation among, 181–82
 styles of, 20–21, 247–48
 of technology, 50–51, 110–12, 273–85
 of technology alliances, 561–73
 See also Leadership; Project management; *specific companies*
Management by Objective, 409
Management by Policy, 409–12
Manufacturing, 29–30
 automated, 385–99
 costs, 234
 cycles, 36, 37
 global competition, 303–4
 new product changes, 144
 of personal stereos, 111, 112, 114–15, 117
 product design and, 125, 126, 386
 product development for, 122, 125, 126
 robotic, 112, 125
 See also Original equipment manufacturers; *specific companies and products*
Margarine products, 455–56
Marginal sectors, 550–51
Market familiarity, 541–43
Marketing, 37
 advertising, 224–25
 of computers, 241–42
 discontinuous innovation and, 353–72
 global, 79–80, 462–63
 lead user research for, 376–84
 of new products, 148, 290
 of personal stereos, 99, 104–6, 111
 project teams and, 429
 quality programs for, 414
 research and testing techniques, 123, 265, 357–58, 368–69
 of VHS vs. Beta videocassette recorders, 75–92
Marks and Spencer, 278, 280, 281
Mary Kay Cosmetics, 206
Mass distribution. *See* Distribution systems
Mass marketing. *See* Marketing
Mass production. *See* Production systems
Matrix management, 247–48, 484–85
Matsushita Electric, 99
 core competencies of, 173, 179, 180
 global innovations and, 458–61, 468, 470
 growth rate, 173
 Home Bakery project, 444
 personal stereos, 112
 and VCR development and standardization, 75, 78, 79, 80, 83, 86, 89, 91–92
 See also JVC; Panasonic
Maurer, Robert, 357
Mazda, 291, 447
MBO system, 201
McGurrin, Frank E., 73, 74
MCI (company), 357, 369, 373n.18, 512, 513
McKesson Drugs, 522–23
McKinsey & Co., 392, 510
Meat packing industry, 535
Mechanical watch, 16, 27, 36, 37, 38–39, 148, 150
Medical and drug industry, 291, 294, 302, 306n.8, 336, 351, 574; *See also specific companies and products*
Mentoring, 510
Merck, (company), 302, 522, 572
Merrill Lynch, 515
Metals industry. *See* Steel and metals industry
Mexico, 304
Microelectronics, 292
Microsoft Corporation, 210, 295, 301, 507, 508, 510
Microwave ovens, 143, 148, 150–51, 152, 155n.23
 Raytheon development of, 527, 528–29, 535, 537
Middle Ages, 133
Middle-level managers, 218, 501
Middle-up-down management, 494–504
Military services, 226
Miniaturization, 178
Minicomputers. *See* Computer industry
Mistakes and failure, 114–15, 206, 207–8, 213, 480
 acceptance of, 222
 of diversification, 545
 by innovators, 288, 354, 362–63, 370, 537
Mitchell, John, 371
MITS (company), 288
Mitsubishi, 84, 86, 334
Mixed modes, 300
Mobile telephones, 373–74n.20
Modesty, public, 206
Monroe. *See* Litton/Monroe
Monsanto, 371–72, 553–54; *See also* Searle
Moore, Gordon, 235, 240, 252, 304, 510
Morgan, Jim, 210
Morita, Akio, 84, 116, 117
Morton, Dean, 241
Motivational approach, 188–89, 403, 406
Motorcycle industry, 224–25
Motorola, 5, 16, 19, 172, 209–10
 cellular phones, 358, 359, 360, 361–62, 369–70, 371, 373–74n.20
 computer processor, 369
 paging products, 423–24
 professional intellect at, 510
 size of, 251, 374n.22
Motorola Semiconductor, 234
Movement manufacturers, 29–30
MS-DOS software, 227, 288, 295, 301
Multinational corporations (MNCs), 333, 452–73
 central innovations, 458–61
 competitive advantage of, 452–58
 feasibility of innovations, 463–68
 innovation management, 468–72
 local innovations, 461–63
Mutual investigation, 448
My First Sony (children's product), 114, 115, 118

Nadler, David, 612
NASA (National Aeronautics and Space Administration), 176, 535
National Cash Register Company (NCR), 279
National Semiconductor, 214, 234, 245, 251
Natural selection, 69, 71
Naval gunnery techniques, 129–40
NEC (company), 173, 176, 177–78, 332, 334, 414
 global businesses, 455, 465, 467
Netherlands, 461–62
Net Present Value, 391
Network alliances, 560–61, 564–67
Network externality, 80–81

Network organization, 325
Network production systems, 24, 25
New Age thinking, 217
New business development, 541–54
  alternative strategies, 543–44
  entry strategy, 541–43, 547–52
  testing proposals, 552–54
  as threat to established firms, 141–54
  types of mechanisms for, 544–47
Newness, 541–42
New process technologies, 53–66
New products, 181–82
  quality programs for, 413–14
  Raytheon's center for, 524–39
  via lead user method, 376–84
  See also Dominant design; Innovation; *specific companies*
Newstream success, 534–36
New Technology Engineering (NTE), 238
Nike, 512
Nippon Data General, 337
Nixon, Richard, 219–20
Nodal networks, 572
No Excuses Management (Rodgers), 215
Noika, 5
Nordstrom, 202, 206, 211, 212
North, Oliver, 225
Norway, 479, 484
Not Invented Here (N.I.H) Syndrome, 195, 405, 474n.6, 529
NovaCare, 517, 521–22
Novellus, 512
Noyce, Robert, 240, 510
NUMMI (company), 201, 202, 212, 560, 569, 570, 579n.3
NutraSweet, 356, 358, 359–60, 361, 362–63, 370, 371–72, 374n.21

Odetics, 209
OEMs. *See* Original equipment manufacturers
Office automation, 413
Oil industry, 513
Olechnowicz, Pawel, 490
Olivetti, 562–63, 564, 579n.8, 580n.17
Olsen, Kenneth, 589
O'Neill, Paul, 607–29, 630n.12
Open system, 161
Operator training, 393
Opportunity gap, 174
Optical fiber industry
  background of, 373n.18
  market testing vs. probe and learn approach, 356, 357, 359, 360–61, 362, 369, 370, 371
  technology alliances, 560
Oral slow release systems, 574–75
Organic architecture, 69
Organic management, 235–36, 249–51
Organizational architecture, 12–16, 18, 161, 170, 631–41
Organizational change, 595–606
  criteria for, 596–97
  implementation problems, 597–98
  implications for management, 598–605
  See also Organizational structure
Organizational control, 598
Organizational evolution, 583–94
  convergence and upheaval patterns, 583–85, 586–91
  executives and, 591–93
  research base for, 585–86
Organizational paradigm, 226
Organizational power. *See* Power issues
Organizational problem-solving. *See* Problem-solving process
Organizational separation, 149–50, 154
Organizational structure, 161–62, 233–53
  change in, 18, 240–45
  charts for, 236–37, 242–43
  effect of research on, 342–52
  information creation and, 498–500
  intellect and, 514–22
  for new product centers, 531–34
  quasi-formal, 245–49, 253
  size of, 217–18, 251–52, 302
  See also Reorganization; *specific companies*
Organization Transition Board, 171
Orient Oil Company, 595–96, 605–6
Original equipment manufacturers (OEMs), 80, 83, 84, 88, 89, 92, 93n.30, 374n.22, 584
Origin of Species (Darwin), 69
Oticon, 3–4, 16
Otsuka Pharmaceuticals Co. Ltd., 332, 334, 336
Output, organizational, 165, 169, 522–23
Owens Illinois, 565

P&G. *See* Proctor and Gamble
Pacemakers, cardiac, 294
Packard, David, 251, 252
Paging products, 423–24
Palo Alto Research Center (PARC), 342–52, 354
Panasonic, 101, 104, 458
Panda thumb principle, 68–74, 274
Papermate pens, 146
Paradigm development, 226
Paradox effect, 266
Parker, Gerry, 237–38
Parker Pen, 152
Parry, Charles W., 608, 617, 622, 626
Participation strategy, 145–53
Parts manufacturing, 29
Pashler, Ted, 359
Patek Philippe, 28
Patents, 289, 302, 558
Patterson, John H., 279
PDCA. *See* Plan-Do-Check-Act
Peer pressure, 510
Pen industry, 146–47, 151
People's Express, 517
PEP Industries, 612–13, 616, 618
Perfection, 509–11
Performance appraisal and evaluation, 165, 193, 207, 339–40, 637–38
Performance gap, 174
Performance issues
  age as factor, 194
  long-term, 146
  for new industries, 151–52
  for product development teams, 436–37
  shortfalls in, 14
  stress or anxiety affecting, 598
  for technology strategy, 282–83
Peripheral technology, 281
Personal computers. *See* Computer industry; *specific companies*
Personnel. *See* Employees
Peters, Tom, 228
Peterson, Donald, 622
Philips, 4, 34, 99
  global business innovations, 452, 456, 458–59, 461–63, 468, 470, 471
  growth rate, 173
  technology alliances, 571–72, 579–80n.11

Philips (*Continued*)
videocassette recorder technology, 78, 79, 81, 84–85, 87, 91
Photocopiers, 344–46, 348–49, 350, 356
Piaget, Jean, 28
Picturephones, 359, 360, 361, 362, 373n.18
Pin lever watch, 39
Pipe hanging system, 378–84
Pirelli, 360
Plan-Do-Check-Act (PDCA), 406, 610–11
Plastics, 356
Platform projects, 106, 425
Poduska, Bill, 584, 593
Poland, joint ventures, 489–91
Polaroid, 583
Political dynamics, 599–602
Pope, Alexander, 68–69
Potential command, 448
Powell, Liz, 113, 114
Power issues, 217–31
ambivalence about, 220–22
decision-making alternatives and, 222–25
implementation difficulties because of, 225–28
in management process, 228–31
of organizational change, 597
for project members, 261, 263–64
Power transformers, 482–83, 484
Pratt & Whitney, 560, 562
Presidential diagnosis, 411
Pressman recorder, 100
Preston Trucking, 211, 613
Pricing. *See* Cost issues
Primary producers, 82–86
Prime Computer, 584, 585, 589, 590, 592
Princeton University, 510
Proactive strategic change, 6
Probe and learn process, 358–63, 368–72
Problem-solving process, 159–71, 191
Process development, 419
Process innovation, 48–49
Process learning, 577–78
Process standards, 413
Process technology, 62, 65–66, 393
Proctor & Gamble (P&G), 207, 209, 547
global businesses of, 455, 456, 457, 470
Product, complementary. *See* Complementary products
Product competition, 101–6, 109, 535
development cycles and, 123, 125
global, 557–58
pricing issues, 181
strategy for new business, 150–51
Product design
automation of, 386, 389
competition in, 47–48
cost of, 118
manufacturing and, 125, 126, 386
in personal stereo industry, 99, 106–10, 111, 112–15, 117–18
problems with, 262
types of, 160
Product development, 419
boundary management in, 433–41
concepts for, 380–83
cycle of, 123–25, 127
discontinuous vs. conventional, 363–68
information creation and, 497–98
joint, 337, 376
market needs and, 459–60
planning for, 3, 36, 100–101, 117
See also Project development
Product differentiation and variety, 86–87, 102–6, 118
Product dominance, 121–22
Product failure. *See* Mistakes and failure
Product families, 10–11, 99–119
customer lifestyle influencing, 112–16
genesis of, 100–101
management of, 110–12
product competition patterns, 101–6
technological evolution of, 106–10
See also specific products
Product innovation. *See* Innovation
Production goals, 65
Production networks, 24, 25
Production systems, 28, 30, 31, 87–89, 414
Productivity dilemma, 46
Product launch, 368
Product-life-cycle shifts, 588–89
Product longevity, 104–6, 119n.3, 453
Product-process relationship, 140, 265–67, 387
Product range, 393
Product substitution, 12, 17–18, 19
Product-technology relationship, 121–28
Product vs. process, 140
Profitability, 291–94, 304
Programming tools, 347, 349
Project charter, 425–26
Project development, 258–59, 411
Japanese approach to, 443–51
organizing teams for, 419–32
See also Product development; Project management
Project evaluation, 391
Project management
attention and effort issues in, 61–66
beginnings of, 59–60
core capabilities and, 256
deadlines and, 63–64
empowerment of members in, 261
leadership for, 427–29, 430
See also Product management; Project development
Promotional vehicles, 408–13
Property rights, 289
Propiconazol (fungicide), 7, 16, 17, 18
Proposal testing, 552–54
Psychological safety, 186
Pygmalion effect, 264–65

QD/QFD. *See* Quality Deployment/Quality Function Deployment
Quality circles, 408–10
Quality Deployment/Quality Function Deployment, 407
Quality programs
at Alcoa, 607–29
at AT&T, 613
automated technology and, 386
automotive industry, 165, 201–2, 407, 411, 613
at Hewlett-Packard, 246–47
in Japan, 201, 404, 618
for marketing and new products, 413–14
in U.S. companies, 402–15
at Xerox, 165, 612, 613–14, 616, 618, 628–29
See also Total Quality Control; Total Quality Management
Quality sweating theory, 403
Quartz watch, 6, 8, 16, 17, 32, 34–35, 37, 39
Quasi-formal organizations, 245–49, 253

QWERTY keyboard (typewriter), 70–74, 274

R&D. *See* Research and development
Radiation Dynamics, Inc., 547
Radio Corporation of America. *See* RCA
Range oven industry, 148
Rapid max strategy, 65
Raytheon Company, 148, 151, 524–39
  background, 526–27
  entepreneurship, 538–39
  new product center, 525–26, 528–34
  newstream sucesses, 534–36
  results of new products, 536–38
  size of, 537
RCA (Radio Corporation of America), 45, 304
  growth rate of, 173
  new industries, 150
  and VCR development and standardization, 78, 79, 81, 83, 88, 89, 90–91, 92
Rechargeable batteries, 103–4, 107
Red bead experiment, 610, 618
Reed, John, 584, 585
Regulatory systems, 27–28
Remington typewriters, 73, 74, 146, 149
Remote interactive communication (RIC), 344
Renaissance, 133
Reorganization, 239–40, 241, 632
Reprographic technology, 16
Research and development (R&D), 127, 154, 307–16
  for banking industry, 279
  corporate reinvention by, 342–52
  cost of, 308, 453, 455, 634, 636
  framework for, 313–14
  for generic technologies, 107
  for global business, 453
  implications for strategy, 302–5
  intrafirm coordination of, 337
  in Japan, 331–41
  lack of, 146
  managers' roles in, 314–16, 319–29
  positive impact of, 312–13
  problem solving by, 197
  profitability and, 292
  strategic perspective on, 177–78, 308–12
  teamwork for, 434–35, 439
  for technology alliances, 574
  for watch industry, 34, 148
  See also Product development; Project development
Resistance pockets, 590
Resource allocation, 499, 500, 501
Resources, organizational, 163, 165
Retailers, 513
Return on Investment (ROI), 307–9, 314
Reward systems, 441, 638–40
RIC. *See* Remote interactive communication
Ricoh Co. Ltd., 332
Rieker, Wayne, 408
RISC/UNIX workstation, 263
Risk taking, 205–7, 213
Robb, Walt, 358, 371
Robotic manufacturing, 112, 125
Rodgers, T. J., 215
Rolex, 28
Rolls Royce, 560, 562
Rollwagen, John, 17, 207
Rolm Corporation, 300, 303, 548
Ronstadt, Robert, 335
Roosevelt, Theodore, 135, 136, 138, 231
Rosenbladt, Peter, 248
Routines, 58, 60–61, 276
Routine users, 382–83
Royal Crown Companies, 287–88
Royal typewriters, 146, 149
Rubbermaid, 203, 206, 207, 209
Rumsfeld, Donald, 361
Russia, 40n.3, 41n.15

Safety issues, 610, 614–15
Samo (Ciba manager), 10, 15, 16, 17, 18
Sanderson, Bryan, 206
Sanford, Charles, 209
Sanyo, 79, 84, 86, 91, 99, 104
SAS (airline), 516–17
Sashimi approach, 445
Satellite telecommunications, 337
Schaeffer pens, 146
Schlatter, James, 374
Schlumberger oil company, 223
Schulmeyer, Gerhard, 479, 486
Science-technology relationship, 122
Scientific conferences, 324
Scientific dominance, 121
SCM Corporation, 148, 149, 151
Scott, Percy, 131–32, 133, 134, 135, 140
Scouting activities, 436, 439
S-curve, 51
Searle, 359–60, 361, 362–63, 370, 374nn.21, 22; *See also* Monsanto
Searle, John, 361
SED system, 496
Seiko watches, 4–5, 6, 17, 30, 34, 37, 40n.9, 108, 152
Self-criticism, 244
Self-designing organizations, 244–45
Semiconductor industry, 121, 122, 148, 149–50, 151
  culture/strategy link in, 214
  dynamic tension in, 240
  organizational structure, 233
  sales in, 414
Senior management teams, 16, 17
Serendipity story, 133
Served market orientation, 181–82
SGT holding company, 40n.5
Shakeout, 50
Shamlin, Scott, 423
Sharp (company), 81, 84, 86, 99, 178, 180
Shearson-Lehman, 209
Sholes, C. L., 72, 73, 74
Siemens, 4, 173, 360, 560, 571–72
Signetics, 579–80n.11
Silicon Graphics, 512
Silicon Valley managers, 204
Simplex algorithm, 289
Sims, William S., 133, 135–36, 137
Sinclair, Don, 584
Singapore, 111, 569
Situational control, 189–93
Skill learning, 578
Skill mix, 394–96
Skills/knowledge dimension, 260, 262–63
SMART system, 112
SMH Group, 37, 40n.6
Smith, Fred, 207
Smith, Roger, 201
Smith-Corona typewriters, 149, 152
SNECMA (company), 560, 561–62, 563, 564, 571, 578, 579nn.2, 5, 6, 7, 8
Soap. *See* Laundry products
Social control, 200–215
Social/cultural design, 160
Socialization efforts, 322–23
Socialization stage, 184–86, 190, 196nn.2, 3, 4
Societal changes, 130, 135, 139–140
SociÇtÇ des Garde Temps, 28

Sociologie du Travail group, 395
Software development, 54, 56, 58–59, 65, 168
  for flexible automation, 392
  joint projects for, 337
  for organizational architecture, 635, 639
  See also specific software
Sony Corporation, 5
  core competencies, 173, 178, 180
  growth rate, 173
  as innovator, 78–79, 83, 100–101, 112–16, 354
  licensees, 83
  management and leadership, 110–12, 116
  manufacturers and, 84, 111, 112, 114–15
  miniaturization processes, 178
  product competition, 101–6
  technological evolution of, 106–10
  videocassette recorder development, 18, 75–92, 93n.30
  See also specific products
Sony Industrial Design Center, 111, 113
Sony Walkman, 10, 99–119, 119n.5
  for children, 114, 115, 118
  customer lifestyle as basis for, 112–16
  evolution of, 100–101, 106–10
  management of, 110–12
  product competition patterns, 101–6
  rechargeable, 103–4, 107
Sophisticated Operating System (SOS), 227
South America, 478
Space industry. *See* Aerospace industry
Speed Queen, 527, 529, 536
Spider's web organization, 518–20
Sporck, Charlie, 245
Sports Walkman, 114, 115, 118
Spreng, Douglas, 242, 243, 244
Springfield ReManufacturing, 210
Spring-powered watch. *See* Mechanical watch
SRAM. *See* Static Random Access memory
SSIH (Swiss watch consortium), 3–5, 28, 34, 37
Stabilization process, 187–88, 196n.1, 396–98
Stack, Jack, 210
Staffing. *See* Employees
Stalker, G. M., 235, 236, 237, 239
Standard procedures, 58
Starburst organization, 517–20
Starkey, 5, 6, 14–15, 16
Static Random Access Memory (SRAM), 237, 238, 571
Statut de l'Horlogerie, 27–28, 37
Steam engines, 144, 149, 293
Steel and metals industry, 54, 64–65, 138–39
Stereos, personal. *See* Sony Walkman
Stock options, 309–12
Strategic alliances. *See* Joint ventures; Technology alliances
Strategic architecture, 175, 176
Strategic behavior, 276
Strategic business unit, 179
Strategic compatibility, 563–64
Strategic competition, 150–51
Strategic innovation, 6, 17–19, 20
Strategic intent, 13, 165, 175–76
Strategic orientation, 19
Strategic partnering, 295
Strategic pitfalls, 153–54
Strategic positioning, 315–16
Strategic task design, 160
Strategy
  business, 608
  innovation/culture link, 214–15
  organizational, 163–65
  quality as, 617
  for R&D, 307–16
Stress. *See* Anxiety and stress
Strîmberg (company), 485–86
Structural change. *See under* Organizational structure
Structural design, 160
Subsidiaries, 458–59, 462–63, 468–72
Substitute product/process technologies, 588
Substitution, era of, 47
Success paradox, 4
Suchman, Lucy, 347
Sun Microsystems, 165, 369
Superconductivity, 126
Suppliers, 393–94, 446–47
Swanson, Robert, 510
Swatch, 28, 41n.16
Sweden, 455, 466, 467, 472, 488
Swiss locomotives, 480
Swiss machinery manufacturers, 376–84
Swiss watch industry, 3–5, 16
  effect of U.S. watch industry on, 26, 28, 30, 31
  global markets compared with, 28–37
  history of, 24–28, 37–40
  organizational structure of, 33–34
  technological changes affecting, 32–37
Swiss Watch Industry Federation, 26, 27, 33
Switzerland, 206, 488
Sylvania, 45, 89
Symbols, change, 602
Symptom identification, 169
Synergy, 590
Syntex, 351–52
System, 161

Tacit knowledge, 289
Taguchi Method, 407
Taiwan, 304
Talent. *See* Intellect
Tamadashi kai (meetings), 496
Tappan (company), 150–51
Task coordination, 436, 438, 439, 440
Task learning, 577
Taub, Louis, 73
TCQ. *See* Total Quality Control
TDIndustries, 212
Team mentality, 250
Teamwork, 205, 207, 213, 222, 409
  building, 194
  environment for, 249–51
  for product development, 433–41
Technical base, 261–62
Technical literacy, 260
Technical systems dimension, 260–61, 263
Technicare, 152, 300
Technological architecture, 168
Technological change and innovation, 4, 45–51, 70
  in gunnery techniques, 136–37
  managing, 391–93
  through organizational structure, 233–53
  in personal stereo industry, 106–10
  value from, 287–306
Technological discontinuity. *See* Discontinuous technology and innovation
Technological evolution, 43
Technological familiarity, 541–43
Technological gatekeepers, 434
Technological guidepost, 48

Technological improvement, 53–66
Technological opportunity, 4
Technology alliances, 556–79
  as learning process, 573–78
  managing, 561–73
Technology cycle, 5, 7–11, 46–51, 369, 375n.35
Technology gap, 304, 453
Technology management, 50–51, 110–12, 331–41
Technology-product relationship, 121–28
Technology-science relationship, 122
Technology strategy, 273–85
  alliances used for, 556–79
  changing, 340–41
  enacting, 279–83
  as evolutionary process, 274–75
  forces shaping, 275–79
Technology threat, 141–54
Technology transfer, 128, 335, 350, 558
Technology vs. competence, 178
Telecommunications industry, 455, 457, 469, 473n.1; *See also specific companies*
Telefunken, 571
Telephones. *See* Cellular telephones; Picturephones; Mobile telephones
Television, 79, 83, 86, 127, 180, 304
  cable market for, 362
  global innovation in, 456, 461, 471
  See also Videocassette recorder industry
Tender Tea Leaf brand, 547
Tennessee Eastman, 618
Texas Instruments, 35, 122, 151, 172, 234, 251, 288
Theobald, Thomas, 209
Third World, 128
Thomson Consumer Electronics, 558, 559–60, 565, 566, 567, 569, 570, 571, 580n.14
Thorn Electrical Industries, 90, 300, 567, 569, 571
3M (company), 203, 205, 206, 211, 212–14, 514, 545, 546, 608
Thurston, Bill, 584
Tiger teams, 422–23
Tilton, 149–50
Time issues, 386
Timex, 28, 31, 40nn.11,12, 148, 150
Top-down management, 494, 499
Topological innovations, 108–9, 112–16, 117
Toshiba
  electronic calculators, 148, 149
  global markets, 334
  growth rate, 173
  innovations, 354
  personal stereos, 99, 101, 104
  technology alliances, 569, 571, 572
  and VCR development and standardization, 79, 84, 86, 91
Total Quality Control (TQC), 246–47, 404
Total Quality Management (TQM), 402–15
Toyota Motor Company, 57, 58, 60, 172, 212
  quality programs at, 407, 611
  technology alliances at, 558, 560, 569, 570, 579n.3
TQC. *See* Total Quality Control
TQM. *See* Total Quality Management
Trade barriers, 304
Trade secrets, 289
Training programs, 64, 323, 338, 349–50
  in English language, 490
  for manufacturing operators, 393
  for professionals, 410
  for quality issues, 408, 413, 617–21
  for technology alliances, 569–71
  See also Learning
Transdermal systems, 574–75
Transformation process, organizational, 165–67
Transistor industry, 121, 144, 149–50, 151
Transition management, 604–5
Transition state, 596
Transnational corporations. *See* Global business; Multinational corporations
Transportation industry, 478–79
Travel, business, 323
Trust building, 448–49
Tuning fork watch, 8, 16, 33, 36, 39
Turkey, 455–56
Tushman, Mike, 611, 612, 628, 629n.1
TV glass business, 565
Typewriters
  keyboard, 69–74, 274
  production, 11, 146, 149
U-Matic, 78–79, 82, 83
Underwood typewriters, 149
Unilever, 455–56, 457
Union Carbide, 295
Union des Branches Annexes de l'Horlogerie, 27
United Kingdom
  engineer training, 323
  global businesses, 461, 464, 471
  naval gunnery techniques, 131–35
  steel industry, 139
  technology alliances, 569, 570
  videocassette recorder industry, 90
  Xerox research lab, 349
  See also specific companies
United States
  competition in, 307
  core competencies in, 255, 557
  electronics industry, 75
  flexible automation, 393
  global businesses, 455, 457, 469, 470
  global competition, 224–25, 245, 304, 305, 332, 353, 354, 389, 391
  industrial development, 126
  innovations, 129–30
  joint ventures with Japan, 409
  leadership loss in, 173
  manufacturing regions, 25
  naval gunnery techniques, 130–40
  organization size, 217
  personal stereo industry, 101–2, 103–4, 105, 115
  quality control activities, 402–15
  scientific strength of, 124
  technological adaptations, 54–56
  videocassette recorder industry, 75, 78, 89, 90, 91
  watch manufacturing, 26, 28, 30, 31, 32
  See also specific companies
Universal Oil Products, 292
University community, 124
UNIX/RISC workstation, 263
Unlearning, 275
Users. *See* Customers

Vacuum tube technology, 48, 144, 149–50, 151
Vadasz, Les, 240, 248, 249
Values
  core capabilities and, 257–58, 261–62, 263–65
  creation of, 174–75

Values (*Continued*)
quality programs and, 621–26
shared, 209
from technology, 281, 287–306
from technology alliances, 563
Variety. *See* Product differentiation and variety
VCRs. *See* Videocassette recorder industry
Vendors. *See* Suppliers
Venture capital, 546, 549
Venture Career Path, 213
Versatec, 547
VHS. *See* Videocassette recorder industry
Vickers (company), 178
Victor Company of Japan. *See* JVC
Victor Comptometer, 150
Videocassette recorder industry, 18, 75–92, 356
complementary products, 80–81, 89–91
core products, 180
global markets, 79–80, 459
inventors, 78–79
mass production and distribution, 87–89
primary producers, 82–86
product differentiation, 86–87
standardization, 77–82
technology alliances, 558, 559–60, 567–68, 569, 572
videotapes, 80–81, 91–92, 350–51
Videocolor, 565
Video Corporation of America, 91
Videotapes, 80–81, 91–92, 350–51
Volatility, 311–12

Wada, Katsutoshi, 503–4
Walkman. *See* Sony Walkman; Sports Walkman
Walleigh, Richard, 246–47
Wal-Mart, 512
Walpole, Horace, 132
Warrick Mill, 619
Watanabe, Hiroo, 495, 496, 497
Watch industry, 3–8, 16, 24–41
global markets, 28–37
history of, 38–40
marketing, 148
pricing issues, 28, 29, 31, 33, 35, 148
technological changes and threats in, 32–37, 143, 144, 150, 152
See also Hong Kong watch industry; Swiss watch industry; *specific kinds of watches and manufacturers*
Waterman, Bob, 203, 209, 211
Weapons systems, 138; *See also* Gunnery techniques
Weisz, Bill, 360
Welch, Jack, 207
Western Electric Studies (1975), 193
Westinghouse, 173, 174, 477
What America Does Right (Waterman), 203
Window of opportunity, 59–60, 61, 62, 63–65
Window on technology, 546
Winning, 175–76
Word processors, 11, 215
Workers. *See* Employees
World markets. *See* Global markets

Xerography. *See* Photocopiers
Xerox, 16, 19
copier redesign by, 348–49
joint projects, 337
new businesses, 547
organizational architecture, 631–41
problem-solving, 159–67, 170–71
product failures, 288
quality programs, 165, 612, 613–14, 616, 618, 628–29
research center, 342–52, 354
technology alliances, 571
See also Fuji-Xerox
X-ray technology, 141–42

Young, John, 239, 241, 242, 247, 250

Zaffaroni, Alex, 574
Zamech. *See* ABB Zamech
Zenith (company), 28, 81, 88
Zurich (Switzerland), 490